The Nature of Mediterranean Europe

An Ecological History

A. T. GROVE

OLIVER RACKHAM

Yale University Press
New Haven and London

To Jan, Jean and Jenny

† Señora Jan López-Gunn, † Dr Jean M. Grove and Dr Jennifer A. Moody

Copyright © 2001 A.T. Grove and Oliver Rackham
Second printing, with corrections, 2003

Edited and designed by Jane Havell
Typeset in Ehrhardt
Printed in China

Library of Congress Control Number: 2003101909
ISBN 0-300-10055-8
Catalogue records for this book are available from
The Library of Congress and The British Library

Frontispiece: a cork-oakery below Burgos,
Sardinia (*see p.* 208).

Contents

Acknowledgements

MEDALUS (*MEditerranean Desertification And Land Use*)

This book is based on work that we did under programmes financed by the European Community through the Department of Geography, Cambridge University. We had both been involved in two early EC projects. The first, in 1988–90, was *Crete and the Aegean Islands: effects of changing climate on the environment* in collaboration with Professor N. Margaris and Evgenia Koutsidou of the University of the Aegean.[1] The second, *West Crete as a Threatened Mediterranean Landscape*, in 1991–3, was in collaboration with Professor V. Papanastasis, M. Karteris, J. Ispikoudhis and A. Kazaklis of the Aristotelian University, Thessaloniki. We remember with gratitude many inspiring discussions with Professor Zev Naveh and the advice of Professors W. Vos and A. H. F. Stortelder of the University of Wageningen.

One of us (Dick Grove) was involved in MEDALUS 1 in 1991–2; both of us took part in MEDALUS 2 in 1993–4.[2] The MEDALUS programmes were under the general direction of Professor John Thornes of King's College, London, to whose work in southern Spain we make many references. In MEDALUS 2, Professor Angelo Aru of the University of Cagliari was coordinator of Project III, *Managing Desertification*, in which we were concerned with the physical, biological and human aspects of environmental change. To all these we are grateful for many kindnesses.

We have differed from many of our colleagues in MEDALUS in our view of the changes taking place in Mediterranean Europe which we present in this volume. We thank Angelo Aru, Ignazio Camarda, Joannis Ispikoudhis, Constantinos Kosmas, Francisco López-Bermúdez, Nikos Margaris, Vasilis Papanastasis, Juan Puigdefábrigas and Maria José Roxo for organizing meetings and leading excursions, and for listening cheerfully to our heretical, though we think well-justified, opinions.

Other contributions to this book

We have drawn on many other research projects, some of them archaeological, with which we have been concerned. Oliver Rackham's interest in the Mediterranean began with Cambridge Botany School excursions to Provence in 1965, Slovenia and Croatia in 1967, and Majorca in 1969; in these (all of which contribute to this book) he is happy to acknowledge his debt to S. Max Walters and Peter Sell. His first visit to Crete, in 1968, was at the invitation of Peter Warren as the expeditionary botanist to the excavation of Myrtos; he repeated his observations on re-visits twenty and thirty years later. He has served on the archaeological surveys of Bœotia (with John Bintliff and Anthony Snodgrass, 1978–81); Cosa and Orbetello, Tuscany (under Dr Martin Jones and Professor Andrea Carandini); Vrókastro, Crete (with Barbara Hayden and Jennifer Moody, 1986–94); Sphakiá (with Lucia Nixon, Jennifer Moody and Simon Price, 1986–95); Grevenά (with Nancy Wilkie and Jennifer Moody); and Atsipάdhes, Crete (under Alan Peatfield, Jennifer Moody and Stavroula Markoulaki). He has also attended surveys and excavations on the islands of Kythera (Dr Cyprian Broodbank) and Pséira (Professor Philip P. Betancourt), at Kavoúsi in E. Crete (Professors Geraldine C. Gesell and William D. Coulson), and at Bozburun, SW Turkey (Dr Fred Hocker). For parallels in Western Australia he is indebted for their hospitality and enthusiasm to Professor Neville Marchant and Mrs Jenny Mills, and in the SE United States to Professor Susan Bratton. He is indebted for grants to the British School at Athens, Corpus Christi College, Cambridge, and the Gladys Krieble Delmas Foundation.

Dick Grove's interest in the Mediterranean was first stimulated when he cycled through the French Midi in 1946, and was maintained in later years by family holidays, travelling overland to the Aegean. He was intrigued by Claudio (now Professor) Vita-Finzi's studies of Mediterranean valleys in the 1960s, and learnt much from Harriet Allen's study of the Kopáis basin in the 1980s. He is proud to have been, at least nominally, their research supervisor and to continue to count them his friends. Until 1980 his own research interests were mainly in Africa, especially in soil erosion, land use and the history of climate change there in the Late Quaternary. In consequence, in the early 1970s he became involved in the question of desertification, and in the early 1980s was invited by Roberto Fantechi, then of the EEC's Directorate-General for Science Research and Development, to become involved in EC studies of this elusive subject in southern Europe.

Our colleagues and helpers

We are deeply indebted for her learning, hospitality, collaboration and encouragement over many years to Dr Jennifer A. Moody (Baylor University, Texas); she has travelled with us to many parts of the Aegean and also in her Texan homeland. Her husband Wick Dossett has been an unfailing friend. Oliver Rackham is much beholden to Dr Diego Moreno

---

[1] Account available as *Stability and Change in the Cretan Landscape*, ed. Dick Grove, J. Moody, O. Rackham: *Petromarula* 1 (available from Dr O. Rackham, Corpus Christi College, Cambridge CB2 1RH, UK).

[2] The final reports of MEDALUS 1 and 2 (1 January 1991 to 30 September 1995) may be had from the MEDALUS office, 20A High Street, Thatcham, Berkshire RG19 3JD, UK. Brandt & Thornes 1996 is a product of MEDALUS 1.

(University of Genoa) for his deep knowledge of the landscape history of the northern Apennines and his kind hospitality and companionship. Dr Jean Grove (Girton College, Cambridge) has been with us on many journeys; we have been greatly influenced by her researches on the Little Ice Age,[3] and by her studies with Annalisa Conterio of its effects on the Mediterranean, especially Crete.

In Portugal we have been guided by Teresa Pinto Correia (Atlantic University, Lisbon) and have profited from her researches into *montado* landscapes. In Spain, Elena López-Gunn (University of Hertford) and her family have given us help and hospitality, especially in connexion with fire and water problems. A birthday party given by Vincenti Font Tullot is especially memorable. Pablo Campos Palacín spared time to discuss *dehesas* with us, the Spanish wood-pastures on which he is an expert. Francisco López-Bermudez escorted us in the Guadalentín valley; Juan Puigdefábrigas and his wife demonstrated the Almería region. Miguel Morey showed us the landscapes of southern Majorca.

Margaret Atherden and Jean Hall (College of Ripon & York St John) have studied pollen cores, especially that from Así Goniá, Crete. Samples from these cores and from fossil trees near Olympia were radiocarbon dated by Roy Switsur (Godwin Laboratory, Cambridge). Tony Carter measured the ring-widths of numerous oak and cypress cores from Crete – without our being able as yet, alas, to make good use of the results.

Alison Grove helped with our survey of the Vidauban area and travelled with us in the Pyrenees. Professor Otmar Seuffert showed us round his instrumented catchment in Sardinia. Professor Gentileschi advised us on the social geography of Sardinia. Dr Gloria Pungetti helped with Sardinian statistics. In south Italy, we have been fortunate to have Annalisa Conterio's company and help. John and Jenny Killingback introduced us to Aliano, where the hospitality of the Scelzi and Contadini households was outstanding (it is no longer the case that 'Christ stops at Eboli'). Other companions in Italy include Charles Watkins and the Rev. Sandro Lagomarsini. Joannis Ispikoudhis and Vasilios Papanastasis took us to Macedonia (Greece). Anna Pyrgaki was largely responsible for the study of Pómbia (Crete) and provided useful information from Gávdhos; she was marooned on that island for several stormy winter weeks in 1993, such was her assiduity. Our many friends and helpers in Crete include Chryssa and Theodhoros Athitakis, Panayiotis Kalomoirakis, and Yanni and Sophia Pappadhakis of Pakhyámmos.

We thank Gottlieb Bosch, Bob Thomas, Alberto Montanari and Letizia Buffoni for supplying meteorological data, and Bill Adams, A. Pulido Bosch, Teresa Pinto Correia, Gaston Demarée, Jean Grove, F. S. Rodrigo, Claudio Vita-Finzi and Dennis Wheeler for reading and commenting on draft chapters. Mike Young drafted or drew several of the diagrams.

We much regret the deaths of Señora Jan López-Gunn in 1998 and of Jean Grove in 2001. We remember them both in our dedication of this book, along with Jenny Moody who, thankfully, is still with us.

OLIVER RACKHAM
DICK GROVE

---

[3] J. M. Grove 1988.

CHAPTER I

Introduction: Ruined Landscapes and the Question of Desertification

Can this be Greece? we asked ourselves. Can it be that Attica is so utterly barren?
The waves of a sparkling sea lapped against a beachless shore of pale grey
limestone, as treeless and bare as the Pyramids. At the top of the cliffs there stood
out, white and clear against the cloudless sky, an exquisite group of marble
columns, a ruined shrine of the God of the Sea. . . .
Can it be that Attica was always thus? In the days of Pericles were its hills so barren,
its soil so thin, and its population, outside the city, so scanty as now? In those days
was the bed of the Ilissos river dry most of the year? Would the old Greek writers
have spoken as they did of the famous grove of the Akademe, where the greatest of
philosophers walked and talked beside the Kephissos river, if then, as now, the
stream had been exhausted by irrigation before reaching the sea . . .?[1]

Historians usually include the natural world in their narratives by making it into
an artefact, by drawing it within the sphere of human influence and diminishing
its natural dimensions.[2]

THE RUINED LANDSCAPE

Nicolas Poussin (1594–1665) was a French painter who spent much of his life in Rome, but never went further into the Mediterranean. His favourite subjects were from Greek mythology, the ancient Hebrews and early Christianity. Like other artists of his time, he set all these in the environments he knew. The Greece of Orion or Achilles, Phocion's Athens, St John's Patmos and even St Jerome's north Africa are minutely depicted as if they were France or middle Italy (Fig. 1.1).

This is the first strand in the theory of the Ruined Landscape or Lost Eden. Renaissance poets and Baroque painters encouraged the belief that the actions of Antiquity took place in lands not too unlike the lush riversides of Normandy or the dramatic wooded badlands of the Papal States. Virgil, their inspirer, had not distinguished harsh Greece from idyllic Italy.[3] When travellers reached the drier and remoter parts of the Mediterranean and compared what they saw with what they expected, they inferred that the landscape had gone to the bad since Classical times.

A second strand comes from the idea that floods are abnormal (rather than extremes of normal behaviour) and that forests, and only forests, prevent them. This apparently comes from Giuseppe Paulini, an elder contemporary of Poussin in Venice. His report on the Venetian Alps in 1608 says that, in ancient times,

both mountains and valleys were full of trees . . . the rains, falling upon these woods, were soon dispersed, and all the water descending directly was almost wholly absorbed by the dead leaves and by the ground itself . . . the snows lying in the shadow of the woods were but gradually liquefied, losing themselves in the soil . . .

whereas now, the mountains having been 'ruined and despoiled of their clothing' by occupational burning and other causes, the rivers rise, 'break the dikes', sweep away buildings and threaten to fill the Lagoon of Venice with debris.[4]

A third strand comes from the fathers of plant physiology. John Woodward (1699) and Stephen Hales (1727) had measured the large quantities of water vapour released into the atmosphere by plants and trees.[5] It became generally accepted that trees increase rainfall by adding moisture to the atmosphere, and that destroying trees decreases rainfall. This idea took hold on colonial and imperial British administrators[6] and was brought back to Europe.

A fourth set of ideas came from the effects of European discovery on remote islands such as Madeira and St Helena. Most of these had never had human or even mammalian inhabitants, and were volcanic; the coming of people, goats and pigs brought disaster to their plants, animals and soils. It was easy to suppose that Mediterranean coasts and islands had suffered a similar fate in the distant past to that of these

---

[1] Lecture by E. Huntington, 1910.

[2] T. Griffiths, 'Secrets of the forest: writing environmental history', *Australia's Ever-Changing Forests II*, ed. J. Dargavel & S. Feay, Centre for Resource & Environmental Studies, Australian National University, Canberra, 1993, pp. 47–50.

[3] 'Virgil's Arcadia is ruled by tender feeling . . . Herdsmen lack the crudeness of the peasant life as well as the over-sophistication of the city. In their rural idyll the peaceful calm of the leisurely evening hours stands out more clearly than the labour of their daily bread, the cool shade is more real than the the harshness of the elements, and the soft turf by the brook plays a larger role than the wild mountain crags'. Bruno Snell, *The Discovery of the Mind*, Blackwell, Oxford, 1953, p. 288.

[4] Quoted in Kittredge 1948, pp. 6 ff, and in excerpts by Perlin 1989, pp. 154 ff. We have not seen the original and cannot vouch for the translation.

[5] J. Woodward, 'Some thoughts and experiments concerning vegetation', *Philosophical Transactions of the Royal Society* 21, 1699, pp. 193–227; S. Hales, *Vegetable Staticks*, Innys & Woodward, London, 1727.

[6] K. Thompson, 'Forests and climatic change in America: some early views', *Climatic Change* 3, 1980, pp. 47–64; R. H. Grove 1995.

Fig. 1.1. *Landscape with the Ashes of Phocion Collected by his Widow,* by Nicolas Poussin, painted in France in 1648 after his return from Rome. The scene is supposed to be Megara, between Athens and Corinth, in 318 BC.

fragile, unstable oceanic islands in the recent past.

From these four strands the theory of the Ruined Land-scape or Lost Eden was woven. Well into historic times Mediterranean lands had been covered with magnificent forests of tall trees: the sort of forests that modern foresters are trained to approve of. Men cut down the forests to make houses or ships or charcoal. The trees failed to grow again, and multitudes of goats devoured the remains. Trees, unlike other vegetation, have a magic power of retaining soil. The trees gone, the soil washed away into the sea or the plains. The land became 'barren', and even the climate got more arid.

Desertification already existed in the fourteenth century in the hellish imagination of Dante:

In mid-sea sits a waste land . . . which is called Crete, under whose king the world was once innocent. A mountain is there which once was happy with water and leaves, which is called Ida; now it is a desert like an obsolete thing.[7]

As a scientific idea it apparently germinated in mid-eighteenth-century writings claiming that the unstable mountains of Provence were turning into desert (p. 241). It is full-grown in the writings of Sonnini, the French traveller of 1777–8, especially about Cyprus, which became the type example of Ruined Landscape. It was popularized by the Abbé Barthélemy in his best-selling reconstructions of Ancient Greece, wherein heroes spear the boar in noble forests and

---

7 Dante Alighieri, *Inferno,* xiv, 94–9.

nymphs swim in crystal fountains, as if they were in Marie-Antoinette's France.[8] Travellers, comparing this with the tangled prickly-oaks and dribbling springs of eighteenth-century Greece, would find the Sonnini theory irresistible, especially if they did not know the country well enough to find the noble forests and crystal fountains that still existed.

An essential part of Ruined Landscape theory is that ruination is cumulative. Damage done in the age of railways was added to that done by Ottoman Turks, Venetians, Arabs and Romans. It is increased by war and misgovernment, especially under political regimes ('excesses of luxury and disorder') of which the author disapproves. Recovery can be achieved only with difficulty, by deliberate and enlightened human intervention.

The 'degradationist hypothesis' went from strength to strength, especially through the great authority of George Perkins Marsh. He was in Italy when most evils, from hail to frost to malaria, were blamed on the supposed destruction of forests. In 1864 he held up 'the fairest and fruitfulest provinces of the Roman Empire' as an example of the 'sterility and physical decrepitude' brought about by 'civil and ecclesiastical tyranny and misrule'.[9] The British took over Cyprus in 1878 partly because they thought they knew better than the Cypriots how to manage the island.[10]

Today, writers on political ecology feel obliged to cite the 'typical' Mediterranean landscape as an example of 'massive ecological degradation'. Scrub and scattered trees are interpreted, without evidence, as 'the debased forms of the forest'.[11] Sir Arthur Evans solemnly stated that the men of Knossós, Crete, took to using gypsum for door- and window-frames because even in the Bronze Age they had run out of trees.[12] We may smile at such naïvety, but within the last ten years scholars have glibly attributed everything from erosion to the decline and fall of the Roman Empire to shortage of trees, without demonstrating either that trees *were* diminishing or that there was no alternative explanation.

Others hold a more scientific version of the myth. Up to the Neolithic, human activity was negligible and effects on vegetation, soils or erosion are attributable to climatic change. From the Neolithic onwards, climate stopped changing and such effects 'must be' due to human action.

One school of environmental determinism claims that wide areas are depopulated, not because people have found better jobs elsewhere, but because the soils and vegetation are no longer usable. This has been most recently expounded by J. R. McNeill.[13] Huge and costly schemes of deliberate, enlightened intervention are intended to halt or reverse these changes and to 'restore' supposedly degraded landscapes. Tree-planting is often – always in Spain – presented as 're-forestation', in some sense restoring a tree cover that is

thought once to have existed and is now missing. (This flatters the vanity of governments, who like to be told they can command even the very trees to grow or not to grow.)

In 1985 the British Broadcasting Corporation Natural History Unit consulted one of us about a television programme on Mediterranean ecological history. They outlined the theory as given above, and asked if we could provide examples of landscapes to illustrate it. We replied that we could, but that the reality was far more complex and we would also provide examples of opposite changes. We heard no more from the BBC, but the programme duly appeared and was a splendid exemplification of Ruined Landscape theory.[14]

Science and the Age of Enlightenment

The climatic part of the theory is a classic story of the Age of Reason. Woodward had said in 1697:

> This so continual an *Emission* and *Detachment* of *Water*, in so great *Plenty* from the *Parts* of *Plants*, affords a manifest reason why *Countries* that *abound* with *Trees* and the *larger Vegetables* especially, should be very obnoxious [that is, exposed] to *Damps*, great *Humidity* in the *Air*, and more frequent *Rains*, than *others* that are more *open* and *free*. The great *Moisture* in the *Air*, was a mighty inconvenience and *annoyance* to those who first settled in *America;* which at that time was much over-grown with *Woods* and *Groves*. But as *these* were burnt and *destroyed* to make way for *Habitation* and *Culture* of the *Earth*, the *Air* mended and *cleared* up apace: changing into a Temper much more *dry* and *serene* than before.[15]

This modest speculation hardened into a scientific principle which nobody thought of denying.[16] That human actions could affect even the climate was a powerful idea in an age which knew nothing of long-term natural changes as an alternative explanation. It was invoked in France to account for 'the year without a summer', 1816, which provoked 'the last great subsistence crisis in the western world', the repeated crop failure the following year and the frosting of the olives in 1819 and 1820. The Ministry of the Interior in 1821 commanded the Académie Royale des Sciences to find reasons for the storms and crop failures, especially to study the part played by the deforestation supposed to have happened over the preceding three decades.[17] (We now know that the real cause was 'nuclear winter' resulting from the eruption of the volcano Tambora, half a world away.)

This climate theory was believed in North America until the 1920s.[18] Even in 1914 deforestation was assumed in West Africa in order to account for an apparent decline of rainfall.

8 C. S. Sonnini, *Voyage en Grèce et en Turquie fait par ordre de Louis XVI*, Paris, 1801; J. J. Barthélemy, *Voyage du jeune Anacharsis en Grèce*, Debure, Paris, 1788.
9 Marsh 1864, chap. 1. Marsh, however, was sceptical of the effect of forest on rainfall. For the higher flights of 'desiccationist' fancy, read R. H. Grove, *Ecology, Climate, and Empire*, White Horse Press, Knapwell, 1997.

10 Thirgood 1987.
11 For example, Braudel 1972–3, p. 239.
12 A. Evans, *Palace of Minos*, Macmillan, New York, 1921– , vol. 2, p. 565.
13 McNeill 1992.
14 *The First Eden: the Mediterranean world and man*, BBC Enterprises, London, 1985.
15 See note 5.
16 R. H. Grove 1995. The belief that loss of trees

diminishes rainfall is often attributed to Theophrastus, but we can find no passage where he says so.
17 J. D. Post, *The Last Great Subsistence Crisis in the Western World*, Baltimore, 1977; V. Bainville & P. Ladoy, 'Préoccupations environmentales au début du XIXe siècle', *La Météorologie* 8, 1995, pp. 88 ff.
18 And in some quarters even later: Saberwal 1998.

But confidence was decreasing: Huntington, having written eloquently of the apparent desertification of Greece, rejected deforestation as a cause, on the strength of American counter-evidence.[19] All this time the meagre original evidence had never been verified by adequate observation, and as far as we know it still never has been. If the effect exists it is evidently very small and difficult to detect among the natural fluctuations of rainfall.[20] There are abundant stories of rainfall increasing after trees have been planted, or decreasing after they have been felled. There are abundant stories of the opposite happening,[21] resulting in the counter-theory that 'rain follows the plow', a belief that lured generations of American and Australian frontiersmen to ruin (p. 120).

This is an example of a repeatable pattern which we shall encounter again. A scientist announces a theory that holds out grave threats or bright promises for mankind. This becomes popularly believed, and commands the attention of governments, who try to make their subjects behave accordingly. The theory develops a life of its own, and it does not greatly matter whether its scientific basis is sound or unsound. This pattern of popular belief and government action based on unconfirmed science or pseudo-science, characteristic of the Age of Enlightenment, gained yet more strength in the twentieth century (a fine rich example is filling the front pages of Europe's newspapers even as we write).

THE MEDITERRANEAN

The Mediterranean region is slowly contracting as the African continental plate burrows under Europe. Mountain building is still in progress; great earthquakes occur every few decades. There are many volcanoes in Italy and the Aegean, with frequent minor eruptions and rare major explosions. California and Chile are likewise precariously placed at the active margins of tectonic plates. The Cape of Good Hope and SW Australia are geologically much older (parts of Gondwanaland) and more stable.

The Mediterranean is characterized by its plants. Among crops, the olive is almost confined to the region; its cold-tolerance is often thought of as defining the limit of Mediterranean climate. Wild plants have strong affinities with middle Europe and SW Asia, and to a lesser extent with Africa. They are unlike those of other mediterraneoid regions (p. 47).

Southern Europe is a region of dramatic variety and glamorous contrast. The great forests of the toe of Italy, with their deceptively Central European appearance, are utterly unlike the burning hills of Sicily, 250 km away. Crete, a splinter of land 250 by 50 km, is a miniature continent with its Alps, its deserts and jungles, its arctic wastes and its tropical gorges, where an afternoon's walk goes from something looking like Wales to a rough equivalent of Morocco. The first objection to Ruined Landscape theory is that it is too generalized. To prove the theory for one side of a mountain does not prove it for the other side (Figs. 1.3, 1.4).

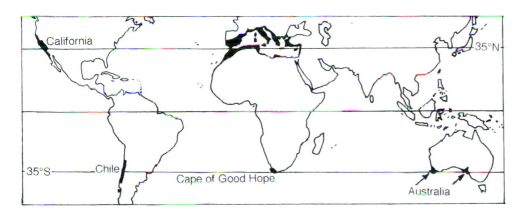

Fig. 1.2. The Mediterranean and the mediterraneoids.

Mediterranean Europe (including south Portugal) is defined by its climate: its warm wet winters and hot dry summers. Winter and spring are the growing season at low altitudes. This regime is confined to about 1 per cent of the earth's land surface, in the six 'mediterraneoid' parts of the world: the Mediterranean (by far the largest, including its Asian and African parts), California, middle Chile, the Cape of Good Hope, south-west Australia and around Adelaide in South Australia (Fig. 1.2). Almost everywhere else has rain either in the hot season or all round the year.

Diversity extends to the inhabitants as well as the environment. Mediterranean plants are often very localized. The cork-oaks of the west Mediterranean, holm-oaks of the middle and prickly-oaks of the east, with their different properties, set the character of their respective landscapes. The animal history of the islands is very unlike the mainland. Human tribes and cultures of extraordinary diversity have created different cultural landscapes out of apparently similar environments: for example, fiercely individualistic Corsica *versus* collectivist Sardinia.

[19] Kittredge 1948, chap. 10; J. Fairhead & M. Leach, 'Reading forest history backwards: the interaction of policy and land use in Guinea's forest-savanna mosaic, 1893–1993', *E & H* 1, 1995, pp. 55–91; Huntington 1910, p. 659.
[20] One undoubted effect is that trees comb droplets out of fog and thereby increase their water supply, presumably at the expense of neighbouring vegetation. This may be significant for savanna trees (p. 199) and for small 'cloud forests' on the tops of island mountains, such as the otherwise arid Asteroúsia in Crete. See Kittredge 1948, chap. 12.
[21] For an early example, Gollut in 1592 blamed deforestation in the Dôle area of France (iron-founders consuming trees faster than they could grow) for increased rainfall in the previous 26 years. Quoted in Braudel 1966.

	Mediterranean		California	Mid Chile	Cape of Good Hope	SW Australia
Area, thousands of sq. km (very approximate)	Europe Asia Africa total	750 150 200 1100	250	70	40	350 (incl. S. Australia)
Latitude	31–45°N		28–44°N	29–40°S	32–35°S	28–37°S
Strength of dry season	+++		++++	++++	+++	++
Tectonics	++++		++++	++++	+	+
Proportion of limestone	++++		+	+	++	+
Soil fertility	+++		+++	+++	+	0
Erosion	+++		+++	+++	+	+
Importance of fire	+++		++++	++	++++	+++++
European settlement history, years	8000		200	450	350	160

Table 1.i. Comparison between the Mediterranean and other regions with a mediterraneoid climate.

Fig. 1.3. The diversity of Mediterranean landscapes: Manouratómylos in Amári district, with the great mountain Psilorítis. *April 1988*

At this point most authors would introduce the Mediterranean 'Triad' of crops: wheat, grapes, olives, barley.[22] These (plus the oft-forgotten legumes) are often held to define Mediterranean agriculture; but they are not present everywhere. The olive may be absent, either because of cold winters or (as apparently in much of Sardinia) for cultural reasons. The Mediterranean is, or has been, full of specializations: sugar-boilers in Motril, acorn-eaters in Estremadura, pig-driers in the Alpujarra, esparto-twisters in SE Spain, palmists in the city of Elx, madder-growers in Provence, cork-cutters in Sardinia, boar-hunters and chestnut-millers in the Apennines, oat-growers on the Macedonian serpentine, cotton-pickers in Bœotia and (formerly) Crete, resin-tappers in Attica, quail-gatherers in the Máni, banana-men in Arvi, ladanum-whippers in one particular spot in north Crete, potatoists in the Lassíthi Plain, distillers of lemon leaves in the Cyclades, shipbuilders in remote nooks of the Aegean, masticators in the southern third of Chios, spongers in the Twelve Islands. With all these local practices and many more, the Mediterranean is no place for facile generalization.

Mediterranean agriculture involves hard work but seldom unrelenting toil; most crops involve weeks of dawn-to-dusk work followed by weeks with little to do. This encourages people to create second jobs, like the merchants and fur-traders of the Píndhos Mountains. The menfolk of remote Kárpathos, renowned stone-masons, used to spend months away from the island building cities like Athens; they now spend years in the Kárpathos colony in New Jersey.

LANDSCAPE

We use 'landscape' in its century-old sense of 'a tract of land with its distinguishing characteristics and features, especially considered as a product of modifying or shaping processes and agents . . .'[23] Landscape is meant as an objective reality, amenable to scientific investigation: gorges, terraces, hedges, pollard trees originated and developed at times and by processes that can be ascertained by observation and analysis. Different observers should, in principle, get approximations to the same answer. People's attitudes to landscape or recognition of landscapes are not our main study: important though they are, they are for more advanced books than this. There is little point in studying them without first ascertaining what it is that people are attitudinizing about.

Nearly all questions of landscape have a historical component, and cannot be understood merely in terms of the

Fig. 1.4a. The desert . . .

Fig. 1.4b. . . . and the jungle.

South and north faces of Mount Kryonerítis, SW Crete *July 1981, April 1988*

[22] A. Sarpaki, 'The palaeoethnobotanical approach: the Mediterranean Triad or is it a quartet?', Wells 1992, pp. 61–76.
[23] *Oxford English Dictionary*, 2nd edn.

present situation. There are four kinds of historical investigation:

(1) Recognising sites and objects of particular importance for their long or unrepeatable processes of development, e.g. savannas with ancient trees.

(2) Understanding what human activities have gone to making a site what it is.

(3) Differentiating changes and problems that are still going on from those due to past events.

(4) Determining the normal dynamics of a site, and thereby defining abnormal processes which lie outside those dynamics. For example, floods may be normal but rare events; absence of burning may be abnormal.

The history of landscape is not the same as that of land-use, nor of country-folk, still less is it only economic history. Most landscapes are produced by human cultures interacting with the natural environment and with plants and animals. They are not necessarily the product of the most recent human culture on the site, nor do all landscape details change with every change of economy and land-use. A twentieth-century person may live in a fourteenth-century house and cultivate fields laid out in the Bronze Age.[24]

Each successive culture keeps some features of its predecessors' landscape, destroys some, abandons some and adds features of its own. Most of the major forest areas in southern Europe in Roman times were still there in the nineteenth century; changes took the form of local subtractions and additions; the complete conversion of wide areas of non-forest to forest, or *vice versa*, was unusual. Property boundaries, roads, canals and even individual olive-trees pass from one culture to another. This is obvious with planned field-systems. Great areas of north Italy still bear the rigid stamp of Roman country planning. Medieval cultivation strips are very visible in Sardinia. Half the Lassíthi Plain in Crete is today divided, as in the sixteenth century, into 193 nearly rectangular compartments, apparently a Venetian imitation of Roman practice.

DESERTIFICATION

Debate about Mediterranean ecology is haunted by memories – real or imagined – of the past. Ruined Landscape theory developed into the idea of desertification.

Desertification properly means the creation or enlargement of a desert, by natural processes such as a change of climate, or by human activities leading to changes such as the loss of soil exposing subsoil or bare rock. The word implies a *change* in environment or vegetation: the making of a desert where there was not one before: a change from a less desert-like past to a more desert-like present and still more desert-like future.

What is a desert?

To the modern English or American reader the answer is clear: an area without, or with very sparse, plant cover, typically because of dryness. This leads to a precise definition of desertification: an example is the southward expansion of the Sahara which occurred five thousand years ago.

However, there is another strand to the idea of deserts. The word has its origin in the Late Latin equivalent of 'deserted'; it originally meant an unpopulated place, which had lost or never had human inhabitants.[25] English writers from the seventh to the eighteenth centuries regularly equate 'desert' either with 'wilderness' or with 'forest', for example Shakespeare:

in this desert inaccessible,
Vnder the shade of melancholly boughes . . .
As You Like It, Act II, sc. vii, 110

The modern sense of 'desert', modelled on the Sahara, begins in the sixteenth century and becomes dominant in English in the eighteenth. The earlier sense, rarely used by English or American writers today, is still active in other European languages. In France *désertion* means 'depopulation'.[26] The German *Wüste* is used for places like the Sahara, but its derivative, *Wüstung*, equivalent to 'desertization', is the normal word for 'deserted medieval village'. In Modern Greek *érimos* can be used interchangeably for 'desert' and 'deserted settlement'. For this linguistic reason there is a tendency, especially in southern Europe, to join the ideas of desertification and depopulation, which English-speakers keep apart. Many depopulations were for reasons that have nothing to do with environmental change, yet are often included in the concept of desertification.

The writings of the Spanish scientist Antonio Ponz, based on his travels in the 1750s, demonstrate how observations of depopulation grow imperceptibly, in a southern European language, into a theory of desertification. Ponz, like ourselves, was fascinated by the contrast between the rich Roman remains of Estremadura, eloquent of past population and prosperity, and its modern state of emptiness and lonely savannas. Unlike the twentieth-century traveller, he found this depressing, an affront to civilization and progress:

How many villages could there not be, ought there not to be, in a land so good and so deserted! On the banks of the Almonte alone and of the other ravines [between Trujillo and Cáceres], in the opinion of very zealous and intelligent persons in Cáceres, there could be a dozen of them. Every day the destruction of this beautiful and fat Province of Extremadura goes further; and if no remedy is applied it will become reduced to a desert. The towns are four, five, and six leagues distant in most places; industry is almost totally destroyed in all the Province; the population is reduced to a shadow of what it was and could be; its wide plains are converted into thick forests of live-oak and cork-oak, and the worst of all into places of *Cistus* and useless shrubs. . . .

I am scandalized to hear that it is all reduced to a hundred thousand inhabitants . . . a province, perhaps the

---

[24] Rackham 1986.

[25] Except for hermits and monks of the Cistercian order, who were required by their Rule to live in desert places.

[26] See, for example, J. Lévy, 'Oser le désert? des pays sans paysans', *Sciences Humaines*, Hors série 4, 1994, pp. 6–11.

most fertile in Spain, and among the biggest in Europe . . . has fewer inhabitants than only one of the chief cities of Europe . . .[27]

This area will be discussed in Chapter 12. Ponz set a precedent by attributing the 'desertification' of Estremadura to mismanagement, especially the depredations of the Mesta, the guild of shepherds (p. 201). Like many later writers, he was over-optimistic about whether an unfamiliar landscape could sustain conventional agriculture.[28]

The idea of desertification, if not yet the word, was a subject of controversy early in the twentieth century. E. Huntington's lecture of 1911 to the Royal Geographical Society, the first of the two quotations at the head of this chapter (p. 8), began a lively debate, in which three positions were taken up:

(1)   The theory of climatic change. Huntington himself claimed that in Classical times the dry, poorly vegetated east side of Greece had been more like the rainy, well-vegetated west side of modern Greece. The change was independent of human activity.

(2)   The traditional degradationist theory.

(3)   The theory that Classical Greece operated in a landscape not, after all, very different from modern Greece.[29]

These three theories are still alive, and it will be our business to ascertain where the truth lies.

The word 'desertification' was introduced in Africa by a British and a French forester. In the 1930s E. P. Stebbing, visiting West Africa in the dry season, saw the supposed encroachment of the Sahara as a threat to the more humid lands to the south.[30] A. Aubréville, in 1949, inferred that in a part of Guinea, West Africa, clearance of deciduous forest had been followed by grass fires, erosion and hardening of the soil, so that the forest could never recover. He speculated that further removal of forest might reduce transpiration, thereby (under Woodward–Hales theory) curtailing rainfall further into the interior. 'La savanisation du continent peut-elle conduire à une désertification réelle de l'ensemble de ce continent?'[31]

In later decades, 'desertification' (or *désertisation* in French) was used by Pabot, Le Houérou and Kassas to denote environmental degradation in the Middle East and around the Sahara: 'reduction of the perennial plant cover, impoverishment of the flora, soil erosion, formation of mobile dunes and establishment of desert pavements'.[32]

'Desertification' still retained its more rigorous meaning in the report of the United Nations Conference on Desertification in 1977.[33] But once institutionalized, the concept broadened. By the time of the Convention on Desertification fourteen years later, the United Nations had come to equate it with 'degradation', defined as 'reduction or loss . . . of the biological or economic productivity and complexity' of natural or artificial vegetation.[34] This has been taken to include changes unrelated to the original meaning, such as problems arising from extracting too much ground-water, or from the disposal of toxic waste. Many of the woes ascribed to desertification in the dry tropics are really the effect of ordinary droughts on human populations that are too numerous to cope with drought in traditional ways and too poor to buy in food.[35]

What does 'degradation' mean and how do we recognise it?

The phrase 'degraded landscape' is often on the lips of our Southern European colleagues. If rightly used, as in the United Nations definition mentioned above, it implies the belief that there has been a change: that the terrain was in some sense better, usually more vegetated, at some time in the past than it is now. (It implies also that the change is permanent or semi-permanent; degradation is not always easy to distinguish from the passing effect of two or three dry years.)

Many users of the phrase find it difficult to explain, from first principles, why they believe a particular landscape to be degraded. If pressed, they appeal to generalities rather than citing evidence for the history of a particular site. Having a preconceived standard ('potential climax') of what Mediterranean vegetation ought to be, they pronounce any landscape to be degraded that now falls short of that standard. They fail to distinguish between a mountain that is treeless because of human misuse, and another that is made of hard limestone and has never been suitable for trees.

The matter is made worse by expectations derived from American studies of succession or soil erosion. The original work was perfectly good, but was done in a part of the world very different in climate, soils and vegetation (not in California, which would have been comparable at least in climate), and should not be made the basis of expectations in the Mediterranean.

The vague term 'degradation' has sometimes seemed to us to include almost any change of which the writer disapproves. It often implies a belief that human action caused the change, unlike 'desertification' which can be due to natural causes. Users of the term often have two types of change in mind: (1) alterations to what they believe had previously been

[27] Ponz 1778, VIII, iii, pp. 16, 60.

[28] Ponz, though he hated forests, was a passionate and ingenious propagandist for tree-planting – as long as trees were kept where they belonged, in hedges and the edges of fields.

[29] Huntington 1910.

[30] E. P. Stebbing, 'The threat of the Sahara', Extra Supplement, *Journal of the Royal African Society*, 1937.

[31] A. Aubréville, *Climats, forêts et désertification de l'Afrique Tropicale*, Société d'Éditions Géographiques, Maritimes et Coloniales, Paris, 1949.

[32] H. Pabot, 'Peut-on arrêter la désertification des régions sèches d'Orient?', *Compte rendu du Colloque de Téhéran sur la conservation et la restauration des sols*, Institut Français de Coopération Technique and Faculté d'Agronomie de l'Université de Téhéran, 1960, pp. 120–25; H. N. Le Houérou, *La désertisation du Sahara septentrional et des steppes limitrophes*, Proceedings of the IBP Hammamet Technical Meeting on the Conservation of Nature, Hammamet, Tunisia, 1968; M. Kassas, 'Desertification versus potential for recovery in circum-Saharan territories', *Arid Lands in Transition*, ed. H. Dregne, American Association for the Advancement of Science, publication 90, Washington, D.C., 1970, pp. 123–42; H. N. Le Houérou, 'North Africa: past, present, future', *ibid.*, pp. 227–78.

[33] United Nations Conference on Desertification, Nairobi, *Desertification: its Causes and Consequences*, Pergamon, Oxford, 1977.

[34] *United Nations Convention to Combat Desertification in those Countries experiencing serious Drought and/or Desertification, particularly in Africa*, Her Majesty's Stationery Office, London, 1995.

[35] Thomas & Middleton 1994.

primaeval forest, and (2) types of change (such as erosion) that cause land to become less suitable for European-type agriculture. The term 'degradation' easily becomes loaded with value-judgements.

Less excusably, some scholars and propagandists use 'degradation' and 'desertification' in situations with no evidence of environmental change at all. The mere existence of a desert, or of something that can be described as one, is taken as evidence of desertification, whereas (for all the writer knows) the desert may have been present throughout history and may even have shrunk. A savanna is categorized as 'degraded forest' without evidence that it is not, in fact, a grassland that has become invaded by trees. In this book we shall try to confine the use of 'degradation' and 'desertification' to examples where it is known that the present state is the result of a change for the worse.

Does desertification (or degradation) lead to deserts?

It is widely believed that human action can enlarge or shrink deserts. The classic example is the High Plains of North America, originally known to explorers as the Great American Desert, and then settled by pioneers in the belief that cultivation and tree-planting would create rainfall. A temporary wet period in the 1870s and 1880s seemed to justify this notion. Then the rains died away and the region became the Dust Bowl of the 1930s, which in turn was supposed to be the result of human misuse of the land. The reality is that in most semi-arid regions rainfall has its ups and downs, regardless of human actions, and it is foolish to mistake unusually wet periods for the normal state.[36]

The experience was repeated in western New South Wales, much of which was well-treed savanna in 1870 and by 1900 had been converted, so far permanently, into treeless semi-desert.[37] This was due to the sudden arrival of European settlers in a relatively wet period; to the introduction of sheep, cattle and rabbits into an ecosystem to which they were not adapted, and which was not adapted to them; to the vegetation being easy to destroy; to the return of drought after a few years; and to the soil and subsoil being exceedingly susceptible to wind erosion.

In Europe such dramatic changes are rare.[38] Land may become unusable for conventional agriculture, but usually remains vegetated. 'Degraded' land is often rich in animal and plant life, strikingly beautiful, loved and valued as an amenity by its inhabitants; an extreme case is the bizarre landscape of the ancient stone-quarries at Syracuse (Sicily). Even land poisoned by heavy metals tends to become vegetated with heavy-metal-tolerant plants. The Arádhena plateau in SW Crete is a tableland of hard limestone which once had soil and is now covered with the walls that enclosed ancient fields. The soil was probably lost through wind and karst erosion (p. 261). Some might regard this as a classic case of desertification through inappropriate agriculture, but the plateau is now covered in pine forest and maquis (Fig. 14.23), and is probably more vegetated than when it was farmed.

Whether government intervention can prevent desertification is uncertain; it can certainly create it. In SW Asia a fatal combination of visionary technologists and corrupt Soviet politicians, setting out to transform the modest traditional economy of the arid lands around the Aral Sea, produced a hell of drought and salinization. In the United States, the government has done its best to rival this achievement by subsidizing the extension of the Great American Desert into savanna rangelands.[39]

Current research on desertification: the scope of MEDALUS

The Sahara has waxed and waned with global climatic changes throughout the Quaternary (Chapter 9). It expanded around 3000 BC, when much of tropical Africa became more arid. In modern times rainfall has varied markedly from decade to decade. Drought in the Sahel, about the time of the United Nations Stockholm Conference on the Environment in 1972, drew attention to the human and environmental problems of semi-arid lands. When the United Nations organized the Nairobi conference on desertification in 1977, Spain (with its extensive desert-like landscapes) was a participant, and was one of the first countries to set up an official body concerned with desertification.[40] Parts of southern Europe were put on world maps of desertification risk.

The matter became more acute in the 1980s through public concern over global warming. Several countries likely to be affected had newly joined the European Community. Dry years used to bring famine to the European Mediterranean, and still bring discomfort: cities such as Athens suffer water shortage, agriculture withers, forest fires turn the sun to blood, and tourists find gypsum in their coffee. If this were to persist year after year the results would be serious, as they well might be if (as early climatic models predicted) rainfall were to diminish in southern Europe over the next half-century.

An attempt to foresee the likely consequences was probably the main stimulus for the series of MEDALUS (MEditerranean Desertification and Land Use) research projects, set up by the European Community from 1991 onwards. MEDALUS has involved some forty institutions all over Europe, most of them university departments, but also official or commercial organizations.[41]

[36] M. H. Glantz, 'Drought, desertification and food production', Glantz 1994, pp. 7–32; Raban 1996.

[37] D. Lunney, 'Review of official attitudes to western New South Wales 1901–93 with particular reference to the fauna', *Future of the Fauna of Western New South Wales*, ed. D. Lunney and others, Royal Zoological Society of New South Wales, Mosman, 1994, pp. 1–26.

[38] See, for example, A. T. Grove, 'Desertification in Southern Europe', *CC* 9, 1986, pp. 49–57.

[39] T. A. Saiko, 'Implications of the disintegration of the former Soviet Union for desertification control', *EMA* 37, 1995, pp. 289–302; K. Hess & J. L. Holechek, 'Policy roots of land degradation in the arid region of the United States: an overview', *ibid.*, pp. 123–41, and 'Government policy influences on rangeland conditions in the United States: a case example', *ibid.*, pp. 179–87.

[40] A. T. Grove, 'Desertification in the African environment', *Drought in Africa 2*, ed. D. Dalby, R. J. Harrison, F. Bezzaz, African Environment Special Report, International African Institute, University of London, 1976, pp. 54–64; A. T. Grove, 'Desertification', *PPG* 1, 1977, pp. 296–310.

[41] For a list of dependent projects see Mairota and others 1998, pp. 6–7.

Most attention has been paid to soil structures and erosion, partly because of the effects on agricultural land, but also because of heavy investment in dams and reservoirs, which get filled up and made useless by sediment resulting from erosion. Models of erosion in relation to rainfall intensity, plant cover, soil and rock conditions, and angle and length of slope are given in the book resulting from MEDALUS I.[42] These researches are further developed in Projects I and II of MEDALUS 2. Project IV of MEDALUS 2 involves studies of portions of three catchments in Italy and Spain where various problems, described as desertification, have arisen or are expected.

Prediction of future changes

Climate is not the only variable. In southern Europe, in contrast to the African and Asian Mediterranean, the total population has stopped growing for the first time in two hundred and fifty years, but wealth and rates of consumption have steadily increased. Tourism, a factor for over a thousand years, has grown so far that the Mediterranean is now the greatest tourist destination in the world. (Even Crete, however, is by no means entirely dependent on tourism.) Tourists and immigrants from the north have greatly increased the populations of cities and coastlands. At the same time, agricultural activity has been concentrating on the plains, where machinery and irrigation can readily be employed. In these areas pressure on resources has been increasing in Mediterranean Europe as fast as in under-developed countries. The demand for water by agriculture and cities is coming to exceed the supply.

In the mountains, human pressure on resources reached a maximum between fifty and a hundred and fifty years ago, and has now declined. In these areas, no longer the most vulnerable to desertification, other threats have arisen. Cultivation has been abandoned, and fewer flocks and herds move from the plains in the winter to the mountains in the summer. Landscapes which once reflected the diversified activities of people extracting a livelihood from a difficult but varied environment are now being reduced to a monotony of trees, shrubs and burnt trees and shrubs.

Over the next few decades the main threat will probably come not from global warming but from technical and economic changes. These are an extension of the powerful and deep-seated forces that have converted Mediterranean Europe within a century from being predominantly poor, over-populated and agricultural to one of the most prosperous regions on earth. New opportunities have been created, but there are penalties. Piped irrigation results in political troubles with distributing limited supplies of water. Labour-saving methods of cultivation increase unemployment and create thousands of tons of rotten plastic. Tourism creates millions of tons of what sometimes are already rotting concrete buildings.

Changes in environment and landscape generally have distant time-horizons. The future political and economic context of 'desertification' in southern Europe is likely to be no more reliably predictable than the climate. Could the changes of the past fifty years have been predicted in 1950? Can the changes of the next fifty years be predicted now?

Many of the problems with which we are concerned can be mitigated by improved management, but not solved: they will not go away. Policies to mitigate 'desertification' are generally unpopular with governments and voters. They are poorly presented, they involve restraint in order to sustain production, they cost money, they restrict profitable activities in the short term in order to secure the resource base for the future, and they reveal that governments are not omnipotent. It is important to select areas and problems that can easily and economically be dealt with, and not to waste resources on trying to cure problems that either are not really problems or that have no solution.

THE WRITING OF THIS BOOK

This book has emerged as a part of Project III of MEDALUS 2, which is concerned with managing desertification, but we emphasize that the views expressed are our own and not necessarily endorsed by our colleagues.

Desertification in this book

We cannot limit this book to 'desertification' in its proper sense of the creation of deserts: if we did there would be no book left. Nor can we cover all the myriad examples of depopulation not linked to environmental change. We try to cover most of the *changes* in the environment and vegetation that have been regarded as desertification. But we insist that it is changes and processes with which we are concerned. The mere existence of a 'desert', without evidence as to how or when it came into existence, does not constitute desertification. Nor does the existence of a badland (a severely gullied area), without knowledge of its history, constitute land degradation. The first problem in dealing with desertification is to identify it.

Methods

The Mediterranean is not easy to study. Many processes are episodic and do not go on all the time. This is not limited to fires, deluges and earthquakes. An area may have trees of about (let us say) 90, 210, 240 and 570 years old, but not of intermediate ages; these represent years when circumstances (such as late spring rains, a preceding fire and disease among the goats) conspired to favour the growth of new trees.

It is difficult to plan studies to be representative in space and time. Somebody studying erosion gets a research grant for three (or, if very lucky, four) years from a body such as MEDALUS. To avoid 'wasting' the grant on nil results, equipment is installed on a site where erosion is already thought to be important. With good management, three rainy seasons' data are collected. However, those seasons are unlikely to include the hundred-year-maximum rainfall which may cause much of the erosion. The measurements thus tend to

---

[42] Brandt & Thornes 1996.

overestimate erosion in space and underestimate it in time – both biases being artefacts resulting from the anthropology of how research is organized.

Historical ecologists have to take their material where they can find it. Written records are limited to times and places where people have been writing things down. Pollen analysis is limited to places with suitable deposits, such as Corsica but, so far, not Sardinia.

Documents

I cannot determine what I ought to transcribe, till I am satisfied how much I ought to believe.[43]

The most important characteristic of an intelligence officer has to be intense scepticism, such that he never believes anything he reads or is told unless there is some other reason for believing it.[44]

Written records are wonderful things, but we must consider what they can and cannot tell us. Taking documents at face value led scholars to construct a pseudo-ecology of Ancient Greece, one strand in Ruined Landscape theory.

To rely on written history has many disadvantages, such as that it cuts us off from knowing what was going on at times when nobody was writing. If the Alpujarra (south Spain) has no records before the Arab period, this does not prove that nothing happened there.

Documents are haphazard in their occurrence: in Graeco-Roman times they have much to say about Athens and Rome, little about Spain or north Greece. This is due to accidents of survival, not because unrecorded areas were unimportant. Archaeology leaves no doubt that the late Roman period was the highest point of Cretan civilization after the Minoan, yet all the writings about Crete at that time could be got on to a postcard.

Ancient Greek authors tell us comparatively little about what Greece looked like; they assumed that their readers would know. Poets, philosophers and dramatists were more interested in getting the metre, philosophy and dramatic conventions right than in handing on accurate incidental details about landscape.[45] The Romans, to judge by surviving writings, were more ecologically minded: for example, Columella wrote about coppice-woods and deer-parks, and Siculus Flaccus and Palladius Rutilius about hedges.

Landscape history is best arrived at from the records of identifiable sites, which can be traced down the centuries in the archives and compared with what is there now. We shall do our best to find a Greek equivalent of Wicken Fen, England, or an Italian equivalent of Harvard Forest, Massachusetts, but we often have to make do with contemporary generalizations and other third-rate sources. Generalizations may or may not have been right. The history of the landscape must not be confused with the history of the things that people have *said* about the landscape.

From the ancient Greeks and Romans, unfortunately, what survives is seldom workaday accounts of actual events ('in year A we cut B acres of the wood-lot called C and sold it to D for E pieces of silver'). Occasionally inscriptions help out the generalized evidence in literary sources. The ancients committed surprisingly mundane regulations and transactions to tablets of stone.

Written evidence has to be verified. What was the status of the story – was it meant as fact, fiction, myth or proverb? Was the author in a position to know what he was writing about? We are suspicious of authors who wrote about places that they had not been to, or times long past. Some, such as the Roman pseudo-zoologist Aelian, mixed fact and fiction and seemed not to care which was which. Crete was a legendary, little-visited land about which even Plato felt at liberty to write nonsense. Everyone 'knew' that Crete has no snakes and no owls, and that any introduced to the island died.

The meaning of words is often not obvious. American authors think they know the difference between 'forest' and 'scrub', but modern Greeks often do not draw that distinction; did ancient Greeks do so, or Spanish Arabs? A landscape which an Arab describes as 'wooded' may not seem wooded to a Finn; what meaning are we to extract from such descriptions in ancient Greek or Roman authors? The Spanish language, like modern Greek, has no certain word for 'forest'. The word *monte* originally meant 'mountain'. Medieval and early-modern Spanish writers use *monte* for almost all rough-land, from forest, savanna and maquis to cistus and esparto-grassland; they seem not to care which is meant.[46] Today, although Spanish foresters usually call forest *monte*, the word has still not lost its other uses, including the original 'mountain'. We shall use words like 'live-oakery' and 'pinery' to mean an area of trees, without committing ourselves to whether it was forest, savanna or maquis.

It is essential to read the original documents and not to trust others' translations and interpretations. As experience in England shows, errors in interpretation are not neutral: they tend to play up the landscape of the past and to play down the landscape of the present. Early references to trees are quoted as implying forests, and re-quoted as 'magnificent forests'; present forests are dismissed as 'scrub'. Translations in this book are ours unless otherwise stated.

A sure route to pseudo-history lies in ignoring the behaviour of plants and animals. Historians gather ancient allusions to people cutting down trees, and assume that these add up to a record of deforestation,[47] as if depleting a forest by cutting down trees were the same as destroying it. In reality, deforestation is felling not balanced by regrowth, and both need to be considered. One does not pronounce a firm to be insolvent without investigating its income.

43 E. Gibbon, *The Decline and Fall of the Roman Empire*, 1776, chapter XVI.

44 F. H. Hinsley, *Intelligence in the Second World War*, Eagle, 1993, p. 40.

45 J. Roy, 'The countryside in classical Greek drama, and isolated farms in dramatic landscapes', Shipley & Salmon 1996, pp. 98–118.

46 Modern Japanese *yama* and Byzantine Greek ὄρος (A. Dunn) mean both 'mountain' and 'forest'.

47 J. D. Hughes, 'How the ancients viewed deforestation', *JFA* 10, 1983, pp. 437–45.

Official statistics

> The pretended exactness of statistical tables is generally little better than an imposture; and those founded not on direct observation by competent observers, but on the report of persons who have no particular interest in knowing, but often have a motive for distorting, the truth – such as census returns – are commonly to be regarded as but vague guesses at the actual fact.[48]

Mediterranean countries have statistical institutes which gather masses and masses of data. If one wants the area of forest in each of the 368 townships in Sardinia (to the nearest hundred square metres), or the number of rabbits in Greece (to the nearest rabbit), or the number of farms in Basilicata of between 3.00 and 4.99 hectares that employ non-family labour, libraries of tables are at one's disposal. How much use are they?

In most countries the ultimate unit is the township (civil parish, 'municipality', 'village community', *comunità*, *commune*, κοινότις, etc.) This is an area of between 2 and 100 sq. km, typically containing one big village, two or three small villages, or a number of hamlets. Every hectare of land normally belongs to one and only one township. Most statistics begin with observations at township level. They are copied, tabulated and summarized by provincial, regional and national institutions, ending, maybe, with the Worldwide Fund for Nature or the United Nations Food and Agriculture Organization. In the process they gain in authority and acceptance, but not in accuracy: however august the sponsoring body, they remain the responsibility of the countryfolk who originally filled in the forms.

A first difficulty is that most Mediterranean countries, and many regions, are not wholly Mediterranean. National statistics for France are of little use in investigating the small part of France that is Mediterranean.

Secondly, statistics hide problems of definition. In England, forests have nice sharp edges, and it is usually easy to decide whether a particular hectare is forest or not. In Mediterranean countries, forest grades into non-forest via maquis (where trees are reduced to the form of shrubs) or savanna (land with scattered trees). A statement of forest area is worthless unless accompanied by definitions of the points at which trees are big enough, and close-set enough, to count as forest. The differences are not trivial: statistics for Crete of the extent of δάσος, usually translated 'forest', range from 4½ per cent of the island in 1981 to 33 per cent of the island (from another official source) in 1992. (Trees did increase between those years, but nowhere near sevenfold.) Spanish forestry statistics are equally useless.[49] Without comparable definitions, official statistics cannot prove that Crete is more, or less, forested than Majorca. If 'forests' covered 24.76 per cent of Ruritania in 1901 but only 15.48 per cent in 1981, we do not know, without further investigation, whether forest has decreased or the definition of forest has become more restrictive. To take another illustration, of what value are

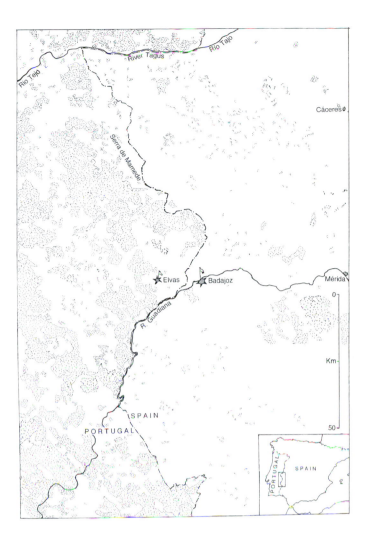

Fig. 1.5. Distribution of 'forest' on the border between Spain and Portugal, after a recent topographical map. The principal vegetation on both sides is savanna, trees widely spaced in grassland (Chapter 12). The Portuguese cartographer considered *montado* to be forest; his Spanish colleague considered *dehesa* to be non-forest. In reality they are identical, and continue uninterrupted across the border. With this example before us, can we believe statistics of 'forest' area?

counts of tractors in a land in which every farm has a grave-yard of tractors in varying states of decrepitude?

Thirdly, statistics can be tendentious. Livestock numbers are doubtful in countries where shepherds regard the number of their animals as confidential and will not reveal it to officials or anyone they suspect of being in league with officials. Sometimes they over-declare their animals to collect subsidies; sometimes they under-declare them because of a long-engrained habit of avoiding taxes. Many studies report far more livestock – which are not starving – than the land is supposed to be capable of carrying. Bureaucracy has devised a certificate system to prevent misdeclaration; but in a wicked world sheep are stolen and goats fall over cliffs, and the wise shepherd keeps a few extra certificates in a dry rockhole.

---

[48] George Perkins Marsh, *Man and Nature*, 1864.   [49] Gutiérrez and others 1985.

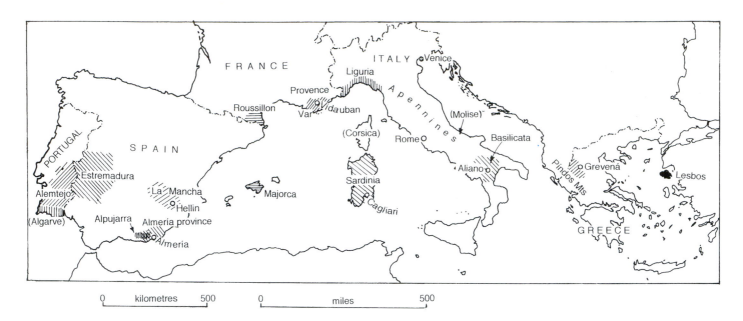

Fig. 1.6a, b. Areas that we discuss in detail. Names in brackets indicate where we do not know the region well ourselves, but draw on the work of others.

Human statistics are not exempt. The United States population was reported to have increased by 11.5 per cent between 1970 and 1980; nearly one-quarter of this increase was due not to more people but to more efficient counting.[50] The census of Britain in 1991 is supposed to have under-counted by 5 per cent because the government had tried to impose a poll tax and people refused to be counted. In countries like Greece, people have second and third homes. Cretans who migrate to Herákleion, or Athens, or even (in one instance known to us) to California have not necessarily given up being mountain farmers. Census figures depend on where people choose to be; many return, like Mary and Joseph, to the ancestral home. Ships and planes are busy around census day. The 'hard facts' of rural depopulation involve guesses about where people really live.

Economists move in a North European world of neatly defined jobs, animals and residences, often at variance with Mediterranean realities. A friend of ours, a typical Cretan, is a farmer, fisherman, restaurateur and hotelier. Is he 'engaged in the agricultural sector?' Do the fish he catches for his guests, or the help he gets from his mother-in-law, count towards the Gross National Product of Greece? How many of his 'jobs' would he have to lose to count as unemployed? If Greek statisticians have solved these problems, can we be sure that they have solved them in the same way as their Italian colleagues or their Greek predecessors in 1960? Statistics from government, and even the Food and Agriculture Organization, are anthropological data and need to be verified. As Paul Halstead remarked at a conference:

Two or three times . . . the argument has been advanced that statistical records can be used to disprove epigraphic

information or travellers' accounts. I think we need to treat statistical records as epigraphic accounts too; they may have their own agenda. This has been very well illustrated in the last ten years as various rural products have gone from being taxed to being subsidized by [the European Union]. The number of olive trees or sheep declared by various places has increased vastly and by margins which exceed the reproductive capacity of olives or sheep.[51]

The importance of fieldwork

Verbis ne temere credes. Non auribus, sed oculis aspice,
Nam aures mendacii, oculos veritatis ostia iure dictitant.

Do not heedlessly believe words. Observe with eyes, not ears; / For they rightly declare that ears are the entrances of falsehood, and eyes of truth.[52]

50 'Overestimated undercount', *SA* 244 (5), 1981, p. 70.
51 Wells 1992, p. 132.
52 Francesco Basilicata, *Cretæ Regnum [The Realm of Crete]*, 1618.

All sources of evidence must be used. It is no use writing a history based entirely on written documents such as Sonnini and Barthélemy had before them, as many authors still do, disregarding all discoveries in other fields in the last two hundred years. Nor is it any use analysing official statistics without verifying them. An argument based on written sources becomes much stronger if corroborated from some independent direction.

There is no substitute for field observation. Even in citing the results of someone's experiment, the site has to be given a context; published reports rarely contain the particulars. The observant traveller, not bound by a preconceived agenda, is forever noting details that other investigators have missed: traces of past fires and deluges, ancient trees and the degree to which their bases have been buried by build-up of soil, the state of vegetation in relation to browsing, etc.

Although landscape itself is objective, different observers notice different things and have different expectations. A non-botanist may overlook the crucial distinction between fire-promoting trees such as pine and fire-hating trees such as fir, or between trees that have or have not survived being cut down. Areas that McNeill describes as deforested we, a few years later, have sometimes remarked as well-forested; this we interpret, not as a real change, but as illustrating how an American has a more exacting standard of what he is prepared to call forest than Europeans.

We make no apology for introducing our own field observations and distinguishing them from those of others (the word 'we' means that one or both of us has been present). The interest of Oliver Rackham in Mediterranean desertification began with an invitation from P. M. Warren, the Cretologist, to the excavation of a Bronze Age site at Myrtos, in arid SE Crete, in 1968. Both of us have been involved in many later studies in Crete, some archaeological, some otherwise, often in collaboration with Dr Jennifer Moody from Texas.

From Crete we have gone out into other countries. We cover most types of Euro-Mediterranean landscape and habitat except urban areas, but do not claim to be either even or random in our choice. In paying attention to Sardinia we do not disparage the glorious and utterly different island of Corsica. To include everything would be a lifetime's work, and would result in a shelf of books.

We have thought it best to select areas small enough to study at first hand (Fig. 1.6). The Mediterranean is more complex than (say) the United States, and one of our criticisms is that other scholars have tried to over-generalize and tend to copy what others have written. We have usually chosen places where colleagues have been working, rather than investigating unknown terrain. In a study of desertification, we have paid more attention to dry regions than to well-watered areas (such as the interior of Tuscany, studied in loving detail by Vos & Stortelder[53]) where the risk of desertification is remote. We have paid special attention to dramatic and special features of the dry Mediterranean, such as badlands, savannas and fiery terrain.

---

53 W. Vos & A. H. F. Stortelder, *Vanishing Tuscan Landscapes: landscape ecology of a submediterranean-montane area (Solano Basin, Tuscany, Italy)*, 2nd edn, PUDOC, Wageningen, 1992.

ON DATES

Radiocarbon dating is based on the fact that a little of the carbon dioxide in the atmosphere contains the radioactive isotope C14 instead of ordinary carbon C12. After a plant has taken up carbon dioxide and has converted it into wood or peat, the C14 disappears by radioactive decay at a constant rate. The proportion of C14 still remaining in the wood or peat is a measure of how much time has elapsed since it was laid down by the plant.

Radiocarbon dates are calibrated by measuring the C14 content of past annual rings of very long-lived trees. For various reasons, the proportion of C14 in the atmosphere has fluctuated, so that the relation between the present C14 content of a sample and the date is complicated (Fig. 1.7).

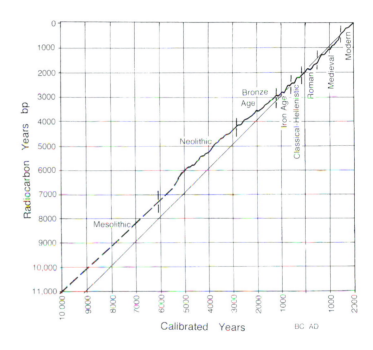

Fig. 1.7. Generalized relation (omitting small-scale details) between 'dates' bp or bc obtained by radiocarbon measurement and real dates BC or AD from tree-rings.

The various sciences on which this book draws have different conventions for expressing dates. For simplicity we shall work wherever possible in terms of calendar years. We adopt these conventions:
• Dates BC, AD or BP represent actual calendar years, derived from documents or from counting the annual rings of trees. BP means years Before Present, that is before 1950 AD. This book is published in the year 2000 AD = -50 BP.
• Dates bc, ad or bp represent calibrated radiocarbon dates, that is radiocarbon dates adjusted for fluctuations in the C14 content of the atmosphere. They are based ultimately on counting the annual rings of trees.
• Dates bp represent unadjusted radiocarbon dates ('radiocarbon years').
• Decades are referred to thus: 'the 1700s' means 1700–9 (not the same as 'the eighteenth century'), 'the 1710s' means 1710–19, etc.

Notes

1. Continuous tree-ring records have been used to construct a calibration curve for the last 9,840 years, and with less certainty back to 11,440 calendar years BP (about 10,000 years bp). Comparisons of uranium/thorium and radiocarbon dating of corals have extended the calibration curve for atmospheric as well as marine samples back to 21,950 bp, with greater uncertainty. The geomagnetic intensity record has been used as a standard to extend the calibration back to 45,000 years BP in a generalized way. An alternative, potentially exact, method, extending back to at least 38,000 BP involves counting the annual layers of lake sediments, and measuring the carbon isotopes of land shells that fell into the lake and are embedded in the layers.[54]

2. Although many scientists think in terms of 'radiocarbon years' bp, these do not represent equal intervals of time as do years BC, and as years BC are intended to do. The interval between 5350 and 5450 bp is not equal to the interval between 5550 and 5650 bp.

3. Dates bp are quoted with confidence limits ('5550 ± 120 years bp') which depend on the precision of measuring the C_{14} content of the sample. For dates BP there is an additional uncertainty deriving from the calibration. Confidence limits for BP dates are (usually) wider than for bp dates, and may be asymmetrical.

4. Where we quote calendar years AD and BC, our practice has been (where possible) to take the original dates bp with their quoted laboratory errors and re-calibrate them by means of the CALIB 3 User's Guide, based on the Stuiver–Reimer calibration. We show the 1σ (1 s.d., one standard deviation) range or ranges.

CHRONOLOGICAL PERIODS

Archaeologists use terms like Neolithic or Iron Age both to indicate a particular degree of human technology and to give an indication of date. For so big an area as the Mediterranean this is difficult. Not only were technologies taken up at different dates, but scholars in different countries have different lists of periods and define them in different ways. Since we cannot avoid the usage, we summarize the information in Fig. 1.8. Further terms are:

Pleistocene: the last two million years, containing the series of powerful climatic fluctuations and glaciations known as the Ice Ages.
Holocene: the last 11,500 years, approximately the interval since the end of the the Younger Dryas, a cold episode that terminated the Last Glaciation.[55]
Quaternary: Pleistocene + Holocene.

Fig. 1.8. Archaeological periods as commonly used in various countries. Dates are approximate: this should not be treated as a definitive reconcilement of the often conflicting data and usages.

[54] M. Stuiver & P. J. Reimer, 'Extended 14C database and revised CALIB radiocarbon calibration program', *Radiocarbon* 35, 1993, pp. 215–30; Van Andel 1998; C. Lal, A. Mazaud & J. C. Duplessy, 'Geomagnetic intensity and 14C intensity in the atmosphere and ocean during the past 50kyr', *Geophysical Research Letters* 23, 1996, pp. 2045–8; H. Kitagawa & J. van der Plicht, 'Atmospheric radiocarbon calibration to 45,000 yr B.p.; Late glacial fluctuations and cosmogenic isotope production', *Science* 279, 1998, pp. 1187–90.
[55] N. Roberts, *The Holocene: an Environmental History*, 2nd edn, Blackwell, Oxford, 1998.

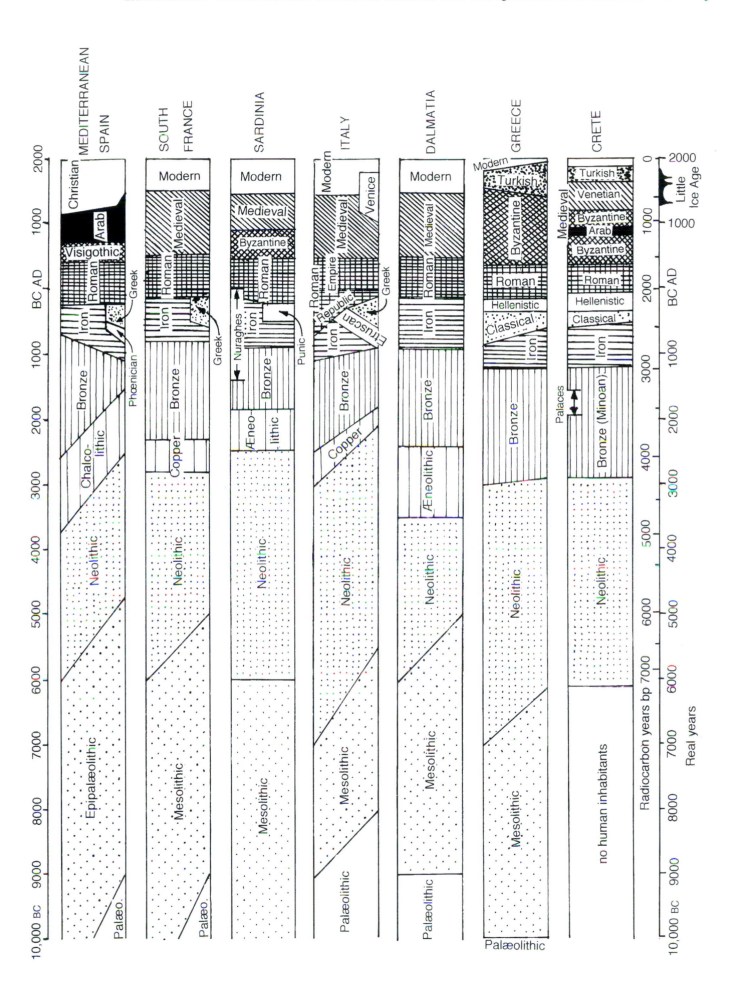

RAINFALL and TEMPERATURE

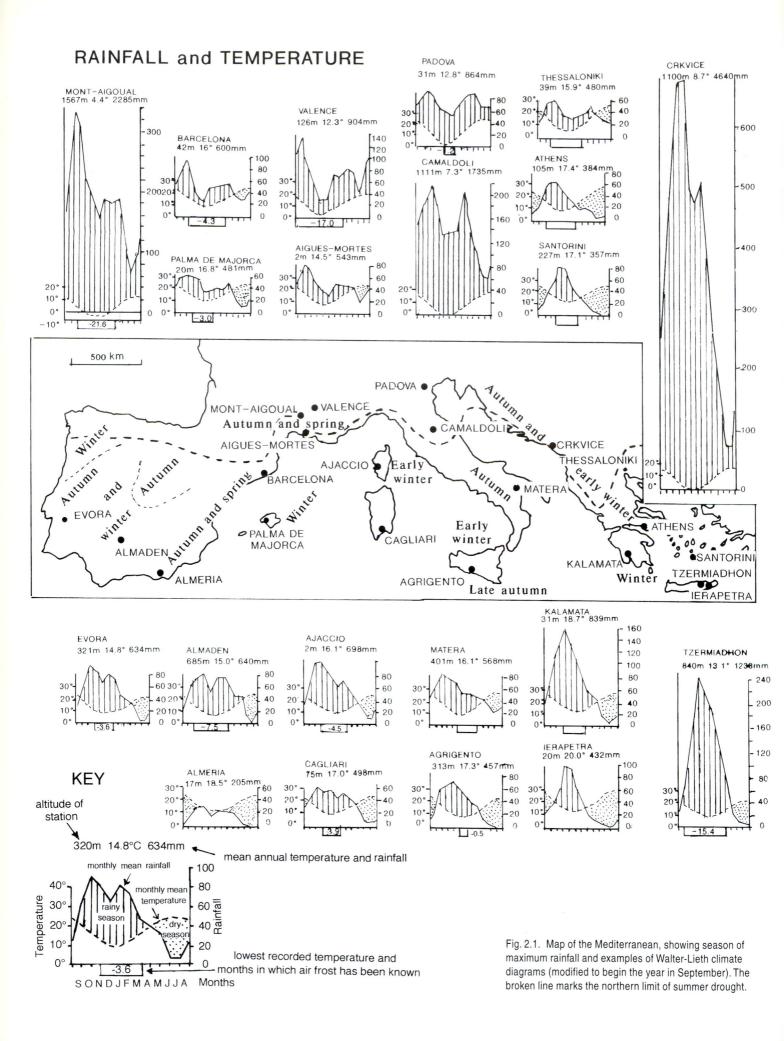

Fig. 2.1. Map of the Mediterranean, showing season of maximum rainfall and examples of Walter-Lieth climate diagrams (modified to begin the year in September). The broken line marks the northern limit of summer drought.

CHAPTER 2

Present Climate and Weather

We cowered watching how primaeval earth was carved. When the veil of rain was
withdrawn at last, I saw the whole face of the landscape changed. The old estuary of
the river existed no more; and a broad and shallow mouth had been opened far to the
north. The bay, which since Spratt's visit in the 'fifties had afforded deep anchorage
close inshore, now shoaled gradually for a mile, and was studded with the toppling
crests of grounded trees . . . Over two-thirds of the plain, where fertile fields and
olive gardens had been, lay sand and stones; and such trees as had held their ground
were buried to mid-trunk. Looking up the river gorge, I saw nothing but naked rock,
where terraced vineyards had clothed the cliff face; while all the ancient tangle of
forest below had vanished to the last shrub, and the sinuous valley-floor, as far as the
eye could follow it, glistened clean as a city pavement after rain.[1]

The Mediterranean year begins in September,[2] when clouds
begin to gather and the sun no longer burns so relentlessly.
The first rains, coming one to ten weeks later, awaken plant
growth for the winter. Frost and snow seldom occur (snow,
however, is abundant in the mountains). Soil moisture typically reaches a maximum in spring; then the rains tail off, the
sun mounts and evaporation increases. Many plants settle
down to their summer sleep, but some – notably the vine – are
less well adapted to the climate and keep going all through the
dry season, their deep roots extracting water from the
bedrock.

However, the Mediterranean eschews dull uniformity
(Fig. 2.1). It is the region of lazy days in the crushing sun of a
summer beach; of the genial warmth of a southern winter; of
people dropping dead from heat in the streets of Athens, or
freezing in the Sierra Nevada, or drowning in an arid Cretan
gorge; of the clammy mists from which the volcano-island
Santoríni draws its moisture; of Crete cut off for days, or
Gávdhos for weeks, by stormy seas; of *diez meses de invierno,
dos meses de infierno* ('ten months of winter, two months of
Hell') on the treeless *meseta* of middle Spain; of the terrible
winter *mistral* blasting through the streets of Arles, or its
sister the *bora* through those of Trieste; of the terrible
summer *sirocco* blasting a fiery south-facing coast; of mules
struggling through the snow in the high passes of Crete in
July, or gondolas crunching through the ice of Venice in
January; of ice-loaded trees plucked off Apennine mountain-
sides in the freezing rain; of the driest place in Europe
(Almería, with 190–200 mm of annual rainfall[3]) and the
wettest place in Europe (Crkvice, 4,640 mm, near the
Montenegro coast).

Walter & Lieth recognise twenty-four different climates in
southern Europe, depending on the duration and tempera-

ture of the dry season (Fig. 2.2), the amount of rainfall in the
wet season and the duration of winter cold if any.[4] At one
extreme, in Almería, Athens, or SE Crete, there is a long, hot,
evaporative dry season and low, variable winter rainfall,
forming a transition to the sub-Saharan climates of the south
Mediterranean. At the other extreme, at the south edge of the
Alps, the dry season is short, rainfall high and winters cold,
forming a transition to the climate of central Europe.

Mountains produce very abrupt local variations. In Crete
there is an immense contrast between the misty, well-
vegetated rain-excess areas on the north sides of high mountains and the arid rain-shadows a few kilometres away on the
south sides (Figs. 1.3, 1.4); there is also a general trend from
the wet west to the dry east of the island.[5] Within Albacete
province, not one of the most mountainous in Spain, eight of
Walter and Lieth's climates have been recognised in an area
no bigger than Crete.[6]

The Mediterranean and the world's weather systems

The Mediterranean lies at the junction of four weather
systems. In summer it comes between the subtropical Azores
High and the Indo-Persian Low. The air in the high-pressure
zone eastwards from the Azores is subsiding and stable. Very
rarely moist air penetrates the basin from the NW or SE to
give a summer storm with tremendous rain and hail.

In autumn, as the zone of high pressure dissipates, the
wave pattern in the upper westerlies changes to form the
North Atlantic Low and the high-pressure belt over north
Africa and south-west Asia. When high-pressure systems
occur over northern Europe they block the westerly
airstream, pushing cold air far to the south. Cool, moist,
unstable air penetrates into the Mediterranean basin to

---

[1] D. G. Hogarth on the Káto Zákro deluge of
15 May 1901, east Crete; from *Accidents of an
Antiquary's Life*, Macmillan, London, 1910,
p. 83.
[2] New Year's Day to most hydrologists is
1 September, as in the Byzantine calendar. The
rainfall for 1997 is counted as the total from
1 September 1996 to 31 August 1997.

[3] Very low rainfalls are so variable that the
average is rarely experienced.
[4] Walter & Lieth 1960– . For a more elaborate
classification of Mediterranean climates in relation to vegetation see: H. N. Le Houérou, 'Vegetation and land-use in the Mediterranean basin by
the year 2050: a prospective study', Jeftic and
others 1992, pp. 175–232. (Note, however, that

the finer details of climatic classification depend
on the particular run of years' data used.)
[5] Rackham & Moody 1996, chapter 4.
[6] J. J. Capel Molina, 'Subregiones fitoclimáticas
(clasificación de Walther y Lieth) en el sudeste de
la Meseta: Provincia de Albacete', *A-B* 23, 1988,
pp. 171–88.

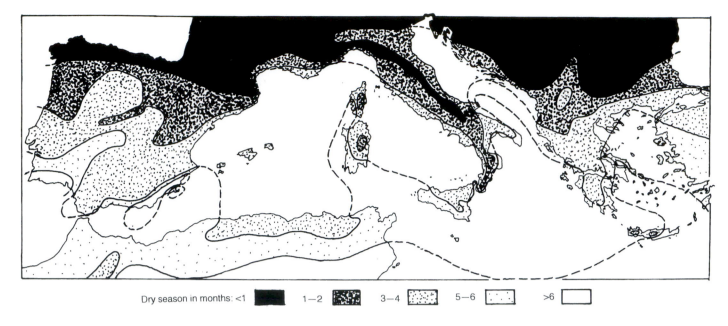

Dry season in months: <1 ■ 1–2 ▦ 3–4 ▦ 5–6 ▦ >6 ☐

Fig. 2.2. Length of the dry season.

generate eastward-moving cyclonic depressions, from which much of the rainfall comes.[7] About half-a-dozen storms a year are Atlantic depressions diverted southwards into the Mediterranean, but most of the seventy or so Mediterranean depressions are generated locally in winter and spring. They move along the northern part of the Sea for a few days before escaping between the main mountain ranges into central Europe, the Black Sea or the Caspian region.[8] Celebrated areas for brewing storms are the Gulf of Genoa and the northern Adriatic in the lee of the Alps, the Gulf of Lions in the lee of the Pyrenees, and the seas west of Crete and near Cyprus (Fig. 2.3).

Rainfall

Much of the rainfall is derived from heavy storms. Where mountains force moist winds to rise, variation over short distances can be very abrupt, with rain-excesses on the windward sides of mountain ranges and rain-shadows on the lee sides. This accounts for the dryness of coastal Almería and Attica, the extreme wetness of the Dinaric Alps in Montenegro and the diversity of Crete.

Precipitation (rain plus the equivalent in snow) increases with altitude at widely differing rates.[9] In Crete the rate is about 0.6 mm for every metre of ascent, at least up to 1,000 metres. On the Sierra de Gádor, west of Almería, the rainfall gradient is estimated at 0.26 mm per metre on the southern slope and 0.31 mm per metre on the northern slope.[10] The Pic du Midi, one of the high summits of the Pyrenees (2,862 m), has about 500 mm more precipitation than the plains to the north in the winter and spring half-year, but the difference in autumn is slight.[11] On and around mountains, mean annual rainfall even at the same altitude varies greatly over short distances with exposure to, or shelter from, rain-bearing winds. Values for individual years, months and days can be even more discordant. Gibraltar, a high isolated rock, illustrates this. Rainfall records for two stations 3.7 km apart, Windmill Hill (0.9 km from the southern tip, altitude c. 116 m) and North Front (near sea-level by the airstrip on the north side), overlap from 1945–8. The mean for Windmill Hill for those three years was 91.4 per cent of the mean at North Front, and for winter 1945–6 it was only 77 per cent.

The season of maximum rainfall varies (Fig. 2.1). In the far south of Europe, with a long dry season, rainfall builds up gradually to a maximum in December or even January. In the north, after a short dry season, rainfall reaches a maximum in October; often there is a second peak in spring, recalling the 'former rain and the latter rain' of the Bible;[12] sometimes the spring rain is the greater. But the rules can change. Portugal used to have a 'latter rain' in March, which disappeared after 1967, causing much distress (Fig. 7.9). In 'deep' Mediterranean countries rain in the dry season is insignificant, although most stations with a long enough record have known it.

[7] T. M. L. Wigley, 'Future climate of the Mediterranean Basin with particular emphasis on changes in precipitation', Jeftic and others 1992, pp. 15–44.
[8] P. Alpert, B. U. Neeman, Y. Shay-el, 'Climatological analysis of Mediterranean cyclones using ECMWF data', Tellus 42A, 1990, pp. 65–77.

[9] In many regions the total precipitation is not well known because of the lack of measuring stations at high altitudes where most of the rain and snow fall.
[10] A. Vallejos and four others, 'Contribution of environmental isotopes to the understanding of complex hydrologic systems . . . Sierra de Gador, SE Spain', ESPL 22, 1997, pp. 1157–68 [quoting

unfinished thesis by W. Martin Rosales, 1996].
[11] J. Dessens & A. Bücher, 'A critical examination of the precipitation records at the Pic du Midi observatory, Pyrenees, France', CC 36, 1997, pp. 345–53.
[12] Deuteronomy 11:14, etc.; in Joel 2:23 the latter rain is placed in the 'first month', late March and early April. See p. 142.

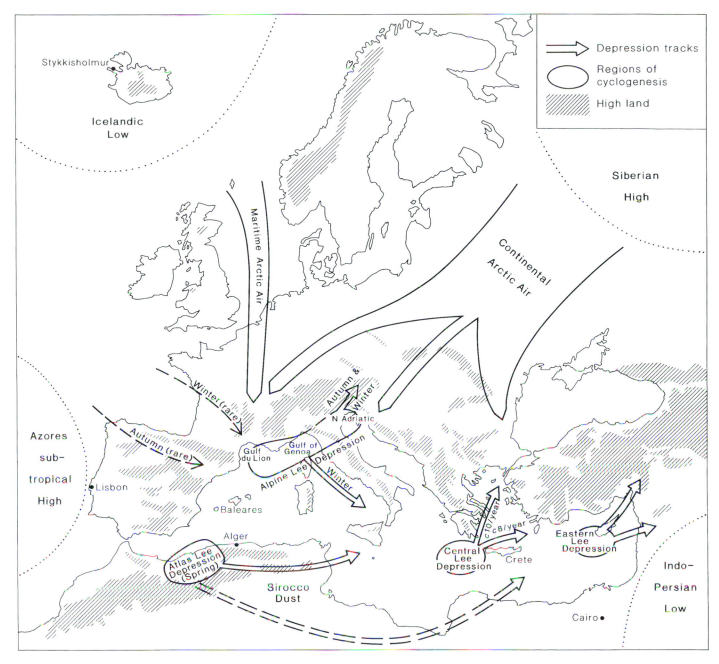

Fig. 2.3. Europe in winter: pressure systems, northerly airstreams, regions of cyclogenesis and principal cyclone tracks. After T. M. L. Wigley.

At any one place rainfall varies widely from year to year;[13] it is not easy to separate changes in climate from ordinary fluctuations of weather. At sites with long records, the driest season on record has typically about one-fifth the rainfall of the wettest season.[14] In Crete the dry season normally runs from early May to mid-September, but can be as long as

[13] The unfortunate practice at some stations of beginning the year in January, thus adding the late rainfall of one season to the early rainfall of the next, exaggerates the variation between years, but less than might be expected. These two examples compare the two methods.
San Fernando (near Cádiz): 179 years of records, mean annual rainfall 593 mm.

Reckoning of year	September–August	January–December
Standard deviation, mm	185	195
Total for driest year, mm	256 (in 1841–2)	233 (in 1994)
Total for wettest year, mm	1229 (in 1871–2)	1262 (in 1855)

Gibraltar 'West Flank': 145 years of records, mean annual rainfall 816 mm.

Reckoning of year	September–August	January–December
Standard deviation, mm	264	272
Total for driest year, mm	283 (in 1984–5)	356 (in 1981)
Total for wettest year, mm	1935 (in 1855–6)	1992 (in 1855)

[14] The lowest year's rainfall recorded in Europe was probably 90 mm in the Guadalentín (SE Spain) in 1993–4.

January to October (1937) or as short as 6 July to 16 August (1995).

Within the twentieth century, wet and dry years have occasionally affected large parts of the Mediterranean at a time, but generally the long-distance correlation is slight. We can detect only a poor year-to-year correlation between Greece and Crete,[15] and none between Greece and Spain. Even within Crete wet and dry seasons are correlated only between some stations (not necessarily those closest together).

San Fernando and Gibraltar are 100 km apart, and have 123 years of overlapping records. The correlation between them is significant but not strong ($r = 0.53$). Of the ten driest years at San Fernando, five were among the ten driest years at Gibraltar; of the ten wettest years at San Fernando, four were among the ten wettest years at Gibraltar. However, the annual rainfall record of San Fernando for the calendar years 1941–85 is quite highly correlated with the average rainfall for the whole of Spain (Fig. 2.4). The coherence of these two sets of values is probably related to the dependence of San Fernando and of much of the rest of Spain on rain brought by strong south-westerly airflows. The Rock of Gibraltar and its surrounding mountains are a disturbing factor absent from San Fernando.

Local variation has tended to encourage trade: anyone whose crops failed could probably get a replacement within 200 km. However, as we shall see in Chapter 8, there is evidence for more synchronous extremes in the historic past.

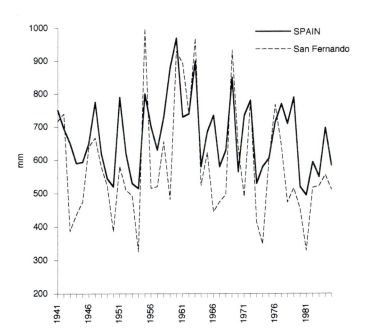

Fig. 2.4. Annual mean precipitation for all Spain compared with the annual rainfall at San Fernando. The correlation is unexpectedly good ($r = 0.71$). Data from Instituto de Meteorología, Madrid; the year begins in January.

Temperature

Temperature decreases with altitude by the usual rate of 0.6°C per 100 m, sometimes more. The highland growing season is curtailed by winter cold as well as summer drought, and may be reduced to little more than two months in spring, as in much of inland north Greece and inland Spain. The boundary between lowland and highland climates is often thought of as marked by the limit of cultivation of the olive, which ranges from 300 m in north Italy to 1,300 m in the Alpujarra; but this depends on risk-taking by the owners as well as the properties of the tree (p. 133). At higher altitudes still, rain is rare; most of the precipitation comes as snow.

About three times in two years a heat-wave occurs: the temperature may exceed the long-term mean by 5°C for a few weeks, or by 7–15°C for a few days. This happens especially in July and August, when the Subtropical Jet Stream is displaced northwards and brings Saharan weather (and fires) to southern Europe.[16]

Sunshine

In the Aegean sunshine is typically over 400 hours per month in June and July, and over 100 hours per month in December and January, about twice the values for London. This gives greater opportunities for generating solar power and for very high crop yields (water permitting) than anywhere else in the world outside tropical deserts.

Frost and snow

Frost limits plant growth. The olive is killed to the ground by a temperature of around -13°C; it sprouts and recovers in a multi-stemmed form as a witness of the event. Tropical fruits such as citrus, banana and kiwi are sensitive even to ordinary frosts. Many native Mediterranean plants, transplanted to Cambridge, do not survive an average winter.

Sea surface temperatures are well above 0°C in the Mediterranean winter, and so frost is rare (but important) near the coast. On clear nights with strong radiation cooling, pools of cold air accumulate in depressions ('frost-hollows'). In March 1980 olives in karst depressions in Bœotia (middle Greece) were killed up to a sharply defined upper limit: a 'tide-mark' of dead foliage sometimes extended halfway up individual trees. Olives in the south of France were devastated by the notorious frost of 1709.

Frost-hollows can have dramatic effects. On 4 May 1990 the oakwoods around Dheskáti (S. Macedonia) were in full leaf at 860 m altitude. On the mountain above, the season came later until the woods were leafless at the oak limit of 1,350 m. In the hollows *below* Dheskáti, the season also became less advanced; in the floor of a big depression at 700 m nightingales were singing and coppice-plants blooming in leafless woods.

[15] Except for the period 1963–85, when Athens and Herákleion had a correlation coefficient (r) of 0.69.

[16] M. Conte, S. Sorani, S. Velletri, 'Heat waves in the central Mediterranean', MEDALUS Working Paper 26, 1994.

Fig. 2.5. Orange-trees damaged by frost between Argos and Návplion (Peloponnese). *March 1998.*

Very sensitive crops such as oranges are confined to the least frosty locations, and may not escape air frost even there (Fig. 2.5). Growers use smoke, fans to stir the air, and fine sprays of water to supply latent heat to reduce the frost hazard. Wholly frost-free spots exist mainly as microclimatic sites rather than whole landscapes. In Crete we have seen oranges growing happily at 450 m on the south-facing slopes of the Lassíthi Mountains. The most frosty place in the inhabited zone is in the low, mountain-girt plain of the eastern Mesará, which also has the highest summer temperatures in the island.

Snow rarely lies for more than a day or two at low altitudes. It can damage trees by breaking boughs, as with olives and cypresses at altitudes down to 300 m in Crete in 1992, the cold winter after the eruption of Mt Pinatubo in the Philippines.

Dust

In spring it rains mud in Crete. Late in the rainy season, when the air becomes unusually murky during a *sirocco* (hot SE wind), a *kokkinovrokhí* occurs, a red rain which stains newly whitewashed houses and turns snowy mountain-tops pink. We set up dust-traps on house roofs and at Herákleion airport in 1988 and 1989 to measure the amount falling on Crete. The results showed a dust-fall of 0.1–1 ton per hectare per year, suggesting that Crete is gaining dust at about 1 mm a century.[17]

Depressions in the lee of the Atlas mountains bring hot dry winds to southern Europe, bearing Saharan dust stirred up by sand-storms. In 1901, 1–4 grams per square metre were spread as far as Germany. In July 1983, 6 grams per square metre of NW African dust were measured at Lannemezan, at the foot of the Pyrenees. In March 1991 an Atlas depression, with winds up to 110 km per hour, dumped tens of thousands of tons of dust over Europe. Thick yellow dust settled over Rome; dust on the Alps contributed to the melting that exposed 'Oetzi', a Bronze Age man buried in ice at over 3,000 m; in far-away Scandinavia people noticed the dust-fall.[18] Autumn and winter dust-falls are rare, but in November 1996, winds round a depression over Portugal transported dust from west Algeria to Genoa, 2,000 km away, where rain deposited 4 grams of mud per square metre.[19] We have experienced red rains in Sardinia and also in middle Texas and in Dubbo, New South Wales, with winds off their respective deserts.

Dustfall has increased in recent decades; there is evidence linking this to the North Atlantic Oscillation (Chapter 7).[20]

Dust from various sources has probably been a feature of Mediterranean countries at least throughout the Pleistocene. It can be a major silt constituent of soils,[21] and may help to explain why Crete is such an unexpectedly fertile island. However, the wind that brought the dust can take it away by wind erosion (p. 244), especially when the soil is disturbed by ploughing.

[17] Z. Antonojiannakis, 'Dust deposition around Crete', *Petromarula* 1, 1992, pp. 70–3; K. Pye, 'Aeolian dust transport and deposition over Crete and adjacent parts of the Mediterranean Sea', *ESPL* 17, 1992, pp. 271–88.

[18] L. G. Frazen and others, 'The Saharan dust episode of south and central Europe and northern Scandinavia, 8–10 March 1991', *Weather* 50, 1995, pp. 313–8.

[19] P. Ozer, M. Erpicum, G. C. Cortemiglia, G. Lucchetti, 'A dustfall in November 1996 in Genoa, Italy', *Weather* 53, 1998, pp. 140–5.

[20] J. Dessens & Pham van Dinh, 'Frequent Saharan dust outbreaks north of the Pyrenees: a sign of a climate change', *Weather* 45, 1990, pp. 327–33; C. Moulin and others, 'Control of atmospheric export of dust from North Africa by the North Atlantic Oscillation', *Nature* 387, 1997,

pp. 691–4.

[21] A. Rapp & T. Nihlén, 'Dust storms and eolian deposits in North Africa and the Mediterranean', *Geoökodynamik* 7, 1986, pp. 41–62; J. Sevink & E. A. Kummer, 'Eolian dust deposits on the Giara di Gesturi basalt plateau, Sardinia', *ESPL* 9, 1984, pp. 357–64; T. Nihlén & S. Olsson, 'Influence of eolian dust on soil formation in the Aegean area', *ZG NF* 39, 1995, pp. 341–61.

A different phenomenon, which no living eye has seen in the Mediterranean, were the 'dry fogs' produced by the dust from certain types of volcanic eruption (Chapter 8).

Deluges

Floods and erosion depend largely on exceptionally intense rains, which happen when warm humid air is carried by strong up-currents to altitudes where the moisture condenses. Ice crystals (or microscopic marine organisms swept up in sea-spray) act as nuclei. This often occurs in autumn, when the sea is warm. A cyclonic vortex, set going by a cold-air intrusion from the north, draws in moist air to form the warm sector of a developing depression. Moving inland and heated from below, the warm, moist air is forced by the pressure distribution to rise over mountain ranges. It becomes unstable where it meets cool northerly air at the cold front. The result is a *deluge* (Catalan *aiguat*, Italian *alluvione*) with tens of millimetres of rain falling in an hour or two, or hundreds of millimetres in a day or two.[22] A deluge is an exciting experience, as D. G. Hogarth (see p. 25) and ourselves have found in Crete (pp. 247–50, Fig. 14.11).

Measuring deluges is not easy. Very high intensities are local, and there may not be a rain-gauge in the area, or if there is it may not be big enough to hold all the rain. (An upturned bucket may serve at a pinch.) Gauges of different design can give different readings for the same event.[23] Strong winds affect the catch.

Some localities are prone to deluges because of their topography. Southern Europe has two deluge capitals. One is Mont-Aigoual (1,565 m), the south-east bastion of the Massif Central towering over Mediterranean France.[24] When southerly winds off the Golfe du Lion, funnelled up the Rhône valley, collide with this mountain, the modest rivers draining it swell, for a day or two, to the volume of the Nile. In one year or another, it has received more than 1,000 mm in every month between September and February (Fig. 2.6). The amount of rain falling on any particular spot depends on vertical currents in individual cumulo-nimbus cloud cells several kilometres high. Horizontal gusts in heavy storms are often very intense, 60–100 km an hour, so a point on a valley floor will sometimes receive as much rain as the peak several kilometres away which created the main updraught. For instance, one September night in 1900, 950 mm fell on Valleraugue, five or six km south of Mont-Aigoual. The other deluge capital is Montenegro, where, too, moist air from the sea strikes abrupt coastal mountains.

Spain and Portugal

In Spain deluges (defined as falls of 200 mm in 24 hours, or 50 mm in one hour) are concentrated on the otherwise dry

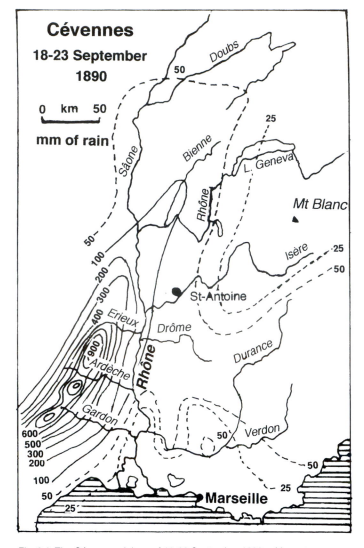

Fig. 2.6. The Cévennes deluge of 18-23 September 1890, with contours of rainfall in mm. After Pardé (1924–5).

south and east coasts, especially where mountains approach the sea (Fig. 2.7). Inland areas are less prone (the highest daily value recorded at Toledo between 1941 and 1970 was only 49 mm) and also SW Spain and Portugal.[25] Two factors are probably involved. The Spanish Mediterranean coast, though protected from depressions that sweep across the Iberian peninsula from the west, is open to 'back-door' storms from the east. The Mediterranean, moreover, is much warmer in autumn than the Atlantic.[26] It has been exceptionally warm in the 1980s, when deluges have been specially frequent in east Spain.[27]

How often can deluges be expected? For Roussillon the return period (the average interval between two such events)

[22] See T. Haiden & three others, 'A refined model of the influence of orography on the mesoscale distribution of extreme precipitation', *Hydrological Sciences Journal* 37, 1992, pp. 417–27.

[23] At Aix-en-Provence on 22 September 1993, different types of gauges a few metres apart measured 222 mm and 185 mm: Météo-France [1995], pp. 12–13.

[24] T. Muxart, C. Cosande, A. Billard, *L'érosion sur les hautes terres du Lingas*, Mémoires et Documents de Géographie, Observatoire Causses-Cévennes, CNRS, Paris, 1990.

[25] However, D. Wheeler tells us of 631 mm at Gibraltar over the four days 11–14 December 1796 (records at MOB).

[26] Martín Vide 1985; M. Millán, M. J. Estrela,

V. Caselles, 'Torrential precipitations on the Spanish east coast: the role of the Mediterranean sea surface temperature', *Atmospheric Research* 36, 1995, pp. 1–16.

[27] J. Querada Sala, [no title], *Weather* 45, 1990, pp. 278–9.

Fig. 2.7. Distribution of intense rainfall in Spain, from Instituto Nacional de Meteorología, Madrid. Right: highest recorded fall in 24 hours. Below: highest recorded fall in one hour.

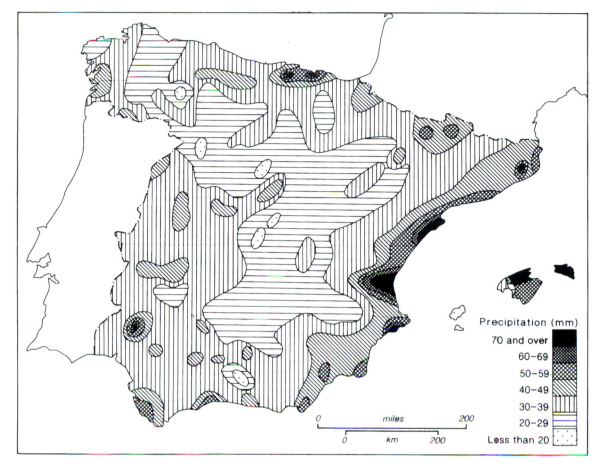

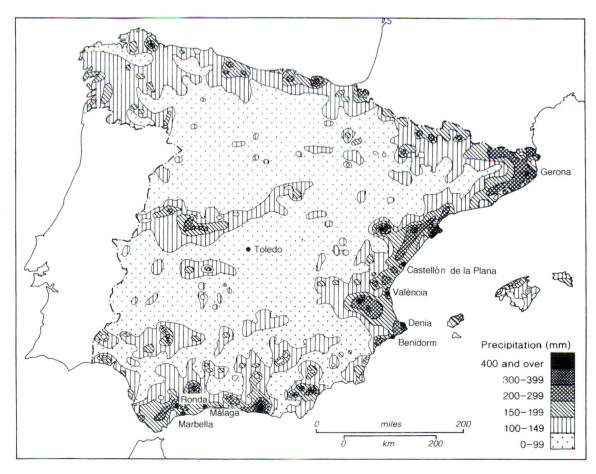

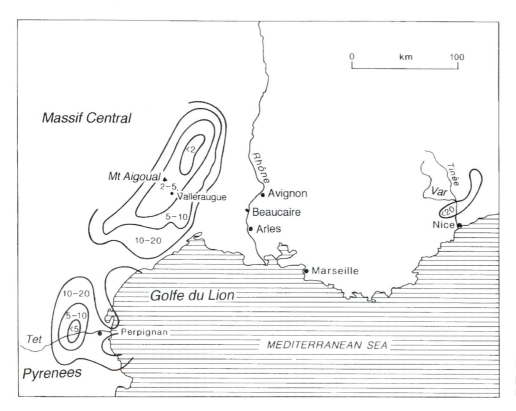

Fig. 2.8. Return period of 200 mm of precipitation in 24 hours, southern France. Based on data in Météo-France [1995].

of a storm of 300 mm in 24 hours is about ten years, and for a 400-mm storm is about fifty years. These figures, however, apply to the whole of Roussillon (4,100 sq. km); the return period at any one point would be much longer. Falls of more than 150 mm in 24 hours can be expected every ten years around Gerona, Castellón de la Plana and Valencia, on the mountains between Denia and Benidorm, and those between Màlaga, Marbella and Ronda.

South of France and NW Italy

The south of France has recently had, unusually, many damaging deluges – occasionally, as on 6–8 November 1982, extending from the Spanish Pyrenees to the Alps. Météo-France produced a collection of maps of the rainfall distribution in deluges between 1958 and 1994 in Mediterranean France and Corsica.[28] In these 37 years there were at least 34 periods of 24 hours in which more than 300 mm of rain fell somewhere in the Languedoc–Roussillon region; on average such a deluge was an annual event! But such very high rainfall intensities were mainly limited to certain mountains; 90 per cent of the Midi never experienced a 300-mm deluge.

These maps provide a means of calculating the return period of deluges of a given magnitude at individual points (Fig. 2.8). Half the total area of some 50,000 sq. km failed to register any deluges of more than 200 mm in 24 hours from 1958 to 1994. But such falls had a return period of less than two years on the 40-km stretch of the high Cévennes northeast from Mont-Aigoual.[29] In a surrounding area of about 5,000 sq. km, constituting the SE corner of the Massif Central, and another area of less than 1,000 sq. km around Mont Canigou in the easternmost Pyrenees, the return period of 200-mm storms was less than ten years. These deluge-prone areas, about one-eighth of the whole area of the south of France, occupy barriers facing winds from the southeast within 100 km of the Gulf of Lions.[30]

The mountains around the Gulf of Genoa also get violent rains. There was a cluster of deluges in the Var valley behind Nice in 1994: approximately 180 mm on 6 January, 200 mm on 26 June, over 100 mm on 22–3 September and 150 mm on 19–20 October. The Var was then caught by the SW edge of the great upper Po deluge of early November: 1,500 sq. km received more than 200 mm in 4 days.[31] Roads and river-banks were still being repaired a year later. There is, however, no sign of a long-term trend; there were six such deluges in the hydrological years 1963–4, 1970–71 and 1977–8, only one in each of the years 1960–1, 1967–8 and 1991–2.

These 1994 storms in the Midi, with falls of 200 mm over 1,000–5,000 sq. km and 300 mm over 100–1,000 sq. km, each delivered something like one-quarter to one cubic kilometre of water in 24 hours. They are comparable with a famous deluge in the eastern Pyrenees in October 1940, supposedly the most violent since 1763. It was of small extent but much more intense (Fig. 2.9), delivering 500 mm of rain to 500 sq. km. The 840 mm at Llau power station on 17 October 1940 is regarded as the highest authentic record of a 24-hour fall for Europe; a fall of 1,000 mm was claimed for St-Laurent

[28] Météo-France [1995].
[29] cf B. Benech and five others, 'La catastrophe de "Vaison-la-Romaine" et les violentes précipitations de septembre 1992: aspects météoro-logiques', *La Météorologie* 8.1, 1993, pp. 72–90.
[30] In the super-rainy Montenegro mountains, any one point experiences a 200-mm deluge almost once a year: Llasat & Rodriguez 1997.
[31] P. Carrega, 'La crue exceptionelle du Var, le 5.11.1994; précipitations et situation météoro-logique', *NSMS* 6–7, 1994, pp. 68–73.

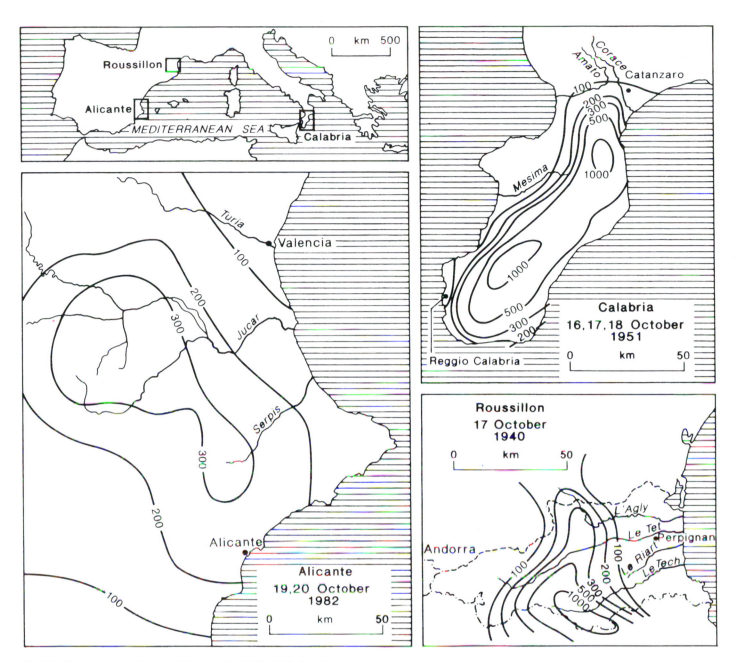

Fig. 2.9. Three great mid-October deluges, with rainfall distributions in mm. After Miro-Granada y Gelabert 1974, Mancini 1986, Pardé 1941.

Fig. 2.10. Sea-level pressure distribution and frontal systems at 1200h on 4 November 1994, after Lionetti.

de Cerdans the same day.[32] The half cubic kilometre of water thus released descended mainly on the upper catchment of the Tech, which flowed at 3,400 cubic metres per second out of the upper 236 sq. km of its catchment at Pas du Loup. 350 people were killed in France and Spain. (For the erosional effects see p. 250–1).

Deluges of more than 200 mm in 24 hours are liable to occur in southern France and Liguria between September

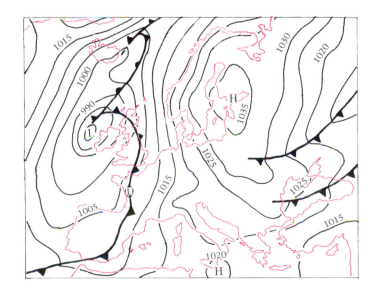

---

[32] Far away in Sardinia, the same day, a discharge of 2,320 cubic metres per second was measured on the Flumendosa at Gadoni, from a catchment area of only 423 sq. km.

	Date	Amount of rain, mm (in 24 hrs unless otherwise stated)	Mean annual rainfall, mm (where known)	Comments
Spain				
Gibraltar	11–14 Dec. 1796	631 in 4 days[1]	c. 800	
Gibraltar	30 Jan. 1959	293	811	
Rambla Honda, Tabernas (p. 315)	28–9 Sep. 1980	150 in c. 5 hours	c. 200	'25–100 years return?'; river last flowed so far in 1964[2]
Velez-Rubio, Almería	16–19 Oct. 1973	300 in 12 hours	c. 200	killed 50 people[3]
Bicorp, 50 km inland from Valencia	19–21 Oct. 1982	c. 650 in 48 hours	c. 500	500 mm on an area of 715 sq. km; massive erosion[4]
Mediterranean France				
Py (E. Pyrenees)	6–7 Nov. 1982	610 in 2 days[5]	c.800	
Easternmost Pyrenees	17 Oct. 1940	840 (1000?) mm	c. 850	highest fully authenticated 24-hour rainfall in Europe
Mt-Aigoual	31 Oct.–1 Nov. 1963	682 in 2 days	2,285	
Mt-Aigoual	24–25 Feb. 1964	702 in 2 days	2,285	
Valleraugue (SE of Mt-Aigoual)	Sep. 1900	950 in a night	c. 1500	
Nîmes	2–3 Oct. 1988	420 mm	700–1,000	killed 11 people
Châteauneuf-du-Pape	30 July 1991	265 in 24 hours (100 in one hour)	720	
Vaison-la-Romaine, Carpentras	21–22 Sep. 1992	448 (72 in ½ hour)	c. 680	killed 34 people
Var basin	9–16 Oct. 1979	up to 300 in a week	c. 1,100	followed by collapse of delta (p. 336)
Corsica				
Cap Corse	23 Sep. 1993	406	530	
Col de Bavella	31 Oct.–1 Nov. 1993	906 in 2 days	c. 1,400	
Italy				
Po valley	7–13 Nov. 1951	600 in Prealps, 500 in Apennines in a week[6]	c.1,000	
Po valley	2–3 Nov. 1968	350 in 2 days at Cengio[7]	c. 700	
Po valley	5–6 Nov. 1994	up to 500; 150 in 6 hours at Ponzone; peak of 55 per hour[8]	c. 700	Po at Turin rose on 6 Nov. to 5.12 m above average level
Arno basin	4 Nov. 1966	190 in Florence city; more in mountains[9]	823	Florence flood, highest for over 800 years
Calabria	16–18 Oct. 1951	> 400 in 3 days; ?1,400 in mountains[9]	c. 1,200	
Crete				
Alikianoú and Zymvrágou (W. Crete)	30 Mar. 1976	298	909	
Exo Potámoi (Lassíthi Mts, E. Crete)	24–25 Oct. 1976	596 in 3 days	1,400	200 mm or more in other places in Crete
Pakhyámmos (E. Crete)	23 Sep. 1986	300 in 36 hours	558	witnessed by one of us (pp.247–50); affected 800 sq. km

Table 2.i. Some notable deluges.

[1] MOB (D. Wheeler); [2] A. M. Harvey, 'Geomorphological response to an extreme flood: a case from southeast Spain', *ESPL* 9, 1984, pp. 267–79;
[3] Miro-Granada & Gelabert 1974; [4] Miro-Granada & Gelabert 1983; [5] French and Corsican records from Météo-France [1995];
[6] M. Rossetti, 'La piena del Po del novembre 1951', *Annali Idrologici* 2, 1958, pp. 3–2-; [7] Lionetti 1996; [8] Lionetti 1996; [9] Mancini 1986

and mid-November; this includes all ten of the greatest floods recorded over the last hundred years in Genoa. They result from the interaction of polar maritime air from the north with warm, wet Mediterranean air. Typically they are associated with a depression over the British Isles and high pressure over eastern Europe.[33]

Italy and Corsica

The deluge on the upper Po on 4–5 November 1994 produced more than 2 cubic km of water. Pressure had been high over mid to east Europe and the Balkans, with a depression over Ireland and a cold front through France (Fig. 2.10). Forecasts such as those of the European Centre for Medium-Range Weather predicted up to 600 mm of rainfall within 4–6 November, which came to pass. A strong, deep flow of warm, very moist air, drawn into northwest Italy from the Gulf of Genoa and the Adriatic, converged on the upper Po valley where it was undercut by cold air from the north and west. About 12,000 sq. km (extending as far as Corsica) received over 200 mm of rain, 1,300 sq. km received 300 mm and 600 sq. km over 400 mm. More than 300 mm of rain in 24 hours were recorded at Oropa, Meugliano and Lanzo in northern Piedmont, their wettest day for over sixty years. The three-day rainfall of 252 mm at Turin was the wettest three-day spell in its 192-year record.[34]

Corsica receives deluges mainly on the east side of its great central mountains. The greatest in 36 years was on 31 October–1 November 1993 and yielded one cubic km on the island alone, with 3,000 sq. km receiving between 300 and 900 mm. Again the setting included a high over central Europe, but this time a flow of air from the south-west was provided by a low-pressure area SW of Portugal.

Southern Italy also attracts deluges, with high mountain ranges near the sea. The toe of Calabria is susceptible to violent rain-bearing winds from both sides. One of the greatest storms known in Europe struck Calabria on 16–18 October 1951, with 400 mm falling over 2,500 sq. km

and 600 mm over 1,500 sq. km. It is claimed (although there were probably no rain-gauges capacious enough to prove it) that 1,100 mm fell on the Aspromonte and Sila mountains (Fig. 2.9).[35]

Crete

The island of Crete forces rain-bearing north-westerlies to rise over its mountains. The heaviest single fall ever measured was probably at Exo Potámoi, on a ridge at the northeast corner of the Lassíthi mountain-plain, on 23, 24 and 25 October 1976 when 596 mm were recorded – more than one-third of the total for the otherwise dry year 1976–7. Falls of more than 200 mm were reported on those same days from scattered places including the west.

The deluge that we witnessed in 1986 (pp. 247–50, Fig. 14.11) was barely noticeable on the weather map. A small depression had appeared off Benghazi on the morning of 22 September (an unusual place for a depression to be at that time of year). It crept slowly east-north-east over the next 2½ days.[36] Hot damp air was sucked south-west from the Aegean and hit the towering cliffs of east Crete with great violence. This affected only 800 sq. km.

Season of deluges

Deluges most often occur in autumn (Table 2.ii), typically a little before the peak rainy season, e.g. at Barcelona, Perpignan and Rome in October, Seville and Athens in December[37] (cf Fig. 2.3). They can occur at any time, even in the dry season, as on 30 July 1991 at Châteauneuf-du-Pape (some of which doubtless got into the wine) and in the Pyrenees in June and July 1996. The eight greatest floods in Florence in the last eight hundred years occurred between August and December, early November being a dangerous time.[38] In east Spain, floods have occurred in all months, but October is by far the commonest season. In Crete, however, deluges occur throughout the rainy season.

Mediterranean France and Corsica: number of *situations* (apparently >190 mm) 1958–94, from Météo-France [1995].
SE Spain: *inundaciónes* 1900–83 mentioned by Font Tullot (1988).
NE Pyrenees: deluges 1820–1940 mentioned by Pardé (1941).
Crete: deluges since 1900, from Cretan records.

	Sep.	Oct.	Nov.	Dec.	Jan.	Feb.	Mar.	Apr.	May	June	July	Aug.	total
France	24	35*	18	5	8	6	2	2	2	5	4	8	119
NE Pyrenees	0	6	1*	1	0	0	1	0	0	1	0	0	11
SE Spain	9	19	12*	8	11	8	10	2	1	3	0	2	85
Corsica	3	11	6*	0	0	3	0	2	0	0	0	0	25
Crete	3	1	1	1	6*	1	2	0	2	0	0	0	17

* Month with highest total mean rainfall.

Table 2.ii. Monthly frequency of deluges in the Mediterranean.

---

33 Russo & Sacchini 1994.
34 L. Mercalli, S. Paludi, F. Dutto, 'Alluvione del 5–6 Novembre 1994 in Italia NW: analisi pluviometrica', *NSMS* 6–7, 1994, pp. 25–32; Lionetti 1996.

35 V. Cantù, 'The climate of Italy', *World Survey of Climatology*, 6: *Climates of central and southern Europe*, ed. C. C. Wallén, Elsevier, Amsterdam, 1977, pp. 127–83; Mancini 1986.
36 Synoptic charts, MOB.

37 Llasat & Rodriguez 1997.
38 F. Nencini, *Florence: the days of the flood*, Unwin, London, 1967; Barriendos Vallvé & Martín-Vide 1998.

Significance of deluges

Deluges, although rare, are important agents of erosion. Losses by evaporation and diversion by infiltration take up a very small part of the total fall, most of which inevitably runs off the surface no matter what the vegetation cover. The runoff scours headwater ravines and valleys, to dump the sediment downstream. The time of year probably matters, in that early deluges (when the soil may be dry and crusted) are expected to promote rills and gullying, whereas late events, when the soil is already soaked by previous rains, may promote slumps and mud-flows.

In coastal mountains, deluges are a rare but normal and recurrent event, like fires and earthquakes. There is no sign of an upper limit to the intensity of rainfall. The alluvial fan at the mouth of any Cretan gorge is built up of layers of sediment deposited and re-worked by millennia of successive deluges, some of them more energetic than any that have been put on record.

People and climate

Reliable summer sunshine is the basis of the region's tourism. However, the capacity of tourists to endure day after day of inexorable heat on an arid beach is only half a century old. In Mediterranean tradition people have avoided 'the burden and heat of the day'; they have distrusted the sun, which brings thirst and weariness and heat-stroke and destroys the complexion, and have loved shade and the summer night.

Cretans and Spaniards are still nocturnal. Although tourism is vulnerable to fashion, economic recession and war, interruptions to growth have so far been short-lived. There has not been much of a shift from the fashionable coast to more glamorous mountain areas, nor from the dry dead summer to the beauty of the Mediterranean spring. How this will change in view of the newly publicized risk of cancer from the sun's ultra-violet radiation remains to be seen.

Mediterranean peoples do not avoid dry areas; they have shown remarkable ingenuity in coping with drought (Chapter 17). At least in the past few centuries, most folk detest winter cold. The upper limit of all-year settlement is seldom much above 1,000 m – well below the upper limit in the Alps – and is often no more than 200 m above the olive limit. (This may be partly due to the dry summers, which make it difficult to make hay, the staple of Alpine high-altitude farming.) Higher altitudes are used by various forms of transhumance and seasonal settlement. In many places all-year settlements have been higher in the past.

Floods seem to cause more damage than they used to, but, as we shall see in Chapter 8, people have been saying this for centuries. It is not necessarily due to there being more deluges, nor even to more publicity. Increasing artificial drainage and canalizing of rivers increase the downstream effects of a given fall of rain. People, too, have become more ready to place lives and property in areas previously shunned as flood-prone – often under the tragic delusion that flood-control technology prevents floods.

CHAPTER 3

Geology and Geomorphology: the Dynamics of a Restless Region

What we can be certain of is the architectural unity of which
the mountains form the 'skeleton': a sprawling, overpowering,
ever-present skeleton whose bones show through the skin.[1]

Nought may endure but Mutability.[2]

Fig. 3.1. The huge fault-scarp, a thousand metres high, which cuts off the east end of Crete. The crack is the great gorge Há (Fig. 3.7). The screes at the base of the cliff, formed by rocks split off from it, are partly loose, partly cemented. *October 1986*

The wondrous variety of Mediterranean landscapes is determined by the rocks and the processes that have shaped the mountains. On these depend soils, vegetation, land-use history, erodibility and even climate.

The northern Mediterranean is the scene of a collision between two of the world's great tectonic plates. Africa and Europe had once been much further apart, separated by a greater Mediterranean known as the Tethys Sea. Some seventy million years ago they began to move together, the leading edge of the African plate being thrust down under the European plate. This movement, producing the Alpine phase of mountain-building, was most active about twenty million years ago. It buckled the strata laid down in Tethys over the previous hundred million years, and even tore up the floor on which they had accumulated. The edge of the European plate was shattered into lesser units, forming islands and peninsulas that move independently of the continents. Geological maps and satellite images display complicated linear patterns related to the fold and fault structures.

Many features result from earth movements within the last two million years. Islands have emerged from the sea. Tracts of land have been lifted or have sunk hundreds of metres (Fig. 3.1), or have been displaced sideways for tens of kilometres as *nappes*. Movement still goes on. It is usually too slow to be

[1] Braudel 1966.   [2] Percy Bysshe Shelley, *Mutability*, 1816.

Fig. 3.2. Fault-scarp at the base of the small, actively uplifting, sacred mountain of Yúktas in Crete. *May 1988*

noticed by the casual observer: 1 or 2 mm per year vertically, or 3 or 4 cm per year horizontally, are fast movements. However, sudden vertical jerks of several metres are not unknown.

Active faults are common in the Mediterranean, especially in the wildly shattered landscapes of Crete. Fault surfaces often look fresh and newly formed (Fig. 3.2), with rock faces that display details of the sliding, crushing and jamming which happen when an irresistible force meets an immovable object.

Stability or instability in general is demonstrated by studies of the shore-line as it was in the Last Interglacial period, 120,000–130,000 years ago, when world sea-level was about 3 metres above the present. In Sardinia, Last Interglacial shorelines are still at about +3 m. Elsewhere around the Mediterranean, they have generally been uplifted at a rate

Fig. 3.3. Kándanos basin, SW Crete, overlooked by steep slopes in limestone and metamorphic rocks. Ancient lake-beds on the floor of the basin are overlain by more than 100 m of coarse sub-rounded gravels in a red sandy matrix. These are exposed in vertical road-cuts. (Heather and arbutus in the middle have been partly grubbed out and replaced by olives.) *November 1994*

of about 0.1 mm per year, reaching about +10 m on the coast of Almería and Murcia, east Corsica and the SW coast of Italy. Parts of Calabria, however, have been uplifted by 120 m since the Last Interglacial.[3]

Erosion has been wearing down the mountains ever since they began to form. The eroded material has been laid down lower on the slopes or in basins to form younger rocks, which in turn have been displaced by the continued mountain-building process and have been exposed to further cycles of erosion and deposition (Fig. 3.3).

RELICS OF AN EARLIER (HERCYNIAN) ERA

The general disturbance spared some fragments of more ancient folding, aged in hundreds rather than tens of millions of years (Fig. 3.4). The biggest is the Meseta, the great table-land of Portugal and west and middle Spain. This Hercynian region is typically underlain by metamorphic rocks, worn down by long slow erosion to rolling plateaus in which rivers such as the Tagus run in gorges. In places there are depressions filled with Mesozoic and Cainozoic sedimentary rocks. These are *dehesa* or *montado* landscapes, with shallow, somewhat acid soils and savannas of evergreen oaks (Chapter 12). They are divided by ancient fault structures, curving across the landscape in a general NW–SE direction from the Atlantic seaboard. These groups of parallel faults, perhaps picked out by ancient wind erosion, show up on satellite images and vegetation maps.[4] Here and there are large or small mountain ranges where the massif has been buckled by later tectonic movements.

Other fragments of Hercynian Europe constitute Sardinia, Corsica and the Massif des Maures in Provence. These are built of ancient granites, gneisses, schists and Palaeozoic sedimentary rocks. They have much the same thin, acid soils and evergreen-oak savannas as Hercynian Spain, but are more rugged; fantastic skylines of granite tors are part of the mysterious landscape of Sardinia. Most of Corsica is a gigantic granite dome whose peaks are the second highest mountains of the Mediterranean islands.

Another Hercynian massif lies in Bulgaria, with metamorphic rocks extending into NE Greece. Metamorphic rocks, not *in situ* but torn from the Hercynian sea-floor and dumped on top of younger limestones, form the distinctive highland landscape of the far west of Crete.

ALPINE MOUNTAIN-BUILDING

Mountains of southern Spain

The Baetic Cordillera extends from Cadiz through the mighty Sierra Nevada to the coast near Valencia, and runs on into the Balearic Islands. Rocks include Palaeozoic metamorphics, especially in the Sierra Nevada, and also limestones and dolomites, especially in the NE.

The Cordillera consists of mountain ranges separated by wide basins, the result of alternate upthrusting and

3 *ZG*, NF 62, 1986, *passim*.
4 For instance, CORINE Land Cover map of the Iberian Peninsula and Southern France 1992, CEC DG XI, European Environment Agency, Brussels.

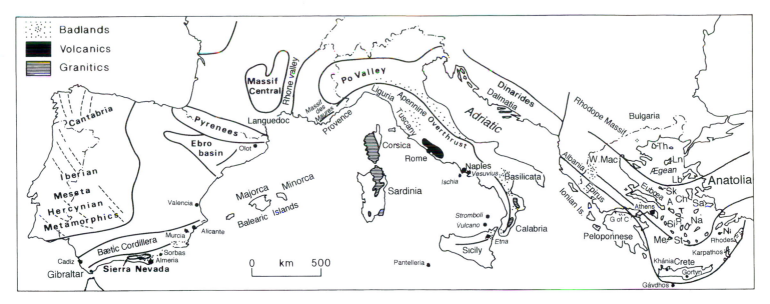

Fig. 3.4. Major geological structures in the European Mediterranean.
W. Mac: western Macedonia.
Aegean islands: A, Andros; Ch, Chíos; G of C, Gulf of Corinth; Lb, Lésbos;
Ln, Lémnos; Me, Mélos: Na, Náxos; Ni, Nisyros; P, Páros; Sa, Sámos;
Si, Síphnos; Sk, Skyros; St, Santoríni; T, Tínos; Th, Thásos.

Fig. 3.5. Mountains and basins of the Wild West-like country of southern
Spain; in the foreground is a pine-covered badland. Sierra del Oró, NW
of Murcia. *April 1994*

subsidence of blocks of the earth's crust. (Thrusting and
overturning movements, however, are less conspicuous than
in the Alps.) The basin floors are covered with great thicknesses of later deposits, laid down both under the sea and on
land. These landscapes resemble the Basin and Range
country of the western USA (Fig. 3.5). Residual tablelands,
with piedmont glacis or pediments[5] at their margins, are
typical of the country between Alicante and Almería. The
glacis are set one into another, as in a river-terrace sequence,
and are presumed to have formed during the Pleistocene.
They are capped by crusts of calcium carbonate (caliche) or
by *raña* (p. 42). Where made of soft Pliocene marls they have
been cut by erosion into glorious badlands (Chapter 15).

Pyrenees and Ebro basin

The Palaeozoic rocks which form the core of the Pyrenees are
flanked to the north by Mesozoic limestones. On the south
flank are inner limestone and outer sandstone ranges with
younger strata in between, including unstable, highly erodible flysch (p. 247).

The great basin of the Ebro was formed by subsidence of
the Hercynian massif during the early Cainozoic, several tens
of millions of years ago. Folding to form a coastal cordillera
closed it off from the sea by the end of the Eocene. Later the
basin was filled, in several episodes, by continental deposits
of sandstones, clays, marls and gypsum. Most of it is now

drained by trench-like valleys, but some of the central parts
are still cut off from the sea: closed depressions were formed
in the mid or early Quaternary by solution of gypsum allied
with wind deflation.

Alps and Apennines

Folded Mesozoic limestones emerge from the plains of
Languedoc and lap on to the southern flanks of the Massif
Central. East of the Rhone, folded ranges at the margins of
the Alps approach the coast in Provence and Liguria. From
the main mass of the continent, the Apennines extend down
the Italian peninsula and into Sicily as a series of linked
mountain ranges and blocks, arranged in longitudinal belts
with intervening valleys and basins. On their western flank is
a volcanic belt.

Some rocks and structures, especially in the Apennines, are
very erodible. The *argille scagliose* of Miocene and later age,
for instance, which occupy one of the longitudinal belts,
consist of clays with complex structures. Embedded in the
clays are irregular sheets of rock which have been displaced

---

5   Alluvial fans and gently sloping erosional
surfaces at the feet of steep scarps that have
retreated backwards.

over long distances; they include mountain-size chunks of limestone, silts, sandstones and even ophiolites. In this rugged, earthquake-prone terrain the clays are unstable and liable to collapse, giving rise to some of the hazards of Campania and the Basilicata badlands in southern Italy (Chapter 15).

Balkans and Aegean

Tectonics in Greece and the Aegean have created island arcs. An outer arc extends south from Albania and Epirus through the Peloponnese and Ionian Islands, curving east through Crete via Kárpathos and Rhodes into southern Anatolia. It includes resistant Triassic and Mesozoic limestones and dolomites, which give mountain ranges and karst landscapes. The intervening synclinal or fault-bounded valleys and basins are floored with flysch or Cainozoic sedimentary rocks; some of these are resistant to erosion and others, including sandstones and ophiolites, are susceptible. Blocks of territory have subsided between faults, to form the *poljes* of the Balkans and their cousins the inland basins of the Peloponnese and the mountain-plains of Crete (Fig. 3.6). Others have been uplifted to form mountainous peninsulas and islands. Where rocks have been subjected to bending movements they have cracked and now form gorges, notably the hundred gorges of Crete (Fig. 3.7).

Fig. 3.7. The world's biggest crack? Há gorge, 500 m deep, splitting at right angles the fault-cliff which divides east Crete. *August 1990*

Fig. 3.6. Askyphou mountain-plain in Sphakiá, Crete. The fault-bounded basin is filled with various deposits, creating a nearly flat surface at 680 m; it grows the best wine in Crete. The surrounding mountains are strongly karstified and pitted with small depressions. *June 1987*

To the north of Crete lies an arc of active or dormant volcano islands. Inner arcs, further north still, include the islands of Síphnos, Náxos and Páros, SE Eubœa, Andros, Tínos and Sámos, consisting largely of gneiss, marble and other metamorphics. Much of Chíos, Skyros, eastern Lésbos and Thásos is also underlain by metamorphic rocks. Acid volcanics form northern and western Lésbos and much of Lémnos; they also outcrop locally in western Macedonia.

EVENTS OF THE LAST 25 MILLION YEARS

Neogene deposits

In Miocene and Pliocene times, river and lake deposits accumulated in inland depressions, often reaching great thicknesses as basin floors subsided beneath their weight. Marine sediments were deposited in shallow seas. Some of these sedimentary rocks are now hard limestones, sandstones and conglomerates forming prominent escarpments. More are pale-coloured marls; others consist of unconsolidated clays, silts, sands and gravel beds. For the most part they lie at less than 500 m above sea-level. These form the traditional lands of dense settlement, at altitudes where cereals and tree crops thrive, yet far enough from wetlands and coasts to escape the worst perils of malaria and pirates.

Messinian Salinity Crisis

The Mediterranean Sea is kept in being by a net inflow of surface water from the Atlantic. Evaporation consumes more than the entire input from rainfall plus the inflow of the Nile, Danube and other rivers. If tectonic action lifts the Straits of Gibraltar above sea-level, the sea dries into a desert basin (a kind of mega-Death-Valley) with salt lakes (a kind of super-Dead-Sea) at the bottom. This last happened in the Messinian Salinity Crisis between 5.7 and 5.4 million years ago. Gypsum and other salts, accumulated from the evaporating water, lie up to 1,500 m thick on the present sea floor. When the Straits of Gibraltar re-formed, the Atlantic, pouring in again, brought a new marine fauna and flora.

The Mediterranean floor continued to be displaced during and after the Messinian event, and its present topography is very different from that of five million years ago. Nevertheless, the consequences persist in submarine and even terrestrial landscapes. The whole continental margin was exposed to erosion. Great torrential rivers, notably the predecessors of the Rhone and the Ebro, cut canyons a thousand metres deep in their lower courses down to the super-Dead-Seas. Rivers draining into the Po canyon are believed to have eroded headwards into the Alps and excavated the valleys, later scooped out by glaciers, that now hold the Italian lakes. Cave systems, penetrating far below present sea-level, date from this time. Massive coral reefs in the Sorbas region of southern Spain, left high and dry when the sea receded, were then covered by evaporites as the waters rose again, and have since been uplifted to form an extraordinary gypsum-karst desert.

Volcanism

The largest volcanic region in southern Europe runs from Tuscany to Sicily and Pantelleria. Activity began in Pliocene times and became more intense in the Pleistocene. It still continues, and affects both weather (p. 143) and landscape. The vast, gently sloping dome of Mount Etna is made up of several overlapping Pleistocene volcanoes; many of the 250 parasitic volcanoes scattered over its slopes date from historic times. There are four apparently active volcanoes around Naples: Vesuvius, the Solfatara of Pozzuoli, the Phlegraean Fields (where Monte Nuovo was born in 1538) and the island of Ischia.[6] Among the islands near Sicily, Stromboli and Vulcano are still active. Volcanoes north and south of Rome erupted 25,000 years ago, their ashes and tufas pushing the River Tiber to the east.

The volcanic island arc of the Aegean has been active all through the Quaternary. Three volcanoes are still active: Méthana across the bay from Athens, Santoríni and Nísyros. Mélos is celebrated as a source of obsidian (volcanic glass) for making stone tools; journeys to fetch it are the earliest known instances of seafaring in Europe. Santoríni is famous for sporadic giant explosions, one of which used to be thought to have caused the downfall of the Minoan civilization of Crete in c. 1450 BC (it actually happened at least a hundred years too early).[7]

In the Quaternary, the eastern Pyrenees were intensely seismic and volcanic, with 44 volcanoes around Olot, now wooded and apparently innocuous.

Pleistocene glaciations

In northern Europe the condition called glaciation has lasted for most of the last two million years, broken by relatively short interglacial episodes such as the present one. In Mediterranean lands ice cover was very limited, but temperatures were several degrees colder than now; climate, vegetation and erosion were profoundly affected.[8]

The Last Glacial Maximum was only 20,000 years ago. Sea-level was lowered by 120–150 m. This generally made little difference to coastlines, except in the shallow north Adriatic, but joined Majorca to Minorca, Sardinia to Corsica, Sicily to Italy, and some of the nearer Aegean islands to the mainland. However, most of the islands remained separate and (so far as we know) inaccessible to mankind, and kept their peculiar animals and plants. The present sea-level was regained only about 7,000 years ago.

Lowered sea-level, though not enough to shut off the Straits of Gibraltar and provoke another Salinity Crisis, restricted exchanges of water between the Mediterranean and Atlantic and between the eastern and western Mediterranean. The sea surface became much colder, especially in the NW Mediterranean. Winter temperatures were probably on average 3–4°C lower than now. Summer surface temperatures around Crete were about 7°C lower than now; off the south of France they may have been 13°C lower.[9]

On land, mean temperatures were typically about 6–8°C lower than now. Winters were much colder, with permafrost extending as far south as Provence and the tree-line there not far above sea-level. These changes were apparently somewhat less in Greece and in southern Anatolia, refugia for warm-temperate plants and animals.[10] Many frost-sensitive plants, including Cretan endemics (p. 155), somehow persisted.

Precipitation was less than it is now, and much of it fell as snow. Snow-lines were about 1,000 m lower than at present. Glaciers on the southern Alps, the Pyrenees and the Sierra Nevada extended for several kilometres; small glaciers occupied cirques in the high Apennines. (The extent of glaciers in Greece is still unsettled.[11])

Mesolithic peoples hunted mammoth and other European animals on the mainland. The islands, lacking human populations until the Holocene, had peculiar, unbalanced faunas of dwarf elephants, dwarf hippopotamuses, deer, antelopes and strange rodents, but usually no effective carnivores.

With little glaciation, Pleistocene sediments are on the whole less voluminous than in northern Europe. However, various forms of erosion were more active then they are now.

6 A. Sestini, 'Apennines and Sicily', Embleton 1984, pp. 341–54.
7 D. M. Pyle, 'The global impact of the Minoan eruption of Santorini, Greece', EG 30, 1997, pp. 59–61.
8 F. Boenzi & G. Palmentola, 'Glacial features and snow-line trend during the last glacial age in the southern Apennines (Italy) and on Albanian and Greek mountains', ZG NF 41, 1997, pp. 21–29.
9 G. R. Rigg, 'An ocean general circulation model of the glacial Mediterranean thermohaline circulation', Palaeooceanography 9, 1994, pp. 705–22.
10 Tzedakis 1993; I. C. Prentice, J. Guiot, S. P. Harrison, 'Mediterranean vegetation, lake levels and palaeoclimate at the Last Glacial Maximum', Nature 360, 1992, pp. 658–60.
11 See, for instance, G. W. Smith, R. D. Nance, A. N. Genes, 'Quaternary glacial history of Mount Olympus, Greece', Geological Society of America Bulletin 109, 1997, pp. 809–924.

Fig. 3.8. Hollow cliff of weak, partly loose conglomerate, case-hardened with redeposited limestone. *Kládhos Gorge, Sphakiá, Crete, August 1995*

Cores from offshore deposits laid down by the Ebro indicate that Late Pleistocene erosion rates in its catchment were two or three times greater than in the first half of the Holocene.[12] Coarse debris, called *Older Fill*, accumulated in valleys and depressions all round the Mediterranean. Frost action generated massive screes below cliffs in faulted regions such as Crete (Fig. 3.1). Strong winds, driven by steep south–north temperature gradients, carried Saharan dust far inland and blew beach sand over low coasts. Gávdhos, the southernmost outpost of Europe, was a desert island covered in sand (now cemented into sandstone).

Cementation

Some erosional deposits remain loose, and thus subject to renewed erosion later; others are consolidated. Percolating water, rich in calcium carbonate, can turn gravels, screes and sand into hard rock. This is the rule in Crete but the exception in Italy. Cementing may occur in bands or only on the outside of the deposit (*case-hardening*); breaches in the cemented layer then allow the loose interior to trickle out, creating the jagged, hollow cliffs so distinctive of Aegean islands (Fig. 3.8). A special kind of cementation occurs where freshwater springs seep out at the shore, gluing shingle into an extremely hard (though very young) conglomerate called *beach-rock*.

A mysterious material called *raña* (in Spain) or '*red-beds*' accumulated during the Quaternary, especially around the Meseta. It consists of angular stones in a slightly cemented, unbedded red matrix. Red-beds appear to be colluvia, derived from Tertiary weathering of resistant rocks from the Hercynian age, which sludged, crept or washed across the lower slopes of pediments or over basin-fill deposits. They commonly form resistant cappings that protect more erodible rocks beneath.[13] In west Crete remnants of such material form the strange spiky red hills SW of Khaniá, and cover the floors of great hollows such as the Omalós, Kándanos and Stylos basins.

Fig. 3.9. Basins in west Crete have been filled by red-beds, sands, gravels and limestone boulders in the course of the Pleistocene. They are now at different stages of dissection. Gullies are extending headwards from sinkholes at the edge of the Omalós mountain-plain. River gorges have broken into the fills of the Kándanos and Stylos basins, but the head of the Samariá Gorge has yet to breach the rim of the Omalós basin.

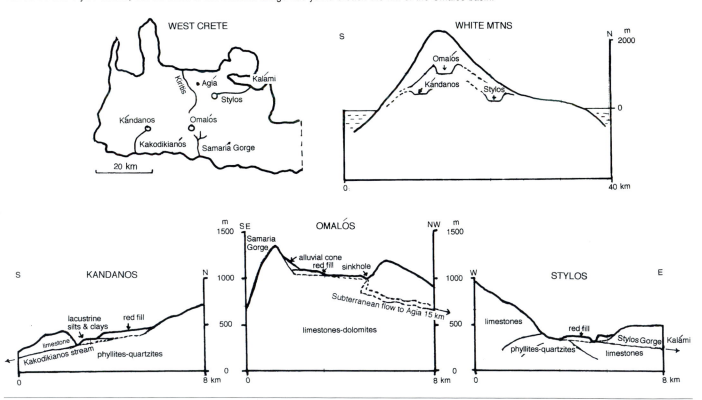

[12] Nelson 1990.
[13] For a discussion of *raña*, see M. Sala, 'Iberian Massif' and 'Pyrenees and Ebro Basin Complex', Embleton 1984, pp. 268–340.

THE HOLOCENE

The cold dry state ended somewhat abruptly about 13,500 BC. It was followed by fluctuations corresponding to the Allerød and Bølling interstadials of northern Europe, separated by cold dry centuries corresponding to the Younger Dryas. The Holocene is the last 11,500 years, beginning about the end of the Younger Dryas, the final cold snap of the Last Glaciation.

Water from melting ice in other parts of the world raised the sea-level to roughly what it is now by about 5000 BC. Mediterranean coasts became very indented: valleys, formed mainly by erosion on the lower courses of rivers in the Pleistocene, were flooded by the rising sea-level to create estuaries ramifying inland.

The Black Sea was then a giant freshwater lake, its surface about 150 m below present sea-level. The lowlands were peopled by Neolithic villagers who made pottery, grew crops and kept animals. In about 6000 BC the rising Mediterranean broke through the straits at the Dardanelles and Bosphorus and rapidly cut down several tens of metres. Water burst into the Black Sea basin at about 50 cubic km a day.[14] Within a few years the sea-level had risen to the present, evicting the inhabitants who thus, it has been suggested, brought Neolithic cultures to the lands north of the Mediterranean.

Since then absolute sea-level has been nearly stable. It is difficult to find a tide-gauge entirely free from tectonic movement, but rises of sea-level in some places seem slightly to outweigh falls in others. Sea-level has been rising in the twentieth century at 1–2 mm per year, as in many other parts of the world, probably through Global Warming.[15]

Changes in general sea-level are greatly outweighed by local movements of the land. The south side of the Gulf of Corinth – that workshop of Poseidon, god of earthquakes – has been been uplifted by 1,800 m in the course of the Quaternary, an average of about 1 mm a year. The coast of the instep of Italy has been uplifted at about one-third of this rate.[16] Such a fall in local relative sea-level is not unusual: NW Scotland is rising more rapidly as a delayed effect of recovery from the weight of Pleistocene ice. However, Scotland is made of hard rocks, and the uplift has been going on for only 15,000 years; whereas the hills of Basilicata are made of weak, highly erodible rocks which have been uplifted several hundred metres and are now unstable.

Local changes of sea-level are recorded by so-called 'wave-notches'. In a tideless sea, when sea-level is constant relative to the land for a few centuries, a notch is dissolved or eaten away by marine organisms at the waterline. If the land is jerked up or down by an earthquake, the notch remains above or below the new waterline, where a new notch begins to form (Figs. 3.10, 3.11). Remains of the waterline organisms afford an opportunity of getting a radiocarbon date – which may,

Fig. 3.10. Notch corresponding to present sea-level within a Roman stone-quarry which has subsided since it was abandoned. *Stavrós, Akrotíri, Crete, April 1990*

however, be erroneously old if the creatures have derived some of their carbon from seawater.

Crete is the classic region for sea-notches, one of which records the best authenticated violent tectonic event of historic times in the Mediterranean: the SW corner of the island was, apparently, suddenly uplifted through nine metres. This is an instructive example of the difficulty of reconciling C14 dates with historical records. Even this tremendous convulsion, glorified as the 'Early Byzantine Paroxysm', is not immediately distinguishable from the ordinary earthquakes (about twenty per century in the east Mediterranean) recorded by miscellaneous writers in a not very well recorded period. Only one possible earthquake, however, is written down for Crete: that which demolished the baths in the capital, Gortyn, in the reign of Theodosius II (408–50 AD).[17]

Y. Thommeret and colleagues, having obtained numerous radiocarbon dates, seven of them from the highest section of the highest shoreline (at Khrysoskalítissa) giving ages between 1550±80 and 1595±70 bp, dated the event to c. 530 AD. They were inclined to identify it with a great earthquake in 438 AD,[18] or less likely one of those recorded in 365, 374 and 448 AD. They attached little importance to the possibility of error from carbon derived from seawater.[19]

D. Kelletat was impressed by the accounts of a tsunami, a freak wave, which ravaged coasts in Egypt, south Greece, Sicily and perhaps Dalmatia on 21 July 365 AD. Referring to the many radiocarbon determinations on notch deposits in Crete that centre on 1530 bp, he concluded that this must have been the date of the uplift. A recalibration of this date (1530±40 bp) by the calibration curve in use in 1986 gave the

[14] W. B. F. Ryan and nine others, 'An abrupt drowning of the Black Sea shelf', *MG* 138, 1997, pp. 119–26.

[15] P. A. Pirazzoli & J. Pluet, *World Atlas of Holocene Sea-level Changes*, Elsevier Amsterdam, 1991.

[16] Brückner 1980.

[17] Guidoboni and others 1994.

[18] According to Guidoboni this happened in 437 by the modern calendar, and is recorded only for Constantinople.

[19] Y. Thommeret, P. A. Pirazzoli and three others, 'Late Holocene shoreline changes and seismo-tectonic displacements in western Crete (Greece)', *ZG* NF Suppl. 40, 1981, pp. 127–49; P. A. Pirazzoli, J. & Y. Thommeret, J. Laborel,

'Crustal block movements from Holocene shorelines: Crete and Antikythira (Greece)', *Tectonophysics* 86, 1982, pp. 27–43; P. A. Pirazzoli, J. Laborel, S. C. Stiros, 'Coastal indicators of rapid uplift and subsidence: exmples from Crete and other Mediterranean sites', *ZG* NF Suppl. 102, 1996, pp. 21–35.

Fig. 3.11. Sea-notches in a cliff – which is an active fault-scarp – near Soúyia, on the south coast of west Crete. The highest, most prominent notch is the most recent. Here, in 1845, Raulin first appreciated the meaning of such notches. *May 1989*

range 341–439 AD with a probability of 0.67, which was more consistent with the date 365 than with 438 AD.[20]

Our own recalibration, using the 1993 calibration program, converts 1530±40 bp to 570±40 AD with a probability of 0.67, and to 530±100 AD with a probability of 0.95. Thommeret's dates, recalibrated, centre on 500 AD. Moreover any error caused by seawater carbon would make the true date even later. The shock could thus have been the earthquake of 537 that brought down the dome of Ayia Sophia in Constantinople, or one of a number of others documented, or (in then remote Crete) an undocumented earthquake.

Whatever its date, this displacement affected all Crete, its island arc and the Anatolian coast. Crete, however, is a very erosion-proof island, and no effects on erosion have yet been recognised.

A last factor is silting. In the last few thousand years Mediterranean coasts have lost much of their complexity. Bold jagged headlands still project into deep water, but shallow bays have been turned into lagoons and then filled in, coastal plains have been extended, deltas created and islets rejoined to the mainland by new deposits. This process has lately been interrupted by the building of dams which intercept the input of sediment, so that the sea is gaining on the land (Chapter 18).

[20] D. Kelletat, 'The 1550 BP tectonic event in the Eastern Mediterranean, as a basis for assuring the intensity of shore processes', *ZG* NF Suppl. 81, 1991, pp. 181–94; D. Kelletat, 'Perspectives in coastal geomorphology of western Crete, Greece', *ZG* NF Suppl. 102, 1996, pp. 1–19.

CHAPTER 4

Plant Life: the Dramatis Personae of Historical Ecology

*For there is hope of a tree, if it be cut down, that it will sprout again,
and that the tender branch thereof will not cease. Though the root
thereof wax old in the earth . . . yet through the scent of water it will
bud, and bring forth boughs like a plant.[1]*

Plants are not just Environment, part of the scenery of the theatre of historical ecology, the passive recipients of whatever destiny mankind's whims inflict upon them. They are actors in the play. Each species of the Mediterranean flora is an independent member of the *dramatis personae*, with its own agenda in life. Even a cursory account of ecological history must begin from a summary of the properties of about forty common trees and shrubs, twenty undershrubs, and forty grasses and herbaceous plants.

It is wrong to accept popular or reductionist misconceptions about the behaviour of plants, such as that forests are 'cleared' merely by cutting down trees, or that vegetation is 'destroyed' by fire, or that 'goats eat everything'. Each plant has its own behaviour in relation to woodcutting and burning: many recover from one or the other or both, others do not. As anyone knows who has watched them, goats are choosy about what they eat: they devour prickly-oak, but dislike the strong flavour of cypress. However, palatability is not simple. Different animals have different tastes: Cretan wild-goats have slightly different gastronomic habits from domestic goats. Starvation can force goats to eat even cypress. Palatability may vary with the time of year or from one individual plant to another: planted cypresses can be less distasteful than wild ones.

FLORA

Mediterranean countries have a richer flora than the rest of Europe. Many common species do not occur north of the Alps; some are of African or Asian affinities.

Plants might be expected to adapt to their environments by a process of Darwinian evolution through natural selection over many generations. An example of such adaptation, or lack of it, is furnished by the Aleppo pine, *Pinus halepensis*. This tree is common on the Greek mainland, but when introduced to Crete somehow fails to prosper – which is fortunate, for Crete is better without this super-flammable tree.

However, European ecology is dominated by environmental change rather than evolution. Most environments have not existed long in evolutionary terms. Plants, except perhaps for annuals which breed every year, have had to make the best of environments into which accidents of climatic and geological history have thrust them, rather than becoming adapted to

some specific environmental niche. The climate of the Mediterranean is of a rare type and has existed for only a few thousand years (Chapter 9). Plants have lived with the present climate for barely longer than they have lived with major human activity. They have yet to become fully adapted to it. Although winter is now the favourable season, only a few trees or shrubs have the trick of losing their leaves in summer; most of them are either evergreen or lose their leaves in winter, a relic of their ancestry in a climate with different seasons.

Endemics

The Mediterranean has many *endemic* plants confined to a particular area. Spanish live-oak, *Quercus rotundifolia*, one of the commonest trees in Spain and Portugal (and Morocco), does not cross the Pyrenees. *Genista balearica* is a species of broom confined to Majorca, where it is one of the commonest plants. *Verbascum spinosum* is an extraordinary spiny mullein confined to Crete, and within Crete almost confined to the region of Sphakiá, where it is one of the commonest plants from sea-level almost to the highest mountains (Figs. 4.1, 4.2). *Petromarula pinnata* is an arrestingly beautiful herbaceous plant, common on cliffs and Venetian palaces and churches all over Crete. *Bupleurum kakiskalae* is confined to just one cliff in Sphakiá.

Some endemics evolved on the spot during the Pleistocene: *Quercus rotundifolia* and *Genista balearica*, closely related to other oaks and brooms, are examples. Many, especially island and mountain endemics, have a longer evolutionary history: they have sat through the Pleistocene climatic upheavals, and appear to be connected (in complex and mysterious ways) with the origins of islands and mountains in the late Tertiary.[2] Examples are *Verbascum spinosum* and *Bupleurum kakiskalae*, which are not closely related to others of their genera, and still more *Petromarula*, a genus on its own. Other endemics are remnants of a wide distribution that failed to survive the climatic upheavals. The elm-like tree *Zelkova* once extended continuously across Asia, Europe and North America; there are now only isolated species in Japan, China, the Caucasus area and high-mountain Crete.[3]

Mediterranean plants, especially endemics, tend to be species of open places; those that like shade (e.g. *Cyclamen creticum*) mostly prefer the shade of rocks to that of trees. In

[1] *Job* 14: 8–9 [5th century BC].
[2] W. Greuter, 'The origins and evolution of island floras as exemplified by the Aegean archipelago', *Plants and Islands*, ed. D. Bramwell,

Academic Press, 1979, pp. 87–106.
[3] D. K. Ferguson, L. Yushang, R. Zetter, 'The paleoendemic plants of East Asia: evidence from the fossil record for changing distribution

patterns', *The changing face of East Asia during the Tertiary and Quaternary*, ed. N. G. Jabelowski, Centre of Asian Studies, University of Hong Kong, 1997, pp. 359–72.

Fig. 4.1 *Verbascum spinosum.* This is the supreme example of a plant adapting to resist browsing. Poisons and distasteful fluff protect all mulleins, but this, being an island endemic, is impenetrably spiny as well. *Anópolis, west Crete, July 1987*

SW Spain, heathlands contain the highest proportion of endemics and the lowest proportion of widespread species, forests being the opposite.[4] Mediterranean forest is a harsh environment, with dense evergreen shade and extreme competition for moisture; it is not a rich habitat.[5] On a geological and evolutionary time-scale, there have been extensive forests in the Mediterranean mainly during interglacials, which occupy about one-fifth of the time; comparatively few plants have adapted to living in forest.

Fig. 4.2 Cretan (and world) distribution of *Verbascum spinosum.* Each dot represents a 2-km square.

Introduced plants

Plants have been introduced from other countries as crops or weeds or by accident. A few are spectacularly abundant: *Oxalis pes-caprae,* brought from South Africa earlier in the twentieth century, happens to flourish under ploughing, and is so abundant in olive-groves as to colour wide tracts of Crete in satellite images. The Chinese tree *Ailanthus altissima* may prove to be the most lasting legacy to Crete of Sir Arthur Evans, the archaeologist. It spreads by suckers and is impossible to contain or get rid of. It has taken over Knossós, where he brought it early in the twentieth century. False-acacia, *Robinia pseudacacia,* an eastern North American tree, has been widely planted in France and Italy, has created woods of its own, and has suckered into native woodland.

Plants from other continents usually remain in or near cultivated areas and do not spread to the landscape at large. Mediterranean wild vegetation has a mysterious resistance to exotic invasions. Much of California, in a similar climate, has been overrun by Mediterranean plants, especially annual grasses;[6] but the Mediterranean is not, in general, overrun by Californian plants.

WILD VEGETATION

Forest, as Scandinavians or Americans know it, is rather limited in the Mediterranean; the trees are often short, crooked and hard. Trees do not imply forest: there are two intermediates between forest and non-forest. The trees may be reduced to the stature of shrubs, constituting *maquis* or *macchia.*[7] The point at which maquis turns into forest or woodland is arbitrary; let us call it 5 m high.[8] Or trees may be scattered among other vegetation, constituting *savanna* (Chapter 12).

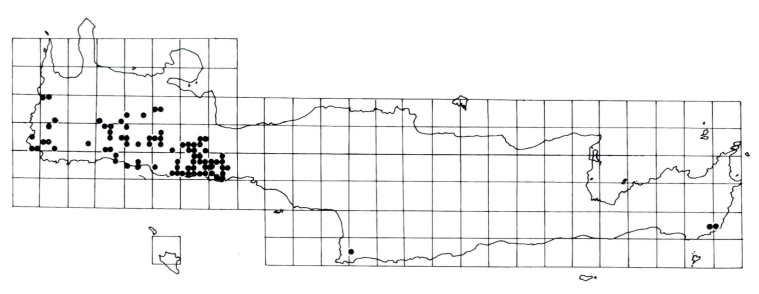

4   F. Ojeda, J. Arroyo, T. Marañon, 'Biodiversity components and conservation of Mediterranean heathlands in southern Spain', *Biological Conservation* 72, 1995, pp. 61–72.
5   Naveh 1994b.
6   L. E. Jackson, 'Ecological origins of California's Mediterranean grasses', *JB* 12, 1985, pp. 349–61.

7   Maquis is a term used in different senses by various authors. We use it to mean shrub vegetation of any height and on any substrate.
8   We do not make the distinction, which some do in Italy, between 'forest' in the sense of vegetation unaffected by human activity, and 'wood-

land' which has been managed by woodcutting, etc. As far as we are concerned, all Mediterranean vegetation (except for *very* inaccessible cliffs, and some very recent types) has been subjected to human activities.

Non-tree vegetation is often characterized by *undershrubs* or 'dwarf shrubs' – for example thyme, sage, many species of broom, species of *Cistus* and *Phlomis,* and the Cretan endemic *Ebenus cretica*. We shall call undershrubby vegetation *phrygana*.[9] Undershrubs are aromatic, and produce the scents and the honey of the dry Mediterranean.

Vegetation characterized by herbaceous (that is, non-woody) plants, including grasses, is called *steppe.*

In savanna the vegetation between the trees may be shrubs (of the same species as the trees or a different species), phrygana, steppe, some combination of these, or cultivated land.

Intermediates make nonsense of any simple attempt to measure forest area, official statistics of which are unreliable or meaningless (p. 19). In vast areas of Greece or Spain it is a matter of opinion whether trees are big enough (in maquis) or close enough (in savanna) to constitute forest; the very languages have no unambiguous word for 'forest'. This leads to mysterious statements such as 'Spain has 25 per cent forest land, but half of this has been deforested'.[10]

Comparison with other Mediterranean-type climates[11]

How has vegetation developed in parts of the world with similar climates but very different biogeographical connexions and human histories?

California is the only such region to have any related native plants. It recalls the Mediterranean in its evergreen and deciduous oaks and evergreen arbutuses, which can be either trees or shrubs. Forests are relatively local; more characteristic are *chaparral* thickets of tall, flammable undershrubs such as *Ceanothus* and *Adenostoma*, resembling (though not related to) the big *Cistus* species of the west Mediterranean. Evergreen-oak savannas recall those of Spain.

In Australia the plants, though utterly unrelated, form plant communities and mosaics looking superficially like those of the Mediterranean: eucalyptus forests, dwarf eucalypts (*mallee*), savannas ('woodland'), tall and dwarf undershrubs (*kwongan*), steppe. Some eucalypts, such as jarrah, can exist in tree or mallee form. However, eucalyptus forests reach 80 m high, which no Mediterranean forest is likely to have done; and the undershrubs, like chaparral but unlike phrygana, are long-lived and sprout after fire.

Chile has short, crooked, drought-resistant evergreen forests of *boldo* (*Peumus boldus*, Monimiaceae) and *quillai* (*Quillaja saponaria*, Rosaceae), as well as chaparral-like undershrubs and maquis-like thorn-thickets. The latter often grow in dispersed, savanna-like formations.

South Africa is least like the Mediterranean, with its *fynbos*, composed of a huge variety of tall and short undershrubs. Native trees are remarkably few.[12]

Mediterraneoid regions have these features in common:
(1) forests are not ubiquitous, often confined to damper areas;
(2) undershrubs are prominent;
(3) savanna is prominent (except in South Africa);
(4) fire is a dominating factor (except perhaps in Chile): most plants are highly fire-adapted and often fire-dependent (p. 218).

Compared with South Africa and Australia, the Mediterranean has relatively fertile soils, the result of tectonics and erosion exposing new material. Exceptions include the stable Hercynian regions of Portugal, Spain, and Sardinia. All the other mediterraneoids adjoin lower-latitude land-masses into which plants sensitive to cold could retreat during glaciations; in the European Mediterranean, however, there was nowhere to go, especially for island endemics.

Mediterraneoid communities are easily invaded by plants introduced from the Mediterranean or elsewhere. *Fynbos*, especially, is threatened by the invasion of exotic trees.

	Mediterranean	California	Mid Chile	Cape of Good Hope	SW Australia
Endemic plants	+++	+++	+++	++++	++++
Evergreen forest	+++	++	+++	0	+++
Summer-deciduous forest	++	0	0	0	0
Savanna	++++	+++	+++	0	++++
Shrubland	++++ (maquis)	+++	++++ (matorral)	++++ (fynbos)	++++ (mallee)
Undershrubs	++++ (phrygana, etc.)	++++ (chaparral)	++++	+++	++++ (kwongan)
Fire	+++	++++	++	++++	++++
Islands	++++	++	0	0	+
Introduced plants in wild vegetation	+	+++	+++	++++	++

Table 4.1. Comparison between the flora and vegetation of Mediterranean and mediterraneoid regions.

[9] We eschew the term *garrigue* as different authors use it inconsistently, sometimes for plant communities containing shrubs as well as undershrubs.
[10] P. Pointereau, E. de Miguel, D. Hickie, *Afforestation of agricultural land – guidelines for environmental assessment*, Commission of the European Communities, 1993, p. 3.
[11] This section is derived from Di Castri & Mooney 1973 and Dallman 1998; also from our own observations in Australia.
[12] R. Cowling, ed., *The ecology of fynbos: nutrients, fire and diversity*, Oxford University Press, 1992.

PROPERTIES OF SOME IMPORTANT PLANTS

Broadleaved evergreen trees and shrubs – maquis

Probably the most versatile European plant is prickly-oak or kermes oak, *Quercus coccifera*. It can take any form from a shrub a few cm high (a *ground-oak*) to a tree 20 m high, with a trunk 4 m in diameter and a thousand years old (Figs. 4.3, 4.5). It is gregarious, growing in large numbers together, not single individuals. At only 60 cm high, prickly-oak performs all the functions of an oak-tree, including producing acorns and pollen. It is not killed by burning, felling or browsing. It burns quite easily but sprouts vigorously after burning (Figs. 13.4, 13.5), sprouts after felling, is moderately palatable to goats but sprouts after browsing. It can be treated by *coppicing* – that is, by being cut down every so many years and allowed to grow again from a permanent base called a *stool*. It can be *pollarded*, cut like a coppice stool but at 3–4 m above ground; or *shredded*, allowed to become a tall tree from which the side-branches are periodically cut for wood or foliage. It is very long-lived in all states, and is seldom seen growing from seed.

If not burnt, felled or browsed for a few years, prickly-oak grows into a columnar shape and then into a tree (Fig. 4.4). A critical stage is the *get-away* height at which a goat can no longer reach the growing apex. The time to reach get-away is determined by site and weather; a couple of wet seasons can help. The ability to grow into a tree increases from west to east. In Spain prickly-oak is rarely seen as more than a shrub, but in Crete it is the principal savanna tree.[13]

Holm-oak, *Quercus ilex* – the evergreen oak of the Balearics, Corsica and Italy, common in Sardinia, uncommon in Greece and Crete – has many of the properties of prickly-oak, but cannot normally exist as a ground-oak.[14] Holm-oak readily recovers (by sprouting) from fire and woodcutting, but is less resistant to browsing than prickly-oak. It is shade-tolerant and more of a woodland than a savanna tree. Holm-oak and prickly-oak are very deep-rooted, and can grow in either rock-crevices or soil.[15]

In Spain and Portugal, instead of holm-oak there is *Q. rotundifolia*, which we shall call 'live-oak' (Spanish *encina*, Portuguese *azinheiro*). This is the commonest savanna tree in Spain – over wide areas it is the only tree. It is often called 'holm-oak' through confusion with *Q. ilex*, but ecologically and in appearance has more in common with *Q. coccifera*. Unlike *ilex* it tends to be shallow-rooted and needs at least a little soil; it appears to prefer non-limestone soils. It is very drought-resistant. It has a strong ground-oak form.[16]

Fig. 4.3. Coppicing, pollarding, shredding and lopping. Each tree is shown before cutting, after cutting, and one year after cutting.

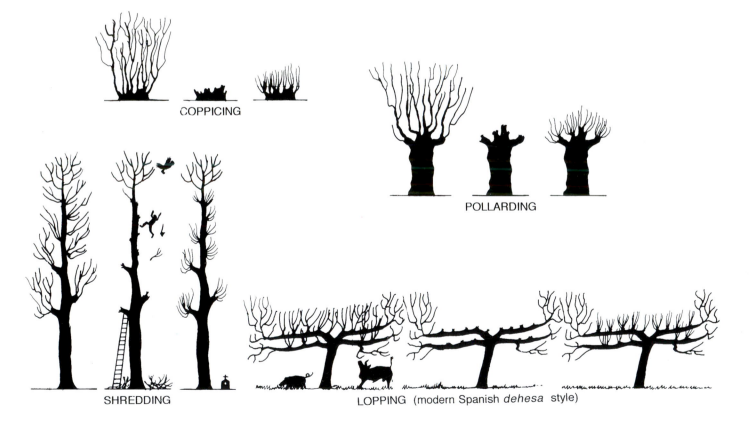

COPPICING

POLLARDING

SHREDDING

LOPPING (modern Spanish *dehesa* style)

[13] East Mediterranean botanists often call the tree-forming prickly-oak a separate species, *Quercus calliprinos*. The idea may be sound (provided it is recognised that the ground-oak in the same locality is also *calliprinos*), but the name is mistaken: the original *calliprinos* was described from Spain.

[14] There are ground-oak forms of holm-oak in Majorca and in one locality in east Crete.

[15] Travellers in Greece and Crete often mistake the tree form of prickly-oak for holm-oak.

[16] European ground-oaks are analogous to the dwarf 'shinnery' oaks of North America – the latter, however, are separate species or subspecies, which cannot turn into full-sized trees as ground-oaks can.

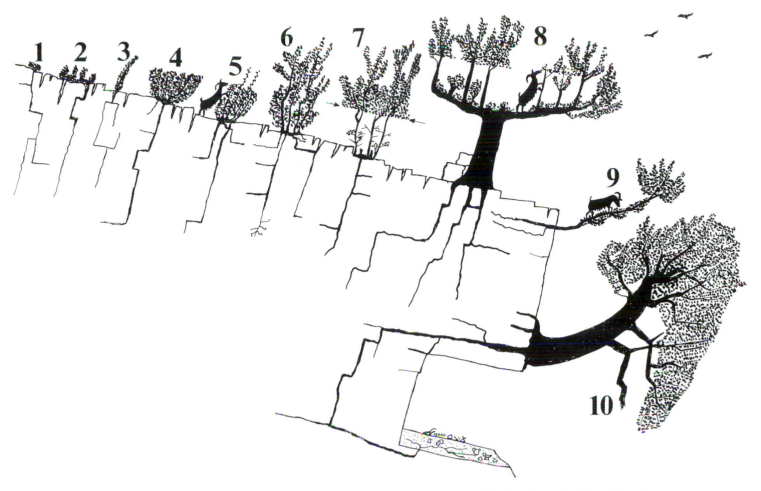

Fig. 4.4. Forms assumed by *Quercus coccifera* under different intensities of browsing. 1 cushion. 2: topiary. 3: columnar. 4: thicket. 5: just after get-away. 6: recently got-away. 7: re-browsed after get-away, showing browse-line. 8: goat-pollard. 9: special form on edge of cliff. 10: never browsed.

Cork-oak, *Quercus suber*, is the typical savanna oak of Portugal and Sardinia, and is locally abundant in Spain. It is less common in forests.[17] It is a very light-demanding tree: in Sardinia it overlaps with holm-oak and is easily displaced by it if the savanna is allowed to turn into forest. It has a ground-oak form. It is confined to relatively acid soils on granites, metamorphic rocks and their weathering products, typically in damper places than live-oak. Cork-oak is only just evergreen; in April it shows 'autumn' colours and loses its old leaves as the new ones emerge. Cork is a product of the trunk and bigger boughs, and is part of the tree's natural adaptation to fire (p. 218). When the tree is about 30 years old the grower strips off the natural cork. The peeled surfaces then grow commercial-quality cork, which is harvested every nine years or so and grows again.

Prickly-oak, live-oak and cork-oak (but not usually holm-oak) can be trees or shrubs depending on circumstances. Many other evergreen trees can be shrubs, collectively forming the thick-leaved evergreen maquis that plant geographers regard as 'typical' Mediterranean vegetation. They include phillyrea (*Phillyrea media*), lentisk (*Pistacia lentiscus*), wild olive (*Olea...sylvestris* [18]), and carob (*Ceratonia siliqua*, an ancient introduction, p. 69). These are long-lived, deep-rooted, relatively palatable and combustible, but not killed by woodcutting, burning or browsing. Tree-heather (*Erica arborea*) and arbutus (*Arbutus unedo*) are similar but shallow-rooted and are confined to water-retaining, usually non-calcareous soils; they are highly flammable.

Coppicing, shredding and pollarding are obvious interactions between trees and human affairs. Some trees play these roles: others, even of related species, refuse. Fire sometimes replicates the effects of coppicing, although many non-flammable trees will coppice. Some trees will sprout from their roots, forming patches of suckers. Each patch is called a *clone*, and is genetically different from other patches. In thickets of prickly-oak maquis the individual clones are often strikingly different in colour in May, when the new leaves and catkins come out (Fig. 4.5c).

---

[17] Occasionally it is planted in pure stands, although evergreen oaks do not transplant well in the Mediterranean climate.

[18] For plant names that are trinomials (genus, species, subspecies) we have invented the abbreviation of replacing the species (middle) name by points: *Olea europaea* subspecies *sylvestris* becomes *Olea...sylvestris*.

Fig. 4.5. Forms of prickly-oak.

(a) *Top left* Bitten within an inch of its life. *Mount Psilorítis, Crete, May 1989*

(b) *Second left* As a shrub ½ m high, overtopped by the grass *Brachypodium retusum* but producing acorns. *Vathy, Sámos, September 1978*

(c) *Left* As near-continuous maquis: the different colours represent different individual clones. *Mount Yúktas, middle Crete, May 1988*

(d) *Above* Almost at the get-away point. This individual has been browsed down by Cretan wild-goats, but is about to grow out of reach. *Samariá Gorge, Crete, May 1989*

(e) *Below left* As a goat-pollard, browsed by climbing goats (Fig. 4.4). *Pass into Katharó Plain, east Crete, August 1986*

(f) *Below* As a big tree. This is one of those select 'pixel trees' that are visible from space on SPOT satellite images. *Near Mourí, Sphakiá, Crete, April 1989*

Fig. 4.6. Strawberry-tree, *Arbutus unedo,* two years after felling; it is already flowering again. *Near Orbetello, Tuscany, August 1980*

Fig. 4.7. Shredded ash-trees, *Fraxinus angustifolia,* a tree very palatable to goats. *Sierra de la Horconera, Andalusia, April 1994*

Conifers

There are two groups of Mediterranean pines. Lowland pines include *Pinus halepensis, P. pinaster*[19] and *P. pinea.* Mountain pines include Corsican pine, *P....laricio,* and its relatives the black pines (*P. nigra*) of the mountains of Spain, Italy and Greece. *P. brutia* is confined in Europe to Crete, where it is the only native pine in both lowlands and mountains (Fig. 17.14). *P. pinea* has edible seeds, and has been cultivated so long that its native limits are uncertain.

Pines exist only as trees (except *P. leucodermis,* a Balkan mountain pine, which has a shrub form). They are gregarious; they are killed by cutting them down, and are not coppiceable; some can be pollarded; they are unpalatable to browsing animals; they grow where there is soil or scree rather than rock.

Pines are very flammable, and fire is part of their ecology (Chapter 13). They react in various ways. *P. halepensis* usually dies when burnt, but grows readily from seed. *P. nigra* survives fire through its thick heat-resistant bark. *P. brutia* is intermediate.[20]

Pines do not only yield timber. They have often been more valued for resin. They produce great quantities of honey, made by bees from the secretions of the tree itself or of insects that attack it.

Other conifers are non-resinous[21] and do not promote fires. *Abies* (fir) is a mountain genus. *A. alba* strays into the Apennines from central Europe; *A. cephalonica* is endemic to Greece. Firs are gregarious, moderately palatable, killed by felling or burning, not coppiceable but can be pollarded or shred. They commonly grow in limestone rock-fissures rather than soil. *A. cephalonica* can live bitten down like a ground-oak (Fig. 4.8).

Cypress, *Cupressus sempervirens,* is the characteristic tree of Crete. It is not native elsewhere in Europe, but a garden variety, the tall narrow 'vertical cypress', has been cultivated

---

[19] For its natural ecology, see L. Gil and 4 others, '*Pinus pinaster* Aiton en el paisaje vegetal de la Península Ibérica', *Ecologia,* fuera de serie 1, 1990, pp. 469–95.

[20] Z. Eron, 'Ecological factors restricting the regeneration of *Pinus brutia* in Turkey', *EM* 13 (4), 1987, pp. 57–65.

[21] Except in France, where all conifers are called *résineux.*

	Abundant in	Coppicing	Suckering	Pollarding	Rooting depth	Palatability	Survival of browsing	Combustibility	Survival of fire	Occurrence on cliffs	Maquis	Savanna	Recent expansion
Evergreen oaks													
Quercus rotundifolia, live-oak	Portugal, Spain	+++	+	++++	++	+++	+++	++	+++	+	++	++++	++
Q. suber, cork-oak	Portugal to Sardinia	++++	+	+++	+++	+++	+++	++	++++	+	++	++++	++
Q. ilex, holm-oak	Sardinia to Greece	++++	–	+++	++++	++++	(+)	+++	++++	++++	++++	+++	0
Q. coccifera, prickly-oak	Greece to Turkey	+++	+	+++	++++	+++	++++	+++	++++	++	++++	++++	++
Other evergreen broadleaves													
Olea...sylvestris, wild-olive	mainly eastern	++	–	+	+++	+++	++++	+++	++++	+++	++++	+	+
Pistacia lentiscus, lentisk	general near sea	+++	+	0	++++	–	?	++	++++	++	++++	+	+
Arbutus unedo, arbutus	general	++++	–	+	++	++	+++	+++	++++	+	++++	–	+
Ceratonia siliqua, carob	widespread near sea	++	–	++++	?	+++	++++	+++	+++	++	+++	++++	0
Conifers													
Pinus halepensis, Aleppo pine	France to Greece	–	–	+	++	–	++	++++	–	–	–	–	++++
P. pinea, umbrella pine	Spain, Italy, S.Greece	–	–	–	+	–	++	++	+	–	–	++	+++
P. brutia, Cretan pine	Crete, E. Aegean	–	–	+	++	–	++	++++	++	–	–	–	+++
Cupressus sempervirens, cypress (wild)	Crete, E. Aegean	++	–	++	+++	+	++	++	+	++	–	++	+++
Wild deciduous broadleaves													
Quercus pubescens and *brachyphylla* oak	Sardinia to Crete	++++	–	++++	+	+++	++	++	+++	–	++	+++	+++
Q. cerris, Turkey-oak	S. France to Greece	++++	–	++++	+	+++	++	++	+++	–	+	+++	++
Q. macrolepis, valonia oak	Greece, Crete, Aegean	+++	–	+++	+	++	?	++	+++	–	–	+	++
Fraxinus spp, ash	Spain to Greece	++++	–	+++	+	++++	++	–	++	++++	++	+	0
Mountain trees													
Fagus sylvatica, beech	Italy, Balkans	++	+	++	+	++	++	–	?	–	–	+	++
Pinus nigra, black pine	Spain to Greece	–	–	+	++	+	+	+++	++	–	–	–	++
Abies spp, fir	Italy to Greece	–	–	–	+	++	+++	–	–	++	–	+	0
Ground-water trees													
Ulmus minor, elm	Spain to Greece	–	++++	+++	?	++++	–	–	+++	+	–	+	+
Alnus glutinosa, alder	Corsica, Sardinia, Italy	++++	–	0	++	++	++	–	?	–	–	0	0
Salix spp, willow	Spain to Crete	+++	–	++++	?	++++	–	–	–	–	–	0	0
Platanus orientalis, plane	Greece, Crete, Aegean	++++	–	++++	?	++	++	+	+++	–	–	+	+
Cultivated trees													
Olea europaea, olive	everywhere exc. Sardinia	+++	+	++++	+	++++	++++	+++	+++	+++	–	+++	++++
Castanea sativa, chestnut	Italy to W. Crete	+++	–	+++	+	++	0	+	++	–	–	++++	+
Juglans regia, walnut	everywhere	0	–	–	?	+++	0	–	++	–	–	–	++
Robinia pseudacacia, false-acacia	S. France, Italy	++	++++	0	++	++	?	+	+++	–	–	–	++++
Eucalyptus camaldulensis, eucalyptus	Portugal to Sardinia	++++	–	0	?	–	0	++++	++++	0	–	–	++++

Table 4.ii. Properties of some commoner trees. 0 means that the property exists but is seldom called into use; in the last column 0 means 'no expansion'.

Fig. 4.8. *Abies cephalonica,* showing various stages from a bitten-down bush to an upstanding fir-tree. *Mount Hélikon, Bœotia, Greece, August 1979*

for ornament (and latterly timber) since ancient times.[22] Wild cypress is gregarious, fire-sensitive and unpalatable (Fig. 4.9), and grows on rocks in preference to soil. Very unusually among conifers, it sprouts if felled when young, and can be regularly coppiced (Fig. 4.10).

Junipers include *Juniperus communis*, a lowland tree in Italy and Sardinia; *J. oxycedrus*, widespread in mountains; and *J. macrocarpa*, the mysterious 'cedar' of remote sand-dunes in Crete and its outlying islets. These are generally gregarious, very sensitive to burning and the competition of taller trees, unpalatable, and have some tendency to sprout.

Fig. 4.10. Coppice stool of cypress in a very remote place. Such a stool results from centuries of felling and regrowth. *Eligiás Gorge, Sphakiá, Crete, July 1989*

Fig. 4.9. (a) *Top* Goat browsing prickly-oak in preference to cypress: she does not mind harsh textures but dislikes strong flavours.

Fig. 4.9. (b) *Above* Prickly-oak bitten back but cypress merely nibbled. *Both from Omalós, west Crete, July 1981*

---

[22] The name *sempervirens* belongs to the cultivar; wild cypress is properly C....*horizontalis.*

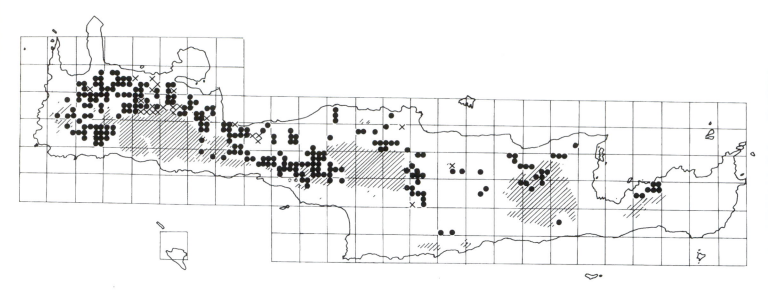

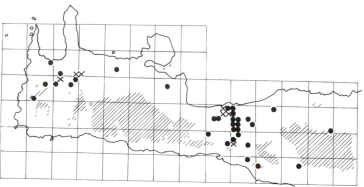

Figs. 4.11a, b. Distributions of deciduous oaks in Crete. Top: *Quercus brachyphylla*. Above: *Q. macrolepis*. Land over 800 m is hatched.

Deciduous trees

The twenty-odd Mediterranean deciduous oaks include *Quercus pyrenaica* (Portugal and Spain, especially at higher altitudes), *Q. cerris* (called Turkey Oak, but more typical of Italy), *Q. macrolepis* (Aegean), *Q. pubescens* (widespread), *Q. frainetto* (mainly Balkans and north Greece), and *Q. brachyphylla* (the common deciduous oak of Crete and south Greece). *Q. macrolepis* (Crete, S. Greece, Cyclades) is the valonia oak, valued for its huge acorn-cups as a source of tannin. These trees are gregarious, moderately palatable, not very flammable; they sprout after felling or after a fire fierce enough to kill them to the ground. They seldom have ground-oak forms, but in north Greece (and a small part of Crete) a non-flowering ground-oak adaptation has evolved in at least two species (Fig. 4.12). Deciduous oaks are shallow-rooted and confined to moisture-retaining soils with good root penetration.

Fig. 4.12a. *Quercus brachyphylla* as tree and as ground-oak (in foreground). *Mesonísi, Amári, Crete, August 1997*

Fig. 4.12b. Ground-oak foliage. *Polydhendhro, Grevená, north Greece, May 1988*

23 W. Greuter, 'L'apport de l'homme à la flore spontanée de la Crète', *Boissiera* 19, 1971, pp. 329–37.

Phytosociologists think of deciduous oaks as an anomaly in low-altitude Mediterranean vegetation,[23] but they are widespread in Greece, Crete and Italy on water-retaining soils – marls, gravels, Pleistocene alluvium, sandstones and some metamorphics. In Crete they are strongly associated with high-rainfall areas (Fig. 4.11). They are relicts of earlier Holocene forests and savannas (Chapter 10).

Other deciduous trees include hornbeam (*Ostrya carpinifolia*), terebinth (*Pistacia terebinthus*) and manna-ash (*Fraxinus ornus*). In general these are less gregarious than oaks, non-flammable, very palatable and readily sprouting; they do not have a bitten-down form.

Ground-water trees

Plane, willow, poplars, oleander and alder grow by streams and torrents and in places that have a water-table. Elms in southern Europe are usually ground-water trees; they sucker and are strongly clonal. In the Sierra de Gádor (SE Spain) there is the remarkable spectacle of a semi-desert, generally too dry for tree growth, in which the only wild trees are occasional elms seeking out small seeps of water in the dolomite.

Northern trees

Trees from middle and north Europe extend into the Mediterranean – for example, beech, lime (*Tilia* species), hornbeam (*Carpinus betulus*), birch (*Betula* species), and elm and alder in Greece. Normally these are at high altitudes, but they formerly occurred in the lowlands (Chapter 10).

Tree management – timber, wood, leaves

In Europe trees do not equal timber. A distinction that pervades European (but not American) vegetation history is between *timber* (*madera, legname, bois d'œuvre, materia,* etc., the stuff of beams, planks and shipbuilding) and *wood* (*leña, legna, bois d'industrie, lignum,* etc., the stuff of firewood, wattlework, charcoal, etc.). Timber is got by felling big trees; wood comes mainly from coppicing and pollarding, and is expected to grow again. The distinction depends partly on the tree: oleander is too slender to be timber, and pine is a poor wood-producer because it refuses to sprout. It is also a matter of cultural preference: Spaniards regard live-oak as a wood-producing tree, although there is nothing to stop them from using it for timber. Wood is also a by-product of cutting timber (branches, offcuts, etc.) and of pruning olives, fruit-trees and vines.

Trees that sprout after felling produce successive crops of wood by coppicing. The sprouts can also be eaten by cattle, sheep and goats; sometimes they are cut for *leaf-fodder* (foliage dried and stored instead of hay) rather than wood. Where animals cannot be kept away, it is commonly the practice instead to manage the trees by pollarding or shredding for crops of wood or leaves (but at some risk to the woodcutter's neck[24]). Shredding keeps the option of felling the tree for timber later.

Fig. 4.13. Group of young shredded deciduous oaks, with stack of leaf-fodder; ground-oak in foreground. *Polydhendhro, Grevená, north Greece, May 1988*

Pollarding and shredding are conservation practices: they enable wood and leaves to be harvested without killing the tree, which often lives longer than it would if left alone. They are more characteristic of savanna than of woodland.

Undershrubs – phrygana

Undershrubs are fundamentally different from shrubs. They are not potential trees; they are short-lived and reproduce by seed; most species are gregarious; most are unpalatable; after a fire most of them die, and are often replaced by a different species of undershrub. They are aromatic and insect-pollinated. Not all undershrubs have all these properties – for example *Ebenus* is long-lived and palatable, and sprouts after fire – but there is rarely any difficulty about what is a shrub and what is an undershrub.

Undershrubs are shallow-rooted and drought-resistant, often growing where there is insufficient root-penetration for trees or shrubs. There are many kinds of undershrubbery, from the dense thickets of *Cistus ladanifera* in Spain and Portugal, 3 m high, covered in sticky flammable resin which catches explorers like ortolans in bird-lime, to the delicate bushlets of *Fumana ericoides*, 10 cm high, on dry islets. The Spanish language is rich in phrygana words: *romeral* for rosemary thickets, *jaral* or *xaral* for *Cistus* thickets, *tomillar* for thyme phrygana, etc. A speciality of the east Mediterranean are 'chicken-wire bushes' such as *Sarcopoterium spinosum*, tangles of spines diverging at 120°.

Steppe

Steppe plants include these diverse types:
● Perennial grasses, for example the tufted *Stipa tenacissima* and *Lygeum spartum* (esparto grasses), and *Ampelodesmus tenax*, the 'elephant-grass' of the west Mediterranean;

---

[24] A statute of Varese (Liguria) in 1628 prohibited shredding 'because often the men who climb on the said trees, some fall and some hit themselves'; Moreno 1990, p. 225.

- Annual grasses, such as various species of *Bromus;*
- Annual clovers and other legumes, such as species of *Medicago* and *Onobrychis;*
- Annual and perennial dandelion-like 'yellow Compositae', especially species of *Taraxacum* and *Crepis;*
- Bulbous and tuberous perennials, such as asphodels, aroids and orchids.

Pasture value lies especially in the softer grasses, yellow Compositae, and legumes. Espartos, asphodels and aroids are tough, distasteful or poisonous. Where steppe is dense enough to burn or grows between flammable shrubs, many plants are encouraged by fire.

Old trees

Pollarding and coppicing prolong the lives of those trees that respond. Ancient pollards and coppice stools, a metre or more in diameter and several centuries old, are to be found in most Mediterranean countries, supremely in Greece and Crete with their ancient evergreen and deciduous oaks, pines, junipers, planes, cypresses and many others. Coppice stools tend to be in woodland;[25] ancient pollards in savanna (plus ancient cultivated olives, chestnuts and carobs). The annual rings of both speak of their history and of what has gone on around them.

Fig. 4.15. Pollard deciduous oak two fields away from Zeus's temple at Dodóna, NW Greece. The Dodonaean oracle spoke in the wind rustling in the boughs of such an ancient oak. Modern commentators tend to deny that there are still oaks at Dodóna: it is easier to copy a previous commentator than to go there and find out. *May 1988*

Fig. 4.14. Giant coppice stool of holm-oak on the Roman site of Sette Finestre near Orbetello, Tuscany. *August 1980*

Wonderful ancient trees are part of the drama and glamour of the remoter Mediterranean; giant chestnuts in the Alpujarra and Corsica, mighty pines and shredded oaks in the Píndos Mountains, the vast cypresses and sweetgums of SW Turkey, and the supreme diversity of west Crete. They are of the greatest importance as a habitat. They are an aspect of wildwood now best represented in savanna. Hollows, redrotted interiors, old dry bark under overhangs, loose bark, etc., are the homes of peculiar insects, spiders, lichens and

fungi. Many birds, butterflies and other animals require some highly specific habitat such as a particular type of tree-hole. A single ancient oak is a whole ecosystem of different habitats which ten thousand middle-aged oaks do not provide at all.[26]

Fieldwork runs into four difficulties. There may be few ancient trees. Their annual rings, especially with evergreen oaks, may be difficult to count. People are often reluctant to cut down trees, or legally inhibited from doing so, so that it is difficult to find stumps to estimate their ages. Mediterranean oaks are exceedingly hard and often hollow, so that it is difficult to take cores for ring-counting with a hollow borer.

Fig. 4.16. Ancient carobs in savanna. *Sykológos, SE Crete, July 1968*

[25] It is said in France that stools become 'exhausted' by repeated felling and need to be replaced from seed. Numerous flourishing ancient stools disprove this notion.
[26] For British examples, see P. T. Harding and F. Rose, *Pasture-Woodlands in Lowland Britain: a review of their importance for wildlife conservation*, Institute of Terrestrial Ecology, Monks Wood, 1986.

ORGANIZATION OF PLANT COMMUNITIES

Plant associations versus mosaics

In the western Mediterranean, maquis, phrygana and steppe tend each to occupy considerable areas, and can be described as separate units. Plant sociologists of the Zürich–Montpellier school have described many such in the south of France and in Catalonia.[27]

The method is less successful farther from Montpellier. As one moves away from the places where phytosociological units were first described, the plant species that define them drop out one by one on reaching their geographical limits, and may be replaced piecemeal by other species; but there is another reason.

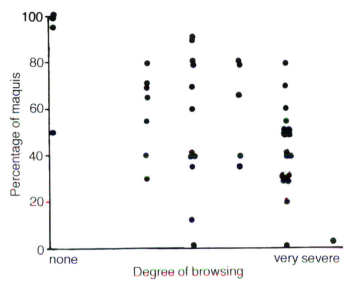

Fig. 4.17. Relationship between (x) the degree to which tracts of vegetation show the effects of browsing (as indicated by the shapes of bushes, Fig. 4.4) and (y) the percentage of each site that is maquis rather than phrygana, steppe or bare rock. From 46 sites in Bœotia, Greece. The extent of maquis is clearly almost independent of the amount of browsing. After Rackham (1983).

In the east Mediterranean vegetation tends to be a mosaic of patches of maquis, phrygana and steppe, rather than continuous areas of one or the other. A typical Greek hillside is mottled with areas of dark-green shrubs, grey-green undershrubs, and yellow (in summer) grasses and herbs (Fig. 4.18a). Each component can vary independently of the others.[28] Spanish savannas can have three components: trees, steppe under the trees, and a different steppe between the trees (p. 199).

Shrubs, undershrubs and herbs have ways of life, sometimes complementary, sometimes competing intensely with each other. Undershrubs have finite life-spans: areas of them exhibit 'colonizing', 'building' and 'declining' phases, often set off by the most recent fire.[29]

Determination of maquis, phrygana and steppe

Ecologists traditionally regard forest as the 'climax' vegetation of the Mediterranean, treating maquis (or savanna) as 'degraded' forest, phrygana as degraded maquis, and steppe as degraded phrygana. This is based on the American theory that any area left to itself, given enough time, will progress through intermediate stages into a stable, climatically determined ecosystem. It has never been wholly convincing: it has rarely been possible to demonstrate that where the three formations adjoin, the steppe has (as a matter of history) received more degradative treatment than the phrygana, and the phrygana more than the maquis (cf Fig. 4.17).[30] Moreover, analogues of savanna, maquis, phrygana and steppe turn up in other mediterraneoid countries, although their human histories as well as their plant species are utterly different. Climax theory is unsuited to the Mediterranean, where the environment is never stable for long enough, at least during interglacials; but its dead hand still weighs on southern European ecologists.[31]

In practice the determinant is likely to be moisture: a combination of (1) rainfall, (2) how much rainfall is retained by the soil and bedrock, and (3) how readily soil and bedrock can be penetrated by roots to get at the moisture. Maquis tends to be more prevalent under high rainfall and on moisture-retaining rocks; also at the foot of slopes, even if browsing is more intense (Fig. 4.18b). Where root penetration is better, for example around the edges of screes, there is more tree and shrub growth (Fig. 12.4). Road-cuts sometimes reveal that a single tree marks some discontinuity, such as an ancient tomb or a geological fault, which lets its roots get into otherwise impenetrable rock.

On rocks such as flysch, with middling to poor water retention but good root penetration, maquis may be almost continuous even under heavy browsing. Limestone exhibits a mosaic of maquis, phrygana and steppe, related to local differences in root penetration and water retention.[32]

We have often observed that, where soil and rock alternate, areas that still have soil tend to be grassland; areas that have lost their soil, or never had any, are treed (Fig. 4.19). Around Mount Parnássos (Bœotia) the soil-filled karst hollows are grassland, while the bare, fissured limestone is fir-wood. In the very different environment of the Gennargentu Mountains in Sardinia, rocky areas are usually treed, whereas areas with deep soils are grassland or savanna.[33]

Figure axis labels: Percentage of maquis (y-axis: 0, 20, 40, 60, 80, 100); Degree of browsing (x-axis: none ... very severe)

[27] For example, J. Braun-Blanquet, N. Roussie, R. Nègre, *Les Groupements Végétaux de la France Méditerranénne*, CNRS [Montpellier], 1951; Folch i Guillèn 1981.

[28] Rackham 1983.

[29] S. C. Clark, J. Puigdefábrigas, I. Woodward, 'Aspects of the ecology of the shrub–winter annual communities of the Mediterranean basin', Mairota and others 1998, pp. 44–7.

[30] In North America, Australia and South Africa, *increase* of woody vegetation at the expense of grassland is regarded as degradation: M. T. Hoffman, W. J. Bond, W. D. Stock, 'Desertification of the Eastern Karoo, South Africa: conflicting paleoecological, historical, and soil evidence', *EMA* 37, 1995, pp. 159–77; W. G. Whitford, G. Martinez-Turanzas, E. Martinez-Meza, 'Persistence of desertified ecosystems: explanations and

implications', *EMA* 37, 1995, pp. 319–32.

[31] See Blumler 1993, Forbes 1997.

[32] Bailey and others 1986.

[33] Explanations advanced for the similar phenomenon in the Near East include greater access to moisture around rocks and reduced competition from herbaceous plants, which flourish on deep soils and may keep out tree seedlings; Blumler 1993.

Fig. 4.19. Karst plateau of Mont Granier, near Grenoble, France. The pines (*Pinus uncinata*) grow on bare rock; areas with loess-like soil have grassland. *September 1970*

Terracing illustrates the importance of root penetration. Marls in dry areas are often inhospitable: they retain moisture, but (unlike hard limestones) are not fissured, and roots cannot pierce the solid rock. One reason for terracing is to create broken rock into which deep-rooted cereals and vines can penetrate.

Maquis and savanna appear to be 'open' plant communities, with large gaps between the trees or bushes. In reality, they may be closed communities below ground: the roots – the part that matters in a dry climate – may be intensely competitive, leaving no gaps.

The meaning of cliffs

Tectonism has given the Mediterranean a profusion of inland cliffs, which are of the greatest significance in reconstructing ecological history. In establishing the effects of man and man's henchmen on vegetation, a first step is to see how places out of reach differ from the rest of the landscape.

Cliffs normally escape cultivation, although old terraces reach surprisingly far down the volcanic ash cliffs of Santoríni. Vertical cliffs escape browsing: a goat climbs rather less well than a person (Figs. 4.20, 4.21). They escape most fires, but not woodcutting: in Crete nearly all trees on cliffs

Left

Fig. 4.18a. Landscape of maquis, phrygana, and steppe. The dark bushes, mainly prickly-oak, constitute maquis; the light bushes between them are phrygana; steppe (now mostly withered) and bare rock form patches among the phrygana. *Mavrommáti, Bœotia, Greece, August 1979*

Fig. 4.18b. Mosaic of maquis, phrygana, and steppe. Maquis (dark bushes) is more continuous towards the foot of the slope, despite the greater browsing from a shepherds' fold. Where deep soil is encountered, however, there are no bushes. *Mavrommáti, Bœotia, Greece, August 1979*

are coppice stools, even where they can only be reached with ropes. Cliffs are thus natural refugia, especially for palatable plants. Yew is very palatable (although poisonous) and in Catalonia, as in much of Europe, is now confined to limestone cliffs.[34] When pressure on land relaxes, these can return to accessible terrain.

Cliffs have been refugia for non-shade-bearing herbs through times when trees have been abundant. Cliffs being very stable, it is not surprising that many endemics are concentrated there. In Crete each of the hundred gorges has its assemblage of plants, often different from those of the next gorge.

Fig. 4.20. A beautiful and sacred cliff: the hermitage of Katholikó on the Akrotíri peninsula, Crete. Records of endemic plants go back to c. 1640; nearly all are still there. *July 1981*

34 Folch i Guillèn 1981, pp. 271 f.

Fig. 4.21. Two of our colleagues studying cliff endemics. *Thérisso Gorge, west Crete, July 1985*

Altitudinal zones

Most ecological books illustrate altitudinal zonation.[35] Evergreen maquis is held to be typical of low altitudes. Above it is a transition zone of *pseudo-maquis* with a mixture of evergreen and deciduous trees and shrubs. Above this is a zone of Mediterranean deciduous trees called by the Croat name *šibljak*. Above this may be a zone of northern deciduous trees, and higher still a zone of mountain conifers, with alpine vegetation above the tree-limit.

This scheme works in some parts of the Mediterranean, such as Croatia and Corsica, but not others. In Tuscany pseudo-maquis, *šibljak*, and northern woodland are juxtaposed in no obvious altitudinal relation. In Spain deciduous oaks such as *Quercus faginea* often form a zone above evergreen oaks, for example in the Sierra de Tejeda in the south,[36] but in the Alpujarra evergreen oaks go higher. In Crete there are only two well-defined zones, namely the alpine zone and the rest: three trees – cypress, pine and prickly-oak – are common from sea-level right to the tree-limit at 1,400–1,800 m.

Crete has a dramatic tree-limit on the western mountains. As the limit is approached, the cypresses steadily diminish in height, increase in thickness, and become sparser. The highest of all are among the oldest trees in Europe, growing into shapes hardly of this world (Fig. 4.22), with cycles of dieback, coppicing and regrowth. Annual rings show that some go back at least to late Roman times (pp. 312–3).

Mediterranean countries often have surprisingly low altitudinal zones. The highest trees in Crete (1,800 m) are lower than in the central Alps. Majorca, though only 1,450 m high, has a treeless alpine zone (Fig. 17.3) extending down to 900 m, barely higher than in Norway. Even Kárpathos at 1,200 m has a tree-limit. These limits have not been depressed by browsing animals, for trees do not go higher on cliffs. All over the world there is the phenomenon of *Massenerhebung*, whereby zones of climate and vegetation are higher in the interior of great mountain massifs than on small isolated mountains. In the Mediterranean it is partly due to the absence, on islands and many mountains, of trees adapted to high-altitude climates. (For alpine deserts see Chapter 17.)

RESILIENCE OF MEDITERRANEAN VEGETATION

Ruined Landscape theory would have it that degradation is a one-way process. Authors from Sonnini to McNeill have stressed the destruction of trees by people cutting them down, the difficulty of recovering forest and maquis, the slow growth of new trees, and the near-impossibility of re-creating eroded soil. The end product, at best, is what is disparagingly called 'scrub'.

Fig. 4.22. Ancient cypress at 1,600 m altitude, above Anópolis, W. Crete. *April 1989*

This theory is entirely at variance with our observations. Mediterranean vegetation is resilient. Maquis very easily reverts to woodland if woodcutting, browsing and burning stop. Shrubs are already there; all they need do is grow into trees. Abandoned farmland turns to steppe within a year, and within five years acquires phrygana. Trees, also, can come quickly. Some are already there as shrubs on terrace walls; others, especially pine, cypress and deciduous oaks, invade both maquis and ex-farmland by seed.

[35] A good summary of traditional lore on how Mediterranean plant communities are related to the environment is H. N. Le Houérou, 'Vegetation and land-use in the Mediterranean basin by the year 2050: a prospective study', Jeftic and others 1992, pp. 175–232.

[36] Gilman & Thornes 1985, p. 156.

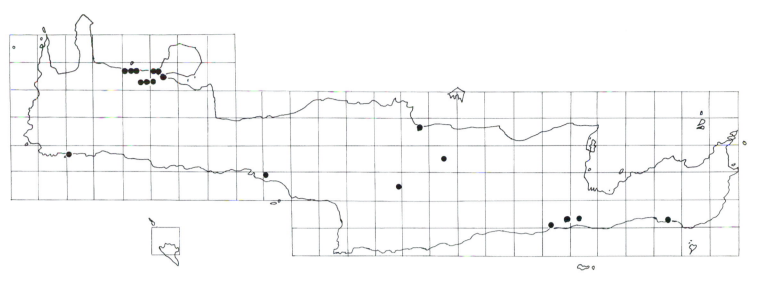

Fig. 4.23. Distribution of the grass *Imperata cylindrica* in Crete.

One reason why tropical forests are easily destroyed appears to be the very vigorous growth of coarse grasses in felled areas, whose competition then interferes with sprouts or new trees. An example is the notorious Alang-alang grass, *Imperata cylindrica*, in Indonesia.[37] This seldom happens in the Mediterranean. Alang-alang occurs in Crete but shows no sign of taking over the island (Fig. 4.23).

Whether soil is present determines which trees grow, but lack of soil – provided the bedrock is fissured – is no obstacle to tree invasion. In Crete, the limestone plateau of Arádhena has become a pinewood since losing its soil by wind erosion (p. 261).

Woodland has very widely increased in the last hundred and fifty years: too much woodland is recognised (e.g. in France) as a conservation problem. The evidence comes partly from photographs, paintings and drawings, often forming the background to a scene taken for some different reason. A photograph illustrating the excavation of Sparta in c. 1909[38] shows only savanna in the middle levels of the Taygetos Mountains, where there are now square kilometres of woodland (Fig. 4.24). Examples can be multiplied. Pashley's and Edward Lear's paintings, the photographs of Diamantopoulos and Gerola in Crete, those by the Wace-Thompson expedition to the Pindos Mountains in c. 1912[39], and the voluminous photographs of the Ligurian Apennines taken by the Italian forestry service in the 1930s show fewer trees than now. The change is evident even on comparing photographs of Crete taken by one of us in 1968 with the same scenes today (Fig. 4.25).[40] Places where trees have increased are common; so are those where there has been no change or no net change; places where trees have declined are rare.

Much well-developed forest, from Majorca to Croatia to Crete, stands on former agricultural terraces (Fig. 4.27). From Portugal to Turkey maquis is turning into woodland

where browsing has declined (it does not always need to stop completely). In all the countries that we have seen, young trees greatly outnumber middle-aged and old. Sometimes old trees, usually pollards, are scattered among them, proving that savanna preceded forest. The savanna pines in the Sparta photograph still stand, hemmed in by their children (Fig. 12.5).

In every European Mediterranean country that we know, woodland has increased by natural processes, following abandonment of terrace cultivation, decline of pasturage, or cessation of woodcutting. The increase is perhaps most obvious in France (where Cézanne painted the first generation of new pines a century ago), least obvious in Spain. The most dramatic increase of all is on Gávdhos, the southernmost islet in Europe, which up to a century ago was renowned for being almost treeless, and is now about 60 per cent pine and juniper woodland (Fig. 4.32).

New woodland is not always continuous. In a mosaic of maquis, phrygana and steppe, we observe that maquis turns into woodland, steppe remains steppe, and phrygana disappears (Fig. 4.33a). This is evidently a matter of root competition. When shrubs turn into trees they use more water, which is abstracted from the gaps between them. The shallow-rooted undershrubs in those gaps are starved of water; being short-lived, they die out and are not replaced.[41] When browsing is renewed, the trees develop a *browse-line*: the foliage suddenly terminates at the height which the animals can reach (Fig. 4.33b).

Most known secondary woodland results from recent land abandonment, but presumably this has happened many times in the past. Ancient secondary woodland can be detected if it contains conspicuous antiquities resulting from past non-woodland activities (Fig. 4.34). Much more would doubtless come to light if archaeologists were able to detect sites buried within existing forests.

Plant communities with no trees whatever are either in very dry climates, as around Almería (Spain), or on rocks with

[37] A. Reid, 'Humans and forests in pre-colonial Southeast Asia', *E&H* 1, 1995, pp. 93–110.

[38] R. M. Dawkins, *The sanctuary of Artemis Orthia at Sparta*, Macmillan, 1929, frontispiece.

[39] Pashley 1837; E. Lear, *The Cretan Journal*, ed. R. Fowler, Denise Harvey, Athens, 1984; Trevor-Battye 1913; G. Gerola, *Monumenti veneti nell' Isola di Creta*, R. Istituto Veneto di Scienze, Lettere ed Arti, Venice, 1905–32; Wace & Thompson 1914.

[40] Rackham 1990b.

[41] Rackham 1983.

Opposite page

Fig. 4.24a, above. Modern Sparta, Greece, in c. 1909. The middle zone of the mountain was then savanna with single trees and patches of woodland; the lower mountains were mainly pasture.

Fig. 4.24a, below. Eighty years on, the savanna has infilled into continuous pine forest, with the original giant pollard pines surrounded and hemmed in by their children (Fig. 12.5). Woods of deciduous oaks and other trees have appeared on the lower mountains. *July 1982*

This page, from top to bottom

Fig. 4.25a. The arid marl country of SE Crete, some twenty years after a decline of browsing: very tenuous phrygana and steppe, with scattered bushes of the unpalatable *Pistacia lentiscus* and a few young *Pinus brutia* trees. In the distance is a long-established pinewood on a north-facing slope. *Myrtos (Foúrnou Korifí), July 1968*

Fig. 25b. The same, twenty years later. The former semi-desert is now thick with esparto grass (*Lygeum spartum*) and the big, palatable, drought-resistant undershrub *Atriplex halimus*, an indicator of not browsing. Pine invasion proceeds apace. (The distant pinewood has been burnt; see Chapter 13.) *July 1988*

Fig. 25c. The same, another ten years on. The scene has been partly burnt: a browsing-dominated landscape has turned into a fire-dominated one. *August 1998*

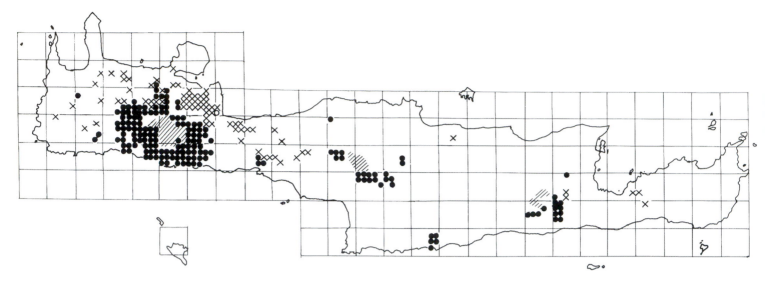

Fig. 4.26, above. Distribution of cypress in Crete. Crosses indicate sites of recent invasion. Cultivated trees are omitted. Land over 1,600 m is hatched.

Fig. 4.27, left. Cypress invading remains of terraces. The original trees are on the soil-less crags behind. *Zoúrva, west Crete, March 1982*

Fig. 4.28, below. Aleppo pine, *Pinus halepensis*, invading abandoned terraces. *Temple of Aphaia, Aegina island, Greece, September 1978*

difficult root penetration, such as the marls of middle Crete. Here there are no upstanding trees, no bitten-down trees, and no trees escaping destruction on cliffs. No amount of not browsing converts such areas into woodland: undershrubs may grow taller (Fig. 4.35), palatable species may move off cliffs, in exclosure experiments the season of flowering may alter,[42] but that is all. It is often asserted that this condition results from intense 'degradation', but this has seldom if ever been demonstrated. There is not enough moisture for trees now, and may never have been (Chapter 17).

Below

Fig. 4.29. Olive-grove invaded by Aleppo pine. *Banyalbufar, Majorca, November 1994*

Fig. 4.30. Landscape of prickly-oak maquis about to turn into woodland. At what point does this count in the statistics as forest area? (The ruin is a thirteenth-century Frankish colonial tower.) *Valley of the Muses, Bœotia, Greece, July 1980*

Fig. 4.31. Deciduous oak (*Quercus brachyphylla*) invading phrygana (from seed) on water-retaining but very stony soil. *Ayios Vasíleios, west Crete, April 1988*

Fig. 4.32. Recent pinewood on the islet of Gávdhos. *May 1992*

Vegetation, released from human activity, does not necessarily revert to what it would have been had that activity never occurred. The new cypress-woods in west Crete (Fig. 4.26) are in places where cypress may never have grown before. In north Europe many plants do not recover lost ground even if the environment is suitable,[43] and this is probably true in the Mediterranean: new deciduous oakwoods in the Peloponnese tend to be of *Quercus brachyphylla*, whereas old woods are of *Q. frainetto*. However, the evidence amply rebuts the belief that Mediterranean ecosystems are fragile and cannot recover. Quite the contrary: they recover more quickly than in England, though not so exceptionally fast as in eastern North America.

[42] E. Bergmeier, 'Zur Bedeutung der Beweidung für die griechische Phrygana', *Berichte der Reinhard-Tüxen-Gesellschaft* 8, 1996, pp. 221–36; E. Bergmeier and U. Matthäs, 'Quantitative studies of phenology and early effects of non-grazing in Cretan phrygana vegetation', *Journal of Vegetation Science* 7, 1996, pp. 229–36.

[43] Rackham 1980, 1990a.

Fig. 4.33a, b. Above: mosaic of woodland and steppe, resulting probably from not browsing a mosaic of maquis, phrygana, and steppe.

Above right: similar mosaic with browse-line resulting from renewed browsing. *Both from Mavrommáti, Bœotia, Greece, August 1979*

Fig. 4.34. Secondary woodland can date from earlier centuries. This deciduous oak (*Quercus pubescens*), said by Professor I. Camarda to be the biggest living tree in Sardinia, grows on a clearance cairn from prehistoric cultivation. *Goceano massif, July 1995*

Fig. 4.35. Effect of not browsing arid phrygana. The long-abandoned enclosure has developed tall dense phrygana (especially the summer-deciduous *Euphorbia dendroides*), but not woodland. *Khálki islet, Rhodes, May 1986*

CULTIVATED PLANTS

Foreign plants are among the determinants of human civilization. Southern Europe is full of edible wild plants, which in some countries (e.g. Crete) have a vast body of folklore.[44] But it has given surprisingly few cultivated plants to the world. Nearly all its crops were once exotic, and some staples of Mediterranean cuisine – tomato, potato, maize, most beans – are American.

Of the five traditional staple crops, olives and some legumes are native to parts of the European Mediterranean. Wheat and barley originated in the Near East, in an environment sufficiently like the Mediterranean to be transferred there in the Neolithic without obvious difficulty.[45] The vine arrived in Crete and Spain by the Neolithic.[46] Its homeland was probably the northern fringes of the Mediterranean. To

---

44 Professor P. M. Warren has pointed this out to us.
45 Oats apparently originated in Crete; J. R. A. Greig, 'A report on plant impressions from Debla, Crete', *BSA* 69, 1974, pp. 341–2.
46 It reached SW Spain by c. 3100 BC: A. C. Stevenson, 'Studies in the vegetational history of S.W. Spain. II. Palynological investigations at Laguna de las Madres . . .', *JB* 12, 1985, pp. 293–314.

judge by surviving wild vines, its ancestors grew in riverine woods. It is ill-adapted to a Mediterranean climate, being leafless throughout the rainy season and depending on deep roots in the dry season.

These five crops grow reasonably well in a Mediterranean environment. They need winter rain, good root penetration, no waterlogging, and no shade or competition. These requirements, uncommon in the world's vegetation as a whole, have become unconsciously accepted as the norm for European-style agriculture. From the Neolithic onwards, farmers have laboured to re-make landscapes in northern Europe, and wherever Europeans have settled, into environments in which crops brought via the Mediterranean will grow.

In dry-land farming the winter rains filter down into the soil and can be used by growing crops well into the spring. This presupposes a depth of soil or soft bedrock into which roots can penetrate; often this has been accumulated by past erosion washing the meagre sediment off rocky hillsides into plains and basins.

The Mediterranean has taken crops from almost everywhere *except the other mediterraneoids*. In Antiquity peaches, carob and fig-mulberry (*Ficus sycomorus*, now probably lost) came from the Near East; white mulberry from the Far East; date-palm from Africa. In the middle ages apples came from central Europe; citrus fruits, rice, sugar-cane, black mulberry and cotton from SE Asia and China. Parts of Italy and Spain became almost as rice-dependent as Japan. Maize came with the discovery of America, and tobacco a little later; tomato, potato, French bean, sunflower and American species of cotton were not widely accepted until the nineteenth century. Eucalyptus followed the discovery of Australia. In the last hundred and fifty years egg-plant and Canary Island pine have come from Africa; banana and kiwi-fruit from SE Asia; *Pinus radiata* and avocado-pear from the Americas. New introductions are still being attempted.

The less well-adapted crops are either grown in special places or require intensive cultivation. Rice needs plenty of irrigation water, as in the Po plain and in east Spain. The vine needs constant tending and pruning, and terracing to give it root-run; without these (unlike olive and carob) it produces no crop and usually dies. Other exotic plants are traditionally grown in irrigated gardens or orchards. Maize, with its high yield, is attractive to irrigation-minded people like the Spaniards.

Irrigation technologies go back at least to the ancient Greeks and Etruscans, using water from natural springs or from wells or tunnels (*qanats, cuniculi*; p. 285) into limestones and other water-bearing rocks. In recent decades there has been increasing use of storage dams and deep boreholes, and of plastic pipes for distribution and of plastic greenhouses since the 1960s. Vegetables and salads are now grown in midwinter in some of the driest coastal plains.

Frost

A frost of -13°C kills the olive to the ground; its base survives but needs to be re-grafted, and takes several years to produce olives again. The altitudinal limit of olive, a frontier of Mediterranean civilization, reaches 1,300 m in the Alpujarra (south Spain) and 750 m in Crete, but in the north declines toward sea-level.[47] Olives avoid frost-hollows (p. 28); they are absent from many inland basins, such as Arkádhia in the middle Peloponnese. The olive limit is set, not only by climate (or rather weather), but by the risks that growers will take. In Tuscany olives are grown up to 450 m, higher – in relation to winter temperatures – than the olive limit in Crete, and are often damaged by frost.

Ancient olives, longest-lived of crop plants, contain a record of historic frosts (p. 133). Olives in Provence were killed to the ground by the great frost of 1709. They gradually recovered, but there were more killing frosts in 1956, in 1963 (-29°C), and on 4 May 1967. Many of the frosted olive-groves – not yet subsidized by the European Community – were abandoned.

Other crops are frost-sensitive: walnut (to late frosts, as in NW Greece in May 1987); oranges and other citrus fruits; bananas (grown out of doors only in SE Crete).

Weeds

Weeds are specialized plants which coexist with cultivation, or latterly with weedkillers. Each type of cultivation has its special weeds. For example, *Gladiolus italicus* is a plant of ploughed land, rare in natural habitats: its corms survive – indeed they almost require – disturbance when dormant.

Weeds have not always been distinguishable from crops, and annual species tend to evolve in parallel with particular crops. Some weeds are thought to have come from the Near Eastern origins of agriculture. The spicy seeds of *Nigella* are now added to the best Greek bread, but originally were probably a contaminant. Other weeds may have been acquired in the Mediterranean from natural habitats. Annuals such as *Mercurialis annua* and species of *Scandix* and *Anagallis*, which in north and middle Europe occur as weeds, grow also in soil pockets on dry limestone cliffs in the Mediterranean. Irrigated crops have a suite of generalized tropical weeds, for example Johnson-grass (*Sorghum halepense*) and species of *Amaranthus*.

The chestnut

This noble tree is of unknown origin, spread by ancient cultivators far outside its native range; it grows unexpectedly well in England. The old claim that it came from Asia Minor is disproved by early finds of pollen and wood from Europe, including Spain.[48] However, we know no localities where it survives as a native: even in Italy it may be a relic of cultivation.

---

47 Olives reach 500 m in the warm Roya valley, on the French–Italian border.

48 Menéndez Amor & Florschütz 1961.

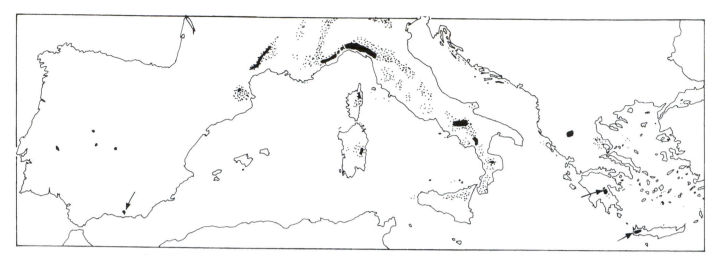

Fig. 4.36. Some principal areas of chestnut cultivation.

Chestnut grows from Portugal to Crete (Fig. 4.36), on or near non-calcareous soils. It is gregarious: where it occurs at all, it is one of the commonest trees. It is much affected by diseases. Ink Disease, caused by *Phytophthora* water-moulds, damages the base of the tree in wet places. Chestnut-growing depends on a narrow range of moisture: if the site is too dry, the trees suffer from drought; if too wet, from disease. In the 1930s the chestnuts of Italy and France (but not outlying areas) were devastated by the fungus *Endothia parasitica*, arriving from nobody knows where. This causes cankers which kill the trunk and boughs. In the 1960s it mysteriously cleared up owing to the appearance of a virus disease of the fungus: *Endothia*, weakened by the virus, still causes cankers, but these do little damage.[49] Chestnut is built to take punishment, and is able to recover even from severe fungus damage.

Chestnut has three functions: nuts, wood and timber. Nut trees are grown in savanna-like orchards. Trees are grafted to special large-nutted varieties – big nuts take less labour to peel. 'Wild' chestnut coppices strongly, and is cut on long or short cycles for different sizes of poles from vine-stakes upward; it is renowned for being durable. Chestnut is a common structural timber in Liguria, Tuscany and Portugal. It makes beams but poor planks, and hence in Italy floorboards and roof-boards are sometimes replaced by specially shaped tiles. It is also a favourite timber for second-class barrels.

Chestnut cultivation is strongest in Italy. Among the Romans, Pliny discusses the different nut-bearing varieties, whose names mostly suggest a south Italian origin. Columella has a detailed account of establishing chestnut coppices, discussing the yield to be expected of poles of various sizes; however, it is not much mentioned as a timber tree.[50]

In the northern Apennines chestnuts have built a whole civilization: they clothe the mountains over hill and vale to the near-exclusion of other trees. Nut-chestnuts are grown in terraced orchards. The nuts are gathered, dried over a slow fire in a building (*albergo*) with a slatted upper floor, and ground into flour to make a delicious and nutritious bread.[51] Chestnut flour tastes better than cereals, and is not so liable to crop failure. Grassland in the orchards need to be carefully maintained for pasture, with distinctive and beautiful plant communities of spring flowers.[52]

This happy symbiosis between people and chestnuts lasted many centuries, and might have gone on for ever; but inferior wheat flour came in, chestnut flour became proletarian and unfashionable, and chestnut-gatherers left the land. *Endothia* was a blow from which the trees recovered, but cultivation did not. Most chestnut orchards have been converted to coppices or are neglected; the mills and *alberghi* are derelict. With land abandonment the woods are increasing, as young chestnuts spring up on semi-derelict pasture. Forest authorities disapprove of occupational burning (necessary to maintain the orchards), and of coppicing, even though chestnut makes excellent poles. Fortunately their disapproval carries little weight. Chestnut woods are still coppiced on a vast scale for poles and firewood.

Chestnut is second only to olive as a long-lived crop tree. Some vast ancient trees are famous, such as the Hundred-Horse Chestnut (*Castagno di Cento Cavalli*) on Mount Etna. Landscapes of ancient chestnuts have an unearthly beauty with their high, bleached dead tops, relics of a defeated *Endothia* or *Phytophthora* attack. They are an excellent habitat, both for creatures that live in the hollow trees and for the special plant communities under them. Our favourites are the giant pollards in remote valleys in west Crete (Fig. 12.29).

[49] S. L. Anagnostakis, 'The American chestnut: new hope for a fallen giant', *Bulletin of Connecticut Agricultural Station, New Haven*, 1978, p. 777; also our observations and informants in Italy.
[50] Pliny, *Natural History* XV. xxv. 92 ff; Columella, *Res rustica* IV. xxxiii. 4.

[51] Which one of us has enjoyed thanks to the kind hospitality of Dr Diego Moreno, to whom we are indebted for much of our chestnut lore.
[52] G. Meriana, *Il castagno*, Cassa di Risparmio di Genova e Imperia, c. 1980; P. Di Stefano, '"Castagneti aggregati a massarie": trasfor-

mazioni nella castagnicoltura a Voltaggio nella seconda metà del '700', *Studi di Storia e Archeologia Forestale*, ed. D. Moreno and M. Quaini, Istituto di Storia Moderna e Contemporanea, Genoa, 1986, pp. 124–37; Moreno 1990.

The carob

The leguminous tree *Ceratonia siliqua* enters into natural vegetation in the hotter parts of the Mediterranean. It is naturalized to such an extent that phytosociologists call arid plant communities Oleo-Ceratonion after it. It is typically a savanna tree: magnificent ancient pollards and bitten-down bushes dot the phrygana and steppe, especially in Crete and the Algarve.

It is a mysterious Oriental tree. No European language has a proper name for it. Some use the Semitic word *harub* or *kharub*. Others have nicknames, such as the Greek κεράτι, 'little horn', whence *Ceratonia*. Names such as 'locust bean' or the German *Johannisbrot* derive from squeamish commentators, who guessed that the 'locusts' that John the Baptist ate with his wild honey were really the pods of this tree.[53]

Although it looks like a native, it is doubtful that carob was known in the European Mediterranean before Classical times. The large pods were used for feeding pigs (and people who lived like pigs[54]). The abundance of its charcoal marks the Roman prosperity of Valencia in Spain.[55] Later, it came up in the world: there was a big trade in carobs in nineteenth-century Crete. Like olives and chestnuts, the trees survive periods of neglect. The carobs of east Crete are still harvested sporadically whenever there is a fashion for carob flour (it makes delicious cakes) or apparently a certain type of photographic emulsion.

Forestry plantations

Forestry means different things in different countries. It may involve managing natural vegetation and developing native traditions of woodmanship (the 'French' style); or it may involve destroying the existing vegetation and making plantations of trees unrelated to what has been before (the 'German' style). Italian forestry is largely a development of traditional woodmanship, although it has a flavour of its own: official forestry became a branch of the Army, and Italian foresters still wear uniforms and tote guns.

Plantation forestry is not obviously suited to countries in which newly planted trees have to survive a hot dry season. Nevertheless, Spanish forestry was influenced by Germany rather than France from the mid-nineteenth century onwards.[56] Vast areas of arid hillsides were (and are) bulldozed and planted with pines and eucalyptus, which is called 're-forestation'. Much the same has happened in Sardinia and Portugal (in the latter from commercial motives). In Cyprus, too, the British introduced German-style forestry in the late nineteenth century, as they had done in India and were later to do in their own country.[57]

Mediterranean foresters persevere in the face of predictable disappointment. Planted trees, torn from their natural environment, are pinched by drought and consumed by goats. If they survive they may not prosper; if they prosper they may be burnt. The bulldozing that goes with planting destroys the archaeology and wildlife, and promotes erosion (even though erosion control is given as a motive). If successful, plantations lack the associated vegetation of natural woodland. At worst they form the typical 'green area' of a Spanish or Greek town: a hectare or two of hot, dusty pines, every tree identical to every other, with nothing beneath them but leaf-litter and human litter. It is an irony of Spain that the official forestry organization should be called the *Istituto para la Conservación de la Naturaleza*, 'Institute for Nature Conservation'.

GRAZING

One component of the desertification story is 'overgrazing'. As scientists have said since the Enlightenment, shepherds are ignorant, and shepherding ought to be planned by scientists and bureaucrats. In an ideal world grazing and browsing would achieve an equilibrium set by the carrying capacity of the vegetation.

Specialists in range management calculate how many animals the landscape will support. If they find more than that number, as they usually do, they infer that the landscape is being degraded. The more nutritious plants are eliminated first, then the tougher ones; forest is converted to savanna, trampling hooves tear away the soil, and desertification supervenes. This happened in North America and Australia, where European livestock and rabbits were introduced to semi-arid ecosystems not adapted to their tastes and habits.

In Europe such an occurrence (if it occurred) lies in the distant past. Virtually all browsable land has been browsed for millennia, though not always at the same intensity. Even in Europe, however, there is much to be said for parts of the story. Individual phases can be observed in operation. Goat-paths can indeed develop into gullies. When browsing declines, palatable species, like *Ebenus* in Crete, come down from their cliff refugia; they often stop at fence-lines (Fig. 4.37).

But it is rare that more than two stages in 'degradation' can be demonstrated in any one place. Nor is it obvious for how much of history ideal conditions were ever fulfilled. Browsing by prehistoric wild animals is one of the hardest factors to quantify in ecological history. There can, however, be little doubt that in Crete and other islands, with their elephants, hippopotamuses and deer, but no predator fiercer than a badger (p. 163), the pre-settlement landscape was heavily browsed by modern standards. On this theory, the normal habitat of *Ebenus* is cliffs; its occurrence on accessible terrain is a freak of abnormally little browsing.

In most carrying-capacity studies in Africa and the Mediterranean, at least twice as many animals are observed as the calculated capacity prescribes that there should be. Rangeland managers conclude that the local shepherds are greedy and short-sighted, rather than that their own calculations are too low. However, Mediterranean shepherds usually survive bad years, and are mostly concerned to keep their animals not merely alive but productive. Milk production falls off at the first hint of under-nourishment.

---

53 Matthew 3:4. Locusts are, of course, perfectly edible; see Leviticus 11:22.

54 Like the Prodigal Son: Luke 15:16.

55 Vernet 1997, p. 180.

56 Bauer Manderscheid 1991, chapter 5, p. 14.

57 Thirgood 1987.

Fig. 4.37. *Ebenus cretica* (crimson) in bloom on accessible ground. The land to the left of the fence has evidently not been browsed for several years. *Zou, SE Crete, May 1990*

Carrying capacity is not simple. Is it to be based on an average year, or the worst year in ten? or a hundred? Are the animals attended and moved around, or can they go where they will? Even in a stable environment the ecological carrying capacity (the number of animals that the landscape will, at a pinch, support) is typically at least 50 per cent more than the economic carrying capacity (the number for maximum sustained yield of produce).[58] In complex countries like Africa or still more the Mediterranean, where animals have access to different types of vegetation within an unfenced landscape and to different landscapes at different times of year, where rainfall fluctuates wildly, and where animal numbers can crash through disease, war or withdrawal of subsidies, the very notion of carrying capacity defies definition.

Carrying capacity is not a precise measurement (nor is the number of livestock, p. 19). The estimator ascertains the productivity of the vegetation in tons of dry matter per hectare per year, multiplies it by a 'proper use' factor (the percentage of those tons that under 'sound resource management' could be eaten), and divides it by the annual consumption per head of livestock. Each of these is problematic. With productivity it is difficult to take account of favoured patches in the vegetation, of good and bad years, and of palatable and unpalatable plants. The proper-use factor (30 per cent? 45 per cent?) is rarely better than guesswork. Consumption per head

---

58 Work of G. Caughley, quoted in: R. H. Behnke and I. Scoones, 'Rethinking range ecology: implications for rangeland management in Africa', Behnke and others 1993, pp. 1–30.

may be derived from improved breeds of livestock in research stations, rather than from local breeds better adapted to the actual vegetation. It is hardly surprising that 'measurements of the same resources have yielded divergent estimates for carrying capacity, while no confidence intervals [statistical limits of accuracy] are ever given'.[59]

Browsing can be self-limiting. Plant communities are adapted to a particular level of browsing.[60] If there are too many animals, they produce less milk or none. If there are too few, vegetation accumulates and leads to wildfires (p. 238). The occupational burning practised by shepherds demonstrates that biomass, or some of it, is being produced faster than animals can consume it.

Transhumance and common land, which scholars dislike and administrators despise (p. 88), increase the economic and ecological carrying capacities, but make them difficult to measure. More animals can be kept with less damage to the vegetation if they can exploit variations in environment and vegetation (seasonality and patchiness) than if they are confined all the year to small, private, fenced pastures.

We find it difficult to measure excessive browsing. It is more convincing to derive qualitative assessments based on the presence of palatable and unpalatable plants, and on the form of the trees and bushes. In Greece the different states of prickly-oak, *Quercus coccifera,* are a guide to the intensity and history of browsing (Figs. 4.4, 4.5, 4.33).

Whether the degree of browsing is excessive depends on how long it continues. Most maquis, phrygana and savanna have an episodic history. Prickly-oaks and other long-lived plants need not regenerate every year, or every fifty years. The trees in savanna are usually of a few ages with long gaps between; the gaps represent centuries of 'over-grazing', punctuated by decades in which circumstances conspired to allow a new generation to grow up. Such a pattern is most difficult to legislate for. Most rangeland managers strive, instead, for an equilibrium model in which roughly the same happens every year. That is not how Mediterranean vegetation works, and would lead to disaster if it could be imposed.

Shepherds, after all, seem to know more about shepherding than administrators. But recent changes warp their judgement and may have counter-productive effects. Discouragement of transhumance and the increase in fencing and private land make it more difficult for shepherds to use the natural variation in the landscape. Where one man owns the winter pasture and another the summer pasture, they use the land less efficiently than if both were communal. Shepherds are encouraged, by subsidized provision of stalls and water-points, to be more sedentary; they over-graze some areas and under-graze others. (The stalls tend to be built to ugly designs devised by non-shepherds.) The provision of roads, often perfunctorily constructed, destroys land and encourages erosion (p. 269). Roads encourage shepherds to be part-timers, commuting to work, leaving the animals to fend for themselves (or, as in one instance we know, delegating the shepherding to a dog) and to devour the palatable plants until they have annihilated them.

More pernicious still are artificial feeding and subsidies per head of livestock, which circumvent natural controls on overgrazing. Herdsmen can graze the summer pastures to destruction, keeping the animals indoors on hay in winter to avoid taking them to winter pastures. They are paid, in part, on the number of animals rather than on the milk or meat. If summer forage runs out they can (they hope) fall back on hay. Cheese becomes the standardized product of a cooperative factory rather than the work of a family of craftsmen. In Crete shepherding is beginning to turn into a mere industry, divorced from the land.

These changes together are bad, but before condemning them out of hand we reflect that if none of them happened the shepherds would give up shepherding and seek an easier way of making a living. Shepherding still holds together landscape and society in remoter parts of the Mediterranean. It deserves encouragement, but after understanding its social and biological complexities; not in the heavy-handed way in which money has been thrown at it in the recent past. Shepherds need to be listened to.

[59]   G. B. Bartels, B. E. Norton, G. K. Perrier, 'An examination of the carrying capacity concept', Behnke and others 1993, pp. 89–103.

[60]   E. Bergmeier, 'Zur Bedeutung der Beweidung für die griechische Phrygana', *Berichte der Reinhard-Tüxen-Gesellschaft* 8, 1996, pp. 221–36.

CHAPTER 5

Aspects of Human History

And Satan stood up against Israel, and provoked David to number Israel. . . . And
God was displeased with this thing . . . And the LORD spake unto Gad, David's seer,
saying, Go and tell David, saying, Thus saith the LORD, I offer thee three *things* . . .
Choose thee either three years' famine; or three months to be destroyed before thy
foes, while the sword of thine enemies overtaketh *thee;* or else three days the sword
of the LORD, even the pestilence, in the land . . .[1]

HUMAN HISTORY AND ECOLOGICAL HISTORY

Human history and archaeology provide the framework of
dates – 'the Hellenistic period' or 'the Punic period' – in
which to set ecological history (Fig. 1.8). Most scholars, from
Plato onwards, go further: they see human history as a more
or less complete explanation of ecological history. Cultural,
economic and social changes are supposed to have their exact
counterparts in changing landscapes. Politics, the supreme
factor, controls all aspects of human and natural activity,
reaching out into the remotest corners of the landscape. Even
the badlands deliver silt, or not, at the politicians' behest.

In reality there are four different stages in the theatre of
history. On one, the *dramatis personae* are the great and the
good and the bad. Time would fail us to tell of Hamilcar and
Metellus, Narses the Eunuch and Basil the Bulgar-slayer, St
Francis and Eleonora the Judgess and Thomas the Albanian-
slayer, Süleyman the Magnificent and Selim the Sot,
Napoleon and Hitler, the World Bank and the European
Commissioner for the Environment. On another, the action
has to be reconstructed from scatters of potsherds, traces of
terracing, ruined chapels, juniper timbers in church walls,
and deserted villages. The third stage is inhabited by people
selling wine and paying rents, by travellers, robbers, diarists
and artists. The actors on a fourth are not human: pollen
grains, elm-trees in a semi-desert, ancient olives and carobs,
dead junipers in forests, burning pinewoods on old terraces,
and oaks in terrace walls.

Some claim that the same play is being performed on the
four stages, but for us it is not so simple. Even human history
may not be self-consistent. It is easy to assume that empires
were efficient and authoritarian, and controlled every detail
that went on in their dominions. This is probably no more
true of historic empires than it was of Stalin's. The Venetian
authorities could not agree even on what villages there were
in Crete; they were forever complaining that they were unable
to enforce Venice's policy of making Cretans grow wheat
instead of vines.

Archaeology and conventional history often tell startlingly
different stories.[2] Historians of Crete leave no doubt that its
conquest by the Turks from the Venetian Empire, from 1645
onwards, was the greatest and most tragic discontinuity in the
island's history. But surviving workaday documents hardly
show that the replacement of a distant Doge by a distant
Sultan did much harm to the average Cretan. Archaeologists,
without knowing the history, would probably be unaware that
anything had happened: pottery styles before and after 1645
are almost indistinguishable, and Creto-Venetian architecture continues well into the eighteenth century.

If this is so for a well-known island, how can we know
whether J. R. McNeill is right in claiming that Epirus
suffered an ecological disaster when it was depopulated by the
Romans in 167 BC? This assumes (1) that the Roman writers
who reported the depopulation did not exaggerate, and
(2) that it had the effects in terms of erosion that McNeill
supposes. Both assumptions need to be verified by detailed
investigation on the ground. He makes a similar claim about
erosion after Philip II of Spain's depopulation of the Alpujarra in 1569.[3] Here, too, it is not obvious that the king
succeeded in removing the 'Moors', nor that (if he did) this
was the reason for the erosion. There are two alternative
explanations: (a) erosion was happening all the time, because
some rocks in the Alpujarra are unstable, regardless of human
action (Chapter 15); (b) it was due to the climate at that stage
of the Little Ice Age (Chapter 16). It is essential to use all
sources of information, and not to rely on written history
alone (p. 18).[4]

INFLUENCES ON MEDITERRANEAN HUMAN ECOLOGY

Peopling the islands

Palaeolithic people spread throughout mainland Europe
during the Pleistocene, and brought about profound ecological changes, especially by exterminating the great
mammals. The islands, meanwhile, had developed peculiar
ecosystems (Chapter 10), having no large carnivores. After
Man and Dog burst upon them in the Holocene, nearly all
the endemic island mammals were sooner or later exterminated and replaced by domestic, feral and wild beasts from
Europe.

---

[1] I Chronicles 21; King David mistakenly
thought pestilence was the least damaging option.
[2] A. M. Snodgrass, *An Archaeology of Greece*,
University of California, Berkeley, 1987, chapter 2.
[3] McNeill 1992, pp. 78, 97 ff.
[4] For an account of methods in the historical
ecology of Italy in relation to the 'degradationist
hypothesis', see Moreno 1990, chapter 1.

	Culture	Date	Overlap with large native mammals
Cephalonia (Ionian Islands)	Middle Palaeolithic	c. 50,000 bc	unknown
Sardinia	Upper Palaeolithic [or much earlier?]	c. .20,000 bp = c. 21,000 BC	long period
Mélos (Cyclades)	Upper Palaeolithic	c. 13,500 BC	no evidence
Cyprus	Proto-Neolithic	10,030 ± 35 bp = 9700–9100 BC	long period
Corsica	Pre-Neolithic	8560 ± 150 bp = 7800–7400 BC	yes
Kythnos (Cyclades)		7875 ± 500 bp = 7500–6200 BC	no evidence
Crete	Pre-pottery Neolithic	7740 ± 180 bp = 6800–6400 BC	no evidence
Balearics	Pre-Neolithic	c. 7000 bp = 5800 BC	long period
Malta	Neolithic	c. 6000 bp = 4900 BC	

Table 5.i. Earliest known evidence of human presence on the islands.

People reached the islands at various dates.[5] There is a Middle Palaeolithic site, from the middle of the last glaciation, on the west Greek island of Cephalonia, then 20 km across the sea. Palaeolithic people were evidently not habitual voyagers, since no further occupation of the island is known until the middle Holocene. In the Upper Palaeolithic they reached Sardinia, where a human finger-bone has been found in an inland cave; the site also contains Mesolithic skull-sherds and tools, as well as abundant Neolithic material.[6] Regular seafaring is implied by finds on mainland Greece of obsidian from Mélos. Sailors who could reach Mélos could have gone on to Crete and, maybe, sampled its tiny elephants and slow deer. But so far the earliest evidence of human presence on Crete is from the Neolithic of Knossós.[7]

Settlement patterns

Mediterranean peoples have created ingeniously different cultural landscapes out of apparently similar natural environments. A powerful fashion in human culture is nucleation: that mysterious influence which at its opposite extremes drives French people to live in tower-blocks and Texans in lonely ranch-houses. Independent of population density overall, settlements can be of various sizes, from isolated farmsteads to *hamlets* (clusters of houses or tens of houses, Fig. 5.2) to *villages* of hundreds of houses and *towns* of thousands of houses.[8] Many Mediterranean 'peasants' are townsfolk, living further from their fields than the three hours' march which anthropologists take as the limit; when work demands, they ride out to stay in *field-houses*, second homes close to their land.

Few contrasts are greater than between Corsica, with thousands of hamlets a kilometre or so apart, and Sardinia, also a mountainous and granite landscape, but with scores of big villages and small towns separated by 20 km or more of empty countryside. Two thousand years ago, Sardinia had thousands of small settlements each clustered round a *nuraghe*, a

tower or castle (pp. 164–5). Sardinia then was more like Corsica now, although Corsica has not had the towers. Some of the desertion is recent: Sardinia is full of deserted medieval villages and hamlets.[9] Sicily, too, is a strongly urbanized island, with people commuting from towns to work an unpopulated, arable countryside.

Much of central Spain is an empty, urbanized landscape: big villages and agricultural towns are surrounded by vast treeless arable plains or long miles of *despoblado* savanna and semi-desert. The great mountains are more intimate: the Alpujarra is a land of rugged fertility with small villages perched above tremendous ravines. In Tuscany and Majorca, besides towns and villages, there are big, aristocratic farmsteads and villas out in their own fields, recalling (and occasionally descended from) the *latifundias* and villas of Imperial Rome. Greece is a land of big villages in an otherwise empty countryside, except for the hamlet-and-castle landscape of the Máni in the extreme south.

There are many curious variants. In Sardinia towns and villages tend to be in twos and threes (as often were *nuraghes* before them). In the Alpujarra villages often consist of several hamlets very close together, each with its plaza, church and its own name. In Crete people may live in villages in winter but migrate into upland hamlets in summer.

In recent centuries there has been a widespread trend towards nucleation: hamlets and small villages have been abandoned or amalgamated in favour of big villages. The motive is often said to be security. The medieval process of *incastellamento*, in which fortified small towns and villages grew up around castles, is familiar to Italian landscape historians, but also occurs in Catalonia and Provence, in Greece with its 'Frankish towers' (Fig. 4.30), and in many small Aegean islands. (In the Máni, however, hamlets were at war with each other rather than with an external foe.)

In Crete both hamlet and village landscapes – and traces of *incastellamento* – still exist. Villages have gained at the expense of hamlets, a change helped by the emergence since

[5]   Most of the evidence is summarized in more detail in J. F. Cherry, 'The first colonization of the Mediterranean islands: a review of recent research', *Journal of Mediterranean Archaeology* 3, 1990, pp. 145–221.

[6]   P. Sondaar and 6 others, 'The human colonization of Sardinia: a Late-Pleistocene human

fossil from Corbeddu cave', *Comptes Rendus de l'Académie des Sciences, Paris* 320 sér. IIa, 1994, pp. 145–50. Claims for a much earlier Palaeolithic culture in Sardinia are based on stone 'tools' of a kind easily confused with naturally fractured stones; for the resulting controversy see articles in Tykot & Andrews 1992.

[7]   C. Broodbank and T. F. Strasser, 'Migrant farmers and the Neolithic colonizaton of Crete', *Antiquity* 65, 1994, pp. 233–45.

[8]   For a more extensive discussion, see Delano Smith 1979, chapter 2.

[9]   Tyndale 1849 *passim;* some of the desertion was attributed to feuds.

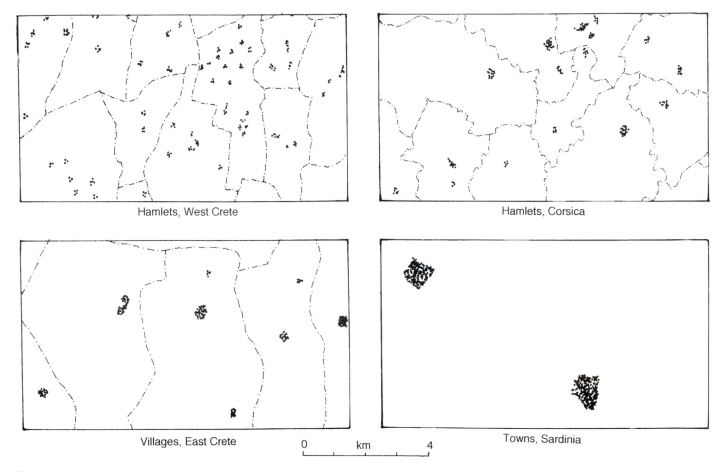

Fig. 5.1. Nucleation: top left, hamlets, west Crete; top right, hamlets, Corsica; above left, villages, east Crete; above right, towns, Sardinia. All are at the same scale.

the seventeenth century of both the coffee-house and the parish church as the cement of village society. There were also upper-class villas in the late Venetian and early Turkish period (the sixteenth to the eighteenth centuries). Crete is full of deserted hamlets and villas. Archaeology shows that the medieval hamlets were preceded by two previous cycles of nucleation and de-nucleation.[10]

Nucleation sets two traps for the unwary scholar. Deserted hamlets and small villages do not necessarily indicate depopulation: the people may have moved to the nearest big village. The other trap concerns village lists and statistics. The unit of recording is normally the *township* or civil parish (p. 19). Ideally a township consists of one village with its fields and roughlands. In practice it often consists of two villages, or a cluster of hamlets or farmsteads. Inconsistency arises about which village or hamlet belongs to which township. A village that disappears from the census record has not necessarily disappeared from the ground: it may have been included in the next village, or been swallowed by the growth of a closely adjacent village, or been missed.

Field-systems

Countries are defined by their land-tenure customs. Crete, whose inhabitants for centuries jealously guarded hundreds of thousands of tiny fields, contrasts with Sardinia, where (until nineteenth-century privatization) folk were casual about ownership and content to let land wander in and out of the public domain.[11]

Most Mediterranean fields are of irregular shape, except in modern schemes of re-allocation or 'reclamation', which are laid out on regular grids. But there are many traces of earlier periods of country planning (Fig. 5.3). The best known is *centuriation*, the division of land into exact squares of 709 m, oriented exactly north and south (or occasionally at 45°). This was fashionable in Roman territories in the latter centuries BC, both in the organized winning of new land and the rearrangement of existing property. Grids were laid out across mountain and gorge (as in Croatia) by planners as rigidly devoted to straight lines as any land office in the American Mid-West or modern ministry official. Field boundaries

[10] J. A. Moody, 'The continuity of settlement and land-use systems in west Crete', *Petromarula* 1, 1990, pp. 52–9; Rackham & Moody 1996, chapter 8.

[11] B. Fois, *Territorio e paesaggio agrario nella Sardegna medioevale*, ETS, Pisa, 1990.

Fig. 5.2. Hamlet landscape of Sássalo in a remote valley of west Crete. It survived through a Muslim period. *January 1988*

being conservative, centuriation remains long after the reasons for it are forgotten, as in the Po plain: the Roman boundaries are joined irregularly by later divisions, and occasionally can be seen to override earlier divisions.[12]

The division of land into strips (typically 200 by 10 m, but often bigger), allocated among owners and often reallocated from time to time, is often thought of as the norm in medieval Europe. It was associated with communal 'open-field' farming practices, and with using ploughland as pasture in a fallow year or after the harvest. On present evidence it appears to have been invented in England in the eighth century, perhaps as a device for making the best use of arable land where there was no roughland into which to expand. Its spread often went with collectivization of agriculture and gathering the inhabitants into villages.[13]

Strip-cultivation penetrated southern Europe haphazardly. Many countries have traces of it: even in Crete some parts of the Mesará Plain (not coinciding with township limits) were covered with strips in 1944 (Fig. 5.4). In Sardinia it was imposed (by whom is not clear) with a rigour unsurpassed even in northern Europe: despite land 'reform' in the last century, it still covers whole townships.

Muslims

Most of southern Europe has been conquered at one time or another by Arabs or Turks or both. Except perhaps in Spain, Muslim empires ruled their European territories, but did not colonize them to any great extent by importing population. Only in Spain and the Balearics are there many Arabic place-names; Turkish place-names in the Balkans and Greece are few except near cities and main roads.

Much of the population took up Islam with vigour and tenacity. They prepared a terrible fate for their remote descendants, who incurred the wrath of Christian regimes and were murdered or expelled: the Alpujarra massacres of 1569, the expulsion of Muslims and ex-Muslims from Spain in 1609, the expulsion of Cretan and Greek Muslims in 1923, and the massacres of Muslims in Yugoslavia in 1994.

It is often supposed that Muslim regimes transformed their respective countries. The high civilization of the Caliphs of Cordova and the kings of Granada, extolled by modern Spanish historians, certainly transformed the place-names: Spanish rivers became wadis – Guadiana, Guadarrama, Guadalquivir, etc. Its effects on the landscape, however,

12 Bradford 1957, chapter 4.   13 Rackham 1986, chapter 8.

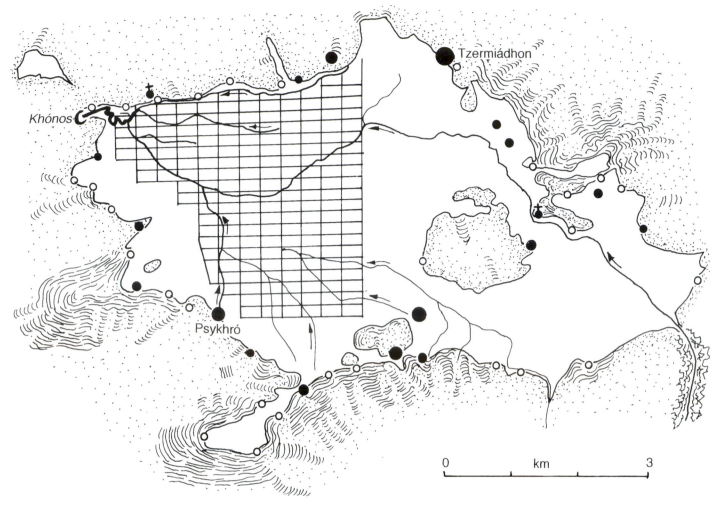

Fig. 5.3. Country planning in the Lassíthi Plain, the largest flat area in Crete, in a karst depression at 880 m altitude. The western half of the plain was laid out by the Venetian State c. 1500 in 193 approximate half-squares of 5.3 ha. (The angles are about 88°, not 90°.) Villages were established around the edge. Each rectangle is now subdivided into about 35 fields. Terraced slopes around the plain are shown, but not individual terraces.

Fig. 5.4. Strip-cultivation in the Mesará, Crete. Parts of the alluvial plain are laid out in curving strips, with no attempt at standardized dimensions. The area shown is 6½ by 4 km. *Townships of Asími, Khárakas and Praitória, German air photograph of 1944 (Unites States National Archives: RG 373:TUGX 2102 SK212)*

can be exaggerated. A recent analysis of Spanish-Arabic irrigation claims that the methods were developed by Arabs out of Roman practices, and some of the actual fabric is Roman.[14] The effects of the four centuries of 'decay' when Greece 'groaned under the Turk', deplored by modern Greek historians, can also be exaggerated. Plague and malaria probably played as large a part as politics; some aspects of civilization (such as Christian monasticism) flourished. The Greek language took up Turkish words (especially technical and kitchen terms, but not place-names) much as the Spanish language adopted Arabic words.

In Crete there were two Muslim periods. The Arabs had the island from 823 to 961. They left no discernible mark on the landscape, except that they established Herákleion as the principal city and called it Handaks, 'Ditch' (*khándaka*, an irrigation canal, is one of the few Arab words still in use in Crete). Apart from a few coins, the Arab period is invisible to archaeologists: artefacts are indistinguishable from those of earlier and later periods. The Ottoman Empire had Crete from 1645 to 1898, but its effects, although so recent, are also

---

[14] Butzer & others 1985.

inconspicuous. Outside the cities with their minarets, the visitor hardly notices Islam, and sees nothing of the Turks except their castles on hilltops. Pre-1645 Christian churches, many with wall-paintings, are no less common in areas with a Muslim history than in those that have always been Christian.

The distinctive flat-roofed architecture of south Spain, with its adaptations to scarcity of trees, is often called 'Moorish', and has parallels in north Africa.[15] But the similar architecture of Crete goes back in principle (though not in detail) to the Bronze Age. Is the 'Arab' attribution of south Spanish architecture merely due to ignorance of the earlier vernacular buildings in Spain?

Piracy

The hydrophobia [fear of water] of the Sardes may be attributed primarily to their want of energy; and, secondly, to their having been from generation to generation the constant prey of corsairs, and having their sea-coast villages attacked and ravaged by the Pisans, Genoese, and Spaniards.[16]

Furat chi benit dae su mare.
He who comes from across the sea is a thief.[17]

Piracy is an obvious way for politics to affect the landscape. Mediterranean peoples have usually feared the sea, bringer of enemies, and have not dared to live by it except in fortified towns.

Pirates, having been executed under the Roman and Byzantine Empires, returned with the decline of the Empire and the growth of warring states. They became institutionalized in the *corsairs*, sent by the Muslim beys of north Africa to fight a perpetual holy war against the infidel. This became a regular branch of the shipping industry. Muslim corsairs had their counterparts in the pirate-monk Knights Hospitallers and the corsairs of the Grand Duke of Tuscany, the Pope and other Christian potentates; but these seldom, save by carelessness, raided European territory, and do not concern us. Corsairs operated mainly against infidel shipping, but when ships were scarce they raided on land. The booty was persons who could be disposed of through the ransom network (or, failing that, the slave trade), treasure and valuable merchandise.[18]

From late-Roman times onwards coastal settlements, such as Soúyia in Crete, were abandoned for sites inland. Most islands with populations of less than 2,000 were deserted. By 1500 the whole Cretan coast, except for fortified towns, had been abandoned, and the entire east end of the island (Fig. 17.12). Great corsairs like Khair-ed-Din of the Red Beard sacked fortified towns such as Réthymnon and raided inland

settlements. Early-warning coastguard posts on mountain-tops were manned all through the summer nights.[19] The kings of Spain, who created a corsair problem for themselves by expelling their Muslim subjects, had to build towers every five km or so along the south coast of Spain. Similar towers in Sardinia, built c. 1590, were manned until 1867;[20] the last great raid was in 1815.

Traditional irrigation

The impact of irrigation was felt most palpably in two fields – technology and law – whose monuments are still visible today in the vast canal network of the Valencian *huertas* (irrigated areas) and in the complex body of customary water law epitomized by the renowned Tribunal of Waters.[21]

Fig. 5.5. Irrigated terrace gardens on a steep slope. *Plokamianá, SW Crete, March 1982*

Irrigation occurs throughout southern Europe. In Crete methods vary from the simple dipping-well (*geráni*) – a bucket raised on a counterweighted lever, pivoted on a forked post – to delicate irrigated gardens on terraces (Fig. 5.5), watered from a nearby spring, with a system for sharing the water among owners: X gets three tankfuls a week, Y gets four, etc.

In Spain irrigation goes back probably to the Copper Age, and was well developed by Roman times.[22] The Arabs brought a further input of water technology and law, and new crops such as rice and sugar-cane. Spain has a strong distinction between irrigated gardens and non-irrigated farmland.[23]

[15] B. Hutton, 'Peasant houses at Mojácar, Spain', *Vernacular Architecture* 17, 1986, pp. 39–50.
[16] J. W. Tyndale, 1849.
[17] Sardinian proverb, from J. Chwaszcza, ed., *Sardinia, Insight Guide*, APA Publications, Hong Kong, 1991, p. 123.
[18] P. Earle, *Corsairs of Malta and Barbary*, Sidgwick and Jackson, London, 1970; J. H. Pryor, *Geography, Technology, and War: studies in the maritime history of the Mediterranean 549–1571*, Cambridge, 1988.
[19] Rackham & Moody 1996, chapter 18.
[20] M. Atzeni, *Amministrazione delle Torri Litoranee in Sardegna*, M.Phil. thesis (ASC: Misc. Sard.).
[21] T. F. Glick, *Levels and Levelers: Surveying Irrigation Canals in Medieval Valencia*, Technology and Culture 9: University of Chicago Press, 1968.

[22] Chapman 1978; Butzer and others 1985.
[23] This goes back at least to the Copper Age, to judge by a recent claim (on the basis of differences in the carbon-13 content of charred seeds) that beans were irrigated but cereals were not; J. L. Araus and 7 others, 'Identification of ancient irrigation practices based on the carbon isotope discrimination of plant seeds: a case study from the south-east Iberian Peninsula', *JAS* 24, 1997, pp. 729–40.

For centuries every Mediterranean Spanish town had its *regadío* gardens. These might be surrounded by a much larger area of *secano* arable and *despoblado* pasture or savanna; but plains with perennial rivers coming from inland mountains might be entirely irrigated, supporting cities like Valencia and Murcia. Vast medieval archives deal with irrigation rights and disputes.

How much of Spanish irrigation is Roman and how much is Arab or Berber is controversial.[24] The Arabs introduced devices like the *noria*, an endless chain of pots for raising water from a well, driven through gears by a donkey going round in circles (this was useful in places with ground-water but no rivers). They brought arrangements for the community (rather than a lord) to maintain dams and canals and to share out the water. Shares in water were measured, to the fraction of an hour, by Arab forms of water-clock. The system and the know-how passed into Christian hands after the conquest: Christian lawyers would appeal to irrigation customs 'as they were in the times of the Saracens'. It coped with droughts and deluges; it supported mills and even navigation as well as agriculture; in the Alpujarra it still operates in some of the highest and most unstable mountains in Europe. Such practices were exported to the Spanish colonies in the Canary Islands, to the Canarian colony in San Antonio, Texas, and even to British India.[25]

Colonies

Outward colonialism – opportunities for people to emigrate – has affected most of the European Mediterranean. It took many forms, according to whether or not the colonies were part of a trading system or empire; how much population was exported; whether emigrants formed distinct colonies or merged with native peoples or other settlers; and whether they retained links with the homeland. Colonization was reversed in the post-Roman period – the Migration Period of middle and north European historians – when peoples from outside came into the Mediterranean. Goths, Slavs, Turks, Vikings or Arabs got into most Mediterranean countries, and left marks on languages, blood-groups, customs and names.

Greece is among the most colonial of nations. By the sixth century BC cities and islands were sending expeditions to settle the coasts of Sicily, Africa and the Black Sea. Sometimes this was a means of getting rid of surplus mouths after a shortage.[26] It did not create empires: most colonies soon broke their links with the mother states, and often sent out colonies of their own.

In the middle ages the Albanians were colonizers: their descendants live from Sicily to the Peloponnese to Samos.

Modern Greece has exported large populations to America and Australia. Greeks tend to form distinct colonies, to keep in touch with the homeland, and to return later in life. For example, the SE Aegean islands of Khálki and Kárpathos have colonies in Tarpon Springs (Florida) and New Jersey; many of the present male islanders have spent years there. Tiny Kastellorízon, far away off the Asian coast, has two Australian colonies.[27] Money sent by colonists to the home community provides private and public amenities (occasionally of a stunningly incongruous character) and upholds many a semi-depopulated village.

Spain had one of the earliest, largest and longest-lived intercontinental empires, and yet colonialism has made surprisingly little difference to the landscape at home. About a quarter of a million Spaniards left Spain for America in the sixteenth century, almost exclusively from the south-western savanna country, which provided nearly all the Spanish cultural traditions of America.[28] Since then, it appears, Spaniards have preferred to endure poverty and overpopulation at home to seeking a new life in New Spain. Nor were the links kept up: Trujillo (Mexico) soon broke with Trujillo (Spain).

Pestilence and fever

. . . the malaria, or *Intemperie* as it is called here, renders [Sardinia], with the exception of the larger towns, uninhabitable for strangers from July to October. Fever, which prevails principally on the low ground, frequently extends its ravages to a considerable height, in consequence of which the mines are deserted during the period above mentioned. The natives, however, appear to be habituated to dangers which would often prove fatal to strangers. The principal precaution they use consists in wearing fleeces . . .[29]

Historians find war exciting and pestilence dull; they exaggerate the effects of the former and play down the latter. In reality, as our own enlightened century has demonstrated, it takes an exceptionally savage war to reduce a population by 10 per cent, whereas bubonic plague can kill three times as many people without attracting much attention. Evidence of disease is elusive: diagnoses from historic written sources are notoriously unreliable, and many diseases kill without leaving archaeological traces on the skeleton. Hence the controversies over such events as the pestilence that crippled Athens in 429 BC, killing Pericles.[30]

Plague could hardly have been possible before the black rat, its vector, came from India in the Roman period. The 'Justinian' plague, which was probably bubonic, ravaged Europe in 542 and several times again in the sixth century,[31] and may plausibly have caused some effects of the decline and fall of the Roman Empire – depopulation (especially of cities), political disorganization and perhaps the increase of trees and forests. On this, as on several other occasions (e.g. Crete in 1595 and 1819), bad weather preceded plague; possibly scarcity resulted in unusual importations of grain and therefore of rats.

24 Glick takes a contrary view from the Butzer school.
25 T. F. Glick, *Irrigation and Hydraulic Technology: medieval Spain and its legacy*, Variorum, Aldershot, 1996.

26 Cf Garnsey 1988, chapter 9.
27 Information from Paul Adam.
28 Butzer 1988.
29 K. Baedeker, *Italy: handbook for travellers: third part*, 1893.
30 J. H. Finley, Jr, *Thucydides*, Ann Arbor, 1963.

31 P. Ziegler, *The Black Death*, Collins, London, 1969; J.-N. Biraben and J. Le Goff, 'La peste au Moyen Age', *Annales Économies, Sociétés, Civilizations* 24, 1969, pp. 1484–1510; D. Panzac, *La Peste dans l'Empire Othoman*, Peeters, Leeuven, 1985.

After a long period of quiescence plague struck Europe again as the Black Death in 1348. Repeated outbreaks resulted in centuries of grief and insecurity on a scale that the modern world can hardly imagine. In Crete there were plagues in at least 1398, 1419, 1456, 1523, 1555, 1580, 1592 (killing 38 per cent of the people), 1655, 1678, 1689, 1703 and 1816. In many areas the population was permanently held down until after the last outbreak: 1730 in Marseilles, 1743 in Sicily and Calabria, 1816 in Crete, 1835 in Epirus.[32] Why plague died out (apart from minor outbreaks in seaports in the twentieth century) is unclear; there are still plenty of black rats in Crete.

Malaria has been present for many human generations. Most Mediterranean populations have genes that protect the bearer against malaria at the cost of incurring disabilities in other directions: favism (a Sicilian gene which when homozygous causes the possessor to be poisoned by broad beans), sickle-cell anaemia, or β-thalassaemia. It is usually supposed that malaria was prevalent in Classical Greece; this is based on medical and other writings, the skeletal evidence being ambiguous.

Malaria comes into Ruined Landscape theory.[33] A familiar argument runs like this. Deforestation causes erosion; erosion brings silt down rivers to choke their mouths and create fens and lagoons; mosquitoes proliferate; malaria wipes out populations; Greek or Roman civilization collapses. This story partly explains why modern administrators hate fens. Parts of it are true. The former bays of Italy and other coasts have turned into mosquito-bearing fens (Chapter 18). English settlers found nineteenth-century Sardinia a 'White Man's Grave'. In the 1920s any medical student studying tertiary malaria would go to Thessalonica, where he would find abundant material among the wretched Christians 'exchanged' from Turkey.

Malaria, however, is a complex disease, with at least three species of malarial parasite and eight species of mosquito. This may explain the inconsistencies in its ecology. Venice and Cagliari flourished in the midst of lagoons, and still swarm with mosquitoes, presumably the brackish-water *Anopheles labranchiae*: did they escape the most dangerous *falciparum* form of malaria? Many travellers remark that malaria was not confined to fens. In Crete it was carried by *Anopheles superpictus*, the mosquito of mountain streams. Carlo Levi, a Milanese doctor banished in the 1930s to political exile in the arid hilltop town of Aliano in Basilicata, found it rotten with malaria, 'a scourge of truly alarming proportions, sparing no-one and lasting a lifetime'. He thought deforestation somehow had a direct effect on malaria.[34]

It is not known what was the incidence of malaria in Antiquity; whether it controlled human populations, whether the deadly *falciparum* parasite was present, and how far ancient civilizations could flourish in spite of malaria. Although the modern ecology is reasonably well known, the scanty and difficult evidence from ancient sources allows scholars to publish very divergent interpretations.[35]

Rising standards of living may reduce malaria as people can afford mosquito nets. They may increase the risk by separating people from domestic livestock: some mosquitoes, which used to feed partly on animals, now have only people to attack. Quinine as a medicine, known from the mid-seventeenth century, was not accessible in large quantities until the late nineteenth. In Italy and Greece the government took responsibility and sold tons of it cheaply to the public.

Malaria has also been combated by environmental measures. In 1770 rice cultivation was banned from certain townships in the plain of Valencia in Spain. After twenty-five years a demographic comparison was published, showing the differences in death-rate and population between these places and townships that still grew rice.[36] In Italy and Albania, malaria control was one reason for the campaign of *bonifica*, the destruction of fens and creation of farmland, in the first half of the twentieth century. Governments naïvely expected that mosquitoes would not breed in drained fenland. In practice, *bonifica* may have increased malaria by attracting farmers to live on drained fenland in daily contact with mosquitoes. (The worst tertiary malaria ever seen by 'Malaria Jones' was in the Lake Kopáis area *after* its draining.[37]) In Albania there was an ingenious scheme to introduce sea-water to the lagoon of Durrës in order that the freshwater mosquito *Anopheles sacharovi* might be replaced by a brackish-water species which was a less efficient vector.[38]

Malaria has been controlled by killing mosquito larvae in their breeding-places by dusting waters with arsenic, creating oil films on the surface, or introducing insectivorous fish. After World War II there were campaigns to 'eradicate' malaria from southern Europe by spraying houses with the newly invented DDT. There was no hope of eliminating mosquitoes; the object was to reduce them temporarily to a level at which the chain of human transmission would be broken. In this way, and by using anti-malarial drugs, the transmission of malaria was ended without using DDT for so long that resistant forms of mosquito were created. The last area with indigenous transmission was Lesbos in the 1970s.

Malaria is supposed to be extinct in the European Mediterranean, although many people aged over fifty must carry the parasite. It also died out in countries like England, where little was done about it. Whether its disappearance will be permanent remains to be seen; every year there are several hundred new cases among travellers from outside Europe.

HISTORICAL CHANGES

It is easy to suppose that the upheavals in Mediterranean countries since 1950 are the effects of technological change on landscapes and peoples that had not known it before. We imagine people leading much the same life from time

[32] Rackham & Moody 1996, pp. 98–100; J. D. Post, *Food Shortage, Climatic Variability and Epidemic Disease in Preindustrial Europe: the mortality peak in the early 1740's*, Cornell University Press, 1985, pp. 27–8; McNeill 1992, p. 167.
[33] Jones 1907.

[34] Allbaugh 1953; Levi 1947.
[35] Bruce-Chwatt & de Zulueta 1980; Sallares 1991, chapter II.7; P. J. Brown, 'Malaria in Nuragic, Punic, and Roman Sardinia: some hypotheses', *Studies in Sardinian Archaeology*, ed. M. S. Balmuth & R. J. Rowland, Jr, University of Michigan Press, Ann Arbor, 1984.

[36] Cavanilles 1795, 1, pp. 153 ff.
[37] See note 33.
[38] Bruce-Chwatt & de Zulueta 1980; Jones 1907, p. 11; G. Harrison, *Mosquitoes, Malaria and Man: a history of the hostilities since 1880*, John Murray, London, 1978.

immemorial as old people remember now: generation after generation of 'peasants' wringing a meagre living from the land by incessant toil, having few dealings with the outside world, and rarely touching money. In reality, the bad times of the early twentieth century followed an earlier wave of technology which had led to overpopulation and poverty, aggravated by war and tyranny; they were not necessarily typical of earlier periods.

The forces behind the modernization of southern Europe were population growth, technological change and the commercial economy. All these go back to prehistory, but the last two centuries involve three new factors: higher populations than ever before, the steam-engine and its successors opening new commercial markets outside the Mediterranean, and the effects of machinery, especially the bulldozer, altering and destabilizing the landscape itself.

However, landscape history is far deeper than a couple of centuries. The last two hundred years may be well documented, but cover at most one-twentieth of the historical time-span. Archaeological survey shows that prehistoric civilization was not confined to good land but had spread to the remotest parts of southern Europe. The discovery of Neolithic and Bronze Age sites in the high White Mountains of Crete – far above modern settlement, and little known to the world outside Crete until well into the twentieth century – is an example.[39] By the late Bronze Age most of the possible land in southern Europe was cultivated (at least intermittently) or grazed. Settlement in some well-favoured sites was as dense as it has ever been. The major changes between the aboriginal landscape and that of the nineteenth century had, where sufficiently known, already taken place by the Iron Age (Chapter 10).

The high cultures and dense populations of southern Europe – late Bronze Age Crete, Iron Age Sardinia, Classical Athens, Imperial Rome, early Byzantine Crete, Arab Spain and Sicily, medieval Italy, Renaissance Crete – were periods of commercial activity, technological advance and urbanization, followed by apparent decline.[40] These earlier cycles provide a scale of reference for the combination of changes we call modernization. Cultural developments that gathered in Renaissance Tuscany were transmitted to the Low Countries and England, to culminate in the industrial revolution which still dominates our environment and continues to transform landscapes.

Stresses imposed by Imperial Rome on environment and vegetation were at least as great as any the Mediterranean experienced before the nineteenth century. North Africa, Sicily and Sardinia were granaries for the imperial mega-city, and Italy was its woodyard. However, it is surprisingly difficult to find evidence for degradation of vegetation or environment that can be dated unequivocally to the Roman period.

Late-medieval adversity

In the twelfth and thirteenth centuries, population again increased; cultivation once more expanded to the limits set by soil and altitude, and to the edge of the sea. This period was brought to an end by the onset of the Little Ice Age, the Black Death and piracy. A possible contributing factor was depletion of the soil according to the Postan theory of 'subsistence crisis': phosphate was accumulating in gardens, middens and graveyards, and was not being returned to the land.[41]

Harvests often failed. In Mediterranean countries it is not easy to starve. The various crops made it less likely that all would fail at once. Except in the hungry season of late summer, people could fall back on snails, wild plants thought to be edible, or acorns.[42] However, the Little Ice Age was approaching, and crops could fail twice or thrice. Famine was often followed by plague: infected rats and fleas possibly came on ships importing corn, especially from Turkey, the traditional supplier of European deficits.

Famine and plague recurred in the 1590s: rats escaping from corn ships could have caused the great plagues of Crete in 1592 and Spain in 1599. In Spain the troubles were compounded by self-inflicted injury, the expulsion of the Muslim and ex-Muslim irrigators on whom food production depended.

Commercialization

Long-distance trade goes back at least to the Bronze Age, as the ubiquitous presence of bronze itself proves. The Roman trade in grain was on a nineteenth-century scale. Commercialization developed in medieval Italian city-states, famously Milan, Florence, Genoa and Venice, but also smaller places such as Prato which specialized (and still specializes) in cloth.

Crete was particularly dependent on trade, although Cretans themselves were seldom shipowners or seafarers. From 1350 to 1700 AD they sent dessert wine to England and Flanders. The people of this densely populated island would grow vines (a valuable, labour-intensive crop), sell the wine, and buy cereals from Turkey. The Venetian authorities tried to make Crete self-sufficient, in order that Cretans should be less willing to live at peace with the Turk. When they forbade the people to grow vines, in order to release the land for corn, Cretans planted olives instead. This prepared the way for the Cretan oil trade to expand in the eighteenth century (p. 137).[43]

Scientific agriculture

Agricultural science is not new in the Mediterranean. Most of the Greek writings have been lost, but those of Roman authors such as Columella and Palladius Rutilius survive.

[39] Rackham & Moody 1996, pp. 161–2, quoting the Sphakiá Survey.
[40] Historians such as O. Spengler and A. Toynbee have constructed theories of the innate cyclical growth and decline of civilizations; O. Spengler, *Der Untergang des Abendlandes* Beck, Munich, 1920; A. Toynbee, *A study of history*,

Royal Institute of International Affairs, London, 1934.
[41] M. M. Postan, *The Medieval Economy and Society*, Weidenfeld & Nicholson, London, 1972, pp. 61–6. The theory is, however, based on experience in England, and even there the frequent but intermittent occurrence of individual years

with low yields is more readily attributable to Little Ice Age weather than to a slow decine in fertility; J. M. Grove, 'The century time-scale', Driver & Chapman 1996, pp. 39–87.
[42] And (in the modern period) rotten oranges; one of us writes from experience.
[43] Triandafyllidou-Baladié 1988.

Scholars in Islamic Spain, notably Abu Zacaria, were fascinated by the theory of agriculture and tree cultivation: they drew on the works of Arab scholars in Africa and Asia, on what they understood of Classical writers, and on their own observations.

The Spanish tradition was continued in the *Agricultura General* of Gabriel Alonso de Herrera, first published in 1513, with many later editions, the foremost book in the western world on the cultivation of crops and trees in its time. In the eighteenth century there came the works of Cavanilles in Spain and Duhamel du Monceau in France. In 1787 a chair of agriculture was created at Palermo. The professor, Paolo Balsamo, spent two years in England studying the latest agricultural practices; he returned full of the ideas of Arthur Young and Adam Smith, and ready to experiment on irrigation, artificial leys, cattle sheds and the rotation of crops.[44] In Sardinia a Royal Society of Agriculture was founded in 1805; under the vigorous direction of the Marchese Manca di Villahermosa, it carried out experiments and introduced crops such as cotton, sumach (for tanning) and *Salsola* (for making soda).[45]

How far this affected actual practice is difficult to say. As in other parts of Europe, there were landowners who tried to put new ideas into effect, but elsewhere 'progress' was slow. Small farmers were either slow to hear of new methods or rejected them as ill-suited to the south European climate. They discovered how to grow new crops (p. 67); but if they had new ideas of their own, small farmers did not write books to tell posterity about them. Only after 1800 were agricultural improvers able to bend the ear of government and give their ideas a currency beyond the power of mere persuasion.

MODERNIZATION

Characteristic of the late-modern period is the new ingredient of energy from fossil fuel and all that depends on it: manufacturing, freight transport, human mobility, urbanization, agricultural productivity and dangerous waste products.

Steamships reached the Mediterranean about 1830 and railways about 1860. No longer was trade a matter of relieving regional shortages and distributing agricultural surpluses: it was now possible for cheap foodstuffs from across the world to destroy a region's agriculture altogether. Products made in far-away factories could undercut the work of rural artisans and craftsmen. Mediterranean cities, however, could set up factories of their own, offering employment to people from the mountains and countryside.

These changes did not happen consistently. Italy acquired railways – and later motorways – much more than Spain. Sardinia was (and still is) an island of railways, but Corsica had few and Crete almost none. As late as 1830 there was barely a single vehicle in Greece: carts were a speciality of south Macedonia, then still Turkish. Crete managed without vehicles – and managed well – even in 1920.

Tractors were not the earliest farming machines. Animal-powered machines were used for raising water and crushing

olives. In many places windpumps, of a semi-American pattern, were used for irrigation: a famous example in the mid-twentieth century were the 14,000 windmills on the Lassíthi mountain-plain in Crete.

Seafaring technology, too, was unevenly applied. Although in the 1830s the Greek navy had the world's most advanced warship, even in the 1950s much Aegean trade was in sailing vessels. Technology applied to sailing ships too: new ship designs made it less dangerous to approach small harbours on lee shores. The development of coastal shipping enabled remote places in Crete to exchange carobs for roof-tiles from Marseilles.

Corsairing almost ended when the French captured Algiers in 1830. Countless coastal villages, extensions of upland settlements, were founded soon after, but bitterly learned habits were not soon forgotten. Even in the 1860s the Cretans dared not cultivate the Frangokástello plain because they feared attack from the sea. Not until late in the century did settlement and commerce creep back to the coasts of Crete; they have still not returned to much of the Sardinian coast.

The 'nuclear winter' of 1816–20 caused by the eruption of Tambora (p. 138) was a bad time for human affairs. Cold and wet caused crop failures even in southern Europe. Spain had been depredated by the Napoleonic Wars and was soon to experience internal troubles. Crete was struck by the last great European plague; Crete and Greece were again to be devastated by war.

Population

The next most significant change is in population. After the ending of plague (but before the ending of malaria) population went up throughout the European Mediterranean, typically at about 1 per cent per year from the late eighteenth century.

The significance of plague is illustrated by Crete. The unsuccessful revolt of the 1820s was accompanied by a fall in population; statistics (for what they are worth) indicate a fall of 39 per cent, far more than wars normally cause. We suspect that the major cause was the coincident plague. From then on the population of Crete rose steadily until 1950. The unsuccessful revolt of the 1860s, by contemporary accounts as nasty as that of the 1820s but not accompanied by plague, is barely noticeable in the population figures.

Rising general population was overlain by rural depopulation, spreading gradually through the Mediterranean from France to Crete (Fig. 5.7). From the early eighteenth century onwards the people of the French Alps began to leave their unstable mountains, driven out (it was said) by the loss of their land through deluges through deforestation, in a classic example of Ruined Landscape syndrome (p. 241).[46] In 1782 the Professor of Botany at Aix remarked that since 1756 the population had forsaken the hills of Provence for the developing port of St Tropez, that seat of wealth and luxury.[47] Maximum population, however, was reached in many rural

44 *Biographie Universelle.*
45 ACCC: Atti delle Advanze periodiche della Società reale d'Agricoltura.
46 de Ribbe and other authorities, cited by Marsh 1864, chapter 3.
47 Quoted by A. N. Brangham, *The Naturalist's Riviera*, John Baker, London, 1962.

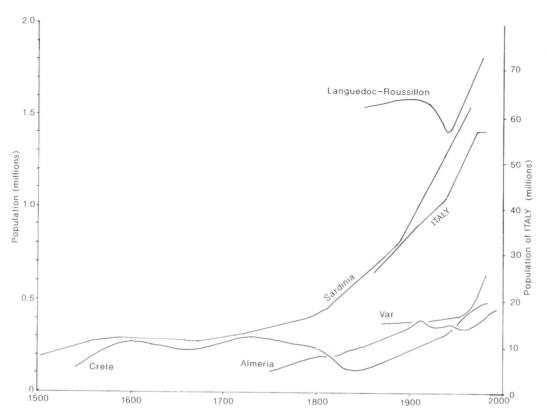

Fig. 5.6. Census population increase in certain countries and provinces.

areas in Mediterranean France between 1850 and 1870, the supposed 'peak of peasant civilization'. Other countries followed, for example Basilicata began to emigrate in 1906.[48] *In Crete the change came late: most villages have the peak of their census population in the 1940s or 1950s (although World War II air photographs, depicting ex-cultivated land, indicate that the real peak was somewhat earlier). Decline began a few years earlier in more mountainous areas (Fig. 5.7).*

Rising population was partly offset by opportunities for emigration. Mountain people might seek seasonal jobs in the nearest lowlands. Or they might emigrate more or less permanently, either (as in France) to non-Mediterranean provinces, or to other countries or continents. With railways and steamships, it became easier to emigrate, to keep in touch with *émigré* relatives, and to return. Trade with distant countries created jobs in those countries.

Mining created distant, often temporary jobs. Some mines, such as the lignite workings of Provence, gave local farmers extra employment in slack periods. Others, as in Sardinia and the Sierra Nevada, brought thousands of people and animals for a few decades to remote places, where food had to be found and lodgings built for them.

Another factor was military conscription. This too was not new: rowing the Venetian Republic's galleys was a hated obligation for sixteenth-century Cretans. But now it took the young men of most south European countries away from their homes, and showed them that there were other countries in the world and other ways of making a living than growing barley on terraces.

Extension of the cultivated area

Rising population outstripped the available capital and land. Scarcity of land persisted until either the coming of foreign produce made land surplus to requirements, or people left to earn their living more easily elsewhere.

When people ventured back to the coastal plains after the end of piracy, they found that the plains themselves had changed. Good new land had arisen since the middle ages in deltas and silted-up bays, such as the Ebro delta and the plain of Thermopylae in Greece (Chapter 18).

Fens had been thought of as under-used and malarious land, or (in revolutionary France) denounced as being somehow aristocratic. They were destroyed wherever possible and made into ordinary farmland. Kopáis, the huge fluctuating lake in Greece, was drained by the British Lake Copais Company in the 1880s. Later the Mussolini regime abolished many marshes in Italy and Sardinia.

Sometimes cultivation reached its limits. Cultivating flood-plains increased the damage done by floods. The steep slopes of the northern peninsula of Corsica were newly terraced and cultivated after 1830. Like many such enterprises, this lasted only a few decades until less strenuous livelihoods arose (p. 116). The Omalós and other mountain-plains in Crete, only partly cultivated today, were uncultivated in the sixteenth century, to the disgust of the Venetian authorities. But remains of terraces show that not only the plains themselves but even the less rocky surrounding slopes were tilled at one time, probably in the eighteenth century (Fig. 6.16).

[48] P. McPhee, *A Social History of France, 1780–1880*, Routledge, London, 1992, p. 232; Douglas N. *Old Calabria*, Secker, London, 1915.

Fig. 5.7. Part of Crete, showing the date at which each settlement reached its maximum census population. (The period of maximum *actual* population was probably a decade or so earlier, p. 20.) Almost everywhere has declined in population in recent decades, except on the coast where settlements are new and still growing. Decline set in later in the bigger settlements. It set in latest in the lowland plain of the Mesará (cf Pómbia, p. 94), but also in some of the uppermost villages round Mount Psilorítis, which confound the theory that mountains are necessarily a difficult environment. Remains of earlier cycles of depopulation are shown by the open symbols for settlements that had higher populations in earlier centuries than this: among these, only those that still had some population in the twentieth century are shown.

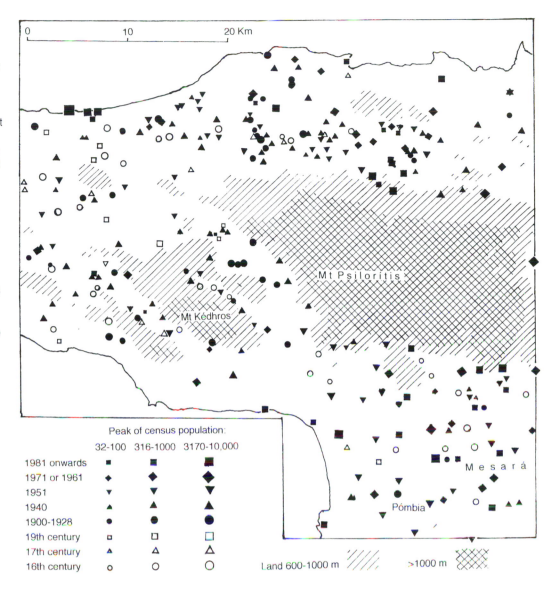

Peak of census population:

	32-100	316-1000	3170-10,000
1981 onwards	▪	■	■
1971 or 1961	◆	◆	◆
1951	▼	▼	▼
1940	▲	▲	▲
1900-1928	●	●	●
19th century	▫	□	□
17th century	▵	△	△
16th century	○	○	○

Land 600-1000 m ⧄   >1000 m ⧈

In some areas the limits of cultivation and settlement were pushed upwards. McNeill attaches great importance to nineteenth-century 'overshoot' as permanently damaging fragile mountain environments.[49] However, we do not agree that mountains are necessarily marginal land nor necessarily fragile, though they may be more prone to erosion than plains. High-altitude Cretan villages such as Máles, Kritsá and Anóyeia have been big and relatively rich since the middle ages (though they are well below the limit of settlement in earlier millennia). In the Peloponnese, too, mountain settlements are neither recent nor impoverished.[50] The claim of Montanari, that mountains afforded greater food security than urbanized or agriculturally developed areas, is at least as plausible as McNeill's.[51] Where limits did rise, this may be as much a response to the ending of the Little Ice Age as to overpopulation.

The fashion for national self-sufficiency in the 1920s brought a new if temporary wave of expansion. Mussolini

Fig. 5.8. Koustogérako, a mountain village in SW Crete, was destroyed in World War II, was rebuilt, and has since lost most of its people by emigration. This shepherd boy is an Albanian refugee. *November 1992*

[49] McNeill 1992, *passim.*
[50] Rackham, 'Laconia', forthcoming.
[51] M. Montanari, *The Culture of Food,* trans. C. Ipsen, Oxford, 1994.

launched the Battle for Grain: marginal lands in Sicily were brought into production by dint of fertilizers, but did not long withstand being sown with wheat year after year. Other potentates too were playing this game, so that Sicily's food exports – olive oil and citrus – fell by a half.

The comic finale came in the dry south of Portugal. There were wheat and sunflower campaigns in the 1930s and 1960s, maquis and savanna being made into permanent arable, sometimes irrigated. When communists came to power in 1974 they introduced high-yielding wheat varieties from Norway; but revolutionary zeal failed to bring enough rain and cold to Alentejo to keep them alive. 1978 was proclaimed the Year of Maize, with the same result.[52] Even more bizarre is the story of the last British overseas colony, founded in 1969 in order to introduce modern English farming to Alentejo. Nemesis struck almost instantly, involving among others the descendant of a noble Norfolk family with a proud history of agricultural innovation, who ought to have known better. One of the few Englishmen to escape ruin married a Portuguese and learnt Portuguese methods of farming.[53]

New crops and crop diseases

New crops appeared and new markets were created. Potatoes, tomatoes and maize from America are grown wherever there is enough water, and have transformed Mediterranean cuisine. The south of France, brought by the railway within two days' journey of the cities of northern Europe, took up fruit and vegetables for export. These crops, however, could be grown only in places favoured by climate, water and transport. More widespread was an increase in the traditional growing of vines, olives and almonds.

From other continents came new crop diseases. Potato blight struck in the 1840s, although less disastrously than in Ireland. Silkworm diseases killed silk production in places like Peyranne in Vaucluse.[54] The chestnut was almost extinguished when *Endothia* blight attacked in the 1920s (p. 68).

Vine-growing has never been the same since America gave Europe powdery mildew in the 1840s and downy mildew in the 1870s. More devastating still was *Phylloxera*, the root-aphid accidentally imported from America by someone trying to improve European vines; first identified in 1863 in Gard, it destroyed almost every single vine in Europe, except on a few islands. All these diseases can be contained, somewhat precariously, but vine-growing now involves grafting and spraying to get a crop at all, and chemical residues alter the wine.

Many vine-growing areas never recovered, but others benefited. Roussillon was not attacked until after means had been found of coping with phylloxera. Before the attack its *viticulteurs* benefited from high prices for wine, and the attack when it came was not long-lasting. At the same time fast trains to Paris, Frankfurt and Cologne made it possible to specialize in peaches, apricots, lettuces and tomatoes. By the

1920s Roussillon was devoted to wine, fruit-growing and market gardening; it avoided industrial development.[55]

Indirect effects of technology

Transport and technology relieve the worst effects of natural bad years, but also link Mediterranean countries to world wars and to the booms and busts of the world economy. The agricultural depression of the 1870s affected especially the south of France; the Great Depression of the 1930s was hard on Crete.

Many technologies are available only to big farmers or people with capital (Fig. 5.9). The mouldboard plough, replacing the ard, made heavy soils easier to cultivate by those who could afford the plough and the extra horses to draw it; its introduction could be more divisive than that of tractors. Imported fertilizers, too, reinforced the division between those who could afford to buy guano from Pacific islands or mummified cats from Egypt and small farmers who had only goat dung or rotten rosemary. Intensification for small farmers could mean over-use of common rights.

Fig. 5.9a. The last horse plough in Aliano, Basilicata. *October 1993*

Technology can destroy people's livelihoods for some trivial reason happening on the other side of the world. In the 1870s the discovery of the dye alizarin suddenly killed the growing of madder (*Rubia tinctoria*), a major crop in the south of France.

These changes can affect the landscape. A year or two of depression or bad weather, such as the frosty winter of 1870–1 which 'killed' the olive-trees in the south of France, can make a viable, though not flourishing, enterprise seem no longer worth the struggle. The people leave for Marseilles, Paris or Algeria, and even if the village survives the terraces are abandoned.

Industrialization brought other effects. In the Sierra Nevada, the demands for timber for construction and wood

52 P. Campos Palacín, 'La degradación de los recursos naturales de la dehesa. Análisis de un modelo de dehesa tradicional', *Agricultura y Sociedad* 26, 1983, pp. 289–381; 'Economía y energía en la dehesa extremeña', Instituto de Estudios Agrarios, Pesqueros y Alimentarios, 1984.

53 Balabanian 1984.

54 L. Wylie, *Village in the Vaucluse*, Harvard University Press, 1974.

55 A. Moulin, *Peasantry and Society in France since 1789*, trans. M. C. and M. F. Cleary, Cambridge University Press, 1991.

for smelting are said to have resulted in excessive woodcutting (pp. 317–8). Conversely, mining of coal or coal-like substances could undercut the demand for firewood and charcoal; this may explain why Sardinia, unlike Italy, has got out of the habit of woodcutting for fuel. The spread of bottle-corks created a new industry, giving a commercial value to a tree previously of only local use. Railways, using many millions of short-lived sleepers, prolonged the market for small, low-grade oaks, previously used for building and fuel. Sleeper-cutting and corking were major employers in nineteenth-century Sardinia (pp. 182, 208).

Government and law

Another change was the growth of central governments. Before 1800 most day-to-day government had been local government at the city-state or parish level. There was usually a rough balance between the activities of the nobility, the Church and the populace. Communities were assigned, by accidents of history, to one or other of a number of European powers, which collected taxes and tried to make laws.

Fig. 5.9b. The remaining mules on the Lassíthi Plain draw ploughs in the morning and carry tourists to the cave of Psykhró in the afternoon. *April 1989*

This varied from one power to another: Turkey, for instance, had a weak nobility and a strong bureaucracy. Only in Spain was there a rigid centralized organization with ideology, policies, ministries and secret policemen, which claimed the power to make its subjects do what the king wanted.

During the nineteenth century governments became fewer and more centralized, as with the unification of Italy. They extended their purview from affairs of state to include more and more aspects of everyday life. Laws came to have the function more of preventing people from doing things, and less of raising revenue from fines. Governments became susceptible to secular ideologies, and encroached on local institutions. Scholars with ideas about how the landscape ought to function or to be subdivided, who in previous centuries had written books, now secured official backing for their theories and imposed them on their fellow-citizens. Countryfolk would discover one day that what yesterday had been a normal, uncontroversial practice, such as goat-keeping or occupational burning, was today a crime by decree of a far-away legislature. (Democratic process had been scrupulously observed: the local representative had protested but had been outvoted.[56]) Attempts to enforce the decree would be inconclusive and would lead to endless ill-will.

One aspect was the regulation of trade. Corn laws were nothing new – they had been prevalent in both Venetian and Turkish Crete – but it now became normal practice for central (rarely local) governments to subsidize, tax or forbid imports or exports. These policies derived from whichever motive (raising revenue, promoting self-sufficiency, encouraging local communities, discouraging luxury) happened to be uppermost in the legislators' minds.

Bureaucracies made work for policemen, inspectors, pen-pushers, taxgatherers, customs officials, etc. This created employment, especially in desperately overpopulated Sicily, but in more subtle ways reinforced the tendency to the growth of cities. In Greece, that very centralized country, social status depends on where one lives: the inhabitants of Athens are upper-class, those of provincial cities are middle-class, and country-dwellers are lower-class.

Land reform, enclosure movements and the Law of Unintended Consequences

The Enlightenment had decided that feudal (especially ecclesiastical) landowners were bad, common-land was bad, and the only proper use of cultivable land (or land thought to be cultivable) was conventional agriculture carried on by private owners in rectangular plots of not less than five hectares or so. For uncultivable land the proper use was timber production organized by the state on the model of either French or German forestry (depending on which state).

These eighteenth-century ideas, a century later, captured the minds of governments, who tried to persuade or force local people to follow them.[57] Traditional tenurial practices, structures and multiple land-uses were overridden, often with results very different from those intended. This began c. 1800 when Napoleon applied French Revolutionary ideas of land-holding to captive Apulia, abolishing the trans-humance rights which had restricted development on the Tavoliere plateau (p. 210). It extended far outside southern Europe: similar changes, especially in regard to forestry, were foisted on native peoples and their landscapes by imperial administrators in other continents.[58] Even statisticians had to make up their minds whether to book a piece of land as 'pasture' or 'forest': it could not be both. Indigenous knowledge was universally despised and marginalized.[59]

56 However, dictators such as Mussolini and Metaxas particularly disapproved of goats. To add to the woes of Aliano in the 1930s (p. 281), the tyrant had taken it into his head that Goats were Bad, and had decided to tax them out of existence; so the peasants had to go without milk, cheese and meat.

57 For philosophical and legal details, see Carletti 1993.

58 For example, R. P. Neumann, 'Forest rights, privileges, and prohibitions: contextualising state forestry policy in colonial Tanganyika', *E&H* 3, 1997, pp. 45–68.

59 R. H. Grove 1995; cf Forbes 1997.

Land over which distant nobles or monasteries had once exercised a tenuous and accommodating overlordship was thus transferred to private farmers, municipalities or state forestry departments, to whom ownership meant the right to keep others out. The quality of the land was consistently exaggerated: often it had been left as common land precisely because it was unsuitable for conventional agriculture. Reorganization was expensive: the beneficiaries had legal expenses to pay and often (as in Spain and Sardinia) had to construct fences or walls round the new boundaries. Capital, supplied by merchants in the cities, became a limiting factor. The change tended to benefit existing large landowners at the expense of small landowners, for whom litigation and wall-building cost relatively more. Those who hitherto had relied on common-rights often got scraps of the worst land (if they could afford to enwall them); transhumant shepherds did worst of all. The south of France, in the hundred years after the French Revolution, was the scene of much discontent.

Reorganization came in a different form to middle and south Greece after the Greek War of Independence. Many of the landowners had fled, being thought to be Turks, leaving their land to be seized by the Greek Crown. The poor state of agriculture in the plains in the 1830s was ascribed to the Crown's inefficiency as landlord.[60]

In Sardinia common-lands were privatized. Dry-stone walls in straight lines, characteristic of 1860s enclosure, march across lonely basalt plateaux full of Iron Age *nuraghe* towers; their construction probably recycled and destroyed the intricate Iron Age field-walls. In savanna the walls run between the pollard oaks in a way that shows that the oaks are older. Field-walls, in turn, are intersected at awkward angles by the slightly later railways. Some landowners acquired great, infertile estates, but the nobleman's country villa is not often a feature of Sardinia. Farmers mostly still lived in towns, where solid courtyard houses of granite, marl or mud-brick give an air of unostentatious prosperity at this period. In the mountains some shepherds became dependent on the new masters, landowners who controlled both dairies and grazing lands.[61] Others defended their grazing rights with sword and musket, and disputes continue to this day.

Sometimes, as in south Italy, new owners showed little agricultural enterprise, or found the land unsuitable; they hired their land out to shepherds who had not previously paid for grazing, and thereby earned their keen resentment.[62] Forests passed into the hands of municipal authorities or the state, and came within the scope of official forestry, in which (in theory) timber production, rather than wood, acorns, etc., was all-important.

Southern Italy, by the end of the century, is commonly portrayed as a land of big estates belonging to absentee landlords, managed by rapacious middlemen, paying subsistence wages to an impoverished peasantry, with the sinister figures of the Men of Honour investing the ultimate proceeds in the vice and crime of Naples and Chicago.[63] The real situation was much more complicated, as shown by the twelve-volume report of a state commission of inquiry in 1909.[64]

The report indicates that the south was not simply a region of large estates worked by wage labour. Census returns set down two-thirds of the rural population as wage workers; but that was not necessarily their main or only livelihood. They had access to land and fruit-trees on terms that varied according to local custom, the individual's needs and opportunities, and the dependability of the particular crop:

> It was common for a peasant to rent grain land from one landowner, and to sharecrop vines from another and olive trees from a third. The peasant lived in his or her own house (often in town) and journeyed to the various plots as required.[65]

Yields of cereals and vines are notoriously variable in south Italy. Olive trees bear in alternate years. Since wine and oil could be stored from year to year – oil was often used in Mediterranean countries as a store of capital – prices did not rise when yields fell. In contrast to irrigated Spain, farmers could not protect themselves much by diversification, because droughts produced bad years in all three crops. Share-croppers could not commit themselves to pay a fixed rent in years when crops might fail; landlords, who often paid for the land to be ploughed and fertilized, wanted their full share of the crops in good years. Complex tenurial arrangements and scattered plots were a well-thought-out means of crop insurance to the benefit of both landowner and tenant.

In Spain privatization had an impetus when the State sold off its own lands to pay war debts and, when this was not enough, stole and sold the Church's lands. Between 1798 and 1862 many existing landowners or newly wealthy people bought up public land at bargain prices, the State having already spent the proceeds. The market was flooded with land which municipalities and noble families were now allowed to sell.[66] It has been alleged ever since that the new owners destroyed forests and that increased floods and erosion were the result.[67] However, detail on the fate of specific areas seems to be lacking, and there is no correlation between provinces where there was little privatization and those that now have forests. Extensive privatization is correlated, if anything, with the incidence of savanna.[68]

There was an attempt to colonize marginal lands. Spain had a Law of Internal Colonization and Emigration, under which the State gave people 3 ha of land each on condition that they cultivated the land themselves and did not rent or divide it. Needless to say, the land given to the hapless colonizers tended to be uncultivable; the schemes won votes but did not create permanent farms.[69]

[60] Rackham 1983.

[61] Schweizer 1988, p. 224.

[62] We acknowledge the help of Paola Mairota in writing about south Italy.

[63] See J. A. Davis, 'The South, the Risorgimento and the origins of the southern problem', *Gramsci and Italy's Passive Revolution*, ed. J. A. Davis, Croom Helm, London, 1979, pp. 67–103.

[64] *Relazione della Commissione d'Inchiesta Parlamentare per Accertare le Condizioni dei Lavoratori della Terra nelle Provincie Meridionali e in Sicilia*, Rome, 1909–10.

[65] F. L. Galassi and J. S. Cohen, 'The economics of tenancy in early twentieth-century southern Italy', *Economic History Review* 47, 1994, pp. 585–600.

[66] A. Shubert, *A Social History of Modern Spain*, Unwin Hyman, London, 1990, chapter 2; A. Diaz Gomez, 'La desamortización en el municipio de Albacete', *A-B* Segunda epoca Ano IV No. 5, 1978, p. 39.

[67] For example Bauer Manderscheid 1991, p. 356.

[68] See Shubert (note 66), Fig. 2.1.

[69] An example appears in G. S. Martínez and

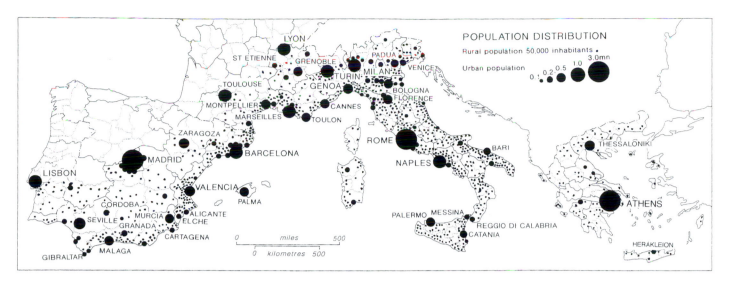

Fig. 5.10. Urban and rural populations of southern Europe.

THE LAST FIFTY YEARS

By the 1950s, southern Europe was recovering from the wars and destruction of the 1930s and 1940s. Health was improving after the campaigns against malaria. But standards of living were low compared with northern Europe. Most regions were overwhelmingly agricultural, farming was not very productive, and other employment was often limited to the bureaucracy.

The period after 1950 saw enormous growth in the economy of northern Europe. Reconstruction and industrial expansion, above all in Germany, sucked in labour to north-European-style jobs from southern Europe. Migrant workers, though cut off from their fields, sent remittances to those left behind, which encouraged others to move.

Population

The population of the European Mediterranean is now about 80 million (Fig. 5.10). Spain and southern France have been filled by ex-colonists expelled from north Africa, and by colonizers from Africa itself.[70] Nevertheless, owing to declining natural increase, the total population rose by only about 5 per cent during the 1980s (the latest available figures). More than half the inhabitants live within a few kilometres of the shore, but many possess second homes inland. Southern Europe attracts pensioners as well as tourists: the emigrants of the 1950s and 1960s will soon retire, and many will return to live on the coast near their former inland homes. The population is now stabilizing, in contrast to the Asian and African Mediterranean, and the average age is rising.

Migration from inland to the coast continued into the 1960s or even later, especially in the far south. However, in areas with good communications such as Provence, population has lately risen up to 50 km inland. Although the biggest cities show signs of stabilizing, small cities continue to grow, as do large and middle-sized towns. Urbanization has become more diffuse, so that a town of the same size now contains fewer people. Outside towns and villages, settlers build villas in pinewoods or ex-cultivated land, each villa surrounded by several hectares of land imprisoned within a fence. This land is not browsed or cropped, and thus accumulates dry vegetation for the next fire. Little of the French Mediterranean coast is still rural; low-density *urbanisación* threatens nearly all the coast of Catalonia; a road girdles almost the entire Italian coast.

Rural depopulation must be seen against excessively high population in the recent past. Even though numbers have shrunk, the remaining population is still high (and very wealthy) by the standards of any century before the nineteenth. Moreover, villages can still grow even if their populations shrink. People live in bigger houses and fewer to a house; they often build an additional, larger house on a new site rather than extending or replacing an old house.

Continuing land 'reform'

Land reform smouldered on in the 1950s, but was being bypassed by capitalist agriculture developed along industrial lines. In south Italy, laws enacted in 1950 redistributed some 7,500 sq. km of land to 113,000 farmers. As social engineering this had little success in the face of out-migration. The new farms were too small. Already by 1962 some 15 per cent of farmers had abandoned them; others were kept going by job-creation in forestry.[71]

Sicily was thought to need more land reform. In 1945 its

A. Estévez, *Parque Natural del Montgó: estudio multidisciplinar*, Generalitat Valenciana, Conselleria d'Administració Publica, Valencia, 1991.
70 For the Mediterranean Sea as a migration frontier, see R. King, 'Migration and development in the Mediterranean region', *Geography* 81, 1996, pp. 3–14.
71 J. Bethemont and J. Pelletier, *Italy: a geographical introduction* Longman, London, 1983; P. Allum, 'Thirty years of southern policy in Italy', *Political Quarterly* 52, 1981, pp. 314–23; E. Pugliese, 'Farm workers in Italy: agricultural working class, landless peasants, or clients of the welfare state?', *Uneven Development in Southern Europe*, ed. R. Hudson and J. Lewis, Methuen, London, 1985, pp. 123–39.

population was approaching five million, having doubled in a century.[72] Agriculture occupied more than half the population; yet one-third of the land was in the hands of only 1,300 owners, mostly under-capitalized arable farmers. In 1948 a new law made criminals of owners of more than 200 ha of contiguous land. Many big landowners sold their estates to avoid confiscation, and poor people bought a few hectares. State agencies were instituted to oversee the distribution of new holdings, providing services such as irrigation and roads at public expense. Despite the reputation of Sicily as an independent and criminous land, the attempt met with some success. Perhaps 3,000 sq. km of new farms (covering one-fifth of Sicily) were created.

In Spain, the Franco regime was promoting irrigation and colonization until the 1970s. Landless 'peasants' were to be settled in dry-farming areas which would be transformed by irrigation. As in previous land reforms, the state's hydrological policies served mainly to increase the capital assets of pre-existing large holdings.[73] Spain has got out of its troubles by higher incomes from abroad, expanding industries, tourism and support from the European Community.

The Food and Agriculture Organization of the United Nations inherited the bureaucratic obsession with modern forestry and detestation of common land. In 1948 it called upon Mediterranean peoples and officials to drop whatever else they were doing and redirect their societies towards 'restoring' the forests.[74]

Disapproval of common land gained a new lease of life with the publication in 1968 of the theory of the 'Tragedy of the Commons': the doctrine that a system of common rights will inevitably break down because each participant pursues his own short-term advantage regardless of the rights of others. This pernicious notion was invented by an American with no experience of how commons actually work. Despite lack of evidence, it has been repeatedly used by governments as an excuse to abolish or oppress commoners, often with results very different from those intended.[75] Real common lands rarely incur Tragedy, either because over-use is self-limiting (if you cut wood too often you merely have to wait longer for the next crop) or because the commoners – the set of people who share the common rights – foresee the problem and agree upon rules to avert it. The communal irrigation arrangements in Spain, working smoothly at a local level for century after century, with no theoretical basis and no government intervention, and surviving even the violent transition from Islam to Christianity, are the most famous of innumerable counter-examples to the Tragedy of the Commons.[76]

Communal arrangements tend to be stable where the people involved have a personal interest in the outcome. They break down as the scale and complexity increase, and especially when governments, politicians or companies intervene. External duties to voters, ideologies or shareholders take precedence over the conservation and just allocation of the resource.

Government assault on common land has often brought about the very disaster that it was intended to avert. Although since 1950 most of the heavy-handed and tragic examples of such intervention have been from lower latitudes, some Southern European governments disapprove of common land and family co-ownership to this day. Not only does private ownership continue to grow, but also the anti-social notion that ownership carries the right to fence land and to keep out persons as well as animals. Robbery of public rights, especially right of access, is keenly resented by local people as well as outsiders.[77]

Agriculture

Agriculture has been affected by five kinds of change:

(1)   Plant and animal breeding, making it possible to grow more crops on the same area of land. For example, between 1949 and 1980 yields of maize in the south of France are said to have increased tenfold and those of wheat $2\frac{1}{2}$-fold.

(2)   Mechanization, which on favourable terrain allows one man to do the work of several men and beasts.

(3)   New materials, especially plastic irrigation pipes, plastic greenhouses and (for olive-growers) plastic nets to catch the olives.

(4)   Transport and refrigeration, which make it no longer essential for any south European country (except Albania) to grow its own food.

(5)   Subsidies and regulations, making agriculture less of a merely economic activity.

Plant breeding, mechanization and transport work to the disadvantage of south Europe, since most crops (except olives and some vegetables) can be more easily grown in other climates. (They would have been even more disadvantageous had Stalin not shot the plant breeders of the Soviet Union, and had eastern Europe been allowed to develop an export trade sooner.) The coming of farm machinery, designed in flattish countries with big fields, has discouraged farming on steep terrain. Agriculture has generally declined and – at least in the statistics – is no longer a major part of the economies of south European countries, except for Albania and Greece.

Southern European agriculture might have fared even worse. It has done very well out of subsidies, tariffs and corn laws, but the effects have not always been as intended.[78] Protection of the European market has kept out the products of other countries which could produce them less intensively and more efficiently. Prices for water, chemicals and fertilizers are kept artificially low, resulting in over-specialization and poor land use, which can be comically counter-productive.

[72]   In the 1900s one-half of Sicily was owned by less than 0.1 per cent of the population; Lorenzoni 1910, quoted by Mack Smith 1968, p. 200.
[73]   D. Garcia Ramon, 'Old and new in Spanish farming', *Geographical Magazine* 57, 1985, pp. 128–33.
[74]   Report of the Director-General (Sir John Boyd Orr), 'Forest Activities in Europe',

Unasylva 2, 1948, pp. 81–2.
[75]   G. Hardin, 'The tragedy of the commons', *Science* 162, 1968, pp. 1243–8; Carletti 1993, *passim*; G. Monbiot, 'The tragedy of enclosure', *SA* 270 (1), 1994, p. 140.
[76]   Delano Smith 1979, chapter 7; E. Ostron, *Governing the Commons: the evolution of institutions for collective action*, Cambridge, 1990.

Hardin (see note 75) later accepted limitations to his theory: G. Hardin, 'The tragedy of the unmanaged commons', *Trends in Evolution & Ecology* 9, 1994, p. 199.
[77]   S. Lagomarsini in Carletti 1993, pp. 155–7.
[78]   C. Stevens, 'The environmental effects of trade', *WE* 16, 1993, pp. 439–51.

Cereals

Mediterranean cereal varieties are low-yielding compared to their equivalents in England and north France, though respectable compared to America and Australia. Much of the product goes into pigs, cattle and chickens. However, in southern Italy the subsidized production of hard wheat (*Triticum durum*) for making pasta encourages the continued cultivation of unstable land which should not be cultivated (Chapter 14).[79] Cereal-growing has virtually disappeared from mountainous countries such as Crete.

Vines

For decades wine has been over-produced: consumption of ordinary wine in southern Europe is declining, and this is only partly offset by increasing consumption, especially of fine wines, in northern Europe and America. The European Union has tried to persuade growers not to replace superannuated vineyards; this has not succeeded because the remaining vineyards yield more wine. The surplus wine lake in Europe is compulsorily purchased and distilled into industrial alcohol.

The south of France became especially a vine region after the collapse of cereal cultivation. In the *département* of Var almost two-thirds of the remaining cultivated area is vineyards; their area diminished from 610 sq. km in 1959 to 400 sq. km in 1986. The production of wine, however, is reported to remain the same, but more of it is fine wine rather than *vin ordinaire*. The region is becoming dominated by big private growers instead of small owners or cooperatives. Not all vineyards have retreated: occasionally, as around Aix-en-Provence, vines have extended into former olive-groves and almond orchards.[80]

In Spain wine-growing has concentrated in La Mancha, Murcia and Estremadura. In Italy production is mostly in Sicily and Apulia, with vineyards shifting to the plains and the abandonment of terraced *coltura promiscua* vines.[81]

Olives

Olive-growing is a paradox. It ought to be precarious, considering the great labour involved in picking olives, erratic cropping, competition from North African olives, the abundance of cheap vegetable oils, and the declining use of the worst oil for soap-making. In practice it flourishes as never before. The European Union has subsidized olives generously, often by payment per tree rather than on production. The effects vary. Sardinians have not overcome their reluctance to grow olives, but we have known Greeks plant olives well above the altitude at which they would survive. Cretans, after four centuries of increases in olive cultivation, have done their best to convert the island into a monoculture of olives. Although there is no immediate sign of an olive oil surplus, this can hardly go on much longer: we hear of dereliction among the olives in Tunisia.

Fig. 5.11. A semi-mechanized method of gathering olives. *Catalonia, October 1995*

The labour of harvesting is sustainable because it happens in winter, when hotels are not providing jobs, and it can be done in school holidays. In Crete, plastic nets are spread under the trees, on to which the olives drop.

There are new varieties of olives and new methods of production. It is fashionable to have large numbers of small trees planted close together – 250 or so to the hectare instead of 80 – and irrigated from plastic pipes. For those who can afford the equipment this has doubled the yield (thus adding to the overproduction) and somewhat reduced the variation between years. The establishment of new olive groves in Crete seems to outrun the labour available to harvest them; it accounts for most of the bulldozing that has damaged the island (p. 269). The canopy of new olives is much sparser than the continuous cover of the maquis that they replace, or of old olive groves in places like Lesbos: they are a less good habitat and fail to protect the soil from rain and sun.

Olives have also increased in Spain. Spanish olives (unlike vines) are rarely irrigated, but in wealthy enterprises they are harvested by mechanical shakers (Fig. 5.11). Although eating-olives are not despised as a proletarian diet, as they are in Greece, they have their ups and downs of fashion. Around Seville, where three-quarters of the olives are destined for the table, production fell for a time and some groves were abandoned, but this use has recovered.[82]

Market gardening

Horticulture flourishes, especially in greenhouses producing fruit, vegetables and flowers. Most suitable areas are coastal, increasing the concentration of people and activities already there or associated with tourism. (So far tourists have been surprisingly tolerant of ugly plastic greenhouses and rotten plastic.) Immigrants from North Africa and Albania provide some of the labour. The communications reach as far as south

[79] P. Sorsa, 'GATT and Environment', *WE* 15, 1992, pp. 115–33.

[80] W. K. Crowley, 'Changes in the French Winescape', *GR* 83, 1993, pp. 252–68; A. de Reparaz, 'Un vignoble méditerranéen français: l'exemple du Var'; C. Durbiano, 'Les arrachages de vigne en Basse-Provence de 1976 à 1985', *Méditerranée* 65 (3/4), 1988, pp. 21–7, 67–70; *Recensement Agricole 1988, Bouches-du-Rhône: Principaux Résultats*, Ministère de l'Agriculture et de la Forêt, Marseille.

[81] I. Tirone, 'Les dynamiques récentes du vignoble italien', *Méditerranée* 83 (1/2), 1996, pp. 87–96.

[82] S. Angle, 'Déclin et renouveau de l'oléiculture dans la province de Séville (Espagne)', *Méditerranée* 76 (3/4), 1992, pp. 27–34.

Crete, where plastic greenhouses came in the 1960s and are now the biggest alternative to olive-growing. In Provence, market gardening, revived by French colonists expelled from Africa, has taken over some of the land vacated by vineyards.

In some areas the market for dried fruits and nuts has long been in decline through competition from countries outside Europe, and because refrigeration now makes 'fresh' fruit available throughout the year. The results have been felt especially in the middle-sized Aegean islands.[83] Conversely, in SE Spain almonds have increased to nearly half the cultivated area, which gives rise to concern, since this is reputedly a very erosion-promoting crop (p. 243).

Market gardening is an objective of the great increase in irrigation through ambitious schemes like the successive canals that have tapped the lower Rhone since the 1960s. The Bas-Rhône–Languedoc irrigation company does many other things: afforesting headwater areas, financing crop-processing factories, building and renovating rural houses, providing marketing and training courses, experimenting with new crops and techniques, mapping and studying soils, and trying to control erosion. In south Spain the plains of Dalías and Níjar are a 'sea of plastic' from greenhouses and irrigated market gardens.

Irrigation has brought great local prosperity, since the people getting the water seldom have to pay its realistic value; they are, however, exposed to floods and vulnerable to the supply being over-used (Chapter 19). Irrigation tends to benefit one area at the expense of another: often it results in silt being trapped in reservoirs instead of being added to deltas (p. 350). It encourages abandonment and depopulation in nearby areas that do not have it. It dries up rivers and spoils wetlands; in some areas of Crete hydrologists think it a reproach to their profession if any water reaches the sea.

Livestock

As in the rest of Europe, arable and pasture husbandry have become disjoined. Cattle have disappeared from many areas, notably Crete. Animal husbandry, especially pigs and poultry, turns into a mere industry, separated from the land. The Mediterranean traveller, except happily in Spain and Portugal, seldom sees a pig, and in many villages never hears a cock.

Sheep and goats are less affected. Numbers of animals reported have generally increased, but the real number is often unascertainable (p. 19). Because of the retreat of cultivation and the declining area browsed by other livestock, the area available has often increased. Over wide areas of the Mediterranean browsing has manifestly declined, but locally it has increased to the point that scholars complain of over-browsing. Another development is the growing of crops specifically to feed sheep.

For example, on the high karst *causses* of Lozère overlooking Mediterranean France, depopulated a century before, there has been a programme to revive sheep-farming by growing winter forage and by grubbing out heathland. By dint of subsidies and machinery, all the land cultivated at the beginning of the nineteenth century was reoccupied to grow fodder. However, this proved expensive to maintain against the return of heath, and remoter areas are not considered worth clearing.[84]

Long-distance transhumance has declined for nearly two centuries. Drove roads are usurped by motor traffic, and overnight grazing-grounds by neighbouring people. However, it goes on, rather secretively, among the Vlachs and Sarakatsani of Greece. The death of Píndos transhumance, predicted nearly a century ago,[85] has not happened. One of us has been in the giant Vlach village of Samarína when it was almost empty; the people and their flocks do not return until 28 May.

In the evening of 30 June 1993, near Castellane in the Provençal Alps, one of us met two flocks of sheep, about 2,000 strong, on their way by night (to avoid the traffic) to the region south of Barcelonnette. The shepherds reckoned they would be able to reach the Alpine grazing grounds on foot for no more than the next five years.[86] However, in Spain, the deserted *cañada* drove roads are being brought back into service; flocks are taken north to the mountains, accompanied by walkers in a new form of agro-tourism.

Agricultural labour

The amount of work done in agriculture is supposed to have diminished throughout Southern Europe, but statistics are deceptive. Farm labour is unquantifiable in a world in which farm work is increasingly part-time and increasingly off-site. 'Workers in the agricultural sector' who make tractor parts or fertilizers in distant factories, or who store, process and transport agricultural produce, are left out of the statistics. However, it is clear that population and agriculture are no longer closely linked. In Provence, and even in remote parts of the Peloponnese, one finds villages with every sign of modest prosperity (such as new building) but virtually no cultivation and often little evidence of pasturage.

Woodcutting

Fossil fuels, expensive to transport even with steamships, were late in displacing wood as the chief rural, even urban, fuel. In many Mediterranean countries in the mid-twentieth century there was a sudden switch from 'excessive' woodcutting – up to the limit set by the growth of trees – to little or no woodcutting. Italy, however, is still a wood-burning land whose coppice-woods are regularly cut even far from a road (Fig. 5.12).

Land abandonment

Much land abandonment, as we have seen, appears to result from easier ways of making a living, from depopulation, and from transport making it unnecessary to grow food locally. These were reinforced in the 1960s by the effects of mechanization and plant breeding. In the early years of the Euro-

[83] N. S. Margaris, 'Effects of recent agricultural changes on the landscapes of Mediterranean Europe', *Petromarula* 1, 1990, pp. 26–8.

[84] F.-E. Petit, C. Cossandry, T. Muxart, 'Défrichements de terre agricoles et risques érosifs: un exemple dans le sud du Massif central français', *Bulletin de la Société Languedocienne de Géographie* 21, 1987, pp. 219–27.

[85] Wace & Thompson 1914.

[86] Cf D. Musset and F.-X. Emery, 'Histoire et actualité de la transhumance en Provence', *Les Alpes de Lumière 95/96* Edisud, Aix-en-Provence, 1986.

Fig. 5.12. Chestnut coppice, Colle di Velvia, Ligurian Apennines. *September 1984*

pean Common Market it became clear that the then reduced area of farmland was still too much for the increased yields. In 1968 the Mansholt Plan proposed to encourage the abandonment of a further five million hectares, mainly in southern France, the Massif Central, Corsica and southern Italy, but for the time being this was thought to be too drastic.

More recently, land in south Europe has been temporarily left uncultivated (though sometimes grazed) under set-aside subsidies. These, however, do not try to identify land that ought not to be cultivated. In Sardinia they have affected not only the basalt plateaux – ploughed up in a fit of enthusiasm only a few years earlier – but even the Campidano, the best land in the island for two thousand years.

The European Union is supposed to keep small farmers on the land. In practice, farm support, like land reform before it, has benefited large farms. Bureaucrats still believe the traditional myth that small farms are inflexible and unable to diversify and modernize. In practice, smallholdings are often a second job for well-educated, resourceful people with modest incomes. They provide extra income, entertainment and insurance against the wages or pension drying up.[87] With the number of retired people in southern Europe outrunning provision for their pensions, the aged should be encouraged to do something useful, pleasant and profitable in tilling smallholdings.

J. R. McNeill has introduced a theory of 'undershoot': the idea that once the population falls below a minimum, the infrastructure – terraces, irrigation works, etc. – can no longer be maintained at all, the ecosystem collapses, 'topsoil' is lost and desertification results. We have never seen this happening, even in the mountains. In our experience, if people can no longer cultivate all the land, they retreat to the easier land.[88] Difficult outliers of the irrigation system may be abandoned, maintenance being concentrated on the more productive parts. Outlying or poor lands become pasture or woodland. The ecosystem reverts to approximately what it would have been before the population reached its maximum. Terraced hillsides revert to the slope they had before terrac-

ing (but see pp. 259–60). Thousands of mountain villages in Crete, Greece and Italy make a perfectly adequate (if now subsidized) living from a fraction of the area cultivated when the population was greater. If this proves to be unsustainable, it will be because it is too dependent on flawed modern technology and economics, not because retrenchment is impossible.

Tourism

Pilgrimage and tourism are nothing new. Ancient Greeks went to the Olympian Games; their oracle at remote Dodóna had a more capacious theatre than any in England today. Pausanias listed the works of art in Greece for the benefit of visiting Romans. Anglo-Saxons thought little of visiting Rome. The devout of all Europe went to the Holy House at Loreto, or passed through Venice on the way to Jerusalem. The humblest south Europeans feasted at the shrine of the local saint.

People have gone to the French Riviera for their health for centuries. In Sardinia, English families seeking the sun built second homes on the south coast – and, lingering too long into the summer, died of malaria. *Murray's Handbooks* and the advertisements therein show that there was some tourist traffic in most of southern Europe in the 1880s; Benidorm, a fishing village six hours by stagecoach from Alicante, welcomed bathers in 1893. After these beginnings, large-scale tourist development was brought to the Aegean by the Italians in Rhodes in the 1920s and 1930s. The magnificent medieval city was heavily restored, and a new city of hotels was built outside it. Good roads, alpine-style hotels, spas and airports were begun all over the island.

Tourism became an industry after 1950 through cheap energy, efficient aircraft, better railways and roads, a newly prosperous northern European work-force entitled to holidays, and the fall of tyrants (Portugal in 1970, Greece in 1974, Spain in 1975). The expansion of the European Community (joined by Britain, Denmark and Ireland 1973, Greece in 1981, Spain and Portugal in 1986) removed some bureaucratic obstacles. The Mediterranean is now said to account for one-third of the world's tourist business.

Numbers of foreign tourists, as officially defined, have increased to about eighty million per year, roughly equal to the resident population. Tourism dominates some of the smaller islands, though even so small and much-visited an island as Corfu does not wholly depend on it. Much tourism, moreover, is internal. Informal, part-time 'bed and breakfast' tourism is probably under-recorded, especially where earnings from it are taxable.

Tourism is still underdeveloped in many areas, especially the mountains of Greece and Crete. In Sardinia there is remarkably little. The NE corner (Costa Smeralda) has been intensively developed, but mainly from outside the island, so that the proceeds do not necessarily remain in Sardinia. On the north coast, tourist settlements are still deserted for much of the year; even in summer they are serviced by parent

---

[87] *The Economist* recently quoted surveys in Italy which revealed that 'part-time' farms fit economists' ideas of dynamism better than full-time, the part-timers being younger, more educated, more mobile and more innovative.

[88] Not necessarily all at once: an individual may still cultivate a tiny isolated vineyard long after the rest of the hillside has been abandoned.

Fig. 5.13. The Sardinian coast east of Cagliari. Most of the development has been orderly and carefully planned, although sometimes precariously close to pollution-sensitive lagoons. *April 1992*

settlements far inland. Elsewhere hotels are few, far between, expensive and good. Sardinia's assets – its bizarre mountains, glorious savannas, prolific antiquities, industrial archaeology, wonderful railways – are little known.

On the coast of Sardinia, holiday homes are increasing; many of them are the second homes of Sardinians and will become their retirement homes.[89] These developments, though often diffuse, show some sign of architectural attention to harmony with their localities (Fig. 5.13): they avoid the grim concrete and harsh greens and blues that disfigure so many Mediterranean developments.

The tourist boom has stimulated investment, generated funds through outsiders buying land, and made work for builders, electricians, painters, waiters, drivers, chambermaids, builders' merchants, boatbuilders, cement makers, surveyors, upholsterers, lawyers, etc., almost exclusively on sandy coasts. This has contributed to depopulation of inland places, although the seasonal work and shorter distance of migration allows people to remain in touch with their homelands, indeed to return and invest their earnings in a hotel of their own. Tourism has also stimulated the production of fruit and vegetables, and restrained the decline of meat and wine.

Industry

Industrialization (not directly related to tourism) has happened in most of the less mountainous parts of southern Europe. Often large-scale industrialization seems to have been skin-deep, industries being owned by extra-Mediterranean companies who took the profits elsewhere. Factories established in the 1960s processed petroleum, iron ore and other raw materials, often imported by sea. Their better-paid workers, too, came from outside. Immense sums were invested in industries that created few local jobs but have required ever-greater support from outside: for example, the petro-chemical plant at Otano in central Sardinia.[90] Many of these are now rusting and polluting; their capital investment is near the end of its life; the jobs they created are precarious. However, especially since the 1970s, middle-sized firms have become established, often with local capital. This has been one of the factors bringing out-migration to an end.

[89] M. L. Gentileschi, 'Tourisme et peuplement de la côte en Sardaigne; les tendances en cours', *Méditerranée* 72 (1), 1991, pp. 43–53.
[90] Schweizer 1988, p. 108.

GENERAL CONCLUSIONS

Mediterranean Europe has adapted to the late-modern world in various ways. The East Aegean islands afford contrasting examples. In Rhodes tourism has contracted since the Italian period, being now concentrated in Rhodes city and the town of Líndos. Much of the island is curiously remote, with subsistence villages linked by excellent, decayed Italian roads, deserted spas and mountain hotels, and ruined airports. Kárpathos, that fierce, hospitable, storm-bound, old-fashioned, ceremonious island (Fig. 5.14), has little tourism, but has every sign of prosperity and strong links with its New Jersey colony. People dress formally on ordinary occasions (men in jacket and tie, women in magnificent national

Fig. 5.14. Olympos, a northern Kárpathos ridge-top town with fourteen disused windmills, some of which are within the town and presumably are older than it. *April 1989*

costume), and cultivate cereals in tiny fields. The island of Chíos is still the world's only producer of mastic, a mysterious constituent of chewing-gum, varnish, etc., as it has been for eight hundred years. The *kámpos* of Chíos city should be the glory of the Aegean: an oasis of beautiful villas and gardens unique in the world, watered by underground *qanats*, with varieties of fruit trees known nowhere else. When we saw it, it was forlorn, unappreciated and encroached upon by decrepit suburbs; half of it had been casually destroyed to make an airport. Common to the three islands is a vast increase in pines and in conflagrations.

As the twentieth century has progressed, the diversity of Mediterranean landscapes has diminished and patterns of land use have coarsened. Distinctive landscapes have turned into non-distinctive grids of fields, pine plantations, excessively browsed pasture, industrial vineyards and burnt pinewoods.

The difference from two hundred years ago is not desertification, but that the land is cut off from the people and even from the animals. Most people no longer belong to a landscape where they spend their working lives and where the pathways, walls, terraces and barns are products of their own or their ancestors' labour. Oddments of the motorized commercial economy have been superimposed on residues of a peasant landscape. Motorways, villas and golf courses are made for tourists from other countries and commuters to other towns, rather than by and for permanent residents. Reduced social diversity and unfinished or decaying concrete structures are as great a cause for concern as threats to the semi-natural environment.

EXAMPLES: PÓMBIA IN CRETE, VIDAUBAN IN PROVENCE, AND THE ALPUJARRA

Pómbia

Repeated information about processes of rural change is very rare. Pómbia was studied by the French geographer Guy Burgel in 1964 and again in 1986–7.[91] We re-investigated it in 1989, with much help from Anna Pyrgaki, Jennifer Moody and the Mayor of Pómbia.

Pómbia is a big village in the south-west of the Mesará, the great dry fertile plain in south Crete. The village, the only settlement in its township, straggles alongside a ravine down the foothills of the arid, treeless Asteroúsia Mountains; it looks out over the plain (a 'sea of olives') to all the high mountains of Crete.

The township covers 15.1 sq. km; it slopes down from the mountain, through the foothill zone, the upper plain, the lower plain (called *Livádha*, 'Meadow', from its probable former use) to the Geropótamos, the main, near-permanent river of the Mesará which forms the northern boundary. By Cretan standards, Pómbia is highly cultivable and has comparatively little pasture. The annual rainfall (1946–85) is 510 mm. There is no sign that Pómbia ever had any forest.

So very desirable a site has been settled since the Bronze Age. A Middle Minoan settlement is reported from Pómbia itself.[92] The south edge of the Mesará is thick with prosperous late-Roman sites;[93] we conjecture that the deserted hamlet of Metókhi Livadhiótis, with its chapel, could represent one of them. A St Pombius was one of the Holy Ten, martyred at Górtyn, the nearby capital of Roman Crete, in the Decian persecution of c. 250. Pómbia is mentioned as a settlement (*casale de bonbea*) in a Venetian perambulation of 1364. In 1393 'Bombea' was a fief of one Cavalarius, formerly Marini. Venetian antiquities include a medieval painted church and two house-towers.[94]

[91] G. Burgel, *Pobia: Étude Géographique d'un Village Crétois*, Centre for Social Sciences, Athens, 1965; 'Vingt ans de modernisations en Messara crétoise' [typescript], Communication au Séminaire international de Recherche 'Collectivités rurales et intégration capitaliste en Méditerranée', Agrinion, 13–15 Novembre 1987.
[92] Σ. Γ. Σπανάκη (S. G. Spanakis), Κρῆτη, A' Sphakianakis, Herákleion, c. 1960, p. 334.
[93] I. F. Sanders, *Roman Crete*, Aris and Phillips, Warminster, 1982.
[94] ASV: Duca di Candia Proclami, *busta* 14bis c.113; E. Santschi, *Régestes des arrêts civils et des mémoriaux, 1353–99, des archives du Duc de Crète*, Bibliographie de l'Institut Hellénique d'Études Byzantines et Post-Byzantines, Venice, 1976; G. Gerola, *Monumenti Veneti nell' Isola di Creta*, R. Istituto di Scienze, Lettere ed Arti, Venice, 1905, 1908, 1917, 1932.

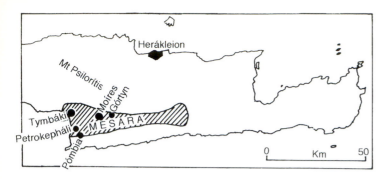

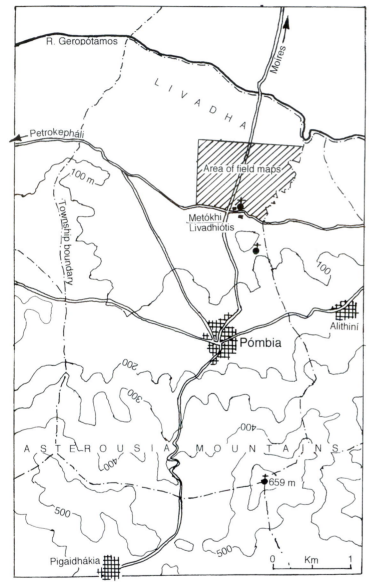

Censuses since 1583 (Fig. 5.15c) show Pómbia consistently among the five most populous of the fifty or so settlements in the Mesará. In relation to its area, and to the area of cultivable land, Pómbia has been almost the most densely populated of all, exceeded only by urban Tymbáki and Moíres, and rivalled by Petrokepháli with its greater area of *livádha*. Being all-Christian, it was unaffected by the exiling of Muslims.

Like all the Mesará, Pómbia suffered severely in the rebellion of the 1820s, but still kept its pre-eminence. The architecture indicates a period of prosperity soon after. Edward Lear, who visited 'Pobia' on 17 May 1864, described 'a large village with no end of archy doors and broken walls', which it still is (cf Fig. 5.16). He sketched the village and its ravine and the magnificent view over the plain of cornfields with patches of trees.[95]

A 1944 air photograph shows ancient field systems, now largely destroyed, which indicate a long and complex history.[96] The foothills had big fields. Parts of the plain had tiny, squarish fields, averaging about 0.1 ha, of a kind which elsewhere in the Mesará are associated with hamlets remaining from Roman farms. There were also rather bigger, elongated fields apparently resulting from the subdivision of a system of strip-fields or strip-meadows of a kind found in other parts of the *livádha*.[97]

Pómbia is not entirely agricultural. Since 1920 it has had a high school, which until 1965 was the only one in the Mesará. The school created jobs and brought income from the feeding and housing of pupils coming from distant villages.

Pómbia in 1963

The peak of census population was reached in 1961, with 1,730 reported inhabitants.[98] Cultivation was in decline before this. The 1944 air photograph shows that all the terraces covering the mountain and some of the fields around the village had gone out of cultivation. Pómbia was on two good roads, but only two or three people had cars. (Nearly all the cars in Crete perished in World War II.)

In Burgel's time the villagers fed on their own produce. Wheat and barley were reaped with sickles, ground at watermills in the next township, and baked once a week in private ovens, mainly into rusks (*paximádhia*) rather than bread.

Burgel in 1963 found about 43,000 olive-trees. The harvest, between October and March, was (then as now) a family occasion. Olive oil was a foodstuff, liberally used. There were several oil-factories; the hundreds of owners of olive-trees would typically give one-tenth of the oil to the oilery owner and workers, consume one-fifth themselves, and sell the rest. Whether there was any oil would depend on frost and biennial cropping.

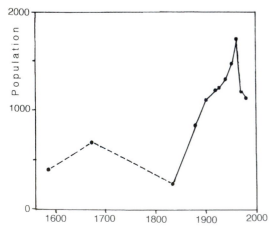

95 R. Fowler, ed., *Edward Lear: the Cretan Journal*, Harvey, Athens, 1984, pp. 82–6.
96 US National Archives, Cartographic Division: TUGX 2102 SK 221.
97 Rackham & Moody 1996, p. 147.
98 The peak is incredibly sharp; we suspect that the census, on this one occasion, included the pupils of the school.

Grapes were grown for wine and brandy (*rakí*) for local consumption, and for sultanas as a minor, labour-intensive cash crop. Another cottage industry, practised in Crete for centuries, was producing silk from silkworms and weaving it into cloth.

Pómbia then used no irrigation except in the *livádha*, where a shallow water table was tapped by means of dipping-wells. Fruit and vegetables were grown mainly for home consumption. Further down the Geropótamos it was possible (perhaps for the only time in Cretan history) to grow rice in a seasonally flooded part of the *livádha*.

Abandoned terraces were browsed by sheep and goats. As throughout Crete, these were divided into house animals and flock animals, of different breeds. The house animals, kept in and near the village, yielded milk and cheese for local consumption. The mountain was grazed mainly by flock

Fig. 5.16. Cistern, Pómbia. *May 1998*

animals brought by transhumant shepherds from Mount Psilorítis across the Mesará (this still continues). They paid grazing rents to the owners of abandoned terraces. Other places with more mountain land had flock sheep of their own; the next township of Pigaidhákia still specializes in sheep and now has a cheese factory. Like all Cretans, the people of Pómbia were not great eaters of meat; only at Easter and Christmas did their animals appear on the table.

Land-holding was complex. A typical family property might consist of 2 ha divided into several scattered plots. This would irritate agricultural writers, but evidently not the people of Pómbia, who were forever buying and selling land, and could have consolidated their holdings if they had wanted to. Ownership of trees could be separated from that of the land on which they grew.

Superficially, Pómbia as described by Burgel might be thought of as exemplifying the 'immemorial' Cretan way of life – of which it had some characteristic features, such as the lack of specialization and the many supplementary activities. The field system, in part, could well have been Roman. However, Pómbia in 1963 was the result of many changes, mostly for the bad. The twentieth century, thus far, had been unkind to Crete. Decades of war, depression, poverty and ill-health had reduced it to a Third World way of life not seen since the 1830s, more dependent on subsistence than was usual in Cretan history. Assuming a true population of 1,500

Fig 5.15d. Fields in part of Pómbia, after a German air photograph of 1944 (Unites States National Archives: RG 373:TUGX 2102 SK221).

Fig 5.15e. The same area in 1963 before reorganization, from the reallocation map. In an area of 1.04 sq. km there are 661 fields, averaging 0.157 ha (0.4 acre) per field. They include traces of an ancient organized system of strip-fields. Changes since 1944 were partly real: the system was not fossilized, and subdivision and amalgamation went on. But discrepancies arise also from uncertainties of air-photo interpretation, and from the official map not exactly corresponding to realities on the ground.

Fig 5.15f. The triumph of geometry: the same area after reorganization. There are supposed to be 273 fields averaging 0.357 ha per field.

and a cultivated area of 10 sq. km, the cultivated area per person was 0.7 ha, dangerously low for a subsistence economy. Pómbia could have cultivated more land, indeed had formerly done so, but at the cost of having little scope for stock-rearing.

Opportunities of escape were developing. People earned seasonal wages harvesting grapes in Malevyzi district to the north. In the 1950s and 1960s they emigrated, first to Athens, then to jobs emerging in Belgium and Germany.

Changes since 1963

Burgel's prognosis was pessimistic. There seemed to be no way for prosperity to return; affairs had got so bad that the people left behind could not afford to buy the land of those who had left.

But times were changing. Funds began to flow back from those who had left, and the post-War regeneration of the European economy spread even here. Although no tourist (save the occasional follower of Edward Lear) ever sees Pómbia, the main tourist area of Crete is 60 km away by a good road and provides a market for its produce.

The next change was the combined reallocation of land and provision of piped irrigation. In 1965, two mainland Greek agronomists arrived and pointed out that consolidating farm plots and providing new roads would make access to fields easier and not waste working hours in moving between plots. These considerations appeal to busy scholars and bureaucrats rather than to Cretan farmers, but in this case, it was claimed, reallocation would make possible the supply of irrigation water. (This was probably true in those days of concrete canals and inflexible pipes.)

A report on Crete by Agridev, an Israeli concern, in 1965–75 pointed out a good aquifer in the Mesará, believed to be replenished by infiltration through river beds and by groundwater entering from the mountains to the north. Water was already coming from a few boreholes, developed with Marshall Aid assistance; concrete canals and earth ditches, and a few pipes, took it to the fields.

Reallocation did not pass unopposed, but the principle was dear to the hearts of the Greek Colonels who had usurped the government. From 1967 onwards most of the western Mesará was given a new grid of field boundaries and roads set out by people who, like the Romans, believed land-holding should be subject to geometry above all else. The axes cross township boundaries, which have been slightly readjusted. An elaborate attempt was made to allocate land and trees according to quality, though doubtless not to everyone's satisfaction. A hectare of grade A land was deemed equivalent to three hectares of Grade D land, or to 83 of the best or 200 of the worst olive trees. One-sixteenth of the area was allocated to roads and ditches. Between 1967 and 1976, 7.2 sq. km was reallocated; the scheme is still not quite complete.

Meanwhile, the state began to increase irrigation supplies for Pómbia and its neighbours. Water from boreholes was pumped up to tanks on high ground; it now flows under gravity to standpipes from which plastic pipes lead to individual plots. This method delivers water direct to the plant by trickle nozzles on the ground. The land does not need to be levelled or reallocated; waste and water erosion are minimized.

Fig. 5.17. Plastic greenhouses near Tymbáki, Mesará. *January 1988*

Greenhouses came to the Mesará in the 1960s. They are concentrated in the Tymbáki area where the frost-free winters and shell-sand soils are favourable (Fig. 5.17). Most of the NW corner of the Mesará is covered with plastic or glass (or its remains), producing tomatoes, cucumbers, eggplant, beans and globe artichokes.

In Pómbia, by 1988, 320 ha of fruit and vegetables were under irrigation in the open air, and 15 ha in greenhouses. Fruit and vegetables have replaced cereals and vines. Potatoes yield twice a year.

Market gardening is labour- and capital-intensive, profitable but risky. Refrigerated lorries and fast ferries bring Pómbia within five days' journey of the markets of western Europe. A hectare typically produces fifty tons of melons or tomatoes per year in the open air, selling at Piraeus (the port of Athens) for at least 15,000 ECU in 1989. The costs of fertilizer, seed, pesticides, plastic and electricity might come to about 750 ECU, and transport to Piraeus about 1,500 ECU. The farmer pays very little for the 3,000 tons or so of water used.

As everywhere in Crete, olive plantations have increased, and are now irrigated. The return in 1989 was about 1–2 million ECU, to which the EC subsidy contributed about 25 per cent; a useful amount, but less than the unsubsidized returns from fruit and vegetables.

Drought no longer matters to irrigated areas, except perhaps to put up the price of the produce. Deluges could still be damaging, but none has occurred since 1928. Repeated droughts, however, as in 1976–7, could lead to further abandonment of unirrigated land. Frost and disease are regular risks.

The high returns have greatly increased the value of land on the sandy upper part of the plain. In contrast, on the *livádha*, formerly the richest land (damp but frosty), the dusty vineyards and great olive trees on cracking soils seem neglected.

Intensive production creates environmental problems. Greenhouse work is hated by the workers, and said to be dangerous owing to the lavish use of pesticides in a confined space. Excess fertilizers are likely to pollute the groundwater. Rotten polythene blows about and accumulates in ravines and on beaches.

Burgel, returning to Pómbia in 1987 after twenty-five years, was impressed by signs of prosperity: new houses, a larger (though less populous) village, electrical equipment, cars, machinery and shops. Though admitting that his pessimism was not altogether justified, he claimed that the future was still threatened. Marriages and births continue to decline; the elementary school was closing; young people were leaving for the towns and were not expected to return until they retired.

Pómbia has passed in thirty years from an unusual state of adversity to an unprecedented state of prosperity. Will this continue? It is still overpopulated. For only a hundred years – less than 3 per cent of its history – has it supported a thousand people on 15 sq. km of land, and it has never before supported them in the style that they now enjoy.

Prosperity came from irrigation plus new markets. (It is doubtful whether land consolidation had much to do with it: had irrigation come a few years later, it would have been done with plastic piping, with which water can easily be supplied to irregular areas and inaccessible places.) The all-sustaining ground-water may be relatively secure: the aquifer seems to be well insulated from the sea, and is probably fed by the snows of Mount Psilorítis, equivalent to more than 2 m of annual rainfall. Markets are more problematical. There are thousands of Pómbias in the Mediterranean, and new ones springing up in other parts of the world, all producing similar goods for which the demand is not infinitely expandable. Pómbia has no control over rising costs of energy and plastic. Its economy is precarious, depending on the whims of people far away who have never heard of Pómbia.

Since 1987, there has been a further change. A Pómbia farmer may have a job in the town of Moíres or the city of Herákleion. People return twice a week to water or gather the crops. They commute in vans and pick-up trucks, which are used as cars but are excused tax as being agricultural vehicles. Such people may appear in the census for Pómbia, but they are more urban than rural. Shepherds can now visit their flocks night and morning to milk them and distribute imported feedstuffs. The distinction between village and city dwellers is becoming unclear.

Vidauban and Chaume [99]

The little town of Vidauban, celebrated for food and wine, lies 20 km inland between Toulon and Cannes in Provence. Its hamlet of Chaume lies 4 km SW of the town. Chaume is a place where (as in New England) land has been cleared, abandoned and resettled, all within two centuries.

The 74 sq. km of the *commune* of Vidauban are divided by the broad valley of the Argens and Aille (Fig. 5.18), in Permian rocks which in places produce fertile soils but in others are sterile, giving rise to an extraordinary landscape of flattish bare rock with scattered seeps of water. To the NW are the Triassic limestone cuestas (tilted plateaux) of Chaume and Astros. To the SE rise the rugged, wooded crystalline hills of the Massif des Maures.

Anciently the area was within the Greek colony of Massilia; it was later traversed by the Roman road from Marseilles to Fréjus. How much of it was settled is unknown. Medieval settlement was dispersed, to judge by the many small castles and *bastides* and the scattered farms or hamlets, mostly on the plain, shown on Cassini's mid-eighteenth-century map.

Sainte-Brigitte, a hilltop hamlet, had fifteen 'hearths' in 1471. Astros was a Knights' Templars' and Hospitallers' estate. Vidauban itself has been added to this pattern, founded by Louis de Villeneuve in 1511 as a *ville neuve* of 72 'families'.

Early-modern Vidauban was, by Mediterranean standards, sparsely populated, infertile and relaxed in its land uses. Cassini shows well over half the area as woodland (or maquis), without noteworthy areas of vineyard. Even on the central plain, over half the area was 'woodland' as late as 1834;

Fig. 5.18. Vidauban: vines on false-terraced fields; pine savanna and woodland with crystalline massifs in the background; *Cistus, Erica* and other undershrubs in the foreground; general invasiveness of pine. *June 1993*

one-sixth was arable land, a proportion that had declined slightly since 1684. The first modern change was the commercialization of vineyards, which expanded from less than 6 ha in 1791 to 535 ha (14 per cent of the plain) in 1834. The annual rings of the umbrella pines on the sandstone suggest that they are relics of an unrecorded afforestation of c. 1848, reduced to a savanna by successive fires.

On the unrewarding siliceous soils of the part of Vidauban in the Massif des Maures, land-use has been remarkably stable for three centuries. Since before 1684 at least, three-quarters have remained woodland and savanna. Better soils along the Aille, converted into ploughland at an early stage, constituted the same 4 per cent of the area in 1684 and 1834. In 1834, 13 per cent was described as *essarts* [probably meaning recently grubbed areas] and 2 per cent as pasture, with trifling areas of olives, vines and chestnuts.

The histories of the limestone uplands differed. Astros has remained an undivided *domaine*. More than 40 per cent of it was still woodland in 1834; the rest was a polyculture of ploughland, olives and vines. Vines and woodland have since increased.

In 1684 only the south-facing slopes of the Chaume hill were a close-knit mosaic of cereals, olives and vines. The plateau was woodland, the subject of common rights of firewood and acorns for pannage of pigs. The growing

[99] Most of our information, outside our own observations, comes from Aubin and others 1980;

J. Galangau, *Carte Phytoécologique de la Commune de Vidauban (Var.)* PhD thesis, Faculté des

Sciences et Techniques St-Jérôme, Université de Droit, Économie et Sciences d'Aix-Marseille, 1984.

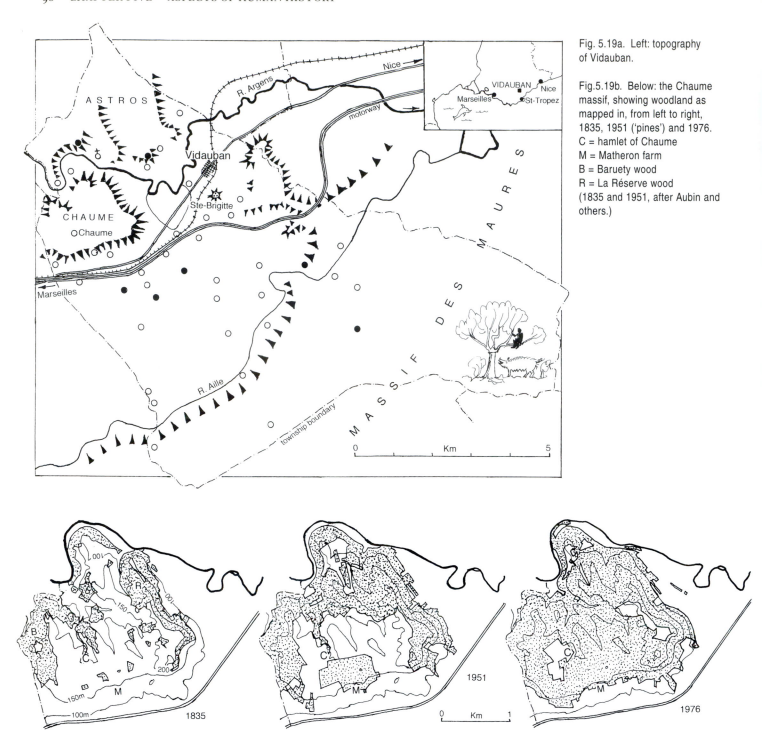

Fig. 5.19a. Left: topography of Vidauban.

Fig.5.19b. Below: the Chaume massif, showing woodland as mapped in, from left to right, 1835, 1951 ('pines') and 1976.
C = hamlet of Chaume
M = Matheron farm
B = Baruety wood
R = La Réserve wood
(1835 and 1951, after Aubin and others.)

population of Vidauban sought permission from the Crown to take over the land and grub out the trees, which they got in 1704, subject to the reservation of about 300 trees 'for His Majesty's service' (presumably shipbuilding). Fires were claimed to have 'destroyed' most of the remaining woodland in 1716 and re-destroyed it in 1764.

Chaume is named on Cassini's map, apparently as a single farm. By 1792 it had become a hamlet; its people had pasturage and woodcutting rights on about 520 ha of remaining oakwood. At the Revolution, they obtained permission to privatize this land and grub out the trees. This arrangement was held to be illegal because the decree for selling common

land excluded woods and forests. In 1810 the community was ordered to put back the trees it had felled. After many years of litigation, the township completed the sale in 1823; the 52 purchasers were supposed to plant their land with vines, olives and figs within the next five years. By 1834, 419 ha were under cultivation, as compared with 103 ha at the southern margins in 1684.

In the 1830s the hamlet counted 65 inhabitants, 85 proprietors and 49 farmers. Nearly all the proprietors were from Vidauban; only fifteen had all their property at Chaume. It had become the stronghold of the Henry family, whose ancestor had established himself in Chaume a century earlier.

Fig. 5.20. Sandstone flats, with stone-pine savanna (a relic of plantation?) and irises in seasonally wet hollows, south of Vidauban. *April 1965*

In general, owners had dispersed plots; most had a parcel of irrigated land for growing vegetables, and also a streamside meadow for fodder.

By 1874, fifty years after the sale, the whole of Chaume was exploited. New terraces had been built, and more were to be added. Woodland survived in steep rocky patches, and also in two large blocks, 'La Réserve' (reserved in 1823 for the continuation of common rights) and 'Baruety'.

Chaume never grew into a village: its population peaked at 85 between 1856 and 1886. They depended on wheat, vines, olives, figs, almonds, vegetables, hunting and collecting wild plants. Swing ploughs were used where possible, but much of the cultivation was by spade and hoe, on terraces where vines grew between olive trees and wheat under the trees. The remaining woodland provided firewood, charcoal and some grazing for sheep. There were two or three milking goats per family, two or three acorn- and fig-fed pigs, some fowls and beehives. They sold scarcely anything except olives, silkworm cocoons and occasionally surplus almonds.

After 1891 Chaume declined, with land cleared around 1823 the first to be abandoned. The population had died of old age by 1936. Fields nearer the hamlet remained under cultivation until World War II. By 1965 the last vineyards were left to be overgrown. Olives never recovered from the frost of 1956, except for two or three ha still maintained in 1980 by a descendant of the original Henry. Most of the site is woodland with scattered ruins; the chestnuts have gone (or are embedded in younger trees); forest fires have increased.

We first knew Vidauban in 1965, and admired the sandstone flats with their umbrella-pine savanna. They were a famous botanical locality, with orchids and winter annuals on thin soils; around winter-wet seeps were rare pre-drainage plant communities with several species of iris, the buttercup *Ranunculus revelieri*, the aquatic stonecrop *Crassula vaillantii*, and the curious, fern-like *Isoetes duriaei* and *Ophioglossum lusitanicum* (Fig. 5.20).[100] Twenty-five years on, this had changed little: although there was a lack of grazing and burning, the shallow, poor soils restricted tree growth.

Vidauban has reverted to almost the same area and distribution of woodland – but not the same species – as in Cassini's time. But it is now near enough by motorway to the concrete jungles of the Côte d'Azur to attract settlers and developers. The population of the township doubled from less than 3,000 in 1975 to more than 6,000 in 1997. The remaining vines were disastrously frosted in spring 1997. Among other changes, a monstrous golf-course appeared under the pretty umbrella pines in former rough pasture. Hundreds of hectares of unkempt woodland and savanna were converted into the

100 Oliver Rackham was introduced to this place by Miss Maybud Campbell and Dr S. M. Walters.

Fig. 5.21. New golf-course, Vidauban. The vegetation is not plastic grass, but looks like it. *June 1993*

over-watered apotheosis of a Scottish sand-dune links (Fig. 5.21). Its boring uniformity is tolerable to someone who never knew and liked the scene as it was, and at least it is scarcely flammable. When we were last there, environmentalists had prevented it from being used, on the grounds of loss of public access.

Recolonization of Chaume by motorized transhumants and commuters began in 1954. By the early 1960s there were seven permanent residents and sixteen second homes. By 1976 there were 25 permanent residents plus 40 second homes scattered through the woods, belonging to people from Marseilles, Toulon and northern Europe. The tourist frontier of the Côte d'Azur had reached Vidauban.

Today a group of reconditioned houses marks the centre of old Chaume. Most of the new houses, with their costly swimming-pools and fashionable patios, are aligned along a grid of new roads, but their grounds incorporate bits and pieces of terraces and field-walls, with remnant olive, almond and fig trees. Buried in pinewoods and dry oakwoods, they await the next conflagration.

Alpujarra

This is the south side of the Sierra Nevada in the south of Spain. This huge mountain range, the second highest in Europe, is separated from the sea by the Sierra de Contraviesa and Sierra de Gádor (Fig. 5.22). The intervening valley contains badlands whose soft ruggedness contrasts with the hard smooth slopes above – for the south side of the Sierra Nevada is a remarkably gentle range for its great height. Most of the settlements form a string of villages just above the badlands. From Trevélez, at 1,400 m the highest village in Spain, they decline eastwards towards Almería and the sea. The zone of villages is abundantly watered and contrasts with the more arid upper and lower slopes.

Alpujarra is a land of majestic beauty. The mountain sweeps gently down from far-away summits flecked with snow, rapidly down through the browns and greens of a zone of encina (Spanish live-oak, *Quercus rotundifolia*) woodland, steeply down through a zone of chestnuts, deciduous oaks and hackberries (*Celtis australis*) on cultivated terraces, headlong down past the villages with their white-cubed houses and olive-groves, ever steeper through wondrous irrigated terraces, plunging into the jagged red and blue depths of the badlands. It has been made famous by two literary writers: Antonio de Alarcón, who travelled in the 1880s, and Gerald Brenan, who lived in Yegen in the 1930s.

The Alpujarra was the last stronghold of the Arabs in Spain. Philip II (he of the Spanish Armada) overcame them in 1570 in a war which 'degenerated into a pandemonium of massacre and pillage',[101] and – in theory – banished the entire population; his successor banished them again in 1610. Nevertheless, scholars and the present inhabitants regard the Alpujarra, above anywhere else in Spain, as the seat of surviving Moorish tradition, place-names, vernacular architecture and agriculture.

Its later history has been uneven. In the nineteenth century it knew periods of brittle prosperity through metal mining. Prosperity has also come from cash crops, especially wine, almonds and grapes, but these too have not proved permanently reliable.

By the late 1980s Alpujarra was in decline, which seemed to J. R. McNeill to be irrevocable. The mines had shut down, cash crops were in retreat, and much of the population had emigrated. Worse still, the environment had been irreversibly damaged by centuries of abuse. The sack of the Alpujarra destroyed terraces and forests and dissipated much of the soil, which was still taking effect four hundred years later. More soil was lost when the remaining forests were destroyed to fuel the mines and smelters, and when cultivation was over-extended to feed the inflated local population, miners and their mules. People have emigrated – McNeill claimed – because the soils can no longer support them. Depopulation has produced undershoot: the remaining inhabitants cannot maintain the terraces and irrigation works, which collapse releasing further soil, so that the carrying capacity of the land is further permanently diminished. For McNeill the Alpujarra was a classic example of a harsh, marginal mountain environment, inhabitable only at the cost of unremitting toil; its mountains had become 'skeletal' and its villages were becoming 'shell villages, home only to the very old and sometimes the very young but to no one in the prime of life'.[102]

Environment

The Alpujarra lies far south and faces southwards. Olives ascend to 1,300 m, probably the highest in Europe. Arable cultivation, too, has gone higher than anywhere in Europe, being reported at 2,400 m in 1940.[103] Rainfall is greatest in the Trevélez area, diminishing eastwards, downwards and towards the sea.

In the geology there are four very distinct regions. The Sierra Nevada is composed of schist, which forms good soils and is not rugged nor very erodible. Water comes out at a strong spring-line along the chain of villages.

[101] Lea 1901.
[102] McNeill 1992, passim.
[103] T. May, 'Human settlement and land use at Trevélez (Sierra Nevada: a historical-geographical approach', *Pirineos* 138, 1991, pp. 53–68.

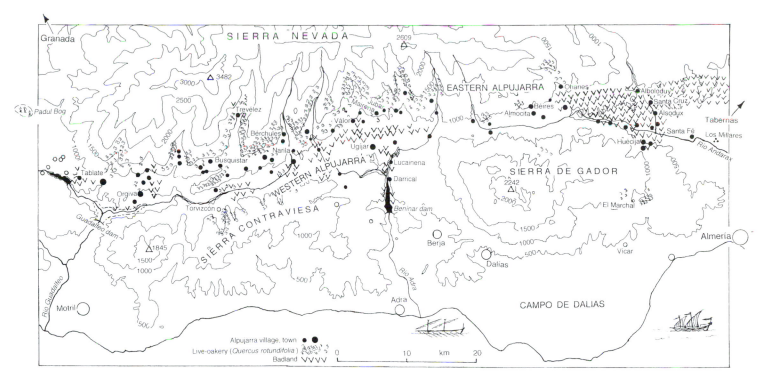

Fig. 5.22. Topography of the Alpujarra.

The Sierra de Gádor is a range of hard dark dolomite rocks, with a great north-facing cliff. Cultivation is mostly confined to pockets of metamorphic rocks. The mountain appears very arid, but has occasional seeps of water. Small aquifers are tapped by water-mines (p. 316). The Sierra de Contraviesa is a schist range, moderately rugged, with crumbly, poorly developed soils and frequent small springs.

Badlands are developed partly in the red and blue metamorphic rocks, and partly in sandstones and marls laid down in Tertiary times. In the east they come to dominate the whole landscape (Chapter 15); even the houses are burrowed into soft cliffs.

Vegetation

The Sierra de Gádor is part of the Almería desert (Chapter 17). Apart from occasional elms by water-seeps, the only sign that it has ever had trees are a few live-oaks on a north-facing slope at 1,000 m, just where we would expect trees to begin with increasing altitude.

The Sierra de Contraviesa presents an arid aspect to the coast. Inland, although highly cultivated, it has remains of encina savanna between the almond-groves. This is probably long-standing. Luis del Marmol Carvajal, writing in the 1570s on the war, mentions great *encinares* (evergreen-oakeries) and much grass for cattle, which suggests oak savanna. Alarcón visited a famous ancient oak, the Encina-Visa, supposedly a relic of the war.[104] This greater degree of vegetation is to be expected from a combination of higher rainfall and more water-retaining soil and rock.

On the Sierra Nevada there are five zones. The zone below the villages has woods of willows and poplars in the barrancas, but is otherwise treeless, lacking even the remains of wild trees. At the main spring-line comes a zone of semi-aquatic trees: elm, hackberry, poplar, ash, willow; these flourish although they are near habitation and are very exposed to woodcutting and browsing.

Above the villages is a zone of chestnuts, evergreen and deciduous oaks, often as coppice-woods mingled with pastures, savanna, and cultivation. (The chestnuts include ancient trees which may date from Moorish times.) Higher on the mountain encinas predominate up to the tree-limit at about 1,700 m. Above them is the alpine zone of grassland and heath.

The zones are most highly developed in the high-rainfall Busquistar area. Eastward the deciduous trees drop out, and then the evergreen oaks. Despite McNeill's claim that there is virtually no forest, a broad belt of woodland extends almost continuously above the villages in the western third of the Alpujarra. It becomes patchy and peters out in the encina savannas above Ohanes. In the eastern third the mountains have virtually no native trees.

There are many endemic herbs and undershrubs adapted to non-shaded habitats, implying that non-forest conditions have dominated evolution far back into the Pleistocene.

Agriculture

Alpujarrine arable farming takes three forms. There is ordinary dry-land *secano* farming of cereals, vines or tree crops, mostly on non-terraced slopes in the Sierra de Contraviesa and (formerly) high in the Sierra Nevada.

Irrigation (*regadio*) is very elaborate, with aqueduct canals (*acequias*) tapping streams high in the mountains, hewn into

---

[104] Carvajal 1600, IV.xii; Alarcón 1892, IV.iv.

Fig. 5.23. In the western Alpujarra, an outlying farm among oaks, chestnuts and poplars above Válor. Remains of dry-farming terraces extend far above. *November 1994*

the faces of cliffs, and distributing the water to the terraces via three orders of branch canals.[105] There is a third form, floodwater farming, in which terraced fields are flooded from a small dam placed to catch flash floods in a dry watercourse, gaining not only water but silt from upstream erosion (pp. 285–7).

The combination of methods depends on the local environment. Floodwater farming occurs mainly in the eastern Alpujarra (and further east), where the rainfall is erratic and where badlands – the essential source of silt – lie above the cultivated zone. Dry-land farming was not expected to produce a cereal harvest every year: the land was not manured, and a crop would be sown only if heavy rain gave promise that it would survive. In the lower Andarax some lands were reported in the eighteenth century as cropped only once in forty years.[106]

Archaeology and history

Two major prehistoric sites lie on the edge of the Alpujarra: Cueva de los Murciélagos and Los Millares. These demonstrate that dryness was no obstacle to Copper Age civilization (p. 316). In the high Alpujarra there is little known pre-Roman archaeology: McNeill claims that settlement began with the Muslims. But absence of evidence probably indicates lack of archaeological activity rather than real absence of settlement. If people lived so early and so well in the arid zone, they would hardly have neglected the far easier environment of the high Alpujarra.

The earliest documents are from the Arab period. The Alpujarra was described as 'densely populated by a very warlike folk' by Ben-Katib-Alcatalami. A ninth-century inscription from Trevélez, the highest, supposedly most inhospitable settlement, records the presence of Christians. In the late ninth century the Alpujarra rebelled under a 'king', Suar Hambun-el-Kaisi, whose head was eventually presented to the Emir of Cordova – a pattern for successive revolts. In the following century Arab writers name Juviles as the principal town and castle, with a silk industry. In the 1230s Mohammed ibn-Hud had a brief and violent career as King of Ugíjar. Later medieval Arabic writers name the *ta'as* or administrative divisions of the Alpujarra and about fifty places in it, most of which can still be identified.[107]

In 1492 Ferdinand and Isabella, Christian sovereigns,

[105] S. Liszewski and A. Suliborski, 'Colonisation et économie en haute montagne: Trevélez (Sierra Nevada; Espagne)', *Méditerranée* 28 (1), 1977, pp. 65–72.

[106] A. Gilman and J. B. Thornes, *Land-use and Prehistory in south-east Spain*, Allen and Unwin, London, 1977, chapter 3.

[107] Alarcón 1892, II.iii; Gómez-Moreno 1951.

Fig. 5.24. The eastern Alpujarra: Huécija among its badlands. *November 1994*

received the surrender of the Islamic Kingdom of Granada. They swore a great oath to give their new subjects 'complete security of goods and property', to 'respect for ever and ever' the Muslim religion, mosques, minarets, services and calls to prayer, and to maintain Muslim buildings, education and the legal system. They appointed a kindly and tolerant first Archbishop in Granada, an ex-Muslim, and all might have been well. But the second Archbishop inaugurated his reign by a public burning of Muslim books. This provoked a cycle of revolts and repressions from 1500 onwards. The Muslims were nominally baptized as Christians (which allowed them to be hounded by the Inquisition), specially taxed, forbidden to speak Arabic, forbidden to wear national costume, forbidden to own slaves, forbidden to slay animals in the Islamic manner, and – the ultimate insult to a religiously clean people – forbidden baths.[108]

Meanwhile the Muslim states regarded the Alpujarra as Christian; the coast was harried by Muslim corsairs and too dangerous to live on. The inland towns of Orgiva and Tabernas were sacked in 1565 and 1566, with the complicity (it was said) of much of the population. People emigrated to North Africa, for example some 250 from Dalías in 1509 and 1512, and 500 from Orgiva in 1565.[109]

On Christmas Eve 1568 the Muslims of Granada rebelled and cut the Christian clergy to pieces. This began two years of bloody sieges, ambushes and battles in the badlands (p. 275). The western Alpujarra was conquered by the Marquis of Mondéjar, a relatively well-organized commander whose object was to rescue Christians. The east was conquered by the Marquis de los Vélez, whose object was plunder.

At the end of the war the king decided to expel all Muslims and ex-Muslims, friend or foe, from the Kingdom of Granada. In theory this was done forthwith, although we are not told how effectively – nor how the Spaniards identified the Muslims. There was a second expulsion of Muslims and ex-Muslims from the whole of Spain in 1610.[110]

What happened to the Moors of the Alpujarra, and who replaced them? In theory they were deported to Estremadura and elsewhere, forbidden to return on pain of death, and in 1610 expelled from the realm. Philip II is supposed to have brought in Christians from Estremadura and other places. If this were so, then Alpujarra should be an Estremaduran colony; it should be full of deserted villages; most of the settlements should be on new sites and have Spanish names; and the dialect, architecture and customs should be Estremaduran, as in much of Latin America.

[108] Alarcón 1892, I.vi; Ortiz & Vincent 1978, chapter 1.

[109] Carvajal 1600, III.xii; Ortiz & Vincent 1978, pp. 30, 86.

[110] Ortiz & Vincent 1978, p. 200.

Fig. 5.25. Huécija. There has been remarkably little change since Brenan photographed this scene in the 1930s, except the building of a school on a bulldozed platform. Even the details of the badland are almost identical. *November 1994*

Manifestly this is not so. Alpujarrine architecture and customs are like those of other ex-Moorish parts of Spain. Virtually all the major place-names, over two hundred of them, are non-Spanish, including the *barrios* or quarters into which villages are divided: for example Bérchules is divided into eleven quarters, all with Arab names.[111] Alpujarra has the densest concentration of 'Arab' place-names in all Spain.[112] Many of the village names, however, are omitted from works on Spanish-Arabic place-names; we suspect that they were as mysterious to the medieval Arabs as they are to modern Spaniards.

Christianity existed in the Alpujarra before the revolt. Most villages had a priest, who was martyred or escaped, and a massive church, in which his parishioners took refuge. Two villages, Santa Cruz and Santa Fé, had Christian names before 1568. The two leaders of the revolt were descendants of the Prophet and were also Spanish Christian grandees. Don Antonio de Valor y Córdoba turned into King Aben-Humeya of Granada; Don Diego Lopez became King Aben-Abóo. Alpujarrine churches today often show evidence of a complex building history and occasionally of fire damage. Alarcón remarks that the peculiar detached church tower at Orgiva, in which the Christians forted up against the rebels, is shown in exactly its modern location in a town plan of 1572. The church of Narila was founded by Philip the Fair and Juana the Mad in 1506, just after the region had in theory been converted to Christianity.[113] Some churches are alleged to be converted mosques.

Censuses and lists before 1568 apparently name 122 villages and hamlets. Of these, only 66 are still villages or hamlets today. It must not be supposed, however, that nearly half the settlements were wiped out in the great revolt. Many were deserted earlier or later: in 1568 (just before the revolt) only 96 seem have been inhabited, of which eighty still were in 1594, and 69 by 1820. Not all this change, however, is real. Some apparently missing places are incorporated as *barrios* of some other place. Jubar, for example, is sometimes listed as part of Mairena and sometimes as a separate place. So far we can identify only fourteen places that are certainly deserted or reduced to a single farm.[114]

The census population of the Alpujarra in 1568 was about 30,000.[115] In 1572 it was proposed to repopulate 54 of the 96 settlements by importing 10,000 settlers. A census of 1586–7 records 68 settlements with a total population of 7,700. The number of settlers, if any, that it was proposed to place in each settlement in 1572 bears no relation to the population as recorded 14 years later. A census of 1594 records nearly 26,000 people in 80 settlements. In the 1610s the Alpujarra, twice depopulated, was faring very badly indeed; many mills and other installations, destroyed in the great revolt, had still not been reinstated.[116]

The Spanish conquest of the Alpujarra, like the Turkish conquest of Crete, would be almost undetectable if there were only archaeological evidence to go on. Place-names and statistics lead us to doubt whether it was quite so terrible as has been made out. Was the deportation or the repopulation

---

[111] Gómez-Moreno 1951.

[112] Lautensasch 1959.

[113] Alarcón 1892, II.iv, VI.iii.

[114] Our figures are for the Alpujarra from Lanjarón to Ragol – excluding the eastern end and the Berja area, for which the data are too

incomplete. Our conclusions differ somewhat from those of McNeill and other scholars, who (we suspect) have sometimes used varying definitions of the Alpujarra. We have added up figures for the same townships each time, and have made estimates for missing data.

[115] The figures add up to 6,064 *vecinos*, a term that in this context means something like 'households' and is conventionally multiplied by five to estimate the population.

[116] Ortiz & Vincent 1978, pp. 207 ff.

Fig. 5.26. Ohanes. The terraces are used for table-grapes or arable crops; the abandoned ones are heavily grassed. There is live-oak savanna in the distance. *November 1994*

fully successful? How many Estremadurans could be persuaded to wrest a living from unfamiliar mountains? How many Moors escaped or returned? What effect did plague have at the same time? Did pain of death deter returned deportees from getting counted on the census?[117] By 1610 it may have been difficult to know who was Moorish and who not. The resulting Christian population spoke Andalusian Spanish, built earth-roofed houses, knew how to irrigate terraces and catch silt, kept silkworms, pronounced place-names like Ugíjar and Huécija, and remembered that their part of the village was called Harat-al-Faguara.

Alpujarra gradually recovered and developed new activities such as growing maize. By 1675 the population is recorded as 14,000, and in 1820 as 64,000. In the early nineteenth century the eastern Alpujarra began to send table grapes to England, using special tough varieties to withstand the arduous journey,[118] and the ham-drying industry (p. 195) is first heard of. Then came the mining boom of the 1820s in the Sierra de Gádor and the high Alpujarra. This brought a huge temporary increase in population and, as McNeill points out, great pressure on land for timber, wood, food and fodder. For two or three decades the Alpujarra was very fully used.

Mining did not last. By 1890 the Alpujarra was back to normal: Alarcón's description gives no hint of a ruined landscape. The Sierra Nevada was 'covered in part with woods which appear embroidered on the sides of the barrancos.' Yegen had a 'magnificent treescape (*arbolado*)' of encinas, hawthorns and willows; above the village there abounded chestnuts, walnuts and encinas. The Sierra de Contraviesa was largely almond and fig orchards and vineyards, with some encinas and (surprisingly) cork-oaks. At Ugíjar there was a large silk factory.[119]

The state in 1994

Entering the Alpujarra from the south, we passed the half-deserted villages of Darrical and Lucainena, rotting above a leaky dam which destroyed their fields. They might confirm the McNeill interpretation, but we soon found that they were not typical. The first shop we came to in Ugíjar sold wedding dresses and the second specialized in First Communion presents, which hardly indicates terminal demographic decline. Most of the villages showed every sign of modest prosperity, and were full of butchers, carpenters and artisans. Nearly every one had a small hotel or restaurant – some of

[117] 'Spanish legislation was apt to defeat itself by its exuberance and violence and its execution to be thwarted by the neglect or cupidity of the officials': Lea 1901.

[118] Gilman & Thornes (see note 105), p.115
[119] Alarcón 1892, III and VI *passim*.

them newly opened by returning émigrés. New schools had opened. There was much civic pride: a new dam was being vigorously opposed. The scars of mining healed long ago. Table grapes are still exported: 'Ohanes grapes' appear on grocers' shelves in Cambridge. Ham-curing flourishes at Trevélez and has spread to neighbouring Juviles. Tree and bush fruits have become a major cash crop.

Cultivation has indeed much declined – but this is a decline from a period of unreasonably intensive land-use.[120] Cereal-growing has almost disappeared. At least half the terraces are abandoned, but the remainder are mostly well maintained. Irrigation systems are fully used – ordinary irrigated terraces, terraced fields and the delicate floodwater irrigation systems. We doubt whether the decline is still continuing.

Erosion has encroached on dry-land fields during deluges,[121] but we could see no sign that it had much to do with the decline. Erosion is a normal fact of life on this geology, but it seems to be concentrated in badlands (and to a small extent in slumps in the forest zone), rather than spread over the whole landscape. The badlands bear no trace of having been cultivated (p. 285). Abandoned terraces gradually return to the original slope and revert to natural steppe, phrygana or woodland, depending on the amount of grazing.[122]

Deforestation has been exaggerated. McNeill does not say why great forests are supposed to have existed in historic times, and we have found no evidence. Our sources imply little if any more tree-land than there is now: the great forests, if they ever existed, had gone by 1550. The trees have been fully used – almost every long-established woodland tree is a coppice stool, felled many times – but not destroyed. Since Alarcón's and Brenan's time trees have increased in the west, but have not spread to the treeless east. Much of this increase is natural. Foresters have also planted great areas of pines, now mostly burnt (Chapter 13).

The Alpujarra has been through spectacularly hard times, but is robust. There has been a remarkable turnround since 1988: our impression is of a place that takes pride in old-fashioned practices[123] and has come to terms with changing circumstances. Undershoot has not happened: people have abandoned the less rewarding part of their cultivation and retreated to the more productive areas. Tourism appears in its rightful place as a modest supplementary activity. The landscape is not physically falling to pieces. The main threat of that comes, as in other places, from too much day-trip tourism leading to large-scale road construction, which – given modern standards of road-making – is likely to be more destructive than any other change.

Maybe the people of the Alpujarra understand their environment and its limitations better than well-meaning outsiders. They know which soils to cultivate and which not. Foresters, dam-builders and road-makers, not the local people, destabilize such a landscape.

[120] T. D. Douglas, S. J. Kirkby, R. W. Critchley, G. J. Park, 'Agricultural terrace abandonment in the Alpujarra, Andalucia, Spain', *Land Degradation & Rehabilitation* 5, 1994, pp. 281–91

[121] J. B. Thornes, *Semi-arid erosional systems: case studies from Spain*, London School of Economics Gegraphical Papers, 1976.

[122] T. Douglas, D. Critchley, G. Park, 'The de-intensification of terraced agricultural land near Trevélez, Sierra Nevada, Spain', *Global Ecology & Biogeography Letters* 5, 1996, pp. 258–70.

[123] 'An atavistic desire to maintain traditional farming methods', as Douglas and colleagues elegantly put it.

CHAPTER 6

Cultivation Terraces

Lakho is prettily situated, with cultivation descending from it on all sides; for, with
remarkable industry, the hill-sides are terraced from top to bottom to support
numberless little strips and plots of soil upon which the olive and vine flourish.
These, in their arrangement resembling a flight of stairs, constitute the chief
portion of the land cultivated by the Lakhiots. By seeing what an amount of
subsistence and fertility can be wrought out of land comparatively so ill adapted
by nature, one learns how large a population the island would support if it were
as perseveringly and industriously cultivated throughout.[1]

Terraces are the most conspicuous features of Mediterranean
cultural landscapes, and among the least understood. Agriculture has been in retreat from terraced terrain for at least a
hundred years. Cereals are now rarely grown, and wild shrubs
and trees have taken their place over vast areas. People still
know how to build terraces, and occasionally make new ones,
but the great majority are nineteenth century or earlier;
nobody now alive remembers when they were not there. They
are a humble feature, usually too insignificant to put on
record: we have to guess when and why they were built, for
those who did the work rarely wrote about it. Archaeologists
and historians have thrown only a dim light on their origins
and likely fate.[2] Yet the future of Mediterranean landscapes
depends to no little extent on what happens to them in the
long term.

Terraces extend northwards into Germany, usually for
growing vines. They are widely distributed in the tropics,
though not in other mediterraneoid climates. Terraced hillsides are scattered across Africa, abundantly in Ethiopia.
They are monumental in the Peruvian Andes; in the
Himalayas they climb vertiginous slopes. In China, Japan,
and SE Asia irrigated mountain rice-terraces are prodigies of
hydraulic engineering.[3]

KINDS OF TERRACES

Traditional terraces are of four types (Fig. 6.1):
• *Step terraces*, which are parallel – either straight, or more
usually curving round the contours. Access is either from the
terrace below (often by steps projecting from the terrace wall)
or by a track intersecting the terraces.
• *Braided terraces*, which zigzag up the slope, being joined by
switchbacks at the ends, so that animals and ploughs can get
up without climbing.
• *Pocket terraces*, crescent-shaped walls providing roothold
for individual olives, chestnuts or fruit-trees.
• *Terraced fields*, squarish fields in which one end is built

up above the hillside and the other end sunk in (typical of
Alpujarra, Spain).

Terraces are usually held up by dry-stone walls, which may
be buried or degraded to give the impression of earth banks.
Earth-banked terraces (*ciglioni*) are, however, a regular practice in NW Italy; the banks are turfed to prevent erosion.[4]
Unwalled terraces cannot always be distinguished from
lynchets formed incidentally by the earth-moving action of
the plough: soil tends to creep down a ploughed slope and pile
up against a hedge or other cross-slope obstruction at the
bottom.

Terraces vary with the material and the skill and time of the
builder.[5] In Liguria (Figs. 6.1j, 6.8) terraces are of the step
type, with walls stylishly built of squared stones set in
courses; the wall does not reach the full height of the terrace
but is continued by a grassy bank which seems to be a deliberate construction rather than a lynchet; many terraces have
only the bank. Irrigated terraces are a speciality of the
Alpujarra and west Crete.

Most terraces were deliberately dug into the hillside; the
boulders were taken out and used to build the retaining wall,
and the spoil piled behind. A layer of stones was often
included to provide drainage and prevent slumping.[6] The
debris fill is thus older than the wall itself, and older material
may cover younger material immediately behind the wall.
The original soil structure was, of course, buried or dispersed, so that the crop grows in subsoil or broken-up
bedrock. If the terrace lasts long enough a new soil may
develop on it.

With *check-dam terraces* the wall was built first and the sediment, washing down the slope, was allowed to accumulate
behind it. The stratigraphy is likely to involve material
younger than the wall, decreasing in age towards the surface. This type is common with terraced fields on the very
erodible terrain of southern La Mancha, Spain (Fig. 14.24).

Terraces vary in the maintenance needed and the effects of
neglect (pp. 259–60). Many terraces in Majorca have been

[1] Lákkoi, west Crete, visited by T. A. B. Spratt
in 1852; T. A. B. Spratt, *Travels and Researches in
Crete*, London, 1865.
[2] They have lately received attention in France:
see Ambroise & others 1989, Blanchemanche
1990, and no. 71 (3/4) of the journal *Méditerranée*
(1990).

[3] J. E. Spencer and G. A. Hale, 'The origin,
nature and distribution of agricultural terracing',
Pacific Viewpoint 1, 1961, pp. 1–40; A. T. Grove
and J. E. G. Sutton, 'Agricultural terracing south
of the Sahara', *Azania* 24, 1989, pp. 113–22;
R. A. Donkin, *Agricultural Terracing in the
Aboriginal New World*, Viking Fund Anthropol-

ogy 56, University of Arizona, Tucson, 1979.
[4] Blanchemanche 1990.
[5] P. Frapa, 'Un patrimoine à valoriser: les
terrasses de culture', *Études Méditerranéennes* 12,
1988, pp. 349–54.
[6] Lehmann 1994.

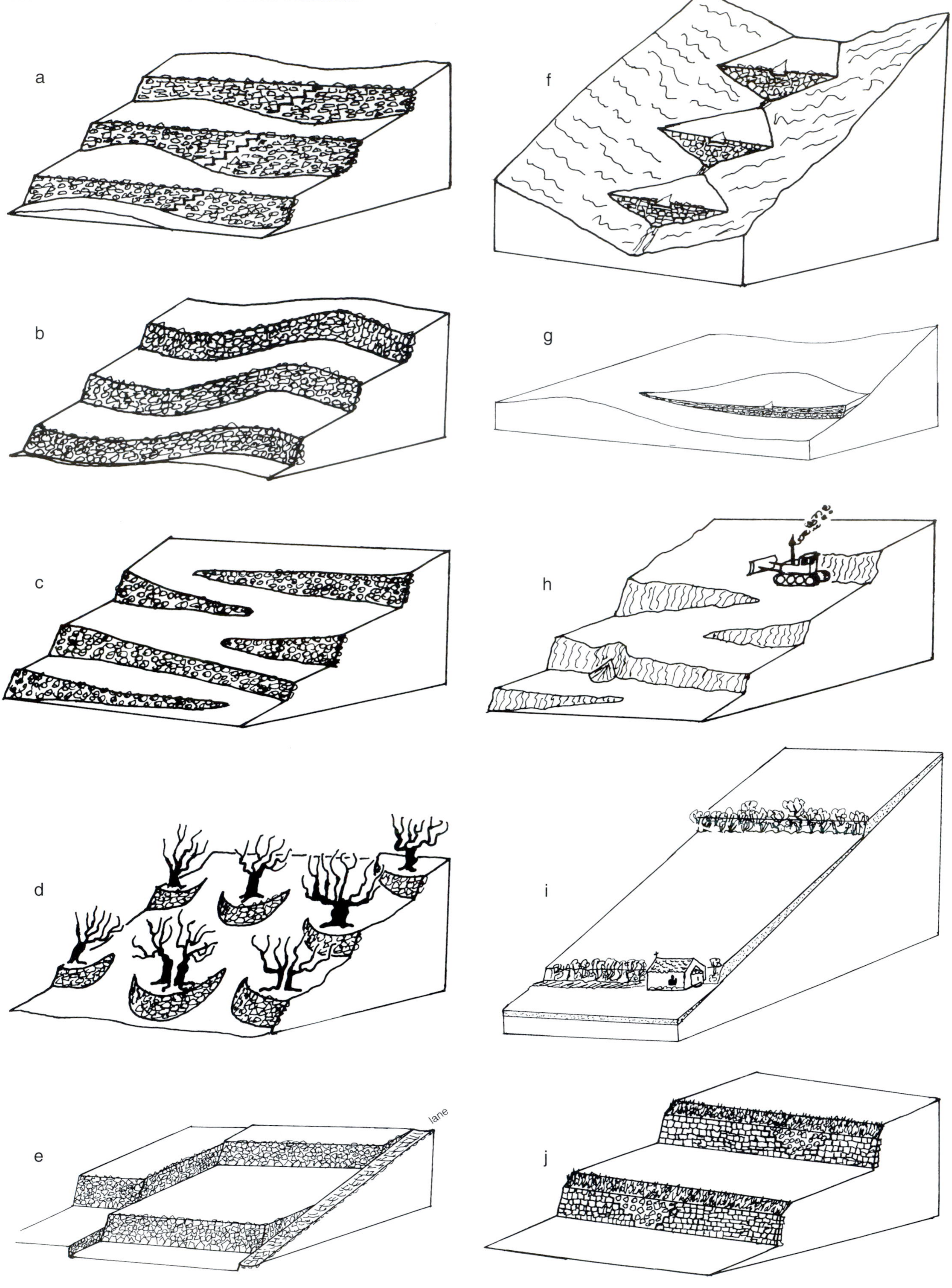
a
b
c
d
e
lane
f
g
h
i
j

Fig. 6.1, left. Kinds of terraces. (a) Step terraces (straight). (b) Step terraces along contours. (c) Braided terraces. (d) Pocket terraces. (e) Terraced fields. (f) Check-dams in gully. (g) Check-dam in gentle terrain. (h) False terraces. (i) Lynchets. (j) Terraces of Ligurian type.

Fig. 6.2, right. Step-terraced vineyards at Thryphtí, east Crete, on slopes of up to 50°. Note the absence of an enclosure-wall. Vines are extremely palatable, and are usually enwalled to keep out sheep and goats; here it is the shepherds' business to control the flocks. *August 1989*

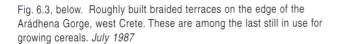

Fig. 6.3, below. Roughly built braided terraces on the edge of the Arádhena Gorge, west Crete. These are among the last still in use for growing cereals. *July 1987*

abandoned for at least fifty years and are still in perfect condition. At the other extreme, terraces disappeared almost completely after eighteen years of disuse in one spot in the Argolid (Greece) (p. 259). In general, terraces carefully built of big, angular blocks need little maintenance; those hastily built of ordinary boulders, especially rounded cobbles, are unstable.

A modern variant, especially on marls and gravels, is to create *false terraces* (French *banquettes*) by digging out the slope with a bulldozer. These are not provided with walls (apart, sometimes, from remains of walls from an earlier phase of terracing); the loose, unvegetated banks stand up, or not, as best they may. The practice, very destructive of the fabric and archaeology of the landscape, is patchy in its occurrence. It is prevalent in Crete in areas reached by piped irrigation (Fig. 6.9), being encouraged by European Union olive grants. The Peloponnese, south Italy and Sardinia get by with far less bulldozing.

Terraces often show several phases of construction: V-shaped patches where walls have collapsed and been rebuilt in a different style; walls heightened in a different style; one terrace split into two, or two terraces run into one; a terrace divided into a series of mini-terraces by parallel walls.

Whether or not to terrace

Most Mediterranean countries have at least a few terraces (Fig. 6.12). The Douro valley in north Portugal is highly terraced, but not south Portugal. In the Alpujarra terracing is a fine art, but in most of Spain it is uncommon. In Sardinia terracing is very local, perhaps because weak rights of ownership discouraged individuals from putting effort into land. The main terraced areas are the densely populated, olive-growing vicinity of Bosa, and Barbagia, the remote mountain interior.[7] The eastern Pyrenees, Provence, Liguria and Croatia are highly terraced; not so Basilicata and the Píndos Mountains. Crete, the Cyclades and Majorca are terraced islands; Sicily and Rhodes on the whole are not.

Terracing occurs on most types of soil and rock, but there is some relation to geology. Terraces are prevalent on hard limestone and marl, less common on schists and clays. On schists they often extend natural outcrops ('foundations indistinguishable from the living rock', as archaeologists used to say). Although angular limestone is the easiest to build with, walls have been made even from rounded volcanic boulders, as on the cliffs of Santoríni. Lack of stone may be a reason for not terracing; in Liguria much of the stone came from quarries rather than from the site itself.

The alternatives to terracing are to cultivate fields on the slope (sometimes ploughing on the contours to reduce soil creep[8]), or not to cultivate. Deciding on whether a slope is too steep, or not steep enough, to terrace was a matter of custom and fashion of which we have no knowledge. In Crete vineyard terraces occasionally cling to slopes of more than 45°, but terraces have been made on the Frangokástello Plain on a slope of less than 2°; unterraced fields in the Amári are ploughed on slopes up to about 15°, and accumulate lynchets. The Alpujarra has a lively sense of which slopes to terrace, which to cultivate on the slope, and which not to cultivate (p. 285).

Further decisions were on the quality of terracing, the distance between walls, and the slope of the finished surface. Well-built terraces often have a nearly level surface. Poor terraces may have only a low, rough wall which somewhat reduces the natural slope of the hillside.

PURPOSES OF TERRACES

When terraces were being built, everybody knew why; now nobody knows, and one has to guess. We can think of six possible reasons.

(1)   To redistribute sediment. This may be the main reason on limestone, where cultivable soil occurs in small pockets.

(2)   To increase root penetration. This may be the main reason on marls, expecially for vines, olives and fruit-trees whose roots cannot penetrate the solid rock. Making a terrace breaks up the bedrock and enables roots to get at stored moisture. The trees are planted in broken-up rock at the front of the terrace.

(3)   To make a less steep slope on which to cultivate.

(4)   To make a wall out of the stones which would interfere with cultivation.

(5)   To increase absorption of water by the soil in heavy rain. Terraces, however, are not more prevalent in semi-desert areas where this would be an advantage.

(6)   To control erosion. This reason appeals to modern agriculturalists, and is believed by peasants who have read their work. Terraces control sheet erosion; they control gully erosion if well designed, or aggravate it if badly designed;[9] they have no effect on wind erosion; they may aggravate slumping by cutting into a slope and by increasing the absorption of moisture to lubricate slip planes within the rock. Erosion control (or taking advantage of erosion) is presumably the motive for check-dam terracing, but we doubt if it was uppermost in the minds of those who built other kinds of terraces. Terracing is far more prevalent on the erosion-resistant islands of Crete and Majorca than in unstable Basilicata and Píndos. It seems to be avoided wherever there is a risk of slumping. False terraces, with their over-steepened, unsupported banks, greatly increase erosion (Chapter 14).

Almost any crop can be grown on terraces; the crop grown within living memory is not necessarily what they were built for. Olives, vines and cereals can be mixed on the same terrace. Pocket terraces are associated with olives and other orchard trees. Well-built, levelled terraces often (but not necessarily) go with vines or with irrigation. Poorly built, braided terraces, especially in remote places and on thin soils, often go with arable cultivation, as proved by threshing-floors nearby.

Fig. 6.4. Braided terraces within enclosure walls. *Vicar, Almería, SE Spain, November 1994*

Fig. 6.5. Pocket terraces on a rocky hillslope protecting the soil around individual olive trees. *Ayiassos, Lesbos, September 1993*

Fig. 6.6. Terraced fields in the Alpujarra. Such wide terraces are relatively uncommon in Mediterranean Europe, but many remain under cultivation here. *Busquistar, November 1992*

Fig. 6.7. Tiny, long-abandoned, terraced fields filling every patch of soil on the desert island of Khálki, Rhodes. The thick walls are a means of getting rid of stones. *May 1986*

Fig. 6.8. Terraces of the Ligurian type, topped with earth banks. These were apparently built in 1522, and are the earliest known to us with a documented date. *Bertignana, Varese, Liguria, March 1996*

Fig. 6.9. False terraces bulldozed out of heathland to grow irrigated olives. *Near Skinés, west Crete, November 1992*

Fig. 6.10. False terraces bulldozed into marly slopes, carefully working round existing almond trees. *Near Laujar de Andarax, Almería, Spain, November 1994*

Fig. 6.11. Terraces poorly maintained but still in use. The very thick walls serve to get rid of the larger stones. As the terrace degrades, cultivation retreats to the front part. *Límnes, Argolid, Greece, May 1989*

---

[7]   It might be thought that the reputed lawlessness of the latter region would have discouraged communal enterprise, but a poster of 'Bandits Exposed to the Public Vengeance' in 1765 shows no concentration of crime in Barbagia (ASC: *Atti governativi e amministrativi* vol. 5 no. 249, p. 270).

[8]   Blanchemanche 1990.

[9]   Vaudour 1962.

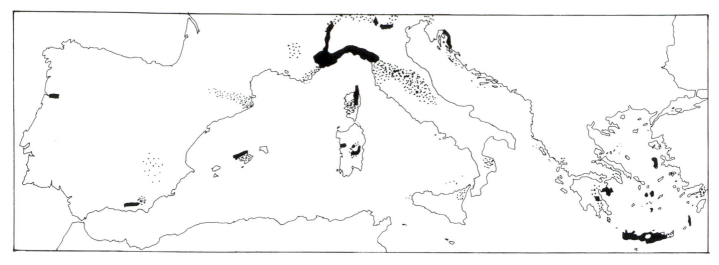

Fig. 6.12. Some of the main terraced areas of the Mediterranean.

HISTORY AND ARCHAEOLOGY OF TERRACES

Terraces in antiquity

Understanding terraces is the key to understanding the chronology and development of many Mediterranean landscapes. We do not have that key. Many scholars have ascribed terraces to various ancient dates on insufficient evidence. Van Andel visualizes soil erosion in the southern Argolid (Greece) as having been controlled by terracing in the late Bronze Age, but admits that no securely dated Mycenaean terrace walls are yet identified.[10]

Terraces are notoriously difficult to date archaeologically: the mere presence of Roman potsherds in a terrace fill does not make it a Roman terrace. They can, however, be dated by careful excavation, especially if farmsteads and other structures are associated with them, or if they occur in places that have not been continuously occupied.

The earliest evidence for terracing seems to be checkdams on the Cretan islet of Pséira, excavated and demonstrated by Julie Clark to be Middle Bronze Age.[11]

The township of Límnes lies near Mycenae in the Peloponnese, on arid hard limestone; its exiguous soils are gathered into terraces with disproportionately massive walls (Fig. 6.11). Although most of the terraces are undatable, some are of two or three phases superimposed. The lowest, earliest walls are of huge boulders, too big for one man to move, in the same style as the road-cut walls that border the ancient roads into Mycenae. It is difficult to resist the conclusion that these are cultivation terraces associated with Bronze Age Mycenae.[12]

There are ancient terraces on the granite islet of Delos in the Cyclades. Delos was a sacred island and a trading centre, but it was also agricultural. Inscriptions indicate at least nine farms in Classical antiquity, growing everything except perhaps olives (usually absent from small islands because of wind). The terraced fields, on a steep slope, are built of polygonal stones over half a metre across, carefully fitted together; earth was brought in from outside. They may have been irrigated from a huge cistern. Careful excavation has dated them to c. 500 BC. Pathways lead to the remains of small farmhouses. The terraces were evidently an investment of professional, probably slave labour. After centuries of use they were covered and sealed by later slope-wash, and thus are clearly differentiated from the rough enclosure walls round the fields of more recent settlers.[13] Similar terrace walls are associated with Classical settlements on Khálki near Rhodes, yet another small, barren, stony island.[14]

On the volcano Methana, across the bay from Athens, terraces have been cautiously dated to the Classical period on grounds of soil development. Brown terrace soils, about half a metre thick, full of Classical and earlier sherds, overlie red, clayey, sherd-free remnants of early Holocene or Pleistocene soils, much eroded before the terraces were made. These terraces are comparatively resistant to erosion after modern abandonment.[15]

Hans Lohmann finds massively built step-terraces associated with Classical farmsteads on very thin limestone soils in south Attica, which he interprets as a reaction to great pressure on land. More ordinary terrace walls have been found attached to several excavated sites of the fourth century BC at Voúla nearby.[16] In Crete, terraces containing Minoan artefacts were noted by one of us on the slopes overlooking the Lassíthi Plain from the SE. However, it is difficult to date terraces so as to satisfy all archaeologists, and many will probably remain controversial.

R. Lehmann, studying erosion on Naxos (Chapter 14), concluded that many of the island's terraces were already

[10] van Andel & Runnels 1987, p. 147.
[11] J. A. Clark, *Soils and land use at an archaeological site: Pseira, Crete*, M.Sc. thesis, Queen's University, Kingston, Ontario, 1990.
[12] Wells & others 1990; E. Zangger, 'Neolithic to present soil erosion in Greece', Bell & Boardman 1992, chapter 12.

[13] M. Brunet, 'Terrasses de cultures antiques: l'exemple de Delos, Cyclades', *Méditerranée* 71 (3/4), 1990, pp. 5–11; M. Brunet and P. Poupet in *AR* 43, 1996–7, p. 95.
[14] Oliver Rackham is indebted to Professor N. Vernicos for demonstrating these.
[15] P. A. James, C. B. Mee, G. J. Taylor, 'Soil

erosion and the archaeological landscape of Methana, Greece', *JFA* 21, 1994, pp. 395–416.
[16] H. Lohmann, 'Agriculture and country life in Classical Attica', Wells 1992, pp. 29–57; *AR*, 1993–4, p. 12.
[17] J.-C. Meffre, 'Habitats augustéens et

Fig. 6.13. Vast ancient olive tree, whose annual rings indicate that it is about 2,300 years old. Like many of the oldest olives in SW Crete, it sits on a pre-existing terrace wall. *Loutró, Sphakiá, July 1987*

abandoned in the middle ages, and constructed at some unknown earlier date. But we suspect that he underestimated the mean erosion rate, through having no information on deluges, and overestimated the volume of fill removed.

Outside the Greek world, terraces on marls and limestones in Provence are claimed on archaeological evidence to go back to the first century BC.[17]

We can also ask the trees. Ancient trees, roughly datable by their size and annual rings, commonly sit on contemporary or earlier terrace walls. In Crete this often applies to olive and plane trees of the Venetian period, occasionally Byzantine, and in one instance probably Hellenistic (Fig. 6.13).[18]

Written records are useless. The whole corpus of Ancient Greek writing mentions no structure that is certainly a terrace, and has no definite word for it, although the words ἀμασιά (normally meaning 'enclosure wall') and ὄφρυς ('eyebrow') *may* occasionally have been used for terraces.[19] Roman land-surveyors fail to mention them: they list ditches, rivers, hedges, roads, etc. as land boundaries, and also *supercilia* ('eyebrows'), defined as 'places which are carried down from the flat in a short slope of up to 30 feet; otherwise it counts as a hill.'[20] This definitely implies a natural feature, such as a small badland gully, not a terrace wall.

Terraces were evidently beneath the notice of poets and philosophers, and still are: the three words for terrace in Modern Greek and the innumerable parallels in Mediterranean dialects are not to be found in most dictionaries. Agronomists, maybe, were writing for more important people than those who had terraces. But it is curious that Hesiod never mentions mending terraces among the tasks of the farm; or that the property of gods or condemned criminals, as listed in inscriptions, should not include them; or that

Pliny the Younger said nothing about terraces in the mid-Apennine landscape (p. 173); or that troops losing a battle never took to the terraces (as they did to the woods) to escape pursuing cavalry; or that the hunted hare, hart or boar never got away by leaping up terraces.[21]

From hints in ancient agricultural writers, Lin Foxhall infers that big slave-owning landowners reserved sloping ground for trees, which were cultivated on the slope by digging and trenching round their bases. Terraces, which did not have to be renewed every year and demanded less labour, were used by peasants.[22]

Medieval and early-Modern terraces

The earliest written mention of terracing of which we are aware is in Boccaccio, who visualizes the Valley of the Ladies thus:

> a little more than half a mile in circumference, surrounded by six little mountains . . . and on top of the summit of each was seen a palace . . . the slopes of these little mountains descended as if stepped toward the plain, as in theatres the tiers are seen in successive order from the top to the bottom, always decreasing their circumference. These slopes . . . on the side of the south . . . were all full of vines, olives, almond-trees, cherry-trees, figs . . .[23]

This is in a work of fiction and does not claim to be any actual place; the terraces are those of formal gardens rather than of farmland. It is more likely that Boccaccio knew about terraced gardens than that he invented the idea. However, the need for careful description suggests that terracing was unusual around fourteenth-century Florence: his readers were more familiar with the terraced seats of Roman theatres than with terraced landscapes.

The next example comes from the dry SE of Spain. A lease of 1528 includes a vineyard in Tarval, near Almería, 'three terraces which contain eight *tahullas* of land, and it has thirteen fig-trees and five mulberries'. We have not seen the original Spanish document, but the word is *bancal*, still used for a terrace in Spanish. Such terms were usual in the 1520s;[24] presumably they meant the terraced fields, irrigated and fertilized by floodwater, still in use today and attributed to the medieval Arabs (Fig. 15.21).

Antonio Cesena, writing a history of his native town of Varese, Liguria, in 1558, says of his vineyard of Bersignana:

> this had been an unenclosed, abandoned field for sowing, in which there were neither vines, nor trees, nor any things wild or tame, nor the slightest trace of management, being pasture for all the animals. In the year of our salvation 1557 I had it all constructed in 56 days, being helped by the men of [nine named hamlets].[25]

'Constructed' is *pastenare*, an obscure word for bringing land

aménagements des versants. Séguret (Vaucluse)', *Méditerranée* 7 (3/4), 1990, pp. 17–21.

[18] Rackham & Moody 1996.

[19] R. Baladié, 'Sur le sens géographique du mot grec "ophrys", de ses dérivés et de son équivalent latin', *Journal des Savants*, 1974, pp. 153–91.

[20] Hyginus, *De generibus controversiarum*. See

also Siculus Flaccus, *De condicionibus agrorum*.

[21] Xenophon, *Kynegetica*.

[22] L. Foxhall, 'Feeling the earth move: cultivation techniques on steep slopes in classical antiquity', Shipley & Salmon 1996, pp. 44–67.

[23] *Decamerone* VI. chiusa.

[24] Cabrillana Ciézar 1977, pp. 472, 458.

[25] F. Carrozzi and S. Lagomarsini, eds., *Relazione dell'origine et successi della terra di Varese descritta dal r.p. Antonio Cesena l'anno 1558*, Accademia Lunigianese di Scienze Giovanni Capellini, 1993, pp. 86 ff. We are indebted to the Rev. Sandro Lagomarsini and Dr D. Moreno for bringing this to our attention.

Fig. 6.14a. One of the earliest pictures of terraces; the monastery of St Michael the Archangel on a peak overlooking Condove, Valle di Susa, west of Turin. *Blaeu,* Theatrum Sabaudiae Ducis, *Amsterdam 1682 (original in Cambridge University Library)*

into cultivation;[26] here, from the great number of man-days of labour, it can hardly mean anything less than the building of a new terrace system or the restoration of a very degraded one. The Bertignana terraces still exist and have vines growing on them (Fig. 6.8). They are built like others in Varese, of coursed stone blocks with a sloping bank on top, with V-shaped patches of different styles of masonry where they have collapsed and been repaired by later hands. Cesena congratulated himself on having achieved a permanent work which his neighbours, 'lazy men' (*huomini pigri*), seldom attempted; but there is nothing to suggest that terracing itself was unusual or an innovation.

Michel de Montaigne, after extensive travels in Italy in 1580, noted at Bagni di Lucca what is still the local method of terrace cultivation in the Garfagnana and Liguria:

the mountains which cover this territory all well cultivated and green up to the summit, grown with chestnut-trees and olive-trees, and elsewhere with vines which they plant around the mountains, and surround them in the form of circles and steps. The edge of the step towards the outside is a little raised, this is vineyard; the hollow of the step, this is wheat.[27]

An early picture of terracing comes from the Piedmontese Alps. In 1609 the military engineer Jean de Beins delineated terraces extending high above the village of Exilles below the Col de Montgenèvre. (These terraces, then used for vines, are now disused and wooded.) In Liguria, a view of Porto Maurizio by P. P. Rizzio in 1622 distinctly shows terraces with trees on them. Similar pictures and views become increasingly abundant from 1650 onwards; it is clear that Ligurian hillsides were fully terraced at least as far back as 1700.[28]

[26] P. Di Stefano, 'Linguaggio e pratiche dell' "agricoltura di villa" nel Genovesato (secc. XVII–XIX)', Còveri & Moreno 1983, pp. 161–71.
[27] M. de Montaigne, *Journal de voyage d'Italie*, ed. F. Garavini, Gallimard, Paris, 1983, p. 268.
[28] Ambroise & others 1989, p. 75; M. Quaini, *Carte e Cartografi in Liguria*, SAGEP, 1986 (especially Pl. 41); S. Lagomarsini, 'Via quotidiana nelle campagne' *Storia illustrata di Genova*, 1994, p. 56.
[29] F. Rebours, 'Versants aménagés et déprise

Fig. 6.14b. The monastery now: the terraced slopes are overgrown with coppiced chestnut and other trees. *October 1995*

On the French–Italian border the narrow and tremendous Roya, most romantic of Riviera valleys, leads up to the Col de Tende. Its phenomenal, well-preserved terraces already existed by 1682, as illustrated in the *Theatrum Sabaudiae Ducis.* Up to heights of 600m about 70 per cent of the slopes were terraced; often the height of the terrace wall is greater than the width of the terrace. Cereals were grown on terraces up to 1,380 m.[29]

In Provence, sixteenth-century (and occasionally earlier) documents refer to *faïsses*, which probably means terraces, although cultivation strips could be meant. The earliest document that expressly mentions a terrace wall is at Barjols in 1620; such records become abundant in the eighteenth century.[30]

Terraces attracted the attention of agronomists and legislators after they had become an established practice. They are commonplace among agricultural writers and in leases and contracts in the south of France between 1720 and 1870. An act of the *parlement* of Provence in 1767 required anyone making a new field on a hillside to subdivide it by 'a wall or bank planted with box or other shrubs for holding up the earth at every *toise* [about 2 m] of slope' on pain of the biggish fine of 3,000 *livres*. Elsewhere the text refers to 'places where the earth has flowed' after making new fields.[31] This appears to be the earliest mention of erosion control as a motive for terracing.

All this evidence (except the first two examples) comes from between Nice and Pisa, and so far we cannot parallel it elsewhere. Among all the early pictures of Crete, we can point only to R. Monanni's view of Merabello (now the town of Ayios Nikoláos) in 1622, which *may* show traces of terraced hillsides.[32] The palatial gardens at Dhiavaidé (near Kastélli Pedhiádhon) and Boutsounária (near Khaniá), described by

Marco Boschini c. 1640, were irrigated and on steep slopes, and therefore must have been terraced, although Boschini, significantly, does not expressly say so.[33]

We suspect that Ligurian artists and cartographers were unusual in taking notice of terraces, rather than that terraces themselves were unusual. Landscape painters generally, such as Lorenzetti in the fourteenth century and Poussin in the seventeenth, are shy of including terraces in their pictures, although they often depict hedges.[34]

Fig. 6.15. Abandoned terraces beginnning to be invaded by pine. Col de Tende above Menton, Provence. *November 1995*

MAKERS AND USERS OF TERRACES

Many observers, such as Spratt in nineteenth-century Crete, stress the enormous burden of labour in making and maintaining terraces. Whether this is true depends on whether it took one or a thousand years, which we do not yet generally know. Was terracing something to fill in idle mornings? Or something a young peasant was expected to do before marrying? Or a job for farmworkers between wheat harvest and vintage? Or something to keep slaves busy?

In our experience, maintenance would usually be no more than replacing fallen stones. Damage arises from goats and deluges, and even this is not too difficult if dealt with at once. However, some flimsy terraces, which rapidly collapse when disused, might be tenable only where there was plenty of labour.

Terracing, over-population and poverty

In a rational world, people would first cultivate flattish land, regarding slopes as a worse option. They would take to the hills when land became scarce through population growth or when plains were denied them by piracy, malaria or occupation by other people. Terraces would be the work of peasant

rurale dans l'est des Alpes-Maritimes', *Méditerranée* 71 (3/4), 1990, pp. 31–6; also our observations in 1965 and 1995; Castex 1980.
[30] Livet 1962, p. 80.
[31] Blanchemanche 1990; Ambroise & others 1989.

[32] BMV: Ital. VII.889 (7798), p. 123.
[33] PBN: Ital. 383, f. 7–10.
[34] Lorenzetti's picture of *The Good City* at Siena, sometimes cited as an early depiction of terraces, really shows contoured rows of vines in sloping,

hedged vineyards (R. Starn, *Ambrogio Lorenzetti: the Palazzo Pubblico, Siena*, Braziller, New York, 1994). *The Adoration of the Magi*, by Domenico Veneziano, c. 1450, doubtfully shows terraced vineyards.

smallholders, pushing cultivation out into remote places. Farmsteads (or field-houses) associated with terraces would be the last settlements to be built, and the first to be abandoned when cultivation retreated. The operation of terracing would be poorly recorded because it would not be done by the sort of people who kept records.

Thus bluntly stated, the theory is a North European point of view applied to another culture with a different set of priorities. However, there is something to be said for it. A. de Reparaz views terracing as an indicator of over-population, as in SE France between 1730 and 1830. (The census population in the eastern Alpes-Maritimes doubled between 1754 and 1848.) The terrace was 'the poor man's field'; abundant labour compensated for lack of land. Often the earliest flights of terraces were constructed around the hill on which a village was perched, sometimes overlooking the fields on the plains originally occupied. By the 1820s peasants were colonizing steeper, more difficult and distant areas.[35] Their isolated houses were to have short lives; the ruins now lie scattered over the hillslopes (unless they have been restored as second homes). The terraces are hidden under rosemary and Spanish broom (to be revealed briefly after fires), or in steeper places have collapsed and been eroded away.

Some terraces were constructed individualistically. At Cabris in Provence, a cadastral map of 1789 shows that it was holders of small parcels who terraced their land, not the big landowners whose lands were often not intensively farmed. Also in the south of France, R. Livet claimed that there are no documents to indicate that the efforts of terrace-builders were concerted; he argues that they worked as individuals rather than being organized from above.[36] But terracing was not always smallholders' work; in Gard, for instance, great landowners are known to have had their land terraced.

Terraces could be constructed quickly if necessity demanded. At Sisco, in the mountainous 'panhandle' of Corsica, slope cultivation was meagre while the people earned their main living at sea. When their livelihood was killed by high technology (steamships) in the mid-nineteenth century they fell back on the land, clearing slopes and making a vast extent of roughly built terraces. At the end of the century they found more rewarding ways of making a living in towns; the terraced slopes were abandoned after less than a lifetime's use.[37]

Terraces often went with poverty. According to de Reparaz, in the Digne area of Provence, smallholders devoted as much as 6 per cent of their farm inputs to repairing and maintaining them for not very productive results. Terraces were often cultivated laboriously with spade or hoe, 'the plough of the poor'. Manure and compost were brought up on wheelbarrows, wives and older children helping.

Terrace cultivation in the south of France (as in Ancient Greece) has thus been attributed mainly to smallholders working for family subsistence rather than for sale. The diversity of crops, the combination of trees and arable land, 'archaic' tools and poor yields are interpreted by outsiders as signs of a self-sufficient economy. However, there is more to terraces than mere subsistence. Smallholders were often wise to grow, not grain, but a more labour-intensive crop – vines or olives or something else – and to exchange the wine or oil for the necessities of life. Sixteenth-century Venetian governors of Crete tried all means (short of paying an adequate price for the product) to compel Cretans to grow grain instead of wine, and failed. The Comte de Villeneuve, statistician of the Marseilles area in the early nineteenth century, regarded terracing as not only a response to increasing population, but also an investment of time and labour to make a profit by planting vines and selling the wine for export.[38] Abandonment may not be due to general depopulation, but to the collapse of a particular crop (e.g. vines through phylloxera) or its market.

Terraces not associated with poverty

De Reparaz's model applies chiefly to the more distant ramifications of terracing, constructed under over-population and poverty. Even here the main body of terraces is documented to an earlier, probably richer, period. They could have been built by comparatively well-to-do people, like Cesena, whose descendants lapsed into poverty and subsistence. Many terraces are the work of professional wall-builders and were the investment of money rather than time. There is nothing amateurish about the regularly planned semi-ashlar terraces of eastern Liguria. Even in France specialist muraillaïres or emparadaïres migrated from Italy, and their skilled work can sometimes be recognised. We know such a skilled wall-builder in Crete.

The late appearance of terraces in the written record may be more to do with the history of record-keeping than with the history of terraces. It is clear from Blanchemanche's researches that agricultural writers expected some of their well-to-do readers to do terracing.

Many step terraces cannot be reached by oxen, and have to be cultivated by hoe. This does not necessarily mean that their builders could not afford a plough (or a share in a plough): the land may have been intended for vines or vegetables which were better cultivated by hand. On the whole, such terraces tend to be professionally built and near settlement. It is the remote, roughly built terraces which are often braided and adapted to ploughing. These could as easily be the work of well-off peasants seeking extra land for cash crops as of the poor seeking somewhere for subsistence. The houses that go with them may always have been second homes.

The terraces celebrated by Spratt (p. 107) are those of Lákkoi in west Crete, a big mountain village, or rather an agglomeration of six hamlets (Fig. 6.16). These terraces presumably date from the thirteenth-century or earlier origins of Lákkoi. The only considerable area of flat land lies far above on the Omalós mountain-plain at 1,080 m. This plain is shared with Ay. Iréne, a group of nine medieval hamlets on the other side of the mountains. Even above the Omalós – as above most of the mountain-plains of Crete (cf Fig. 6.18) – limestone slopes bear degraded remains of terraces.

---

35 de Reparaz 1990.
36 Castex 1980; R. Livet 1962.

37 A. de Reparaz, 'Diversité et évolution des structures agraires corses', Méditerranée 1961 (4), pp. 39–61; 1962 (1), pp. 51–72.

38 de Reparaz 1990.

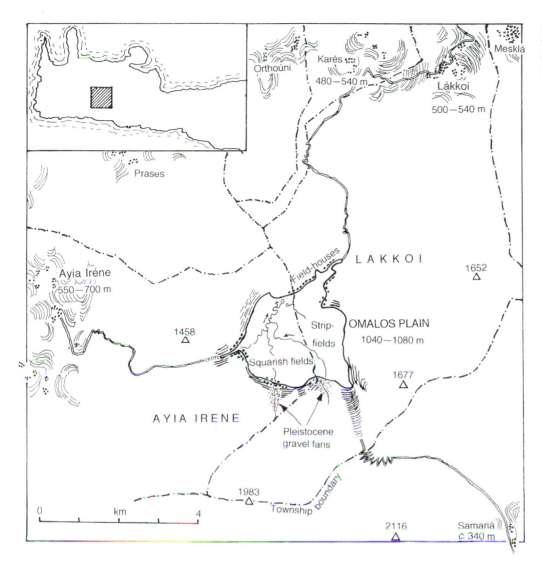

Fig. 6.16. Omalós Plain, west Crete, its field-systems and surrounding terraces, and the townships of Lákkoi and Ayia Iréne. Terraced slopes are shown, but not individual terraces.

Lákkoi and Ayia Iréne presumably began by cultivating their local terraces. They set up field systems on the Omalós – strip-fields in the Lákkoi half, squarish fields in the Ay. Iréne half – and cultivated them whenever the Little Ice Age and the claims of their warlike neighbours of Sphakiá allowed. The Venetians reported that Omalós was uncultivated in the sixteenth century, but by the nineteenth it was fully cultivated. The terraces above Omalós seem to date from the eighteenth century and to have been abandoned by Spratt's time. There are many well-built (though now decrepit) field-houses around the Omalós, whose owners live in Lákkoi and Ayia Iréne. The peak extent of terrace cultivation thus appears to have been in the late eighteenth century. Population, if we can believe the censuses, reached a maximum at other times: in Lákkoi in 1900 and in Ayia Iréne as early as 1583. This remote terracing seems to be the work of people who had second homes.

Terraces often go with areas of hamlet settlement where there is no sign that villages ever existed. West Crete has a multitude of tiny mountain settlements (Fig. 5.2) and imposing flights of terraces. Hundreds of hamlets are known, either from Venetian records or from their churches, to go back at least to the thirteenth century, and could hardly have existed without terraces because there is nowhere else to cultivate.

The terrace walls can occasionally be dated to that period from ancient trees on them. The medieval churches, with carved stonework, embedded Turkish pots, and wall paintings by celebrated artists, do not suggest poverty. The same is true of the mountain villages (with their massive churches) and extinct hamlets in the Alpujarra. In Liguria virtually an entire province depended on terrace cultivation.

As with so much else in the landscape, the easily accessible nineteenth-century evidence must not mislead us into supposing that it tells the whole story. Terraced slopes are not always the worst land, nor are mountain settlements necessarily late and poor. Terracing may have been widespread in antiquity, the evidence surviving on remote islets because these have been less disturbed by later cultivation. It was well established by the middle ages, when many hamlets and villages and some whole landscapes – including relatively rich settlements – depended on it. It entered the written record in the sixteenth century, but not copiously because much of the better terracing had already been built. With the rise of population, terraces were pushed out to the limits of marginal land. In the twentieth century terrace cultivation has shrunk until only the more rewarding, usually older, terraces are still in use.

Fig. 6.17. Top: abandoned terraces on a debris cone (formed during the last glaciation but one, p. 291) on the south side of the Omalós mountain-plain, Crete. *April 1988*

Fig. 6.18. Above: karst sinkhole in SW Turkey. The flat bed is surrounded by a massive enclosure-wall, indicating that at one time it was the only cultivated land. Terraces over the surrounding hills show that cultivation later expanded far beyond. It has now retreated to little more than the basin floor; most of the terraces are abandoned to woodland. This situation is repeated with many mountain-plains in Crete. *Near Bozburun, July 1996*

CHAPTER 7

Climate in the Last 150 Years: the Period of Instrumental Measurements

We are very imperfectly acquainted with the present mean and extreme
temperature, or the precipitation and the evaporation of any extensive region,
even in countries most densely peopled and best supplied with instruments and
observers. The progress of science is constantly detecting errors of method in
older observations, and many laboriously constructed tables of meteorological
phenomena are now thrown aside as fallacious . . . because some condition
necessary to secure accuracy of result was neglected . . .[1]

Questions of desertification, past or future, depend partly on climate. Diminished rainfall will affect natural vegetation, especially that which is already at the desert margin. It will affect ground-water, irrigation, hydroelectricity, city water supplies and the frequency of fires. Higher temperatures will have similar effects, through increasing the rate at which water evaporates from plants and from water surfaces, and increasing the demands for water and electricity (especially for air-conditioning).

Temperatures in the world as a whole are expected to rise during the next century, owing to the continuing increase in the output of carbon dioxide through rising fuel consumption, especially in hitherto underdeveloped countries. Temperatures over land are supposed to have risen already by about 0.6°C over the last hundred years[2] – a difference roughly equivalent (in the Mediterranean) to descending 100 m in altitude, or moving 100 km southward – and still continue to climb (Fig. 7.1).

A change of a fraction of a degree in the mean temperature – less than the difference between one day's weather and the next – can make a noticeable difference to climate. Such small changes, however, are not easily measured: they are readily masked by changes in the balance between day and night or summer and winter temperatures. Future Global Warming is unlikely to be distributed evenly over the planet. Southern Europe may get more or less than its share of warming: it may even get cooler, being balanced by greater warming somewhere else. Experience of the Little Ice Age (p. 135) shows that changes in mean temperature can be accompanied by changes in variability and in the occurrence of exceptional heat or cold. Global warming should also affect rainfall, although this is even more difficult to predict.

Climatic change is not an idle threat. There has been a calamitous decline in rainfall in the Sahel-Sudan zone on the south side of the Sahara. The mean for the thirty years 1961–90 was 25 per cent lower than the mean for 1931–60.[3] This is roughly equivalent to the southern edge of the Sahara advancing by 120 km. It is not necessarily related to Global Warming: southward advances and retreats of the Sahara

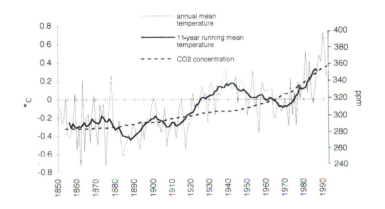

Fig. 7.1. Mean temperature over land in the Northern Hemisphere, compared with carbon dioxide concentration. Temperatures are expressed as means of individual years, and as 11-year running means (the mean of the year in question and the five years before and after), both in terms of deviation from the 1951–70 average.[1] Some of the change, estimated to be no more than 0.1°C, is an artefact due to the local growth of the cities in which many of the stations are sited. CO_2 concentrations are from the Antarctic Siple Station ice core (1854–1959) and from Mauna Loa, Hawaii (1960–92).[2]

[1] Parker and others, 'Global and regional climate in 1995', *Weather* 51 (Feb.), 1996, pp. 202–10.

[2] *A Compendium of Data on Global Change*, Carbon Dioxide Information Analysis Center, Oak Ridge, Tennessee, 1994, pp. 12–19.

have happened from time to time throughout the Quaternary.[4] Could a comparable decline in precipitation in Mediterranean Europe lead to a northward expansion of the Sahara?

Here we examine climates during the period of instrumental recording, to see if this gives any basis for extrapolation into the future. The record extends over a mere century or two, going back for a few stations to the tail of the Little Ice Age. As Chapter 8 will show, climate in earlier centuries shows greater extremes than those in the instrumental period.

[1] George Perkins Marsh, *Man and Nature*, 1864.

[2] C. K. Folland & others 1990; N. Nichols & others, 'Observed climatic variability and change', *Climate Change 1995 . . . contribution of Working Group 1 to the second assessment report of the Intergovernmental Panel on Climate Change*, Cambridge, 1996, pp. 130–92.

[3] The mean rainfall of the Northern Hemisphere tropics since 1970 has been about 10 per cent lower than the century mean; M. Hulme, 'Estimating global changes in precipitation', *Weather*, February 1995, pp. 34–42.

[4] M. A. J. Williams and others, *Quaternary Environments*, Edward Arnold, London, 1993.

Expectation and reality

People adjust their affairs to the climate that their parents remembered, and are caught out if this happens to have been an unusually cool or wet period. Again and again, settlers in unfamiliar climates thought every run of dry years was an abnormality for which some physical or spiritual human failing was responsible.[5] Misjudgement had tragic consequences in the American Dust-Bowl and inland New South Wales, when settlers, arriving during an unusually wet period, organized their affairs on the basis that such rainfall would continue for ever (p. 16).[6] This has often been repeated on a small scale in southern Europe, when farmers, coming to land made newly available under 'land reform', mistook a period of high rainfall for the norm, and found out too late that the land would better have been left as common pasture.

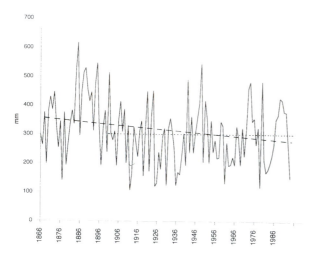

Fig. 7.2. Annual rainfall for Murcia. It shows a downward trend (broken line) since records began in 1866. This trend depends on high rainfall in the 1880s and 1890s; it disappears for the twentieth century alone (dotted line).

Even if a decline in rainfall appears to be supported by measurements, such a trend can still be an artefact. A new rain-gauge is carefully sited in a wide open space: as the years pass new trees and new buildings arise, intercept rainfall in driving winds, and reduce the catch.

If this can be excluded, trends in rainfall (or temperature) may depend entirely on which particular run of years is selected. Climatological writers can easily repeat the Dust-Bowl error. For instance, if annual totals for Murcia (Spain) since 1860 are plotted and the linear trend is calculated, a long-term decline in rainfall appears (Fig. 7.2). Closer examination shows that the trend depends entirely on unusually rainy years between 1880 and 1895. If we take rainfall values for Murcia in the twentieth century only, no significant trend, up or down, can be detected. At Athens, too, high rainfall in the 1880s gives the impression of a general long-term decline; this disappears if the entire record since 1858 is considered. In south-central Italy it has been asserted that drought has set

in since 1940, with two years out of three having less than the mean annual rainfall for the period 1810–1990;[7] however, that mean was inflated by unusually wet decades around 1900.

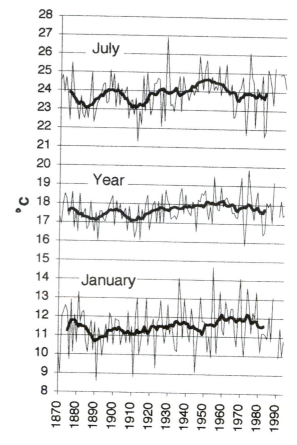

Fig. 7.3. Long-term records of January, July and annual temperature for San Fernando (near Cadiz, SW Spain), 1870–1994, expressed as annual means and 11-year running means.

TEMPERATURE

Most stations with a long temperature record show upward or downward trends of at least 1°C, well outside the effects of chance fluctuation (Fig. 7.3); but they are not consistent with each other nor with the world as a whole. In Italy there has apparently been a warming trend for more than a century, as Global Warming would predict. In the west and east Mediterranean there is as much evidence of cooling as of warming. The balance of the seasons has changed: for example, the steady fall in mean annual temperature in Crete over sixty years is largely due to cooler autumns.

Some instrumental records document the end of the Little Ice Age. In Milan, the winters of 1766 and 1767 were among the coldest ever recorded there; January means of -2.5° and -4.7°C compare with a 1930–60 January mean of +1.79°C.[8] The 1810s were a generally cold decade, famously 1816 (p. 138). In both Milan and Rome 1845–54 was the coldest decade since their instrumental records began, with mean temperatures 0.9° and 1.1° below the 1930–60 means.

5   R. H. Grove 1995.
6   For how to do this, see Raban 1996.
7   Conte & Sorani 1993.
8   Buffoni & others 1996.
9   Barriendos Vallvé & Martin-Vide 1998.

	Lisbon	Palma	Marseilles	Milan	Sassari	Rome
1891	8.4	7.7	3.6	-1.5	6.0	4.7
1892	10.1	10.4	7.8	1.1	9.2	8.1
1893	9.8	8.6	3.0	-2.7	6.4	4.2
1894	10.0	8.9	6.2	-0.4	7.9	6.4
1895	9.7	8.7	3.6	-0.5	6.7	6.9
Mean 1866–1921	10.3	10.5	6.8	1.3	8.1	7.0
Difference between mean 1891–95 and mean 1866–1921	-0.7	-1.6	-2.0	-2.5	-0.9	-0.9

	Lisbon	Palma	Marseilles	Milan	Sassari	Rome
1891	15.2	17.1	13.8	12.3	15.3	14.9
1892	15.8	17.2	14.3	12.7	16.2	15.9
1893	16.4	17.8	14.6	13.0	16.4	15.8
1894	15.4	17.0	14.3	13.0	15.3	15.5
1895	16.3	17.7	14.4	12.5	16.0	15.5
Mean 1866–1921	15.7	17.4	14.1	12.3	15.5	15.5
Difference between mean 1891–95 and mean 1866–1921	+0.1	0.0	+0.2	-0.4	+0.3	0.0

Table 7.i. Top: January mean temperatures in the 1890s, degrees C. Above: annual mean temperatures in the 1890s, degrees C.

Minimum temperatures were especially low: the mean daily minimum temperature for the decade in Rome was 9.8°C (8.5°C in 1844 and 1845), compared to 11.6°C in 1930–60.

A cold spell throughout the northern hemisphere was 1891–5. In the west Mediterranean this produced cold winters, with January temperatures 0.9–2.5°C lower than the long-term mean; but mean temperatures for the year were close to average (Table 7.i).

We quote these figures for what they are worth. All the stations are urban, and tend to get warmer even if the climate is stable. A station set up conveniently near a city is later engulfed by the growth of the city and its warm urban micro-climate. It may be moved to another site either warmer or cooler than the original; if left *in situ* it makes a false contribution to statistics of warming. This 'urban heat island' effect is almost impossible to measure, because there are no long-term control records from stations that have remained rural.[9]

Most of the real warming has been at night, perhaps because of increased cloud cover; daytime temperatures have changed less, and in some places have decreased.[10] Warming may therefore not have been detected at stations such as those in Greece, where it is not the custom to measure temperatures at night.

West Mediterranean stations generally show a cooling from the 1860s to the 1910s, followed by warming until the 1940s, then cooling, and then a rise in the 1980s to the high temperatures recorded in the 1990s.[11] Southern Portugal got cooler (especially in spring) by 2–3°C from the 1940s to the early 1970s. In the late 1980s, temperatures returned to those of the 1940s. Warming in southern Spain between 1885 and 1985 amounted to 0.5–1.0°C; it was most marked in winter.

In the Pyrenees, the mountain-top observatory at 2,862 m on the Pic du Midi de Bigorre (Fig. 7.4) recorded an increase in mean annual temperature of 0.83°C over the period

Fig. 7.4. The observatory on the Pic du Midi de Bigorre, seen from the Pyrenean foothills on the north side. *Bigorre, June 1998.*

9 For attempts at estimating the effect, see P. D. Jones and others, 'The effect of urban warming on the Northern Hemisphere temperature average', *JC* 2, 1989, pp. 285–90; P. Bacci and M. Maugeri, 'The urban heat island of Milan', *Il Nuovo Cimento* 15C, 1992, pp. 417–24.

10 Folland & others 1990.

11 P. Maheras, 'Principal component analysis of western Mediterranean air temperature variations 1866–1985', *Theoretical and Applied Climatology* 39, 1989, pp. 137–45.

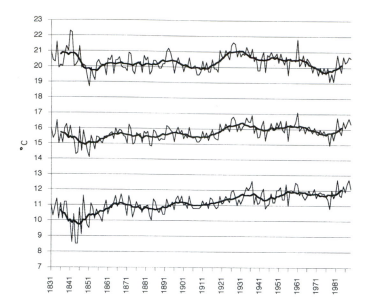

Fig. 7.5. Mean annual temperatures, and 11-year running means, at the Collegio Romano, central Rome, since 1831. (Records go back to 1782, but these are omitted as being too difficult to interpret.) Part of the increase is due to the warming microclimate of the city, for which no correction is made.

[M. Colacino and A. Rovelli, 'Principal component analysis of western Mediterranean air temperature variations 1866–1985', *Theoretical and Applied Climatology* 39, pp. 137–45]

1878–1984. The increase mainly involved warmer nights; the mean daily range diminished by 2.5°C, possibly on account of cloud cover increasing by about 15 per cent.[12] In the Gulf of Lions, mean sea temperatures cooled from nearly 15°C in 1949 to below 14°C in 1977.[13]

Italian stations became warmer from 1870 to 1970. This trend, most marked in November, went with a decrease in Adriatic anticyclones bringing cold *bora* winds from central Europe.[14] At Rome, after the cool 1840s, mean temperatures fluctuated around 15.5°C between 1860 and 1910, and then increased by about 0.8°C up to 1935 (Fig. 7.5). There followed a slight cooling in the 1960s and 1970s, and then a warming of about 0.3°C in the 1980s and 1990s. Warming, especially in spring and early summer, has been detected in southern Italy, Sicily and Sardinia (and elsewhere in Europe).

Warming in the eastern Mediteranean has been less than in the west. In Greece, mean temperatures in the twentieth century peaked in the 1920s or 1930s, and then fell by about 1°C over the next half-century. The temperature of Crete, which had risen by about 1°C from 1910 to 1927, then fell slowly to its previous level. There has been somewhat of a rise in the late 1980s in Greece and Crete owing to unusually hot summers. The east Aegean has shown a marked cooling trend in the last twenty years.[15]

RAINFALL

In Fig. 7.6 we present 11-year means of precipitation for stations with a long record. We also give the figures for individual years, beginning of course on 1 September.[16]

Problems of interpretation

As with temperature, long-term records are seldom homogeneous. Tracing the history of recording and making corrections is not simple. The rain-gauge has been moved, its environment has altered, or an old-fashioned gauge has been replaced by an up-to-date one with different properties.

Fig. 7.6a. Mean annual precipitation for southern Europe, much generalized at this scale.

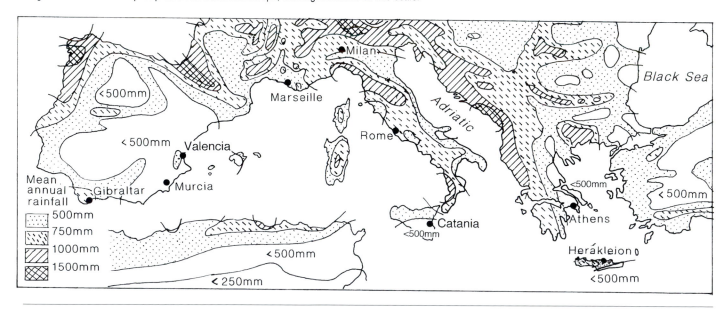

[12] A. Bücher and J. Dessens, 'Secular trend surface temperature at an elevated observatory in the Pyrenees', *Journal of Climate* 4, 1991, pp. 859–68.

[13] Jeftic & others 1992, p. 355.

[14] Palmieri & others 1991.

[15] C. J. E. Schuurmans, 'Climate variability and its time changes in European countries, based on instrumental observations', Flohn & Fantechi 1984, pp. 65–101; B. D. Giles and A. A. Flocas, 'Air temperature variations in Greece, Part 1. Persistence, trend and fluctuations', *JC* 4, 1984, pp. 531–39; D. A. Metaxas, A. Bartzokas, A. Vitsas, 'Temperature fluctuations in the Mediterranean area during the last 120 years', *IJC* 11, 1991, pp. 897–908; B. D. Giles and C. J. Balafoutis, 'Greek heat waves of 1987 and 1988', *JC* 10, 1990, pp. 505–18; M. Türkes, U. M. Sümer, G. Kiliç, 'Variations and trends in annual mean air temperatures in Turkey with respect to climatic variability', *IJC* 15, 1995, pp. 557–69.

[16] 1 October in Portugal.

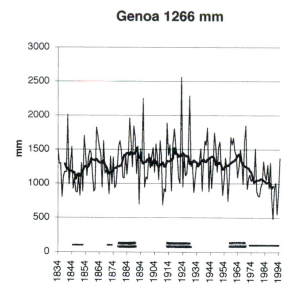

Genoa 1266 mm

Milan 985 mm

Valencia 450 mm

Rome 788 mm

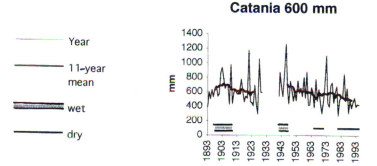

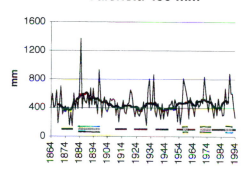

Murcia 308 mm

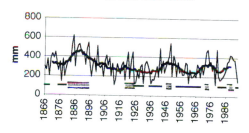

Year

11-year mean

wet

dry

Catania 600 mm

Gibraltar 811 mm

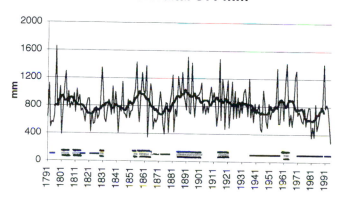

Athens 398 mm

Herákleion 499 mm

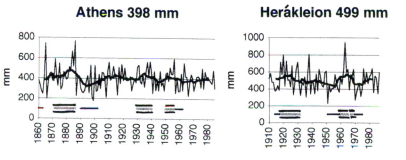

Fig. 7.6b. Long-term precipitation records (yearly totals and 11-year running means) for nine stations. The bars indicate runs of years well above or below the long-term mean.

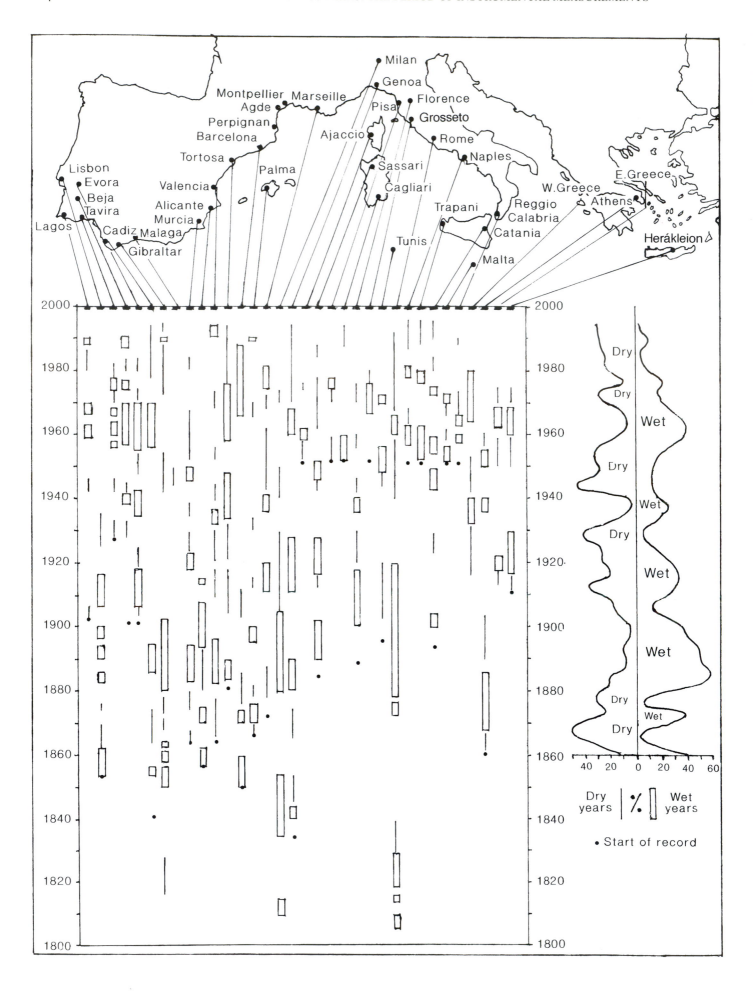

Left: Fig. 7.7. Runs of wetter-than-average and drier-than-average years at individual stations. The graph at the right gives a general percentage of the number of stations experiencing wet and dry periods over Southern Europe as a whole.

At Athens three types of gauge were used between 1858 and 1893, and the site was changed eight times between 1858 and 1890. At the Pic du Midi de Bigorre the published annual precipitation fell from 1,705 mm in 1891–1930 to 925 mm in 1951–80, whereas low-altitude stations generally recorded differences of less than 10 per cent; this was probably due to new buildings near the meteorological platform.[17] Future recording is threatened by the fashion for modern automated gauges, which are unreliable: by the time a breakdown has been noticed and someone has travelled from the capital to mend it, three weeks' data can be lost.[18]

Rainfall varies widely from year to year, and also shows runs of wet and dry years. At Gibraltar the wettest year (1856 with 1,959 mm) in two hundred years of observations had 5½ times the rainfall of the driest year (1981 with 353 mm).[19] It helped to boost mean rainfall over the fifteen years 1852–66 to 1,049 mm, compared with the 1791–1980 average of about 800 mm.[20] Rome's mean precipitation for 1900–10 reached 1,000 mm, compared with 600 mm for 1942–52. Genoa's mean rainfall is 1,266 mm: in 1923 it was twice that amount, c. 2,550 mm, and in 1992 less than half, 500 mm: the eleven-year mean was 1,500 mm for 1910–20 and less than 1,000 for 1984–94.

Trends in particular regions

In Fig. 7.7 we show runs of wet and dry years (related to the local long-term mean) at thirty-two stations with reasonably long records.

Atlantic Iberia
From Lisbon to Gibraltar there is a general correspondence in annual rainfall. There were wet years around 1860, in the 1880s and 1890s, around 1910 and in the 1960s; dry periods around 1870, in the 1900s, around 1930, in the 1940s and early 1980s.

The long record for Gibraltar shows wet periods in the 1800s and 1850s; dry periods in the 1790s, 1820s, 1870s, 1940s and 1980s. Most remarkable is a general decline in rainfall from 1,000 mm in 1890 to 600 mm in 1980, especially in March; this seems to be real and not the result of the rain-gauge being moved (Fig. 7.8). However, in 1989 and again in 1995, there was a very wet autumn and early winter.[21]

Southern Portugal lost its spring rainfall peak. From 1931 to 1963, there were generally two seasons of maximum rain-

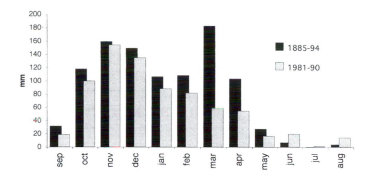

Fig. 7.8. Drying of Gibraltar: mean monthly rainfall in the decades 1885–94 and 1981–90, showing the much lower values in the latter decade, especially for March.

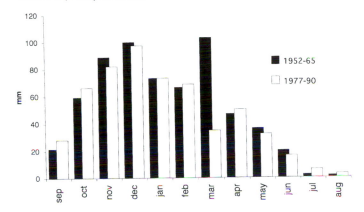

Fig. 7.9. Disappearance of the March rainfall peak in Alentejo. The values plotted here are the average monthly means for Beja, Serpa and Evora for the years 1952–65 and 1977–90.

fall, winter and spring. In the next thirty years there was usually only one, in winter. Successive droughts struck Alentejo, in springs 1968–73 (March rain 30 per cent short), autumns 1970–75, winters 1979–83 and springs 1982–88. Average of rainfall for March in Alentejo between 1968 and 1973 was about 30 per cent below the long-term mean (Fig. 7.9).[22] The March rainfall was especially missed by cereals, grasses and other plants.

Malaga to Valencia
As in SW Iberia, wet years were characteristic of the late nineteenth century in SE Spain. Between 1880 and 1895 annual rainfall in several years was more than 40 per cent above the long-term mean. Murcia had 2½ times its average rainfall in 1883; four years out of the following ten had twice the average. Wet Septembers contributed much of the excess (Fig. 7.10). Mean discharge of the River Guadalentín (then with little abstraction for irrigation, cf Chapter 19) was three times the

[17] Katzoulis & Kabetzidis 1989; J. Dessens and A. Bücher, 'A critical examination of the precipitation records at the Pic du Midi observatory, Pyrenees, France', *CC* 36, 1997, pp. 345–53.
[18] Based on our experience in Crete, where fifty stations operated during 1953–83, and there is now automated equipment at a smaller number of stations. A better solution to the problem of automation would be a simple gauge that collects and stores a month's rainfall at once.

[19] The year begins on the previous 1 September.
[20] Even greater extremes are shown by the record of San Fernando, 100 km from Gibraltar. The driest complete run of ten years in nearly two centuries was 1917–27, with a mean of 487 mm, barely half the mean (921 mm) in the wettest ten years (1863–73): the wettest year in 1917–27 had less rainfall than the driest year in 1863–73. However, these two decades were not abnormal at

Gibraltar, and Cádiz, only 8 km from San Fernando, shows 1863–73 as near average (we do not have the Cadiz record for 1917–27). We suspect that the San Fernando record is faulty.
[21] D. G. Tout, 'A very wet autumn–early winter in southern Spain in 1989', *Weather* 46, Jan. 1991, pp. 11–15.
[22] Casimiro Mendes 1993; Palutikof & others 1996.

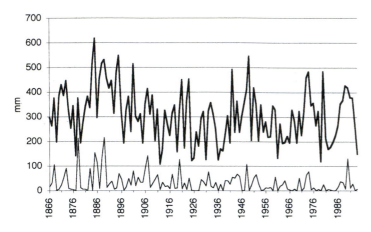

Fig. 7.10. Murcia rainfall: annual and September totals.

long-term mean since 1895. There was then a general decline until the 1940s, with several totals less than 200 mm; between 1923 and 1928 Murcia had no rain in September. Rainfall returned to average or higher from 1956 to 1980; then the early 1980s were exceptionally dry.[23]

Tortosa to Marseilles
Southern France and Catalonia contrast with SE Spain. At Agde and Marseilles, rainfall was high from 1910 to 1918 and also in the early 1970s; it was low around 1870, 1950 and 1980. The 1930s were generally wet, with 27 per cent above average rainfall at Perpignan. In SE France there was a downward trend between 1941 and 1979, with moister intervals in 1958–63 and 1975–79. On the plains at the foot of the Pyrenees there was no significant trend in annual rainfall from 1882 to 1992.[24]

Italy, Corsica, Sardinia, Sicily
Italian meteorologists differ in their interpretation of the trends. Palmieri's analysis of all available data fails to show any long-term trend for Italy as a whole, except that April rainfall has decreased in recent decades; the spring rainy season (where it exists) has become weaker. Conte and Sorani, however, assert that drought has become endemic in south-central Italy since 1940. Droughts were especially serious in south-central Italy in the 1940s and early 1950s and again (as in Gibraltar to Valencia) in the 1980s. Italy did not share in the wet period of the late 1980s in the west Mediterranean.[25]

Claims of long-term drying tend to be made on the basis that the period 1880–1920 was normal, which it was not. In the long record for Rome the years 1895–1915 had a mean rainfall of 960 mm, the wettest twenty years on record. The first half of the nineteenth century was generally dry (mean rainfall 780 mm). The 1970s and 1980s may seem perilously

dry with a mean rainfall of 680 mm, but this is not abnormal compared with the 1830s (675 mm) and 1940s (620 mm).[26]

Greece and Crete
There is little consistency between west and east Greece. The one long record, for Athens, shows greater fluctuations in the nineteenth than the twentieth century; the 1890s were specially dry.[27]

At Herákleion there has been no consistent long-term trend over the twentieth century: the usual clusters of wet and dry years have shown some consistency with mainland Greece.[28] In the late 1980s there was a marked drying, confirmed by records from other Cretan stations, but this is unlikely to persist; there was a much more dramatic, short-lived decrease in rainfall from 1929 to 1934.

Middle and west Turkey
Rainfall records from 1930 to 1993 show cycles of 10–20 years superimposed on a slight downward trend. Dry periods came to most of Anatolia in 1962–69, 1982–87 and 1988–93. Rainfall tended to shift from winter to spring.[29]

General trends

Before 1850 records are too few to draw general conclusions. Since then *there is no evidence for a general wetting or drying trend*. Most stations have experienced runs of wet or dry years as well as random fluctuations. There is some consistency between stations within each region, rather less over the west Mediterranean as a whole, and much less consistency when SE Europe is included.

The 1850s tended to be wet and the late 1860s dry. There followed a generally wet period from 1880 to 1900. From then until 1925 no strong pattern emerges. The late 1920s and early 1930s were dry, the late 1930s wet, and the mid-1940s dry. The 1960s were generally wet. The early 1970s, 1980s and early 1990s were generally dry. From what we hear, 1996–98 has generally been wet.

One somewhat consistent change is a weakening of the 'former and latter rain' pattern (Fig. 2.1) typical of the north Mediterranean and Iberia.

ACID RAIN AND OZONE

Sulphur pollution from the industrial world is not an obvious problem in the rural Mediterranean. On the contrary, lichens – generally indicators of non-acid rain – flourish to a degree which we, coming from polluted eastern England, find surprising. They include the large leafy and shrubby species that are very sensitive to acidity and flourish in areas of moderate to high rainfall,[30] despite the summer drought. We

[23] Font Tullot 1988, pp. 130–34.
[24] J. Dessens and A. Bücher, 'A critical examination of the precipitation records at the Pic du Midi observatory, Pyrenees, France', CC 36, 1997, pp. 345–53.
[25] Palmieri & others 1991; Conte & Sorani 1993.
[26] H. Flohn, 'Rainfall and water budget', Flohn & Fantechi 1984, pp. 106 ff.

[27] P. Maheras and F. Kolyva-Machira, 'Temporal and spatial characteristics of annual precipitation over the Balkans in the twentieth century', JC 10, 1990, pp. 495–504; G. T. Amanatidis, A. G. Paliatsos, C. C. Repapis and J. G. Bartzis, 'Decreasing precipitation trend in the Marathon area, Greece', IJC 13, 1993, pp. 191–201; Katsoulis & Kambetzidis 1989.

[28] D. A. Metaxas, 'Climatic fluctuations in Crete during the twentieth century', Petromarula 1, 1990, pp. 21–5.
[29] M. Türkes, 'Spatial and temporal analysis of annual rainfall variations in Turkey', IJC 16, 1996, pp. 1057–76.
[30] D. L. Hawksworth and F. Rose, Lichens as pollution monitors, Edward Arnold, London, 1976.

have seen one of the most sensitive, *Lobaria pulmonaria*, in the Apennines only 15 km from the urbanized Riviera coast and 45 km east of the city of Genoa.

Acidification of soils is seldom a problem. Geology and climate make Mediterranean soils alkaline, except at high altitudes. Saharan dust, much of which consists of limestone particles, falls over much of southern Europe every year and helps to neutralize sulphuric acid in the rain.

However, there are many local sources of pollution. Most Mediterranean cities have at least one decaying factory emitting noxious fumes. Rust-bucket smelters and refineries are sited at random in the countryside, each with its halo of olive trees crusted with nameless dust. In places there are reports of effects on vegetation, as in the Po plain and among the stone-pines of Tuscany.[31] These are probably due to ozone pollution deriving from modern vehicle engines, which should be aggravated in the Mediterranean climate.

Athens is one of the most notoriously polluted cities in the world. Acid rain and chemicals in the atmosphere have dissolved limestone and marble monuments, including those of the Parthenon sculptures that Lord Elgin had not made off with to England. In summer the city is a hell of humidity and smog which can kill hundreds of people in a hot year. The vegetation, however, is surprisingly resistant. It is a mystery to us that pinewoods flourish within 20 km of the centre of Athens, and many trees and plants remain alive in the very middle of the city.

ATMOSPHERIC PRESSURE AND MECHANISMS

Many attempts have been made to link fluctuations to a wider pattern. South-east Spain, for example, shows some similarities to the Sahel-Sudan zone where the late nineteenth century was wet and the years around 1913 and 1984 exceptionally dry.[32] The cool dry springs in south Portugal began in the late 1960s, when drought began in the African Sahel. They have sometimes been attributed to a 'Great Salinity Anomaly' with cooler, less salty water in the north Atlantic.[33]

Some have detected cycles and periodicities, especially in rainfall. In the mid-western Mediterranean, superimposed on a downward trend between 1900 and 1990, there has been an oscillation with a periodicity of 15–20 years. At Agde and Marseilles, there are signs of a 20-year cycle, with peaks in the 1910s and early 1970s, and lows around 1870, 1950 and 1980.[34] Should the cycle be of 22 years, a link with the 11-year sunspot cycle comes to mind. At Rome, a negative correlation has been shown between rainfall and sunspot numbers.[35] However, there are many patterns of Mediterranean climate to choose from, and many possible outside influences to compare them with; with less than two hundred years' data to go on, some correlations would be expected by chance.

Climatic fluctuations in southern Europe have some coherence in space and time, but the pattern in the west differs from that in the Aegean region. This must have an explanation in terms of movements of air masses and the fronts separating them, which are largely controlled, at any one time, by two of the world's great pressure systems (p. 25).

A major influence on winter weather in the Mediterranean is the variation in strength and location of the *Azores High Pressure* system. This is part of two large-scale atmospheric seesaws. It sits at the southern end of the North Atlantic Oscillation and at the western end of the less well-recognised Mediterranean Oscillation.

The strength of the North Atlantic Oscillation can be measured by the difference of sea-level atmospheric pressure in winter between Lisbon (representing the Azores High) and Stykkisholmur in Iceland (the Icelandic Low).[36] Records go back to 1870. When this index is high, as it was from 1900 to 1915 and 1920 to 1930 and again in the early 1980s and early 1990s, there are stronger-than-average westerlies in middle latitudes, with warming across Europe, less snow, and retreating glaciers in the Alps. Less moisture is transported in the atmosphere, which produces drought in parts of southern Europe, the west Mediterranean and north Africa (Fig. 7.11). When the index is low, as in the 1960s and 1995–7, glaciers begin to advance, more moisture is transported over southwest Europe, especially Iberia, and more rain falls there. As might be expected, the negative correlation between the index and precipitation is strongest in the west and decreases south-eastward (Table 7.ii).

Table 7.ii. Variation in the strength of the correlation between the North Atlantic Oscillation index and Mediterranean rainfall.

Station	Number of winters on record	Mean daily precipitation, December to March, mm	Difference of daily rainfall between winters with North Atlantic Oscillation index above and below 1.0, in mm	Correlation coefficient between winter rainfall and North Atlantic Oscillation index
Lisbon	130	3.0	-1.8	-0.64
Madrid	129	1.3	-1.0	-0.69
Marseilles	120	1.5	-0.5	-0.32
Ajaccio	42	2.2	-1.1	-0.48
Milan	130	2.2	-0.8	-0.35
Rome	119	2.6	-0.8	-0.37
Athens	98	1.7	-0.1	-0.11
Istanbul	64	2.7	-0.7	-0.36

[31] A. Bottacci and 8 others, *Inquinamento ambientale e deperimento del bosco in Toscana*, Società Botanica Italiana, Florence, 1988.

[32] A. T. Grove, 'A note on the remarkably low rainfall of the Sudan zone in 1913', *Savanna* 2, 1973, pp. 133–38.

[33] M. C. Serreze, J. A. Maslanik, R. G. Barry, T. L. Demaria, 'Winter atmospheric circulation in the Arctic basin and possible relationships to the Great Salinity Anomaly in the northern North Atlantic', *Geophysical Research Letters* 19, 1992, pp. 293–6; Casimiro Mendes 1993; Palutikof & others 1996; Font Tullot 1988, pp. 124–28.

[34] Conte & Sorani 1993, pp. 68–70; Jeftic & others 1992, p. 359.

[35] Thomas 1993. Normalized accumulated departures from the mean annual precipitation correlate negatively with sunspot numbers over 1882–1989 ($r = -0.94$).

[36] J. W. Hurrell, 'Decadal trends in the North Atlantic Oscillation: regional temperatures and precipitation', *Science* 269, 1995, pp. 676–79; J. W. Hurrell and H. van Loon, 'Decadal variations in climate associated with the North Atlantic Oscillation', *CC* 36, 1997, pp. 301–26.

A Mediterranean Oscillation has been identified in MEDALUS and other studies. The pressure distribution over the Mediterranean can be depicted by plotting the height at which atmospheric pressure falls below 500 millibars (mb). (This is at an altitude of roughly 5,000 m above sea-level). Mounds in the 500-mb surface indicate centres of high pressure in the lower troposphere, and hollows indicate centres of low pressure. Information is available over about fifty years for certain stations, notably Algiers and Cairo, where radiosonde balloons are regularly released to make meteorological observations through the thickness of the atmosphere. They indicate a seesaw in atmospheric pressure about an axis passing from Tunisia north-east through Crete.[37]

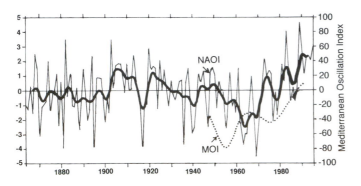

Fig. 7.11. North Atlantic Oscillation in winter 1864–94 (after Hurrell & van Loon 1997, Fig. 5) and Mediterranean Oscillation 1946–89 (after Palutikof & others 1996, Figs 4.12, 4.14).

The North Atlantic Oscillation index increased from the 1960s to reach higher positive values in 1994 than at any time since 1870. About the same time, from 1955 until 1990, mean annual pressure over the mid-west Mediterranean (between Algiers and Sardinia) increased relative to that in the easternmost Mediterranean (Fig. 7.11). The Azores High was becoming more powerful. It produced more frequent Saharan air and less rainfall in the western Mediterranean south of 38°N, and accentuated cool northerly flows over the Aegean. Since 1994 the High has diminished: rainfall in most parts of Mediterranean Europe was well above average in 1995–6.[38]

The periodicity of these seesaws, and what causes them to rock and tilt back every decade or two, is uncertain. The Oscillations are changes over decades in mean seasonal or annual values: the pressure distribution and wind flow pattern at any particular time bears little resemblance to the annual mean. However, the Oscillations might be expected to be associated with changes in decadal mean temperatures and precipitation – the changes differing in amount, and possibly direction, from the west to the far east Mediterranean.

Another investigation has studied changes in seasonality since 1946. In the summer quarter-year, the pressure gradient from the west down towards the east has steepened intermittently, and winds from the north and west in the eastern Mediterranean have accordingly strengthened. In the winter quarter-year, when conditions are complicated by cold airstreams sporadically entering the basin from northern Europe, wind speeds throughout most of the Mediterranean have slackened somewhat, except in the east where the pressure gradient has steepened and given stronger northerly winds.[39]

The eleven-year sunspot cycle is often suspected of affecting the climate. Sunspots affect the sun's output of radiation only very slightly, but this could conceivably have cumulative effects on sea-surface temperature, and consequently on the temperature and moisture content of air passing over the sea. At Rome, rainfall is correlated (negatively) with sea surface temperature in the Bay of Biscay as well as with sunspot numbers.[40] This would be worth pursuing at other stations with long records.

CONCLUSIONS AND THE FUTURE

The instrumental records of temperature in Mediterranean Europe show no constant trend over the twentieth century. By themselves such records afford no firm basis for predicting the next century. However, many of them, especially in the west and centre, conform with the general temperature record for the land areas of the Northern Hemisphere. Such warming as has been measured in southern Europe was mainly in the period 1900–40, followed by a cooling until about 1980. From then until the late 1990s, temperatures have risen by about 0.5°C, as they have in most of the Northern Hemisphere.

It is possible that the eastern Mediterranean has been shielded by atmospheric pollution. Sulphate aerosols from industrial sources elsewhere, especially power stations in Europe, act as nuclei for cloud droplets which reflect solar radiation back into space and curtail warming. A great plume of sulphate aerosols has been depicted as curving southwards over the eastern Mediterranean.[41] The relative stability of the region's climate may depend on northern and eastern Europe continuing to pollute the atmosphere.

A world-wide restraint on high temperatures in the early 1990s was the sulphate blasted into the upper atmosphere in 1990 by the Pacific volcano Pinatubo. Over the following two or three years there was a cooling of about 0.4°C (Fig. 7.12). Between 1993 and 1998, a prolonged El Niño is supposed to have boosted temperatures in Mediterranean Europe by 0.6°C. Mediterranean temperature change is likely to continue to respond to this type of global disturbance.

[37] Palutikof & others 1996; M. Conte and M. Colacino, 'Notes on the climate of the Mediterranean and future scenarios', *Desertification in a European context: physical and socio-economic aspects*, ed. R Fantechi and others. European Commission Directorate-General Science, Research and Development, Brussels, 1995, pp. 79–109.

[38] Parker & others, 'Global and regional climate in 1995', *Weather* 51 (Feb.), 1996, pp. 202–10.

[39] J. M. Reddaway and G. R. Bigg, 'Climatic change over the Mediterranean and links to the more general circulation', *IJC* 16, 1996, pp. 651–61.

[40] Thomas 1993.

[41] K. E. Taylor and J. E. Penner, 'Response of the climate system to atmospheric aerosols and greenhouse gases', *Nature* 369, 1994, pp. 734–37.

Although rainfall figures show a decline from 1950 to 1990, over a longer period they provide no more support than the temperature record to encourage the belief that the Sahara is expanding into Europe. There is no consistent trend towards lower rainfall at stations that ought to be specially exposed to desertification. The nearest such change is the reduction of the spring rains in south Portugal, but even this seemed to have come to an end by the late 1990s.

There is, however, a very considerable risk of a run of years with rainfall totals 20–30 per cent below the mean. This could be part of the normal fluctuation of climate, as happened for instance in southern Spain between 1925 and 1940 and again in the early 1980s and 1990s. It would hardly be desertification: the local vegetation has coped with such fluctuations in the past and can do so again. It would, however, bring great distress to agriculture, industry and even domestic consumers in regions where the population has increased, and has become more dependent on water, since the last time such an event occurred. In south Spain great efforts have been made to exploit every drop of water supplies, indeed more than every drop (Chapter 19), leaving nothing in reserve for a dry decade.

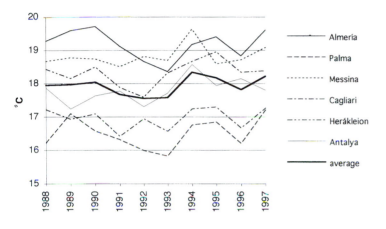

Fig. 7.12. Mean annual temperatures at five stations in southernmost Europe and Antalya in south Turkey, 1988–1997.

CHAPTER 8

The Little Ice Age: Extreme Weather in Historic Times

The farmers cannot sow the ground [on 10 December] because it is too hard . . .
The month of November was hotter than of August; and now [22 January] it is
hotter than in March, and there was no rain until the 6th of December old style . . .
If this drought continues [people] will die of hunger . . . Meat is not to be found
because from the past drought the animals are dead of hunger . . .[1]

The year 1102 [Muslim reckoning] was a vine year, the olive-trees did not bear
fruit, the year was barren, the beneficial rains, by the divine will, were very meagre;
in consequence of this our crops and cereals did not turn out.[2]

Between the beginning of the fifteenth century and the year 1761, thirty-one
destructive floods of the Arno are recorded; between 1761 . . . and 1835, not one.[3]

Instrumental records reveal modest but not unimportant changes in the Mediterranean climate over the last 150 years. In the previous few centuries, the climate of the rest of Europe was far from stable. Even without instrumental records, it is possible to recognise a Medieval Warm Period followed by a Little Ice Age, a period characterized by several phases of cool summers and cold, snowy winters. Glaciers advanced far down Alpine valleys; food and fuel were scarce and expensive in England; famines occurred in much of Europe, culminating in those disastrous decades for human affairs, the 1310s and 1320s, 1590s and 1600s, and the 1680s and 1690s.[4] Glaciers advanced also in other continents, showing that the Little Ice Age was not a local fluctuation but was related to changes in world weather patterns.[5] How was it manifested in the Mediterranean?

THE DATA

Before instrumental records, we mostly lack a continuous record of the climate of normal years. People had little opportunity of recording slow fluctuations in climate, and show little interest in doing so. They rarely tell us that one century was warmer than the next, and when they do we do not know whether to believe them. There are two other types of data: qualitative records of weather rather than climate, usually of damaging events such as exceptional floods and frosts; and proxy records, such as data from tree-rings.

Qualitative data

Unusual events were put on record because they attracted interest in themselves, or they affected human life, death or commerce. It is important to verify the record, and to treat with suspicion second-hand data, especially those set down by people living far away or long after the event.[6] Unfortunately many compilations, such as the indefatigable labours of Font Tullot in Spain,[7] do not give full references to original sources; without verification the reader does not know how much to believe. Some studies withhold the original data altogether, presenting only an arithmetically transformed version of them which is difficult to relate to what actually happened. (There are also the usual difficulties in collating dates of events from diverse historical sources; for example, the date of a new year varied by nearly 24 months between different Italian cities.[8])

Pierre Alexandre has gathered and assessed reports of weather in western and middle Europe in the period 1000–1425 AD, including records from Languedoc, Provence, Liguria, Tuscany, Umbria, Abruzzi and Latium. Rodrigo and others have reconstructed the rainfall regime from the archives of Andalusia and Granada from 1500 to 1700, comparing the results with proxy data from other parts of Spain. Barriendos Vallvé and Martín-Vide have compiled lists of floods near the Mediterranean coast of Spain between the fourteenth and nineteenth centuries. Camuffo's studies involve, besides early Italian instrumental records, accounts of floods and other disasters going back to 414 BC.[9]

---

[1] Reports by Venetians on the 1595–6 season in Crete; ASV: PTM filza 762: Cavalli, 10 December 1595; Onorio Belli, published in G. Morgan, 'The Canea earthquake of 1595', Κρῆτικα Χρωνικά 9, 1955, p. 79.
[2] Report on the 1690–1 season by the Turkish governor of Crete; Stavrinidhos 1985–6 no. 1397.
[3] George Perkins Marsh, *Man and Nature*, 1864, chapter 4. Marsh thought this was because part of the Arno catchment had been diverted into

the Tiber in 1761.
[4] J. M. Grove, 'The century timescale', Driver & Chapman 1996, pp. 29–87.
[5] Grove 1988.
[6] D. Camuffo and S. Enzi, 'Reconstructing the climate of northern Italy from archive sources', Bradley & Jones 1992, pp. 143–54; M. J. Ingram, D. J. Underhill and G. Farmer, 'The use of documentary sources for the study of past climates', Wigley, Ingram, Farmer 1981, pp. 180–213.

[7] Font Tullot 1988.
[8] C. R. Cheney, *Handbook of Dates*, Royal Historical Society, London, 1978.
[9] P. Alexandre, *Le Climat en Europe au Moyen Age*, École des Hautes Études en Sciences Sociales, Paris, 1987; Rodrigo & others 1994, 1995; Barriendos Vallvé & Martín-Vide 1998; D. Camuffo and S. Enzi, 'Climatic features during the Spörer and Maunder minima', *PKF* 16, 1995, pp. 106–24.

Weather records must be assessed in relation to the general quantity of records surviving from the area and period in question. Most scholars ignore this task, and few attempt it in any but the vaguest terms, yet without it the reader does not know whether the data represent the number of actual events or merely the extent and survival of the records.[10] There are few countries from which any but the most fearful events will be recorded before 1200 AD. The fifteenth century is, in general, less well documented than the fourteenth or the sixteenth. There is virtually nothing yet known for Crete between 1760 and 1810, so even the most tremendous happenings in that island will be unknown.

Crop failures are widely recorded, but a cause was not always assigned. Crops can fail because of drought, but also because owing to war or pestilence there are not enough labourers to harvest them; or through diseases or locusts, which may or may not be linked to some aspect of weather.

Many good data are from places only marginal to the Mediterranean. Eustache Piémond kept a valuable meteorological diary from 1579 to 1600 – a critical period for the Little Ice Age – but he lived in St Antoine in Dauphiné where, for example, frosts and summer rain are not unusual.[11] Floods on the Ebro reflect Pyrenean snowmelt as well as local deluges. There is a copious record of floods on the Rhone, but care must be taken to discount those of non-Mediterranean origin. The season of flooding on the lowermost Rhone indicates the source of the floodwater: June to August floods almost invariably originate in Switzerland or the Isère in the French Alps (June and July); autumn floods are usually the result of heavy rain in the Cévennes and Durance.

Events in border regions are not irrelevant; they contribute to pictures of large-scale weather patterns affecting the Mediterranean. The greater precipitation and lower temperatures which gave glacier advances in the Alps and Pyrenees are likely to have involved blocking[12] over NW Europe, greater north–south movements of air masses, more intense cyclogenesis in the Mediterranean, and thus more frequent extreme weather and floods in southern Europe.

Data on floods of wholly Mediterranean rivers might be thought to be objective, consisting as they often do of levels incised on bridges, or records of the depth reached in particular places. However, floods are made by human activities as well as by rain. Flood protection works tend to aggravate flooding if they fail, or to shift it downstream if they hold. The great floods of medieval Florence were blamed on people keeping stacks of wood and timber near the river, which floated off and jammed the arches of bridges, creating dams.[13] Floods on many rivers have become rarer because of dams constructed in the twentieth century. Dams, whatever their purpose, tend to even out the flow and reduce small and moderate floods.

Frost is another reasonably objective kind of record, especially when lakes freeze or olives are 'killed', which requires

Fig. 8.1. Basilica of St Titus, excavated from the sediments that buried it. Note the exposed roots of the old olive-tree growing on that sediment above the modern retaining wall. *May 1995*

a well-defined degree of frost (p. 28). However, credulity is strained when vines and almonds too are reported as 'killed'. These are much more frost-resistant than olives, and are grown in places with much colder winters than in the Mediterranean (e.g. vines in Ontario). Any frost capable of killing vines or almonds, even locally, would wipe out olives for decades over much of the Mediterranean. Since this clearly did not happen, our informants must have been reporting frost damage only to the almond or vine blossom.

Proxy data

Deluges and archaeology

Deluges leave a mark in their earth-moving activities – soil erosion, gullying, changes in river channels, flood-plains and deltas – which can be approximately dated. These are dealt with in Chapters 16 and 18. Here we give some data from buried churches in Crete.

The basilica of St Titus on the edge of Górtyn, the Roman capital of Crete, stands at the foot of a slope of marly sandstone. It has been excavated from the sediments that covered it, about 1.8 m above the original floor level on the up-slope side and 0.9 m on the lower side. On both sides there stand old olive-trees whose exposed roots (p. 252) indicate the loss of 0.6 m of sediment. At some time between the construction of the building (c. 500 AD) and the origin of the olives (c. 1600 AD) the site was buried in 1.5–2.4 m of sediment brought down by a nearby torrent. During the lifetime of the olives another event took away 0.6 m of sediment (Fig. 8.1).

The plain of Kouphós in west Crete is covered with sediment apparently brought by floods from the surrounding hills. On it stands the magnificent church of Aikyrgiánnis, of three phases: one attributed to the eleventh century; an

10 For an attempt to deal with this problem, see Grove & Conterio 1995.

11 Piémond recorded three unusually frosty or snowy winters or springs, nine rainy summers, and seven droughts or heat-waves in the seventeen years 1581–96; M. Greengrass, 'The later

wars of religion in the French Midi', Clark 1985, pp. 126 ff.

12 Blocking happens when a system of high atmospheric pressure settles over northern Europe and diverts westerly airstreams from the Atlantic south towards the Mediterranean. It is

more likely when the zonal circulation is relatively slow and the jet stream meanders widely, giving more north–south air movement.

13 F. Nencini, *Florence: the days of the flood*, Allen and Unwin, London, 1967.

Fig. 8.2. Church of St George, Axéndi, Crete. Note the unweathered stonework lately exhumed from the fill that had buried it (up to the arrows). *May 1995*

question of whether earlier deluges were more extreme than those of the instrumental period. As far as we can judge, they were: earth-moving on this scale does not happen in Crete now. The Pakhyámmos event of 1986, for example (pp. 247–50), buried no churches, even though it was assisted by bulldozing. Venetian reports mention destruction of buildings, which did not happen in 1986.

The snow trade

This is an obvious manifestation of the Little Ice Age. Before refrigerators, snow was gathered on mountains, stored in pits, and brought down on insulated donkeys to cities. In Crete, snow from the high White Mountains (up to 2,400 m) cooled the sherbet of the burghers of Khaniá. In 1683, after a snowy winter (p. 137), the populace complained to the Sultan that the authorities had been imposing unlawful taxes, including that 'during the days of the heat-wave in summer, they compelled each group to carry 6,300 loads of snow'; this would not be possible now. Tournefort found snow on the mountains throughout his visit from 3 May to 28 July 1700, and claimed that it never disappeared.[14] Although this might have been an exaggeration, the existence of a summer snow trade implies that years (like 1992) in which snow lasts into August were then regular rather than exceptional. Raulin in 1847 found snow still being brought down to Khaniá and shipped to Alexandria. Spratt, however, in 1851–2 noted that the mountains 'after midsummer appear bald and grey', as they do now.[15]

Sardinia had a well-documented snow trade. In 1636 three entrepreneurs got the king's licence to establish snowpits on Mount Gennargentu at some 1,700 m altitude, and snow from them was sent nearly 100 km to Cagliari. Cold drinks and ices were a normal branch of cuisine, not an unusual luxury. The trade was still going strong in the 1840s.[16] The pits are still to be seen.

Snow was sent to Málaga and Algeciras from the Serranía de Ronda, the southernmost mountains in Spain, only 1,900 m high.

Trees

In Mediterranean countries the widths of the annual rings of old living trees or timbers in ancient buildings might be expected to provide indications of weather, especially of dry years, as in the SW United States. Unfortunately much archaeological timber was brought down from high altitudes, where growth is limited also by cold summers or short growing seasons, or (as with larch in Venice and Marseilles) from outside the Mediterranean altogether. The wide-ranging studies of P. I. Kuniholm, sampling vast numbers of conifers and archaeological oak timbers around the Aegean, have identified a well-marked period with narrow tree-rings between 1585 and 1600 AD, the second peak of the Little Ice Age.[17]

Creus and Puigdefábrigas interpret the ring-widths of certain Spanish pine-trees as responding to rainfall in the current summer and preceding autumn. They infer from the

extension, also Byzantine; and an inserted Gothic doorway. The church, originally at ground level, is now three steps below ground. The original and the extension have been buried, but the Gothic threshold is at the present ground level. The extra material was therefore deposited between, roughly, 1200 and 1400 AD.

At the deserted village of Axéndi, near Ayia Varvára in middle Crete, is the church of St George, in the bottom of a shallow valley. The original building is in a Gothic style, probably of c. 1300. The original porch was partially blocked by a wall which bears the date AΧΙΕ = 1615. The church has been buried by up to 1.5 m of sediment all round (Fig. 8.2). The burying extends over the blocked porch and is therefore later than 1615, but probably not much later, since the buried stonework is little weathered.

These and similar examples indicate a number of localized events, separated in time, but often pointing to a date between 1200 and 1700 AD. The later ones corroborate written reports of destructive floods in Crete. They answer the

[14] Stavrinidhos 1985–6, no. 773; Tournefort 1717/1741.

[15] Raulin 1858, p. 426; Spratt 1865, p. 149.

[16] Pinna 1903, p. 223; La Marmora 1860, 1, p. 434.

[17] P. I. Kuniholm and C. L. Striker, 'Dendro-chronological investigations in the Aegean and neighbouring regions', *JFA* 10, 1987, pp. 411–20; P. I. Kuniholm, 'Archaeological evidence and non-evidence for climatic change', *Philosophical Transactions of the Royal Society of London* A330, 1990, pp. 645–55.

tree-ring record that the Little Ice Age in southern Spain was characterized by increased wetness, with rainfall anomalies predominating over those of temperature. It has been claimed that certain trees near Rome respond to summer temperatures, which were 1–2°C lower between 1685 and 1705 than the subsequent long-term mean.[18]

We have taken cores from oaks and cypresses in Crete, but so far have found them unhelpful. Rings are often indistinct, double, or variable around the circumference; and in an island full of rain-excesses and rain-shadows it is difficult to get long enough climate records from sites near the trees.

Upper or lower tree-limits are informative. In Crete, the highest cypresses in the White Mountains are much older than the Little Ice Age (Fig. 4.22). Their bizarre shapes tell a story of centuries of warfare against the elements. They appear to represent a stable, natural tree-limit at 1,400–1,800 m, unaffected by browsing animals. The tree-limit cannot have been any lower, and there is no sign that it has been more than 50 m higher, within the lifetimes of the present trees (p. 60).

A limit to the possible severity of Little Ice Age frosts is set by the upper limit of ancient olive-trees. In Crete these are frequent almost throughout the zone of present olive-growing, up to 700 m (occasionally 800 m) altitude. Although some of the higher ones have several big stems, suggesting Little Ice Age frost damage (p. 137), single-stemmed ancient trees occur up to over 600 m, having survived the winter of 1709 which 'killed' the trees in the south of France. Above 400 m in Tuscany, in contrast, ancient olive-trees either do not occur or are multi-stemmed through frost damage – Tuscans take greater risks with their olives.

Perennial rivers

In Venice there is a list, dating from 1625, of Cretan rivers 'abounding in good water'.[19] Venetian galleys then carried huge crews and needed to fill up their drinking-water every few days. Galleys operated mainly in late summer, when rivers were at their lowest. Nevertheless, the list names twenty-eight rivers (including six in the arid east of the island), only four of which abound in good water today. These rivers no longer flow partly because modern irrigation works have drunk them dry. This is not the whole explanation, since Raulin, who visited Crete in 1847, included only five of the twenty-eight in a list of rivers that reached the sea all the year.[20] Maybe forests and wild vegetation either diminish or (as some claim, p. 257) promote the year-round flow of water-courses. This too can hardly be the cause in Crete, where

forests and wild vegetation diminished down to Raulin's time but are now better developed than in 1625. The most likely explanation is that at the height of the Little Ice Age rainfall was less strongly seasonal (or possibly greater), and snow-lie more copious, than it is now.

High-altitude cultivation

Another Cretan indication lies in the Lassíthi Plain. This large flat mountain-plain, 850 m high (Fig. 5.3), was intermittently inhabited from Neolithic to Roman times. After centuries of trying to exclude people from the plain, the Venetian authorities tried to revive farming there; in 1548 they brought a quantity of refugees from mainland wars and told them to grow wheat. By 1583 the experiment had failed. The Venetians made renewed attempts, but Lassíthi was not successfully cultivated until the eighteenth century.[21]

There is no difficulty in growing wheat in Lassíthi now: why not in c. 1600? Likely reasons include the absence of varieties adapted to high altitudes, and lack of knowledge and enthusiasm on the part of the town-bred refugees. The reasons given at the time were bad weather, a condition called *sirica*[22] and especially lack of drainage. Rainwater and snow-melt from the plain and its surrounding mountains make their exit underground through the Khónos cave in a corner of the plain. The Khónos was not big enough to take the water away, and it backed up and flooded the crops. (Another mountain-plain, the Omalós, was sometimes mapped as a lake.) This is no longer a problem: we have found the Khónos half-choked with polythene bags, and nobody seemed to care. We infer that there was more precipitation in a normal year in the Little Ice Age than there is now.

RECORDS OF EXTREME WEATHER

Different types of extreme weather often go together. Periods of cold winters tend also to have floods, droughts and out-of-season rain. Hot winters and cold summers attract less attention, but also tend to go with the other extremes of climate. Times of variable weather are clustered in four main phases (Fig. 8.3).

Deluges, cold and drought in the fourteenth century

After a minor period of floods and cold winters in 1200–20, the first major period of extremes began in the 1280s, reached a peak in 1305–55, and continued until the 1380s. It was expressed in deluges and cold winters, but also in summer

[18] J. Creus and J. Puigdefábrigas, 'Climatología histórica del Pinus uncinata Ramond', *Cuadernos de Investigación del Colegio Univers. de Logroño* 2, 1976, pp. 17–30; 'Climatología histórica y dendrocronología del Pinus nigra Arnold', *Avances sobre la Investigación en Bioclimatología*, ed. A. Blanco de Pablos, VIII Reunión de Climatología, CSIC-Universidad de Salamanca, 1983, pp. 121–28; F. Serre-Bachet, J. Guiot, L. Tessier, 'Dendroclimatic evidence from southwestern Europe and northwestern Africa', Bradley & Jones 1992, pp. 349–65.

[19] BMV: Ital. 340/5750.

[20] Raulin 1858, 1860.

[21] Rackham & Moody 1996, pp. 95, 150.

[22] Our researches, with the assistance of J. M. Grove, have failed to establish what *sirica* was. *Sirica* was prevalent from 1574 to 1603, mainly but not exclusively in the Lassíthi. The Venetian word should mean 'bunt' (Modern Greek σείρικα), a cereal disease caused by the fungus *Tilletia caries*. The characteristic blackening of the ears is mentioned, but not the fishy smell that bunt should have. Smut, the cereal disease caused by *Ustilago* species, is an alternative, but like bunt it should have affected barley as well as wheat,

which *sirica* apparently did not. *Sirica* was sometimes associated with fog – whether ordinary fog, Saharan dust haze or volcanic 'dry fog' – shortly before the time for harvesting. Bunt and smut are systemic fungi and ought not to be affected by weather at harvest time. ASV: PTM filza 734: Michiel, 25 July 1574; PTM filza 733: Foscarini, 18 October and 16 November 1574; PTM filza 741: Foscarini, 9 October 1577; PTM filza 743: Captain-General of Candia, 18 August 1579; PTM filza 753: Mocenigo, 26 June 1586; PTM filza 755: Giustinian, 28 November 1590; Sen. Dis.Rett.1, Priuli, 7 July 1603.

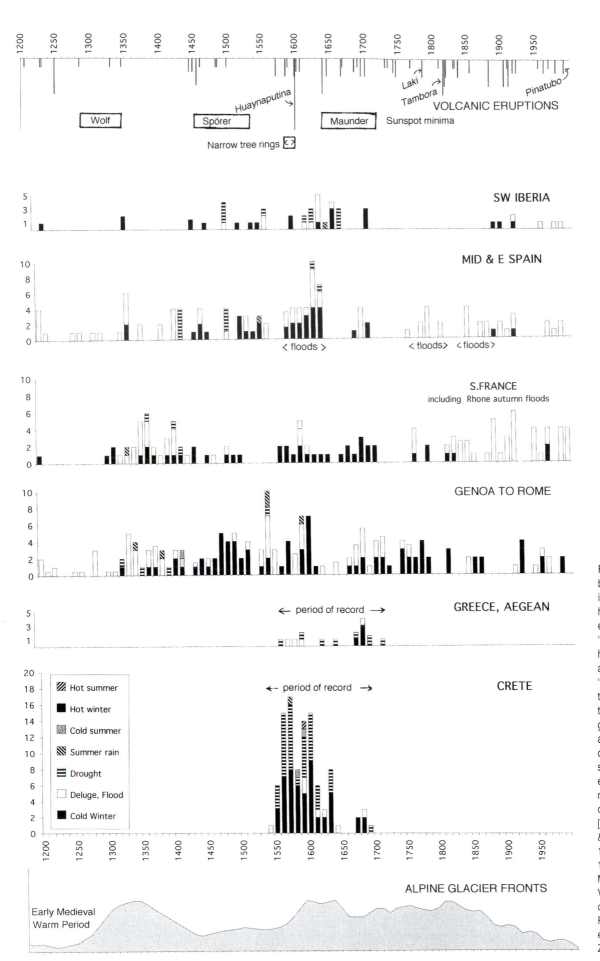

Fig. 8.3. Extreme events by decade. Where information is copious we have tried to categorize extremes: in Florence a 'medium' flood counts as half a unit, a 'large' flood as one unit, and an 'exceptional' flood as two units. Peaks and troughs in the Alpine glacier curve represent advances and recessions of the tongues. Also shown are great volcanic eruptions, sunspot minima and periods of narrow tree-rings. [Principal sources: Billi & Rinaldi 1997; Font Tullot 1988; Grove & Conterio 1994, 1995; Marsh 1864; Martin-Vide & Barríendos Vallvé 1996; Rodrigo & others 1994, 1995. Relevant volcanic eruptions are added after Zielinski (1995)]

rain and occasional drought, chiefly in Tuscany and the south of France. Dearth or famine is recorded in Tuscany in 26 seasons out of the 138 between 1282 and 1421.[23] As Giovanni Villani put it:

in the months of October and November 1345 at the time of sowing there was abundant rain so that the seed rotted, and then in April and May and June . . . it did not stop raining, and storms at the same time, and so in the same way the seed of the *biade minute* was lost, and the sowing was totally lost and the same happened in many parts of Tuscany and of Italy, and in Provence and in Burgundy and in France, from which was born a great famine and dearth in these countries, and at Genoa and in Avignon.[24]

1383–92 was a ghastly decade, despite fewer mouths to feed after the Black Death.

Floods are well recorded on the River Arno, involving Florence, Pisa and Pistoia (Fig. 8.4c). Then as now, they were commonest in autumn. For example, on 3 November 1333 there were floods all over middle Italy; a bridge then swept away in Florence was replaced by the Ponte Vecchio, which has withstood all subsequent floods. In 1328–9 there was no rain at Siena until 20 April. Great frosts included the Rhone freezing in 1364 and the Arno in 1368. Processions praying for rain early in 1368 were followed by floods in the following season. In 1371–2 processions prayed for rain at Florence in December, followed by processions praying for the rain to stop in May.

Records in Spain are more scanty, but continue into the fifteenth century. Villani mentions great rains in Seville (and in Cyprus) late in 1330. In Crete we know of no suitable records, but debris burying churches would be compatible with similar deluges there.

Weather disasters around the 1590s and 1600s

A minor run of events – mainly cold winters – centres on the 1470s to 1490s. For example, the burghers of Tortosa walked across the frozen Ebro in 1442, 1447 and 1506. There was a minor cluster of deluges, droughts, and out-of-season rain around the 1540s: the exceptional 'Wrath of God' flood struck Florence unexpectedly in 1547.

The next and greatest cluster of extremes began in the 1560s, reached a peak in the 1590s, and tailed off in the 1640s. Droughts, floods, cold winters and other exceptional events appear wherever we have records.

In Spain, a recent study regards 1580–1630 as the first of three catastrophic periods of flooding of the south-eastern rivers.[25] Font Tullot records many unusual snowfalls. 1617 was a 'deluge year' throughout eastern Spain, with intense rain in October; this was repeated early in 1626. Rodrigo and colleagues find this to be a generally wet period in south Spain, punctuated by occasional droughts such as that of 1604–5.

Records of flooding and freezing at Arles on the lower Rhone since 1500 show a good correspondence between freezing of the river and the frequency of floods over four metres (Fig. 8.4b). The 1590s were the first of a number of periods when floods were more frequent than in the twentieth century.[26]

In south France and Tuscany the 1590s were marked by crop failures, attributed variously to drought, cold and out-of-season rain; these led to wheat being imported, unusually, from northern Europe (which was itself in trouble). Floods, though not numerous, could be severe; in Rome they included that of December 1598, when the ancient Aemilian Bridge was carried away and 1,400 people were drowned.[27] 1601 was a cloudy and frosty year.

In Crete, for the period 1548–1648, weather is better documented than elsewhere in the Mediterranean. Venetian administrators sent home a mass of reports and letters mentioning, among other things, droughts, cold winters, great heat and out-of-season rainfall.[28] These are contemporary, precise, first-hand descriptions from observers present at the time.

Between 1585 and 1605 six winters were 'very severe' and a further five were 'severe'; two were dry. (Winters are rated as 'severe' if much damage was caused by rain to buildings, particularly public buildings, or if there were unusual difficulties with transport or many mentions of 'cold'. 'Very severe' winters are those with exceptional falls of snow, prolonged abnormal cold, no rain at all, or rain so excessive as to delay sowing of cereals until late spring.) The vines, of course, survived and Cretan wine was in much demand, for grape harvests in southern and especially central Europe had suffered badly.[29]

A 'very severe' winter was that of 1590–1. On 14 October a deluge hit the city of Réthymnon which, being on a promontory, is not ordinarily susceptible to flooding:

Most of the inhabitants of Retimo fled out of their houses and sheltered themselves as well as they could, especially in some of the churches. All the rivers flooded the town of Retimo, and the countryside, destroying a lot of buildings and causing high water. Eighty old buildings have fallen down . . . A lot of lands have been completely ruined.[30]

Throughout the winter excessive rain, storms and strong winds caused damage, trees and even vines being uprooted, buildings collapsing and people being killed. Around Canea (Khaniá) too, the winter had been very cold, with structures damaged by continuous and extraordinary rains. The following winter was merely 'severe': 'although the winter was cold and too long, and the season began with unusual snows, nevertheless the farmers have cultivated and sown wheat.'[31]

[23] C. de la Roncière, *Prix et Salaires à Florence au XIVe Siècle (1280–1380)*, Collection de l'École Française de Rome, 1982; G. Pinto, 'I livelli di vita dei salariati fiorentini (1380–1430)', *Toscana medievale: paesaggi e realtà sociali*, ed. G. Pinto, Le Lettere, Florence, 1993, pp. 113–49.
[24] G. Villani, *Nuova Cronaca* XIII.73, ed. G. Porto, Fondazione Pietro Bembo, Parma, 1990–1 (kindly brought to our attention and

translated by Dr P. Spufford).
[25] Barriendos Vallvé & Martín-Vide 1998.
[26] Pichard 1995.
[27] Braudel 1966, p. 273; Pichard 1995; N. S. Davidson, 'Northern Italy in the 1590s', Clark 1985, pp. 157–76.
[28] Grove & Conterio 1992a, 1992b, 1994, 1995.
[29] Was it a coincidence that the big late-medieval trade in malmsey wine between Crete and

England arose shortly after the bad weather of the fourteenth century? See Rackham & Moody 1996, pp. 78 ff.
[30] ASV: PTM filza 755: Nicolò Priuli, Rector of Retimo, 22 November 1590.
[31] ASV: PTM filza 755: Francesco Malipiero, Rector of Canea, 10 February 1591; PTM filza 756: Regiment & Captain, 16 February 1592.

Winter 1589–90 was dry. The Governor of Candia (Herákleion) reported that it had started very harsh and cold. By May, however, a strong south wind had been blowing for six months and had 'burnt all our lands'. Then on 1 June there was a thunderstorm, it rained day and night, and 'a lot of lands were ruined', which may imply that sediment was shifted by a deluge following drought. The greater drought of 1595–6, though coming to a population diminished by the 1592 plague, led to famine over much of the Aegean. Bad as things were in Crete, refugees fled there from Mélos and other islands because of food shortage.[32]

Cold could be a hazard, although frosts were not severe enough to kill olive trees. In 1602, for instance, many animals

died, and there were even deaths amongst the unfortunate conscripts manning the galleys because of 'the bad season of very bitter cold'.[33]

Between 1548 and 1645, twenty-one 'very severe' winters are recorded, of which thirteen were associated with drought, three with exceptional snowfall, and the rest with heavy rains. In addition there were twelve 'severe' winters with cold or rain or both. Since instrumental records began early in the twentieth century there have not been three continuous months of complete drought in winter.[34]

Out-of-season rain was reported, although we cannot determine whether it was more common than during the present century. At least one such storm had a more severe and more widespread effect than any in recent years. In 1576 the Governor-General of Crete reported that

> the very heavy and unusual rains we have had this year in June, July and August have greatly ruined the vintage . . . also destroyed the salt-works of Suda and Spinalonga.[35]

These places, where sea-water was evaporated in the open air, are 150 km apart.

The disastrous 1690s and 1700s

After a period of quiescence in the mid-seventeenth century, the extreme events in and around the 1590s were repeated a century later. The records are best from the east Mediterranean in 1675–1715.[36]

Rhone floods of Mediterranean origin at Beaucaire

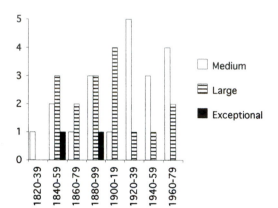

Number of months per decade when there was ice on the lower Rhone and autumn floods at Arles reached over 4 m

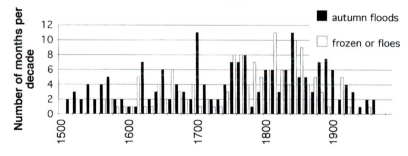

Arno flood frequency at Florence

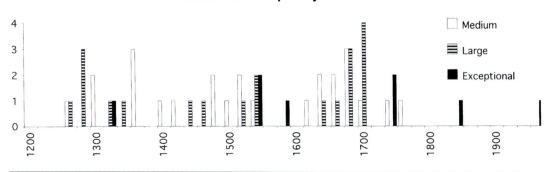

Fig. 8.4a. Top left: floods on the Rhone at Beaucaire, at the head of the delta.

Fig 8.4b. Floods and ice on the Rhone at Arles. The greater numbers in the eighteenth and nineteenth as compared with earlier centuries may reflect fuller records. Until the 1890s there is a good correspondence between autumn floods on the one hand and following cold winters on the other – with the notable exception of 1700–10. Based on Pichard 1995, Fig. 3.

Fig. 8.4c. Floods on the River Arno.

[32] ASV: PTM filza 763: Donado, 16 May 1596.
[33] ASV: PTM filza 770: Priuli, 10 March 1602; Cà' Taiapiero, 24 March 1602.

[34] Grove, Moody, Rackham 1991; rainfall data in appendix.

[35] ASV: PTM filza 738: Foscarini, 14 September 1576.
[36] Grove & Conterio 1994.

Trouble began with drought in the Ionian Islands in 1671–2. In Cephalonia

> the most deplorable misery was to be seen, since there was not bread for two [days' consumption] . . . the poor people had been eating the wild plants on the island, while a severe drought had deprived them even of this nourishment.

There were similar reports of drought from Zante (Zákynthos) and Corfu. On the smallish Aegean island of Tinos the Venetian authorities reported that winter 1682–3 was one of strong winds, continuous rain, heavy snow and 'very intense cold'; 480 cattle and 3,427 sheep and goats perished. The Turkish authorities in Crete reported that torrential rain had caused serious damage to aqueducts.[37]

Matters were worse in the 1690s. In July 1692 the Governor-General of the remaining Venetian islands reported that

> the scantiness of a very thin crop of wheat having increased the worry that this current year of famine is almost universal, so that there is very little help from the mainland and from the kingdom of Morea [Peloponnese].

There was similar trouble in 1694.[38]

In Sicily there was a run of disastrous harvests. In 1691, 1692 and 1694 wheat stored in a depot at Termini Imerese was spoilt in summer, either because of rain during harvest or because out-of-season rain got into the warehouses. In 1693 and 1694 wheat and barley crops were scanty. In 1700 the Duke of Gampilerio noted that even less wheat had been brought down from the mountains to the various stores than in the past few years 'even though the crop then was poor'; he speculated that this was due to the climate having become colder and wetter. In 1692 heavy snowfall was reported from Castroreale.[39]

Reports from the west Mediterranean are less copious. The 1700s again saw unusual floods on the lower Rhone. Mediterranean Spain, in contrast, was free of 'catastrophic' floods between 1690 and 1720. The 1690s there were probably the coldest decade of the century; 1693–4 was the coldest winter of that decade, with ice on the Ebro at Tortosa three metres thick (or so it was said). 1694–5 was a very snowy and frosty winter also in middle and east Spain, and 1696–7 in Majorca; snow fell at Turin in late April 1713 and on 1 May 1714.[40]

Frost in the west Mediterranean caused the olive oil crop to fail (over and above its biennial tendency), and encouraged the oil merchants and soap-boilers of Marseilles to look for sources out of reach of severe frost. In 1700 the botanist Tournefort remarked that the olive-trees 'never die in *Candia* [Crete], because it never freezes there.' The French Consul

> assur'd us, that in the Year 1699, the Island yielded 300,000 Measures [somewhat over 3 million litres] of Oil; of which the *French* bought 200,000 . . . The Crop of Oils failed that year in *Provence*, and the ports of *Candia* were crouded with Ships from *Marseilles*, to fetch Supplies for the Soap-makers there.[41]

This Consul reported in 1701 that merchants from Marseilles and ports nearby had set up establishments in Crete. Cretan prosperity was confirmed by the great winter of 1709, ruining olives altogether in the south of France and north Italy.[42]

A noteworthy drought in the eastern Mediterranean was that of 1713–14. The wheat crop was scanty in south Greece, and the Turks forbade trade in food. In Cerigo (Kythera) 'because of the very severe drought which occurred during the winter, the watermills could not be used because of lack of water'.[43]

THE END OF THE LITTLE ICE AGE

In the Mediterranean the third phase of unusual events died out in the 1710s. After a last cluster of floods and hard winters in the 1740s and 1750s a period of 'normality' appears to set in, with only a background level of unusual events, some of them confirmed by the emerging instrumental record. The 1760s were particularly cold in north Italy (p. 120) and the south of France. In 1789 the Lombardy rivers froze, and snow is said to have lain for three months.[44] South-east Spain experienced a second period of inundations in the last quarter of the eighteenth century.[45]

Among the unusual events of the late eighteenth century were so-called 'dry fogs', when dust reduced visibility, a phenomenon which has not been seen in living memory. Unlike loess rains (p. 29) they occurred over wide areas and could persist for weeks. They were commonest in spring, when the sea was cooler than the land. The most important in the last seven hundred years was that in summer 1783, observed from Portugal to Anatolia, and attributed at the time to a violent earthquake in Sicily and Calabria. The Italian volcanoes Stromboli, Vulcano and Vesuvius, known to produce dry fogs, were active at the time, but the main source of the dust was almost certainly Iceland. A new volcanic island erupted there early in the year, followed by a colossal eruption of Lakagígar, one of the vents of the great volcano

37 ASV: Sen. Dis. Ret. Zante filza 3: Priani, 12 April 1672; PTM filza 941: Valier, 25 April 1672; PTM filza 900: Marcello, 28 April 1683; Stavrinidhos 1985–6, no. 813.

38 ASV: PTM filza 1175: Vendramin, 28 July 1692; PTM filza 1176: Molin, 26 December 1694.

39 Grove & Conterio 1994; ASPG: VPT: Lettere e ordini 1690–94, busta 150: Duke of Gampilerio, 8 October 1691; busta 150: Errante, 22 September 1691; busta 150: Miceli, 29 September 1692; ASPTPR: Mete di Frumenti reg.43, Giurati da Menfi, 17 October 1693; ASPG: VTLO: 1693–1701, busta 150, Micheli, 14 September 1694;

busta 150, Duke of Gampilerio, 7 October 1700; ASPGTPR: Consulte, busta 124, Munallio, 22 August 1692.

40 Pichard 1995; Barriendos Vallvé & Martín-Vide 1998; R. O'Callaghan, *Anales de Tortosa e Historia de la Santa Cinta*, Tortosa, 1886, p. 89 (quoted by J. M. Fontana Tarrats, *Historia del Clima en Cataluña Noticias*, Madrid (mimeo), 1976, p. 214); Font Tullot 1988, p. 83; Leporati & Mercalli 1994.

41 Tournefort 1717/1741.

42 Triandafyllidou-Baladié 1988.

43 ASV: PTM filza 957: Sagredo, 17 March,

28 April and 18 July 1714; PTM filza 959: Dolfin, 22 September 1714.

44 Not confirmed by the Milan thermometer, which recorded a moderately cold winter (Buffoni & others 1996).

45 Delano Smith 1979, p. 213; Barriendos Vallvé & Martín-Vide 1998; C. Pfister, G. Schwarz-Zanetti, F. Hochstrasse, M. Wegmann, 'The most severe winters of the fourteenth century in Central Europe compared to some analogues in the more recent past', *PKF* 15, 1998, pp. 45–61; 'Winter severity in Europe: the fourteenth century', *CC* 34, 1996, pp. 91–108.

Laki. A brimstone-smelling haze reached Oslo two weeks later, and was observed from Labrador to China. This was followed by two hard, long winters in Italy (with a record snowfall of 233 cm in Turin in 1784–5), but had little if any long-term effect.[46]

The 1810s were cold, as the instrumental records for Cadiz and Milan show.[47] There were great winters in north Italy in 1812 and 1814, followed by the 'year without a summer' in 1816, famous for its physical and political repercussions in the south of France (p. 10), and then frosty winters in the south of France and floods on the Rhone.[48] The mighty eruption of Tambora in Java in 1815 is held responsible for cold weather throughout the northern hemisphere, but no dry fog was reported in the Mediterranean.

Turin had very snowy winters in 1844–5 and 1847–8. On the lower Rhone there were two more clusters of floods: four between 1836 and 1846, six between 1882 and 1891 (Fig. 8.4a).[49] These floods were all between September and February and of Mediterranean origin, mainly from heavy precipitation in the Cévennes and Durance valley. On the Ebro there were five great floods between 1888 and 1895, and no more until 1926.[50]

For the last sixty years or so, to take the records at face value, extremes, especially deluges, have again increased. In four single years – 1940, 1966, 1982 and 1994 – flooding in the western Mediterranean has been severe and widespread, despite the effects of dams. It would be easy to collect reports to indicate that the 1980s and 1990s are a period of extremes, comparable with the 1690s. Is this really so?

The twentieth century was peculiar in other respects. In the past, deluges and other localized events were rarely recorded unless they affected well-known places. The Florence flood of November 1966 would have attracted attention in any century since the eleventh, but newspapers now notice similar floods in remote places. Many modern activities cause a given event to do more damage than it used to. Rivers have been narrowed and canalized, and braided river-beds reduced to single channels. An ordinary deluge turns into a disaster when roads, bridges, flood protection walls, etc. leave the water with nowhere to go. People have moved homes and factories on to flood-plains, abandoning their self-protective caution on the plea that modern technology has made these places safe; too late they discover that it protects them against middle-sized but not great floods. They also take more risks with sensitive crops.

The apparent cluster of extreme events in the last few decades is thus at least partly an artefact of information technology and a consequence of incautious modern development.

POSSIBLE INFLUENCES ON EXTREME EVENTS

The events we have considered show definite clusters. The cluster in the late sixteenth century was apparently the most intense in terms of the number of extreme events, though not necessarily of their individual severity. This is only partly a consequence of record-keeping being better then than earlier or later: evidence comes from tree-rings and alluviation as well as documents. The fourteenth-century cluster was apparently also prolonged; the events were severe but apparently less numerous. The cluster of events around 1690 (well documented from Sicily) was apparently of shorter duration, with relatively few events but some of them very severe. A fourth cluster of extreme events came in the 1810s.

Relation to glaciers and weather patterns

Extreme weather in the Mediterranean does not occur in isolation. Even during the Medieval Warm Period, intervals of cold winters, such as that of the 1200s, may have corresponded to minor advances of Alpine glaciers as known from studies of moraines and buried trees.[51]

The first phase of the Little Ice Age manifested itself throughout Europe, but not at quite the same date. The disasters of the 1310s in England were comparable to those of the 1330s and 1340s in southern Europe. Several Alpine glaciers are known to have advanced: the Allalin glacier blocked the Saas valley in the thirteenth century; the Gurglerferner advanced in the 1310s and the Lower Grindelwald in 1338; the Großer Aletsch, best known of Swiss glaciers, had blocked a major irrigation canal before 1385.[52]

The 1590s phase went with a generally colder and wetter, as well as more variable, climate in the Mediterranean. This coincides with the second peak of the Little Ice Age. Cold or wet seasons are recorded all over north and middle Europe. Glaciers such as the Großer Aletsch were advancing in the Alps, reaching a peak some fifteen years later than the exceptional weather in the Mediterranean. .

The 1690s phase, also accompanied by evidence for cold and wetness, coincides with the third peak of the Little Ice Age, going on into the eighteenth century. It too came some fifteen years before the peak of an advance of Alpine glaciers. During the 1810–23 cluster of climatic extremes, many European glaciers again extended down-valley to positions that they have never since exceeded.

Some more recent clusters of events are linked to minor advances of glaciers. The groups of autumn and winter floods on the Rhone in the 1840s and 1890s, although not directly of Alpine origin, were each followed, a few years later, by

[46] D. Camuffo and S. Enzi, 'Impact of the clouds of volcanic aerosols in Italy during the last 7 centuries', *Natural Hazards* 11, 1995, pp. 135–61; R. B. Stothers, 'The Great Dry Fog of 1783', *CC* 32, 1996, pp. 79–89; G. R. Demarée, A. E. J. Ogilvie, De'er Zhang, 'Further documentary evidence of Northern Hemisphere coverage of the Great Dry Fog of 1783', *CC* 34, 1998, pp. 1–4; Leporati & Mercalli 1994.

[47] D. Wheeler, 'Early instrumental weather data from Cadiz: a study of late eighteenth and early nineteenth century records', *IJC* 15, 1995, pp. 801–10.

[48] Pichard 1995.

[49] Leporati & Mercalli 1994; M. Pardé, *Le régime du Rhône, étude hydrologique*, Allier, Grenoble, 1924–5.

[50] A. F. Gellatly, J. M. Grove, A. Bücher, R. Latham, W. B. Whalley, 'Recent historical fluctuations of the Glacier du Taillon, Pyrénées', *Physical Geography* 15, 1995, pp. 399–413.

[51] For general information on glaciers, see Grove 1988.

[52] H. Holzhauser, 'Zur Geschichte der Aletschgletscher und des Fieschergletschers', *Physische Geographie (Geographisches Institut der Universität Zürich)* 13, 1984, pp. 1–148; 'Rekonstruktion von Gletscherschwankungen mit Hilfe fossiler Hölzer', *Geographica Helvetica* 39, 1994, pp. 3–15; J. M. Grove and R. Switsur, 'The Medieval Warm Period', *CC* 30, 1994, pp. 1–17; Holzhauser & Zumbühl 1996; G. Patzelt, 'Former lakes dammed by glaciers and resulting floods in the Ötztal, Tyrol', *Mountain Research and Development* 16, 1996, pp. 298–301.

glaciers in the Alps advancing to reach forward positions in the 1850s and 1890s. The Ebro floods of 1888–95 overlapped with one of the last glacier advances in the Alps and Pyrenees. (After a minor readvance around 1970 most of the glaciers have retreated to their positions at the end of the Medieval Warm Period.)

Extreme weather, of various kinds, in the Mediterranean is thus linked in time to cold periods in the rest of Europe. There is an obvious link between glacier advances and some types of extreme close by in the Mediterranean. Glaciers advance when a succession of snowy winters is reinforced by a later onset of melting in unusually cool late springs.[53] Weather systems that gave heavy autumn snowfall and contributed to glacier advances in the Alps would also account for much of the flooding in southern Europe. Extreme hydrological, glacial and erosional events may thus have occurred simultaneously in the Alps and the north Mediterranean. However, such events also occurred in regions like Crete which are remote from the weather systems of the Alps.

Possible causes: volcanoes

The dust and especially the sulphuric acid blasted into the upper atmosphere by certain types of volcano might be expected to result in unusual weather, especially cold, for a few years, as happened with the eruptions of Mount St Helen's in 1980 and Pinatubo in 1991 (pp. 29, 128). A single eruption would probably not be enough to set off a phase of the Little Ice Age, but a cluster of eruptions could reinforce a tendency already apparent in the global circulation.

The 1810s, even before Tambora erupted, had already been cool. This can plausibly be ascribed to a cluster of eruptions: Etna in 1809, Vesuvius in 1806, 1811 and 1812, Sabrina (Azores) and Mayon (Philippines) in 1811, and then Tambora in 1815.

To extend this correlation backward runs into the problem that even huge volcanic eruptions, affecting weather over half the globe, escape the notice of historians if they happen in remote places. The eruption of Huaynaputina, Peru, in 1600 has only recently been recognised among the greatest of historic times, turning the summer of 1601 into one of the coldest of the past six hundred years in the Northern Hemisphere.[54] A more complete record of this type of eruption is beginning to emerge from fallout stratified into the ice layers of Greenland (this we have added to Fig. 8.3).

Global weather patterns

H. H. Lamb, discoverer of the Medieval Warm Period in northern Europe, surmised that the early medieval period may have been wetter in the Mediterranean, including rain in summer, because the anticyclonic belt was then frequently displaced somewhat to the north. Changes to the colder climate that followed were likely to be due to shifts of the main features of global circulation.[55] In cold phases of the Little Ice Age, wind patterns over at least the European part of the globe would have been more southerly than now, with cold air more frequently entering the Mediterranean basin from the north. This would have stimulated the formation of cyclonic depressions and increased the frequency of deluges. One might expect to find erosional features attributable to weather events in these periods.

A modern analogue for the Medieval Warm Period is provided by the synoptic situation prevailing in summer 1976. A high-pressure system, sitting firmly over NW Europe, caused atmospheric blocking, resulting in exceptionally severe drought in Ireland, Britain and northern France. Airstreams coming from the west were diverted south towards a low-pressure trough extending into the central and eastern Mediterranean; this produced a cool summer with out-of-season rainfall. Rain fell in Greece, south Turkey and Algeria; there was a deluge in Valencia on 28 August. A similar anomaly in August 1956 produced out-of-season rainfall in Cyprus and Greece.[56]

Similar relationships are to be found in other parts of the world. In the United States, extreme flood events over the last five thousand years are not random in time, but are tied to regional and global climatic variations.[57]

Solar activity

The ultimate cause of the Little Ice Age and its Mediterranean manifestations is often held to lie in solar activity. Two of the major periods of unstable weather roughly correspond with the Wolf and Maunder minima of sunspots in 1282–1342 and 1645–1715. The Spörer sunspot minimum, 1450–1534, coincides with a minor period of unstable weather.[58] However, the second (c. 1600) and fourth (c. 1815) phases of the Little Ice Age were not at sunspot minima. Sunspots alone probably have had some influence, but are not a complete explanation for climatic fluctuations of Little Ice Age type.[59]

53 C. Pfister, 'Switzerland: the time of icy winters and chilly springs', *PKF* 13, 1994, pp. 205–24.
54 S. L. De Silva and G. A. Zielinski, 'Global influence of the AD 1600 eruption of Huaynaputina, Peru', *Nature* 393, 1998, pp. 455–8; K. R. Briffa, P. D. Jones, F. H. Schweingruber, T. J. Osborn, 'Influence of volcanic eruptions on Northern Hemisphere summer temperature over the past 600 years', *Nature* 393, 1998, pp. 450–55.
55 H. H. Lamb, 'The early medieval warm epoch and its sequel', *PPP* 1, 1965, pp. 13–37.

56 A. H. Perry, 'Mediterranean downpours in 1976', *Journal of Meteorology* 2, 1976, pp. 10–11.
57 L. L. Ely, 'Response of extreme floods in the southwestern United States to climatic variations in the late Holocene', *Geomorphology* 19, 1997, pp. 175–201.
58 J. A. Eddy, 'Climate and the role of the sun', *Journal of Interdisciplinary History* 10, 1980, pp. 725–47; N.-A. Morner, 'The Maunder Minimum', *PKF* 13, 1994, pp. 1–8; A. Nesje and P. Johannessen, 'What were the primary forcing

mechanisms of high frequency Holocene climate and glacier variations?', *The Holocene* 2, 1992, pp. 79–84; M. Stuiver and T. F. Braziunas, 'Evidence of solar variation', Bradley & Jones 1992, pp. 593–605; 'Sun, ocean and climate and atmospheric 14C: an evaluation of causal and spectral relationships', *The Holocene* 3, 1993, pp. 289–305.
59 This will be discussed at greater length in the forthcoming second edition of Grove 1988.

CONCLUSIONS

There is arguably some correlation between unusual weather and each of the three factors shown in Table 8.i – volcanoes, sunspots and glacial advances. A single volcano is seldom enough to make itself felt: for example the giant eruption in 1259, perhaps of El Chichón (Mexico), failed to produce any known events in Mediterranean weather, as far as records go. However, the cold winters of the 1810s and the deluges of the 1910s can each be explained by a cluster of eruptions, even though the sunspots were not in favour.

Of the four phases of the Little Ice Age, two were associated with volcanoes and two with sunspot minima. The greatest in its Mediterranean effects was the second phase, centred on the 1590s. Glacial advance coincided with a cluster of great eruptions far away, including Billy Mitchell (Bougainville Island near New Guinea) in c. 1580, Kelut (Java) in 1586, Raung (Java) in 1593, Hekla (Iceland) and Ruiz (Colombia) in 1599, and Huaynaputina (Peru) in 1600. Solar activity, however, seems not to have contributed. The same is true of the fourth phase, centred on 1815. The first and third phases, centred on the 1340s and 1690s, were apparently less severe (though also less well recorded); each time a glacial advance coincided with a sunspot minimum but with only two big eruptions.

	Type of unusual weather	Areas affected	Volcanoes	Sunspots	Glacial advances
1980s, 1990s	deluges, hot summers	general	Mt St Helen's, El Chichón, Pinatubo		
1920s	cold winters	N. Italy	1		minor
1900s, 1910s	mainly deluges	Spain, S. France	Katmai (Alaska) and 6 others		minor in 1890s
1880s	deluges	S. France	Krakatau		
1830–70	deluges, cold winters	Spain, S. France, N. Italy	5		minor in 1850s
1810s	COLD WINTERS, COLD SUMMER	S. France, N. Italy	Tambora, Mt St Helen's and 5 others		major in 1810s
1740–1800	deluges, cold winters	Spain, N. Italy	2		minor in 1760s
1680–1720	DROUGHT, COLD WINTERS, rain	general	Tongkoko (Indonesia) and Fuji (Japan)	latter part of Maunder Minimum	minor
1560–1620	DELUGES, DROUGHT, SUMMER RAIN, COLD WINTERS	general	Huaynaputina, Mount Billy Mitchell, and 5 others	none	major
1540s	esp. deluges	Spain, N. Italy	none		
1460–1520	esp. cold winters	Spain, N. Italy	5 between 1460 and 1484	Spörer Minimum	
1320–70	DELUGES, COLD WINTERS, SUMMER RAIN	probably general	2	latter part of Wolf Minimum	major
1200–30	deluges, cold winters	general?	2		

Table 8.i. Periods of frequent unusual weather and correlation with possible influences. Capital letters indicate major clusters of events. Volcanic eruptions are those thought to affect weather, mainly from the Greenland ice record (Zielinski 1995).

CHAPTER 9

Climate in Early Historic and Prehistoric Times

The Cretans among others say that nowadays the winters are more severe and
more snow falls, adducing as evidence that the mountains were settled in olden times
and bore grain and [tree-]fruit as the land was planted and tilled. For there are
extensive plains among the mountains of Ida and in the other mountains, none of
which are worked now because they do not bear. Whereas in those days . . . they
were settled . . . because then the rains were generous, and wintry weather did not
often occur. If then what they say is true, the Etesian winds must be more
numerous (today).[1]

Climate has attracted attention from scholars looking for environmental change as a cause of the 'Neolithic Revolution' and later events. For example, the fall of Minoan civilization has been attributed to a supposed 'Mycenaean drought'.[2] Other scholars infer that lack of obvious evidence for change implies lack of change. As the last two chapters have shown, climate history is likely to be complex, and the data are fragmentary.

Southern Europe is susceptible to decades of unusual climate or violent weather, but if recent experience is typical, it is somewhat less prone to longer-term climatic variations than lands to the north or south. Only parts of southern Europe, not the whole Mediterranean, have had experiences comparable to the 1968–93 drought in the Sahel–Sudan zone.

During the Holocene and the end of the Pleistocene, on the south side of the Sahara, there were dramatic changes in lake levels; in what is now desert the degree of aridity fluctuated strongly. In Alpine Europe glacier fronts repeatedly advanced and retreated. Compared to these northern and southern regions, the climate of the Mediterranean seems to have been more stable, though we may have this impression because of a lack of sensitive indicators.

There are hardly any glaciers in southernmost Europe to record cooling phases during the Holocene, though the Calderone glacier, perched on the Gran Sasso 110 km NE of Rome, may yet tell its story.[3] The rarity of bogs not only limits the data on vegetation as a proxy record of climate, but has made it difficult to infer climate changes from the behaviour of the bog itself, as is possible in Britain.[4] Lakes providing records of changing levels are few, and are mainly in limestone depressions where fluctuations in level can be caused either by climatic fluctuations, or by subterranean outlets becoming blocked with debris or opened by dissolving of the limestone or by earthquakes.

Tree-ring studies (pp. 132–3) may turn out to yield useful information, but no known living trees in southern Europe are half as old as the bristlecone pines of the western United States, and subfossil wood is scarce. The nearest equivalent, the 1500-year-old cypresses near the Cretan tree-limit (p. 60), are difficult for dendroclimatologists because the trees have been coppiced every few hundred years, and growth is uneven: individual rings have a disconcerting tendency to merge or subdivide.

Computer modelling of climates is of little help. It is a coarse-grained procedure; it may be of use in the vast expanses of the Sahara, but cannot cope with the towering coastal mountains of the Mediterranean.

Apart from fluctuations of global weather patterns, there may have been shorter-lived events through volcanic eruptions: directly through ash-fall, dust and toxic fumes, or indirectly through alteration of the chemistry of the upper atmosphere.

EARLY HISTORICAL RECORDS

The climate of the early historical Mediterranean cannot have been very different from what it is today. Most of the apparent indications to the contrary are at best ambiguous.[5] The White Mountains (Montes Leuci) of Crete may be named after their once permanent snow, but more likely after their white crystalline limestone rocks. As Lucia Nixon points out, the fact that many of the rivers of Antiquity now often do not have water in them tells us more about how the ancients defined a river than about changes in the rivers themselves.

For the ancient Greeks and Romans, as in the early-Modern period, harvests fluctuated. Food crisis was common, but could usually be met by importing food or by eating something unusual; famine was rare.[6]

Some Greeks and Romans believed that climate had changed,[7] but the tantalizing habit of ancient philosophers of theorizing from insufficient data usually makes it impossible to verify the facts. Saserna, the first-century BC agricultural writer, said that Italy used to be frostier; he was not

[1] Theophrastus, *On Winds*, fourth century BC, trans. revised after V. Coutant and V. L. Eichenlaub.

[2] R. Carpenter, *Discontinuity in Greek Civilization*, Cambridge, 1966.

[3] A. F. Gellatley, C. Miraglia, J. M. Grove, R. Latham, 'Recent variations of the Ghiacciaio del Calderone, Abruzzi, Italy', *Journal of Glaciology* 40, 1994, pp. 486–90.

[4] J. H. Tallis, 'Climate and erosion signals in British blanket peats: the significance of *Rhacomitrium lanuginosum* remains', *JE* 83, 1995,

pp. 1021–30.

[5] Sallares 1991.

[6] Garnsey 1988; see p. 80.

[7] J. Neumann, 'Climatic change as a topic in the Classical Greek and Roman literature', *CC* 7, 1985, pp. 441–54.

interested in this information for its own sake, but immediately embarked on an astronomical explanation of the phenomenon, without first making sure that there was a phenomenon to explain.[8] The few scraps of definite information relate mostly to local microclimates. Theophrastus (fourth century BC) says that Larissa in Thessaly had become a frost-hollow since its surrounding plain had been drained.[9]

A few indications in ancient writers are not compatible with the present climate. God sent the ancient Hebrews 'the first rain and the latter rain',[10] as he did in the twentieth century to parts of the west Mediterranean (Fig. 2.1). In Palestine there is now only one rainy season, in late autumn.[11] In Sicily, in Theophrastus's time, 'the spring rain is abundant and gentle, the winter rain is sparse', which was good for cereals.[12] This suggests a late maximum rainfall, much as in the modern Ebro basin. In Catania and other places in Sicily today, spring rain sometimes happens, but rarely exceeds that of January and February. A spring as well as a late autumn rainfall peak was characteristic of SW Iberia before 1965, but not after (Figs. 7.8, 7.9); similar changes could reasonably have occurred in earlier times elsewhere in the Mediterranean.

Ancient authors give a definite impression that winter snow was less rare in inhabited places than it is today. Biblical writers were familiar with snow, though it seldom enters into the action of the Old Testament. Xenophon the hunter (early fourth century BC) expected to go out after hare or boar in time of snow,[13] which a modern Greek would not.

This is corroborated for Italy by one definite piece of proxy evidence. At Pliny the Younger's villa above the Città de Castello basin, on the upper River Tiber,

> The climate in winter is cold and frosty; it rejects and expels myrtle, olives, and other things which flourish in favourable weather; however, it allows laurel and even brings it forth luxuriantly, but sometimes kills it, though not more often than around [Rome].[14]

The winters therefore were too cold for myrtle and olive, and 'killed' laurel – only to the ground – every so often (which exactly matches Cambridge today). Pliny's home was at 370 m; the plain, which would probably be a frost-hollow, is at c. 300 m. The olive-limit today in this part of Italy is about 460 m. It is possible that Roman growers, like modern Cretans, were more cautious with their olives than modern Italians and did not push them so far into the frost zone. The laurel evidence, however, seems conclusive: laurel-killing frosts then were not uncommon even in Rome. Winters in the first century AD were colder than today by the equivalent of at least 100 m extra altitude. As we have seen (p. 122), winters at Rome have varied by several degrees in the last hundred and fifty years: occasional frosts have damaged, but have not expelled, olives near the altitudinal limit.

Another unexpectedly high olive-limit is implied in the account of a battle of the second century BC, fought on Mount Aphrodite 'planted round with olives'; this was somewhere in middle Spain, where olives are marginal today.[15]

Theophrastus remarks that the plains of Latium, near Rome, produced beech of exceptional size.[16] If this is true – if Theophrastus was correctly informed – it means that beech then extended several hundred metres lower than now: the summers would have to have been less arid.

Ancient seafarers divided the year into a safe sailing season from March to October and a close season when they avoided travelling. To take a famous example, St Paul correctly predicted the shipwreck that would happen to travellers who stayed at sea too late.[17] The seasons seem not to be so clearly demarcated now, and summer gales may be more frequent. However, modern ship design and rigging, as well as modern climate, have reduced the terrors of jagged capes and lee shores.

For a change of climate within the Graeco-Roman period, the passage by Theophrastus on page 141 is the most definite reference. It implies that winters had become more severe in the fourth century than previously. It seems to apply to the Nídha Plain in Crete, which at 1,400 m is well above the present limit of settlement, and also to the Lassíthi, Omalós and possibly Askyphou Plains. So far the abandonment of these plains has not been corroborated by archaeology.[18]

Was not North Africa the 'granary of Rome'?

It has long been believed that the climate of North Africa at the time of the Roman Empire was wetter than at present. Lepcis Magna and Timgad, ruined cities now in the midst of 'desolation', seem to demonstrate this. Grain was exported, but more from Tunisia, Cyrenaica and Egypt than from the interior of the Maghreb.

B. D. Shaw found most of the climatic inferences from fauna, water resources and water-control systems to be indecisive. Hippopotamus and rhinoceros once existed in north Africa, but the rock pictures of them date back to the Neolithic when the climate was indeed wetter than now. Elephant and lion, though preyed upon by Roman amphitheatres, persisted into medieval times; lion and leopard into the nineteenth century. Many Roman wells, cisterns and aqueducts are still usable. It is erosion, not a drier climate, that makes it difficult to restore some of the irrigation systems. Shaw estimates the grain exports of the Maghreb to Rome at 'somewhat less than 5 million bushels [about 200,000 tons] annually', with much more going to other parts of the Mediterranean. Such exports continued into modern times; they ceased only when other sources of grain displaced them and growing North African populations increased local demand. Most of the literary and archaeological arguments for conditions having been generally wetter are

8 Quoted by Columella, *Res Rustica* I.14–5.
9 *De Causis Plantarum* V. xiv. 2.
10 There are many allusions to two rainy seasons, from Deuteronomy 11:14 (seventh century BC) onwards. The last is James 5:7 (c. 100 AD), but by then it may have been merely proverbial.
11 Walter & Lieth 1960–.
12 *Historia Plantarum* VIII.vi.6.
13 *Kynegetica*, 4.9.
14 Pliny the Younger, *Letters* V.vi.7.
15 Appian, *Histories: Iberia* VI.ix.45.
16 *Historia Plantarum*, V.viii.3.
17 Acts of the Apostles 27:9, 10.
18 This passage has been misinterpreted as showing that the ancients were 'aware' that felling forests changes the climate (p. 8), although it mentions neither forests nor felling; J. D. Hughes, 'Forestry and forest economy in the Mediterranean region in the time of the Roman empire in the light of historical sources', *PKF* 10, 1994, pp. 1–14.

unsubstantiated and should be discarded.[19] Nevertheless, it would be surprising if in the six centuries of the Roman empire there were not wetter and drier decades at least as extreme as those recorded by rain-gauges in the last two centuries.

Floods, droughts, frosts

References to extreme weather in Antiquity are of the same general character as in the middle ages. The ancient Greeks believed in Deukalion's Flood, which for some was not merely mythological, but was an event in about 720 BC from which other events could be dated: the Olympian Games were founded in 776 by Klymenos, a Cretan, who might have known it.[20] The River Tiber, then as now, had occasional floods. In 241 BC

> an overflow of the Tiber, swollen by unusual rains, beyond imagination both for magnitude and duration, destroyed all the buildings of Rome sited in the plain.[21]

Drought was not unusual. In Athens there were several food shortages from 360 to 330 BC, from which some have inferred a thirty-year drought. However, not all the shortages were necessarily due to drought, and this may have been a well-documented rather than an unusual period.[22] Theophrastus writes of heavy rainfall causing Lake Kopáis to rise exceptionally high before the battle of Khaironéia in 338 BC. Later in the century, wells in the Agora in Athens were abandoned, and it has been argued that they failed because of low rainfall. They could equally have fallen out of use because a new method of waterproofing cisterns had been invented.[23]

Some of these fluctuations may indicate changes of climate similar to those of the last thousand years; but there are not enough reliable data to identify periods of extreme events in Antiquity comparable to those in the last chapter.

Volcanoes

The explosive eruptions of Vesuvius in 79 and 472 AD, Etna in 44–43 BC, and Santoríni in the seventeenth century BC were world-class events. Santoríni's colossal eruption is said to have affected tree-rings all round the northern hemisphere in 1629–28 BC. However, other means of dating the eruption are approximate and conflicting, and there is nothing in the tree-rings themselves to show that the anomaly was caused by a volcanic event; a recent analysis of volcanic ash in the Greenland ice appears to show that if a volcano was the cause it was not Santoríni.[24] The Etna eruption had hemisphere-wide effects, to judge by an ash-fall dated in the Greenland

ice at 50±5 BC and a 'frost ring' at 43 BC in the bristle-cone pines of Arizona. In 472 AD

> Vesuvius . . . vomited forth its bowels, with darkness of night oppressing the day, covered all the face of Europe with minute dust. They celebrate the memory of this fearful ash at Byzantium on [6] November.[25]

In Portugal, that year was exceptionally cold.[26]

A distant volcano (or was it a meteorite crash?) apparently produced a chain of events in 536–7 AD. As with Tambora in 1816, years of cold, drought or narrow tree-rings are reported all over Eurasia. In Italy there were cold, drought, famine and a 'dry fog' like those attributed to volcanic air pollution in modern times. Procopius, the historian, witnessed it in North Africa or Sicily and wrote:

> A dreadful portent occurred. For the sun was without brightness of its rays, like the moon, during all this year, and it seemed very much like an eclipse; its rays were not clear like those it usually makes. And beginning from when this happened, people lacked neither war nor pestilence nor anything else bringing death. The time was when Justinian was in the tenth year of his reign.[27]

An anonymous contemporary writer, probably Zachariah, Bishop of Mitylene, left this account:

> The earth [at Constantinople] with all that is upon it quaked; and the sun began to be darkened by day and the moon by night, while the ocean was tumultuous with spray [?] from the 24th March in this year till the 24th June in the following year.
>   . . . and, as the winter [in Mesopotamia] was a severe one, so much so that from the large and unwonted quantity of snow the birds perished . . . there was distress . . . among men . . . from the evil things.[28]

Six centuries later, an annalist was still recounting this fearful eighteen months in which – he says – the sun shone faintly for only four hours a day and 'the wine tasted like what comes from sour grapes'.[29]

Although some of these details echo the *Book of Revelation*,[30] the event was not confined to Christendom. China had frosts and snow in July and August. Pines in northern Europe and oaks in Ireland, Britain and Germany put on remarkably narrow rings from 536 to 541. This is suspected to come from the eruption of some unknown volcano – possibly Rabaul in New Britain, radiocarbon-dated to 1410±90 BP = 450–630 AD.[31] The event coincided, at least approximately,

[19] B. D. Shaw, 'Climate, environment and history: the case of Roman North Africa', Wigley, Ingram, Farmer 1981, pp. 379–403.

[20] Pausanias V. viii. 1

[21] Orosius 4.11 (written some seven hundred years later).

[22] Garnsey 1988, chapter 9; P. D. A. Garnsey and C. R. Whittaker, eds., 'Trade and famine in classical antiquity', *Proceedings of the Cambridge Philosophical Society* suppl. 8, 1983.

[23] Theophrastus, *Historia Plantarum* IV.xi.3, VIII.vi.6 ff.; Aristotle, *Politics* 1330b 4–7;

Plutarch, *Solon* XIII.5; Sallares 1991, pp. 392 f.

[24] P. I. Kuniholm and 5 others, 'Anatolian tree rings and the absolute chronology of the eastern Mediterranean, 2220–718 BC', *Nature* 381, 1996, pp. 780–83; G.A. Zielinski and M. S. Germani, 'New ice-core evidence challenges the 1620s BC age for the Santorini (Minoan) eruption', *JAS* 25, 1998, pp. 279–89; Hardy and others 1990 *passim*.

[25] Marcellinus Comes, *Commentaries* 90 24, 472 AD.

[26] R. B. Stothers and M. R. Rampino, 'Volcanic eruptions in the Mediterranean before AD 630

from written and archaeological sources', *JGR* 88:B8, 1983, pp. 6357–71.

[27] Procopius of Caesarea, *Vandalic War* IV.xiv.5–6.

[28] F. J. Hamilton and E. W. Brooks, *The Syriac Chronicle known as that of Zachariah of Mitylene*, Methuen, London, 1899, pp. 267, 298 [editors' translation].

[29] *Chronique de Michel le Syrien*, trans. into French by J.-B. Chabot, Leroux, Paris, 1901, vol. II, p. 220.

[30] Book of Revelation 8.

with the sudden uplift of west Crete by up to 9 m (pp. 43–4). It preceded a great drought in Italy in 539 and the coming of plague in 542 (p. 78). With these Apocalypse-like disasters, all in the reign of Justinian, the great Byzantine Emperor (527–65), southern Europe declined into the Dark Ages.

CLIMATE IN THE EARLY TO MIDDLE HOLOCENE

Ice cores and deep-sea cores demonstrate that Quaternary glacials and interglacials fluctuated markedly over short periods. However, in northern and middle Europe the Atlantic Period, 6000–3000 BC, has been widely regarded as a long era of stable climate somewhat warmer and wetter than the present, in which the various trees completed their migrations to form the plant communities of the pre-Neolithic wildwood. Fluctuations of climate, if any, were not prolonged or intense enough to leave a clear mark on the pollen record.

Long-term changes in the inclination of the earth's axis in the early Holocene resulted in the heat received from the sun 10,000 to 8,000 years ago being 7 per cent greater in summer, and 7 per cent less in winter, in Mediterranean latitudes than now. One might expect this to have given frostier winters and hotter summers. However, if a stronger monsoon brought more rain to the interior of northern Africa in summer, it is possible that moist air reached as far as southernmost Europe.

In the Younger Dryas period (11,200–9600 bp, c. 10,400–8,700 BC), ice-caps were melting: fresh water poured into the North Atlantic and altered its pattern of circulation. The melt-water, being light, stayed on the surface, halting the 'Atlantic Conveyor' which normally takes cold, heavy, salty water to the bottom of the ocean. The result was widespread cooling around the North Atlantic; icebergs and sea-ice were plentiful as far south as Portugal. The western Mediterranean ought to have been cooler, and SW Europe ought to have been swept by cold north-westerly winds for much of each winter and spring.[32]

Fig. 9.1. Lime (*Tilia rubra*) and wild-olive growing together in the Acheron gorge, Epirus, as they did in the mid-Holocene in Crete, 500 km south. *May 1988*

The Holocene in southern Europe

The pollen record is dealt with in detail in Chapter 10. For the first half of the Holocene, what is now thought of as 'typical' Mediterranean vegetation, dominated by evergreens, was decidedly local. Pollen deposits were dominated by deciduous trees. Some of these, such as deciduous oaks, extend far into the Mediterranean today. Others are 'north European' trees – alder, ash, birch, elm, hazel, hornbeam (*Carpinus*) and lime – which regularly occur in pollen diagrams well to the south of their present southern limits. These trees were then normal constituents of ordinary lowland landscapes in southern Europe; today they occur only rarely and in special habitats.[33]

For example, there are two prehistoric pollen sites in Crete. Ayia Galíni – a hot, dry area near the south coast – had

deciduous oak, hazel, alder, elm and lime. Tersaná, on the NW coast, also now a dry site, had oaks, lime, hazel and *Ostrya*.[34] Hazel, alder and lime are now extinct in Crete and rare in most of Greece. The southern limits of vegetation have, in effect, moved at least 500 km northwards since the early Holocene (Fig. 9.1).

Most of the evidence is from Greece (Table 10.i), but there are western examples too. Neolithic sites near Xàbia, between Alicante and Valencia, contain charcoal of deciduous and evergreen oak and ash, indicating a difference from this arid corner of Spain today.[35]

Why did these trees become extinct? This can be inferred from the nearest places where they just survive. The nearest lime to Crete is a group of ancient coppiced and pollarded

[31] R. B. Stothers, 'Mystery cloud of AD 536', *Nature* 307, 1984, pp. 344–5; M. G. L. Baillie , 'Dendrochronology raises questions about the nature of the AD 536 dust-veil event', *The Holocene* 4, 1994, pp. 212–7.

[32] W. S. Broecker and 6 others, 'Routeing of meltwater from the Laurentide ice sheet during the Younger Dryas cold episode', *Nature* 341, 1989, pp. 318–21.

[33] Any possibility that long-distance transport of pollen may then have been more active than now is disproved by the nature of the pollens. Lime, whose pollen is sparse and does not travel far, appears regularly in these diagrams. Pine, whose copious pollen blows for long distances, is rare or absent at many sites.

[34] Bottema 1980; Moody & others 1996.

[35] Work of Badal Garcia and others, summarized by Vernet 1997, pp. 129–40.

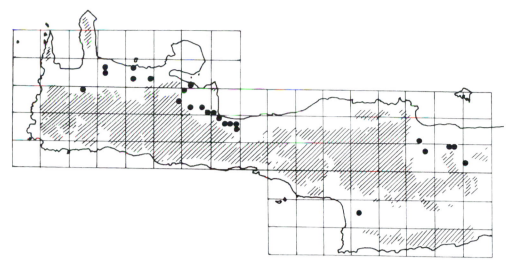

Fig. 9.2. Present distribution of elm (*Ulmus minor*) in Crete, by 2-km squares. Hatched areas are over 500 m.

trees on a north-facing cliff high in the Langádha pass of the Taygetos mountains west of Sparta; it is accompanied by *Ostrya*, ash and a Central European wild pear. Lime also survives on north-facing cliffs high on Mount Hélikon in middle Greece.[36] The nearest alder to Crete is a cluster of trees by the River Eurotas near Sparta; this is a permanent river and very cold in summer. Elm is rare in Crete, but consistently present on the wet north sides of the mountains (Fig. 9.2). All these sites are the coldest or wettest parts of their respective landscapes. A secondary factor is that some of them are inaccessible to animals but not to people.

The disappearance of north European trees is thus explained by a change to a hotter or drier climate. Explanations in terms of human activity, however, need to be considered. Woodcutting could not exterminate them, for they all tolerate it well and are extensively coppiced further north. Nor could burning, for they are not combustible. If they had been grubbed out because they grew on cultivable land, they would survive as field trees, in ravines, in ancient woodland (p. 185) and on rugged ground among cultivation, as deciduous oaks do. This is not their pattern of survival.

What about browsing? These trees are palatable, and browsing is partly responsible for the decline of lime in England.[37] However, if it had exterminated them in Mediterranean Europe, they should survive on cliffs in general (like, for example, *Pistacia terebinthus*). In reality, where they survive on cliffs, they survive on *specially* cold cliffs.

The vegetation, therefore, unequivocally implies a long period of less arid climate. To judge by the plants' behaviour, it was wetter and less seasonal than now.[38] The difference increases south-eastward. Deciduous oaks and lime then flourished – to take an extreme example – at Ayia Galíni, where there is now not even evergreen oak, resistant though it is to woodcutting, burning and browsing.

When was the change to something like the present climate? Pollen evidence indicates a gradual change – the *Aridization* – over the Neolithic and Bronze Age; in Crete

lime and hazel diminish over a long period, finally disappearing from the pollen record after 2500 BC. This is confirmed by a tendency for lake levels in Spain and Portugal, and also in Greece, to have fallen between 6000 and 5000 bp = 5000–4000 BC.[39]

Ludwig Hempel has brought together the results of nine botanical studies (excluding Tersaná) and nine studies of other types (e.g. sedimentation, animal remains) from Crete, Greece and Croatia, all of which indicate a date for the onset of a 'Younger Aridization Phase'. The range of dates is from 4800 to 2400 BC, with a concentration around 3400 BC. Hempel has scantier, non-botanical evidence for an 'Older Aridization Phase' around 6000 BC.[40]

In Santoríni the change was apparently still incomplete by the seventeenth century BC (p. 321). A similar change is reported from a study of olive cultivation in SE Spain.[41] Analyses of the carbon-13 content of carbonized prehistoric cereals from NE and south Spain – which is taken to indicate the rainfall in spring, when the grains are developing – have indicated increasing aridity between 4000 and 1000 BC.[42]

Aridization affected the east and west Mediterranean more than the middle. We, like the ancient Romans, are impressed by the difference between verdant Italy and dry Spain and Greece. Northern trees are still abundant in Italy; yew, alder and holly survive in Sardinia and even Sicily.

Claudio Vita-Finzi made a pioneer attempt to explain phases of erosion, alluviation and the downcutting of rivers in terms of changing climates (Chapter 16). In 1992 an international conference at Cambridge brought this area of study up to date. There is some consistency in the results. A wet period is inferred to have affected the whole eastern Mediterranean down to 5000 bp; a roughly contemporary wet period laid down the principal Holocene valley fills of the southern French Alps (and in NW Africa). However, in Provence erosional features of 8000 bp are interpreted as being 'related to a relatively dry Mediterranean ecosystem'.[43]

[36] Rackham 1982.
[37] Rackham 1986, 1990a.
[38] Rackham & Moody 1996.
[39] Summarized in S. P. Harrison and G. Digerfeldt, 'European lakes as palaeohydrological and palaeoclimatic indicators', *QSR* 12, 1993, pp. 233–48. Most of the Greek examples are karst

lakes, whereas the Iberian ones have surface outlets.
[40] Hempel 1990.
[41] Terral & Arnold-Simard 1996.
[42] J. L. Arans and 7 others, 'Changes in carbon isotope discrimination in grain cereals from different regions of the western Mediterranean basin during the past seven millennia. Palaeo-

environmental evidence of a differential change in aridity during the late Holocene', *Global Change Biology* 3, 1997, pp. 107–18.
[43] Lewin & others 1995 (especially the articles by Collier & others, Ballais and Provansal); J. Vaudour, 'Évolution Holocène des travertins de vallée dans le Midi méditerranéen français', *Géographie Physique et Quaternaire* 48, 1994, pp. 315–26.

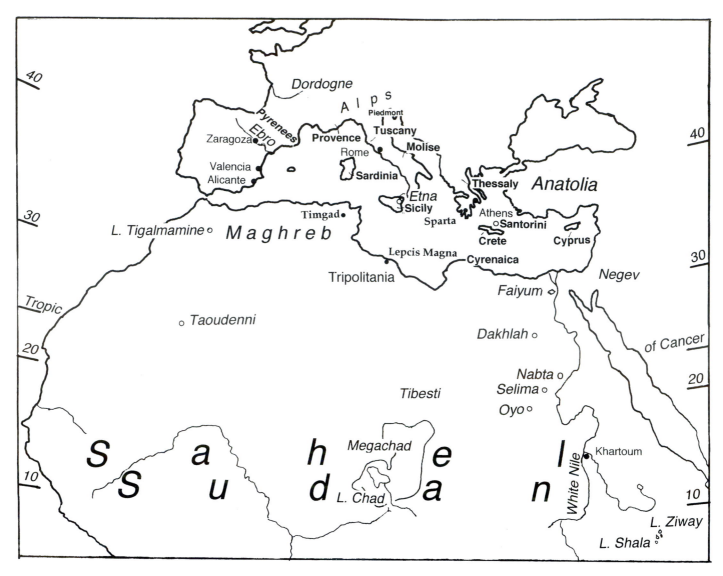

Fig. 9.3. African lakes with a Holocene record.

*The southern parallel: Holocene climates of
north and middle Africa*

Hempel's Aridization of 3400 BC corresponds with the date for aridization in Egypt. Further evidence comes from fluctuations in the shore-lines of lakes south of the Sahara (and one in the mountains of Morocco) which have no outlet, whose level depends on the balance between input (from rainfall, runoff and ground-water) and evaporation (Fig. 9.3).

Around 12,000 bp, about the time of the Allerød warming in NW Europe, Lake Chad was fifteen times as extensive as in the twentieth century. Further east, a long narrow lake formed in the valley of the White Nile south of Khartoum. This wet phase on the south side of the Sahara was followed by a dry period contemporary with the Younger Dryas. Chad and other tropical lakes shrank between 11,200 and 10,700 bp and again between 10,300 and 9800. Then levels of lakes in inter-tropical Africa rose, and the lakelet at Tigalmamine in Morocco was created. Further changes are listed in Table 9.i.

At Tigalmamine the lake fluctuated, but the pollen record – so far as it is published – does not show any clear linkage between climate and the surrounding oak and cedar vegetation.

Climatic modelling is here of some value. It indicates that, in the early and middle Holocene, the sun's greater radiation to the northern hemisphere and the greater heating of the Saharan region would have increased south-westerly monsoonal inflows of moist air to West Africa and northern India.[44] Such warm incursions – interrupted by the cooling of the Younger Dryas – evidently brought rain to the Sahara. They could have touched Morocco and southern Europe, fuelling thunderstorms to relieve late summer aridity, and giving more cloud to temper the effects of the hotter sun.

44 J. E. Kutzbach and F. A. Street-Perrott, 'Milankovitch forcing of fluctuations in the level of tropical lakes from 18 to 0 kyr bp', *Nature* 317, 1985, pp. 130–4.

Table 9.i. Wet and dry periods in and around the Sahara.
[A. T. Grove, 'Africa's climate in the Holocene', T. Shaw and others, eds., *The Archaeology of Africa*, Routledge, London, 1993, pp. 32–42; H. F. Lamb and 7 others, 'Relation between century-scale Holocene arid intervals in tropical and temperate zones', *Nature* 373, 1995, pp. 134–37; Z. Garba, A. Durand, J. Lang, 'Enregistrement sédimentaire des variations de la dynamique éolienne pendant la transition Tardiglaciaire/Holocene à la limite Sahara/Sahel (Termit, Bassin du lac Tchad)' , *ZG*, NF Suppl. 10, 1996, pp. 159–78]

Date bp	Date BC	Locality	Observation	Remarks
2350–1700	400 BC–380 AD	Tigalmamine (Morocco)	Small shallow lake	Arid interval
3050–2650	1300–800	Tigalmamine	Shallow lake	Arid interval
c. 3500	c. 1820	Chad basin, NW Africa		Last wettish phase
8300–3840	7350–2300	Taoudenni, mid Sahara	At least four sequences of flooding and regression of lakes	Wetter phases from time to time in Sahara and Maghreb
4550–4050	3300–2540	Tigalmamine	Shallow lake	Arid interval
7200–5900	5900–4800	Ziway–Shala (Ethiopia)	Deep lakes	Wet phase
> 7400–6250	> 6200–5200	Tigalmamine	Shallow saline lake	Arid interval
8000–7000 (–6100)	6850–5840 (–5000)	Oyo and Selima (Egypt)	Deep lakes	Wet phase
10,550–9200	10,530–8280	Tigalmamine	Shallow but variable lake	Interrupted arid period
11,200–9600	12,400–10,700	Chad and other tropical lakes	Shrunken lakes	Dry phase
12,000–11,200	13,400–12,400	Chad and other tropical lakes	Very extensive lakes	Wet phase

Date bp	Date AD/BC	Name of advance	Data from	Remarks
c. 950 c. 1300 c. 1650	1070 AD 700 AD 410 AD	Göschenen II	W. Alps	Großer Aletsch and Lower Grindelwald glaciers
2820–2280	980–370 BC[1]	Göschenen I	E. and W. Alps	Fernauferner (Tyrol) and Lower Grindelwald (Bernese Oberland)
3600–3000	1930–1250 BC	Löbben	Many glaciers throughout Alps	Sometimes more extensive than Little Ice Age
4600–4200	3350–2700 BC	Rotmoos	E. Alps	Advances of several glaciers comparable with Little Ice Age
6600–6000	5700–5250 BC	Frosnitz	Throughout Alps	Oberaar glacier reached its maximum Holocene extent; advance in Pyrenees more extensive and prolonged
7600–7400	6400–6200 BC	*unnamed*	Alps and elsewhere	Coincides with dry, dusty, cold period in Greenland ice core
8700–8000	7700–6900 BC	Venediger	E. and W. Alps	Double advance, reaching up to 400 metres beyond Little Ice Age extent
c. 9300	c. 8400 BC	Schlaten	E. Alps	Poorly dated from wood

Table 9.ii. Advance of glaciers in the Alps before the Little Ice Age.
[1] Calibration problems arise for 14C dates between 2300 and 2900 bp.

[M. Gamper and J. Suter, 'Postglaziale Klimageschichte der Schweizer Alpen', *Geographica Helvetica* 37, 1982, pp. 105–14; A. F. Gellatly, J. M. Grove and V. R. Switsur, 'Mid-Holocene glacial activity in the Pyrenees', *The Holocene* 2, 1992, pp. 266–70; Grove 1988; H. Holzhauser and H. J. Zumbühl, 'Methoden zur Rekonstruktion von Gletscherschwankungen – Großer Aletschgletscher', *Die Alpen* 64(3), 1988, entire issue; Holzhauser & Zumbühl 1996; F. Mayr, 'Postglacial glacial fluctuations and correlative phenomena in the Stubai Mountains, eastern Alps, Tyrol', *Glaciation of the Alps* [INQUA 1965], ed. G. M. Richmond, University of Colorado Series in Earth Science 7, 1968, pp. 143–65; F. Mayr, 'Die postglazialen Gletscherschwankungen des Mont Blanc-Gebietes', *ZG* Suppl.8, 1969, pp. 31–57; G. Patzelt, 'Die neuzeitlichen Gletscherschwankungen in der Venedigergruppe (Hohe Tauern, Ostalpen)', *Zeitschrift für Gletscherkunde und Glazialgeologie* 9, 1973, pp. 5–57; G. Patzelt, 'Holocene variations of glaciers in the Alps', *Les méthodes quantitatives d'étude des variations du climat au cours du Pleistocene*, 1974, pp. 51–59, Colloques Internationaux du Centre de la Recherche Scientifique 219; Röthlisberger 1986.]

The northern parallel: glaciers in the Alps

In Holocene prehistory there have been at least eight periods of advance and retreat among Alpine glaciers (Table 9.ii). These were comparable to those of historic times: sometimes less extensive, sometimes rather more, than those in the Little Ice Age. As with the Little Ice Age, the advances were probably caused by mean temperatures no more than 1°C cooler, combined with greater snowfall.

Glacial advances thus came roughly every thousand years, but it would be misleading to read a regular cycle into them. The Frosnitz advance, if correctly carbon-dated, lasted into the Atlantic period, when glaciers might be expected to have retreated high up their valleys. Indeed, the glacier record does not correspond closely with the pollen sequence for NW Europe or even for the Alps themselves. The changes that produced ice advances were evidently not severe or prolonged enough to have much effect on the vegetation at

Table 9.iii. Summary of Holocene chronology in the Mediterranean and other relevant areas.

Calendar date	Radio-carbon date bp	Historical or archaeological period	Periods in NW Europe	Glacier advances in the Alps	African lake levels	Deluges and non-badland alluviations (Chapter 16)
AD 1800–1998	—		Little Ice Age IV	1970s 1920s 1885–95 1850–60 1815–20	low since 1970; high in 1960s; low in 1910s; high in 1870s	Provence, Molise, Tuscany
1600–1800	—		Little Ice Age III c.1695; sunspot minimum	1780		Provence, Tuscany, Crete
1400–1600	480–250	Late Medieval, early Modern	Little Ice Age II c.1595; earlier sunspot minimum	1580–1650		Provence, SE Spain
1200–1400	830–480	High Medieval	Little Ice Age I c.1320; sunspot minimum	1250–1350		Crete, Tuscany
1000–1200	1020–830		Medieval Warm Period			Tuscany, SE Spain
800–1000	1160–1020	Byzantine		800–900 AD Göschenen II; sunspot minimum		
600–800	1430–1160	Arab invasions				Dordogne Provence, Crete
400–600	1630–1430	Christian Roman	Sub-atlantic		Tigalmamine small and shallow	Provence
200–400	1780–1630	Imperial Roman				
1–200	2010–1780					
BC 200–1	2140–2010	Republican Roman			Modern climates established in Sahara and Sahel	Molise; Bœotia
400–200	2320–2140	Hellenistic				Argolid; Italy; Spain
600–400	2480–2320	Classical Greek				Sicily; Argolid
800–600	2660–2480			Göschenen I	NW Sahara very dry	Eubœa
1000–800	2800–2660					
1200–1000	2980–2800	LATE			Tigalmamine a shallow lake	
1400–1200	3100–2980					
1600–1400	3290–3100					
1800–1600	3440–3290	BRONZE		Löbben		Cyprus
2000–1800	3610–3440				Lake Chad high	Argolid; Crete
2200–2000	3760–3610					Argolid
2400–2200	3890–3760	EARLY		Intermittent warm period 4000–3600 bp		
2600–2400	4030–3890		Sub-boreal			
2800–2600	4180–4030	end of Aridization				
		LATE			Tigalmamine	
3000–2800	4280–4180	NEOLITHIC			small and shallow	
3200–3000	4480–4280				DRYING	

Calendar date	Radio-carbon date bp	Historical or archaeological period	Periods in NW Europe	Glacier advances in the Alps	African lake levels	Deluges and non-badland alluviations (Chapter 16)
3400–3200	4630–4480		Sub-boreal	Rotmoos		Provence
3600–3400	4830–4630				Drying of Oyo and Nabta and lakes in Ethiopia and Taoudenni	
		NEOLITHIC				
3800–3600	5040–4830			Intermittent warm period 6000–4600 bp	lake levels falling	
4000–3800	5200–5040					
4200–4000	5470–5200					
4400–4200	5570–5470					
4600–4400	5750–5570	EARLY				
4800–4600	5870–5750	start of Aridization			Lake levels high in central and eastern Sahara and Ethiopia	
5000–4800	6020–5870		Atlantic			
5200–5000	6190–6020			Frosnitz		
5400–5200	6570–6190					Cyprus
5600–5400	6820–6570	sea-level reaches present			arid interval: Nabta playa in SW Egypt dry; Tigalmamine shallow, saline; E Saharan lakes falling	
5800–5600	7030–6820					
6000–5800	7200–7030	Older Aridization?				
6200–6000	7400–7200					
6400–6200	7600–7400			unnamed ice advance		
6600–6400	7800–7600				deep lakes in SW Egypt and NW Sudan Rep.	
6800–6600	7930–7800					
7000–6800	8070–7930					
7200–7000	8200–8070		Boreal		Tigalmamine Sahara, Sahel (MegaChad) and Ethiopia	Dordogne S. France
7400–7200	8360–8200			Venediger		
7600–7400	8600–8360					
7800–7600	8800–8600					
8000–7800	9000–8800	MESOLITHIC			Lakes begin to rise in E Sahara	
8200–8000	9200–9000					
8400–8200	9400–9200			Schlaten		
8600–8400	9530–9400					Portugal
8800–8600	9670–9530					
9000–8800	9800–9670		Pre- Boreal		Tigalmamine shallow but variable	
9500–9000	10,000–9800					
10,000–9,500	10,200–10,150					
10,500–10,000	10,500–10,200		Younger Dryas cold	Younger Dryas cold	Saharan lakes dry	
11,000–10,500	11,000–10,500					
11,500–12,000	11,500–12,000		Allerød warm	Allerød warm	Saharan lakes high	
12,000–12,500	12,000–12,300					
12,500–13,000	12,300–12,700				'MegaChad'	

low altitudes. As in the Little Ice Age, individual trees could live through several advances and retreats of the glacier tongues beneath them. However, the upper tree-limits are known to have shifted up and down a hundred metres or more.

During Göschenen II people were beginning to write down sunspots observed with the naked eye. The records go far enough back to reveal a minimum in the eleventh century (the Oort Minimum) and another in the seventh and early eighth centuries, which could be related to the last two phases of glacier advance.

The eastern parallel: Holocene climate in Israel

A proxy record comes from G. A. Goodfriend's studies of carbon and oxygen isotopes in the subfossil shells of land snails in the Negev desert.[45] The ratio of C_{13} to C_{12} is held to be determined by whether the snails were eating C-3 or C-4 species of plants, which is a measure of the degree of aridity. Results indicate that between 6500 and 3000 bp the mean annual rainfall was at least 300 mm, instead of 150 mm as it is now. The findings are supported by the oxygen-18 values of carbonates in the shells, some 2 per cent less in those dating from 6500–6000 bp than in modern snails; from this, by a lengthy argument, Goodfriend infers a change in the atmospheric circulation, with more frequent storms reaching the region from NE Africa. By 3500 bp (c. 2000 BC) oxygen-18 values had reached modern levels, indicating that a relatively stable pattern of atmospheric circulation has existed since that time.

CONCLUSIONS

'Typical Mediterranean vegetation', and the summer-arid climate to which it is supposed to be adapted, have been typical for only about five thousand years since the Aridization. This is roughly the period over which El Niño (which is related to the North Atlantic Oscillation, pp. 127–8) seems to have been a feature of global circulation.[46] The early and middle Holocene had a less arid climate.

The evidence is clearest from vegetation, although even here there are gaps in the story: it has seldom been observed exactly what happens when plants die out through climatic change. Much of the evidence of river behaviour, of snails in Israel, and – in the most general terms – of African lakes supports this conclusion. Estimates of the date of the change vary by a thousand years or more. This may be an effect of the different proxies employed: trees (especially the long-lived lime) could easily have lived on and shed pollen for centuries after a change of climate made it impossible to reproduce their kind.

The written evidence, for what it is worth, suggests that in the Classical to Roman period the climate was a little cooler and less arid than now, and bigger areas had a double rainy season.

Evidence for shorter-term fluctuations is less consistent. This may be partly because of real variations between regions within the Mediterranean, and partly because different proxies respond to different aspects of climate and over different time-scales. As the Little Ice Age demonstrates, changes in Alpine glaciers may correspond to conspicuous short-term events in the Mediterranean, but without affecting the pollen record. As modern deluges demonstrate, a single great storm may move more sediment than centuries of slightly increased average rainfall.

Glaciers demonstrate, rather than long-term changes, many short-term changes of the Little Ice Age kind; they are roughly backed up by the details of lake levels in Africa. By analogy with the Little Ice Age, we would expect each ice advance in the Alps to have been associated with unstable, stormy, erosive weather in southern Europe, rather than with a major change of climate.

[45] G. A. Goodfriend, 'Rainfall in the Negev desert during the Middle Holocene, based on 13C of organic matter in land snail shells', QR 34, 1990, pp. 186–97; 'Holocene trends in 18O in land snails from the Negev desert and their implications for changes in rainfall source areas', QR 35, 1991, pp. 417–26.

[46] J. Shulmeister and B. G. Lees, 'Pollen evidence from tropical Australia for the onset of an ENSO-dominated climate at c. 4000 BP', The Holocene 5, 1995, pp. 10–18.

CHAPTER 10

Vegetation in Prehistory

To anyone who is familiar with the present-day landscape around the
Mediterranean Sea, a vision of almost unbroken forest from the
water's edge up to the crests of all but the highest mountains may
seem too fantastic for credence.[1]

SCOPE AND SOURCES OF MEDITERRANEAN VEGETATION HISTORY

Vegetation history holds one of the keys to the desertification
question. We need to know when there was a golden age
of magnificent forests throughout the Mediterranean – if it
ever existed outside the imaginations of Ruined Landscape
scholars – and when and how it came to an end. We ask:

(1)   What is the origin of the present Mediterranean-type
vegetation?

(2)   Was there ever the idealized aboriginal vegetation from
which degradationist theory begins?

(3)   If it existed, what exactly did it consist of?

(4)   At what date did it cease to exist?

(5)   Which changes in vegetation have been due to human
activity, and which to natural changes in climate and soil?

(6)   Were these changes progressive or reversible?

(7)   Which landscapes, now considered as degraded, are in
fact so, and which have never been any 'better'?

(8)   Are the changes of the last fifty years unique, or are they
a repetition of something that has happened before?

   This chapter and the next are largely and inevitably
concerned with trees and tree-land (forest, savanna, maquis).
Most trees produce more abundant pollen than most other
plants; timber and wood have some chance of leaving an
archaeological record; trees appear more often than other
vegetation in most types of historical record.

Pollen analysis

Pollen preserved in lakes and bogs is the main source of information about vegetation before the Neolithic period; it
continues, though sites become fewer, to the present. Many
trees and other plants produce great quantities of pollen
grains which, to some extent, are identifiable under the
microscope. Pollen grains are exceedingly durable if kept
permanently wet. The investigator looks for a wetland with a
stratified deposit, in which each year's pollen fallout is incorporated in a layer of peat or mud built up on top of the previous year's. A core is taken, the sediment sampled centimetre
by centimetre, and the pollen extracted, identified and
counted (Fig. 10.1). Samples containing plenty of organic
matter are set aside for radiocarbon dating.

Mediterranean vegetation history will probably never be
so exact a science as north of the Alps. Palynology encounters
greater difficulties. Owing to the annual dry season and lack
of acid soils, bogs and lakes that preserve pollen are rare. The
pollen is often in poor condition. Wetlands have been so
persecuted in the last hundred years that only the smallest
pockets of fen or bog still retain their top layers. There is often
a lack of datable organic material.

   Pollen sites (Fig. 10.2) are far too few and too unevenly
scattered to represent the complexities of Mediterranean
vegetation. They tend to be in mountains and on the fringes
of the Mediterranean area, for example middle and north
Spain,[2] Corsica and north Greece. The amount of evidence
for the pre-Neolithic Holocene for all Mediterranean Europe
is no greater than for the one English county of Norfolk,
yet the wildwood of Norfolk is by no means exhaustively
understood.[3]

   Pollen deposits, occurring as they do in wetlands in an
otherwise semi-arid environment, tend to portray vegetation
as less arid than it was in reality.

   Pollen preserved in undersea deposits is more difficult to
interpret. It will have come from a vast, undefined area,
including anything from the coast to high mountains; salt-
marsh plants are likely to be over-represented. The pollen
may not be contemporary with the deposit, but washed out of
earlier deposits by river erosion. Nevertheless, treated with
caution, submarine pollen deposits eke out the sparse material from on land.

   Palynology depends very largely on wind-pollinated trees
and other plants that shed abundant pollen. In the Mediterranean, whole categories of vegetation are insect-pollinated,
using their pollen more 'efficiently' and leaving very little to
provide a pollen record. These include nearly all undershrubs
(Cistus, Labiatae, etc.) and most endemics. Trees and shrubs
can produce identical pollens: in Greece there is no telling
whether an evergreen oak pollen grain came from a tree 20 m
high or a shrub 0.6 m high.

   There are also general problems of pollen analysis. Related
species may produce indistinguishable pollens. Of the
twenty-odd species of oak in the Mediterranean, palyn-
ologists can identify at best three types: 'evergreen', comprising Quercus ilex, coccifera and rotundifolia; 'intermediate',
comprising semi-evergreen oaks such as cork-oak (Q. suber),

[1]   S. R. Eyre, 1963; the vision is indeed incredi-
ble to us, but was not to Eyre.

[2]   P. López García, 'Resultados polínicos del

Holoceno en la Península Ibérica', Trabajos de
Prehistoria 35, 1978, pp. 9–44.

[3]   O. Rackham, 'The ancient woods of Norfolk

[England]', Transactions of Norfolk & Norwich
Naturalists' Society 27, 1986, pp. 161–77.

Fig. 10.1a, b. Top: coring a peat-bog at Así Goniá, west Crete. *July 1985* Above: The core in an aluminium tube, opened out and about to be sampled by our colleagues. *July 1985*

Q. cerris, macrolepis and *trojana;* and 'deciduous', comprising the remainder. Several hundred grass species reduce to a handful of pollen types – 'cereal', 'rye', 'reed' and 'other grasses'. Pollen of certain species may be blown long distances and may not represent the local presence of the tree:

thus stray grains of cedar (*Cedrus*) pollen, from NW Africa, have been found in Spain, Corsica and France.

Another limitation is artificial. A pollen diagram is a large, unwieldy piece of paper. Palynologists are tempted to simplify it, including only those aspects that seem to be relevant to the question they happen to be studying. The rest of the information is too often left unpublished. Someone else, interested in (let us say) *Tilia* pollen – an uncommon and very significant type – may never know whether its absence from the diagram is because it did not occur or because it was a 'minor' type not thought worth publishing. This is disastrous when pollen deposits are wasting and the opportunity to sample them may not recur.[4]

Scholars, confronted with an incomplete record, tend to interpret ambiguities – however unconsciously – in favour of a preconceived theory. To a north European or North American, it comes naturally to interpret tree pollen as implying forest. In Mediterranean countries there are two other possibilities. Tree pollen may come from trees reduced to the state of shrubs and forming maquis. Or it may come from savanna trees scattered in some other, often insect-pollinated, vegetation. Only if some discriminating characteristic can be found can forest be properly inferred.

Important and little-known matters are the actions of fire (Chapter 13) and of browsing wild animals, and the balance between them. The fact that many sites now 'revert' to forest if protected from both fire and browsing cannot be taken as evidence of past vegetation.

4 Electronic 'publication' is too impermanent to
be an adequate record.

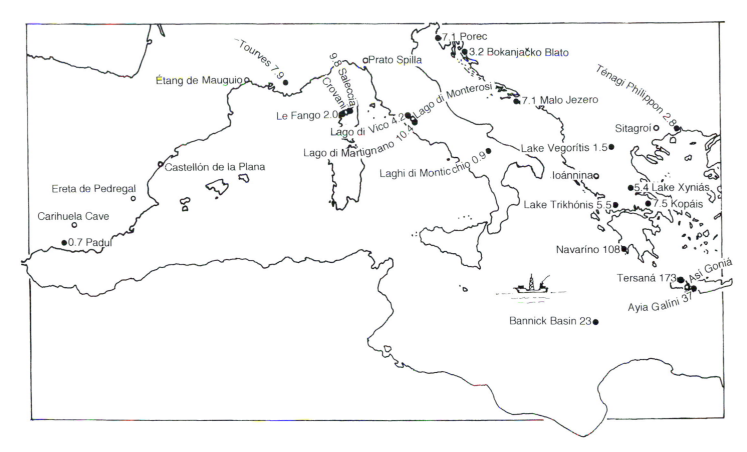

Fig. 10.2. Some pollen sites. Black spots are sites used in reconstructing wildwood in Table 10.i; numbers are values of the Non-Forest Index for each site.

Interpreting the pollen record

Pollen reaching a wetland site may come from various plant communities, including forest or maquis; phrygana (under-shrubby vegetation); steppe (non-shade-bearing herbs); and the fen vegetation of the wetland itself. Some pollen types can be recognised as coming from one or other. *Cistus* or *Helianthemum* pollen can only come from phrygana; *Artemisia* (wind-pollinated) or *Asphodelus* (insect-pollinated) must mean steppe; *Sparganium* pollen comes from wetland. Other types are ambiguous. Compositae Tubuliflorae (thistles, etc.) may come from steppe plants, or from marsh thistles on the wetland itself. The biggest uncertainty concerns non-distinctive grasses. Grass pollen may come from steppe, but also from reed (*Phragmites*) and the many other grasses that are abundant in fenland. We therefore leave grass pollen out of our discussion.[5]

The tree to non-tree pollen ratio, often used by palyn-ologists to estimate the extent of forest, is of little use in Mediterranean countries. A high proportion of non-tree pollen may mean that the landscape as a whole was not forested; but it may also mean that the coring site happened to be surrounded by a grass-dominated wetland, the rest of the landscape being forest.

Plant geography

Much can be inferred from the ecological characteristics of endemic plants and those with disjunct distributions (those that are native in widely separated countries). Endemics, by definition, lack long-distance dispersal, and cannot migrate in the face of climatic change; island endemics have nowhere to migrate to. Mostly they are long-lived perennials and cannot evolve in the face of rapid climatic change. Hence they tell us about the possible limits of the vicissitudes to which they have been subjected by climate and – in this interglacial – by human activity.

One does not, on the whole, find endemics in forests. They are plants of open ground – cliffs, high mountains, phrygana and savanna. For instance, of the 250 or so endemics of Crete and neighbouring islands, only twenty-one normally grow in shade, most of these in the shade of rocks; only eight can be classed as woodland plants, of which two are limited to woods. This is to be expected in a land where forests have been abundant only during relatively brief periods in interglacials. Even these periods could not have been wholly hostile to light-demanding endemics; this casts doubt on any theory that the normal state of wildwood was trees upon trees upon trees, from high mountains to the seashore. On the contrary, there were abundant and permanent gaps in the forest where endemics could persist.

---

5 *Phytoliths* (silica bodies of distinctive shapes, formed between plant cells, which persist when the plant decays) offer a hope of distinguishing between grasses.

Plants with disjunct distributions (native in widely separated countries) tell a similar story. The well-developed desert flora of SE Spain, like the different flora of the desert of SE Crete, show that areas too dry for forest persisted through both the early Holocene and previous interglacials.[6]

Archaeology

Excavated material sometimes indicates vegetation. Seeds can be preserved in a charred form, or in cesspits, or as impressions on the undersides of pots. Buildings and equipment – mills, storage jars, etc. – reveal much about cultivation and land-use, less about wild vegetation. Building construction may reveal the sizes of trees available. For example, Sir Arthur Evans found traces of great timbers in the Cretan palace of Knossós, from which it has (mistakenly) been inferred that cypress trees grew bigger in Bronze Age than in modern Crete. Plant remains, preserved in waterlogged deposits, have seldom been studied in Mediterranean archaeology.

In archaeological survey the object is to search the landscape for traces, especially scatters of potsherds in ploughed fields, from which the past pattern of settlement and land use can be reconstructed. This can be an excellent guide to the intensity and character of past human activity. However, it is very laborious, and is typically applied to areas no larger than a few tens of square kilometres. The distribution of conspicuous sites – cities, palaces, villas – may be a poor guide to the total extent of activity.

Archaeological survey is not free from bias. Sites can be removed by erosion or buried by sedimentation (Chapter 16). Sites are difficult to find in areas that are now grassland, and especially in woodland: people choosing areas to survey tend to avoid such terrain. The archaeologist must be on guard against the preconception that areas now wooded have always been woodland. This may be untrue, and yet surveys may fail to dispel it.

Charcoal

The study of charcoal, especially from archaeological sites, complements pollen analysis, although less has been done. Charcoals can be identified to genus and sometimes to species. As with pollen, there is a long chain of circumstances between the plant producing charcoal and the scientist identifying it. For example, people decide to cut one tree and not another; some trees grow again and produce a further crop; some woods carbonize more easily than others, and thus are more likely to be preserved as charcoal; some charcoals are easily shattered into unidentifiable particles. Charcoals are often poorly published: archaeologists who draw, photograph and store thousands of pots never think to tell posterity how they identified their charcoal, or to preserve the material for future verification.

Nevertheless anthracology has much to commend it. There are many more charcoal than pollen deposits. They do not record only the wettest spots in the landscape. Charcoals sieved out from an archaeological deposit are a sample of the firewood that people were using, for which selection, though not negligible, was less rigorous than for timber. A school of anthracology at Montpellier has developed methods of sampling and evaluating this material.[7] Charcoal shares some difficulties with pollen: herbaceous plants are poorly represented, and there is usually no means of distinguishing between trees in the form of forest, maquis and savanna.

Archaeological charcoal can also provide evidence about rates of growth and woodland management.[8] In the wider landscape it tells us about charcoal manufacture and forest or heath fires.

The present landscape

In interpreting any vegetation record one must ask how the ancient vegetation differed from that *of the same area* today – or, rather, of the same area 'yesterday' (as Anthony Snodgrass calls it), before twentieth-century changes.

North Europeans and North Americans must allow for the different ecological behaviour of many genera of plants in the Mediterranean. Scholars trained in the forestry traditions of the last two hundred and fifty years believe that Mediterranean forests of the past, unlike many of those today, were of tall trees. This is, for example, often asserted by J. R. McNeill.[9] It is probably why the Council of the European Communities included only 'forests of tall trees' in its list of 'natural habitat types' requiring 'special areas of conservation'[10] – excluding forests (however remarkable) of short trees, such as the ancient high-altitude cypresses of Crete (Fig. 4.22). In reality, there is no evidence whatever in the pollen record for heights or shapes of trees; there is seldom evidence from any source except on rare occasions when timbers, or traces of timbers, survive in ancient buildings.

It is important to study places such as cliffs where vegetation has escaped some of the effects of browsing, woodcutting and burning. The theory that these influences have caused deforestation can be tested. If it is true, then trees (and other sensitive plants) will survive on cliffs – as they do, for example, in southern Greece. Where there is no trace of trees even on cliffs, as in the very dry areas of SE Crete and SE Spain, we infer that the present climate is too dry for trees.

ORIGINS AND SURVIVAL OF THE MEDITERRANEAN FLORA

Some Mediterranean plants, especially annuals, are well adapted to the climate, growing in winter and shrivelling away in summer. Many perennials, however, struggle to remain active during the harsh summer. Deciduous trees, especially, are dormant in what seems to be the favourable time of year and active in the unfavourable time. This is a reminder that the Mediterranean climate has existed for far too short a time for plants, especially slow-reproducing trees, to become fully adapted to it by Darwinian evolution (p. 45).

6 M. Costa Tenorio, M. García Antón, C. Morla Juaristi, H. Sainz Ollero, 'La evolución de los bosques de la Península Ibérica: una inter-pretación basada en datos paleobiogeográficos', *Ecología* fuera serie 1, 1990, pp. 31–58.
7 See Chabal 1997 (for methods) and Vernet 1997 (for results).
8 Rackham 1972, 1986b.
9 McNeill 1992.
10 Council Directive 92/43/EEC, Annex 1.

Fig. 10.3. *Phœnix theophrasti*, an endemic palm. *Phœnix* in *Séllia, S coast of Crete (a locality already so named in Classical times), April 1982*

Many plants are still bound by their ancestry in a climate with rainy summers and cold winters.

Accounts of the origin of the Mediterranean flora generally begin with the very different environment of the Pliocene, when the Mediterranean was an arm of the Indian Ocean, and its shores were mantled with forests – or what are assumed to have been forests – of laurel, sassafras, tulip-tree and avocado pear: rain-forests reminiscent of the present SE United States, Caribbean and Japan, with a curious outlier in the Canary Islands.[11] In the Mediterranean the rustle and scent of these 'laurisylvan' forests still faintly echo and breathe in the mountain bay-laurel woods of Apokórona in Crete, but otherwise they belong to the geological past. Only isolated species have struggled through cold, drought and fire to the present: for example, the mountain maples, Cretan *Zelkova* and the fern *Woodwardia*, the oleander fringing streams and many an endemic.

Most of the last two million years has consisted of glaciations; interglacials as warm as the present were relatively brief, 20,000 years each at the most. Glaciations were punctuated by less cold periods (*interstadials*) of a few centuries each.[12] Glaciations took a different form in the Mediterranean from northern Europe. Although trees and other pollen-bearing plants retreated southwards, indicating a generally lower temperature, the effects of dryness are more conspicuous than those of cold. Glaciers were confined to high mountains.

Glaciated regions such as north Europe have few, poorly differentiated endemics. The Mediterranean has many, often very distinct and confined to some particular island, mountain or gorge. How many endemics perished during glaciations we shall probably never know, but many survived, evidently close to their present localities. Frost was not severe enough to exterminate, for example, the palm *Phœnix theophrasti*. This Tertiary relict, formerly more widespread,

contracted during successive glaciations to its present few sites in the warmest parts of Crete and the SE Aegean (Fig. 10.3). To take three other Cretan examples, *Petromarula pinnata*, confined to the less dry lowlands, *Ebenus cretica*, a plant of the drier lowlands, and *Anchusa cespitosa*, confined to the western high mountains, all somehow got through the glaciations. So did the many shrubby pinks (*Dianthus* species), each confined to its particular gorges and islets.[13]

Glaciations had more effect on vegetation than on flora. The characteristic vegetation of southern Europe during glaciations was steppe. The commonest pollen type, sometimes making up one-third of the total pollen, is the wind-pollinated herb *Artemisia*. Pollen of grasses and of Chenopodiaceae, also plants of non-shaded habitats, is abundant. Tree pollen, mainly pine and oak, is sometimes nearly absent, sometimes present in quantities indicating savanna.[14]

The longest record, from Lake Ioánnina in the mountains of Epirus, goes back 420,000 years, covering eleven vegetation cycles. In each cycle an 'interglacial' period of forest of oak, pine, or hornbeam, with less abundant hazel, fir (*Abies*) and beech, alternated with a 'glacial' period of 'open vegetation' with abundant *Artemisia* and Chenopodiaceae but also some pine and oak.[15] The forests were very like those of NW Greece today (although beech has recently been increasing); the open vegetation can be paralleled now in the pine savannas and oak savannas of the same region (p. 210–2), though their steppe component no longer contains *Artemisia*.

In lowland north Greece, too, periods of steppe alternated with phases of 'forest steppe', savanna with scattered trees or patches mainly of pine. The savanna included also *Pistacia*, which presumably was terebinth; this tree of relatively warm climates persisted even on this northern fringe of the Mediterranean.[16] The savanna periods presumably correspond to interstadial and interglacial warm periods as known in other parts of the world, but their dates are not well enough established to be sure of the exact relationship.

In south Italy, the record of the Monticchio lakes extends back some 75,000 years, covering much of the last glaciation. For this long period the pollen is overwhelmingly that of grasses, *Artemisia* and Chenopodiaceae. There are small but fluctuating amounts of tree pollen. The commonest was generally pine, accompanied by birch and juniper, which indicate forest or trees of a relatively arctic character. However, oak was almost continuously present, mixed at times with beech, fir, hazel, elm, lime and ash, all of them north-temperate trees. In this far-southern region, therefore, the interglacial tree flora apparently persisted as pockets of forest in favourable sites. There is little or no evidence for the survival of strictly Mediterranean plants, but the elevation (650 m) might be too high for them.[17]

The Palaeolithic site of L'Hortus, near Montpellier, from a warmer interval during the last glaciation, has charcoal of deciduous and evergreen oaks, phillyrea and the deciduous trees of the *šibljak* zone of Mediterranean mountains.[18]

[11] Vernet 1997.
[12] Van Andel 1998.
[13] Similar arguments apply to island mammal populations which, once they arrived, evolved independently on their various islands and often persisted until people arrived in the Holocene (pp. 72–3). Half the land molluscs of Crete are endemic: K. Βαρδινογιάννη, *Βιογεωγραφία των χερσαίων μαλακίων στο νότιο νησιωτικό αιγαιακό τόξο* (K. Vardinogiannis, *Biogeography of the land molluscs of the south Aegean island arc*), Ph.D. dissertation, University of Athens, 1994.

[14] Wijmstra & Smit 1976 regard between 30 and 55 per cent of tree pollen as indicating 'forest steppe'.
[15] Tzedakis 1993.
[16] Wijmstra & Smit 1976.
[17] Watts, Allen, Huntley, Fritz 1996.
[18] Vernet 1997, p. 77.

In Corsica there is evidence from the tail of the last glaciation for two endemic trees, *Pinus...laricio*, Corsican pine and *Alnus...suaveolens*, the Corsican alpine alder, which have recognisably distinct pollens.[19] At the Carihuela Cave and at Padul, both in the mountains of south Spain, the latter part of the last glaciation exhibits steppe with *Artemisia* and grasses, and patches or scattered trees of pine; the pine and steppe pollens fluctuate against each other. In Spain, occasional records of Cistaceae pollen (including *Cistus ladanifera*) show that the modern tall phrygana had its antecedents in glacial-period vegetation.[20]

The last glaciation was punctuated by short intervals of less cold, as measured by changes in the ratio of the isotopes of oxygen. An undersea core from off Sicily shows periods with some oak pollen, correlated with these intervals, which lasted long enough for oaks to increase.[21]

Much of the Mediterranean thus incurred deforestation (as it would now be called) many times during the Pleistocene. Trees survived the glaciations – where they did survive – as small populations in favourable spots (refugia) in the regions where they now occur. These would be the warmest and wettest parts of the landscape, especially valleys in the various mountain ranges where water from glaciers drained southward to the sea. This explains why, for example, the evergreen oak of Spain and Portugal is not *Quercus ilex*, as known in the rest of the Mediterranean, but the peculiar Iberian species *Q. rotundifolia*. A remarkable survival is walnut, whose pollen comes from various glacial periods at Carihuela; it is probably extinct as a Mediterranean native, surviving only in cultivation. Horse-chestnut is still endemic (as a native tree) to Albania and NW Greece which were evidently its glacial refugium. It is disjunct – it has relatives in America and the Far East – and seems to have lost its means of dispersal.

The main variation in glacial vegetation was between treeless steppe and treed steppe or savanna. The steppe was a plant community which apparently no longer occurs in the European Mediterranean, even at high altitudes. *Artemisia* is seldom abundant and often absent in Mediterranean countries; Chenopodiaceae[22] are characteristic of arable land and the seaside. They were sometimes accompanied by cocklebur (*Xanthium*), now a universal weed of the subtropics. Imagination fails to reconstruct what this plant community looked like.

The end of the Pleistocene

The end of the last glaciation can first be detected at around 12,000 bp. On the fringes, as in the south of France and the Pyrenees, the steppe was first colonized by relatively arctic trees such as birch and willow, then by more temperate species such as oak and elm.[23] In warmer regions the climate became less arid. In general, by 10,000 bp, deciduous oaks had become the commonest trees in most of the European Mediterranean, followed by pines.

A setback to near-glacial conditions, corresponding to the 'Younger Dryas' minor glaciation (p. 22), can be discerned mainly in submarine cores. A corresponding fluctuation at Monticchio lasted from 11,500 to 10,300 BC.[24]

Evergreen trees and shrubs expanded typically from about 8000 bp onwards. They spread mainly from the south-east: evergreen oak from south Greece and Crete, *Pistacia* (lentisk, but also terebinth which is deciduous) from the southern Adriatic, *Phillyrea* from the middle Adriatic.[25] These spreads probably represent, not so much migration of these trees towards the west, as their expansion in a westward sequence from a number of different, specially favoured places where they had survived. The means of spread will have varied. Oaks are dispersed by birds, especially jays; most pines by wind. Sea currents were probably unimportant, to judge by the rarity of land plants colonizing drift-lines today.

Comparisons with mediterraneoid regions

In California it is claimed that glacial periods were wet as well as cold, giving a different relationship to the present flora. In South Africa and Western Australia we are told that climatic changes – much as in Crete – were not severe enough to disturb the complex pattern of locally endemic plants.[26]

WILDWOOD – PALAEOLITHIC AND MESOLITHIC[27]

The relative consistency of Mediterranean Europe during de-glaciation breaks down in the Holocene. Changes led to regional and local differences in *wildwood* – natural forest, maquis or savanna, not altered by civilized human activities. In some regions wildwood resembled natural vegetation as it remains today; in others it was very different.

[19] Reille 1992.

[20] J. S. Carrión and P. Sanchez-Gómez, 'Palynological data in support of the survival of walnut (*Juglans regia* L.) in the western Mediterranean area during last glacial times', *JB* 19, 1992, pp. 623–30; Pons & Reille 1988.

[21] M. Rossignol-Strick and N. Planchais, 'Climate patterns revealed by pollen and oxygen isotope records of a Tyrrhenian sea core', *Nature*, London 342, 1989, pp. 413–46.

[22] It is usually impossible to distinguish the pollen of genera within Chenopodiaceae, or to separate that family from Amaranthaceae, which have a similar ecology. Electron microscopy could help: A. Smit and T. A. Wijmstra, 'Application of

transmission electron microscope analysis to the reconstruction of former vegetation' *ABN* 19, 1970, pp. 867–76.

[23] J. Guiot and A. Pons, 'Une méthode de reconstruction quantitative du climat à partir de chroniques pollenanalytiques: le climat de la France depuis 15000 ans', *Académie des Sciences, Comptes Rendus*, Paris 302, 1986, pp. 911–16.

[24] M. Rossignol-Strick, 'Sea-land correlation of pollen records in the Eastern Mediterranean for the glacial-interglacial transition: biostratigraphy versus radiometric time-scale', *QSR* 14, 1995, pp. 893–915; W. A. Watts, J. R. M. Allen, B. Huntley, 'Vegetation history and palaeoclimate of the last glacial period at Lago Grande di

Monticchio, southern Italy', *QSR* 15, 1996, pp. 133–53.

[25] Huntley & Birks 1983.

[26] D. I. Axelrod, 'History of the Mediterranean ecosystem in California', *Mediterranean Type Ecosystems: origin and structure*, ed. F. di Castri and H. A. Mooney, Chapman and Hall, London, 1973, pp. 225–77; S. D. Hopper, 'Patterns of plant diversity at the population and species levels in south-west Australian mediterranean ecosystems', *Biodiversity of Mediterranean ecosystems in Australia*, ed. R. J. Hobbs, Surrey Beatty, 1992, pp. 27–46; (also correspondence with Dr Hopper).

[27] This section is based in part on Huntley & Birks 1983.

Two sample sites

At Padul, a peat-bog in the Sierra Nevada, the end of the last glaciation was marked by a change from pine savanna directly to evergreen oak. Pine pollen greatly diminished, replaced by evergreen oak with a proportion (about one-third) of deciduous oak (plus the occasional grain of cork-oak type). For about two thousand years *Artemisia* and Chenopodiaceae held out, indicating oak savanna (and therefore presumably a dry climate) rather than forest, but they become scarce about 9500 bp.[28] The appearance of *Pistacia* confirms that the immediate post-glacial vegetation was what would now be regarded as typically Mediterranean – indeed more so than much of the Sierra Nevada is today. Cork-oak type became a significant minority of the oak pollen from 6800 BC; this suggests an increasing frequency of fire (p. 218). Shortly afterwards, the appearance of quantities of olive and cereal pollen indicates the beginnings of agriculture.

In Corsica which, like the Sierra Nevada, is a non-limestone country, a very different Mediterranean-type forest developed early in the Holocene. Tree-heather and *Arbutus* became dominant. There was some *Pistacia*, *Phillyrea* and deciduous oak, and also trees with a northerly distribution, such as lime, elm and hazel. Evergreen oak was rare except (surprisingly for a Mediterranean evergreen) in the mountains. The mountains were also the home of Corsican pine, Corsican alder and locally of birch.[29] Something like the characteristic maquis of modern Corsica, but without evergreen oak, thus goes back well beyond human civilization. The maquis or forest was apparently dense, with little room for non-shade-bearing plants, although occasional grains of plantain, ligulate Compositae, etc. indicate patches of steppe-like vegetation. At two coastal sites (Saleccia and Crovani) there is a small but continuous presence of *Cistus* pollen, which (considering that *Cistus* is insect-pollinated) means substantial phrygana-filled gaps in the maquis.

Summary for the European Mediterranean

By 7000 BC there were trees in all parts of the Mediterranean that have a pollen record. Usually oaks predominated, but in Corsica the predominant tree was tree-heather, and in parts of Spain, south Greece and Crete it was pine. Often there was more, sometimes much more, deciduous oak than evergreen. There is generally evidence for other typical Mediterranean trees and shrubs: *Pistacia* (lentisk or terebinth), *Phillyrea*, *Arbutus*, plane, olive. There were also northern trees: alder, birch, elm, hazel, hornbeams (*Carpinus* and *Ostrya*), lime (particularly significant because it is insect-pollinated), etc.

The last entry in Table 10.i is from a deep-sea core between Crete and Sicily; it would contain pollen from much of the Mediterranean, depending on winds and currents at the time. The investigators considered that most of the pollen came from southern Europe.

Was the vegetation forest?

Tree pollen does not imply forest: isolated trees produce at least as much pollen as forest trees.

A landscape consisting of, let us say, 80 per cent oak forest and 20 per cent asphodel steppe would produce much less than 20 per cent of asphodel pollen. Asphodel would be under-represented through being insect-pollinated and also through its low stature: an asphodel pollen grain would be less likely to reach a wetland than a wind-blown oak pollen grain. Each asphodel grain recorded therefore counts for much more than an oak grain. Similar under-representation applies also to most undershrub pollens.

As a rough criterion we propose to infer a substantial phrygana element in the landscape if the total undershrub pollen exceeds 1 per cent of the tree pollen. If the wind-pollinated non-shade-bearing herb pollen (excluding grasses and sedges) exceeds 10 per cent of the tree pollen, we infer that there was a substantial steppe element; likewise if the insect-pollinated non-shade-bearing herb pollen (e.g. asphodel) exceeds 1 per cent of the tree pollen.[30]

Combining these criteria, we introduce the *Non-Forest Index* (NFI):

$$\frac{\text{Undershrub pollen} + \frac{1}{10} \text{ of wind-pollinated non-shade-bearing herb pollen} + \text{insect-pollinated non-shade-bearing herb pollen}}{\text{Total tree-and-shrub pollen}}$$

Provided that the predominant trees are good pollen producers such as oak and pine, this gives a rough measure of the ratio of open ground to trees. It will probably tend to underestimate open ground and overestimate forest, for three reasons:

(1) trees would tend to congregate near the damper parts of dry landscapes where pollen is preserved;
(2) the index excludes steppe grasses, whose pollen is indistinguishable from that of wetland grasses;
(3) many abundant undershrubs leave little or no pollen record.

At Ténagi Phílippon, the NFI varied from about 4 per cent during interglacials (regarded as forest) to over 300 per cent during glaciations (regarded as steppe).[31] At Monticchio, typical values for warm and cold periods within the last glaciation are 16 per cent and 36 per cent respectively. For wildwoods of the middle Holocene, the NFI varies from 0.7 per cent to 173 per cent (Table 10.ii).

Other less fully studied sites provide further data. At Castellón de la Plana (coastal NE Spain), pine was by far the commonest tree, but the researchers remark that 'The percentages of Gramineae, Cyperaceae and "varia" [grasses, sedges and uninteresting types] on some occasions are so high that the woods must have been very open' – in our terms constituting savanna, not forest. Charcoal deposits in this area indicate more evergreen than deciduous oak.[32] Further

---

[28] Pons & Reille 1988.
[29] Reille 1992.
[30] It is conceivable that non-shade-bearing herbs could really belong to forest, not steppe, being awakened from dormancy every time the forest is burnt. If so, a high proportion of such herbs would indicate frequently burnt forest rather than discontinuous forest. However, asphodel (one of the most widespread pollen types in this category) does not behave thus today. We leave this possibility to be explored.
[31] Computed from information taken from Wijmstra & Smit 1976.
[32] For example, Vernet 1997, p. 128.

Sites are chosen to avoid high altitudes and the fringes of the Mediterranean region. The figures are approximate, estimated by measuring from published pollen diagrams. Trees marked * in the third column contribute a significant fraction of the total pollen.

Intermediate oak is counted as evergreen in Spain and Corsica (where it is likely to be *Quercus suber*); as deciduous in the east Mediterranean (*Q. cerris* or *trojana*).

Principal undershrub pollen types are *Cistus*, *Ephedra*, Ericaceae (unless identified as *Erica arborea* or *Arbutus*, which are trees), *Helianthemum*, Labiatae, *Poterium*.

Principal pollen types counted as wind-pollinated non-shade-bearing herbs are *Artemisia*, Chenopodiaceae, *Plantago*.

Principal pollen types counted as insect-pollinated non-shade-bearing herbs are *Asphodelus*, Compositae Tubuliflorae, Cruciferae, Umbelliferae.

Site and altitude	Principal trees & shrubs	N. European trees	Deciduous oak / Total oak %	Undershrubs / Trees & shrubs %	Wind-poll. herbs / Trees & shrubs %	Insect-poll. herbs / Trees & shrubs %
Spain						
Padul (785 m)	oak > *Pistacia* > pine (cedar, juniper, olive, *Phillyrea*)	(alder, birch, hazel)	27	< 0.1	1.5	0.5
Provence						
Tourves (298 m)	oak > juniper > *Phillyrea*, pine	*Abies*, ash*, elm, hazel,* lime,* maple,* yew* (alder, beech, birch, *Carpinus*)	73	2	8	5
Corsica						
Le Fango (10 m)	tree-heather > pine > oak > *Pistacia* > arbutus > *Phillyrea* (olive)	*Abies*, alder*, birch, elm, hazel*, lime, yew* (box, *Ostrya*, *Viburnum*)	70	0.6	0.7	1.3
Saleccia (15 m)	tree-heather > oak > pine > arbutus (juniper)	*Abies*, alder*, beech, elm, hazel*, lime (birch, *Carpinus*, *Ostrya*, willow)	c. 99	4.6	7.8	4.4
Italy						
Lago di Vico (507 m)	oak > beech, lime, hazel > *Phillyrea* > pine, elm	alder*, beech*, *Carpinus**, elm*, hazel*, lime*, *Ostrya*, *Picea*	?	1.7	0.9	2.4
Lago di Martignano (200 m)	oak > beech, *Ostrya* > *Phillyrea*, ash	*Abies*, alder*, ash, beech*, birch, elm, hazel, hornbeam, *Ostrya**	90	2.8	9.2	6.7
Lago di Monticchio (660 m)	oak > ash, *Ostrya* > hazel, beech, elm, lime	*Abies*, alder, ash*, beech*, elm*, hazel*, lime*, hornbeam, *Ostrya**	'rare to absent'	nil	0.6	0.8
NE Adriatic						
Porec (-28 m)	oak > pine > *Phillyrea* (*Pistacia*)	*Abies**, alder*, beech*, birch, elm*, hazel*, hornbeam, lime*, spruce (chestnut, walnut)	81	2.4	4.8	4.2
Bokanjačko Blato (c. 0 m)	beech ≥ oak > juniper, *Fraxinus ornus* > hazel, elm	ash, beech*, birch, *Abies*, elm*, hazel*, hornbeam, lime, yew	85	0.8	3.3	2.1
Malo Jezero (0 m)	oak > *Phillyrea* > pine > *Pistacia* > juniper	alder, ash, beech, hornbeam, elm*, hazel*, lime (*Picea*, yew)	99 +	insignificant	3.4	1.7
Greece						
Tenági Philíppon (40 m)	oak > pine, juniper > *Pistacia*	hazel*, lime*, ash, birch, *Abies*, *Carpinus*	mainly evergreeen	1.6	4.3	1.2
Lake Vegorítis (570 m)	pine > oak (heather, olive, *Phillyrea*, *Pistacia*)	alder, ash, beech, hazel, hornbeam*, lime* (birch)	99 +	0.1	1.1	1.3
Lake Xyniás (500 m)	oak > pine > juniper, *Pistacia* > *Phillyrea* (olive, plane)	alder, birch, elm, hazel*, hornbeam*, lime	91	3.9	1.9	4.3 (19 pollen types)
Lake Trikhónis (20 m)	oak > heather > pine > olive, *Pistacia*, *Phillyrea*	*Abies*, alder, ash, birch, hornbeam*, hazel, lime (beech, elm)	81	2.4	11.3	2.0
Kopáis (30 m)	oak > lime, juniper > elm	*Abies*, beech, elm*, lime*	mainly deciduous, becoming more evergreen	0.9	2.0	4.6
Navarino (0 m)	pine > oak > olive > juniper, *Pistacia*	beech, elm, hornbeam* (alder, willow)	not known	3.2	9.6	104
Crete						
Tersaná (0 m)	oak > pine > cypress > *Pistacia*	hornbeam, lime	mainly deciduous	88	51	80
Ayia Galíni (1 m)	pine > oak > plane	alder, elm, hazel, lime	69	2.4	4.2	34
Euro-Mediterranean						
Bannick Basin (-3,400 m)	oak > ash, olive > hazel, lentisk	alder, ash*, hazel*, *Ostrya*	unknown	13	36	44

Table 10.i. Summary of Mediterranean pre-Neolithic wildwood characteristics as represented in certain pollen diagrams.

Abstracted from the following parts of each profile:

Padul	2.04–2.75 m	Pons & Reille 1988.
Tourves	7.5–9.0 m	Nicol-Pichard 1987.
Le Fango	5.95–7.00 m	Reille 1992.
Saleccia	6.05–8.10 m	Reille 1992.
Lago di Vico	0.70–1.20 m	Frank 1969.
Martignano	2.80–3.80 m	Kelly & Huntley 1991.
Monticchio	3.00–3.50 m	Watts & others 1996.
Porec	Po-1	Beug 1977.
Bokanja˘cko Blato	Zone 1	Grüger 1996.
Malo Jezero	phase A1	Beug 1967.
Tenági Philíppon	2.00–3.50 m	Wijmstra 1969.
Lake Vegorítis	core 8, bottom 6 samples	Bottema 1985.
Lake Xynias	core I, phase Z3–Z4	Bottema 1979.
Lake Trikhónis	bottom 6 samples	Bottema 1985.
Kopáis	300–376 cm and 4–6 m	Mean of two cores: Turner & Greig 1975 and Allen 1997.
Navarino	phase A3	Kraft and others 1980.
Tersaná	bottom	Moody and others 1996.
Ayia Galíni	samples 3–7	Bottema 1980.
Bannick	2.5–2.9 m	Cheddadi and others 1991.

inland behind Valencia, pine and oak were dominant, but there was a good representation of non-forest types including the undershrubs *Ephedra* and *Helianthemum*. In west Majorca, although pine was again predominant, there was some oak; again there was a high percentage of non-forest types.[33] On the south coast of France, deciduous oaks predominated, with a local presence of pine, and abundant north European trees, especially hazel; but many Mediterranean evergreens were present, including wild-olive and evergreen oak.[34] In the extreme south of Spain, at Nerja, umbrella pine (*Pinus pinea*) occurred, and people roasted the cones and ate the nuts from far back in the last glacial until at least the Neolithic.[35]

Mediterranean wildwood – the natural vegetation in about 7000 BC – was thus not much less varied than the vegetation today. Some of the present variation can already be discerned. Italy stands out as less arid than Spain or Greece; forests like those of central Europe reached far south at a modest altitude. The variation was much too great to be sampled adequately from twenty or so sites, especially since drier types of vegetation are likely to be under-sampled.

Trees were present everywhere, but there was a general trend for forest to diminish eastward, southward and on the coast. There is some correlation with diminishing rainfall, complicated no doubt by the effects of water retention, root penetration and ground-water.

The present 'typical' Mediterranean forest, maquis or savanna of thick-leaved evergreen trees or shrubs was not then ubiquitous. Deciduous oaks, hornbeams and even beech were dominant much more widely than they are today.

Northern trees were present everywhere, even far south, and were sometimes dominant, especially in Italy.

Mediterranean evergreen trees were always present, but only in parts of Spain (with its own distinctive evergreen oak) and Corsica were they dominant. Normally they formed either a minor constituent of the deciduous oakwoods or patches of special vegetation within them (e.g. on limestone outcrops). There is no discernible correlation between these trees and higher NFI or greater aridity. Besides evergreen oaks and tree-heather, there was also *Phillyrea*, widespread but not often abundant today. Pine was generally less abundant than now, but has always been characteristic of Spain.

Where forest or maquis was discontinuous, vegetation in the gaps was normally steppe (non-shade-bearing herbs). Phrygana – bearing in mind the difficulty of recognising its pollen – was apparently far less common than today, abundant mainly in some coastal locations.

The undersea core, which samples pollen fallout from a larger area than those on land, appears more arid than most of the others, with a high non-forest component, though even here there is little evidence for undershrubs other than *Ephedra*. This confirms the view that land pollen cores tend to record the wetter parts of a dry landscape. However, this core probably contains pollen from Africa, and reworked pollen from glacial times, as well as from mid-Holocene Europe.

At high altitudes wildwood was more like that of Central Europe. For example, at Prato Spilla, at 1,550 m in the Ligurian Apennines, the pre-Neolithic vegetation was forest of fir, oak and lime, with some elm and hazel. There were some gaps, as shown by pollen of Ericaceae and Compositae, but juniper, now prevalent in the area, had died out in the early Holocene and did not return until modern times.[36]

Differences from the present

Differences from present vegetation could be due to four factors.

(1) Lapse of time: changes after the last glaciation, such as the migration of plants from refugia, may still have been incomplete after five thousand years, continuing into the Neolithic period and beyond.

(2) Change of climate.

(3) Human activity:
(a) grubbing out vegetation for cultivation;
(b) exterminating native mammals (especially on islands, where the first effect of settlement may have been a reduction of browsing);
(c) browsing (by domestic livestock) and wood-cutting;
(d) altering the fire cycle (Chapter 13);
(e) introducing foreign plants (Chapter 4) and animals.

(4) Appearance of new species and varieties of plants with different ecological behaviours (Chapter 4).

[33] J. Menéndez Amor y F. Florschütz, 'La concordancia entre la composición de la vegetación durante la segunda mitad del Holoceno en la costa de Levante (Castellón de la Plana) y en la costa W. de Mallorca', *Boletín de la Real Sociedad Española de Historia Natural*, Geologica 59, 1961, pp. 97–100; Menéndez Amor y Florschütz 1961.

[34] N. Planchais and I. Parra Vergara, 'Analyses polliniques de sédiments lagunaires et côtiers en Languedoc, en Roussillon et dans la province de Castellon (Espagne); bioclimatologie', *Bulletin de la Société Botanique de France, Actualités Bota-niques* 131, 1984, pp. 97–105; cf also charcoal evidence in Vernet 1997.

[35] Vernet 1997, pp. 147 ff., quoting work of Badal Garcia.

[36] Lowe & others 1994.

Site and NFI	Altitude, m	Modern rainfall, mm	Forest, savanna, or maquis as inferred from pollen	Mediterranean trees	Northern trees	Non-tree vegetation
Padul (Spain) 0.7	785	558 ***	continuous evergreen oak forest or maquis	++++	+	no appreciable undershrubs or steppe
Monticchio (Italy) 0.9	660	815 *	forest of deciduous oak, ash, Ostrya	+	++++	some insect-pollinated herbs
Lake Vegorítis (N. Greece) 1.5	570	c. 650 *	forest of pine and deciduous oak	++	+++	little steppe, no phrygana
Le Fango (Corsica) 2.0	10 coastal	c. 600 **	tree-heather maquis with some deciduous and a little evergreen oak	++++	+++	no appreciable undershrubs; very little steppe
Malo Jezero (Croatia) 2.0	0 island	c. 780 *	forest of deciduous oak and hazel	+++	+++	no appreciable undershrubs, very little steppe
Tenági Philíppon (NE Greece) 2.8	40	c. 640 **	forest of evergreen oak and hornbeam	+	+++	some steppe, little phrygana
Bokanjačko Blato (Croatia) 3.2	25	915 *	discontinuous forests of beech, oak etc.	++	++++	few undershrubs, some steppe
Lago di Vico (Italy) 4.2	507	c. 800 *	savanna or discontinuous forest of deciduous oak, beech, lime, hazel	++	++++	no appreciable undershrubs, little steppe
Lake Trikhónis (W. Greece) 5.5	20	c. 850 ***	savanna of mainly deciduous oak	+++	+++	steppe abundant; some phrygana
Porec (Slovenia) 7.1	-28 undersea	c. 800 *	savanna or maquis of deciduous oak and hazel, later turning into oak with some pine	++	+++	some undershrubs; steppe well represented
Lake Kopáis (mid Greece) 7.5	30	c. 450 ***	savanna of mainly deciduous oak	+++	++	steppe abundant, little evidence of phrygana
Tourves (Provence) 7.9	300	c. 870 *	savanna of mainly deciduous oak with juniper	++	+++	steppe abundant, little evidence of phrygana
Lake Xyniás (mid Greece) 8.4	500	c. 600 ***	savanna of mainly deciduous oak	++	+++	steppe well represented; some phrygana
Saleccia (Corsica) 9.8	15 coastal	c. 600 **	tree-heather maquis with some deciduous oak	++++	+++	some undershrubs and steppe
Lago di Martignano (Italy) 10.4	200	c. 700 *	savanna or patchy forest of oak, beech, Ostrya	++	++++	undershrubs and herbs well represented
Bannick Basin (deep-sea) 23	–	–	savanna, principally oak	+++	+++	undershrubs poorly represented; abundant steppe
Ayia Galíni (Crete) 37	1 near coast	600 ***	savanna or patchy maquis of deciduous and evergreen oak, pine forest at a distance	++	+++	abundant steppe; some phrygana
Navarino (S. Greece) 108	0 coastal	c. 800 ***	savanna of pine and oak, later replaced by pinewood	+++	+++	steppe abundant; some phrygana
Tersaná (Crete) 173	0 wind-exposed coast	c. 550 ***	scattered maquis or savanna of mainly deciduous oak	++	++	phrygana and steppe

Table 10.ii. Vegetation at pre-Neolithic sites, displayed in ascending order of Non-Forest Index.
* weak dry season   ** moderate   *** severe dry season

Compared with today, wildwood forest was more extensive but by no means complete. The climate, although wetter than now, was evidently not wet enough for continuous forest, at least in the presence of browsing wild animals and lightning fires. As in modern savannas, trees may have formed a closed community below ground, but with gaps above ground. Maquis, savanna and steppe have since extended partly through people burning, woodcutting and keeping livestock in the remaining natural vegetation. Nevertheless, the drying climate would have resulted in some decline of forest even if these activities had never happened.

The increasing predominance of evergreen vegetation in Mediterranean countries is explained by the different behaviour of deciduous and evergreen trees, especially oaks. Evergreen vegetation is mainly on thin soils and bare rock. Holm-oak and prickly-oak now grow mainly on fissured limestones, live-oak on thin granite and other non-limestone soils. Deciduous oaks are the trees of potentially cultivable soils. Being shallow-rooted, they are unsuited to a fissured substrate, and are rare on hard limestone even in rainy areas or where protected from browsing; it is hard to believe that they have ever been dominant on it. In recent centuries deciduous oaks have been confined to occasional uncultivated spots, but as cultivation retreats they are now increasing (pp. 61–5). Moreover, deciduous oaks, except where they have evolved a ground-oak form (p. 54), are less adapted to browsing than prickly- or live-oak.

In wildwood, therefore, deciduous oaks would predominate in pollen samples because they would grow on the sort of soils that occur around pollen sites. Agriculturalists would later destroy these oaks, leaving evergreen vegetation on more distant, uncultivable soils to supply the tree pollen. Hard limestones in wildwood times, as now, were probably treed mainly with holm-oak and prickly-oak, sometimes with pine and fir (*Abies*). The anomalously early abundance of evergreen oak in the Sierra Nevada may be because Spain's peculiar live-oak can compete with deciduous oaks on non-calcareous soils.

North European trees have disappeared except from specially cold, damp relict sites. Although some of these are protected from browsing, the pattern of survival strongly indicates climatic change (the Aridization) rather than human activity as the main factor in their decline (p. 145). This change is more marked in Spain and Greece than in Italy.

Another change from the pre-Neolithic is the appearance of phrygana. Communities of undershrubs had been specialized, probably confined to places where tree growth was difficult because of wind exposure or lack of root penetration (p. 57). Today they tend to be maintained by browsing, and often disappear if browsing ceases. The rise of undershrubs is largely a consequence of civilization, although the drying of the climate may have helped.

THE COMING OF AGRICULTURE AND DOMESTICATED ANIMALS

Long before the Holocene, Palaeolithic peoples on the mainland, although few in number, may have had profound effects on the ecology – as did their contemporaries in other continents – by exterminating the great mammals (especially elephants, the only European mammal without an effective predator other than man), and also by altering the fire regime.[37] The pollen record is not detailed enough to show up these influences. The wildwood period ends when the beginning of the Neolithic brought about alterations through cultivation or stock-rearing.

Recognising the Neolithic

The Neolithic is marked by the invention of most of the things that make up civilization as we know it: cultivated plants, domesticated animals, permanent settlement, buildings, pottery, graveyards and temples. In north Europe most of these arrived together, and quickly had an impact on the landscape. In the Mediterranean the Neolithic is less recognisable. Inventions built up gradually over an immensely long period: the Neolithic lasted 2,500 to 4,000 years.

It is difficult to correlate palynology and archaeology. In north Europe, dated Neolithic artefacts are often found within or near dated pollen-bearing deposits. In south Europe, Neolithic material is often scarce and poorly preserved. Intensive archaeological survey is needed to find sites, and areas chosen for survey seldom contain pollen deposits going back so far. It is thus seldom possible to relate changes in pollen fallout to known Neolithic activities nearby. More often a poorly dated pollen diagram has to be compared with a vague and poorly dated archaeological record.

This might not matter if Neolithic activity had some distinctive effect on the vegetation. In north and middle Europe it let loose Dutch Elm Disease:[38] a sudden drop in elm pollen production appears in almost every pollen diagram, and can be used to mark the onset of the Neolithic even in profiles for which there is no other date. In southern Europe the Elm Decline can just be detected on the Mediterranean fringes,[39] but can seldom be used to define the Neolithic.[40]

Human activity can be inferred from pollens of crops and of weeds (such as *Plantago*, *Mercurialis annua* and *Centaurea solstitialis*) north of the Alps, where crops and weeds were introduced together. In the Mediterranean, the distinction between crops, weeds and plants of natural habitats is less clear (p. 67). Cereal pollen, distinctive in north Europe, cannot be distinguished from that of wild Mediterranean grasses. In Crete, the three plants mentioned above are abundant not only in farmland but also in semi-natural steppe and on cliffs. A closely related *Centaurea idaea* is endemic and presumably grew in Crete long before agriculture. All we can infer from such weed pollens is that there was not continuous forest.

[37] Z. Naveh, 'Ancient man's impact on the Mediterranean landscape in Israel – Ecological and evolutionary perspectives', Bottema & others 1990, pp. 43–52.

[38] Rackham 1980, chapter 16.
[39] M. Reille and J. J. Lowe, 'A re-evaluation of the vegetation history of the eastern Pyrenees (France) from the end of the last glacial to the

present', *QSR* 12, 1993, pp. 47–77; Lowe & others 1994; Huntley & Birks 1993, p. 412.
[40] Elm Disease exists today in Spain, Italy, Sicily, Greece and Crete, but is not obviously a product of civilization.

Changes in the middle Holocene are gradual and not very consistent. Tree pollen tends to decrease and non-forest pollen to increase. Oak pollen changes from mainly deciduous to mainly evergreen. This is accompanied by some expansion of *Pistacia* and *Phillyrea*. North European trees decline, but usually continue into the Bronze Age. These changes form a transition to the 'classic' Mediterranean vegetation as understood today. On the north-western fringes, however, beech advanced into the Mediterranean.[41]

Olive cultivation began to spread. Charcoal and pollen prove wild-olive to be native in south France, SE Spain and Crete, and thus probably throughout the European Mediterranean. It began to be domesticated in late-Neolithic Crete and also – if wood identifications are reliable – from the early Neolithic in Spain.[42]

In the south of France the Neolithic is reasonably well understood from pollen and charcoal. From a slow beginning changes in the landscape gradually accumulated, until, by the third millennium BC, large areas of forest had been modified into coppice-woods and savanna, or replaced by steppe and phrygana. In the charcoal as well as the pollen, deciduous oak was replaced by evergreen oak at different dates. Box – a characteristic shrub of sub-Mediterranean France today – appears, also at different dates: it was already abundant at one site in the Mesolithic.[43] Around the Languedoc lagoons, close to the edge of Mediterranean vegetation, north European trees lingered: lime into the late Bronze Age, beech, hazel and ash until the Roman period. As in Greece, the present Mediterranean climate appears to have set in about 4000 BP; this is confirmed by the mollusc assemblages.[44]

It is claimed that the shift from deciduous to evergreen oak and the appearance of box represent 'disturbance' to the remaining forest, presumably by woodcutting. This may be true in the south of France, where holm-oak often occurs on the same soils as deciduous oak, and is favoured by repeated cutting.[45] Woodcutting may cause box to flourish and produce pollen by removing the shade of taller trees.[46] In Greece, however, deciduous and evergreen oaks grow on different soils and seldom compete; the change probably represents the subtraction of deciduous oak from soils that were worth cultivating.

At Sítagroi in NE Greece, pollen evidence shows that the extensive plain remained wooded with deciduous oak throughout the Neolithic and well into the Bronze Age; the scattered settlements had only local effects. Charcoal identifications show oak as the predominant type, but ash, elm and other trees were also present, even occasional birch. There is slight evidence of coppicing. There are a few records of undershrubs.[47]

Did Neolithic activity transform the landscape?

In most of the European Mediterranean the Neolithic impact is still only vaguely known from palynology or archaeology. There is uncertainty on what to regard as 'anthropogenic indicators'.[48] At Lake Kopáis, the proportion of tree pollen rapidly and permanently fell by half at 3255 bc = 4050 BC (mid to late Neolithic). This has been attributed to 'forest clearance', but since the site is a karst lake a plausible alternative might be a change in the behaviour of the lake, causing more grasses to grow round its edge and dilute the tree pollen.

The extent of Neolithic activity is often an open question. Archaeological survey seldom reveals many sites, but this may be because they are poorly preserved and difficult to recognise. The lack of a detailed Neolithic chronology makes it difficult to decide how long activity continued at any one site. However, Neolithic peoples reached every part of the Mediterranean, and did not confine their activities to favourable spots. For example, there is Final Neolithic material in the high mountains of west Crete, at twice the altitude of permanent settlement today. In the Apennines, where pre-Neolithic material is already fairly abundant and goes up to 2000 m, pollen evidence shows that conversion of wildwood to subalpine pasture began in the Neolithic. Small occupation sites, too high for year-round settlement to be likely, with a high proportion of sheep bones, were presumably forerunners of the mountain transhumance of historic times.[49]

Most commentators assume that changes in vegetation before the Neolithic are due to climate whereas, after the Neolithic, climate stops operating and changes must be due to human activity. It is tempting to see the beginnings of a 'progressive degradation' which has continued at an ever-accelerating rate. This may be partly true, but not all the explanations offered are convincing. We doubt, for example, that the expansion of evergreen oaks resulted from 'opening' of the deciduous oak forest and the creation of temporary agricultural clearings into which evergreen oaks could spread. In our experience, most deciduous oaks very easily colonize ex-agricultural land, whereas *Q. ilex* and *Q. coccifera* do not easily start from seed.

Alternative explanations need to be considered. Changes of climate are more than likely, especially over such a very long time-span as the Neolithic. Natural changes resulting from the end of the last glaciation may still have been in progress when the Neolithic began.

[41] Lowe & others 1994.

[42] Vernet 1997; Moody 1987; Terral & Arnold-Simard 1996.

[43] Vernet 1997, especially p. 113; S. Thiébault, 'Les premières déforestations', *De Lascaux au Grand Louvre: archéologie et histoire en France*, ed. C. Goudineau and J. Guilaine, Errance, Paris, 1989, pp. 48–55; Guilaine 1993.

[44] Planchais 1982; Guilaine 1993.

[45] Chabal 1997.

[46] At Box Hill, one of the few box sites in England, box flourished after the beech trees that shaded it were blown down in the 1987 hurricane.

[47] J. Turner and J. R. A. Greig, 'Vegetational history', Renfrew & others 1986, pp. 41–54; Rackham 1986b.

[48] K.-E. Behre, 'Some reflections on anthropogenic indicators and the record of prehistoric occupation phases in pollen diagrams from the

Near East', Bottema & others 1990, pp. 219–31; Bottema & Woldring 1990.

[49] L. Nixon, J. Moody, V. Niniou-Kindelis, S. Price, O. Rackham, 'Archaeological survey in Sphakia, Crete', *Classical Views* 34, 1990, pp. 213–20; Lowe & others 1994; G. W. W. Barker, 'Prehistoric territories and economies in central Italy', *Palaeoeconomy*, ed. E. S. Higgs, Cambridge, 1992, pp. 111–76.

Settling the islands

Mediterranean islands had a strange set of aboriginal inhabitants: elephants the size of calves, pig-sized long-legged non-aquatic hippopotamuses, deer that could not run, antelopes, a bewildering variety of rodents, but no carnivore fiercer than a badger or a giant owl, nor any cattle or pigs.[50] These were repeated with variations throughout the large and middle-sized islands, including Malta and even Tílos close to Turkey.[51] We infer that island vegetation was (by modern standards) over-browsed during the Pleistocene. It is no accident that island floras, especially in Crete, are rich in endemic plants that live on cliffs or have extraordinary adaptations to resist browsing.[52]

Even if people only occasionally visited islands such as Crete, they could have had a profound effect by hunting and by leaving dogs behind. Island mammals, unused to predators, could have been slaughtered by hunters walking up to them and wringing their necks. The first effect of human contact could well have been a reduction of browsing and an increase of trees.

Not all the island mammals were exterminated at once.[53] On Cyprus and Sardinia they coexisted with the settlers for a long time. The endemic rabbit of Sardinia and the endemic chamois of Majorca[54] probably lasted into the Bronze Age; our own eyes have beheld the spiny-mouse of Crete. However, neither in Crete nor Corsica – islands for which there is pollen evidence but, at this time, no zoological record – is there a vegetation change different in character from changes on the mainland. In Crete there are the beginnings of olive cultivation and a general increase in non-woodland plants, especially those associated with burning. In lowland Corsica, during the probable period of the Neolithic, tree-heather continued to predominate, but evergreen oak became abundant and undershrubs such as *Cistus* became common.

A Neolithic legacy to some islands was wild descendants of domesticated mammals. The *mouflon* of Corsica and Sardinia is descended from sheep intoduced in the mid-Neolithic period.[55] The *agrími* of Crete is derived from a Neolithic domestic goat; its browsing preferences are not quite the same as those of modern goats.

BRONZE AND IRON AGES

Archaeology now begins to furnish, not only a general chronology, but indications of the amount of human activity in an area. In Crete, for a brief period, there are written records, not of vegetation as such, but of activities affecting it. Most of the evidence is available for only one or a few countries. In Corsica there is a good pollen record but the archaeology is problematic. In Sardinia the archaeology is spectacular but there is no pollen record.

The Bronze Age marks the final disappearance of north European trees from Crete and south Greece. On the margins of the Mediterranean the forest often changed in character: for example, in the Ligurian Apennines beech (the dominant tree today) began to increase at the expense of lime and hazel.[56] How far this was due to drying of the climate, to human activities (and if so which activities), or to the passage of time is still a matter of speculation.

Cultivation and 'anthropogenic indicators'

By the standards of north Europe or North America, the conversion of an area from forest to farmland should be recorded as a rapid change from mainly tree to mainly non-tree pollen – reversing the fall in the Non-Forest Index at the beginning of the Holocene.[57] In the Mediterranean, even where pollen deposits continue into later prehistory, the change is seldom well defined, probably because 'forest' and 'farmland' themselves were less well-defined. Roman farmers tolerated trees and bushes growing in their fields,[58] and in many parts of the modern Mediterranean there are hedges and wild trees on farmland. In SW Turkey we have seen farmers ploughing round bushes and trees. The civilizing (or degrading) of the Mediterranean will often have taken the form of converting savanna – steppe with trees and bushes – to farmland with trees and bushes, a change that would not show up well in palynology.

The crops most easily detected are not cereals (whose pollen is poorly defined and dispersed) but vine, olive, walnut and chestnut. All four have a pre-Neolithic record; their native limits in Europe are very uncertain. However, the quantities of pollen in late prehistory imply cultivation, for example of olive in Bronze Age Crete. Chestnut, throughout historic times a staple food tree in Italy, is doubtfully native there but has been cultivated since the sixth millennium BC.

The plane tree is sometimes considered an indicator of land-use. *Platanus orientalis* is a wild tree in SE Europe and Asia Minor. In several pollen diagrams in Greece and Turkey it expands at the same time as vine, chestnut, and walnut. From this it has been inferred that

> its sudden appearance in many pollen diagrams . . . points to deliberate planting of plane trees. . . . It cannot be excluded that *Platanus* profited from changes in habitat caused by human activity.[59]

To this it may be objected that the plane tree is tough, dry, dusty and uneatable; it serves no human purpose for which other trees will not do equally well. Elsewhere it has been

50 Mediterraneoid islands off California also had dwarf elephants but no carnivore fiercer than a fox; D. L. Johnson, 'The origin of island mammoths and the Quaternary land bridge', *QR* 10, 1978, pp. 204–25.

51 L. Caloi, T. Kotsakis, M. R. Palombo, 'La fauna a vertebrati terrestri del Pleistocene delle isole del Mediterraneo', *Bulletin Écologique* 19, 1988, pp. 131–51.

52 Adaptations against elephant browsing do not necessarily protect plants against goats; G. C. Stuart-Hill, 'Effects of elephants and goats on the Kaffrarian succulent thicket of the eastern Cape, South Africa', *JAE* 29, 1992, pp. 699–710.

53 E. Lax and T. F. Strasser, 'Early Holocene extinctions on Crete: the search for the cause', *Journal of Mediterranean Archaeology* 5, 1992, pp. 203–24.

54 R. Burleigh and J. Clutton-Brock, 'The survival of *Myotragus balearicus* . . . into the Neolithic on Mallorca', *JAS* 7, 1980, pp. 385–88.

55 J.-D. Vigne, 'Les premiers pasteurs', Bottema & others 1990, pp. 80–83.

56 Lowe & others 1994.

57 Such a change is visible in the Roman period or later at the Étang de Mauguio, Languedoc; Planchais 1982.

58 Palladius Rutilius, *De condicionibus agrorum*.

59 Bottema & Woldring 1990.

conjectured that prehistoric people destroyed the mountain forests, leading to erosion, leading to the pebbly torrent-beds that plane is supposed to favour.[60] This is a long, unverified chain of inference, and it is unlikely that all the links are true.

The present ecology of plane (apart from street trees) is not really related to human affairs. In Crete (where its record goes back beyond the Neolithic) it is the universal tree of places with a water-table, going far into the mountains along every little ravine. It flourishes where the rest of the landscape is occupationally burnt, but can hardly be regarded as fire-dependent. Its story raises the philosophical question, repeatedly encountered in Mediterranean historical ecology, of whether a change can properly be attributed to human activity on the sole evidence of coincidence in time, without knowing what the activity was or whether it could cause the change.

The rise of plane is presumably connected with the decline of alder, the alternative riverine tree. This can hardly be a mere matter of competition, for alder has completely disappeared from Crete and is very rare in south Greece. Nor can human activity be hostile to alder in some unknown way, for in Italy (where it has always been commoner than in the east or west Mediterranean) it still flourishes. Alder appears to require more continuously flowing water than plane, and drying of the climate may have to do with its disappearance from Greece and Crete (p. 145).

Population and settlement

In parts of the Mediterranean settlement and population were very dense. In Bronze Age Crete archaeological surveys reveal a dense hierarchy of hamlets, villages and towns, culminating in the great 'palaces'. For a few centuries the rural population may have been rather greater than it is now. The tablets at Knossós enumerate at least eighty thousand sheep, nearly all males kept for wool. These alone imply at least a hundred and sixty thousand sheep in Crete, compared to at least half a million in modern times, but it is most unlikely that the tablets record all the sheep that there were. The pollen record of Ay. Galíni shows an abrupt change at the beginning of the Bronze Age: steppe and phrygana dominated the landscape for the first time.[61]

In Sardinia the Iron Age is dramatically visible: in much of the island the traveller is never out of sight of it. The distinctive monument is the *nuraghe*, a dry-stone round tower built of huge stone blocks (sometimes not local stone). These represent fortified manor-houses or big farms; around many of them the remains of houses and fields can be seen. They were used for centuries; the more important ones were added to and turned into castles. At least four and a half thousand *nuraghes*[62] are known, about eight times the number of modern towns and villages, and even these do not represent all the settlement that there was. Sardinia had a very different settlement pattern from the widely spaced small towns and big villages of today.

These mighty towers, gilded with lichens and wreathed with wild fig (Fig. 10.4), are a Sardinian invention, unknown in Corsica or even on Sardinia's islets. They extend from the fever-haunted lagoons by Oristano to the brigand-haunted peaks of Barbagia (up to 1,350 m); they are perched on projections of the cliffs edging the *giaras* (basalt tablelands); they rise out of the maquis of granite Sarrabus and crown the marl mesas of Logudoro.

Over much of the island *nuraghes* are less than 2 km apart (Fig. 10.5), and represent at least as great an intensity of land-use as there ever has been. In general they are not on the 'best' land. They are commoner on basalt plateaux, marginal agricultural land in historic times. One of the densest concentrations is in the volcanic and metamorphic mountains of Marghine and Goceano. This is now deep forest, containing rare trees such as yew, yet full of megalithic field-walls. A deciduous oak 8.45 m in girth stands on what appears to be an ancient clearance-cairn (Fig. 4.34). Huge holm-oaks grow on the *nuraghes* themselves.[63]

Nuraghes are few in Gallura, the granite north-east, although abundant in other granite areas. They avoid the great wheat-belt of Campidano and other fertile plains. Not that these areas were still untouched wildwood. Place-names like *Naragu* and *Nuraxi* suggest that *nuraghes* have been destroyed, probably by later people recycling the stones. Moreover, not all nuragic settlements possessed a *nuraghe*, and there were two other contemporary cultures – Phœnician and Roman – to be found room for on the island.

We infer that in the last millennium BC all parts of Sardinia had at least some habitation, and most were fully used. The island was probably no more wooded – in places less wooded – than it is today, with at most small tracts of wildwood.

Corsica, as ever, contrasts with Sardinia. No great change is visible in the lowland pollen record. Evergreen oak continued to increase; there was some decline in heather (even though tree-heather can be frequently cut or burnt and yet produce pollen).

[60] Bottema 1985.
[61] Rackham & Moody 1996; Bottema 1980.
[62] Books on Sardinia usually say seven thousand, but we suspect this is an exaggeration. The modern 1:25,000 map of the island depicts some three thousand *nuraghes* and places named after them. Sample areas which have been archaeologically surveyed turn out to have about 1½ times as many *nuraghes* as the map shows (A. Moravetti, 'Gli insediamenti antichi', *Montagne di Sardegna*, ed. I. Camarda, Carlo Delfino, Sassari, 1988). The figure of seven thousand is probably an estimate of the total number of nuragic settlements, some of which had no tower.
[63] We are indebted to Professor Ignazio Camarda for an excursion to this mysterious area.

Fig. 10.4. Left: remains of deserted settlements: a lonely church on one bluff and a *nuraghe* on the other. *Near Perfugas, Sardinia, April 1992*

Fig. 10.5. Right: surviving Sardinian *nuraghes.* Circles represent those recorded only as place-names. Each spot, representing a *nuraghe,* is about 1 km in diameter: where the spots join up the *nuraghes* are less than 1 km apart. Note the association with basalts and Pliocene marls, and the avoidance of (or non-survival on) alluvial plains.

C = Cagliari, G = Goceano massif, M = Marghine massif, O = Oristano, S = Sassari.

From the 1:25,000 map, supplemented by our own observations and a few from other sources.

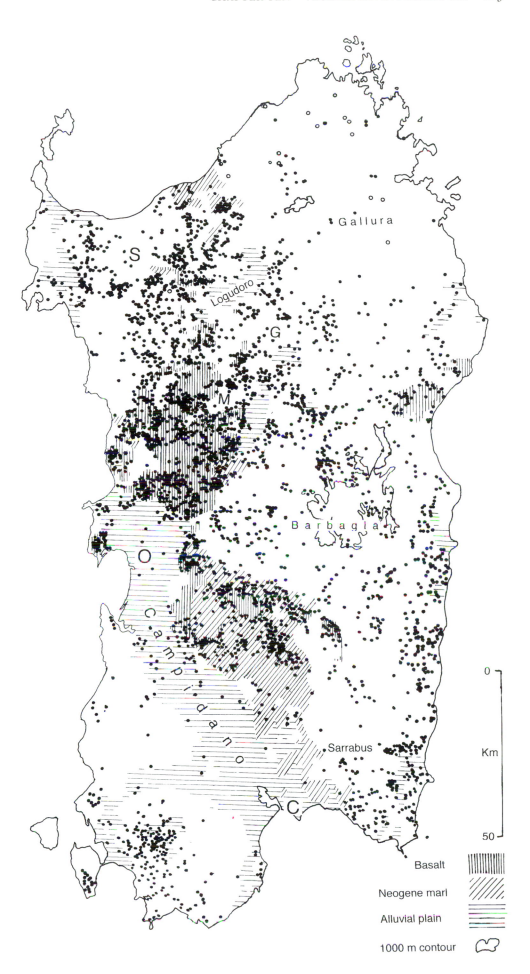

Declines of human activity

Progress was not forever. In Crete the density of settlement suddenly decreased after the Bronze Age (late first millennium BC), and did not recover for fifteen hundred years. Although this cannot be translated exactly into amount of land-use – the survivors, for example, may have tilled less land but kept more animals – there can be little doubt that vegetation would have recovered. Cores from Lake Trikhónis and other sites in south Greece reveal an increase in tree pollen after the Bronze Age. These fluctuations were not, of course, synchronous throughout the Mediterranean. The heyday of Sardinia and Classical Greece was the mid-first millennium BC, when Crete was underdeveloped.

Returning forest, then as now, was not necessarily the same as the original wildwood. In SW Turkey, on several occasions, forest returned after periods of Neolithic and Bronze Age farming. The wildwood had been of oak, pine and juniper; the secondary forest was of pine only.[64]

CONCLUSIONS

In general there were two critical events: the aridization of the climate from the fifth millennium BC onwards, and the advance of agriculture and pasturage in the Bronze Age. In much of the Mediterranean these two changes converted the 'aboriginal' landscape to something like the present landscape in the third or second millennium. The processes we have enumerated – expansion of phrygana and steppe, decline of deciduous oak, disappearance of northern trees, rise of plane and (in places) of pine – were mostly completed at about this time.

In Crete the intensity of land-use in the late Bronze Age could hardly have been less than it is now; in Iron Age Sardinia it was apparently more intensive than today; but in Corsica people still had only a modest impact on the pollen record. Another area of Bronze Age activity was south Greece, where in the Argolid cultivation was extensive enough for widespread erosion to be attributed to it (the circularity of this argument will be discussed in Chapter 16).

Did any 'remote' areas escape human influence until much later? Claims to this effect have a habit of being disproved once a proper survey is done. Absence of evidence for settlement is easily mistaken for absence of settlement. Even the great Pendlebury, who knew Crete better than anyone not a native,[65] thought that much of west Crete was untrodden wildwood until the late Bronze Age, a mistake thoroughly disproved by later work.

---

[64] N. Roberts, 'Human-induced landscape change in South and Southwest Turkey during the later Holocene', Bottema & others 1990, pp. 53–67.

[65] J. D. S. Pendlebury, *The Archaeology of Crete*, Methuen, London, 1939.

CHAPTER 11

Natural Vegetation in Historic Times

... The angel of repentance ... took me away to Arkadia ... and showed me a great plain, and round the plain twelve mountains, and the mountains differed in appearance from one another. The first was black like soot. The second was bare: it had no plants. The third was thorny and full of thistles. The fourth had half-dry plants, the top of the plants green, but towards the roots dry; part of the plants became dry whenever the sun singed them. The fifth mountain had green plants and was rough. The sixth mountain was all full of ravines ... the ravines had plants, but the plants were not too well-nourished, but rather shrivelled-like. The seventh mountain had happy plants, and all the mountain was flourishing, and every kind of cattle and birds pastured upon that mountain; and as much as the cattle and the fowls fed, more and more the plants of that mountain sprouted. The eighth mountain was full of springs, and every kind of creature of the Lord drank of the springs of that mountain. The ninth mountain had no water at all and was all deserted; and it had in it beasts and deadly reptiles destroying men. The tenth mountain had very big trees, and was all overshadowed, and under the shade lay sheep resting and ruminating. The eleventh mountain was very thick with trees, and the trees on it were very fruitful, adorned with different fruits, so that one seeing them would want to eat of their fruits. The twelfth mountain was all white ...[1]

The historic period, by definition, is when written records are available. Greek and Roman writings and inscriptions go back to about the eighth century BC. Before this, there was a brief flash of enlightenment in the second millennium BC, when Linear B tablets record rustic activities in Crete.

Writings do not supersede other information. Palynology continues into the modern period, although sites are fewer because many have lost their top layers, and dating becomes unsatisfactory. Archaeology increases in importance as, for example, timbers survive in more recent buildings. More information comes from the present landscape as we pass into periods from which trees survive.[2]

Historical landscapes have been divided into cultivation and roughland. Roughland is land not ploughed or sown, with natural vegetation (forest, maquis, phrygana and especially wetland); but it is part of the cultural landscape and was used. It provided pasturage, fuel, snails and other edible animals, edible and medicinal herbs.[3] Roughland was part of the assets of most communities. It was not essential, but townships that were all farmland had to buy in roughland produce. Its present under-used state, especially the lack of wood-cutting, is not typical of even its recent history.

Some scholars, where direct written evidence of vegetation fails, argue from proxy evidence such as hunting. They should beware of preconceptions about the behaviour of huntable beasts. Those who infer 'noble forests' from hunting wild boar do not know their Homer[4] or their ecology. The wild pig of the south of France today, or of Sardinia (where he is introduced), as in Homer's Greece, lives in dense, low, evergreen maquis. Or in less than maquis: in La Mancha (Spain) we visited mountains covered with esparto-grass and sparse undershrubs, and our Spanish colleagues assured us that it was excellent pig country.

'Using up trees'

Shipbuilding and metal-working are often cited as proxy evidence for the existence, or depletion, or (more rashly) destruction of forests in antiquity. One hears it claimed that in Minoan times the Cretans had a sea empire; they therefore possessed a fleet; they therefore built ships; they therefore had plenty of timber; Crete was therefore well wooded; excessive consumption through shipbuilding contributed to the (supposed) disappearance of trees from the island. None of the links in this chain of argument has been tested; some of them are reasonable, but it is highly unlikely that they are *all* simultaneously true. Shipbuilding and metal-working must be set in a context.

Shipbuilding is evidence for the existence of trees, but not necessarily near dockyards: shipwrights can very well bring their material from a distance. In the Mediterranean (where most trees are crooked) there was little difficulty in getting curved timbers for shaping hulls, but long straight trees for keels and masts might have had to come from special sources.[5] Maritime history may show what was considered an outsize tree and where such trees came from.

Did shipbuilding consume timber faster than the growth of trees could replace it? This depends on its scale. In eighteenth-century England the merchant and naval fleets grew exponentially, through intercontinental trade and the arms race, to a size greater then the world had ever seen. England had only about 6 per cent of woodland, yet only at

---

[1] *Shepherd of Hermas*, Parable ix; Hermas was a first- or second-century Christian divine, who probably came from Arkadia in the Peloponnese.

[2] For an account of methods in Italian historical ecology in relation to the 'degradationist hypo-thesis', see Moreno 1990, chapter 1.

[3] H. Forbes, 'The uses of the uncultivated land-scape in modern Greece: a pointer to the value of the wilderness in antiquity?', Shipley & Salmon 1996, pp. 68–97.

[4] *Odyssey* XIX. 439–43.

[5] For the early middle ages, see M. Lombard, *Espaces et réseaux du haut Moyen Age*, Mouton, Paris, 1972, chapters 6–7. This work, however, makes rather more of the woodland of the time than the sparse written sources really say.

Fig. 11.1. Building a timber ship. *Near Bozburun, SW Turkey, July 1996*

the end of that century did shipbuilding begin to dominate the market for trees.[6] Mediterranean shipbuilding, with rare exceptions, was on a far smaller scale. Ships were not built to withstand ocean storms nor to carry great guns, nor did they make voyages to India. Minoan ships are more likely to have been the size of the longboat of Nelson's *Victory* than of the *Victory* herself. Most Mediterranean shipbuilding was comparable in scale with the building of caiques on Greek islands today.

An old shipwright wrote an account of building fishing-drifters at Lowestoft (England) in the 1910s.[7] The shipyards produced about ten vessels a year, each some 30 m long and 6 m wide – not quite as long as an Ancient Greek trireme but much wider and stouter, about one-third the size of the Roman cargo ship excavated at La Mandrague de Giens (Provence), and a little smaller than the Byzantine ship of Yası Ada (Turkey). Most of the hull timbers came from within about 30 km radius SW of Lowestoft, a much less wooded area than most Mediterranean countries today, though with plenty of trees in hedges. Even in this unpromising area, there was apparently no difficulty in getting timber, and no suggestion that the dockyard used up the supply.

Timber shipbuilding still flourishes in SW Turkey, which one of us investigated in 1996. Ships are not built in docks, but on any convenient piece of ground; they leave almost no archaeological traces. A typical ship, 25–31 m long (Fig. 11.1), is built of pine (*Pinus brutia*) with some elm and mulberry; other Turkish shipwrights use oak. The ship's frames are made from short, crooked, savanna-type pines, some fifty years old and 50 cm in diameter. Each frame consists of about twelve parts bolted together; a typical pine-tree produces about eighteen of these parts. The forty-five or so frames of a ship thus represent some thirty trees, which would have grown on about 0.4 ha of savanna. If we multiply this figure by three to allow for decks, planks and other parts of the ship, each ship represents about 1.2 ha of pines at fifty years' growth. The industry around Bozburun turns out some thirty ships a year, i.e. 1,500 ships in fifty years, and would thus be in equilibrium with the growth of pines on about 1,800 ha or 18 sq. km of pine savanna. This is much less than

the actual area of pineries on the Bozburun peninsula. If this is representative, it does not justify scholars' obsession with shipbuilding as the great consumer of trees.[8] One big fire in 1996 (p. 221) killed in a few days as many pines as the shipyards in Bozburun would use in a century.

Metallurgy and other fuel-using industries have been the object of modern tirades against 'deforestation'. Some authors believe that ancient civilizations depended on timber and wood to such an extent that they used up the supply and then went out of business.[9] Supporting evidence has been gathered with remorseless diligence. Every detail of life – using gypsum in Minoan palaces, recycling bronze and glass, using iron instead of copper, not roasting sulphide ores, re-locating industries, orientation of buildings, cooking practices – has been ruthlessly interpreted as a reaction to scarcity of trees. Even charcoal-burning hearths (Fig. 11.2) have been blamed for sterilizing the soil and causing erosion, although they occupy only a tiny area and were re-used on successive fellings.

Fig. 11.2. Charcoal, made in traditional earth-covered stacks, is an important cooking fuel in Crete. It is made almost entirely from olive logs, and has no effect on woodland. *Mourniés near Khaniá, July 1981*

For Britain this thesis was overturned thirty years ago. Charcoal and wood fuel were renewable. They came mainly from coppice-woods; to use timber incurred the extra cost of chopping it up. Industries often lasted a century or more, which implies that their fuel had grown and regrown and regrown during the industry's life. Their solvency depended not on the cost of trees but on other, bigger costs, including the labour-intensive processing of the trees. Industries that spread to other areas were expanding their production, not abandoning 'exhausted' forests. Industrialists had every reason to conserve their fuel supplies. Throughout Britain and Ireland, medieval woodland survives *more* often in places where there was an industry than where there was not.[10]

These considerations apply equally to the ancient world. A fuel-using industry was an investment in mines, buildings, equipment, animals and slaves and arrangements for feeding them, not to be jeopardized by allowing the fuel supply to run

6   Rackham 1980, chapters 10, 11.
7   T. Frost, *From tree to sea: the building of a wooden steam drifter*, Dalton, Lavenham, 1985.
8   Modern ships are at least as heavily timbered as ancient ones.
9   e.g. Perlin 1989.
10 M. W. Flinn, 'Timber and the advance of technology: a reconsideration', *Annals of Science* 15, 1959, pp. 109–20; G. Hammersley, 'The charcoal iron industry and its fuel', *Economic History Review* 2nd ser. 26, 1973, pp. 593–613; Rackham 1990, chapter 4.

out. Technologies that had the effect of economizing fuel *may* result from a shortage of trees, but may equally result from the supply of trees remaining constant but the demand increasing, or from shortage of labour to fell, chark and transport the trees; or from some reason that had nothing to do with trees.

A wood-burning industry tells us little unless we know how much wood was used, over how long a period, and how long the trees took to grow. The first may be estimated, working from the volume of slag left behind to the amount of metal that this represents and the wood needed to produce this metal. This is the basis of statements such as that the Roman Spanish silver mines consumed more than five hundred million trees during the four hundred years of their operation. This does not mean the conversion of 500,000,000 trees' worth of forest to non-forest. Trees consumed at the end of the 400 years are unlikely to have been already alive at the beginning. To estimate the area needed to support the industry we need to know over how many square kilometres the wood laid down by a year's growth of trees equals the wood consumed in a year at the mine. We shall try to do this later; but for most of the ancient world there is no way to know how fuel was organized.

To refute all the nonsense written about deforestation, point by point, would take up more space than the claims themselves. We warn against four general errors:
(1)   Forgetting that scarcity may be due to increased demand rather than reduced supply.
(2)   Disregarding alternative explanations. Scarcity of charcoal does not prove lack of trees.
(3)   Building an inverted pyramid of argument on surviving scraps of evidence.[11]
(4)   The all-too-common error of allowing enthusiasm to read into the documents statements that simply are not there in the original.

Forestry legislation

At conferences on forest history about two-thirds of the papers tend to be on the history of forest legislation and one-third on the history of forests. Many countries have centuries of forestry regulations. Study of these is a way to earn a Ph.D., but is of limited value for the history of forests. Modern legislators tend to be ill-informed and to have a high opinion of legislation: they pass regulations, to satisfy voters or the European Community, which there is little hope of enforcing. Earlier legislators, though not so bound by public opinion, must not be taken at face value.

Regulations are some indication of what people were in the habit of doing. A law forbidding pollarding is evidence that pollarding occurred. It is not evidence that pollarding ceased: indeed the object may not have been to stop pollarding, but to raise revenue from fines. Regulations are of most value at the lowest level (village bye-laws are more significant than national statutes). The regulations themselves are of less value than evidence of enforcement: how common were prosecutions? were the fines deterrent?

THE GREEK AND ROMAN PERIOD

Ancient Greece

Homer introduces maquis in his pigsticking scene (p. 167), but otherwise is not read for his landscapes. Homeric epithets, like 'woody Samothrake' and 'craggy Samos', are sometimes treated as one-word descriptions. It would be nice if they were – it might imply that the island of Samos was less wooded than it is now – but we do not know whether Homer intended them as such, or whether they were meant for contemporary Greece or for a vaguely remembered heroic past. The ancient Greeks, to judge by their surviving writings, knew less ecology than the Romans and Hebrews; like all the ancients, they were hampered by the difficulty of identifying plants before the medieval invention of wood-block printing made it possible to reproduce illustrations accurately.[12]

We know that the ancient Greeks had woodland management from farm leases written on tables of stone on the island of Chios. The farms had woodland (ἄλσος) which was expected to produce so much wood each year for ever, either for home consumption or by way of rent. The produce was measured by weight, showing that it was wood, not timber; we are tempted to render ἄλσος as 'coppice'. The quantities, a ton or two per year, are quite small. One clause is said to mean that the lessee was forbidden to let livestock browse on growing trees, but the text is damaged and the meaning uncertain.[13]

Ancient writers give scraps of information specific enough to compare with what is there now. Probably the earliest plant records in Europe are to be found in Hesiod, the writer of the seventh century BC, who lived in the Valley of the Muses in Bœotia, and made his farm implements out of prickly-oak, deciduous oak, laurel and elm.[14] All these still grow there. Several other authors, notably the fourth-century botanist Theophrastus, give particulars for Bœotia and Attica;[15] other parts of Greece are less well served.

The most useful topographer is Pausanias, the guidebook writer of the second century AD, who mentions notable vegetation. A learned commentary was written by Frazer, who went over the same ground in 1890–95;[16] one can compare Pausanias with Frazer and with what is there now. Pausanias mentions any woods that he encountered in terms that imply that they were somewhat of a rarity. Another late source is Hermas's vision of the twelve contrasted mountains of Arkadia (see p. 167). To anyone who knows Arkadia his vision

[11]   If the Emperor wrote to the proconsul of Africa, granting certain concessions to 'African shippers transporting wood suitable for public uses and needs', this does not mean that Italy had exhausted its forests and was having to turn to North Africa for supplies: *Codex Theodosianus* XIII.5.10; cf Meiggs 1982, pp. 258 ff.; Perlin 1989, p. 128.

[12]   O. Rackham, 'Ecology and pseudo-ecology: the example of Ancient Greece', Shipley & Salmon 1996, pp. 16–43.
[13]   A. Plassat and C. Picard, 'Inscriptions d'Éolide et d'Ionie', *BCH* 37, 1913, pp. 154–246; B. Haussoullier, 'Inscriptions de Chio', *BCH* 3, 1879, pp. 45–58, 230–55. Ancient leases some-

times forbid tenants to cut down trees, but all the others appear to allude to olive and fruit trees and are not evidence of concern to prevent deforestation.
[14]   *Works & Days*, pp. 435–6.
[15]   Rackham 1982.
[16]   Frazer 1898.

Fig. 11.3. Greece, showing places mentioned in the text.

vividly evokes the harshly romantic geology and vegetation, the limestone and flysch, the savannas and badlands and oakwoods, of the inland Peloponnese today – so unlike the politely romantic literary Arcadia.[17]

Soldiers and hunters might be useful sources, for their lives depend on having an eye for landscape.[18] In *Cynegetica*, Xenophon (or pseudo-Xenophon) hunts the hare or, on grander occasions, hart or boar, in a land divided into mountains and tillage, a land of rocks, thickets and occasional woods. He looks for boar in 'oakeries, depressions, roughs, meadows, fens and waters' – as hunters do in Spain today.

Fig. 11.4. Oakwood on the site of ancient Skotítas in the Peloponnese. The tree dead at the top is a sign of the intense competition for water in Greek deciduous oakwoods where coppicing has been long delayed. *July 1985*

A synthesis, comparing ancient authors point by point with nineteenth-century writers and with the present vegetation, has been published for Bœotia;[19] others are being prepared for parts of the Peloponnese. Where Pausanias or some other ancient author mentions a wood, sometimes there is a wood today, sometimes not. On the borders of Laconia and Arkadia he found 'the whole place full of oaks' sacred to Zeus Skotita, above the shrine of Artemis and the Nymphs at Káryai.[20] The site is now a great coppice-wood of deciduous oak (Fig. 11.4). Skotítas is probably the oldest named wood-lot still in existence, and the only surviving European wood harbouring its own god. Káryai (meaning 'nuts') is now the village of Arákhova (meaning 'nuts' in Slavonic); Artemis and the Nymphs may be represented by a chapel with plane-trees 11 metres in girth.

Are there woods today that were not there in Ancient Greece? The magnificent forests now on Mount Taygetos in

the Peloponnese appear not to be mentioned in any ancient author. They would surely have been noteworthy as a source of shipbuilding timber for Sparta and later for other states. Travellers' accounts, early photographs, and field evidence make it certain that they were of much less extent in the nineteenth century (Fig. 12.5); they may hardly have existed at all in Classical and Roman times.[21]

Comparing ancient with modern Greece, surprisingly little has changed, except for coasts, deltas and the absence of written mention of terraces (Chapter 6). Xenophon would instantly recognise most of nineteenth-century Greece, before the recent decline of cultivation and expansion of woodland in the mountains. However, in modern Greece one misses fens. In ancient times wetlands were abundant and probably useful as pasture, but nearly all have been destroyed in the last 150 years (p. 82).[22]

In Classical and Hellenistic Greece forests were uncommon, mainly in the mountains and the north. Modern scholars quote this as evidence of deforestation, but that is not what the ancient writers say. It is never expressly stated that a wood had existed earlier and was not there in the writer's time.[23] For the ancient Greeks it was normal that woods should be rare, remarkable, and often sacred to gods and nymphs.[24] They had no record of any different state of affairs, although Plato may have had a vague idea that woods had been more general in remote antiquity (p. 288).

Trees, timber and wood

Athens, giant of Greek cities, was not self-sufficient in food or timber. She had woodland of pine and evergreen broadleaves on Mounts Párnes and Kíthairon and in the Isthmus. Woodland could be private property: a list, graven in stone in 414 BC, of the lands of condemned criminals includes an oakery and probably a pinery attached to a house.[25] Athens was apparently self-sufficient in wood and charcoal, supplied by professional woodcutters. Demosthenes mentions a farm with six full-time donkeys carrying wood to market. Woodland was probably less extensive and certainly less well-grown than it is today. Wood was regularly brought to woodless islands like Delos.[26]

Timber for shipbuilding or big construction came from far away. According to Theophrastus, the sources were Macedonia (the best quality), the Black Sea, Thessaly, Arkadia, Eubœa and Mount Parnássos (the worst).[27] Other documents and inscriptions record minor supplies from all over the Aegean and even Sicily. The Athenians' foreign policy meant that they had to maintain access to the timber of Macedonia: if they failed in this, they had to do without a fleet. The quantities were not great, for a trireme was a huge, lightly

[17] 'Arcadia' is mainly a Renaissance development, based on the *Eclogues* of Virgil, who never saw the real Arcadia; see J. P. Mahaffy, *Rambles in Greece*, 2nd edn, London, 1907, pp. 290–93.

[18] R. L. Fox, 'Ancient hunting: from Homer to Polybios', Shipley & Salmon 1992, pp. 118–53.

[19] Rackham 1982.

[20] Pausanias III.x.6.

[21] O. Rackham, 'Laconia', forthcoming.

[22] O. Rackham, 'Ancient landscapes', *The Greek*

city from Homer to Alexander, ed. O. Murray and S. Price, Oxford, 1990, pp. 85–112.

[23] A single doubtful instance, from the northern fringe of the Greek world, is Theophrastus's statement (*De Causis Plantarum* V. xiv. 6) that at Philippi 'the whole plain was full of trees' when the Thracians lived in it (fourth century BC), implying that it was no longer so. 'Trees' can mean savanna, orchard or forest.

[24] As in other countries including Christian

Greece, *sacred* groves were specially protected.

[25] ΔΡΥΙΝΟΝ Κ[ΑΙ ΠΙΤ]ΥΙΝΟΝ: the first word means 'place of deciduous oaks', and the third, if the editor has correctly restored the damaged text, means 'place of pines': W. K. Pritchett, 'The Attic Stelai Part II', *Hesperia* 22, 1953, pp. 288 ff., pl. 84.

[26] S. D. Olson, 'Firewood and charcoal in Classical Athens', *Hesperia* 60, 1991, pp. 411–20.

[27] *History of Plants* V.ii.1.

built racing rowing-boat;[28] but specially long timbers were needed for keels and masts.

Attic vegetation fuelled the silver-mines at Láurion, whose output paid for imports of corn and timber. Theodore Wertime, in support of his thesis that fuel-using industries were the prime factor in deforesting Mediterranean lands, worked out that all the silver produced at Láurion would have called for one million tons of charcoal, implying that this is an unreasonably large amount and proves deforestation.[29] (He apparently did not realize that industrial Attica was less deforested than non-industrial Bœotia.) Such a figure is useless without knowing how long it took the wood to grow. A million tons of charcoal over five hundred years implies about 14,000 tons of wood a year. Even if one ha of maquis produced only 1 ton of wood a year, Láurion could have kept going for ever on only 14,000 ha of land. Maybe we should double this area, to allow for most of the smelting being done over a shorter period (c. 480–300 BC). This means that the Athenians were devoting one-seventh of their land area to a fuel supply for what was by far their biggest industry.

Pines in ancient Israel

In Israel pines (*P. brutia* and *P. halepensis*) have long been thought to be relics of aboriginal, or at least Biblical, vegetation; but a recent study shows that they have much increased and may not even be native. *Brutia* charcoal is rare, and mostly occurs in archaeological contexts with imported timber. *Halepensis*, whose charcoal can be distinguished from that of *brutia*, is not found at all.[30]

Crete[31]

Pausanias never visited Crete. All we have are scraps of evidence, fairly copious but unreliable, which imply a landscape not very different from the present. Theophrastus and Pliny knew of it as the homeland of cypress, then as now the principal wild tree of the island, especially in the western mountains. Cypress was exported as a special and precious timber, but we are not told how common it was. Plato (who was ignorant of Crete) says the island had little shipbuilding timber. Inscriptions defining boundaries record a landscape of mountains, rocks and caves, no more wooded than today. Crete, with its non-shade-bearing endemic plants, was a renowned source of drugs, exported by professional pharmacists and prescribed even to Caesar himself.

The Classical, Hellenistic and Roman periods, to judge by the archaeology, were a time of recession in human affairs, but probably not enough to give rise to a great increase in woodland. In the late Roman period, however, Crete recovered to a density and prosperity of settlement not seen since the Bronze Age, and was again very thoroughly used.

Roman Italy

Italy was less arid, more populated, and more vegetated than Greece. Russell Meiggs has summarized the abundant evidence for woodland.[32] Woodland was very extensive, mainly (but not exclusively) in the mountains. There was probably more than in the nineteenth century, but to prove this would involve site-by-site comparisons of which few have been done. Much of what Theophrastus, the Greek, says about trees and woods comes from Italy. Wetlands too were abundant and an asset especially of early Rome itself.[33]

Some ancient Italian landscapes

Among the volcanic craters 50 km NW of Rome lay the Ciminian Wood, scene of a famous espionage operation against the Etruscans in 310 BC. Three centuries later it was remembered by Livy the historian as a place 'more pathless and horrifying than the wilds of Germany recently; up to that time no trader had entered the wood'.[34] However, there were two Etruscan towns nearby, and the Roman army, once a way had been found, easily crossed the mountains by night. The massif is still wooded: maps of the 1950s show chestnut, coppice and beechwood symbols, with vineyards on the lower slopes.

The Ciminian Wood is the home of Italian pollen analysis: four cores have been taken from the crater lakes. None of these confirms the story of dense wildwood persisting into the Roman period and then turning into a landscape of woods, vineyards, chestnut groves and farmland. The wildwood itself, some six thousand years before, appears to have been distinctly patchy (Non-Forest Index 4 to 10). At Lago di Vico, in the heart of the massif, the core is poorly dated. In its top layers there are successive maxima of oak; hazel + elm + olive family; hornbeam + alder; olive family again; and finally chestnut. At Martignano, probably outside the Ciminian Wood, the core continues into Roman times with a modest fall in oak pollen, a rise in beech, and rises in some plants of open ground such as docks and horsetails, but no dramatic change. At Lago di Monterosi, now surrounded by farmland, there is a gradual decline in oak since the Roman period, but much tree pollen continues to the most recent layer. At Lago di Baccano, oak and pine continued to dominate until deposition was terminated by Roman drainage activity.[35] The area thus appears to have changed gradually from dense savanna to well-treed farmland, with no easily recognisable deforestation. We suspect that the pathlessness and the horror, unless they arose in Livy's vivid imagination, were due more to gullies in the soft volcanic rocks than to the vegetation.

Somewhat earlier than this are leases written on a table of bronze – the *Tabula Heraclea* – found near Pisticci in Basilicata in 1732. Dionysos, god of wine and of the Greek colony of Heraklea, founded in 432 BC near the coast, was

---

[28] The empty hull of the Athens trireme replica weighs only 25 tons: J. S. Morrison and J. F. Coates, *The Athenian Trireme: the history and reconstruction of an ancient Greek warship*, Cambridge University Press, 1986.

[29] T. Wertime, 'The furnace versus the goat: the pyrotechnologic industries and Mediterranean deforestation in Antiquity', *JFA* 10, 1983, pp. 445–52.

[30] G. Biger and N. Liphschitz, 'The recent distribution of *Pinus brutia*: a reassessment based on dendroarchaeological and dendrohistorical evidence from Israel', *The Holocene* 1, 1991, pp. 157–61.

[31] Rackham & Moody 1996; cf A Chaniotis, 'Problems of "pastoralism" and "transhumance" in Classical and Hellenistic Crete', *Orbis*

Terrarum 1, 1995, pp. 39–83.

[32] Meiggs 1982.

[33] N. Purcell, 'Rome and the management of water: environment, culture and power', Shipley & Salmon 1996, pp. 180–212.

[34] Livy IX.xxxvi.1 ff.

[35] Bonatti 1970; Frank 1969; Kelly & Huntley 1991.

letting out his land somewhere on the Agri, one of the badland rivers. His surveyors had divided it up, regardless of the terrain, into a grid of rectangles parallel to the river, a division not unlike that which is still archaeologically visible behind Metaponto nearby (pp. 280–1). The lots were not of identical size (had the surveyors been worshipping their employer?) and were typically of around 80 ha. About one-third of the land was arable; the rest was described as σκίρος and ἀρρήτος and δρυμός. The second word, 'unbroken', presumably means land that might be ploughed but was not. Δρυμός, 'oakery', is a common Greek word for woodland; we suspect it really means 'coppice', which many Greek oakwoods are. The first word, σκίρος, is a technical term for what looks like some form of coppice or maquis.

The lessees were required to build houses and barns, and to plant vineyards with olives among them. They were

> not to sell nor cut nor burn the wood in the oakeries, nor in the σκίροις . . . except wood for domestic purposes in building and [as stakes] in vineyards . . . or for use in their houses from the oakeries and σκίροις on each lessee's own lot.

The amount of roughland seems too much for domestic wood-lots; we are not told whether it was used also for grazing, or whether Dionysos was utilizing the remaining wood for something else.[36]

The lower Agri is not very different today. Arable land is more extensive on the broad silty plain, which tends to gullies. On the footslopes, where old badlands have eaten into the hills, we noted maquis on the north-facing slopes and patches of deciduous oakwood on south aspects.

A letter – a *belle-lettre* rather than ordinary correspondence – of Pliny the Younger about his beloved villa contains one of the few descriptions of an ancient Italian landscape:

> Imagine an immense amphitheatre . . . a broad and outspread plain is girdled with mountains; the mountains on their top part have tall groves and ancient. Abundant and varied is the game. From thence coppice woods run down with the mountain itself. . . . below these on every side vineyards extend . . . on whose lower edge bushes grow like a border.[37]

He goes on to mention the cornfields, flowery meadows, clover, and boats in winter on the infant river Tiber. The site was east of Arezzo. Everything he mentions, including the conifers on the mountains and the coppices clothing lower slopes, is characteristic of this part of the Apennines today, except that according to Pliny there were then no olives (p. 142).

Rome and its supplies
Rome was much bigger than Athens. There was a very high density of rural settlement over much, but not all, of the surrounding lowlands.[38] In comparing, say, Liguria with south Etruria (both well studied archaeologically) one gets a sense of division into mainly woodland regions and mainly farmland regions. (But sites are difficult to detect in modern woodland, p. 154.)

By the second century BC Rome outran the local farmland and became dependent on imported food. Rome probably consumed more timber and wood per head of population than Athens. Romans were lavish users of building timber; they indulged vastly in bricks, mortar, baths, iron, glass, funerary pyres, hypocausts and throwaway pottery, all of which required fuel. There were special city markets for timber and wood.[39] Unlike food, these were adequately supplied from within Italy, mostly from within 200 km of Rome itself, and seldom brought from outside the peninsula. Authors such as Vitruvius and Pliny discuss building timber. A difference from later centuries was the lack of chestnut, the common building timber of medieval and modern Tuscany: the Romans used it for nuts and wood, but rarely mention it as timber.

To supply Rome with half a million tons of African and Sardinian corn a year became the chief objective of Roman shipping. It called for a fleet comparable in size with that of seventeenth-century England. The ships were very big by ancient standards, some over 1,000 tons. Little seems to be known of where they were built or how long they lasted. (The wreck of Mahdia, Tunisia, dating from the first century BC, was 40 m long by 14 m in the beam, and carried a cargo of 200 tons of column-drums.)[40]

The iron-mines on Elba were a big fuel-using industry. In the early Empire iron was smelted on the island and sent to the mainland as blooms for further processing. This has sometimes been attributed to 'exhaustion of the forests' on Elba, making it impossible to complete the working on the island. Even Meiggs claimed this as one of the few records of diminution of forests in Italy. But this is not what Strabo and Diodorus Siculus say: they explain that smelting and fining were different industries, run by different groups of people, one on Elba and one on the mainland. We are not told why: it could have been due to lack of fuel on the island, but also to lack of labour or to some quite different cause.[41]

There is only one ancient mention of scarcity of metallurgical fuel. Pliny says that the re-heating of Campanian copper is done with charcoal 'because of the scarcity of wood'.[42] This does not make sense (unless the word *carbo* is to be read as mineral coal), because charcoal produces less heat than the wood it is made from.

Did the woods of Italy diminish during the Roman period?
It has been claimed that ancient authors in several places record diminishing forests. Only two of these withstand scrutiny.[43] Strabo says that the slopes of Avernus (an entrance to the Underworld), formerly 'wild, great-treed, impassable

36 R. Dareste and others, *Recueil des inscriptions juridiques grecques*, Leroux, Paris, 1913, XII.1.
37 Pliny the Younger, *Letters* V.vi.7 f.
38 Potter 1979, p. 120.
39 Meiggs 1982, chapter 8.
40 G. Rickman, *Roman Granaries and Store Buildings*, Cambridge, 1971; L. Casson, *Ships and Seafaring in Ancient Times*, British Museum,

London, 1994; A. J. Parker, *Ancient Shipwrecks of the Mediterranean and the Roman Provinces: BARIS*, 1992, p. 580. Bigger still were the two mysterious ships excavated on Lake Nemi. Although confined to a small lake, one of them was 73 x 24 m, one of the biggest timber ships ever built, but flimsy.
41 Meiggs 1982, p. 379; Strabo, *Geography* V.ii.6;

Diodorus V.18.
42 *Natural History* xxxiv.95–6.
43 Some of the others depend on misreading the text: Strabo's account of Pisa (*Geography* V.ii.5), given as an example by Perlin (1989, p. 116), does not say that the woods of Pisa had diminished, merely that they had once been used for Etruscan shipbuilding and were now being used for buildings near Rome.

wood' had lately been brought by the toil of men into cultivation at the instance of the great Agrippa. Diodorus says that the tyrant Dionysos brought timber from Mount Etna 'in those days full of sumptuous[44] firs and pines', implying that it was no longer so in Diodorus's own time.[45]

If ever an ancient people ran out of timber or wood through over-use, the Romans should have done so; but the evidence points in the opposite direction.[46] The Rome of the late Emperors and early Popes was no less prodigal with timber and wood than the Rome of Julius Caesar.[47] On archaeological evidence, ship design in the later Roman Empire was constrained by rising labour costs, timber being relatively cheap. Meiggs could find only uncertain scraps of evidence of shortage of wood (some of which are over-stated), and none of timber. One of the greatest Roman shipbuilding operations came after the end of the Empire, when Theodoric built a thousand large ships on the Adriatic coast, and did not have to scrounge every last tree to do so.[48]

The Romans were practical people, and wrote learned accounts of coppicing.[49] Why should they not have practised woodland conservation? Modern Italy (which never had a coal age) is still a wood-burning nation, and its woods are increasing. This could easily have been true of Roman Italy.

Mediterranean France

On the Languedoc plain, despite an increasing population, considerable tree cover – perhaps as savanna or wood-lots among farmland – remained in late prehistory. The trees were mainly deciduous and evergreen oaks; arbutus and box (regarded as disturbance indicators) increased. There was a curious local increase of elms, ashes and other moisture-loving, palatable trees, attributed to rivers spreading silt over the land and thus increasing the area of suitable soils. In the Iron Age and Roman times there was a transformation to something like the early-modern Mediterranean cultural landscape with cypress or juniper, olive and chestnut.[50]

Spain

Although Spain was an important Roman province, only Pliny and Strabo say much about it. Like dry Australia today, it produced wheat and wine. Olives were exported in pickled form, as witness amphorae found in eastern England.[51] Julius Caesar, though his *Spanish War* reveals disappointingly little

about the terrain, encountered the predecessors of the present olive-groves SE of Cordova.[52]

Spain was famous for esparto-grass, the fibre of which was used for ropes, cloth and many other things. It grew in

the plain called Spartarian, which means Rope-bearing. This is vast and treeless, grown with the rope-twisting *sparton*, which is exported to all places, but mainly Italy.[53]

Already in the Neolithic esparto had been much used in south Spain; even clothing was made of it.[54] Pliny says that the industrial source was an area thirty miles wide and a hundred miles long, extending inland from Cartagena. Even the mountains were covered with esparto. It could have been fetched from a wider area but for the difficulty of transport.[55] From Strabo's and Pliny's accounts it appears that the Plain of Esparto extended far inland, comprising most of Murcia and La Mancha and some of the coastal mountains. Esparto also grew in the extreme NE of Spain, but was not used. Another indication of semi-desert was the 'arid plain' along the Guadiana river south of Mérida.[56]

Woods and trees receive little mention. Turdetania (the Atlantic coast of Andalusia), one of the few good agricultural regions in Spain, built ships from local timber. (However, there were so few timber trees within six miles of Osuna that Pompey's men were able to cut them all down and take them into the town to deprive Caesar's army of them.) The mountains west of Málaga had 'thick and great-treed woods'. The hills east of the Plain of Esparto, south of Valencia, were bare, but in the mountains to the west there was a great oakwood. The south side of the Pyrenees was very wooded, including evergreen trees, but the north side is described as bare – a difference that still exists.[57]

Minor tree-products included edible acorns, presumably of *Quercus rotundifolia*. Eating acorn bread is mentioned as a proof of Spanish barbarism (although Italians cooked with chestnut flour). Acorns, however, then as now, could appear on polite tables. The Mérida area is described as the best source of scarlet-grain, the dried bodies of the insect *Kermococcus vermilio* used as dye; this is surprising, as the insect is supposed to live on *Quercus coccifera*, which we did not see there.[58]

The impression is of a distinctly more arid Spain than today, at least in the south and east, the best known parts. To the ancients its sterile mountains were very different from fertile, wooded Italy. Larger areas than now were occupied by

[44] Πολυτελοῦς, which may mean 'general purpose'.
[45] Strabo, *Geography* V.iv.5; Diodorus XIV.xlii.4.
[46] Within 200 km of Rome there are about 60,000 sq. km of land. If one-third of this was tree-covered it would be equivalent to 2 million ha of forest. If this produced 5 tons per ha per year (more productive than Attica, owing to higher rainfall) it comes to 10 million tons of wood and timber a year. Among a population of 1 million for Rome itself and ½ million for the rest of the area, this would make 7 tons per person per year. This very rough calculation shows that Rome was not obviously out of balance with its wood and timber supply.

[47] If people burnt branches, vine-roots, pine-cones and scrap timber, this does not indicate a 'distressing' fuel situation (Perlin 1989, p. 128). Why go to the trouble of cutting, logging, transporting and drying trees when Rome, like all cities, produced excellent fuel from its own gardens and building sites? (Cf Fig. 11.2.)
[48] K. Muckleroy, *Maritime Archaeology*, Cambridge, 1978; Meiggs 1982, pp. 152–3.
[49] Columella, *Res rustica* IV.xxxiii.4.
[50] Planchais 1982; Chabal 1997.
[51] J. J. van Nostrand, 'Roman Spain', *An Economic Survey of Ancient Rome*, ed. T. Frank, Johns Hopkins, Baltimore, 1937, vol. 3, pp. 119–224; P. R. Sealey and P. A. Tyers, 'Olives from

Roman Spain: a unique amphora find in British waters', *Antiquaries' Journal* 59, 1989, pp. 53–72 (we are indebted to Dr B. Juniper for this reference).
[52] Caesar, *Spanish War* 27.
[53] Strabo, *Geography* III.iv.9.
[54] Brenan 1957.
[55] Pliny, *Natural History* XIX.viii.30.
[56] Strabo, *Geography* III.ii.3. Ancient authors ignore the spectacular Roman buildings in Mérida city.
[57] Strabo, *Geography* III.ii.6; III.iv.2; III.iv.10–1; Caesar, *Spanish War* 42.
[58] Strabo, *Geography* III.iii.7; Pliny, *Natural History* XVI.vi.15; Strabo, *Geography* III.ii.3.

esparto, a semi-desert plant. Then as now, some of the mountains were woodless; it was not the esparto trade that kept them so, for more esparto grew than could be used. Even the semi-desert regions, however, were populated (like those in north Africa) and had towns and a network of Roman roads.

Did mining and metal-working in Spain 'use up' the woods? Were they responsible for the 'desertified' state of the country? At this point most modern commentators loose the reins of conjecture, unhindered by lack of evidence. It is claimed that the Roman Empire itself tottered when the emperors put base metal into the coinage, having run out of silver because of lack of wood to smelt it.[59] In reality, ancient authors describe the mines of Iberia in some detail,[60] but the only word surviving about their wood supplies concerns the pit-head baths.[61] It is not to be assumed that the Romans, masters of technology, somehow neglected to secure a permanent fuel supply; nor that monetary inflation necessarily happened because the trees on which the money grew had stopped growing.

The Plain of Esparto and other semi-deserts were not associated with mines, although some of the wooded areas were. Although the mines had been worked since the Bronze Age, they had not yet reached their heyday in Strabo's and Pliny's times, so they can hardly already have run out of fuel. Where they got their fuel is unknown. For all we know, the Spanish silver mines may have burnt 'aromatic shrubs', as mining districts are reported to have used c. 1830.[62]

The esparto areas are now largely agricultural plains, singularly lacking in trees except along watercourses. Much of the agriculture depends on irrigation, but not all (Chapter 17). It looks as though Roman Spain was drier than it is now, to the point where lack of moisture restricted agriculture as well as tree growth.

BYZANTINE, MEDIEVAL AND RENAISSANCE

During the decline and fall of the Roman Empire (400–1450 AD) the story becomes complex and obscure. Populations diminished, especially through plague (pp. 78–9), but immigrants came from outside. Written records and archaeology often become sparse or non-existent. For those who believe in a link between political history and vegetation history, this ought to have been a time of recovery and increase of woodland. (There is some charcoal evidence for this in the south of France.[63])

This impression, however, is partly due to the nature of the record. Populous and wealthy civilizations, likely to have made big demands on the production of the land, still flourished: in Crete down to at least 800 AD, later in Arab Spain and Sicily, and in much of Italy. Where the archaeology is particularly obscure, as in Crete from 800 to 1200 AD, this may be because artefacts are not distinctive enough to separate them from those of other periods.

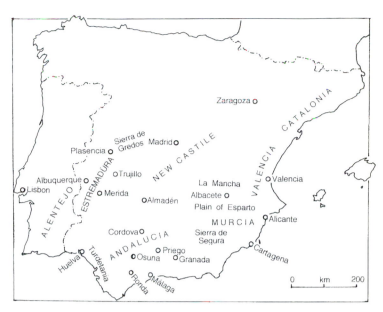

Fig. 11.5. Spain and Portugal, showing places mentioned in the text.

Spain

The Arabs conquered Mediterranean Spain in 711–14 AD. Their words for vegetation, as for many other things, have left a mark on the language and place-names: *alcornoque*, 'cork-oak'; *acebuche*, 'wild-olive'; *jara, xara*, 'cistus', etc. The town of *Albuquerque*, now in the oak savannas on the Portuguese border, is said to mean 'land of oaks'. Scholars draw a distinction between *algaida*, 'woodland' (there is a place so called in Majorca), and *moeda*, 'maquis' – but who is to say where Spanish Arabs drew the line?[64]

Medieval Arab geographers, such as Al-Edrīsi in the twelfth century,[65] continue to give the impression of an arid Spain and Portugal, possibly more wooded than in Roman times, but less wooded than later. The Plain of Esparto still existed in La Mancha and around Albacete, both of which place-names are said to mean 'plain'. Esparto was exported from Alicante to 'all maritime countries'. The plain, however, was far from uninhabited; most of the present towns have an Arab phase.

Sources of timber in Mediterranean Spain and Portugal were the Catalan mountains, the Sierra de Segura, the mountains between Granada and Algeciras, the coast near Huelva, and the coast south of Lisbon (Fig. 11.6). Pine, the principal timber, could be transported long distances: the roof-timbers of the Grand Mosque of Cordova, although not specially large, were said by Al-Edrīsi to have come from the pineries of Catalonia.[66] All these, except perhaps at Huelva, continued down the centuries and are the main areas of non-oak forest in Mediterranean Spain and Portugal today. Spaniards, then as later, did not regard live-oak as a timber tree.

[59] Perlin 1989, p. 126.
[60] E.g. Polybius, *History* XXXIV.9.
[61] An inscription forbids the bath-keeper to sell the mine's wood 'except the off-cuts of branches which are unsuitable for fuel'; C. G. Bruns, *Fontes*

Iuris Romani Antiqui, 7th edn, Mohr, Tübingen, 1909, pp. 289 f.
[62] Cook 1834, p. 226.
[63] Chabal 1997.
[64] Lautensasch 1959.

[65] Al-Edrīsi, 1901 edn.
[66] Later travellers claimed that the roofs were of a mysterious conifer (*Tetraclinis articulata*?) from north Africa. Modern restorers have thrown away nearly all the ancient timber.

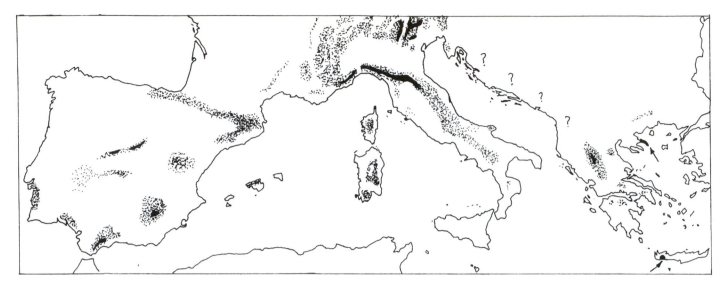

Fig. 11.6. Principal forested areas in medieval southern Europe.

Alfonso XI, king of Spain 1312–50, compiled a great *Book of Hunting*.[67] He was the mightiest hunter of all European sovereigns; he stuck great boars and smote raging bears with his royal hand. Wild boar was the commonest big game in medieval as in modern Spain. Bear was to be found in fewer sites all over Spain. Deer were rare – they are possibly more a woodland animal than the others.

Alfonso had an eye for practical detail, especially in veterinary medicine and field surgery. His third volume is a list of hundreds of *montes* throughout the then Christian kingdom of Spain, with notes of what beasts to hunt and when. For him the word *monte* meant not a mountain, nor yet a forest, but a place of hunting (*montería*). Each *monte* seems to have been a few square kilometres in extent. The author knew the localities, and reminisces about thrilling hunts: about 'where We slew two bears that were big enough for six' or when, between Priego and Alcála, SE of Cordova:

> One day We slew here a pig that slew two hunters and two alaunts [a fierce hound like a rottweiler] and one mule and tore a horse.

Vegetation is described in words such as *enzinar* (live-oakery), *xaral* (*Cistus* heath), *madroñal* (stand of arbutus) and *alcornocosa* (cork-oakery), in which Spanish is so rich. It is only occasionally stated whether trees comprised forest, maquis or savanna (p. 19). Several localities were called Colmenar, a place of beehives. Swine and bear were to be hunted in time of *belota* (live-oak acorns), *panes* (pannage, that is deciduous-oak acorns) or *madroñera* (tree strawberries, *Arbutus*).

In the late middle ages particulars of the extreme south are recorded in the context of the wars of Granada. As in earlier and later periods, the main forest area was the pineries and live-oakeries of the mountains around Ronda; some of these have later become live-oak savanna, recently invaded by pine. Much of the higher mountains was covered with cistus.[68]

Another documented area is the northern Sierra de Gredos in the middle of the Meseta. Here the forests of *Pinus sylvestris*, often thought to be an anomalous occurrence of a northern tree in a supposedly Mediterranean climate, go back at least to the fourteenth century. In the Little Ice Age they were more extensive than now, and supplied the best building timber to the country 150 km around. Much of the pine has been replaced by deciduous oak.[69]

Fernando Colón, said to be the son of Columbus the Explorer, explored Spain from 1517 to c. 1520, giving distances and the terrain between thousands of places.[70] At 83 points in Spain south of Madrid and on the Valencia coast (outside the main forested areas) we have compared his description with what is there now, mainly from our own observation. The results are shown in Table 11.i.

The broad outlines of the modern Spanish landscape already existed in the middle ages: the pineries of eastern Spain, live-oakeries of western Spain, cornlands around Madrid, esparto-lands of the SE, etc. Later changes have been in the density and local balance of vegetation. No widespread medieval types of vegetation have disappeared. Among new land-uses – pine plantations, eucalyptus plantations, greenhouses, citrus-groves – only pine plantations occupy any large fraction of Spain.

The mountain forests of medieval Mediterranean Spain were probably less extensive and less continuous than now; we have not checked this in detail, although the information is there in Alfonso and Colón for anyone willing to do plenty of fieldwork.

Outside the forested sierras the most notable changes since the 1510s are increases of cultivation (arable and olives) and increases of wild trees, especially live-oak and pine. (The pine increases are mainly the expansion of existing pine areas.) These have been at the expense of non-tree vegetation, especially cistus, rosemary and esparto. Olives and almonds have increased, partly at the expense of existing arable.

[67] Alfonso XI, 1983 edn.
[68] J. E. L. de Coca Castañar, *La tierra de Malaga* *a fines del Siglo XV*, Universidad de Granada, 1977, p. 42.
[69] Mancebo, Molina, Camino 1993.
[70] Colón 1517– .

Change	Number of instances	Localities
Wild trees		
Live-oak: increase	15–20	mainly NW New Castile
decrease	8	scattered
change from 'live-oakery'		
(*enzinar*) to live-oak savanna	17	mainly Estremadura
Cork-oak: increase	4	
Decidous oak: decrease	3	
Pine: increase (other than forestry)	7	mainly Murcia, Valencia
decrease	1–4	
Elm: increase	2	
Lentisk: decrease	>1	Estremadura
Other wild vegetation		
Cistus (or 'heath'): decrease	8	Estremadura, Andalucia
Rosemary: decrease	6	E. Spain
Esparto: increase	1	
decrease	8	mainly La Mancha
Cultivation		
Arable: increase	14	scattered, especially La Mancha
decrease	5	scattered, usually by change to other cultivation
unchanged	13	especially SE New Castile
Olives: increase	8	especially Granada
decrease	1	
Vines: increase	2	
decrease	3	
Citruses: increase	3	
Almonds: increase	4	
Pine plantations	5	E. Spain

Table 11.i. Principal changes in the landscape of the southern half of Spain between the 1510s and 1990s.

Live-oak has increased in regions where it was already extensive, and decreased in some outlying areas. Another common change (if it is a change) is from medieval (or early-modern) 'live-oakery' (*enzinar, monte de enzinas*) to what is now live-oak savanna, sometimes with cork-oak as well (Chapter 12).

Medieval Spain was apparently less polarized than it is today into western live-oak and eastern pine regions. There were more live-oakeries in the east than now, and occasional natural pinewoods in the west. Several other trees have declined, such as lentisk inland, and the arbutus whose fruits fed Alfonso's bears. Deciduous oak was less rare in Mediterranean Spain than it is now; probably it grew on deep soils which were later cultivated.[71] Alfonso and Colón often mention *texeda*, apparently meaning a wood or savanna of yew.

Sardinia

Sardinia was one of the most treed parts of the Mediterranean, but not uniformly. Around the cities of Cagliari and Sassari there was little woodland, and many charters from the fourteenth century onwards regulate timber- and wood-getting. The wood-rights of Cagliari lay in places extending fifty kilometres across the lagoon to the SW, and also in an area lying fifty kilometres across the mountains to the NE. Timber was less accessible than wood, and penalties for stealing it could theoretically extend to the loss of one hand. By a statute of 1429, Sassari's wood-rights extended thirty miles around the city.[72] The wood supply lasted more than five centuries, but transport was problematic: the cost of carting and guarding the wood exceeded that of cutting it.[73]

Italy

Italy declined after the Roman period, with lessening population and the abandonment of parts of the hills and of the more difficult lowlands.[74] This was succeeded by medieval recovery, the growth of nucleated villages, and later the growth of cities on almost the scale of ancient Rome – some of them, such as Venice and Genoa, in difficult terrain.

Presumably woodland increased, although this has seldom been archaeologically demonstrated. There was also an ill-documented shift towards chestnut becoming important as building timber and food (p. 68). For example, at Vico (Piedmont) in 1329 a property comprised a chestnut-bakery (*castanearum pistarum*) and a chestnut-grove (*castagnetum*).[75]

At least by the late middle ages, woodland conservation was well established. Woods and savannas were divided into various management categories according to whether they grew timber or underwood or chestnuts, and whether there was shredding or grazing.[76] The supply of wood as urban fuel was well organized. Probably in the fourteenth century a mountain canal more than twelve kilometres long was dug to supply Bologna with wood from the Apennines.[77]

Genoa was a large, flourishing city which had shipbuilding (naval and merchant) and iron-making as well as the ordinary urban uses of timber and wood. All these were supported out of the mountain woods of the small (5000 sq. km), densely populated territory of Liguria. These woods shared the mountains with wide areas of grassland (partly of prehistoric origin) at the same altitudes.

Venice

Venice was smaller than ancient Rome, and had a far smaller territory, extending into the Alps. Trees were used for the thousands of piles that underlie every building in Venice; for building timbers; for shipbuilding; and for wood (including brickmaking fuel). House timbers, from our observation, are

71 Deciduous oak was later disapproved of in the Sierra Morena because brood mares that ate it were supposed to miscarry; Cook 1834, p. 250.
72 Pinna 1903, *passim;* ASC, Ballero 2, pp. 26 ff.; ASC, vol 402 I, 403 V; cf p. 181.
73 Sardinia, unlike many countries, always had vehicles and roads.

74 N. Christie, 'Barren fields? Landscapes and settlements in late Roman and post-Roman Italy', Shipley & Salmon 1996, pp. 254–83; Potter 1979, pp. 120 ff.
75 G. Comino, 'Sfruttamento e ridistribuzione di risorse collettive: il caso delle Confrarie dello Spirito Santo nel Monregalense dei secoli XIII–

XVIII', *QS* n.s. 81, 1992, pp. 687–702.
76 For example, D. Moreno, 'The making and fall of an intensive pastoral land-use-system. Eastern Liguria, 16–19th centuries', *Rivista di Studi Liguri* A56, 1990, pp. 193–217.
77 G. Filippi, 'Un canale per la fluitazione nell' Appennino bolognese. Primi rilievi', *QS* n.s. 49, 1982, pp. 137–47.

larch and fir which must have come from the Alps; the piles are uninvestigated.

Shipbuilding is well documented through the researches of F. C. Lane. Hulls were of oak; masts of Alpine fir; oars (up to 15 m long) of beech. The total amount of shipping existing at any one time in the fifteenth and sixteenth centuries appears to have been between 50,000 and 100,000 tons burden. Since a ship lasted about twelve years, less than 8,000 tons were built a year (excluding boats and barges), which makes shipbuilding a relatively minor user of timber, although it called for large sizes and special shapes. Oak and beech came mostly from the hills west of Venice and the foothills of the Alps. At first the Arsenals (the State dockyard), as well as private shipbuilders, bought timber on the open market, but in the sixteenth century the State bought, stole and managed woods of its own.

Much of the timber came from outside Venetian territory, and in the later sixteenth century whole ships were increasingly bought from elsewhere, especially Holland. This has been interpreted as meaning that timber was the limiting factor and the local supply was being used up. This would not be surprising: Venice had an exceptional concentration of users of wood and timber on a small, though very wooded, territory. The Senate, moreover, had legislated harshly against the growth of suitable trees. Anyone who let his trees reach shipbuilding size risked having his woodland banned to the State by sound of trumpet: the hapless owner had his trees and their successors confiscated, and might have to provide free labour to transport them off his land. Nevertheless, the evidence that trees did decline is inconclusive. Censuses of oaks show an increase up to 1660,[78] and the Venetian navy seems never to have been short of shipping. Shipbuilding, more than any other industry, is international: shipowners were forever evading regulations forbidding them to buy from overseas. The Dutch, with far less local timber, could out-compete Venice with her old-fashioned ship design, poor dockyard layout, corrupt workforce and inefficient management.[79]

Greece

The surviving documents of the Byzantine Empire contain many scraps of evidence for natural vegetation; although these hardly add up to a coherent story, they enumerate activities in medieval Greece, especially Macedonia, from the seventh to the fourteenth centuries. There were various kinds of maquis and woodland, imperial, monastic or private property, managed under the direction of woodwards (οροφυλάκες). There was a regular trade in wood and charcoal to industries and cities (such as Thessalonica), and in timber for shipbuilding and other purposes – some of it (illegally) to the Muslims. The monks of Mount Athos lived from trading in timber, boards, wood and pitch. Other products included carobs, acorns, chestnuts, acorn-cups (for tanning), scarlet-grain (the dried bodies of the prickly-oak

scale-insect *Kermococcus vermilio*, used as a dye) and umbrella-pine nuts (from the Khalkídhike coast).[80]

There is little evidence for the extent of these activities, still less of increase or decline, but the impression given is of extensive roughland as well as cultivation, of maquis, savanna and forest, and of various and apparently stable methods of management. Nothing is known about fire or terracing.

Crete

For the post-Roman dark age in Crete there is one good piece of evidence, a pollen core from the southernmost peat-bog in Europe at Así Goniá.[81] This is now a remote spot in the western mountains close to the upper limit of settlement (800 m). It is surrounded by phrygana and a little maquis, with plane-trees around the bog itself, ribbons of woodland along the watercourses, and a thin scatter of big trees, mainly evergreen oaks (Fig. 10.1a).

The bottom of the core, dating probably to the sixth century AD, depicts a landscape of evergreen and deciduous oak, with plane and occasional olive. This could represent either woodland or maquis with occasional trees. Hornbeam and alder still occasionally survived. At this date there is little sign of human activity. Woodland was stable or increasing, and evidence for burning (in terms of charcoal fragments in the sediments) is low compared to later periods.

Between 900 and 1100 AD there is a sudden drop in tree pollen, especially evergreen oak; this is not accompanied by charcoal, and probably represents felling rather than burning. Next comes a brief period with little tree pollen, after which trees rise again to moderate levels about 1300. It is tempting to attribute the subsequent drop in tree pollen to a rise in population and prosperity in the early Venetian period, when innumerable chapels in west Crete were built.

Copious written records from 1200 AD onwards show that something like the present settlement pattern had already arisen. Medieval Crete was a fully developed cultural landscape. The island was covered with hamlets every kilometre or so, or with villages every 4 km or so; there was already a tendency for hamlets to disappear in favour of villages (Chapter 5). Settlement was cut off at the altitude of 800 m, above which there was mainly mountain grazing. The coast, exposed to raids by corsairs, was becoming depopulated.[82]

Cretan documents mention trees but rarely woodland. The traveller Buondelmonti, who perambulated the island in 1415, made particular mention of woods of cypress and pine, which then as now were extensive on the middle elevations of the western mountains. There was a trade in cypress boards, a precious timber for making chests and furniture, of which Crete was the only source in Europe. This trade is documented in the records of the Venetian authorities (p. 230). The chests themselves still exist in churches and as collectors' pieces as far away as England. Crete otherwise had rather less woodland, and far fewer trees, than now. The Cretans found local trees, as best they might, for ordinary construction, and

[78] A. Di Berenger, *Saggio storico della legislazione veneta forestale*, quoted in Lane 1934.
[79] Information in this and the preceding paragraph from Lane 1934.

[80] A. Dunn, 'The exploitation and control of woodland and scrubland in the Byzantine world', *Byzantine & Modern Greek Studies* 6, 1992, pp. 235–98.

[81] Atherden & Hall 1999.
[82] Rackham & Moody 1996.

used the bark of deciduous-oak roots for tanning leather. Ships for the navy were built in Venice of Venetian timber, and timber for public works was usually sent from Venice.

Existing trees, many of which date back to the Venetian period and a few beyond, show that savanna with pollards was more widely distributed than woodland. There were also olives and other trees among fields.

EARLY MODERN

Modernization upset the relations between people and vegetation. Rising population increased pressure on land, extended cultivation on to worse land, and increased the use made of the remaining wild vegetation. Shipbuilding reached its maximum consumption of timber in the nineteenth century, and railways succeeded it as a consumer of poor-quality timber trees.

Pressures were aggravated in some countries by Enlightenment attitudes to land-use and the growing interference of governments. Graver in its consequences than mere tree-felling was the Age of Reason's insistence that common land was bad and that trees and farming should be separated. The belief, often spectacularly ill-founded, got about that governments knew better than local populations how land should be managed. Professional foresters interpreted savanna and maquis as forest degraded by pasturage and woodcutting. They approved of harvesting the trunk of a tree, but disapproved (as some still do) of harvesting the branches by pollarding or shredding. The consequence was that legal or customary constraints on grubbing out trees were removed at a time of maximum population.

These destructive influences, however, often had not time to take full effect before depopulation set in and vegetation could recover.

France

Mediterranean France provides the classic example. The south of France had been very wooded; in Provence more than half the land area was describable as forest. Before the French Revolution in 1789 the larger 'forests', in general, had been common land; they belonged to landowners, institutions or the Crown, whose rights over them were limited by custom; local inhabitants had rights of pasturage, woodcutting, cork-cutting, etc. The importance of pasturage indicates that the 'forests' could not have been continuous but would have had some savanna character. A balanced and regulated system of wood-pasture, in which each party acted as a check on the excesses of the others, had apparently grown up, much as in the ancient and stable wood-pastures of England.[83]

Revolutionary bureaucrats hated multiple land-uses. Many forests passed under the control either of the state or of municipalities, who sought to manage the land for public profit and according to the doctrines of their forestry advisers. The new owners, in effect, stole the common rights and used the profit from growing trees to finance either the state's

wars or the public amenities of towns. This caused great ill-feeling: commoners refused to accept that bureaucrats could criminalize woodcutting and pasturage. Petty 'crime' of the early nineteenth century was dominated by actions, previously normal, that now constituted offences against forestry regulations.[84]

Similar conflicts between commoners, landowners, 'scientific' foresters and men in suits fill the proceedings of conferences on forest history. Disputes were widespread in France and Italy; they also occurred in Spain and (more recently) in Cyprus, Greece and Turkey.[85] Usually the foresters won, if not always completely. The effects on the forests are much less well known. The incentive to grow trees was increasing, not only for timber but for wood, charcoal (an industrial fuel where there was no coal, and also a big source of employment), and oak bark for tanning. This 'de-savannization' will be pursued in Chapter 12.

Spain and Portugal

Spain was an underdeveloped but also sparsely vegetated country, in which five thousand people constituted a city. It was beset by disasters both external (Little Ice Age, plague, Napoleonic Wars) and self-inflicted (expulsion of the Moors and ex-Moors).

In the 1570s there were reports on the state of agriculture and settlement, township by township. These establish that most of Spain had no local timber and often no wood. For example, in Old Castile most villages built houses of mud-brick, with pine timber brought from distant sierras; some had local firewood, but many had to buy it or use vine-prunings.[86]

Sources of timber were much as in the middle ages. The great survey of the mid-eighteenth century, the Catastro de Ensenada, lists trees, often to a suspicious degree of precision (there were 513,313,072 pines in the former Kingdom of Murcia). In most forest areas pines were predominant. In what are now savanna regions live-oaks and cork-oaks predominated. There is some evidence that deciduous oaks were more abundant than now. What are called 'new' trees, especially pines, were overwhelmingly predominant: this suggests either a period of increase or a rapid turnover.

Spaniards commonly believe that shipbuilding for the royal navy, especially the Armada of 1588, destroyed or depleted the forests of Spain; they have told us that Ferdinand and Isabella (1479–1504) pruned the whole of Spain, while the rest of the oak was cut for the Invincible Armada. In reality, the Armada was scrounged from all over western Europe except France. Of 132 ships of known origin, only 22 came from Mediterranean Spain.[87] Spanish shipbuilding was concentrated on the Atlantic coast. The Spaniards later had the sense to build warships, including the giant *Santísima Trinidad*, of mahogany in Cuba. Shipbuilding had less effect on timber supplies in Mediterranean Spain than it did in France or England.

Attempts were made to increase vegetation by legislation,

83   Rackham 1989.
84   Agulhon, 1982 edn, chapter 2.
85   M. Özdönmez, 'Les droits d'usage dans les forêts de l'État en Turquie', *Actes du symposium*

international d'histoire forestière [à Nancy, 1979], École Nationale du Génie Rural, des Eaux, et des Forêts, Nancy, 1982, pp. 213–15.
86   Bauer Manderscheid 1991. Vine-prunings are

an excellent fuel for ovens.
87   M. J. Rodríguez-Salgado and others, *Armada 1588–1988*, National Maritime Museum, Greenwich, London, 1988, p. 161.

for example by the proclamations of Zaragoza (1518) and of Plasencia (1567). Mostly this amounted to little more than commanding trees to grow.[88] Simultaneously, as in Venice, trees were (in effect) commanded not to grow through 'an ancient law by which the king was proprietor of every tree which his officers judged fit for any purpose of naval construction'.[89] In the 1790s Cavanilles, the state botanist, drew up a more practical proposal:

> It would be appropriate to divide the uncultivated territory of each [township] into six parts, designating five for pasture and wood, and reserving the sixth for plantations and woods, in which on no account should livestock be allowed nor wood be cut by the space of eight years, until the trees and shrubs should have regained enough strength. After this time cutting underwood, pruning and thinning the trees would be allowed, supervised by experts nominated by the local council. The livestock could be allowed in again, and this sixth part of the territory could now remain free, enclosing another by repetition for an equal number of years . . . and so successively in the remaining ones. In this way in half a century the whole realm could be planted. To ensure this operation it would be appropriate that the Mayor and Council should exact fines from infringers, and should pay when the delinquents could not be identified.[90]

Not even Spain, however, was authoritarian enough for this. Cavanilles did not foresee that his proposal would result in an increase of fires. However, he may have originated the unfortunate doctrine, which persists to this day, that the whole of Spain was once forested and ought to be forest again, regardless of climate and soil.

What did this amount to on the ground? Travellers in Spain included Bowles the Irishman in c. 1753, the learned and indefatigable Antonio Ponz from 1755 onwards, Swinburne in 1775–6, Cavanilles in c. 1793, Link the botanist in 1797–8, and Borrow the Gypsy scholar in 1835–6. The last two also visited south Portugal. They tended to cover the same ground; several travellers were fascinated by the quicksilver mine at Almadén (Estremadura), with its coppicewoods (then seldom recorded in Mediterranean Spain) that assured a permanent supply of fuel.[91] They tell a consistent story, which we summarize, from these six travellers, in Table 11.ii. The big changes – expansion of live-oaks and pines, decline of heathland, expansion of arable, change from arable to other crops – evidently happened mainly since these travellers wrote. Iberia two hundred years ago was more like medieval Iberia than like Iberia today. Even in the 1830s Borrow could write of Alentejo: 'the greater part consists of heaths . . and gloomy dingles, and forests of stunted pine; these places are infested with banditti.'[92] This area is now mainly oak savanna, with well-grown pinewoods in the west; heaths without trees are almost as rare as banditti. At the same time Le Play found around Almadén cistus 'forests' at least six metres tall.

Change	since 1750–1790 (out of 36 points)	since 1790–1836 (out of 57 points)	Localities
Wild trees			
Live-oak: increase	5–7	16–17	Alentejo, NW New Castile, Estremadura
decrease	2	2	
change from 'forest' or 'wood' to live-oak savanna	3	3	
Cork-oak: increase	2	13	mainly Alentejo
decrease	–	3	
Pine: increase (other than forestry)	–	8	mainly Valencia, Alentejo
decrease	3–4	2	
Elm: increase	2	–	
Other wild vegetation			
Cistus (or 'heath'): decrease	–	17	mainly Alentejo, Estremadura
Cultivation			
Arable: increase	4–5	4	scattered
decrease	2–3	6	especially Portugal
unchanged	4	4	
Olives: increase	4	1	
Pine plantations	7	7	scattered, especially Alentejo
Eucalyptus plantations	–	7	especially Portugal

Table 11.ii. Principal changes in the landscape of the southern half of Spain and Portugal between the 1750s and 1990s.

Timbers of buildings corroborate this story. From the middle ages to the nineteenth century, the building timbers of most substantial structures (including the massive doors of churches) that we have seen in south and middle Spain are pine, even where pine must have been carried 200 km. For humble buildings, timbers are poplar, which may have been local (although the Ensenada does not record much of it), and chestnut in Portugal. Spain avoids using boards: for example upstair floors are made of pine joists on which are laid bricks or special curved tiles to hold up an earthen floor.

Balearic Islands

These were supposedly one of the most wooded parts of Spain. A survey of Majorca in 1745 reported 2.38 million holm-oaks, 4.71 million pines, and about 100,000 other trees.[93] Their distribution among townships is illustrated in Fig. 11.7. However inaccurate the counts, the eighteenth-century woods were certainly either less extensive or sparser than they are now. Pine and holm-oak are still the commonest trees, and their relative distribution has not much changed; but pine has increased at least sixfold and oak roughly twofold.

In Majorca holm-oak, unlike Spanish live-oak, was sometimes used for big timbers. We have seen a pair of olive-

88 Among other examples of disobedience to such commands, Ponz noted that the trees supposed to have been planted along a new road where it entered villages 'could not be defended from the livestock'; Ponz 1778, IV.x.21.

89 Cook 1834, p. 223.
90 Cavanilles 1795, p. 228.
91 For example Le Play 1834. Mercury was needed in extracting New World silver; much of the historic output of Almadén is doubtless now

polluting Mexican and Peruvian rivers.
92 G. Borrow, *The Bible in Spain*, Murray, London, 1842 (many later editions), chapter 1.
93 Bauer Manderscheid 1991, pp. 142 ff.

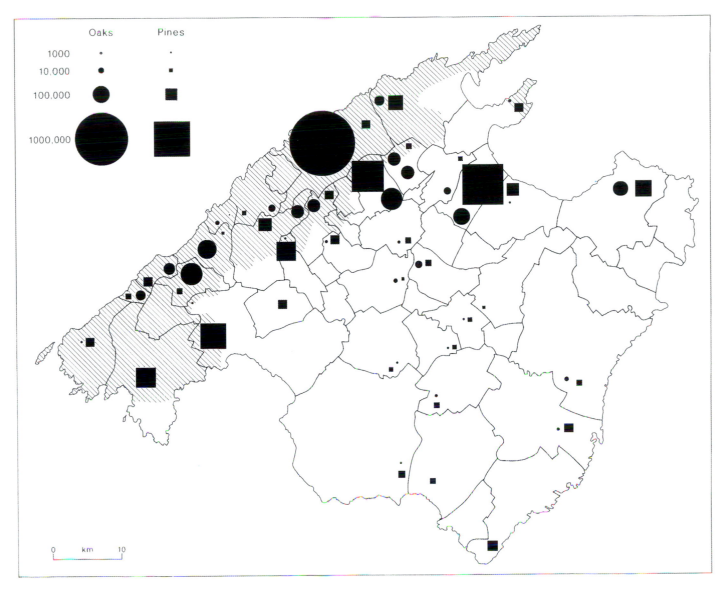

Fig. 11.7. Pine and oak in Majorca, 1745. Numbers of trees are converted to squares and circles on the basis that pines grew 5 m apart (40,000 to the sq. km) and oaks 6.8 m apart (22,000 to the sq. km). Pinewood now covers most of the northern mountains (hatched); much of it, especially in the west, is on old terraces.

presses, worked by levers pivoting on stone posts in the Ancient Roman manner,[94] in which each lever is made from two nearly straight holm-oak logs 13 m long.

Sardinia

Medieval arrangements for fuelling cities from distant coppices continued to operate. In 1717 the villages within 30 km of Cagliari, most of which had no wood of their own, were required (as a tenurial obligation) to provide 1,290 carts to take wood to the city from the woodland to the SW.[95]

Pietro Ballero wrote an account in c. 1805. Although the island was very wooded, agricultural tracts and areas round the cities lacked woodland; he claimed that timber had to be imported from Corsica and even Switzerland. He thought the woods were diminishing, despite recent statutes protecting

them, and brandished an eloquent pen against woodcutting, burning and browsing:

> Whoever tours La Nurra [the NW corner], whoever passes into the Barbagias [the mid-eastern mountains] and other parts of the Island cannot but be filled with grief at the sight of such great destruction of trees, of forests, of woods...[96]

J. W. Tyndale estimated that one-fifth of Sardinia in the 1840s was 'forest', especially of cork-, evergreen, and deciduous oak and chestnut. About eighteen thousand tree-trunks were exported annually to Marseilles and Toulon, which he thought excessive: 'a few years continuance of the present system will deprive the island of its great abundance of trees and superiority over every other part of the Mediterranean'. He claimed, nevertheless, that in the small district of Cuglieri

94 D. J. Mattingly, 'The olive in the Roman world', Shipley & Salmon 1996, pp. 213–53.

95 ASC: vol 402.I, 'Provviste Legno 1550–1792'.

96 ASC: MS Ballero 2, pp. 67 ff. and elsewhere.

there were a million trees, mainly deciduous oak, 'very good for shipbuilding'.[97]

At this time the Sardinian Crown, far away in Turin, was claiming ownership of many of the trees and inviting British entrepreneurs to exploit them. Concessions were granted to the infant cork industry (p. 208), and for timber. In 1839, for instance, Signori 'Grants Balfourt' and Domenico Giordano were allocated four thousand deciduous oaks in Sarcidano, in the remote middle of the island, if they paid more than other speculators offered.[98]

A survey of 1845 reported 110,800 to 124,800 saleable holm-oaks, some of them bigger trees than there are in the localities now. At Gutturumannu and Guttureddu, in the mountains across the lagoon from Cagliari, where wood had been cut for the city for centuries, there were 18,000 to 20,000 holm-oaks 2–3 m in circumference and 7–9 m long. At Capoterra holm-oaks had useful lengths of up to 12 m, suggesting that they were densely crowded, following some long-forgotten fire or logging.[99]

The great La Marmora – halfway through his career from political prisoner to governor-general – protested against the concession of a hundred thousand oaks in the middle of the island: repeated fellings, he said, had left so few deciduous oaks that the number had to be made up with the less valuable holm-oak.[100] Most of the localities are forests or savannas of deciduous oak and other trees today.

Italian railways used millions of oaken sleepers, some from Sardinia. In 1847, for example, it was proposed to sell 80,000 holm-oaks from the limited woodland of Bosa. The average tree contained 1.2 cubic metres, equivalent to about eight sleepers.[101] After 1867 Sardinia acquired railways of its own. The railway companies themselves negotiated timber concessions.

Fattening pigs was a lower-class use of woods and *saltos* (savannas), little mentioned by general writers, but filling twelve mighty boxes in Cagliari archives with letters on 'Estimates, Acorn-Bearers, 1838–1857'. Bureaucrats wrestled with common rights of pannage, payment and competition with other uses on lands claimed by the Crown. Nearly every township was involved. In 1850 Laconi, with its extensive mountain woods, had about 6,980 great pigs called *mardiedu*, plus 54 yearlings, belonging to about a hundred owners. Capoterra, whose woodland kept Cagliari in wood, had 811 pigs assigned to 18 localities, plus 215 from places without woodland up to 30 km away. Villamassargia divided its woods into 56 localities, each with a Sard name; 39 of these had from 5 to 15 pigs each; the other 17 wood-lots had no pigs (was this because they had recently been felled?). Returns were sent in, year by year as the acorn crop varied, for each of the roughly three-quarters of townships that had woodland.[102]

Travellers continued to comment on what they thought was the destruction of the forests, and the evils they expected to follow. For example Tennant, while asserting that reports of destruction were exaggerated, drew the conventional Ruined Landscape conclusions:

Large areas have been completely denuded of vegetation and consequently the meteoric waters run off the lands more rapidly than before; the rivers become more swollen in the rainy season, and are dry at an earlier stage in the summer, and the slopes of the hills are more easily denuded of soil; . . . the earthy matter, brought down by the streams when in flood, gradually raises the beds of the rivers and obstructs their channels, occasioning more frequent overflows; and the water lies for a longer time on the lowlands, creating new swamps and increasing the malaria on the arrival of the hot weather.[103]

Destruction continued into the 1900s, according to a colourful account by D. Lovisanto. Despite the de-wooding, there was still a big wood and charcoal trade. Down the Barbagia mountain railway came eight trains a day, each bearing 18 tons of charcoal – which would represent over a year the annual growth of about 700 sq. km of trees.[104]

Building timbers in Sardinia tell a rather different story. Timber was apparently scarcer and smaller than we would expect from the archives. Juniper – which cannot have come from forests – is unexpectedly common (p. 209). Holm-oak is abundant, usually small and crooked. Biggish timbers are sometimes yew; *longufresu*, one of the Sard words for yew, means 'long beam'.[105] Imported sawn boards of pine are abundant. In a remote nineteenth-century farm, set among huge oaks, we found pine timbers 10 m long, which must have come from outside Sardinia.

Sardinia had come to terms with its woods from the middle ages onwards. Pressure on land was never very great. Mining areas, as in the SW, had regular supplies of timber and wood, and were still well-treed in the nineteenth century. There have been interchanges between forest, savanna, coppice and maquis, and between holm-oak, cork-oak and deciduous oak. In the late eighteenth century, for some reason, timber trees seem to have increased, and were then depleted by nineteenth-century fellings. Woodcutting collapsed in the twentieth century and the trees grew back.

In recent decades bulldozing and planting have taken their toll, and fire has taken over some of the functions of woodcutting. However, Sardinia is still increasingly wooded. Most of the great wooded areas are still recognisable. Often they contain *nuraghes*, megalithic field-walls and stone-piles, proving that they were not formed out of virgin wildwood. The woods that supplied Cagliari are now vast tracts of maquis and woodland, not cut for several decades; giant ancient coppice stools and charcoal hearths testify to centuries of fairly intensive use. Although impenetrable with arbutus and tree-heather, they appear to hide earlier antiquities. They are little used today, apart from cutting cork from scattered cork-oaks. Sardinia now imports wood from Italy, and we had great difficulty in finding stumps on which to count tree-rings.

[97] Tyndale 1849.
[98] ASC: Regio Demanio, Boschi e Selve, b.14.
[99] *Ibid*.
[100] La Marmora 1849, 1860.

[101] ASC: Boschi e Selve, b.14, 'Tentative di Vendita Elci per Strade ferrate'.
[102] ASC: Regio Demanio, Boschi e Selve, Estimo Ghiandiferi, b.4

[103] R. Tennant, *Sardinia and its Resources*, Spithöver, Rome, 1885, pp. 95–108.
[104] D. Lovisato, *Sul disboscamento della Sardegna*, Cagliari, '1900' [but mentioning events in 1902].
[105] Information from Professor I. Camarda.

ITALY

In Italy the histories of individual wood-lots can sometimes be followed, comparing the documents with what is there on the ground. The results are similar to those well known in England.[106] For example, the Bosco della Pianora near Pisa is a wood of about 5 sq. km, documented since the middle ages; its complex boundaries – defined in part by hedges of maple (*Acer campestre*) – have changed little for at least two hundred and fifty years. It was a coppice-wood since at least the sixteenth century, felled (according to the evidence of living coppice stools) every ten years or so. Originally it was mainly of deciduous oak and chestnut (including nut-bearing varieties), with some elms. Since the eighteenth century it has been invaded by pines, and more recently by *Robinia*, the suckering North American tree. These changes were helped by outbreaks of chestnut-blight (*Endothia parasitica*, also an American introduction) and 'Dutch' elm disease. Increasing pines have brought increasing fires, which help *Robinia* to spread.[107]

Ligurian Apennines

In parts of Italy woodmanship developed to a level of detail seldom equalled anywhere in Europe. Around Genoa, since the seventeenth century if not earlier, it was apparently the practice to train oaks into curved or forked shapes specifically for shipbuilding. This is also reported from around Toulon and Venice: evidently the shipyards of those ports had gained some control over the woods, at least by paying extra for specially shaped trees (as shipyards in England did not). These practices are described in a manuscript by D. G. Pizzorno, a landowner whose woods were managed in several different ways but whose main interest was in fuelling iron-works.[108]

Industrial fuel production was well organized. Ironworks were not the rapacious, short-lived enterprises of popular and scholarly myth. They belonged to long-established families who owned fuel supplies in woods and savannas. Far from destroying or depleting their property, they managed the trees to combine fuel-growing with an increasing output of chestnuts for food (for their farm tenants and for sale, p. 68), and sometimes with timber production. They did not like to use timber for fuel – few industrialists do – but preferred 'charcoal rods', coppice or pollard poles of 7 cm diameter. These were produced in various ingenious ways from trees that also yielded nuts of cultivated varieties. The investigators conclude that 'no evidence has been found that the tree cover of this area was degraded before the nineteenth century'; indeed, many of the mighty chestnut stools have withstood *Endothia* blight and are still there.[109]

The Bosco Ramasso, a non-chestnut wood in Liguria, was the scene of two curious practices. Part of it was maquis, cut on a short coppice cycle, with timber trees of pine. The latter were shredded to yield branches for industrial fuel, and later felled for timber. (Shredding would also reduce the fire risk.) Another part was grassland with stools of deciduous trees of ash, hornbeam (*Ostrya*) and alder. The grasses and coppice shoots were cut every year and made into hay. Likewise in Friuli, on the north edge of Mediterranean Italy, alders in meadows were supposed to increase grass production, probably through the nitrogen-fixing action of root-nodule bacteria.[110]

In the sixteenth to eighteenth centuries stock-farming and even agriculture expanded into what had earlier been great woods. For example, the Bosco di Savona in east Liguria, once a near-continuous mountain wood of some 6 sq. km, had by the end of the eighteenth century been eaten away by farms so that only the steeper slopes remained wooded. In contrast, the rather earlier settlement in the western half of the Boschi d'Ovada, in west Liguria, did not, mostly, destroy the wood, but exploited it more intensively by woodcutting and chest-nut-cultivation.[111]

There were intermediates between woodland and agriculture. A curious practice called *ronco* in the mountain beech-woods, established by the seventeenth century, was to fell the trees, burn the branches, sow cereals between the stools for two years, and then allow the trees to grow up again. The fire was supposed to encourage sprouting. This coexisted uneasily with the intention of the Republic of Genoa, as owner, to use the beeches for growing oars – as in ancient Athens, oars for state galleys were an important speciality. The sites now have huge, ancient coppice stools of beech.[112]

The many early maps for Liguria show the relation between woodland and grassland. There was a modest loss of woodland down to the nineteenth century, attributable less to industrial over-exploitation than to over-population and grubbing out for farmland. Field evidence indicates that cartographers, then as now, had no way of showing intermediates between woodland and non-woodland; some of what they depicted as woodland was really savanna. In the twentieth century the ancient grasslands, rich in wildflowers, have greatly declined through under-use and invasion by trees. This is abundantly clear from field evidence and the excellent photographic record. In less than a century grassland has passed from its historic maximum to its least extent for two thousand years.[113]

[106] Rackham 1990.

[107] P. Piussi and S. Stiavelli, 'Dal documento al terreno: archeologia del Bosco delle Pianora (Colline delle Cerbaie, Pisa)', *QS* n.s. 62, 1986, pp. 445–66.

[108] D. Moreno, 'Querce come olivi. Sulla roveri-coltura in Liguria tra XVIII e XIX secolo', *QS* n.s. 49, 1982, pp. 109–36.

[109] Di Stefano 1986, '"Castagneti aggregati a massarie": trasformazioni nella castagnicoltura a Voltaggio nella seconda metà del '700', *Studi di Storia e Archeologia Forestale: studi in memoria di Teofilo Ossian de Negri*, ed. D. Moreno and

M. Quaini, Istituto di Storia Moderna e Contemporanea, Genova, 1982, pp. 124–37; Baraldi & others 1992.

[110] Moreno & others 1993; M. Guidi and P. Piussi, 'The influence of old rural land-management practices on the natural regeneration of woodland in abandoned farmland in the Prealps of Friuli, Italy', Watkins 1993, pp. 57–67.

[111] D. Levi, '"Una memoria oscura ed incerta". La toponomastica del "Bosco di Savona" in una fonte settecentesca', *Studi di etnografia e dialettologia ligure in memoria di Hugo Plomteux*, ed. L. Còveri and D. Moreno, SAGEP, Genoa, 1983,

pp. 85–105; D. Moreno, 'La colonizzazione dei "Boschi d'Ovada" nei secoli XVI–XVII', *QS* 24, 1973, pp. 977–1016.

[112] D. Moreno 1990, chapter 5.

[113] M. Quaini, 'I boschi della Liguria e la loro utilizzazione per i cantieri navali: note di geografia storica', *Rivista Geografica Italiana* 75, 1968, pp. 3–32; D. Moreno, 'Une source pour l'histoire et l'archéologie des ressources végétales: les cartes topographiques de la montagne ligure (Italie)', *L'œil du cartographe et la représentation géographique du Moyen Age à nos jours*, ed. C. Bousquet-Bressolier, Paris, 1995, pp. 175–98.

Venice

Most of the big sources of Venetian timber are still forest today. An example is Montello, the karst massif 45 km north of Venice. In 1471, that fearful year when the Turks depopulated east Crete, the Senate had confiscated the oaks in the forest, estimated as enough for more than a hundred galleys, which is surprisingly little for an area of over 50 sq. km. The forest was successfully maintained by the State, used as a source for crafts and activities, as long as the Republic lasted and even beyond, down to 1848. Only after the unification of Italy were common rights abolished and preparations made to destroy the forest, described (of course) as degraded. An account published in 1905 relates how legal obstacles had been overcome, roads and check-dams built, and half of it allotted to peasants in 2½-ha plots.[114] This attempt to destroy Montello was short-lived; by 1971 about two-thirds of it was mapped as forest again.

Molise

This province illustrates how in the nineteenth century wood-pastures in lowland Italy were liable to be privatized, made into farms, and removed from the map. Almost all the lowland 'forests' of Molise thus disappeared, those of the mountains being less affected.[115]

Basilicata

Works on Basilicata begin by claiming that it was very forested in Antiquity; this appears to be based on the notion that its Roman name, Lucania, is derived from the Latin *lucus*, 'grove'.[116] There are occasional complaints of, or prohibitions on, destruction of forests from the sixteenth century onwards; it is not known what lies behind them.

From 1826 onwards the Kingdom of Naples kept meticulous records and maps of the extent and composition of woodland from the modern forester's point of view. Although coppice-woods and pasture rights were well recorded, there is little evidence of savanna, more likely because there was no official category for it than because it did not exist. Between 1826 and 1950 about one-third of the woodland of Basilicata became arable land or pasture (Fig. 15.11); this is attributed to rising population, shortage of land, and the privatization of commons removing legal restraints. Much of the total, however, was due to the grubbing for arable of certain very large private woods, such as those that covered most of the township of Montemilone in the north of the province. A minor cause was pasturage following a fire.[117]

Greece

The main sources of evidence are travellers' accounts, together with the story told by the vegetation itself, especially old trees. Written sources emphasize cultivation and woodland; maquis and phrygana were taken for granted and are less often described.

Attica and Bœotia[118]

Numerous travellers' accounts can be compared, point by point, with what was visible in the 1980s. Bœotia was a land of cultivated plains, with maquis and phrygana forming mosaics on the hills, small sacred groves around chapels, and woods, mainly of pine and fir, confined to the mountains. Attica had some bare hills, such as Hyméttos famous for its honey, and more extensive mountain and lowland pinewoods. There was not very much terrace cultivation.

These variations, existing since Classical times, were familiar to the travellers. For example, Wheler in 1675 wrote of a lonely khan on Mount Parnes:

> there is a very curious Fountain hard by it, where the Wolves, and Bears, and Wild Boars come to drink; to which this Mountain is yet a great Covert. For it is indeed almost covered with Pine-trees; of which we made a good Fire to keep us from the cold; and stopped up the entrance into the *Kan*, to secure us from the Assaults of Wild Beasts.[119]

The beasts vanished in the nineteenth century, but the pinewoods remain. Industries in Attica included sawing boards and making pitch and resin. Du Loir found the Attic hills in 1641

> covered with old trees, particularly Terebinths, Lentisks, Firs, Myrtles, and Pines which have a hole in the base, from which resin comes out.[120]

Resin-tapping has much declined (more retsina wine is now made than ever, but most of it contains little resin), but the scars that it leaves are familiar on Aleppo pines all over lowland Greece.

In contrast Leake, travelling in 1804–5, remarks of Bœotia:

> The great level [of Thebes], as well as the long slopes of the hills to the right [south] of it, is for the most part a cultivated corn-field, without a single fence . . . as well cultivated a district as any in Europe.[121]

In Bœotia was the wonderful fluctuating Lake Kopáis. Its unique vegetation, aquatic animals, and craft of musical instrument making had been celebrated by Theophrastus, Pliny and Plutarch.[122] This vast basin, some 18 by 11 km, filled itself every spring from melting snow in the mountains and dried up in autumn. Excess water overflowed through caves at the edges of the lake. Kopáis was valuable grazing-land, the source of fish, peat fuel, wildfowl and wild swine. The lake was destroyed by the British Lake Copais Company in the late nineteenth century. It was made (with difficulty) into ordinary farmland on lake marls and shrunken peat, not unlike the English Fenland.

Apart from the destruction of the lake, the great change of the last three hundred years has been the advance of woodland, especially pine spreading out from the edges of existing pinewoods. On the Attica–Bœotia border, pines have steadily marched out from the base of Mount Parnes at the rate of

[114] P. Bertolini, 'Il Montello: storia e colonizzazione', *Nuova Antologia* 200, 1905, pp. 68–94.
[115] Di Martino 1996.
[116] The ancient Lucanians are unlikely to have spoken Latin!
[117] Tichy 1962.
[118] Rackham 1982.
[119] G. Wheler, *A Journey into Greece*, London, 1682.
[120] Du Loir, *Voyage du Sieur du Loir, contenu en*
plusieurs lettres écrites du Leuant, Paris, 1654.
[121] Leake 1835.
[122] Theophrastus, *History of Plants* IV.x.7 to IV.xi.11; Pliny, *Natural History* XVI.164–71; Plutarch, *Life of Pelopidas* XVI.3–4, *Life of Sulla* XX.4–5.

about 100 m a year for three centuries. Forest fires have marched behind the pines (Chapter 13). Mountain fir-woods have also increased, and recently there has been a huge outgrowing of pricky-oak maquis into woodland (Fig. 4.30).

Peloponnese

The Peloponnese is a grand and wonderful land. It has fir-woods on erosion-slotted limestone, oakwoods on marls and metamorphics, plunging badlands mantled in pinewoods, peaks soaring far above the tree-limit, internal basins of farm-land and steppe, deserts of crumbly schist, barren gravel river-beds and populous mountains of rugged fertility. For the ancients, beyond fashionable Corinth and sporty Nemea lay the mysterious, fearsome land of Pan and the Nymphs, the Erymanthian boar, Lake Stymphalos of the terrible birds, Sparta the city of terrible men, Mount Lykaon where Zeus's priests liturgically turned into wolves, and the river Styx of hydrofluoric acid.[123]

Theophrastus gives details of many kinds of Peloponnesian tree, but without giving the impression that forests domi-nated the landscape. Strabo says that Arkadia had been much depopulated by wars and was famous for horses and asses.[124] If Hermas (p. 167) is to be believed, the landscape in the Roman period was very like that of the nineteenth century.

Travellers in the Peloponnese included Leake the British agent in 1805, the Bory de St Vincent expedition in 1830, and Philippson the German in 1888. There had been Venetian maps and censuses in the 1690s.[125] We refer in other chapters to recent surveys in the Argolid. We have taken part in the Laconia and Nemea surveys, and have visited the scene of the Olympia studies (pp. 291–5).

Throughout modern times the Peloponnese has been settled mainly in the wetter west, with most of the present villages already in existence. The highest densities of popula-tion have been on some of the rugged metamorphic moun-tains, especially Mount Párnon.[126]

Modern changes, as elsewhere, include the drainage of the fenny depressions, although Lake Stymphalos continues to hold out. Travellers report a somewhat more arid-looking, much less vegetated landscape than today, with occupational burning but apparently less wildfire. Browsing has since declined: in 1984 we could find none in much of Laconia.

In most of the Peloponnese there is copious evidence for increase of trees. Views and photographs of the deserted Byzantine city of Mistra, from 1831 onwards, chronicle the invasion of the site by trees. A better example still is Mount Taygetos, where Philippson's remarks about the limited (and supposedly diminishing) extent of woodland in 1888 are corroborated by photographic evidence of its subsequent increase (Figs. 4.25, 12.5).

In many parts of the Peloponnese young woods of pine and deciduous oak are now springing up on heathland, terraces

and badland. Maquis also turns into woodland. The pine landscapes are becoming fire-dominated; fires in oakwoods are less hot and destructive.

Such recent woods can be distinguished from the ancient woods mentioned by travellers, e.g. Skotítas rediscovered by the 1830 expedition. Some of the latter have expanded within the last hundred and fifty years and are thus partly recent woodland. Ancient woodland does not have boundary earth-works as it would in NW Europe, but does have big coppice stools (the effect of centuries of felling and regrowth) and distinctive plants.[127] The dominant oak, *Quercus frainetto*, is reluctant to produce acorns and has very little power of spreading, unlike other deciduous oaks. The ancient woods also have a distinctive flora, often of plants outlying from NW Europe, such as wild strawberry (*Fragaria vesca*), primrose (*P. vulgaris*), and cowslip (*P. veris*); even the fungi have more in common with middle Europe than with the local environs.[128]

The Peloponnese, like many other regions, is recovering from a period of great pressure on land. Phenomena that might be termed 'degradation', however, are correlated more with geology than with the degree of pressure. Some of the best-vegetated areas, including Skotítas, are in rugged, fertile mountains with a history of dense population. The badlands and semi-deserts (pp. 283–4, 326–7) are on particular types of marls, schists and unconsolidated sands and gravels with below-average human activity.

Píndos Mountains[129]

The province of Grevená is a borderland, very untypical of Greece. Mediterranean-type vegetation is confined to low altitudes (< 400 m) or to limestone: olive, vine, fig, prickly-oak, lentisk, thyme and all their fellows are rare or absent. The main cultivation is arable. Terraces are rarely seen: every-thing too steep to plough is pasture or woodland. Hundreds of square kilometres of deciduous oakwood and oak savanna give way to pine and beech in the mountains. The soft, erodi-ble rocks form splendid badlands (p. 283). Permanent settle-ment, in the form of widely spaced big villages, ends at about 1,200 m, above which is a belt of large transhumant villages inhabited for half the year by descendants of the medieval Vlach nation, made famous as the 'Nomads of the Balkans' by the Wace-Thompson expedition of c. 1910.[130]

Grevená gives the impression of much more relaxed historic use than Bœotia or Laconia. However, every scrap of land has certainly been used for something. Even the vast woods have been thoroughly coppiced and pollarded. How people found the time and energy to keep up with the growth of trees is partly explained by the climate, with its very short growing season: Grevená has the ferocious winters of the Balkans and the ferocious summers of the Mediterranean. People would thus have spent much time cutting firewood

[123] Pliny, *Natural History* VIII.81; Pausanias, VIII.ii.4, xviii.5.
[124] *Geography* VIII.viii.4
[125] Leake 1830; Bory de St Vincent 1836; Philipp-son 1892; Sauerwein 1969.
[126] P. Topping, 'Premodern Peloponnesus: the

land and the people under Venetian rule (1685–1715)', *Annals of New York Academy of Science* 268, 1976, pp. 92–108; Sauerwein 1969.
[127] cf Rackham 1990.
[128] These are not the localities where north Euro-pean trees survive from before the Aridization, p. 145.

[129] This summarizes observations made by Dr Jennifer Moody and Oliver Rackham when serving with the Grevená Archaeological Survey at the kind invitation of Professor Nancy Wilkie.
[130] Wace & Thompson 1914. Vlach is a language close to Rumanian.

and leaf-fodder; also the oaks grow very slowly, taking 30–75 years even to produce wood. Oaks are used for many purposes, including acorn bread as food in times of famine (p. 195).

The area lacks conspicuous antiquities; even the churches are seldom earlier than 1700. Archaeology seems to be thinly scattered, but except in ploughland sites are difficult to find, because of erosion and dense vegetation. Most of the villages have names in the Slav language, suggesting that they were founded or re-named in the middle ages; the names often refer to trees, especially willow.[131] One of the transhumant villages, however, has a Greek name, Perivóli, 'garden', although it is a Vlach settlement. Perivóli – at 1,200 m one of the highest and remotest settlements – thus presumably goes back to early Byzantine times, before Slavs and Vlachs came. *Fermans* of the Sultans record its existence as a village since at least the late sixteenth century.[132]

The tree-rings of ancient savanna oaks and pines, pollards and coppice stools, show that there has been a long period of relative stability in the mountains; wood-pasture has existed for at least six hundred years (p. 211).

The travels of Pouqueville and Leake in the 1800s can be followed and compared with what is there now.[133] The plains were then mainly grassland, and their villages impoverished; their clay soils, drought and cold winters were inhospitable to farmers who (though they had vehicles) lacked machinery. The mountains were more cultivated and more prosperous, with more varied crops than there are now, as well as opportunities for trade. For example, Leake found

. . . Kraniá, a Vlakhiote village of fifty neat cottages, pleasantly situated in an opening of the forest, amidst fields of maize and other corn, fenced with a well-made palisading. . . . The corn after deducting the Vezír's portion, suffices only for a small part of the consumption of the inhabitants, whose means of subsistence are chiefly derived from the cheese of their sheep and goats; from the wood which they cut in the forest and transport to Ioánnina and other towns; and from the profit of their horses and mules, which are let to traders and travellers. The master of the house in which I lodge, who possesses two horses and two oxen, formerly kept a shop at Smyrna, Constantinople, and Saloníka; and now employs himself in transporting wood to Ioánnina, Grevená, Lárissa, and Tríkkala . . .[134]

Pinewoods now come up to this mountain village; maize, often mentioned by Leake, has now disappeared almost throughout the province, and vines are much reduced. In 1987 palisades were still to be seen.

With the coming of tractors and bulldozers, cultivation spread in the plains and retreated in the mountains. Little but cereals, walnuts and tobacco is now grown. With cultivation has come stubble-burning and resulting wildfires: the lowland is becoming fire-dominated.

We have found many examples of increased trees, especially in the mountains. Savannas have infilled; oaks have invaded steppe and arable; pines have invaded farmland. Much of this happened when World War II and the ensuing civil war destroyed men, property and livestock. Although annual rings demonstrate that the increase of trees began well before the war and went on well after it, many of the new trees are from this period. Human tragedy, here as in many other places, favoured trees.[135] The Greek Forest Service, moreover, has been very active in the mountains. When we were there, very unusually among forest bureaucracies, they were not at war with the shepherds; they sought to attract shepherds into certain areas, and away from others, by providing waterholes and other amenities.

There have also been changes from one kind of wild vegetation to another. Beech has increased into pine areas, and fir (*Abies*) has declined. The pine zone has tended to move downward into oakwood.

Erosion seems not often to threaten houses or cultivation, although these are often fitted in cunningly between gullies. Wace and Thompson relate that the people of the giant transhumant village of Samarína, fearing for their safety, decreed that certain slopes should be allowed to grow up to woodland. We found, 77 years on, that the surroundings are indeed more wooded than in Wace and Thompson's photographs (though the explanation offered to us for the change was startlingly different from what they heard), but gullying continues.

These observations so far are consistent with the kinds of change that we have seen in other environments in Greece. But they are at variance with McNeill's conclusions regarding the Píndos Mountains, that there has been general deforestation leading to erosion and loss of soil, and to settle-ments becoming untenable.[136] Possibly both may be right. McNeill's study area was different from ours; but where they overlap, we cannot always reconcile the differences.

In a few places we do record a loss of trees. North of Métzovo, Leake found 'Mavro-vúni, which is a long mountain covered with pines' – the Greek name, 'black mount', implies as much. Pouqueville, never at a loss for words, called it a 'crête sublime couronnée de pins'.[137] There is now only a thin scatter of old and fallen pines on the Métzovo side, except at the east end, where a patch of pinewood is mingled with beech. The altitude, 1,700–2,000 m, is near the upper limit of trees. This is probably part of a general retreat of trees at high altitudes. On the Bótsa Pass the upper slopes (1,800–2,000 m) have scattered giant pines; for every hundred living trees there are about twenty standing dead and a hundred fallen. The dead trees appear to be the accumulation of at least a hundred years. At these elevations recruitment of new trees is evidently a rare event, happening every two hundred years or so when favourable weather coincides with reduced browsing.

---

[131] M. Vasmer, *Die Slaven in Griechenland*, Akademie der Wissenschaften, Berlin, 1941. (Many villages have been re-named in modern Greek, replacing place-names in politically incorrect languages.)

[132] Θ. Κ. Π. Σαράντη (T. C. P. Sarandi), 'Τό χωριο Περιβόλι–Γρεβενῶν', Athens, 1977.

[133] Leake was British Resident in Ioánnina when Byron rode through.

[134] Leake 1835, vol. 4, p. 300.

[135] Wace & Thompson vividly describe the perils of shepherding close to the then Greek–Turkish frontier (1914, p. 77).

[136] McNeill 1992.

[137] Leake 1835, vol. 6, p. 296; Pouqueville 1826, vol. 2, p. 232.

We are prepared to agree with McNeill that Mount Mitsikéli in Epirus may have lost its trees, though his informants had not personally seen them. But this is a special case – the mountain closest to the capital city Ioánnina, very accessible to woodcutters and shepherds. We do not agree that Mount Olytzika (Tomáros), also in Epirus, has been permanently deforested: our notes and photographs of 1988 show plenty of maquis, fir and young pines. The Zygos Pass, east of Métzovo, had in places in 1805 'a dense forest of large pines'; these are now scattered, but the site is not deforested; they have been replaced by coppice-like beech. Much of what McNeill interprets as diminution of forest betrays a difference between European and American standards of what constitutes forest: what Leake or Pouqueville or ourselves would call woodland he dismisses as 'scrub'.

Our part of the Píndos Mountains may have been through a period of reduced vegetation in the last century, but has fully recovered. The character of the forest has often changed (for reasons that are not clear), but the area is now nearly as forested as any part of Greece could be. We do not believe that loss of forest has had much effect on either erosion or farming. Gullies eat into woodland and non-woodland alike (Fig. 15.3); unstable terrain tends to have been left as woodland. Most of the abandoned farmland still has soil.

Crete

At times in the last four hundred years Crete has been very fully used. Terraces reach to surprisingly remote places (Figs. 6.2, 6.16, 6.17). In general, the rural population reached a maximum in c. 1940 before depopulation set in (pp. 81–3). Some travellers' accounts suggest even greater pressure on land in the seventeenth or eighteenth century. The rural population was not then so great, but much of the best land was unused because it lay near the sea and was exposed to corsairs (p. 77). There is no sign that overpopulation effects were particularly severe at high altitudes.

Nearly all that was not cultivated was browsed and often burnt. Photographs and pictures indicate that nineteenth-century Crete was far less vegetated than now. The Venetian walls of the cities rose out of a desert-like landscape. According to Raulin in 1846, Crete imported two-thirds of its wood and timber; the remainder was supplied by coppice-woods and by small sawmills in the White Mountains. Imported pine beams and boards are still to be seen in houses, even far inland.

In the period of overpopulation, wild trees, where there were any, were used to the full. Every house, and even field-houses, had a fireplace. Almost every wild tree of that age, except (in part) the timber species cypress, pine and deciduous oak, is a coppice stool or a pollard – even on cliffs. This does not mean that trees necessarily were diminishing. Although, for example, most of the deciduous oaks to the south of Réthymnon are less than sixty years old, the occurrence of older trees suggests that oaks were already increasing, although slowly, from the eighteenth century onwards.

Woodcutting ended, to judge by the annual rings, in the 1950s. This was not wholly due to the coming of electricity and paraffin. In the 1930s it must have been fashionable to be cold: the then modernistic architecture, only now going out of fashion, does not provide chimneys. Moreover, the increase of olives and fruit and nut trees now generates wood. The groves resound with chainsaws; a myriad wood-stoves, cunningly adapted out of gas cylinders, are stoked with logs of olive and the finest walnut; wild trees are rarely cut at all; we have seen firewood exported to Athens; and wood is so plentiful that nobody troubles to cut trees charred in wild-fires.

Crete is now more wooded (and vastly more treed) than at any time for at least seven hundred years. However, there is concern at local concentrations of livestock. The north side of Mount Psilorítis is apparently going through a period of over-use. People are taking excessive advantage of European Union subsidies, and access roads enable them to bring in hay and thus to keep more animals than the terrain will support.[138] (Roads, perfunctorily constructed, also do great damage in themselves, p. 269.) Other areas are hardly browsed at all, and become dominated instead by wildfires (p. 238).

Crete and the Aegean furnish some remarkable examples of continuity, especially among endemic plants (p. 45–6). At nearly every place called Phœnix or Vaï the rare relict palm *Phœnix theophrasti* still occurs. It suckers and is extremely tenacious of life. It still grows at the two places in Crete called Phœnix in Classical times (Fig. 10.3), at one of which St Paul could have seen it; and on Santoríni, where eruptions of the volcano bury it in red-hot ash.[139]

PLANTATION FORESTRY

A new encroachment on wild vegetation is tree-planting, sometimes of native species, more often of exotics. The earliest example on a large scale was the afforestation of the Karst of Slovenia in the mid-nineteenth century. Plantations usually replace heath or phrygana, less often savanna or farmland, seldom (as in Britain or Germany) existing woodland.

Plantation forestry is a northern idea which does not transfer well to the Mediterranean – especially the rigorous German style as practised in Spain. Summer drought is hostile to transplanted trees, and most of the native trees are too short and crooked to produce valuable timber; this is no doubt why plantations came late. Within the twentieth century pines, poplars and eucalyptuses have been tried. Poplars grow well where there is water, but compete with modern agriculture and have an uncertain market. Six species of planted pine have been fashionable, four native (*Pinus halepensis, pinaster, pinea, nigra*), one from the Canary Islands (*P. canariensis*), and one mediterraneoid (*P. radiata* from California). They seldom produce good timber, and are extremely combustible (Chapter 13).

Eucalyptuses introduced into Europe are a curiously random selection from the thousand species of *Eucalyptus* that cover Australia. The earliest, *E. camaldulensis*, River

[138] Lyrintzis 1996.
[139] Rackham & Moody 1996, pp. 68–9;

O. Rackham, 'Observations on the historical ecology of Santorini', Hardy & others 1990, vol. 2, pp. 384–91.

Red-Gum, makes a magnificent street tree where there is ground-water, but is not well adapted to the Mediterranean climate. It survives drought, but like most eucalypts responds by dying back, which foresters hate. It is usually coppiced on a cycle of twenty years and ground up for pulpwood. It sprouts vigorously but does not invade; like all eucalypts it is fire-dependent and burns like the blazes (Fig. 13.20). Of the other ten eucalypts widely grown in Mediterranean Europe, only two come from mediterraneoid Australia; none of the great timber eucalypts of Western Australia has been introduced.[140]

Eucalypts were fashionable in the 1950s in Portugal, Spain and Sardinia. They work well as shelter-belts in agricultural plains, but have been disappointing timber trees; they form a very bad habitat, conducting chemical warfare against European plants and animals; they appear to promote erosion (p. 258). They are now banned in some areas, but are extremely difficult to kill. As so often, Mediterranean Europe has not learned from California's mistake:

> The bubble began to burst in 1910, when it was finally learned that eucalyptus is a generic name applied to over 700 species, that the trees transplanted to California were commercially worthless, and that they constituted a serious fire hazard. By 1920 the eucalyptus craze had evaporated into that nirvana of wornout California enthusiasms, but it left behind millions of closely stocked eucalypts. More than once that legacy in Alameda County has threatened to burn Berkeley to the ground.[141]

Plantation forestry, and its accompanying bulldozing, has destroyed wild vegetation, soils and antiquities, especially in Spain (p. 269); it has seldom lived up to its promise of restoring original forest cover or of conserving soils. Plantations often burn before they fulfil their destiny. However, as Italian experience shows, a plantation after a fire (*provided it be neglected*) can turn into something like a natural wood, with native trees among the remaining conifers.

CONCLUSIONS

Despite regional and local differences, certain generalities may be allowed. Much of the making of the landscape had already happened before written records begin. Changes during prehistory were followed by less drastic and less extensive changes in the historic period. Classical Greece and Italy were more like nineteenth-century than early prehistoric Greece and Italy. On the whole, the major areas of forest, pasture and cultivation were already determined in antiquity. There is little question of the destruction of 'old-growth' wildwood: Mediterranean forests and savannas are, in American terms, umpteenth-growth. Changes took the form of

depletion and regrowth more often than of the progressive conversion of large areas of forest to non-forest or vice versa. Some historic forest areas, notably in Sardinia, were not forest in prehistory.

A recent symposium on Roman deforestation, including archaeological surveys and palynology, produced very little evidence for decrease of trees definitely dated to the Roman period. Such evidence as there is comes mainly from the fringes of the Mediterranean – north Spain, southern Alps, parts of the north of modern Greece, Bulgaria; there is virtually none from ancient Greece.[142] Whether or not trees decreased in the Mediterranean proper is an open question.

Structures and compositions of vegetation changed, such as between forest, savanna, maquis and coppice. The vegetation seems to have become slightly more 'Mediterranean-type' as deciduous trees declined further, especially in Spain, following their previous decline in prehistory. A notable decline is of yew, once relatively common in Spain, Corsica, Italy and Sardinia; it is now rare and a relict all over Europe except in Britain.[143]

Travellers of the nineteenth century were often dismayed by the apparent destruction of the 'remaining' woods of Mediterranean countries. Often they exaggerated, not distinguishing between cutting down trees and destroying woodland. At this time exceptional population was combined with privatization, improved transport, and a period of good markets for poor-quality trees. The decline would not be suspected by anyone revisiting these places today: their woods are flourishing. The Taygetos woods, whose apparent destruction was noted by Philippson in 1888, and the Cretan pinewoods, where Trevor-Battye deplored the cutting of telegraph-poles in 1909,[144] have greatly increased. Most of the present trees had not begun to grow when the travellers visited the sites. All round the north Mediterranean, from the Alpujarra to Croatia to Turkey, woods are increasing at the expense of grassland and ex-cultivated land. Modern forestry has been only a minor contributor to this increase in trees, and even then the trees that grew were often not the trees that were planted.[145]

Over-exploitation of woodland was succeeded by at least fifty years of under-exploitation. Fossil fuels, although expensive to transport, in time displaced wood as the chief urban, even rural, fuel. In the mid-twentieth century many Mediterranean countries switched from 'excessive' wood-cutting – up to the limit set by the growth-rate of trees – to little or no woodcutting. Pollarding and lopping, however, continue on a reduced scale in Spain; and coppicing by the square kilometre in the Píndos Mountains.

Italy remains to a large extent a wood-burning nation; despite nearly two hundred years of foresters' disapproval, it is the greatest stronghold of coppicing in Europe. Woods of

[140] N. A. Burges in *Flora Europaea*, Cambridge, 1968, vol. 2, pp. 304–5.

[141] S. J. Pyne, *Fire in America: a cultural history of wildland and rural fire*, Princeton University Press, 1982, p. 188.

[142] B. Frenzel, ed., *Evaluation of Land Surfaces Cleared from Forests in the Mediterranean Region during the Time of the Roman Empire: PKF* 10,

1994 (whole volume).

[143] Tens of thousands of yew staves were exported to make longbows, the favourite weapon of fifteenth- and sixteenth-century England (where yew was then rare). Transport of bows from Venice was linked to that of wine from Crete; see, for example, *Calendar of State Papers Venetian*, 1511, 1518. This could have depleted stands of

yew but hardly have destroyed them, since only the best timber would have been suitable.

[144] A. Trevor-Battye, *Camping in Crete*, Witherby, London, 1913.

[145] See, for example, G. F. Croce and D. Moreno, 'Storia e archeologia delle risorse ambientali: il "Bosco Ramasso" [Liguria]', *Bollettino Ligustico* 3, 1991, pp. 69–87.

oak, chestnut, hornbeam and arbutus are regularly felled (Fig. 5.12), including on mountainous terrain far from roads. Woodstacks – sometimes of thousands of cubic metres of billets – must be well worth stealing, to judge by the precautions taken to lock them up. Even so, growth of trees (on the now increased area of woodland) has been gaining on woodcutting.

The appearance of stability may be exaggerated. The ups and downs of the last two hundred years could easily have happened on many previous occasions for which the record is less copious. However, the theory that Mediterranean vegetation has been relentlessly and progressively destroyed is not upheld. On the contrary, most of it is robust and resilient; attempts to destroy it have usually been short-lived, and attempts to 'restore' it often have different results from those intended. Only in the twentieth century have bulldozers, fertilizers and large dams given farmers and foresters a power to destroy landscapes and vegetation unknown in the days of mattocks and goats. Wetlands, coasts and non-limestone gorges are more at risk than forests.

CHAPTER 12

Mediterranean Savanna: Trees without Forests

Egressi sunt duo ursi de saltu, et lacerauerunt ex eis quadraginta duos pueros . . .
There came two bears out of the savanna, and tore up forty-two of the boys
[for being rude to the prophet Elisha].[1]

In a region absolutely covered with trees, human life could not long be
sustained . . . The depths of the forest seldom furnish either bulb or fruit suited to
the nourishment of man; and the fowls and beasts on which he feeds are scarcely
seen except upon the margin of the wood, for here only grow the shrubs and
grasses, and here only are found the seeds and insects, which form the sustenance
of the non-carnivorous birds and quadrupeds.[2]

Savanna, like the old English 'wood-pasture', consists of trees scattered among some other kind of vegetation such as grassland or heath. Usually the trees are wild, but orchards, olive-groves and 'agro-forestry' intercropping can be regarded as artificial savanna. Often the trees are pollarded, lopped or shred (Fig. 4.3) to provide wood or leaves on which to feed animals.

The word, of American-Indian origin, was first applied to various kinds of North American and Caribbean vegetation, and later to the treed grasslands of Africa, Asia and Australia. Savanna or 'orchard-bush' is sometimes thought of as particularly African,[3] but examples occur in most European countries.

In north Europe savanna has much declined. In England it was associated with the husbandry of deer in parks and royal

Forests.[4] Hence the word 'parkland' used by English-speaking writers, which we avoid because savannas lack the boundary fence which is the defining feature of a park. In middle Europe savanna was widespread, being thought of as a kind of forest (*Hudewald*), until the nineteenth century.[5] In the Alps the transition between subalpine forests and alpine heathland often takes the form of scattered larches or pines. Savanna, under the vernacular name *montado*, still covers at least one-sixth of Portugal; under the academic name *dehesa* it covers about one-eighth of Spain; it covers much of Sardinia, northern Greece and Crete, but has largely disappeared from mainland Italy (Fig. 12.1). The savannas of pre-European North America included the 'oak openings' (prairie with scattered big deciduous oaks) of Michigan and the 'motts' (clumps of evergreen oaks or elms) of middle Texas.[6]

Fig. 12.1. Mediterranean savannas, with their principal tree species.

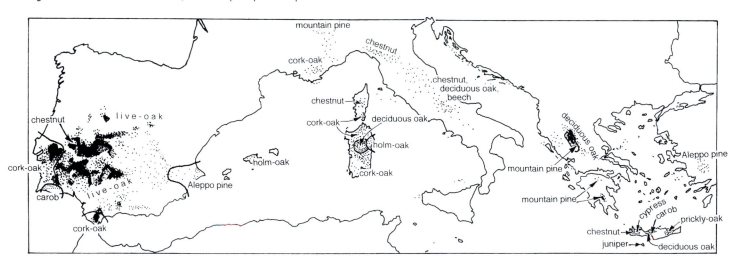

[1] IV Kings 2:24 (in the Vulgate Latin of St Jerome, c. 400 AD).

[2] George Perkins Marsh, *Man and Nature*, 1864 (based on North America).

[3] Some of our European colleagues restrict 'savanna' to tropical ecosystems, in which (they suppose) the balance between the two components is determined only by climate, independent of human activities. They limit the term in differ-

ent ways. Some use it – despite its New World origin – only for the African ecosystem, in which the trees are *Acacia* species and the ground vegetation is perennial grasses. These restrictions are very recent: *Der Große Brockhaus* of 1933 and the *Enciclopedia Italiana* of 1936 give varied examples from Africa, Asia, Australia and both Americas. We keep to the earlier, more general usage.

[4] Rackham 1989, 1990a.

[5] Behre 1981; R. Pott, 'Anthropogene Einflüsse auf Kalkbuchenwälder am Beispiel der Niederholzwirtschaft und anderer extensiver Bewirtschaftungsformen', *Allgemeine Forstzeitschrift* 23, 1981, pp. 569–71; R. Pott, 'Geschichte der Hude- und Schneitelwirtschaft in Nordwestdeutschland und ihre Auswirkung auf die Vegetation', *Oldenburger Jahrbuch* 83, 1983, pp. 357–76.

[6] Rackham 1998.

Fig. 12.2. Savanna and infilled savanna. Note the roots.
(a) Forest (b) Savanna formed by removing trees from forest (a), here termed 'subtraction savanna', otherwise 'degraded forest'.
(c) Savanna not formed by removing trees from forest. (d) Savanna, not formed from forest, with ancient trees.
(e), (f) Forests formed from savannas (c) and (d) by infilling; note what happens to the original trees.

Speakers at conferences on forest history allude to historic 'forests' used as pasture for cattle, sheep and goats. If one asks what the animals ate, one gets an evasive answer. Most shade-bearing grasses and herbs are non-nutritious, distasteful or poisonous. Domestic livestock and deer (except roedeer) are not really forest animals.[7] They love tree leaves, but cannot climb for them. In forest they eat all the leaves within reach, creating a 'browse-line' beneath which nothing edible is allowed to grow. When historical documents mention grazing in 'forests' they cannot, therefore, mean continuous forest as modern foresters understand it: they imply at least a tendency to savanna. Fernand Braudel was wrong in claiming that medieval forests 'were thicker than they are today.'[8] On the contrary, many of them were sparse enough to let light-demanding edible plants grow between the trees.

Savanna is an affront to the ideal that vegetation ought to be in a state of climax, either forest or grassland but not something between. Foresters regard it as not proper forest, agronomists and officialdom as not proper agriculture, ecologists as a non-natural ecosystem beneath their notice, and writers on orchards as not proper orchard. Much has been done to remedy this state of ignorance in Africa,[9] but Europe remains backward.

HOW SAVANNA WORKS

Some savanna-like ecosystems are highly artificial, for example an olive-grove intercropped with barley. Others are stable and semi-natural, for example the wood-pasture of the New Forest, England. The trees may be artificial and the pasture natural, or vice versa. Very often a natural savanna owes its structure but not its composition to human activities: burning, woodcutting and especially grazing. If grazing stops, new trees arise between the old ones, so that the savanna turns into a wood. Most North American savannas depended on fire (partly set by native peoples for their land management) and were lost or greatly altered (local ecologists say 'degraded') where European settlers prevented fires.[10]

7 Bailey & others 6.
8 Braudel 1972–3, 1 42.
9 For example, R. H. Behnke, I. Scoones, C. Kerven, eds., *Ra Ecology at Disequilibrium:*

new models of natural variability and pastoral adaptation in African savannas, Overseas Development Institute, London, 1993.
10 A. J. Rebertus and B. R. Burns, 'The impor-

tance of gap processes in the development and maintenance of oak savannas and dry forests', *JE* 85, 1997, pp. 635–45.

Importance of roots

Trees and ground vegetation appear to conflict: the shade of the trees is bad for the pasture (and still worse for arable crops), and animals grazing the pasture will eat any new trees. Instead of having 100 ha of pasture of which one-fifth is shaded by trees, why not have 80 ha of pure pasture and 20 ha with only trees? In an enlightened world, would this not produce better pasture and better foresters' trees? The answer appears to lie underground (Fig. 12.2, 12.3).

Roots do two jobs: they hold the tree up against wind, and supply it with water and minerals.[11] In climates without a dry season, much more root is needed to hold up the tree than to supply it. The great storm of 16 October 1987 uprooted many millions of trees in England. Most of these survived two subsequent dry summers, although only one-quarter of their roots remained in the ground. The trees most often uprooted were those crowded together in woodland, especially in artificial plantations – not free-standing trees, despite their greater exposure to wind. Root development in woods is limited by the competition of neighbouring trees, which restricts each tree's roots to the area beneath its own branches. Isolated trees can spread their branches widely, and can develop roots more widely still.

In dry climates – especially if the dry season comes when the trees are in leaf – trees depend on moisture stored in the soil. The extent of root system needed to extract moisture is

Fig. 12.3a. Shallow, spreading root system of deciduous oak, filling the space between the trees. *Kypouréio, Grevená, N. Greece, August 1987*

Fig. 12.3b. Oak stool, left after savanna has been grubbed out, showing zone of drought around the tree. *Kívotos, Grevená, May 1988*

greater, relative to the size of the tree above ground, than in England. More root may be needed to supply the tree than to hold it up. A tree can survive if it has space to develop a root system unconstrained by its neighbours' roots.[12]

A region too arid for forest may thus have scattered trees. The tendency for trees to be scattered is reinforced by burning, woodcutting and especially browsing; these maintain savanna more readily where forest growth is already marginal because of moisture.

In the Mediterranean even small evergreen shrubs

---

[11] The second function requires the help of mycorrhizal fungi which inhabit the roots.

[12] In Estremadura the truffle-like fungus *Terfezia* (p. 200), which is probably mycorrhizal on the fine roots of oaks, occurs between, as well as under, the trees.

(lentisk, prickly-oak) can have roots penetrating at least 6 m into limestone fissures. We have found the shallow roots of *Juniperus macrocarpa* extending in sand-dunes 20 m beyond the spread of the branches.[13] Deciduous oaks, too, have shallow roots outrunning the tree's canopy (Fig. 12.3). We have rarely known trees uprooted by storms.

Savanna as transitional between forest and non-forest

Savanna occurs where drought, browsing, or cold allow trees to grow but not forests. It was apparently the main form of tree-land in the dry, cold climates of the glaciations (Chapter 10). It may still come at the altitudinal upper limit of forest, as in the Alps, Balkans and Crete. Most Mediterranean savannas are in areas where individual trees grow well but forests are limited by drought. Savannas occur in all mediterraneoid regions except South Africa.

	Rainfall transition between forest and savanna, mm	Rainfall transition between savanna and steppe or desert, mm	Principal savanna trees
European Mediterranean	700	400	evergreen oaks, pines
West Africa (summer rain)	1400	400	acacias
Texas (all-year rain)	1000	400	evergreen oaks, elms
SW Australia (mediterraneoid climate)	600	250	specialized eucalypts

Table 12.i. Mediterranean and other savannas.

Mediterranean savannas fall within similar rainfall limits to those in other continents (Table 12.1). Not all dry-climate savanna is wholly natural in origin. Many Mediterranean, African, American and Australian savannas owe their origin to human activities. But in a semi-arid climate savanna turns less easily into woodland after the activities that created it have ceased. If the existing trees monopolize the water-supply, it is difficult for new trees to arise between them.

Details depend on the local environment: on rainfall, water retention and root penetration (p. 57–9). The competitiveness (especially below ground), growing seasons, and drought resistance of different trees, and of different vegetation between the trees, may influence the determination between forest, savanna, maquis and steppe.

Savanna, or something like it, may be determined by topography. As we have seen, grassland tends to grow on soil and trees in rock fissures (p. 57–9). In arid parts of Crete, where undershrubs are dominant and root penetration appears to limit tree growth, anything that increases penetration, such as a geological fault, an ancient tomb, or even a large boulder fallen from above, is marked by a tree or two. The biggest trees in the mountains are often on the edges of screes, which retain moisture and have excellent root penetration (Fig. 12.4).

Fig. 12.4. Scattered *Quercus coccifera* on scree and limestone rock. The biggest trees are at the edges of the screes. *Near Kavoúsi, east Crete, August 1989*

Prehistory and early history

The earliest evidence for cultural savanna is from pollen cores near the SW coast of Spain, where Stevenson & Harrison find the earlier oak and pine forests replaced in the Copper Age (about 2500 BC) by scattered oaks among herbaceous vegetation. They interpret this as the beginnings of medieval and modern *dehesa*, managed by intermittent cultivation, grazing and burning.[14] This may well be correct, although the identifiable herbs are mainly weeds of cultivation rather than specifically *dehesa* plants. We would point out that the change came during the Aridization change of climate (pp. 145–6). In Italy, too, abundant grass pollen in prehistoric tree-land has been held to indicate 'parkland'.[15]

Savanna is a favourite subject with artists, including even the medieval Flemings who would not have been familiar with it. Pictures of savanna go back to the beginnings of landscape painting. The Miniature Fresco on Santoríni, c. 1700 BC, shows a mountain landscape with scattered trees in an unknown country. The trees are unidentifiable, but have the spreading habit that distinguishes them from forest trees.[16]

The tomb of Philip II of Macedon, who died in 336 BC, is adorned with a fresco showing the deceased lancing a lion in a mountainous savanna. It is by a great artist, who could depict recognisable trees. There are four free-standing beeches, at least two of which are dead, and a grove of trees including two more dead ones.[17] The beeches are ancient, knobbly, spreading, non-pollard trees, quite unlike the towering beeches of Macedonian forests.[18]

The Romans had two contrasting words for tree-land. *Silva* was the normal word for woodland, forest or coppice, places with trees only. *Saltus* usually appears in the context of pasture as well as trees. It had some of the historical range of

[13] Savanna junipers and pines in New Mexico depend on catching rain falling between, as well as on, the tree canopies; D. D. Breshears and 4 others, 'Differential use of spatially heterogeneous soil moisture by two semiarid woody species: *Pinus edulis* and *Juniperus monosperma*', *JE* 85, 1997, pp. 289–99.

[14] Stevenson & Harrison 1992.

[15] Potter 1979.

[16] C. Doumas, *The Wall-paintings of Thera*, Thera Foundation, Athens, 1992.

[17] Many artists like dead trees, perhaps because they are easier to paint than live ones.

[18] M. Andronicos, *Vergina: the Royal Tombs and the Ancient City*, Ekdotike Athenon, Athens, 1984.

Fig. 12.5. Infilled savanna: a big spreading *Pinus nigra* tree surrounded by its close-set children. *Mount Katoúnes, Taygetos, August 1984*

Fig. 12.6. Savanna of deciduous oak, slowly infilling. The dead tops of these small trees indicate that they are outrunning the moisture supply. *Kivotós, Grevená, N. Greece, August 1987*

meanings of 'savanna'; in north Africa it could mean treeless pasture; occasionally a site, like the Ciminian Wood (p. 172), could be alluded to as both *silva* and *saltus*. It is hard to resist the conclusion that the Romans thought of wood-pasture as different from woodland, and although they sometimes confused them they had different words for them.

Influences on savanna

The Mediterranean mosaic of maquis, phrygana and steppe (p. 57) is a sort of savanna in which the trees have been reduced to shrubs by browsing, burning and woodcutting. If brows-ing is reduced, the shrubs revert to trees and the phrygana declines, producing a mosaic of herbaceous plants and thickets of trees (Fig. 4.33).

In savanna, if browsing is reduced, new trees may spring up between the old trees to form a distinctive *infilled savanna* (Fig. 12.5). Often the trees grow slowly because of intense competition for moisture (Fig. 12.6). Many hillsides have evidently been through alternations of maquis–phrygana–steppe and savanna. If they are now in the maquis–phrygana–steppe stage, they often retain a few big old trees; in the savanna stage they usually retain some maquis as well.

In savanna trees are related to fire differently from in forests (Chapter 13). Being scattered, the trees cannot control their environment to produce the particular fire régime to which they are adapted. The dominant partners in fire-raising are usually the grasses and undershrubs.

A warning

So far we have assumed that trees dominate, the non-tree partners making do with whatever the trees leave unused. However, foresters and fruit-growers know that grasses are thirsty plants and can interfere with the growth of trees, especially newly planted trees that have impaired root-systems. In some African and North American savannas, grasses are dominant; trees arise when occasional periods of severe

grazing, by wild or domestic ungulates, prevent the grasses from taking all the moisture.[19] To be certain of understanding how a savanna functions would call for detailed study of how the roots of the partners are distributed in relation to moisture. Until this has been done for a Mediterranean savanna our interpretation cannot be definitive.

SAVANNA TREES

The commonest Mediterranean savanna trees are oaks and pines. These are not very attractive to livestock, and can grow up in the face of considerable browsing. Oaks, but not pines, can survive being cut down, and grow again from the base. Some oaks have a further adaptation: under sufficient browsing they turn into a dwarf ground-oak form (Chapter 4). This remains alive indefinitely until a period of less browsing enables it to grow back into a tree.

Evergreen oaks – live-oak, cork-oak, holm-oak, prickly-oak – all form savannas: holm-oak is less well adapted than the others. So do many deciduous oaks. Chestnut savannas are characteristic of Alpujarra, Italy, Corsica and west Crete. Beech savannas occur in Italy, Corsica and north Greece; carob in the Algarve (Portugal) and Crete.

Wood-pastures are seldom well documented: their documents tend to be a record of land-use rather than landscape. The student must therefore ask the trees. Wood-pasture trees tend to live longer than in woodland: periodic cutting slows the ageing process and prolongs their lives. Most ancient trees are in savanna, and are a record of its history.

USES OF SAVANNA

Savanna is an 'agro-silvo-pastoral system' producing inter-related products.[20] In Spain pigs, sheep, grain (from shifting cultivation), acorns and wood are the five interlocking products, often thought of as indispensable to each other.

Livestock need to eat in the long dead months of summer and (at high altitudes) of winter. Sometimes they use the

[19] B. H. Walker, D. Ludwig, C. S. Holling, R. M. Peterman, 'Stability of semi-arid savanna grazing systems', *JE* 69, 1981, pp. 473–98.
[20] T. Pinto Correia, 'Threatened landscape in Alentejo, Portugal: the "montado" and other "agro-silvo-pastoral" systems', *LUP* 24, 1993, pp. 43–8.

stubble of arable fields. Most savanna areas have a tradition of transhumance, with the animals marched into seasonal pastures elsewhere. (Pigs are marched off into sausages and hams.)

Savanna trees are usually too short and crooked for timber. They may be pollarded for repeated crops of wood, or shredded for leaves to feed livestock. In the Píndos Mountains in Greece, where grass grows for only a few months in the year, leaves shredded from deciduous oaks are dried and stored to feed non-transhumant sheep (Fig. 4.13). In Crete, as well as cutting browsewood for livestock in the summer after a dry winter, it is the custom to make evergreen oaks into 'goat-pollards': the tree is pollarded so as to create a platform of shrubby foliage into which a goat is encouraged to climb and browse (Fig. 4.5e).

Fig. 12.7. Iberian pigs in savanna of youngish oaks. *Near Trujillo, April 1994*

Acorns for pigs and people

Savanna trees fruit more abundantly than forest trees, and the nuts or acorns are easier to gather. Contrary to popular belief, pigs are best fattened in savanna, not woodland: they need grass (for extra protein) as well as acorns. The savanna practice of pannage is known from England to Crete, but is best documented in Spain and Portugal. Acorns were supposed in 1957 to amount to one-sixth by value of all 'forest products' in Spain.[21]

Spaniards are serious eaters of real pigs: fine hams, bacon and sausages from 'viper-fed' pigs of the black and red Iberian breeds are a popular, very expensive, delicacy. These noble beasts feed in the western savannas (Fig. 12.7); their hams and sausages are processed in other places, such as Trevélez high in the Alpujarra with its special ham-drying climate.[22] Acorn-fed pigs are too fat to be fashionable for butchers' meat.

Piglets, born in winter or in late summer, eat grass until the acorns fall. In the Serra de Mamede, we were told that the season (*montanera*) opens in September, running through four successive species of oak until February. A pig gains about 60 kg live weight by consuming 500–600 kg of acorns – from fifty or so live-oak trees – and 100 kg of grass.[23] There is a lively trade between pig-breeders and oak-owners.

Among live-oaks the acorn crop can vary from half the average in a bad year to twice the average in a good year; we were told that acorns could be smoked and stockpiled. Live-oak is semi-diœcious, and trees tend to have sexes; male-ish trees produce few acorns. Cork-oaks are less productive; the acorns are less nutritious, though shed over a longer period.

People, too, eat live-oak acorns, *belotas*, not only in times of famine. Grandees of Pliny's time had roasted acorns with their dessert wine, as did Don Quixote among his goatherd friends.[24] We have seen bottles of *belota*-flavoured spirit. Acorns as human food appear in south Spain as far back as the Neolithic.

Other acorns can be made edible by removing the tannin. Holm-oak acorns are reported from Neolithic to Bronze Age excavations in Corsica, Sardinia and the Po plain. Charred acorns are found with remains of equipment for preparing them. Acorn bread was known in Sardinia and Corsica until modern times.[25] In north Greece Turkey-oak acorns were used to eke out maize and rye in World War II. Although it was a famine food, our informant enjoyed acorn bread and regretted that it was no longer made.

Cork

Cork, the most distinctive savanna product, is produced in Portugal, Spain and Sardinia, and a little in France, Corsica and formerly Italy. It is the fire-resistant outer bark of the cork-oak, which can be peeled off at a separation layer (Fig. 12.8) and is replaced by the tree.[26] Cork is harvested usually on a nine-year cycle. The first 'virgin' harvest from the tree is of poor quality and is often thrown away. In Spain all the cork on a tree is taken at once. In Portugal parts of a big oak may be corked in different years; each tree (or part of a tree) is labelled with the year of harvesting.

Fig. 12.8. Newly corked cork-oaks. *Abbasanta, Sardinia, July 1995*

[21] Balabanian 1984, pp. 171 ff.; Parsons 1962.
[22] Parsons 1962.
[23] Joffre & others 1988.

[24] Pliny, *Natural History* XVI.vi.15; *Don Quixote* I.xi.
[25] J. G. Lewthwaite, 'Acorns for the ancestors: the prehistoric exploitation of woodland in the west Mediterranean', Bell & Limbrey 1982, pp. 217–30; Tyndale 1849, vol. 2, p. 239.
[26] Cork-oak does not shed its bark spontaneously after a fire, as do many Australian eucalypts.

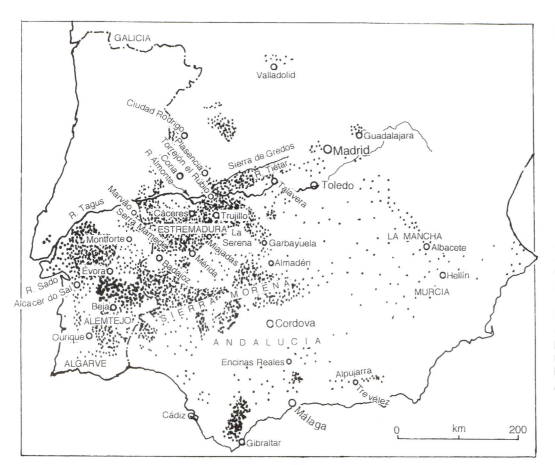

Fig. 12.9. *Left Dehesa* and *montado* areas in Spain and Portugal.

Fig. 12.10a. *Right* Typical live-oak *dehesa* on thin, uncultivable soils with granite rock coming through. *Esparragalejo, Mérida, April 1994*

Fig. 12.10b. *Far right* Live-oak *montado* (Portuguese *dehesa*), pruned and lopped in the classic manner, with a cereal crop between the trees. *Arronches, Alentejo, May 1993*

Cork-oak, the most light-demanding oak, is the ideal savanna tree. Its abundance implies many generations of something like the present savannas. In Sardinia savanna turns into cork-oak forest if browsing decreases, but does not regenerate: the next generation of trees, growing up in shade, is holm-oak.

Cork was well known to the ancient Romans but not to the ancient Greeks.[27] It has been used for floats for anglers and fishing-nets and children learning to swim, for beehives, roof-insulation, as roof-tiles[28] and in footwear. The English word 'cork' originally meant a cork over-boot ('galoches or corkes'[29]). A cork trade from Portugal to north Europe was already 'immemorial' in the fifteenth century. In Ciudad Rodrigo in Spain fifteenth-century documents refer to cork as a valuable export to Burgundy.[30]

Cork acquired a new use with the rise of bottled drinks. This is first heard of in England in 1530 ('stoppe the bottell with a corke'[31]). Cork stoppers and bungs are frequent in Shakespeare. By the late eighteenth century France was deriving much of its wine-bottle cork from NE Spain; some 1,400 tons were exported annually from Barcelona to make bottle-corks in Paris.[32] In the nineteenth century came other uses of cork, such as linoleum for floor-covering, making an ever-rising demand for a wild product of which the supply could increase only slowly. By 1900 virtually every tree was being corked.

Thirty years ago cork was expected to be wholly replaced by plastic stoppers for wine-bottles. Now, on the contrary, there is a vast and increasing export to wine-growers in North America, Chile, South Africa and Australia, none of which grows cork themselves. In Australia, all but the cheapest wines have corks; bottlers cope with the finite production of Mediterranean cork-oaks by using composite corks, in which a sliver of good cork is in contact with the wine, the rest of the cork being made from inferior bark.[33]

SAVANNA IN SPAIN AND PORTUGAL

In Portugal savanna (*montado*) extends from the Algarve mountains to the river Tagus. In Spain *dehesa* is abundant throughout the SW, with outliers that may be relics of a once wider distribution (Fig. 12.9). This impression of retreat is confirmed by place-names such as Encinas Reales south of

[27] There is a puzzling mention by Pausanias of cork-oak in the Peloponnese: VIII.xii.1; S. Amigues, 'Le témoignage de l'Antiquité classique sur des espèces en régression', *Revue Forestière Française* 43, 1991, pp. 47–58.
[28] Tyndale 1849, vol. 1, p. 311, reports this use in Sardinia at Bortigiadas, deriving the place-name from *oltiju*, the north Sard word for cork.
[29] *Oxford English Dictionary*.
[30] D. Stanislawski, 'The Monchique of southern Portugal' *GR* 52, 1962, pp. 36–55, n. 23, quoting Portuguese secondary sources; J. Viera Nativi-
dad, cited by Balabanian 1984, p. 174, n. 226.
[31] *Oxford English Dictionary*.
[32] Swinburne 1779; Ponz 1788, XIV.iii.
[33] Oliver Rackham's observations on corks in Australia.

Cordova, where no live-oaks (*encinas*) are seen today. The Málaga oak-savanna is a long-established outlier, now much invaded by (or planted with) pines.[34]

In *dehesa* or *montado*, every 15–20 m, there is a biggish, low, spreading live-oak or cork-oak with four great horizontal boughs about 3 m above ground. These march over hill and valley, over pasture, arable or even roads, as far as the eye can see. In spring the pasture is brilliant with hundreds of species of flowers, the arable with pre-EU weeds. The oakery, empty of habitation, resounds to the cry of the hoopoe, most characteristic of savanna birds. In summer the flowers suddenly wither, the pasture disappears, but the tough evergreen trees keep going. With the autumn rains the grass returns but, on the high Meseta, winter cold delays the growing season until March and April.

Montado and *dehesa* are typical of granites and metamorphic rocks, where soil is thin or absent or consists of sand or rotten granite (Fig. 12.10). Areas of deep or clayey or better-watered soils are arable or vineyards, in which trees are eliminated with extreme thoroughness. The landscape is typically rolling or flattish; savanna can extend on to slopes steep enough for people to worry about erosion, and occasionally goes high into the Sierra Nevada and other mountains.

Savannas are set in an urbanized, otherwise empty, region. Towns and large villages, on cultivable land, are 20–40 km apart in a countryside otherwise populated by occasional ranch-houses. Some townships are huge: Cáceres covers 1,768 sq. km, bigger than the entire province of Réthymnon

in Crete with 145 townships. There are also tiny townships of less than 10 sq. km, sometimes quite surrounded by big ones.

Not all savannas are on this grand scale. In the Serra de Mamede we encountered *montados* with deciduous oak (*Q. pyrenaica*) mixed with cork-oak, both shredded in the spreading style, and also pollard chestnuts (supposed to be a Roman introduction). These are set in an intricate landscape of hamlets and single farms.

Pigs are now the typical savanna animals. Goats used also to be characteristic, as Don Quixote knew. They are well adapted to savanna (but suffer from Malta fever), but many landowners now disapprove of them. Sheep are normally of the merino breed, kept for wool only; they are sent out of the *montado* in summer. Cattle, traditionally of small breeds like the Alentejo, now include beef animals and 'brave' cattle for bullfighting. Sheep and cattle can be fed on surplus acorns.[35]

---

34 J. E. López de Coca Castañar, *La tierra de Malaga a fines del Siglo XV*, Universidad de Granada, 1977, pp. 42 ff.

35 Balabanian 1984. Encina acorns are evidently not poisonous to cattle, as are those of the English oak if eaten to excess.

Fig. 12.11. Variation in live-oak. The nearest tree (too young to have been lopped) is semi-male. *Talaván, Plasencia, Spain, April 1994*

Fig. 12.12. An unusual savanna, with a new generation of live-oaks coming on. *Near Trujillo, April 1994*

Fig. 12.13. Effects of improper ploughing in live-oak savanna: invasion by undershrubs (lavender and cistus), erosion. *Talaván, Plasencia, Spain, April 1994*

The trees

Spanish and Portuguese savannas usually have a formal, almost monotonous appearance. The trees are about 80 per cent live-oak in Spain, 60 per cent in Portugal. Cork-oaks are chiefly in west Portugal and south Spain. Generally the trees are all the same species and all the same (usually youngish) age, regularly spaced though not in straight lines. Live-oaks are typically about 30 to the ha, covering about one-fifth of the ground.[36] Young trees, old trees, dead trees and stumps[37] are usually absent. We have seen all variations from near-forest (mainly cork-oaks in Portugal) to steppe with only the occasional tree.

Live-oaks are wild trees and genetically variable: apart from clonal suckers, no two are exactly alike. From the four boughs spring tufts of branches and foliage (*chupones*), which are cut at intervals supposedly of nine years. This cutting, which we shall term *lopping* (Fig. 4.3), yields wood, and is held to increase the yield of acorns and to reduce the shade on what grows beneath. Traces of earlier fashions in pollarding and shredding are to be seen in older trees. Live-oak-style lopping is practised less rigorously on cork-oak, and occasionally on other trees, like the elm at the public pillory in Marvão town. This style of lopping is less prevalent in outlying savanna areas.

The main product of lopping is charcoal. For centuries live-oak has been valued for its hard charcoal, which can be transported without turning to dust.[38] Charcoal-hearths are common: in one Estremadura *dehesa* we found a huge one, 22 m in diameter. Commercial lopping was often supposed to be excessive and harmful to the oaks. Lopping is now regarded as a conservation practice benefiting the pasture and the acorn crop; the wood is given to the loppers. A by-product used to be live-oak bark for tanning the famous shoe-leather of Cordova. Livestock are not fed on leaves: the main lopping season is March and April, when there is little need for leaf-fodder.

In formal savannas it might be inferred that live-oak has been selected or even cultivated for its fruit quality. Authorities disagree on this point. Balabanian says that plantations are rare and fail unless expensively watered (as expected of a tree that makes great demands on its root-system); González Bernáldez cites instances to the contrary.[39] Our experience indicates that cultivated *dehesas* are unusual. We rarely saw live-oaks in straight rows where planted, and we never saw grafted trees. Live-oak savannas are not consistently of productive or well-flavoured acorn trees; many trees are semi-male (Fig. 12.11). The steppe is clearly a semi-natural component. Cork-oaks grow in well-defined habitats, in valleys and on granite tors,[40] and not wherever the whim of owners has put them. Occasionally we found old ash-trees, again in well-defined habitats along streams.

In Spain, we never saw *dehesa* trees felled, other than for roadworks. Where we found stumps, these appeared to come

[36] Our colleagues in the Centro de la Dehesa, Torrejón el Rúbio regard a hundred trees to the hectare as desirable, but these would be smaller trees than usual.

[37] Lack of stumps might be due to termites, which consume stumps in less than the hundred years which an oak stump would otherwise last.

[38] Cf G. de Herrera, *Agricultura general*, Madrid, 1645 edn, p. 65.

[39] Balabanian 1984; González Bernáldez 1991. De Herrera recommended planting live-oak.

[40] Outcrops that have weathered into piles of great boulders.

from thinning the current generation of oaks rather than removing a previous generation.

In the less tidy savannas there are patches of live-oak bushes, from which the next generation of trees will come. New trees arise easily unless prevented by de-bushing. If browsing is not too intense the ground-oaks grow up into a columnar form and get away. At the next lopping, they are pruned into the conventional four-branched form. Occasionally we have seen new savannas formed in this way out of a maquis-steppe mosaic (Fig. 12.12); the trees first produce acorns at some 4 m high.

In Portugal the artificial appearance is less marked. Live-oaks and cork-oaks are mixed over a wide area. Huge wild pears are lopped in the same style. New savannas are not uncommon, especially between Évora and Beja. Cork-oak (now the more valuable tree) tends to be promoted rather than live-oak. Young cork-oak plantations occur locally around Évora.

Ploughing under savannas is resorted to in order to get rid of the undershrubs, as well as to get an arable crop. It is suspected of damaging the tree-roots; this may be why Iberian savannas seldom involve shallow-rooted trees such as deciduous oak or chestnut. It probably encourages a new generation of undershrubs, as well as discouraging a new generation of trees (Fig. 12.13).

Pine savannas are probably rare. We saw umbrella-pineries, partly arable, partly pasture, on the small sierras between Albacete and Hellín. There are, or were, 800 sq. km of pineries near Valladolid, 'in open stands, scattered as isolated trees, often shredded'.[41]

Chestnut savanna is abundant in part of the western Alpujarra; it includes huge old pollards, as in Italy and Crete. Where pasturage is neglected the groves change into woodland of other species containing relict chestnuts.

Ground vegetation

Steppe in *montado* can be remarkably rich, with 30 species of clover, medick and other plants on as little as 20 sq. cm, and more than a hundred species on a few square metres. Many of these are annuals a few centimetres high, quick-growing and nutritious. Their seeds lie through the summer, and germinate with the first rain. They vary from year to year with the abundance and timing of the rains.[42] They depend on grazing being severe enough; without grazing the smaller finer, plants are suppressed by taller, coarser species.

In Spanish savannas there are often two distinct types of grassland, under the trees and between them.[43] The presence of two grasslands, which develop at different rates (Fig. 12.14), is said to be important in prolonging the grazing season. Examples we found are shown in Table 12.ii.

Note the many annual plants, including annual members of mainly perennial genera. Despite the semi-artificial nature of savanna, there are several endemics.

An oak affects grassland in various ways: (1) its roots reduce the moisture available to the grasses and herbs; (2) the

		Under trees	Between trees	Thin soils at edge of granite rock
a	Plantago lagopus	co-dominant, tall	abundant but tiny	abundant but tiny
a	Silene gallica	locally abundant	abundant	
a?	Medicago ?lupulina	co-dominant		
p?	Leucanthemum vulgare	abundant	+	
aL	Lathyrus angulatus	frequent	+	
a•	Silene scabriflora	locally abundant		
a	Cerastium vulgatum	+		
p?g	Dactylis sp.	+		
ag	Lamarckia aurea	edge of shade zone		
p•	Armeria hirta		abundant	
aL	Ornithopus compressus		abundant	
a	Rumex bucephalophorus		abundant	
pX	Asphodelus microcarpus		locally abundant	
a	Leontodon taraxacoides ssp longirostris		locally abundant	
p	Serapias sp. (orchid)		locally abundant	
a?	Plantago coronopus		occasional	
a•	Linaria spartea		local	
a	Erodium botrys		+	
a	Tuberaria guttata		+	
pX	Urginea maritima		+	
ag	Vulpia sp.		+	
a•	Sedum arenarium			abundant

Savanna on granite 14 km NW of Mérida, 5 April 1994
a annual p perennial g grass L legume X inedible • endemic

Table 12.ii. Herbaceous plants under and between live-oaks.

tree increases the effective rainfall immediately beneath, by trapping driving rain and combing moisture droplets out of fog; (3) its shade reduces the light for grasses and herbaceous plants; (4) its deep roots bring up minerals from the bedrock, which get into the soil via fallen leaves; (5) its canopy protects the ground against slight frosts; (6) cattle and sheep assemble in its shade and deposit their dung. Each of these effects applies either to the area directly beneath the tree, or to where its shadow falls, or to where its leaf-litter lies, or to the area occupied by its roots. Our impression is that effects associated with shadow predominate.

Measurements of soil moisture (by neutron probe) at three Andalusian sites showed consistently more moisture under the trees than between them, the difference being greatest in the rainy season and least in late summer. This was explained on the hypothesis that the tree roots are deeper than those of the herbs and intercept moisture that would otherwise drain away.[44]

Mixed with steppe, or instead of it, are undershrubs, especially rosemary, lavender and *Cistus ladanifera*, which are unpalatable and very combustible. In cork-oak savannas

[41] Cavaillès 1905.
[42] T. Espigares and B. Peco, 'Mediterranean annual pasture dynamics: impact of autumn drought', *JE* 83, 1995, pp. 135–42.

[43] T. Marañon, 'Plant species richness and canopy effect in the savanna-like "dehesa" of S.W. Spain', *EM* 12 (1/2), 1986, pp. 131–41.
[44] R. Joffre and S. Rambal, 'How tree cover influ-

ences the water balance of Mediterranean rangelands', *Ecology* 74, 1993, pp. 570–82. (Several variables were uninvestigated, including the real distribution of roots and the differences in rainfall under or between the trees.)

Fig. 12.14. Live-oak savanna. Note the lusher grass under the tree. *Esparragalejo, Mérida, April 1994*

undershrubs are often the normal vegetation, and have to be grubbed out from time to time to restore the pasture and reduce the risk of fire. This may be a consequence of more moisture. Among live-oaks grassland is more stable, but can be invaded by undershrubs especially if ploughed, eroded or severely grazed.

On schist near Reguengo (Serra de Mamede) we saw a distinctive plant community, rich in small undershrubs (e.g. *Genista triacantha, Erica umbellata, Lithodora diffusa, Tuberaria lignosa)* as well as herbs and grasses. We were shown this as an area 'degraded' by goat-browsing, but it appears to be a stable, light-demanding plant community on thin soils. Even slight shade inhibits the distinctive plants and results in a community with fewer species.

If there is enough soil the land may be cultivated with wheat, barley or oats, either regularly or sporadically. Land that has once been cultivated usually has grassland of fewer species, with traces of earth-moving round the tree bases.

Shifting cultivation has provided many habitats for specialized flora and fauna, especially butterflies and other insects. Meadows in damp depressions between cultivated fields are habitats for red partridge, sand-grouse and bustard. Warblers and other migratory birds from northern Europe find winter quarters.

There are many minor products and activities. Honey comes from the steppe flowers. In spring the steppe produces a delicious underground fungus, a species of *Terfezia;* it is so abundant that experienced gatherers hunt it without the aid of hound or pig.

Ownerships and boundaries

The apparently uniform savanna hides many tenurial and social practices. On big estates (sometimes thousands of hectares), the acorns, grazing and woodcutting are the absolute property of one owner. Even here, however, owners are inhibited by laws or customs (which appear to be effective) from felling trees or cutting main boughs. Other savannas belong to small owners or municipalities, or are common land. The soil, trees and pasturage may each belong to a different owner.

Some savannas are surrounded by walls of stone or mudbrick or mud, others by fences or by no fixed demarcation. Hedges are rare: they are distrusted because they are supposedly susceptible to fire (in Spain hedges do not easily arise spontaneously).

Where a road through savanna is delimited by walls, the road verges, up to 100 m wide, are also savanna. Grazing is usually neglected, but the trees on public land have often been lopped in the same style, but not on the same cycle, as those on the private land beyond the wall. The public trees often overhang the wall and drop acorns on to private land.

Prehistory of montado *and* dehesa

Savanna areas have a distinguished archaeology. A contribution to Ruined Landscape theory in the eighteenth century (Chapter 1) was the contrast between the magnificent Roman antiquities and the present vast, empty, richly vegetated *despoblados*.

The Évora district is full of chamber tombs, cromlechs, stone circles, cup-marked rocks and other monuments vaguely dated to the Neolithic and Chalcolithic periods. The Bronze Age, though present, is less visible. In the Iron Age there were widely spaced towns, much as now, in most of the savanna region; the extent of settlement between them is uncertain.[45]

In Roman times Mérida was a great city, 'the Rome of Spain in respect of stupendous and well preserved monuments of antiquity', as Ford put it in 1845. Ponz was impressed by the Roman bridges and ruins of Trujillo and other places. Roman antiquities extend into Portugal, for example the irrigated farmstead at São Cucufate north of Beja. Évora was another great Roman city; sites around it indicate olive cultivation among other agriculture.[46]

Early written sources are vague. There is a hint of savanna in the high plains of Andalusia, 'great-treed and well-pastured' according to Strabo, the ancient geographer.[47]

The archaeological magnificence is still not fully explained. Ponz's theory of environmental collapse through the mouths of medieval sheep has little to be said for it. Although there are claylands with good arable cultivation, it is hard to imagine that the sands and granites have ever been fertile. The Romans evidently made better use of a difficult environment than their successors, probably by skilled irrigation. Most scholars have attributed the change to a catastrophic expulsion of Muslims in the thirteenth century: 'making a solitude and calling it pacification', as Ford said. The medieval rural archaeology is not well enough known to confirm this.

Montado *and* dehesa *country in the middle ages*

The twelfth-century geographer Al-Edrīsi mentions a district transliterated as Fahs al-Ballut and interpreted as 'plain of live-oaks' – the word *belota*, 'acorn', having passed into Arabic. It lay in the Sierra Morena, and it would be nice to see an early allusion to the savannas of this area. However, the location is only vaguely identifiable, and the word could mean something quite different.[48]

The Portuguese word *montado* now means a pasture set with oaks. Dictionaries quote examples of the word back to the Knights Templars in 1261, but it was then used for a pasturage rent and does not necessarily imply savanna. The ancient but rare Spanish term *monte hueco*, 'empty forest',

unequivocally means savanna, defined as 'land in which there are live-oaks and other trees and in which, looking low down, one can see to a distance'.[49]

The modern word among learned Spanish and foreign writers is *dehesa*. It originally meant 'fenced pasture', from Late Latin *defensus*, 'fenced'. It came to mean 'pasture' in general, its normal meaning in Modern Spanish. As late as 1872 a dictionary cites *dehesa*, in the restricted sense of treed pasture, as a provincialism from Galicia.[50] Some of our Spanish colleagues use *dehesa* in very precise senses – for one of them it must be a savanna used in a specific way and of more than a certain extent – while others still use the word for any kind of pasture. The place-name *Dehesa* occurs all over Spain except the north coast and NE of Madrid; its distribution coincides roughly with the wider distribution of savanna.

Spanish savanna is haunted by memories of the Mesta, the guild of transhumant shepherds, scarcely less formidable a public institution than the Inquisition. The Mesta's flocks wintered in the south – Andalusia to Murcia – and travelled along designated roads, *cañadas* (Fig. 12.15), to summer pastures in north Spain and the mountains. The present savannas were winter or spring pasture. This pattern apparently arose in Arab times, survived the Christian conquest, and was institutionalized in the thirteenth century.[51]

The Mesta was a cooperative of shepherds, kings, municipalities, landowners and monasteries, all owning merino sheep and supplying fine wool to the cloth industries of England and Flanders. It had rights of usage over most land that was not cultivated, including areas abandoned after the Black Death. The kings of Spain found in the transprovincial activities of the Mesta a reason for their own existence and a source of revenue.

Mesta sheep could have had a great ecological influence, though their numbers have sometimes been exaggerated. In the heyday of the guild (1490–1550) around 2.7 million sheep paid toll each year – fewer than there are in a single province today. In a wintering area of something like 60,000 sq. km, this works out at about one sheep per 2 ha. This is quite a lot of sheep, given that some of the 60,000 sq. km would have been cultivated land, and it would also have to support cattle, goats, sheep that evaded the toll, and the many non-transhumant sheep. In the eighteenth century the Mesta claimed that their activities discouraged 'undergrowth' which might harbour two- and four-legged wolves. The populace quarrelled with the Mesta, claiming that shepherds' fires consumed savanna trees.[52] However, the numbers involved are insufficient to justify the claim that Mesta sheep and shepherds destroyed the great forests supposedly existing in the non-mountainous parts of medieval Spain.

Mesta archives, as reported in Julius Klein's work, draw a distinction between *bosques* and *montes*. *Bosque*, like its equivalents in other Romance languages, meant 'woodland'. *Monte*

[45] Burgess 1987; J. C. Edmondson, 'Creating a provincial landscape: Roman imperialism and rural change in Lusitania', *Historia Antigua [Salamanca]* 10, 1992, pp. 13–30.
[46] S. Willis, Appendix 2 to Burgess 1987.
[47] *Geography* III.ii.6. Butzer 1988 has tried to

make sense of the scraps of Roman written evidence for livestock.
[48] Transliterations of European place-names from Arabic are uncertain because short vowels are omitted.
[49] M. Lemos, *Encyclopedia portugueza*, Lemos,

Porto, c. 1900; E. Zerolo, *Diccionario enciclopedico de la lengua castellana*, 9th edn, Garnier Hermanos, Paris, c. 1920.
[50] *Diccionario enciclopédio de la lengua española . . .* Gaspar y Roig, Madrid, 1872.
[51] Butzer 1988.
[52] Klein 1920, p. 75; Parsons 1962.

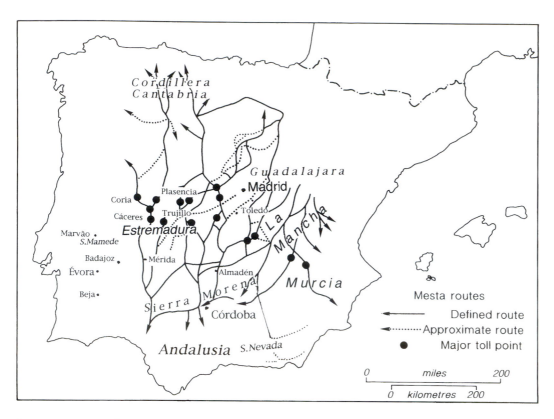

Fig. 12.15. Main transhumance *cañadas* of Spain in the time of the Mesta, as recognised by Klein. In reality the *cañadas* extended into Aragon in the east, Granada in the south, and Portugal.

here meant 'not forests, but rolling country with scattered trees', evidently savanna. (Unfortunately Klein does not illustrate these uses by quotations from documents.) From *monte* he derives *montazgo*, evidently the Spanish spelling of *montado*, which was a widespread term from the ninth century onwards for a tax or rent imposed on users of *montes*. The verb *ramonear*, 'to branch [a tree]', was the name of a woodcutting right documented in shepherding contexts from the 1270s onwards. The wood was used for corrals, fences, cabins, fuel, dairy implements, fodder and tanbark. As an established right this implies, not casual cutting of boughs of trees in forests, but savanna woodmanship by pollarding, shredding or lopping. Neither of these kinds of evidence is as yet wholly watertight, but together they indicate that savanna (unlike forest, p. •) was widespread and familiar in medieval Spain.[53]

Many of Alfonso IX's fourteenth-century hunting 'forests' (p. •) were in savanna country around Trujillo and Badajoz – the land from which, two centuries later, Pizarro was to leave his pigs and grab an empire.[54] Without identifying the hundreds of localities we cannot be sure how many of them coincide with savannas today. The place-names include occurrences of 'La Dehesa', but without indicating what it meant. In one place there is an explicit account of savanna:

La Madroñera [near Trujillo] . . . it is a flat *monte*, and it is an *encinar hueco* [a live-oakery with a clear space under the trees] in which a man can go on horseback because it is raised up, and for those who keep the hounds it is a good *monte* to walk.
[Chapter XVIII(1)]

Montado *and* dehesa *country in the sixteenth and seventeenth centuries*

Fernando Colón, in the 1510s, often specifies the vegetation bordering roads. Very common terms are *monte de enzinas* and *enzinar*, 'live-oakery'; it is tempting to interpret these as a distinction between woodland and savanna. Both occupied definite areas ('the distance between A and B is 4 *leguas* of which the middle 1½ *leguas* are *enzinar*'). They occurred throughout the present Spanish savanna region and its outliers. The actual sites of present savannas were often reported as *monte de enzinas* or *enzinar* by Colón, but often also as cistus heath or other non-treed roughland (Table 11.i).[55]

Friction developed between Mesta shepherds and local townspeople over pasture rights; the tide of litigation gradually set in favour of the townspeople. Sometimes, however, local people encouraged Mesta sheep to feed on cultivated land in autumn, benefiting from the manure.

Acorn-fed hams from the Sierra Morena are said to have been well established by the early sixteenth century, being specified on the manifests of ships sailing to America. The archives of Jérez de los Caballeros in 1554 record fattening of swine on roughly the same scale as in 1960.[56]

In Portugal, municipal ordinances in the Serra de Mamede make regulations for what seems to be savanna on common lands. In 1512 we find an early reference to cork as a product:

The felling of timber and wood shall be both forbidden, in the same way as the drawing of cork and bark from the cork-trees. Infringement of cutting of timber or a tree

[53] Klein 1920; see his p. 321n. for secondary sources.

[54] Alfonso XI, 1983 edn.
[55] Colón 1517–.

[56] Parsons 1962.

felled by the foot is punished with a fine of 60 *reaes*, with 108 per load of cork and bark, and with at least 40 *reaes* for a single load of wood.

This emphasizes the distinction between timber (*madeira*) and wood (*lenha*). The fine for cutting a tree by the foot implies that pollarding was the proper practice, and therefore that the regulations are for savanna, not forest. An ordinance of 1647 imposed a new fine for felling fruit- and wood-bearing trees, including chestnuts.[57] (Chestnuts are still a feature of the Mamede; they are the commonest timber in old houses and old furniture.)

In the sixteenth century cork-oak is much the commonest tree mentioned in *Don Quixote*, which is set mainly in parts of Spain where it is not now abundant. Colón, however, mentions *alcornocal* (cork-oakery) less often than *enzinar*; perhaps Cervantes picked on it as the most recognisable Spanish tree. His reclusive Lucinda dwelt in a hollow cork-tree; today she would not easily find one big enough (but see Fig. 12.20).[58]

Savannas were not large commercial producers of wood. According to Klein, the late-medieval carters' guild transported charcoal to the potteries of Talavera from the woodland of Toledo. Talavera is about 50 km from the wooded mountains of Toledo, and now has savannas near it. Another regular journey was taking 'wood' (fuel or pit-props?) from Coria and Plasencia to the quicksilver mine at Almadén, 200 km away. This would now be unnecessary: Coria and Plasencia have extensive savannas, but so does most of the country between them and Almadén.

Modern history of montado *and* dehesa

William Bowles met with savanna on his visit to Almadén in 1752:

> The greater part of these hills are covered with live-oaks, now hollow, because they have their trunks and boughs lopped [Italian *potati*]; they yield, however, infinite acorns for the pigs, which are all black in this country. The income of the owners there consists in pastures, in acorns and in berries.

He mentions cattle and horses. He wrote this of the hills around Garbayuela, 110 km east of Mérida; then as now, this treed sierra contrasted with the vast steppe in the basin of La Serena. Hollow trees fascinated him, and he wrote a chapter speculating on why some pollards went hollow and others did not.[59]

The Mesta was still a power in the land. The flocks wintered in Estremadura, where pollarding was done by the shepherds:

> The shepherds . . . build themselves huts with branches of trees and clay, for which purpose, and to make a fire, the law permits them to cut a branch from every tree, which in my opinion is the reason why all the trees of the environs, where the Merinos feed, are rotten and hollow.[60]

Bowles and Antonio Ponz (a few years later) found live-oakery (*encinar*) distributed roughly within its present limits, north to the Sierra de Gredos, east to Talavera and Almadén, south to the Sierra Morena, and west into Portugal. *Encinar* was interspersed with large cultivated plains and wild heaths. The higher mountains, as around Guadalupe, had mixed forests including deciduous oak. Patches of live-oakery occurred outside these limits, roughly (sometimes precisely) where they still do: east to Madrid and beyond to the border of La Mancha, south to Cádiz and Málaga.

Although vague words such as *monte* are used, Ponz often specifies savanna. South of Badajoz he passed for four leagues [about 40 km] of 'empty *monte*, that is to say of trees without density'; this he regarded as the remains of a forest. Elsewhere the terrain varied from treeless *xaral* (*Cistus* heath) to 'dense and stately *monte* of live-oak'. Ponz distrusted a country where 'passers-by risk money and life at the hands of highwaymen' and a horseman could get lost in the thickets.

Around Trujillo, says Ponz, the people lived from pigs. The *dehesas* had landowners, but were not subdivided – the traveller had no roadside walls to guide him – and were common swine-pasture. They were patrolled by Keepers of the Vert (*Guardas del verde*), whose task was to oppress the swineherds and to raid their sausages and hams. A decline from formerly more intensive land-use is suggested by terrible places where, he says, cistus and arbutus were so high and thick that cattle and even dogs could hardly get through.[61]

Link the botanist, travelling in 1797–8, regarded savanna as a kind of forest (rather than of grassland), but describes it unequivocally:

> This tree [*Quercus rotundifolia*] seldom grows high, generally about the size of a moderate pear-tree; the stem is thick, and covered with a thin fissated bark, with a head formed by short branches crowded together . . . The trees generally stand at a distance from each other, so that their tops do not touch . . . The short thick stems often afford an appearance of great age, the curled leaves have a very thirsty appearance . . . the soil is parched and bare, and there is scarcely enough shade to render even a German summer tolerable, much less that of Spain.

Savanna then was less extensive than it is today; much of the ground was arable land, heath and retama. It was used for sheep pasture, feeding swine, cork-gathering, acorns 'to be roasted for the use of man' and charcoal. Link was fascinated by the wild savannas and heaths, the *despoblados* between the towns, where the traveller trusted in his pistols.

Link thought the tract of cork-oaks east of the Sado estuary was the most extensive in Portugal. (It probably still is, though few of the present trees are old enough to have been seen by him.) Elsewhere south of the Tagus there was savanna both of cork-oak and live-oak, but also vast areas of 'heath' and *Cistus*. In Algarve the carob savanna was less extensive than it is now.

Link was not aware of the integration of land uses in savanna. In Spain he mentions transhumant sheep, and

57 C. Videira, *Memoria historica da muito notável villa de Castello de Vide*, Lisbon, 1908; *Relacão de successos historias . . . da notável vila de Castelo de Vide*, Castelo de Vide, 1955, p. 75.
58 *Don Quixote* I.xi; I.xxvii.
59 bowles 1783, vol. 1, p. 226; vol. 2, p. 123 ff.

60 W. Bowles, *Tratado sobre el Ganado Merino*, Rodd, London, 1811.
61 Ponz 1778.

blames overgrazing for what then seems to have been an impoverished flora. Swine-feeding in Portugal was well developed: the people knew that cork-oak acorns were two-thirds as valuable as live-oak. Around one lonely roadhouse the savanna had not been managed recently:

> Venta del despoblado . . . stands in the middle of an extensive forest of evergreen oaks, which in some parts are rendered almost impassable by the thickness and closeness of the cistus.

He says little of how savanna trees originated. In one place he records 'creeping-oaks (*Quercus humilis* Lam.)' which might have been ground-oak forms of *Q. rotundifolia* or *suber*.[62] According to Link, oak was little used for timber, which agrees with our observations.[63]

The savanna country was fought over in the Peninsular War, and took long to recover. Richard Ford, who travelled in the 1830s,[64] believed that conventional agriculture had once prospered here, and was deceived into thinking it could do so again:

> . . . the finest wheat might be raised here in inexhaustible quantities, and under the Romans and Moors this province was both a granary and a garden. It is still called by the gipsies *Chin del Manro*, 'the land of corn' . . . The lonely *dehesas y despoblados* . . . are absolute preserves for the botanist and sportsman: everything displays the exuberant vigour of the soil, teeming with life and food, and neglected, as it were, out of pure abundance.

Ford, like so many others, failed to appreciate that luxuriant vegetation can go with thin, poor soils.

From the villages, 'coalitions of pigsties', scattered through 'these parklike scenes', pigs were every morning 'turned out in legions' and galloped back at night, each to its own sty. Indeed

> Estremenian man is quite a secondary formation, and was created to tend herds of these swine, who lead the once happy life of the Toledian cathedral dignitaries, with the additional advantage of becoming more valuable when dead.

Ford praised the local hams. 'Parklike' is the term that an English writer of the period would have used for savanna. Where Ford's travels coincide with Link's the extent of savanna seems to have been much the same.

He disapproved of the Mesta, but misunderstood it and exaggerated the numbers of sheep:

> They quit their highland summer quarters, *Agostaderos*, about October, and then come down to their winter quarters, *Invernadores*, in the warm plains. . . By the laws of the mesta the king is the Merino Mayor; his deputies,

wolves in Merino clothing, compel landed proprietors to leave a Cañada de Paso, or free sheep-walk, 90 paces wide, on each side of the highway, which entirely prevents enclosure and good husbandry.

(*Cañadas* were, of course, specifically intended *not* to prevent enclosure: their purpose, as with the drove roads which Ford should have known in England,[65] was to allow sheep to get through enclosed areas.)

Le Play, writing about Estremadura in 1834, remarks on the wide-spaced trees – the inhabitants had no word for 'woodland' or 'forest' – the pigs, the giant cistus that hemmed in the roads, and the cereals grown in places between the trees.[66]

The Mesta was dissolved in 1836, but has left its mark on the landscape in the *cañada* drove-roads, each with its name.[67] Originally these were supposed to be 6 *sogas*, 75 m, wide, or a fraction of that width for minor *cañadas*. This width was defined wherever they passed between cultivated lands and enclosed pastures. Like all roads, *cañadas* were forever nibbled away by neighbouring farmers, and there were *entre-gadores* charged with preventing such encroachments. *Cañadas* were not defined where they passed through common land. As commons became privatized in the eighteenth and nineteenth centuries, so the defined *cañadas* were extended. Today they run, usually between walls, through almost continuous private savanna. Some are paved. They are still public property and legal rights of way, though the authorities are lazy about protecting them; many have been interrupted by reservoirs or stolen by farmers.[68]

Widdrington's travels (1829–32) reveal the privatizing and formalizing of savanna, which he likened to the 'improvement' of common land then fashionable in England. In the western Sierra Morena, he recognised three types of land: (1) 'dehesa or open waste, covered with a few cistus and other shrubs, with remnants of better trees which once clothed the soil'; (2) land covered with 'scrub', meaning evidently young oaks rather than *Cistus*; (3) land 'already covered with good trees'. Landowners were sowing pines on the first type of land; on the second they employed peasants to 'clear out' the 'useless scrub', leaving the better trees to be trimmed up.

> Independently of the wood for charcoal, which is the grand object, no less than three successions of animals are turned in . . . In the spring and summer bullocks occupy the ground, to the enormous benefit of themselves and their owners as well as the proprietors of the land; in the autumn droves of pigs are driven in to eat up the belotas and other fruits . . . ; and in the winter they are succeeded by sheep . . . ; for each of these occupations a rent is paid cheerfully and liberally . . .

---

[62] *Flora Europaea*, however, now regards *Q. humilis* as a separate species.

[63] Alexandre de Laborde (*Itinéraire Déscriptif de l'Espagne*, Nicolla, Paris, 1808) says that 'the greater part' of Estremadura 'is converted to pasturages . . . absolutely deprived of trees [except] a few tufts of elms, poplars and willows along . . . the rivers . . .' This claim, contradicted by other travellers and by ages of trees still alive,

should not be taken at face value. Laborde took the viewpoint of the conquering French; he disparaged Spain, with its commons and Mesta and pigs, for not joining the Age of Reason. For him, trees generally meant fruit-trees.

[64] References to the Mesta prove that Ford was writing of the time before 1836, although his work was published in 1845.

[65] Rackham 1986, chapter 12.

[66] Le Play 1834.

[67] At just this time the equivalent of *cañadas* were being established, along with merino sheep, in inland New South Wales. Today these 'long paddocks' are still public land, and survive as strips of savanna in areas where the private land has been grubbed out and made into arable.

[68] H. Villavilla Asenjo, *En Defensa de las Vías Pecuarias*, Asociación Ecologista de Defensa de la Naturaleza, Madrid, c. 1994.

Goats, 'the delight of the peasantry', were rigorously expelled. Only once did he hear of trouble, when peasants burnt a piece of land which had been privatized and on which they could no longer keep goats and donkeys.[69]

At a rough estimate, savanna has increased in the last two hundred years by about one-third in Spain between Madrid and Badajoz. In Portugal south of the Tagus it has increased about threefold since Borrow's travels in 1835. Cork-oaks have increased more than live-oaks. The increase has been at the expense of heath and *Cistus*. It has not been uniform: around Miajadas (Estremadura) the 'forest' has partly become arable land; around Alcacer do Sal what Link records as heaths are now cork-oakeries infilled with pines.

The first certain use of the modern term *dehesa* for savanna is by the French writer De la Laurencie in 1889 in Estremadura:

> The bushy *dehesas* with a few trees, or an even more peculiar type, the *vacuos* (hollow forests), vast areas 'covered with scattered trees of holm-oaks [live-oaks], under which there extend either pastures or farm crops, a sort of wooded meadow' . . .[70]

According to Balabanian, much Portuguese *montado* is roughly a century old, and was derived from a mosaic of maquis and steppe called *charneca* (cf Fig. 12.16). Large-scale maps of the Beja area in the 1880s indeed show more *charneca* than *montado*. Statistics of various Alentejo townships in 1902, repeated in 1957, show fluctuations in the area of *montado*, ranging from small increases to big decreases.[71] We do not press these figures, for the definition of *montado* is unlikely to have remained steady for 55 years; but the large areas of *montado* with oaks monotonously about 50 cm in diameter could well date from the high-rainfall period (p. 120) of the late nineteenth century.

Ages of trees[72]

In a 'healthy' savanna trees are supposed to be of all ages, but in practice there are seldom more than two ages and often only one. Frequently, in otherwise even-aged Portuguese *montados*, there are single giant old trees up to about 1.5 m diameter. These tend not to be lopped in the spreading four-bough style, but pollarded into a more compact shape, often with several points of cutting. Like most ancient trees, they are hollow; one or more limbs can disappear, leaving the remainder healthy.

In Spain old trees are generally rare, but some savannas contain single veterans. Two such trees near Monroy are about 3.0 m in girth, one on a road and the other on private land as if they date from before the demarcation of the road. On the bottomland of the Río Tiétar, below Plasencia, we saw among lopped cork-oaks a group of four ancient pollarded

Fig. 12.16. Live-oak savanna extending into steppe via ground-oak. *Talavan, Plasencia (Spain), May 1994*

ashes, 1.5 m diameter, among the biggest trees we have seen in Spain. (Younger ashes are lopped in a high style similar to cork-oaks.) Old trees are commoner as the savannas thin out eastward.

An extensive tract of old oaks, reminiscent of the wood-pastures of England or Greece, lies on the road from Trujillo to Plasencia. Ford held up this lonely country between the Ríos Almonte and Tozo as an example of 'what a *despoblado* and *dehesa* mean'. For at least 10 km ancient hollow live-oaks still predominate, with outliers to the town of Jaraicejo and beyond. They are wonderfully variable: some pot-bellied, some sny (with a gradual curve), some tall and straight, some crooked (Fig. 12.17). Most have a history of pollarding at varying heights, with attempts at conversion later to the modern lopping style. Ground vegetation varies between steppe, *Cistus* and arable. In general they are in good condition,[73] though dead trees are slightly more prevalent than usual, especially where the pasture is neglected. We estimate these to be at least three hundred years old, some much older. They show that the youngish appearance of other *dehesas* is not an illusion. Live-oaks, like other oaks, have a long and indefinite life-span, and can flourish to a much greater age than most *dehesa* trees today. The general rarity of ancient live-oaks is due to management history, not to their limited life-span.[74]

Old trees (or the lack of them) amplify our inference that from 1870 to 1920 there was a phase of renewal and creation of new savannas. The current style of lopping then became fashionable, replacing an earlier style less distinct from other European forms of pollarding. This new style was, of course, introduced mainly on trees being lopped for the first time.

The Almonte–Tozo tract is, as far as we know, the historic heartland of savanna. Ford thought these were 'the oak woods in which Pizarro fed his pigs'. He could well be right: Colón mentions many *enzinares* in this area, and it is close to Alfonso XI's savanna called La Madroñera.

[69] Cooke 1834, vol. 2, p. 246, etc.

[70] Cavaillès 1905, quoting L. de la Laurencie, 'Les forêts de l'Espagne', *Revue des Eaux et Forêts* 28, 1889, pp. 481–96.

[71] Balabanian 1984, pp. 109 ff.

[72] Our observations on stumps: Cerca, Alentejo, Portugal, cork-oak: 46 cm diam., c. 100 years. Monfraguë, Estremadura, Spain: live-oak, c. 16 cm diam. under bark, 28 years. Between

Peñarroya and Azuaga, Estremadura, live-oaks grubbed for road-widening: one 48 cm diam., c. 137 years (rather bigger than the average *dehesa* live-oak); one c. 92 cm diam., probably c. 300 years (live-oaks of this size are uncommon in Spain). Near Torrejón el Rúbio, coppice shoots of live-oak, 19 cm diam., c. 40 years; 23 cm, c. 45 years.

[73] Our Spanish colleagues, however, subscribe to

the notion that hollow trees are diseased.

[74] The inhabitants believe their trees to be much older than our estimates, pointing out the dry climate and short growing season as reasons for their slow growth. This is not impossible: a tree that is lopped and also produces a heavy crop of acorns may have little substance left over for the annual increase in girth. However, local people appear to have had no better success than we in counting annual rings.

Fig. 12.17a. *Above left* Old oaks in savanna; note the nineteenth-century roadside wall.

Fig. 12.17b. *Above* Savanna with ancient oaks, pollarded in the ancient manner. *Both near Jaraicejo, Trujillo, April 1994*

Fig. 12.18. *Left Montado* of ancient pollard chestnuts. *Reguengo, Serra Mamede, May 1993*

The last fifty years

Savanna last expanded in Spain in the 1930s and 1940s, when (according to Balabanian) population was at a maximum and labour was available to grub out ground-oak and formalize the more promising trees. This would produce oaks about 30 cm in diameter, the youngest widespread generation today.

A photograph of Montforte in Alentejo, taken probably c. 1950, depicts live-oak *montado*. The scene, which we revisited in 1994, shows increased trees, partly through planting pines, but also through savanna beoming denser. One patch of savanna is now sparser, perhaps through burning; a new olive-plantation has appeared.[75]

Cork production increased in the second quarter of the twentieth century. Official statistics show a rapid upward trend, especially in Portugal.[76] If the figures represent an increasing crop (rather than increased efficiency in collecting the data) they mean that the c. 1900 generation of trees was coming into production, and was an addition rather than a replacement of earlier trees.

Disaster struck in the 1960s with African swine fever. Between 1955 and 1974, the numbers of Iberian sows officially counted in Spain diminished from 567,000 to 77,000.[77] The disease is now controllable; Iberian pigs have largely recovered in Spain, but less so in Portugal, where they were replaced by 'improved' pigs which do not flourish in *montado* nor produce fine meat.

At about this time the local people began to escape from poverty. About one-third of the workers emigrated, and the wages of the remainder doubled in real terms during the 1960s.

Sheep and goats have been partly replaced by beef cattle, which require less attention. There is a tendency to 'improve' breeds, which increases yields, but begins to separate the animals from the land: the improved animals, less adapted to the environment, are fed on imported feedstuffs or on sown clovers and grasses. Being heavier, they are held to damage the soil. Most of the cereals now grown in savanna are fed to cattle.

In the 1950s there was talk of 'improving' savanna, with 'rationalization' of oaks and pigs in order to increase yields. The only practical effect was the spraying of nearly two-thirds of the Spanish *dehesa* with DDT to prevent defoliation by caterpillars. It was claimed that this more than doubled pork production; history does not relate what it did to the ecosystem or the pork-eaters.[78] Many savannas, particularly in Portugal, were grubbed out (often illegally) and replaced by pine and eucalyptus plantations.

Savanna has suffered many attempts to force it into conventional European agriculture, often at the hands of vainglorious, isolationist and optimistic regimes (p. 84). Governments up to the 1970s disapproved of trees; the area of wood-pasture in Estremadura (for what the figures are worth) is said to have declined by about 40 per cent between 1955 and 1978. The Évora archaeological survey reported a further increase in arable and eucalyptus cultivation since 1987.[79]

[75] H. V. Beamish, *Hills of Alentejo*, Geoffrey Bles, London, 1958.

[76] G. B. Cook, *Cork and the Cork Tree*, Pergamon, Oxford, 1961.

[77] Balabanian 1984, p. 140.
[78] Parsons 1962.
[79] Burgess 1987.

Conclusions

Savanna in Spain and Portugal goes back at least seven hundred years, if not to prehistory (Chapter 10). Formal *montado* and *dehesa*, with their interlocking system of land-uses, are more recent. In the eighteenth century savanna was less extensive and less formal than now; more of the trees were old and hollow; and they were less exclusively evergreen oak – there were also arbutus and big wild-olive trees.

Montado and *dehesa* became a way of life in parts of Spain and Portugal where much of the population had left for America and Africa. The effects of the Mesta should not be exaggerated: savannas, and the heaths that preceded them, were no less well developed in Portugal, where the Mesta had much less influence than in Spain.

Formal savannas are probably a nineteenth-century intro-duction using cheap labour – the only agricultural improve-ment in Alentejo and Estremadura to have lasted more than a decade or two. The present live-oaks and cork-oaks are prob-ably the first generation of trees to be formalized. This was the last large-scale revival of wood-pasture, then unfashion-able and declining in all the rest of Europe; it attracted little attention among agricultural writers.

This revival could be due to the devastation of the Napoleonic Wars in 1806–14, which allowed oaks to invade former pasture, but the time interval is rather long. It may result from the privatizing of common lands in the 1830s and 1850s, which allowed landowners to do what they liked with trees – and also, by depriving the poor of common rights, may have created the necessary casual labourers.[80]

Formal *montado–dehesa* is superimposed on the remains of a much older type of savanna. This involved different forms of pollarding and shredding, still to be seen on pre-nineteenth-century oaks, on trees of other species, and outside the central *dehesa* area. Practices were adaptable, and could respond to new influences such as the rise of commercial cork.

Savanna at present functions relatively well. The most profitable operation is gathering cork. Cork-cutting is a well paid job. Workers (we were told) can draw the equivalent of full pay for doing two months' work a year, the balance being paid by the European Community. This solves the problem of finding workers for seasonal jobs. At a modest cost to the Community, it prevents the poverty resulting from seasonal unemployment, which has given the savanna parts of Spain such a tragic and bloody modern history.[81]

We were repeatedly told in Spain that to provide a living a *dehesa* has to be at least 400 ha in extent. Some families went to the trouble of marrying close relatives to keep the land in the family and in one piece. However, in practice areas of savanna can be much smaller: they do not have to give a full-time livelihood.

Fully developed savanna is labour-intensive, especially in lopping and gathering acorns. It seems to the outsider a curious land-use in an area with almost no rural population: workmen would often have had to commute 20 km from a town. Spaniards, unlike Greeks, seldom have field-houses for living away from home, but there was some use of temporary huts. In its classic form *montado* and *dehesa* may be a relic of the nineteenth and early twentieth centuries, with swarms of ill-paid labourers. However, it would be perfectly possible to maintain a rougher, less formal savanna in which labour-intensive operations were not done over the whole territory.

Savanna is protected by bureaucratic laws which prohibit felling trees or cutting major boughs without licence, and which lay down the manner and even the season of lopping. These seem to be observed, despite the difficulty of enforcing them in a country with few public roads. They probably helped to preserve savanna through its unfashionable period in the 1960s.

BALEARICS

The woods of Majorca have holm-oak, not live-oak or cork-oak. At the tree-limit in the mountains, holm-oak woods peter out into grassland with scattered trees. Savanna was once widespread in the island when oaks were fewer than now (Fig. 11.7): in the northern cordillera some oakwoods have the structure of infilled savanna. There is a partial survival of something like the Spanish pig culture, involving pink pigs with black hair.

SOUTH OF FRANCE

The 'forests' of *ancien régime* France, to judge by their uses (Chapter 11), were partly in the nature of savannas. Some-thing of their character lingers in the maquis with scattered cork-oaks around La Garde-Freinet. The cork trade became industrialized in Provence around 1830. It was the main employer in a region where the population peaked in about 1880; it was noted for bad labour relations and propensity to Red revolution.[82]

Long-distance transhumance developed on the lower slopes of the Pyrenees and the Alps and in the southern Massif Central, with drove roads leading from the coast to summer pastures on the highlands of the interior.[83] It reached its peak in the mid-eighteenth century, and lingers today (p. 90).

SARDINIA

In Sardinia the Latin word *saltus* is still current (as *saltu* or *salto* in Sard), and savannas like Salto de Quirra play a part in Sardinian history. They involve cork-oak, holm-oak and deciduous oaks. Professor I. Camarda points out that much of the savanna-like countryside of present Sardinia is recent and unstable, the result of people grubbing out young forests to make pasture, leaving some of the trees. This often follows an earlier phase of cultivation followed by abandonment. This is certainly true, but there are also areas of long-standing savanna with old trees.

[80] A point of view expressed by Cavaillès 1905.
[81] Information from Dr Pablo Campos Palacin,

1994.
[82] Agulhon 1982 edn, chapter 2.

[83] H. Isnard, *Pays et paysages méditerranéens*, Paris, 1973.

Medieval saltus

A royal charter in 1388, following a fire in Cagliari, mentions the 'scarcity of . . . timber for the use of the city', and assigns to Cagliari certain *saltos* 'with the little woods therein . . . without damaging private interests'. The places named were evidently savannas, for other charters, from 1397 onwards, allow the city's butchers to pasture beasts awaiting slaughter on them.[84]

Cork-oak savanna and the cork trade

Cork-oak, the commonest tree in Sardinia, forms denser, taller, less formal savannas than in Spain. The oaks grow into the natural shape of a free-standing tree (Fig. 12.19), which in windy Sardinia is often lopsided through wind-pruning. The extent of cork-oak savanna is impossible to define. It merges into cork forests in which the trees – for one generation only – grow close together, and into woodland in which cork-oaks form a scattered over-storey over holm-oak coppice-wood.

Young trees arise, as in Spain, by the growing up of oaks that were previously coppiced or browsed.

We did occasionally find big pollard cork-oaks which (to judge by the known rate of growth) ought to be at least two hundred years old (Fig. 12.20). They produce cork, but not all is usable: the cork-cutters harvest only the smoother areas of the trunk. These old – but hardly ancient – trees appear to be relics of an earlier phase of pollarded or shredded savanna. La Marmora in 1849 mentions shepherds who shred (*rimondano*) trees to feed their animals. He regarded this as an abuse, on the ground that the purpose of trees (even cork-oak) is to produce timber, and trees so 'mutilated' became rotten and unusable.[85]

Our researches in Cagliari archives show that cork as an industry began in the 1830s. English and French entrepreneurs bought leases on cork-trees from the Crown. Their first task was to 'cultivate' the trees by removing the natural, useless, bark. They then waited until the commercially usable cork had grown. Unless they were lucky enough to find trees already 'cultivated' by someone else it was seven years before they got any return on their investment. They had difficulties

Fig. 12.19. A rather sparse cork-oakery. *Below Burgos, Sardinia, July 1995*

Fig. 12.20. An exceptional giant pollard cork-oak. *Nurri, Sarcidano, Sardinia, April 1992*

Cork-oak country in Sardinia, as in Spain, is an empty, urbanized land, with towns and big villages widely spaced. It was very different in the Iron Age (pp. 164–5).

Most cork-oakeries appear to be semi-natural: the trees, though even-aged, are not evenly spaced nor in rows. In the NE we saw cork-groves ploughed beneath, rather like olives. Pruning is frequent but casual; only occasionally did we find traces of formal lopping in a variant of the four-branch style of Spain. Most cork-oak groves have been burnt within the last ten years.

We are puzzled by the age-distribution of cork-oaks. The great majority appear to be between sixty and a hundred years old. Writers assert that the tree goes on producing cork until about a hundred and fifty years, but we rarely encountered trees so old. It is not that they are cut down on becoming superannuated: stumps of any age are exceedingly rare.

with shepherds and others who were already using the trees for leaf-fodder or other purposes, and who 'stole' the cork or disputed the Crown's title to the trees, often in the forcible manner customary in Sardinia.[86]

Cork-gathering was still not a fully developed industry in the 1840s. Not all the cork that grew was used, and little processing was done within Sardinia. Trees were leased for twenty-one years, which provided for two corkings.

Many of the trees are said to have attained their fourth century; an immense quantity of young ones are coming forward, to be barked when eight to ten years old.[87]

What happened to these trees is somewhat of a mystery.

The industry uses almost every cork tree in the island; even in remote places we rarely saw trees that had never been corked. Trees now abandoned are mainly in dense maquis;

[84] Pinna 1903, §223, 238.

[85] La Marmora 1849; La Marmora 1860, vol. 2, pp. 565–6.

[86] ASC: Regio Demanio, Boschi e Selve, b.4, b.14, *passim.*

[87] Tyndale 1849, vol. 2, p. 3.

probably corking and woodcutting went together, the oaks being corked when they became accessible after each felling of the underwood.

The establishment of the cork trade depended on previous land-uses having generated vast numbers of cork-trees which the entrepreneurs could take over. This points to the great importance of fire in early-modern Sardinia (Chapter 13).

Other oak savannas

The holm-oak of Sardinia, *Quercus ilex*, is much more a forest tree than the live-oak of Spain. Cork-oak savannas readily turn into holm-oak forests. As a savanna tree holm-oak is mainly in the limestone mountains of the east, where there is no cork-oak. Here we found spectacular ancient shredded trees.

Deciduous oaks (known to Italian writers as *Quercus bianca*) include at least two species: the widespread *Quercus*

Fig. 12.21. Savanna of deciduous oaks (*Quercus pubescens*), some of them ancient pollards. *Fonni, Barbagia Ollolai, Sardinia, April 1992*

Fig. 12.22. Ancient pollard yews. Note (i) the strong browse-line, (ii) the absence of a ring of corpses (yew is reputedly an instant mortal poison). Why should anyone pollard yew? *Badde Sálighes, Goceano, Sardinia, July 1995*

pubescens and the endemic *Q. conferta*. In the mountainous interior, deciduous oaks occur among fields, pastures, phrygana and recent woodland, and are pollarded or shred in various styles. There are often ancient oaks (1–2 m in diameter), with younger pollards between them (Fig. 12.21). They are rarely felled, and therefore difficult to date; but one stump, of a tree intermediate in age, dated from c. 1750 and had been cut five times between c. 1827 and 1955. If this tree is representative, the older pollards would be of various dates before c. 1700; the younger would be nineteenth century. Most pollards appear to pre-date the nineteenth-century enclosure walls. Pollarding mostly ceased between 1930 and 1950; the numerous young oaks from about that period have never been cut.

Sardinian juniper

Indirect evidence about past savanna comes from juniper timbers in buildings. Juniper is prevalent from the thirteenth to the eighteenth or nineteenth centuries, and is found in lengths of up to 5 m, much bigger than junipers now

Fig. 12.23. Medieval Sardinian floor of juniper (note the distinctive hollows in the logs). *Portoscalas Gate, Cagliari, July 1995*

(Fig. 12.23). Many countries prefer juniper as a building timber – its small size is outweighed by its durability – but these indicate an environment in Sardinia much more favourable to juniper than today.

Most junipers are non-forest trees, even more light-demanding than cork-oak. The only one locally abundant in Sardinia today, the coastal *Juniperus phœnicea*, is seldom big enough to be carpentry timber. Something was happening in medieval and early-modern Sardinia to produce a steady supply of juniper for at least five hundred years. Place-names like Zinnibiri and Su Enapru, which occur well inland, afford a clue. We suspect that the medieval and early-modern *saltus* of Sardinia contained large tracts of savanna,[88] in which juniper established itself more or less regularly and cork-oak was abundant enough to attract a later cork industry.

---

88 For juniper as a savanna indicator, see Behre 1981.

CORSICA

We know of three types of savanna in Corsica. There are remains of extensive groves of pollard chestnuts much as in Italy. A few of the trees are of vast size, up to 8 m in diameter, and well over a thousand years old. In the mountains the beechwoods contain scattered pollards, probably relicts of beech savanna. In the south of the island are cork-oak savannas resembling those of Sardinia.

ITALY

The tree- and grazing-land called *saltus* was widespread in the mountains of Roman Italy. Since Roman cattle and sheep could not (as far as we know) climb trees to eat leaves, there would have been at least a tendency towards savanna, with grassland between the trees. Cicero and Varro refer to summer *saltus* in the mountains and winter *saltus* in the lowlands. This apparently increased – on pollen and place-name evidence – in the Apennines from the sixth century AD onwards.[89]

Among ancient representations of ancient trees is a pair of Roman silver bowls of the time of Augustus; each depicts a vast old plane-tree with the stubs of two huge boughs near the base.[90]

Shepherding transhumance was described by Varro, the scholar-farmer of the second century BC:

> Those who pasture in savannas (*saltibus*) and are far away from roofs carry hurdles or nets with them, with which they make corrals (*cohortes*) in the wilderness, and other equipment. They habitually pasture far and wide in different places, and often the winter pastures are many miles distant from the summer. I know this myself, I say: for my flocks used to winter in Apulia, which summered in the Reatine mountains, and between these two places . . . public lanes join the distant pastures.[91]

From Apulia to the mountains behind Rieti is at least 210 km of rough going. To Varro transhumance was a normal, established custom, but his words imply that it was so unfamiliar to his readers that he might be disbelieved.

Transhumance was later institutionalized. The Mesta-like *Dogana della Mene delle Pecore di Puglia* controlled the movement of flocks along generally similar routes to Varro's. It had the king's authority and provided revenue. The number of animals is supposed to have reached a peak of 5½ million in 1684, twice as many as in Spain on about one-quarter the area.[92] Corresponding to the *cañada* drove-roads of Spain were *tratturi*, sixty paces wide between hedges, leading from the mountains to the Murge and Tavoliere. This interior plateau of Apulia, a vast sheep-run of some 3,000 sq. km, was protected from development partly by the Dogana and partly by inhospitable limestone or calcrete soils.

Savanna was far more extensive than one would suppose from its remnants in Italy today. It often appears in Paolo Di Martino's study of the ecological history of Molise province. From at least the eighteenth century there were oakeries specializing in pigs, others (apparently with more maquis-like vegetation) in cattle and sheep. These were remarkably like the later formal *dehesas* of Spain; some were even called La Difesa.[93] Acorns and charcoal were produced.

Leaf-fodder, with its many technical terms, has been investigated by Diego Moreno and his colleagues from twentieth-century survivals. Shredded and pollarded trees stood in the less mountainous parts of woodland, but also scattered as savanna (*runchi*) in pasture, and among arable land and in hedges. Turkey-oak was shredded every three or four years; beech, ash, alder, poplar and hornbeam less often; elms only every ten years. Trees were shredded in the 'classic' manner, cutting the side-branches and sometimes leaving a tuft at the top (Fig. 4.3). Leafy boughs were stored to feed stalled cattle. In beech savannas, unlike the oak grasslands of Spain and Portugal, fodder production is mainly from the trees, also from grassland between the trees; the shaded land *under* the trees is unproductive.[94]

The forest of Montedimezzo, in the high Apennines, involved wood-pastures and coppice-woods of oaks and other trees. These can be traced back to the seventeenth century and possibly the middle ages. The pasture was served by a *tratturo*. An echo of Spain is the frequent mention of acorn-fed pigs.[95]

Chestnut-groves form intermediates between savanna, orchard and woodland; since the Iron Age they have provided nuts, flour, wood and timber (p. 177).[96]

Various tools were used for pollarding and shredding, from the Roman tree-sickle (*falx arborea*) onward. The mysterious *houlette*, a kind of sharp-ended spoon on a long, beribboned staff, which is the tool of the aristocratic shepherdess as portrayed by artists in Marie-Antoinette's France, originated in real shepherding practice in Liguria and elsewhere; she would use it to cut leafage from a shredded tree without risking her neck.

Wood-pasture survives less in Italy than in other countries, partly because modern forestry has taken a greater hold, and partly because *Endothia* blight damaged the pollard chestnuts between 1920 and 1970. However, ancient savanna trees are still to be found, often embedded in younger timberwoods or coppice-woods. Great pollard beeches sometimes survive in ex-savannas near the upper tree-limit in the Apennines.[97]

GREECE

Deciduous-oak savannas and wood-pastures[98]

In the mountains of south Macedonia oak savanna exists on a scale second only to Iberia. For at least 60 km great oaks are

[89] Moreno & Poggi 1996.
[90] *Apollo*, July 1995, pp. 32 ff. The objects are in the Metropolitan Museum, New York. The tree is wrongly described as an oak.
[91] *Rerum Rusticarum* II.ii.9. For a wider discussion, see H. F. Pelham, *Essays*, Clarendon,

Oxford, 1911, chapter 14.
[92] Delano Smith 1979, chapter 7.
[93] Di Martino 1993, 1996.
[94] Moreno & Poggi 1996; G. Salvi, 'Alberi da foraggio: foglia e stalla a Bertassi (1880–1980)', Còveri & Moreno 1983, pp. 85–105.

[95] P. Di Martino, '"Pascoli boscosi del Molise". Pratiche silvo-pastorali nella Foresta di Montedimezzo (XVII–XIX secolo)', *QS* n.s. 62, 1986, pp. 467–89.
[96] Moreno 1990b; Baraldi & others 1992.
[97] D. Moreno and O. Raggio, 'The making and

Fig. 12.24a. Deciduous oaks, Píndos Mountains, showing different styles of shredding. *Between Polynéri and Ætiá, Grevená, August 1987*

Fig. 12.24b. Oaks similar to the one that produced a five-hundred-year ring-count. *Above Polynéri, May 1988*

scattered among steppe, pollarded and shredded in different styles (Fig. 12.24). There are also continuous oak coppices, ground-oak forming a pasture in itself, and pollards among fields (p. 185). A whole cultural landscape depends on various forms of at least seven species of deciduous oaks; the people called themselves *Kupatshari*, the Oak People.[99]

Five treatments (ground-oak, pollarding, shredding, coppicing, timber) are applied to the various deciduous oaks with little discrimination, except that not all species play the ground-oak game. Different treatments overlap, and are probably not stable. Ground-oak and coppice often have pollards intermingled, suggesting that they developed out of savanna. Sheep or goats sometimes browse the shoots of coppice as well as the ground-oak.

The geology involves sandstones, ophiolites and their redeposition products, which form shallow, very erodible non-calcareous soils (Chapter 15). Oak savanna is mainly at 800–1,100 m, at the upper limit of oak and below the beech-conifer zone. Oak coppices are at 600–900 m, interspersed with cultivation. In the Grevená plain, at 400–700 m, most of the land is cultivated, with oak pollards which appear to be relics of former savanna.

The grassland is of several types, depending on altitude and moisture. The dominant grasses are perennial species of *Festuca* and *Poa*. Among them are perennial herbs such as *Acinos...alpinus*, *Erysimum linariifolium* (a Balkan endemic), *Euphorbia myrsinites;* legumes such as *Astragalus monspeliensis*, *Lathyrus digitatus* (a Balkan speciality), *Trifolium physodes;* bulbs such as *Muscari neglectum*. Annuals, less prominent than in Spain, include *Cerastium...pindigenum* (endemic), *Evax pygmaea*, *Lamium...balcanicum* (Balkan endemic, new to Greece) and *Scleranthus uncinatus*.

South Macedonia is higher than the savannas of Spain, with an even shorter growing season, limited by winter and spring cold and summer drought. It is too cold for deep-rooting evergreen oaks. Deciduous oaks are shallow-rooted, and because of drought they perform badly as forest trees. Oaks in woods are extremely slow-growing and often dead at the top (Figs. 11.4, 12.6); they compete severely with each other. They can take seventy years to reach 15 cm in diameter. Free-standing oaks, having only grassland to compete with, grow faster. Oaks between fields, in presumably better-watered places, can put on annual rings of 5 mm or more.

Away from the mountains, at the lower limit of the coppice zone, rainfall is probably too limited for deciduous oaks to form continuous forest. The woods are patchy coppices with many gaps: a sort of coppiced savanna, with steppe between the patches of trees, very rich in plant and bird life.

Livestock are mainly sheep, with some cattle. The object of woodcutting is leaves, which are stored in small ricks, like hay (Fig. 4.13), and are used to feed sheep in winter.[100]

Annual rings prove that pollarding has been established at least since the Middle Ages. For example, in a wood-pasture in Ætiá we found the stump of an oak which began life in 1497 and was first pollarded in 1556; it was again pollarded in 1578, 1617, 1636, 1706, 1754, 1814 and 1850, and died in 1950. Another oak began life in 1742, was shredded in 1760, 1777, 1845, 1851, 1863, 1884, 1918, 1933 and 1948, and died in 1960.[101] In this particular, very browsed, wood-pasture most of the oaks appear to date either from the mid-eighteenth century or from c. 1500 (Fig. 12.24b). They were pollarded or shredded at very irregular intervals, usually on a cycle of ten to twenty-five years, but sometimes at up to seventy-five years.

In less remote places, oaks may be shredded more often; we found cycles of six to eleven years between Amygdhaléi and Syndhendhron, and four to seven years in Méga Seiríni. The oldest datable oak that we found was a stump by a chapel in Mavronéi, with rings from which we extrapolated an age of seven hundred years.

fall of an intensive pastoral land-use-system. Eastern Liguria, 16–19th centuries', *Rivista di Studi Liguri* A56, 1990, pp. 193–217; Moreno & Poggi 1996.

98 Observations gathered on the Grevená Archaeological Survey (p. 185n).
99 From Vlach *kupatshu*, 'oak' (Wace & Thomson 1914).

100 Cf P. Halstead and J. Tierney, 'Leafy hay: an ethnoarchaeological study in NW Greece', *Environmental Archaeology* 1, 1998, pp. 71–80.
101 Dates, based on field counts, are approximate.

Fig. 12.25. Savanna of *Pinus nigra*. Smíxi, Grevená, August 1987

Pine savannas

Pines in south Macedonia form a higher zone, from 1,000 to 2,100 m, of pinewood and pine wood-pasture. Pinewoods are usually young (up to a hundred and fifty years), and are felled for timber.

In wood-pasture there are big old pines, *Pinus nigra* or *leucodermis* according to elevation – less often *P. heldreichii* – set in grassland or bracken (the fern *Pteridium aquilinum*, Fig. 12.25). Occasionally they are pollarded, but mostly they develop flat tops and descending branches (*nigra*) or big horizontal upturning boughs (*leucodermis*). They make a very romantic landscape, often with the pine variant of mistletoe, *Viscum album*.

Ancient pines are commonly 30–40 m high and 1.5–2.5 m in diameter. Their ages vary with altitude and growth-rate: thus a big black pine at 1,200 m on the way up to Smíxi is 195 years old; one of similar size at 1,400 m is about 420 years old; those on the pass to Dhístraton (1,750 m) are 470–550 years old; *leucodermis* pines at Lákkos in Vovoússa (1,760 m), up to 1.4 m diameter, are 630–700 years old. Their shapes indicate that they have always been free-standing – at least since they were very young – and are not the remains of a continuous wood. About half the ancient pines have been struck by lightning. Repeated strikes burn a great scar down one side of the tree without killing it (Fig. 13.7).

Grassland in pine wood-pastures is dominated by species of *Festuca* and other perennial grasses. There is a rich flora of mainly perennial herbs, such as *Arabis subflava* (endemic), *Doronicum orientale*, *Geranium macrostylum*, *Valeriana tuberosa*, *Vicia lathyroides*.

Grazing is transhumant in the pine zone: the Vlach shepherds inhabit their villages in summer, spending the winter near the coast (p. 90). Pockets of deeper soil lack trees and have traces of cultivation; we were told they were last tilled after World War II.

The principal product of pines, besides timber, was resin. A rectangle of bark, 1 m by 30 cm, was removed, and the wood underneath hacked or charred to stimulate the flow of resin. Resin accumulates for many years at such wounds, and may

be set on fire by lightning, sometimes burning a hole through the tree, which survives such treatment. Scars can be dated back for several centuries. Resining has been discontinued for at least fifty years.

Pine wood-pasture is more artificial than oak. It probably results from centuries of more severe browsing than there is now. Often the steppe is replaced by bracken, which is unpalatable. Some savannas, invaded by beech, are now coppice-like beechwoods with relict pines towering above (pp. 186–7).

In the Peloponnese, savannas of *Pinus nigra* occur on mountains such as Taygetos. These pines, which appear to date from the eighteenth century, are pollards. With the general increase of pine since that date, most of the savannas have turned into pinewoods by infilling (Figs. 4.24, 12.5).

CRETE

Crete has the most varied savannas in Europe, with at least seventeen kinds of tree. The most widespread, especially in the eastern limestone mountains, are of prickly-oak, less often holm-oak or phillyrea. The trees are commonly made into 'goat-pollards' (Figs. 4.4, 4.5e). If browsing declines, ground-oak grows up into maquis and then into infilled savanna. Holm-oak pollards by themselves sometimes punctuate the luxuriant arbutus maquis of the phyllite mountains of west Crete.[102]

In 1968 one of us visited the oak savanna above Máles which had impressed the French botanist Tournefort in 1700.[103] The immense prickly-oak pollards have stems up to 3 m in diameter and up to a thousand years old (Fig. 12.26). They grow out of fissures in bare rock; the ground vegetation is commonly also of prickly-oak and phillyrea, bitten into a ground-oak form, sometimes with phrygana and steppe.

Wild-olive, carob and occasionally terebinth form savanna in the same style, often also on hard limestone with little soil – for example, wild-olive in the Akrotíri Peninsula, carob in the hills behind Réthymnon. It is often difficult to distinguish natural savannas from derelict olive-groves and carob plantations (Fig. 4.16).

Deciduous-oak savannas (Fig. 12.27) are usually on phyllite with continuous soil. The trees (*Quercus brachyphylla* and *Q. macrolepis*), although large, are not always pollarded and may be young; they tower above the tree-heather maquis of the western mountains. An alternative in west Crete are giant trees of plane (*Platanus orientalis*), pollards up to 5 m in diameter and over a thousand years old. This moisture-loving tree, scattered among rather arid phrygana, marks the small springs characteristic of a different variant of phyllite-quartzite rocks.

Other savannas are of wild pear (*Pyrus amygdaliformis*) and hawthorn (*Crataegus* species) on the limestone rubble and loess that fill mountain-plains (Fig. 12.28). The pears, often huge ancient pollards, are grafted to domestic pear varieties.

The chestnut-groves of west Crete are still actively used. Although some travellers report chestnut 'forests', all those

[102] For middle Crete, see G. Lyrintzis and V. Papanastasis, 'Human activities and their impact on land degradation – Psilorites mountain in Crete; a historical perspective', *Land Degradation & Rehabilitation* 6, 1995, pp. 79–93.
[103] P. de Tournefort, *Rélation d'un Voyage au Levant, fait par ordre du Roy*, Anisson et Posuel, Lyon, 1717. (In 1998 there was little change.)

Fig. 12.26. Subalpine savanna of prickly-oak, pollards with stems up to 3 metres thick and a thousand years old. The shrubs are also partly prickly-oak. *Selákanon, Lassíthi Mountains, Crete, July 1968*

Fig. 12.27. Infilled savanna of deciduous oak (*Quercus brachyphylla*) with ancient pollard. *Tsikalarió, Ayios Vasíleios, Crete, August 1997*

Fig. 12.28. Savanna of hawthorn and wild pear. *Omalós Plain, west Crete, May 1992*

Fig. 12.29. Ancient pollard chestnuts with *Lecockia cretica*. Each of these famous trees, up to seven hundred years old, can be divided sectorially among several owners. *Noúlia in Flória, west Crete, April 1988*

we have seen are in the nature of orchards (Fig. 12.29). Their ground vegetation is dominated by the remarkable umbellifer *Lecockia cretica*, otherwise known only in Asia, together with moisture-loving Atlantic plants, rare in the Mediterranean, such as *Primula vulgaris* (primrose), *Osmunda regalis* (royal fern) and *Anagallis tenella*.

One of the world's rarest trees, the endemic *Zelkova abelicea*, is almost confined to subalpine savanna. It is adapted to an island with a Pleistocene history of over-browsing (p. 163). Like elm (to which it is related), it pollards, suckers and is very palatable. However, it has a unique defence mechanism. Its suckers grow into a permanent ground-oak-like state: the first time they are browsed, the sharp-ended woody cores of the twigs are left behind as spines that arm the shoots against a second attack. The trees, confined to a few mountain localities, comprise ancient pollards, youngish trees that got away the last time browsing declined, and various states of suckers. This tree, commonly called *ambelitsiá*, plays its part in Cretan civilization: all the shepherds' crooks in west Crete are made from its suckers. For cypress, also, the alpine transition is via savanna with the most extraordinary ancient trees (p. 60).

On Gávdhos, the southernmost island in Europe, the sand-dune juniper *Juniperus macrocarpa* forms savannas on spray-swept marls – often infilled by the general increase of trees on the island.

ORIGINS OF SAVANNA

Savanna is customarily interpreted as 'degraded forest' rather than treed grassland – even though, by most measures, Mediterranean savanna is a richer habitat than forest.[104] European ecologists contrast the 'artificial' wood-pastures of Europe with the similar but supposedly natural wood-pastures of other continents. This distinction cannot be upheld on environmental grounds. These ecosystems occupy regions with much the same rainfall (Table 12.i), and other mediterraneoid regions have savanna too.

Ecologists on other continents, however, often regard their savannas as cultural. There is an entrenched belief that 'savannization' of forest is part of the process of desertification in the tropics; in West Africa this notion has been held by generations of French, British and post-imperial administrators, despite evidence that trees have been stable or increasing for at least a century.[105] In Europe, too, there has been a similar lack of verification of the change from forest to savanna. Learned visitors, seeing scattered trees, are irresistibly tempted to interpret them as remains of a forest: in reality they may be trees that have sprung up out of maquis, heath, or cultivated land, or may be the latest of several generations of savanna trees.

If savanna had a prehistory as a wholly natural ecosystem, it should have species of plants and animals with special adaptations. Savannas, as in Spain and north Greece, are indeed very rich in plant and animal life, with many endemic species peculiar to those countries. Mediterranean forests, in contrast, are poor in shade-bearing plants, and especially poor in endemics (p. 46). Most of the endemics are not well enough known to exclude the possibility that they evolved in some different ecosystem – such as steppe or cliffs – and only by chance flourish in savanna. However, the cork oak is not only typical of savanna today but is uniquely adapted to savanna life. Its corky bark can hardly be other than an adaptation to withstand fire, yet the tree itself is barely combustible. It appears to be adapted to the sort of fire that would occur in a savanna where the non-tree component is more flammable than the oak itself, for example *Cistus* undershrubs or dry grass.

As was shown in Chapter 10, many pollen samples from the early to middle Holocene contain enough pollen of non-shade-bearing herbs to show that the aboriginal wildwood was patchy, not continuous forest. Before civilization it already had at least some characteristics of savanna. The proportion of the pollen of non-shade-bearing herbs is generally greatest in the dry eastern and southern Mediterranean (Fig. 10.2). Although the climate as a whole was then less arid than now, this observation supports our hypothesis that (other things being equal) a drier climate promotes savanna *versus* continuous forest.

We have remarked on the richness of savanna in birds. It is conventional to regard most European birds, especially tits and thrushes and crows and other common birds of farmland and gardens, as originally forest birds which have made a home for themselves in non-forest. However, most of them nest in trees but feed on shrubs, herbaceous plants and on the ground: they would not prosper if there were only trees. Unless they have changed their behaviour, savanna seems to be the habitat to which they are adapted.

There have probably always been savannas at the various borders between forest and non-forest. In the Pleistocene they would have been encouraged both by climate and by the great herbivores, especially on islands (p. 163). The early Holocene was exceptionally favourable to forest, not only because of climate but because for the first time the giant tree-breaking beasts were extinct. Even then, there was undoubtedly some savanna where moisture allowed trees but not forests. In the later Holocene the balance swung towards savanna, as people and livestock to some extent replaced the missing elephants. In the last two hundred years the balance has swung towards forest again.

The making of individual savannas

'Degraded forest' presumably means forest from which some trees have been subtracted. This happens in Sardinia: somebody wanting a field or pasture may partially grub out a wood, leaving scattered cork-oaks as a 'subtraction savanna'.

Oaks in the open have a spreading habit, which contrasts with their narrow habit when growing among competing trees in a forest (Fig. 12.2). This architecture is determined early in the life of the tree, and persists if the surrounding trees are later removed. It is thus possible to detect trees of forest origin in a savanna, or trees of savanna origin in a forest.

The shape of cork-oaks and live-oaks in Spain and Portugal usually excludes any possibility that they were forest trees. The savannas have not been forest since before the lifetimes of the present trees. As we have seen, many documents report heathland where there is now savanna. In other countries, too, such as Crete, the present wood-pastures are definitely not subtraction savannas. Some instances in south Macedonia, however, are inconclusive.

THE FUTURE

Conservationists are obsessed with what are (often wrongly) regarded as 'undisturbed' ecosystems, and neglect the merits of cultural ecosystems. Savanna – whether natural or cultural – forms some of Europe's most beautiful landscapes, and supports many human activities. With its juxtaposition of many habitats and freedom from poisons and pollution, it is one of the best ecosystems in Europe. In Spain and Crete it forms the last strongholds of vultures. Oak savanna, not forest, is the preferred habitat for the young of the imperial eagle, 'the most endangered bird of prey in Europe'.[106] Particularly important are savannas that have old grassland or ancient trees. The juxtaposition of abundant insect food, nectar from flowers, and nesting cavities in tree-holes sustains complex food-chains. On the scale of conservation values, savanna should come at least as high as forest.

Savanna has declined towards extinction in most of north and middle Europe (although England has some of the best examples, such as Hatfield Forest[107] and the New Forest). In the south, despite official neglect, it is relatively flourishing, except in Italy and the Balearics.

Threats of destruction have receded but still exist.[108] Although most savanna farmers are content to cultivate between the trees (a combine harvester is very manoeuvrable), some still seek occasion to grub them out. Where cereals are grown too often, grassland is destroyed and ploughing damages shallow tree-roots. Conversion to irrigated arable also goes on, but lack of water has protected Spanish and Portuguese savannas from major extensions of irrigation. Enthusiasm for pine and eucalyptus plantations is waning.

[105] J. Fairhead and M. Leach, 'Reading forest history backwards: the interaction of policy and local land use in Guinea's forest-savanna mosaic, 1893–1993', *E&H* 1, 1995, pp. 55–91; K. de Selincourt, 'Demon farmers and other myths', *New Scientist* 27, April 1996, pp. 36–39.

[106] M. Ferrer and M. Harte, 'Habitat selection by immature Spanish imperial eagles during the dispersal period', *JE* 34 , 1997, pp. 1359–64.

[107] Rackham 1989.

[108] For the situation a decade ago, see Gutiérrez & others 1985.

Neglect

Different aspects may be neglected. In Spain, woodcutting has diminished since World War II, perhaps because of better-paid alternative jobs: the oaks are lopped, if at all, at longer intervals. Lopping is less neglected in Portugal: we often found young oaks shredded for the first time.

Browsing is often neglected through the fashion for 'improved' breeds of animals which need special fodder. This tendency reinforces the decline of transhumance: the remaining animals often stay in the savanna all the year and are fed artificially when vegetation is not growing.

The consequences of not browsing depend on the environment and the plants. In places well within the climatic limit of forest, savanna turns into forest by infilling, as with the oak savannas of Sardinia, the black-pine savanna of Mount Taygetos and the prickly-oak savanna of east Crete. In south Macedonia there are several kinds of infilled savanna, including deciduous-oak savannas infilled with oak, or pine savannas infilled with beech. In many cork-oakeries in Portugal the older trees are more spreading than the younger, indicating that a formerly open *montado* has got denser. In drier places, savanna is more stable, as on thin soils in south Macedonia and on limestones in Crete.

In Spain and Portugal, although infill can happen by sprouts, the outcome is often determined by tall, aggressive, very combustible grasses and undershrubs. The coarse perennial grasses *Stipa lagascae*, *S. gigantea* and *Agrostis castellana* suppress annuals. Undershrubs, especially *Cistus ladanifera*, lavender and rosemary, invade either directly or after a fire in the tussock-grasses. These form dense thickets, competing against tree seedlings and even tree sprouts. Retama (*Lygos sphaerocarpa*) may invade on more silty soils in drier climates. Especially if there is arbutus or eucalyptus as well, this creates a fire-dominated landscape, as around Monchique in the Algarve mountains (Fig. 13.20). Infilled pine savannas are peculiarly flammable in SW Turkey (Fig. 13.9).

Inappropriate use

While some areas are neglected, overstocking becomes possible where road access has been made easy, so that feedstuffs can be imported in bulk and shepherds do not need to live with their animals. At worst, savanna which is neither destroyed nor neglected can degenerate either into 'parking lots' for artificially fed livestock, or into a combination of sown clover and poor-quality grassland. Trees are neglected, lop-wood is wasted, undershrubs are disc-ploughed with heavy equipment, and no provision is made for regenerating the trees.

This happens in Spain and Portugal, but an example is developing on Mount Psilorítis in Crete, where more animals are kept than there is pasturage to sustain (p. 187).[109] However, periods of severe browsing are not necessarily beyond the normal dynamics of the landscape. The age-distribution of trees indicates periods of 'excessive' and 'insufficient' browsing in the past. Over-browsing is harmful only if it goes on too long. Attempts to prevent it risk the collapse of shepherding, the consequences of which could easily be worse (p. 71).

Disease

Many ecologists in Spain, Portugal and Sardinia consider unspecified 'disease' or 'live-oak decline'[110] to be a problem among oaks. This is ascribed vaguely to live-oaks and cork-oaks growing on soils too wet or too infertile for them; to old age; to 'excess of biomass' through too little pollarding; and to too much pollarding.[111] The first cause is possible, although it seems to be based on theories of what climax vegetation should be, rather than on the known history of savanna sites. The second is disproved by the existence of healthy trees well over twice the average age. The third may have some truth: lopping, like pollarding, can rejuvenate a tree – although we doubt whether many of the oaks are old enough, or have been long enough unlopped, to decline from this cause. A cause that invites investigation is ploughing damage.[112]

In our travels most oaks appeared healthy, with unexpectedly few dead trees. We encountered, however, a cork-oak disease. Branches wither suddenly one at a time, leading to the death of whole limbs. This is locally attributed to an insect called *coubrilha* (*Corebus bifasciatus*) which is evidently a phloem-borer, specific to cork-oak, well-known and long-standing in Portugal. Trees also die completely, but this seems to be a different phenomenon, preceded by a general loss of vitality.

Oaks suffer caterpillar defoliation from time to time.[113] This used to be regarded as an abnormality and a problem, and in the carefree 1960s led to broadcasting insecticides (p. 206). There are reports of three fungi: the twig canker *Diplodia quercus*, the acorn parasite *Taphrina kruckii*, and the root parasite *Armillaria* sp.[114] None of this need be abnormal: the world's oaks are famous for the wildlife they support, including fungi and periodic or cyclical caterpillar attacks. Claims of 'forest decline' are often based on the unspoken, and therefore untested, assumption that the normal state of any tree is perfect health.

Erosion

Erosion is thought to be a hazard, but we have seen little evidence, although the trees make it easily detectable (p. 252).

[109] Joffre & others 1988; González Bernaldez 1991; Lyrintzis 1996.
[110] Not the condition so called in the southern United States, where evergreen oaks are attacked by the fungus *Ceratocystis fagacearum*.
[111] A. M. Delgado Gil, 'Las podas, un factor de destrucción del encinar', *Quercus* 15, 1994, pp. 16–19.

[112] A recent study implicates the root-attacking fungus *Phytophthora cinnamomi*, doing damage under alternate drought and wet: F. J. Gallego, A. Perez de Algaba, R. Fernandez-Escobar, 'Etiology of oak decline in Spain', *European Journal of Forest Pathology* 29, 1999, pp. 17–27.
[113] A dramatic example is recounted by Trevor-

Battye 1913, who in 1909 rode through the prickly-oak savannas of Apokórona (Crete), leafless from gipsy-moth caterpillars. They have completely recovered.
[114] A. Rodriguez Martin, 'El vareo de las encinas transmite enfermedades del arból', *Quercus* 2, 1986, p. 21.

In Spain, Portugal and Sardinia even rills are very local. We seldom saw more than a few centimetres of empedestalment, even in ploughland. Olive-groves show much more: continued ploughing can remove something like 2 mm of soil a year.[115]

North Greek savannas are often on badlands, and erode as do forests and pasture. Old pollards often stand on long-dormant erosion surfaces.

Conclusions

Formal savanna management will probably go on declining in a world of expensive labour. However, management in this particular way does not seem to us to be central to the conservation of savanna.

Neglect leads to either woodland or maquis, depending on the type of neglect and the part played by fire. Sometimes the resulting vegetation may resemble, at least superficially, that out of which savanna was formed in the last century. We do not expect desertification; but there is is a loss to conservation when a distinctive savanna turns into a non-distinctive forest and loses its grassland and old trees.

Many Spanish and Portuguese savannas are owned (as commonly in the past) by absentee landlords.[116] Their interest is game: wild swine, red partridge, hare, rabbit and wood-pigeon. A game habitat involves maintaining a balance between trees, shrubs, undershrubs and steppe, although often in a simplified form with a moderate degree of neglect. The landscape benefits from the willingness of townspeople to invest in it for pleasure and status. This does not seem to be an improper use of a fragile landscape, and might well be encouraged.

A difficulty in conserving savanna is its inherently episodic character. It is all very well to decide on an ideal management regime and to get it applied year after year after year, but that is not how savannas have evolved. Their character depends on historic fluctuations – decades of over-browsing alternating with years of recovery – which depend on the ups and downs of human affairs. This may occur in the future as in the past, but it is difficult to legislate for.

The conservation of savanna as a whole deserves attention, but areas of especial conservation value should be identified. Two types, in our view, are specially worthy of study and protection: those with old grassland rich in species, and those with ancient trees – and supremely that minority of savannas with both. These include the Almonte–Tozo savannas in Spain, the deciduous-oak grasslands of Sardinia and the Píndos, certain ancient beech and chestnut sites in the Apennines, and the ancient mountain prickly-oaks, *ambelitsiá* and cypresses of Crete.

---

[115] Field-boxes have been set up to measure erosion in the Almonte heartland of *dehesa;* preliminary results show that (as in most erosion studies) there is extreme local variation: D. Gomez Amelia and S. Schnabel, 'Procesos sedimentologicos e hidrologicos en una pequeña cuenca bajo explotación de dehesa en Extremadura', *Estudios de Geomorfología en España*, 1992, pp. 55–63.
[116] González Bernaldez 1991.

CHAPTER 13

Fire: Misfortune or Adaptation?

Deny anthropogenic fire and you deny humanity and many of its ancient allies
a legitimate place on the planet.[1]

All agree that the suppression of every fire only leads to fuel accumulations of such
magnitude that suppression becomes impossible.[2]

Myrtos [Crete] is . . . liable to change from the grazing-dominated landscape of the
past to a fire-dominated landscape in the future.[3]

Firebreaks work great unless there's a fire burning.[4]

Fires are thought of as a misfortune of Mediterranean lands, a tragedy self-inflicted by human malice, greed or carelessness. As we write, Mediterranean newspapers are recounting the area of forests and other lands 'destroyed' in conflagrations.[5] Summer by summer, ignition is attributed to shepherds, arsonists, beekeepers smoking swarms (or scaring badgers), madmen, developers, speculators buying the timber of burnt trees at bargain prices, foreign agents, smugglers distracting preventive officers, or right-wing politicians taking revenge against the estates of rich Communists. Editors wax rhetorical on the 'struggle against fire' as though it were an enemy that could be overthrown. The late Prime Minister of Greece was called upon to resign for failing to prevent the pines around Athens from burning, and escaped by promising never to neglect this 'duty' again. (He died, and the next big fires occurred in July 1998.)

Scientific writers regard fire as part of the 'degradation' of natural ecosystems that progressively converts forest into desert;[6] but (as with other links in the chain) there is a shortage of examples of deserts that are known, as a matter of history, to have been created in this way.

The British in India tried to prevent all forest fires in order to 'protect' sal and other valuable timber trees. This was unpopular, for occupational burning was an ancient and necessary practice in the local ecology. Where the British succeeded they got no more sal, a tree whose behaviour often depends on fire.[7] Similar folly once prevailed in North America. What Pyne calls 'Europe's pyrophobia' is entrenched in popular, academic, legal and political habits of thought. The inevitability of fire in fire-adapted vegetation; the idea of fire as a cyclical process with a definite recurrence interval; the fuel characteristics of particular types, ages and structures of vegetation; the need to understand fire in traditional land management; the idea that the intensity of a fire is

proportional to the time elapsed since the previous fire – all these are now well known in America (where pyrophobic officials strove for half a century to suppress that knowledge[8]) and Australia. In Brazil it is claimed that fire is merely a tool, the behaviour and consequences of which can be confidently predicted by managers of roughland.[9] This wheel has been reinvented at least four times, but not yet in the European Mediterranean.

Firefighting and fire-protection schemes are often drawn up without considering what makes the landscape combustible, or how the actions proposed would in practice affect the intensity and frequency of fires. Some schemes do not even state a clear objective of what pattern of fires it is hoped to achieve. It is vaguely assumed that fire is a bad thing, and that fewer fires mean less fire damage. But fire is not like a pollutant, whose effects can be eliminated by removing the source. 'Fire control' may mean reducing the frequency of the sort of fires that can be controlled, while increasing the severity of big, uncontrollable fires. Is this the intended objective?

We distinguish three types of fires:
(1) *Natural fires* ignited usually by lightning.
(2) *Occupational fires* used deliberately as a means of improving or controlling natural vegetation. (Occupational fires ignited by command of scientists or foresters are termed *prescribed fires.*)
(3) *Wildfires*, caused by human accident (e.g. burning rubbish or farmland after harvest) or malice, or by occupational fires getting out of control.

Fires, like floods, come in different sizes. Just as most flood damage is caused by exceptional floods called deluges, so most fire damage is caused by exceptionally big fires which we shall call *conflagrations* in contrast to ordinary fires.

[1] S. J. Pyne, 'Keeper of the flame: a survey of anthropogenic fire', 1992, in Crutzen & Goldhammer 1993, pp. 245–66.
[2] R. B. Keiter and M. S. Boyce, *The Greater Yellowstone Ecosystem: redefining America's wilderness heritage*, Yale University Press, New Haven, 1991.
[3] Oliver Rackham, 1989, in Rackham 1990b.

This prediction was fulfilled within six years.
[4] A sceptical American friend, on beholding yet another instance where a fire leapt effortlessly across a main road.
[5] The tragic aspects of fire are recounted with great (but uncritical) eloquence and ghastly photographs by O. Olita, *Sardegna in fiamme: prospettiva il deserto?*, STEF [Sardinia], 1992.

[6] E.g. Tomaselli 1977.
[7] K. Sivaramakrishnan, 'The politics of fire and forest regeneration in colonial Bengal', *E&H* 2, 1996, pp. 145–94.
[8] Schiff 1962.
[9] V. R. Pivello and G. A. Norton, 'FIRETOOL: an expert system for the use of prescribed fires in Brazilian savannas', *JAE* 33, 1996, pp. 348–56.

NATURAL HISTORY OF FIRE

Fire is essential to most mediterraneoid ecosystems.[10] In South Africa, thousands of plant species refuse to reproduce unless burnt. In Australia, nine trees out of ten bear the marks of fire; they are adapted not only to fire in general but to particular frequencies and intensities. *Eucalyptus regnans*, one of the world's tallest trees, reproduces after the world's grandest conflagrations. Even in Chile, with its apparently weaker fire history, vegetation is combustible and fire-adapted.[11] The seeds of many plants in California, SW Australia and South Africa need heat, smoke or charred wood to induce them to germinate;[12] it is not yet known how many do in the Mediterranean.

In southern Europe fire was normal long before Mediterranean vegetation existed. The cork-oak has an extraordinary, insulating bark (like that of several unrelated Australian trees). Cork-oak is not very combustible in itself, but grows among flammable grasses or undershrubs, encouraged by its light shade (Fig. 13.1). An ordinary fire leaves it unscathed; a fierce fire burns the foliage and twigs, leaving the trunk and branches which sprout and generate new foliage. If repeatedly burnt, cork-trees become deeply scarred but are seldom killed (Fig. 13.2). Such an adaptation, in a slowly reproducing tree, implies hundreds of thousands of years' exposure to fire. Cork-growers dislike fire because it damages the cork, and because harvesting the cork makes the tree vulnerable; but the cork-oakeries of Portugal are supposed to have been created by artificial fire,[13] and most of those in Sardinia show signs of having been burnt at least once.[14] Without fire this weakly competitive tree is squeezed out by more aggressive species.

Trees and shrubs in Mediterranean-type climates have three types of fire-adaptation. They may have insulating bark as in the cork tree. They may be killed to ground level but sprout, as do holm-oak, prickly-oak (Figs. 13.3–13.5),

Fig. 13.1. *Above left* Cork-oak savanna overrrun with *Cistus ladanifera*. In commercial cork-growing this is a sign of neglect, but it roughly represents the circumstances to which the cork tree is naturally adapted. *SW Alentejo, Portugal, May 1993*

Fig. 13.2. *Top* Cork-oaks sprouting after a severe crown fire. These trees have been burnt more often than is good for them. *Iglesias, Sardinia, July 1995*

Fig. 13.3. *Above* Crown fire in prickly-oak maquis about 1 metre high. *Nemea, Corinth, Greece, July 1985*

arbutus and eucalyptus. Or they may be killed completely and start again from seed, sometimes stimulated to germinate by fire. Trees damaged by fire develop a characteristic scar at the base, usually on the uphill side. A fire-scar gets overgrown and embedded in the annual rings, and can be dated when the tree is felled. Repeated fires enlarge the original scar into a cavity.

Among undershrubs, some are killed but stimulated to germinate. Several species of *Cistus* produce two kinds of

[10] Naveh 1994a.

[11] Chile is supposed not to have lightning fires in the present climate, but has volcanoes as sources of ignition and a Pleistocene history of fire; E. R. Fuentes, A. M. Segura, M. Holmgren, 'Are the responses of matorral shrubs different from those in an ecosystem with a reputed fire history?', Moreno & Oechel 1994, pp. 16–25.

[12] R. Cowling, ed., *The Ecology of Fynbos: nutrients, fire and diversity*, Oxford University Press, 1992; C. A. Thanos and P. W. Rundel, 'Fire-followers in chaparral: nitrogenous compounds trigger seed germination', *JE* 83, 1995, pp. 207–16; C. M. Tyler, 'Factors contributing to postfire seedling establishment in chaparral: direct and indirect effects of fire', *JE* 83, 1995, pp. 1009–20; K. W. Dixon, S. Roche, J. S. Pate,

'The promotive effect of smoke derived from burnt native vegetation on seed germination of Western Australian plants', *Oecologia* 101, 1995, pp. 185–92; W. J. Bond and B. W. van Wilgen, *Fire and plants*, Chapman & Hall, London, 1996, p. 212.

[13] Mabberley & Placito 1993, p. 222.

[14] We are indebted to Professor Ignazio Camarda for pointing out the more subtle signs.

Fig. 13.4. Prickly-oak sprouting after a fire. *Bœotla, Greece, August 1979*

Fig. 13.5. Prickly-oak maquis 1¾ years after fire. *Mount Yúktas, middle Crete, May 1988*

Fig. 13.6. *Juniperus phœnicea* burnt four years ago. The middle of the bush was somewhat sheltered and survived. Note the feeble sprouting. A juniper as badly burnt as this seldom lives. *Vasilikí, E. Crete, August 1990*

seed: a hard-coated seed that germinates after heating, and also (lest fire should fail) a soft seed that needs only to be wetted. Different species require longer or shorter fire cycles.[15] Less often, the undershrub is killed to ground level but sprouts, for example the long-lived Cretan endemic *Ebenus cretica*. This may depend on the temperature of the fire. As we shall see, herbaceous plants are stimulated by fire too.

Some Mediterranean plants dislike fire. *Juniperus phœnicea* has little power of sprouting (Fig. 13.6). Many undershrubs are fire-sensitive, including the spiny *Calicotome spinosa*. Most forms of this, in Crete and south Greece, are indicators of areas that have not recently been burnt. However, Professor V. Papanastasis assures us that in north Greece it recovers from fire. Fire-sensitivity can thus depend on ecotypic variation within a species.

Why do trees burn?

Plants are not combustible by misfortune but by adaptation. They make flammable chemicals such as resins and essential oils, or have other provisions to promote fire. It is their business in life to burn from time to time and to set back their less fire-adapted competitors.

The two common Greek pines are examples. *Pinus nigra* is a thick-barked pine with a loose, combustible leaf-litter; it attracts lightning (Fig. 13.7). In the Peloponnese it competes with *Abies alba*, a non-resinous, thin-barked conifer which burns with difficulty and is sensitive to fire. Pines are typical of metamorphic rocks, *Abies* of limestone. *Abies* can, however, grow on metamorphics, where (being shade-tolerant) it would take over from pines if they did not burn it out. The pinewoods depend on periodic fires for their continued existence.

Pinus halepensis is thin-barked. It is killed by all but the mildest fires, but grows from seeds released from cones remaining on the burnt trees. In south Greece we have seen it maintain itself on a fifteen-year cycle of burning and re-establishment (Fig. 13.8a, b). Its loose fallen leaves act as 'fine fuel'; they burn hot and allow a ground fire to spread quickly. The light shade allows undershrubs to grow beneath the trees, which catch the dead needles and prevent them from

[15] C. A. Thanos and K. Georghiou, 'Ecophysiology of fire-stimulated seed germination in *Cistus…creticus . . .* and *C. salvifolius . . .* ', *Plant, Cell & Environment* 11, 1988, pp. 841–49; Mabberley & Placito 1993, p. 70.

Fig. 13.7. *Pinus nigra,* several hundred years old, burnt to a chimney by repeated lightning strikes, but still alive. *Smíxi, Grevená, N. Greece, August 1987*

Fig. 13.8a. Young pinewood (*Pinus halepensis*), burnt a year ago. The pines are dead but retain their cones, producing seed for the next stage in the cycle. The arbutus and other maquis, which the pine had invaded, has sprouted.

Fig. 13.8b. *Pinus halepensis* seedling on the burnt area.

Both at Nemea, Corinth, Greece, July 1986

compacting into a less flammable mat (Fig. 13.9). The Spanish forest service claims, to justify pine plantations, that pine is a pioneer tree, and will in time be replaced by evergreen oak or arbutus growing up beneath it. However, *halepensis* pine has provided against this contingency by being flammable. When it burns this does not kill the oaks and arbutus (which are fire-adapted by sprouting) but prevents them from getting the upper hand.[16] We know of no examples where *halepensis* pines have remained unburnt for long enough to let successor trees become established.[17]

Ericaceae (heathers, arbutus) and palms also make flammable chemicals. Perennial grasses, especially such vigorously tussocky species as *Ampelodesmos tenax,* the 'elephant-grass' of the western Mediterranean, are fire-adapted. Their loose airy structure makes them an effective fine fuel, but fire does not penetrate the tight-packed base of the tussock, which survives and grows more fuel for the next fire.

Plants sensitive to fire do not make fire-promoting chemicals, and are difficult to burn unless mixed with flammable species. Deciduous oaks will not burn unless very dry, but their leaf-litter can support a weak fire, which the oaks survive.

Combustibility is related to age. In the mediterraneoid *chaparral* of California the hazard increases as fuel accumulates; a stand more than thirty years old is very combustible indeed. The historic pattern was a small-scale mosaic of fires which burned older stands but went out when they reached younger stands. Where fires are suppressed, this creates entire landscapes of thirty-year-old chaparral which explode into gigantic conflagrations.[18] This example has failed to inspire Mediterranean administrators with caution when

[16] B. G. Williamson and E. W. Black, 'High temperatures of forest fires as a selective advantage over oaks', *Nature* 293, 1981, pp. 643–4 [American species of pine and oak].

[17] Invasion by fire-sensitive trees through lack of fire is a conservation problem with the short-lived pines of the mountains of the SE United States.

[18] R. A. Minnich, 'Fire behavior in Southern California chaparral before fire control: the Mount Wilson burn at the turn of the century', *AAAG* 77, 1987, pp. 599–618.

Fig. 13.9a. Maquis on old terraces invaded by *Pinus brutia*. The mixture of sizes and species makes this an extreme fire risk. *Island off Bozburun Peninsula, SW Turkey, July 1996*

Fig. 13.9b. A conflagration occurred on similar terrain, 25 km away, four days later. *Near Marmaris, SW Turkey, July 1996*

Fig. 13.10. Hollow *Pinus brutia* set on fire – without, as it happened, igniting anything else. Lightning was the cause, as shown by fresh lightning scars on the surrounding trees. *Near Arádhena, Sphakiá, Crete, July 1987*

suppressing ordinary fires in shrublands; we shall see the results later.

The natural pattern of fires

What would be the natural pattern of fires if people did not ignite them? This is a hypothetical question, since people have been igniting fires in Mediterranean and mediterraneoid countries since before the present climate and vegetation became established.

There was a fire regime before people invented fire, in some ecosystems at least as far back into geological time as the Jurassic. In the boreal forest of Canada, closer to wildwood than anything in the Mediterranean today, any area can expect to burn roughly once in a hundred years; lightning is supposed to account for 90 per cent of the area burnt. There are many studies of fire before human inhabitants, or before

European contact, in other ecosystems, especially through charcoal preserved in pollen deposits.[19] Mediterranean-type ecosystems are less well known.

Lightning fires in America, Australia and probably Europe start most often when a standing dead or hollow tree is struck (Fig. 13.10). In Neolithic times or earlier, when the Mediterranean was more forested, such trees were probably commoner. Trees killed by a lightning fire would remain standing dead for many years, ready to ignite the next fire.

It is difficult to conjecture how large an area would burn at once – this would depend on weather, topography, plant species, patchiness of vegetation, intensity of natural browsing and, especially, the pattern of areas previously burnt. A fire started by the last thunderstorm of the rainy season might be relatively small and cool. A fire started by the first thunderstorm after the dry season could be hotter and more extensive than the biggest conflagrations today.

[19] J. S. Clark and J. Robinson, 'Paleoecology of fire', Crutzen & Goldhammer 1993, pp. 193–214; E. A. Johnson, *Fire and Vegetation Dynamics: studies from the North American boreal forest*, Cambridge, 1992; J. S. Clark and P. D. Royall, 'Local and regional sediment charcoal evidence for fire regimes in presettlement north-eastern North America', *JE* 84, 1996, pp. 365–82.

The natural frequency of fire in any one place with the present climate and species makeup would probably be of the order of decades more often than of centuries. The rarer fires were, the hotter and more damaging, and the greater the distance they would burn before stopping at an area previously burnt.[20]

Relations between fire and human activities are often not straightforward. When European livestock were let loose on North American savannas, they ate up the grasses which acted as fuel, causing fires to become less frequent: trees increased to the detriment of the pasture.[21] One reason for the modern destruction of tropical forests is the letting loose of giant grasses that invade felled areas, promoting fire and preventing trees from coming back.

Effects of fire

A big, late-season fire, as above Monastiráki (Ierápetra, Crete) in 1987, is an awesome phenomenon. When stoked with several square kilometres of young pines, fallen needles,

Fig. 13.11. Landscape of burnt pines (*Pinus brutia*) after a great fire. *Monastiráki, E. Crete, July 1988*

sage and spiny-broom, it releases as much energy as a small atomic bomb (Fig. 13.11). Gusts of red-hot air sear plants on cliffs, and jump gorges to ignite phrygana on the other side. The very rocks are calcined at the edges. Burning brands take to the air and start fires in advance of the main front. Islands of non-flammable vegetation, such as vineyards and riverine woods, are burnt; we have seen pears in an orchard roasted on the tree. Isolated houses are consumed, but villages, with their buffer zones of gardens, usually escape.

However, fires are part of the natural order, and even the fiercest does not sterilize the land for long. This is what happens after an ordinary summer fire in a maquis–phrygana–steppe mosaic (p. 57).

Almost at once lizards and ants emerge from their holes. Within days underground plants are stimulated to emerge

and bloom (Fig. 13.12). Over the next winter, the ground is carpeted with herbaceous plants. *More species grow than before the fire.*[22] They benefit from increased light and moisture, and from minerals, especially phosphate, recycled in the ashes. This can be dramatic if the previous vegetation was so densely shading that nothing would live under the trees. It resembles the 'coppicing plants' which are stimulated each time an English wood is felled.[23]

Post-fire plants have several adaptations. Many perennials, already present, sprout and flower in greater abundance: these include bulbs and tubers such as yellow asphodel (*Asphodeline lutea*), grasses and dandelion-like plants. Others were not apparently present before the fire, for example the hollyhock *Lavatera bryoniifolia*, *Psoralea bituminosa* and thistles. Some have good means of dispersal, moving around the landscape from one fire site to another; others come from a seed-bank in the soil, from seeds shed by their predecessors after the previous fire.

Without grazing, maquis shrubs quickly recover. In one year *Quercus coccifera* or *Arbutus unedo* are often a metre high

Fig. 13.12. Squill (*Urginea maritima*) in bloom three days after fire. The plants are nearly a metre high. *Vasilikí, E. Crete, September 1986*

(Figs. 13.5, 13.8). Undershrubs, having mostly to start again from seed, recover more slowly. The number of species is sometimes greatest in the second year after the fire, later diminishing as the maquis canopy closes in and by its shade and water abstraction out-competes the undershrubs and herbaceous plants.[24]

Post-fire plants can be vigorous competitors. In the Sierra de Alcaraz (SE Spain) thin-barked pines (*halepensis* and *pinaster*) are often slow in returning to burnt areas because of the rapid regrowth and severe competition of shrubs (*Quercus coccifera*), tall *Cistus* and even the grass *Brachypodium*

---

[20] For an attempt to estimate the prehistoric pattern of fires in California, see F. W. Davis and D. A. Burrows, 'Spatial simulation of fire regime in Mediterranean-climate landscapes', Moreno & Oechel 1994, pp. 117–39.

[21] K. Hess and J. L. Holechek, 'Policy roots of land degradation in the arid region of the United States: an overview'. *EMA* 37, 1995, pp. 123–41.

[22] This is commonplace from our own observa-

tions. See also P. Sanroque, J. L. Rubio, J. Mansanet, 'Efectos de los incendios forestales en las propiedades del suelo, en la composición florística y en la erosión hídrica de zonas forestales de Valencia (España)', *Revue de l'Écologie et Biologie du Sol* 22, 1985, pp. 131–47.

[23] Rackham 1990a.

[24] M. de Lillis and M. Testi. 'Fire disturbance and vegetation dynamics in a mediterranean

maquis of central Italy', *EM* 18, 1992, pp. 55–68; L. Trabaud and J. Lepart, 'Diversity and stability in garrigue ecosystems after fire', *Vegetatio* 43, 1980, pp. 49–57; L. Trabaud, 'Dynamics after fire of sclerophyllous plant communities in the Mediterranean basin', *EM* 13, 1987, pp. 25–37; L. Trabaud, 'Postfire plant community dynamics in the Mediterranean basin', Moreno & Oechel 1994, pp. 1–15.

Table 13.i. Changes in vegetation 1, 2, 4, 5, 8 and 10 years after the Vasilikí (Ierápetra, Crete) fire of 1986.

1 very rare   2 rare   3 occasional   4 frequent   5 abundant   6 very abundant.   * implies the word 'locally', thus 4* means frequent over part of the area.   db   dieback
sl seedling.   Legumes are shown in bold. 'Additional rare species' : the number of further species (not listed) which were in categories 1 and 2 and were seen in only one year.

	before fire	1987 Sep.	1988 Aug.	1990 Aug.	1991 May	1994 May	1996 Aug.
Trees and shrubs							
Ceratonia siliqua *(height)*	2.1m	1.2m + sl	1.4m	1.8m	1.1m (db)	1.1m	1.2m
Ephedra campylopoda	3*	4*	2*	2*	–	–	1*
Juniperus phœnicea	4*	dead	–	–	–	–	–
Olea...sylvestris *(height)*	0.9m	0.5m	0.6m	0.6m	0.5m	0.9m	0.9m
Pistacia lentiscus *(height)*	1.1m	0.8m	0.9m	0.8m	0.8m (db)	0.6m	0.7m
Prunus webbii	1	1	1	1	1	1	–
Rhamnus...oleoides *(height)*	0.3m	0.1m	0.15m	0.25m	0.3m	0.3m	0.3m
Undershrubs							
Anthyllis hermanniae	–	–	–	2	3	1	–
Asparagus aphyllus	3	3	4	4	3	2	2–3
Asperula rigida	–	–	–	4	5*	2	–
Calicotome spinosa	6	1 (sl)	2	2	2	2	4
Cistus...creticus	–	–	–	–	2	2	1
C. salvifolius	–	–	2	2	–	–	3*
Capparis spinosa	–	–	1	–	–	1	–
Cuscuta palaestina *(parasite on undershrubs)*	1	2	–	–	4	3	2
Ebenus cretica	–	–	–	–	3*	3*	3*
Fumana thymifolia & arabica *(flowers from seed in 1 year)*	2	6	5	6*	5	5	5
Genista acanthoclada	4–5	–	–	2	3	2	2
Helichrysum...barrelieri	2	–	4	3	4	2	3
Phagnalon graecum	2	–	3	3	4	2	3
Phlomis lanata	2	2	3	3	5*	4	4
Salvia triloba	–	1 (sl)	2	3*	3	3	3
Sarcopoterium spinosum	5?	2–3 (sl)	5*	5*	5*	4 & dead	4
Stachys spinosa	3?	4*	3	4*	1	2	2
Teucrium alpestre	–	–	–	–	–	3	–
T. divaricatum & microphyllum	3	–	–	–	3	2–3	2–3
T. polium *(flowers from seed in 1 year)*	2?	4	3	4	4	3	2
Thymus capitatus	6	–	3	5	5	5 & dead	5
Bulbous and tuberous perennials							
Allium rubrovittatum	–	–	–	–	–	2*	4*
A. tardans	–	–	–	–	–	–	3
Asphodelus aestivus	1	4*	5	6	5*	3	3
A. fistulosus	1	–	–	–	1	1	–
Muscari spreitzenhoferi	–	–	–	–	1	1	3
Ophrys sp. (orchid)	1	–	–	–	–	1	1
Orchis anatolica	1	–	–	–	–	1	–
Scilla autumnalis	–	–	2	–	–	–	–
Serapias sp. (orchid)	–	–	–	1	–	–	2
Urginea maritima	3	4	4	5*	2	2	4*
Additional rare species	[0]	[0]	[0]	[0]	[0]	[1]	[3]

(continued on p. 224)

	before fire	1987 Sep.	1988 Aug.	1990 Aug.	1991 May	1994 May	1996 Aug.
Other perennial herbs							
Carlina corymbosa	–	–	–	–	–	1	1
Lavatera bryoniifolia	–	–	–	2	2	–	1
Onosma erecta	–	–	–	–	–	–	2–3
Prasium majus	2?	–	2	–	3	3*	3
Psoralea bituminosa	–	2	6	–	2	1	4*
Reichardia picroides	–	–	–	–	2	1	2
Rubia tenuifolia	–	–	–	2	–	2*	2*
Selaginella denticulata (crust)	4*?	–	–	–	–	–	–
Steptorhamphus tuberosus	–	–	–	–	1	1	–
Additional rare species	[0]	[0]	[0]	[2]	[0]	[2]	[5]
Perennial grasses							
Brachypodium retusum	4	5*	5*	2	5*	4	2
Dactylis hispanica	2?	–	2	5*	–	1	5*
Hyparrhenia hirta	3?	5*	2	6	2	4	4
Piptatherum miliaceum	–	3	6	6	5	3	4
Stipa bromoides	2?	–	–	2*	2*	1	2
Additional rare species	[0]	[1]	[0]	[0]	[0]	[0]	[1]
Annual and biennial herbs							
Anagallis fœmina	–	–	–	–	4*	–	2
Anthyllis tetraphylla	–	2	–	2*	–	–	–
Atractylis cancellata	–	4	–	–	2	3	1
Bupleurum gracile	1	4	5*	–	5*	1	3–5
Campanula ?sprunerana	–	–	–	–	–	3*	–
Carlina lanata	–	–	3	2	–	–	3
Centaurea idaea	–	–	–	–	–	–	3*
C. solstitialis	–	–	–	2	–	2	–
Crepis cretica	–	–	–	–	–	2	3*–5
Crucianella imbricata & latifolia	–	4	–	1	2	1	2
Crupina crupinastrum	–	2	3	3*	4	1	3–4
Galium setaceum	–	5*	–	–	–	–	3
Helianthemum salicifolium	–	–	–	4*	–	3*	–
Hippocrepis unisiliquosa	–	–	–	1	–	1	1
Lagœcia cyminoides	2?	–	–	2	3	–	1–3*
Linum strictum	–	4	–	–	2	–	2
Lotus halophilus	–	2	–	–	–	1	3*
Misopates orontium	–	–	–	–	–	–	3*
Onobrychis aequidentata	–	–	–	–	–	1	4*
Ononis reclinata	–	5*	–	–	–	3*	3
Pallenis spinosa	1	3	5	3	5*	2	2–3
Pimpinella cretica	–	–	–	–	5*	1	2
Pseudorlaya pumila	–	–	–	–	–	1	2–3
Rhagadiolus stellatus	–	–	–	–	–	2	2
Scandix . . . brachycarpa	–	–	–	–	–	–	3*

	before fire	1987 Sep.	1988 Aug.	1990 Aug.	1991 May	1994 May	1996 Aug.
Sedum rubens	–	–	–	–	4*	–	–
Sideritis curvidens	–	–	–	–	–	–	3*
Tragopogon . . . australis	1	–	–	–	4*	2	2
T. hybridus	–	–	–	–	3*	1	–
Trifolium campestre	**–**	**–**	**–**	**–**	**4***	**1**	**2**
Valantia hispida	–	–	–	–	2	1	2
Additional rare species	[0]	[1]	[0]	[1]	[0]	[5]	[7]
Annual grasses							
Aegilops dichasians	–	–	–	–	2	3*	–
Ae. triuncialis or umbellulata	–	–	–	–	2	1	–
Avena sp.	1–2	3*	4*	5*	4	2	3–4
Brachypodium distachyon	–	–	–	–	2	1	–
Briza maxima	–	–	2	–	–	2*	1
Bromus fasciculatus	3?	4	5*	–	2	–	4–5
B. hordeaceus	–	3	–	–	–	2	3–4
Desmazeria rigida	1?	–	–	–	4*	1	2
Gastridium ventricosum	–	–	–	–	5	1	–
Lolium rigidum	–	–	–	–	–	–	2
Lophochloa cristata	–	–	5	–		3*	2–3
Stipa capensis	–	–	–	3*	5*	1	–
S. capillata	–	3	–	–	–	1	2
Plant cover							
Trees and shrubs	20%	8%	10%	10%	15%	15%	15%
Undershrubs	45%	10%	30%	35%	40%	40%	60%
Herbs and grasses	20%	12%	30%	30%	30%	20%	18%
TOTAL (exc. bryophytes & lichens)	c. 85%	30%	70%	75%	85%	80%	90%
Total number of species							
Trees and shrubs	7	6	6	6	5	5	5
Undershrubs	13	9	13	15	18	20	17
Perennial herbs	7	3	5	7	9	16	22
Perennial grasses	4	4	4	5	4	5	6
Annual herbs	4	11	4	10	14	25	33
Annual grasses	3	4	4	2	8	11	8
TOTAL	38	37	36	45	58	82	91
Number of common species (categories 4 to 6)							
Trees and shrubs	5	4	4	4	4	4	4
Undershrubs	4	3	4	7	9	4	5
Perennial herbs	1	2	3	2	2	0	0
Perennial grasses	1	2	2	3	2	2	3
Annual herbs	0	6	2	1	7	0	2
Annual grasses	0	1	2	1	4	0	1
TOTAL	11	18	17	18	28	10	15

retusum. For this reason pines prosper on south-facing slopes despite the greater aridity.[25]

Table 13.i gives ten years' observations after an ordinary wildfire. The site is typical of the drier parts of Crete. A marl slope rises to a plateau bounded by low cliffs (which were not burnt). It has a shallow soil, with long-abandoned cultivation terraces in places. The rainfall is not especially low (about 550 mm per year), but the vegetation is arid, probably because of poor root penetration. There has been little browsing except by hares, either before the fire or since. The fire, in September 1986, covered about 1 sq. km, of which these observations cover some 30 ha.

The area had been a mosaic of maquis, phrygana and steppe. Phrygana was predominant, dominated by spiny-broom (*Calicotome spinosa*), thyme (*Thymus capitatus*) and spiny-burnet (*Sarcopoterium spinosum*). Maquis patches were low and scattered, composed of carob, wild-olive and lentisk. The site borders the juniper-woods of the Skinávria Mountains, and outlying junipers (*Juniperus phœnicea*) extended into it. To judge by the junipers, there had not been a fire for at least forty years.

Juniper is exceedingly fire-sensitive. Every juniper more than three-quarters burnt died, and there have been almost no seedlings. (Junipers often escaped burning where isolated by rocky ground.)

Apart from juniper, every individual tree and shrub survived, being killed to ground level but sprouting. Within four years they were back to their normal drought-bitten state. Since then they have fluctuated, probably because of dry winters. Although there were a few seedlings, it is unlikely that any new trees or shrubs were established: all the available sites for them were occupied.

The fire especially affected undershrubs. Every individual of the dominant undershrubs appears to have been killed, and a new generation of different undershrubs has arisen from seed. *Calicotome*, which formed spiny thickets before the fire, took ten years to become again a significant part of the vegetation. *Genista acanthoclada*, also spiny, had still not recovered after ten years. *Sarcopoterium* and *Thymus* returned slowly, and by the fourth year had recovered much of their abundance. They have since declined: we suspect that their lifespan (always finite) has been shortened by drought. In place of the original undershrubs has arisen a more numerous group of species in which no one is dominant. The dwarf undershrubs *Asperula rigida*, *Fumana* and *Teucrium* species have been (temporarily) abundant or locally dominant despite their small stature. New undershrubs continued to arise at least five years after the fire.

There was a rapid increase in annual herbs. Perennial grasses then became dominant, fluctuating from year to year. The number of species of herbaceous plants went on increasing for at least ten years. Many were not recorded every year, especially tiny annuals such as *Ononis reclinata* and *Galium setaceum*.

This fire was typical in that trees and shrubs, except juniper, recovered rapidly. Great changes in undershrubs are probably also typical, although the increasing number of species was unexpected. Undershrubs are often the least stable component of roughland: even if not burnt, most are short-lived, and the species change from decade to decade according to no discernible pattern.[26] The increase in grasses and other herbs, most of them species invisible before the fire, is also usual. It was unexpected that the number of species increased for as long as ten years: it probably took so long for herbs and grasses to fill the void left by the missing undershrubs. The wet winter of 1996, following several dry winters, may have been responsible for the unusually numerous species that year.

Maquis has returned to the same state as before the fire, but it is doubtful if the other vegetation will ever do so. The previous generation of undershrubs would have grown up through changes following the previous fire; it is determined by a history of weather and browsing that cannot be repeated.

Less typical are the changes still continuing after ten years. This is not because the maquis was slow to recover, but because there was not enough of it to dominate the ecosystem. On a less arid site, post-fire plants might reach a peak in the second year and then decline as maquis regrowth closed in. Here, after ten years, successional changes after the fire are merging with general changes common to the whole area. Drought years have taken their toll. A general decline in browsing has resulted in the spread of the palatable undershrubs *Ebenus* and *Teucrium*.

Had this been an occupational fire, it would be judged successful. It severely set back *Calicotome* and *Genista*, and partly set back *Thymus* and *Sarcopoterium*, four unpalatable undershrubs. Potential pasture includes grasses, legumes and other herbs, sprouts of maquis, and some palatable undershrubs. Even after ten years without browsing, there is still plenty of potential pasture. Another occupational fire would not be needed for at least five years.

It would be hard to maintain that this fire did ecological damage or was a step towards desertification. On the contrary, it created richer and more diverse plant communities. This and other fires created the habitats of several rare plants including *Aristolochia cretica*, *Convolvulus oleifolius*, *Eryngium amorginum* and *Euphorbia dimorphocaulon*.

FIRE IN THE CULTURAL LANDSCAPE

Human activity encourages fires in various ways, deliberate and accidental. It would discourage fires in less obvious ways (Table 13.ii).

Periodic burning favours fire-promoting, fire-dependent trees, shrubs and grasses over fire-sensitive species, and thus makes the landscape more combustible. This is well established in aboriginal times in Australia and North America. In Australia, Aborigines have been credited with using fire to hold back rain-forest from advancing over the landscape when the climate grew warmer in the Holocene.[27] In the Mediterranean, matters are complicated by aridization, since drought-resistant species are often also fire-dependent.

[25] J.-M. Herranz, J. de las Heras-Ibáñez, J.-J. Martínez-Sánchez, 'Efecto de la orientación sobre la recuperación de la vegetación natural tras el fuego en el valle del Río Tus (Yeste, Albacete)', *Ecología* 5, 1991, pp. 111–23; J.-J. Martínez-Sánchez, J.-M. Herranz, J. Guerra, L. Trabaud, 'Natural recolonization of *Pinus halepensis* . . . and *Pinus pinaster* . . . in burnt forests of the Sierra de Alcaraz in Segura mountain system (SE Spain)', *EM* 22, 1996, pp. 17–24.

[26] Rackham 1990b.

[27] P. Adam, *Australian Rainforests*, Oxford University Press, 1994, chapter 9.

Tend to increase fire	Tend to reduce or postpone fire
Additional sources of ignition	Removal of big dead and hollow trees (reduces lightning fires)
Use of fire for driving hunted animals?	
Exterminating native herbivorous mammals, thus reducing browsing (especially on islands)	Browsing by goats and sheep (reduces fuel)
Occupational burning	Cultivation (breaks up large continuous areas of flammable vegetation)
Woodcutting (results in thickets of young stems)	
Decline of agriculture (replaced by flammable pioneer vegetation)	
Modern forestry (flammable trees)	

Table 13.ii. Human activities and fire.

Australian Aborigines and non-agricultural American Indians burnt the landscape for purposes connected with hunting and gathering, especially to improve the pasture of deer or kangaroos. The ethnographic evidence is copious and varied. Many tribes did not regard fire as an enemy, and did not extinguish it when it had done its work.[28] It would be surprising if Palaeolithic and Mesolithic peoples of Europe were different. This would lead to occupational burning by shepherds in favour of domestic livestock.

Occupational burning

With certain exceptions, such as the alpine pastures of west Crete, most Southern European pasturage is regularly burnt. This can be a skilled art: the area burnt can range from a single bush (to keep the shepherd warm in winter) to a whole hill-side. The interval between fires can be anything down to three years (grassland is sometimes burnt every year). A study in Corsica found that land was seldom burnt more often than once in five years.[29]

One of us has studied occupational burning in the Ligurian Apennines under the guidance of Dr Diego Moreno. This tradition was partly responsible for the (now much diminished) flower-rich grasslands of the low and middle mountains. In late March 1996 the hills around Varese were lit up every evening by fires in grassland and *Erica arborea* maquis. Grass banks (*Brachypodium pinnatum*) at the tops of terraces (p. 107) were neatly blackened. Occasionally women were gathering and burning dead leaves in chestnut groves to promote the pasture under the trees (Fig. 13.13), part of the historic management of this chestnut-and-pasture landscape. At the end of the rainy season the fires are relatively cool and controllable. Yet all the perpetrators are now arsonists in the eyes of the law. The Forest Service was busily scurrying around pursuing these desperate crooks.

We studied a fire of about 1 sq. km in grassland and *Erica arborea*, with patches of chestnut and other trees, on the steep, badlandish slopes of Monte Porcile. The fire was very sensitive to topography and vegetation. Tree-heather and dead

Fig. 13.13. Burning chestnut leaves, a necessary part of the management of chestnut wood-pasture. *Val di Vara, Liguria, March 1996*

bracken burnt fiercely. The fire burnt into grassland, but went out on encountering *Erica carnea* (less flammable than *E. arborea*) or on reaching a ridgetop. Patches of chestnuts, alders and aspens formed unburnt islands (Fig. 13.14a): the trees did not burn themselves, and their shade had prevented flammable vegetation from accumulating under them. Conversely, patches of briar (*Rosa canina*) formed burnt peninsulas at the edges of the fire: briar is flammable itself, and protects the grasses under it from browsing animals, building up fuel (Fig. 13.14b). On serpentine rock, gullies were the most burnt parts of the landscape: the extra moisture encouraged bracken to accumulate in them. On adjacent limestone, gullies escaped burning because willows and other incombustible trees replaced bracken. Scattered hawthorns in grassland displayed scars of up to four previous fires on the up-slope sides of their trunks (Fig. 13.14c). Juniper (*Juniperus communis*) was confined to rocks, ridge-tops and other protected areas; some of these junipers, to judge by scars, had just escaped both the present and previous fires which had evidently followed the same pattern (Fig. 13.14d).

Fire hindered, but did not prevent, invasion by trees. Typically, a new chestnut tree would start from a bird-dropped nut. After ten years or so, a fire would kill it to the ground, and it would sprout. After two or three fire cycles, the chestnut – by now multi-stemmed – would be big enough to shade out the vegetation underneath and to create its own fire-break island, protecting itself from further fires (Fig. 13.14e).

This fire was the first for about fifteen years. We interpret it as a normal and proper example of occupational burning, dependent on damp weather and on not too much accumulated fuel. Had it been postponed later in the season or to another year, it would have become a hot, uncontrollable wildfire, overwhelming the natural fire-breaks to sweep through chestnuts, junipers and all.

[28] Pyne 1982, chapter 2; S. J. Pyne, *Burning Bush: a fires history of Australia*, Henry Holt, New York, 1991.

[29] Rackham & Moody 1996; H. A. Forbes and H. A. Koster, 'Fire, axe, and plow: human influence on local plant communities in the southern Argolid', *Annals of the New York Academy of Sciences* 268, 1976, pp. 109–26; Conventi 1993.

Fig. 13.14

(a) *Top* The Monte Porcile fire of March 1996 burnt round, but not into, a coppiced chestnut-wood.

(b) *Top right* Burnt peninsula under briar (*Rosa*) bush, surrounded by weakly combustible grass.

(c) *Above* Base of isolated hawthorn, scarred by four previous fires. The scale is 10 cm long.

(d) *Above right* Juniper which just escaped this fire and is scarred by two previous fires.

(e) *Right* Unburnt island under an isolated chestnut-tree, surrounded by burnt tree-heather. The tree is multi-stemmed because previous fires burnt it to the ground when it was smaller. *Val di Vara, Liguria, March 1996*

Occupational burning is pronounced a 'short-term strategy' in land management by scholars who have not investigated its history. In reality it is often as necessary to shepherding as ploughing is to farming. The object is to promote edible plants: grasses, legumes, yellow Compositae and other herbs, and the young growth of shrubs. Shepherds need to get rid of spiny or distasteful undershrubs, to favour the young growth of grasses, and to remove unpalatable dead leaves. Sheep-grazing immediately after burning delays the regrowth of the shrubs and prolongs the useful early stages. Goats feed more on shrubs – they too dislike undershrubs – and require a longer burning cycle. However, if the area is never burnt, edible shrubs grow up out of reach, developing into tall maquis with a browse-line (p. 66) which provides little sustenance for goats.

Occupational fires can turn into wildfires, which is one reason why many scholars and agronomists disapprove of shepherding. This can result from laws forbidding burning, which compel shepherds to do it furtively. It may happen when shepherds move into unfamiliar areas. We understand that in Sardinia the age-old feud between shepherds and cultivators has taken a new turn, as mountain shepherds buy abandoned farmland in the lowlands for winter pasture, and set fire to it, sometimes with unfortunate results.

Part of the anthropology of fire is that conservationists and administrators prohibit occupational burning whenever they get the chance. This is done routinely in National Parks, in the belief that fire will be prevented through forbidding it. In practice, occupational fires are replaced by wildfires. Fuel accumulates over long years without burning, and when a fire does come it is late in the season and therefore hotter. We hear that exactly this happened in the Doñana, SW Spain. The traditional burning of maquis on a 25–30-year cycle, producing a mosaic of small areas of different stages of regrowth, was stopped as 'unnatural' when the Doñana National Park was set up in 1969. This created uniform tracts of vegetation with enormous accumulations of fuel, and then, in the 1980s, the conflagrations began.[30] The history of California repeated itself. Replacing controlled fires by wildfires has not benefited the habitat.

FIRES IN HISTORY

Fires are seldom recorded in ancient and medieval writings. Charcoal evidence leaves little doubt that they happened, but probably only big fires near densely populated areas were regarded as abnormal and put on record. Occupational burning, like terracing, was too commonplace to write about – just as general books about modern Scotland seldom mention occupational burning of heather.

Prehistory

The early Holocene abundance of cork-oak – if its pollen can rightly be distinguished – in the western Mediterranean implies that fire was a factor in the Upper Palaeolithic. Direct evidence from later periods comes from charcoal dust preserved in pollen deposits.[31] Fragments, big enough to identify the species that burnt, can be preserved in soils, and sometimes indicate a vegetation different from the present,[32] but have not been much studied.

Charcoal is abundant in Holocene sediments in the Algarve, and indicates 'very extensive burning' at least since the Bronze Age. There is archaeological evidence for fires in the northern Apennines during the Copper and Bronze Ages, attributed by the excavators to 'forest fires', although on pollen evidence the forests, of beech, oak and fir, would barely have been combustible. In Catalonia, charcoal in soils is reported to increase after c. 430 BC, with peaks between about 600 and 1200 AD and again around 1650 AD.[33]

We are sceptical of the theory that prehistoric people 'cleared forest by fire' to make farmland. Even if a fire kills the trees and bushes, it does not remove them, and there remains the labour of disposing of the charred stems and digging up the roots.

Classical authors

Homer twice uses forest fires in mountains as a simile.[34] There are occasional minor allusions in the Bible. A few other Classical writers allude in general terms to forest fires and ignition by lightning. It was widely believed that fires began by the wind rubbing two sticks together – a factoid apparently put into circulation by Thucydides. Fires were ignited, though seldom, during battles, either by accident or for tactical purposes. The pinewoods around Pylos in the Peloponnese burnt during the battle of 425 BC – much as those around Dubrovnik did in the war of 1991. The ancients in general, however, found floods more impressive than fires.[35]

Medieval

Sardinian towns, fearing wildfires, had the custom of making a *doha* or firebreak, a zone from which dry grass or thorns were removed or burnt. This was institutionalized in the *Carta de Logu*, the famous law-code of 1386, Article 49 of which relates to *faghere sa doha pro guardia de sa fogu*, 'making the *doha* for protection from fire'. Any citizen neglecting to work on the *doha* before Petertide (29 June, i.e. 7 July in the modern calendar) was fined and held liable for any fire damage resulting.[36] Later Sard custom allowed burning before that date except on farmland.[37]

[30] Information from M. Martin Vicente and F. Garcia Novo.

[31] Atherden & Hall 1999; Moody & others 1996.

[32] A. Pons and M. Thinon, 'The role of fire from palaeoecological data', *EM* 13 (4), 1987, pp. 3–11.

[33] P. A. James and D. K. Chester, 'Soils of Quaternary river sediments in the Algarve', Lewin & others 1995, pp. 245–62; J. J. Lowe, C. Davite, D. Moreno, R. Maggi, 'Holocene pollen stratigraphy and human interference in the woodlands of the northern Apennines, Italy', *The Holocene* 4, 1994, pp. 153–64.

[34] *Iliad* XIV.396–7; XX.490–4.

[35] Judges 9.15; Lucretius V.1241; Thucydides II.lxxvii.4; III.xcviii; IV.viii.6.

[36] ASC: MS Ballero 2, p. 79. The *Carta de Logu* is sometimes supposed to have institutionalized occupational burning. The actual text fines or imprisons anyone burning *before* 8 September (i.e. 17 September), so probably refers to stubble-burning which, if done earlier, might threaten neighbours' standing crops.

[37] ASC: MS Ballero 2, pp. 69 ff. The deliberate fire-raiser was liable to pay for one quarter of the damage and to forfeit £50 and a right hand or, in graver cases, to be burnt himself.

In Spain the habit of occupational burning has been inferred from statutes forbidding it (or legitimizing it, as in 1491). Pedro the Cruel in 1362 is said to have decreed that anyone starting a fire should be cast into it. Despite the royal favours shown to sheep-drovers (Chapter 12), Philip II in the 1550s discouraged pasturage in certain provinces on the grounds that evils were caused by fires.[38]

In Crete we hear of occupational burning from a decree of the Venetian Senate in 1414 concerning the decline of cypress, a fire-sensitive tree. The document shows unusual ecological insight:

Since the cypress timber of the Island of Crete is consumed from day to day, and is diminishing because of the great quantity which is continually being exported from the Island . . . and also because of the fires which are set by countryfolk in the groves and mountains where all the cypresses are; And the said cypresses are now coming to such a state that the mules which go to load up such wood to convey it to the city can hardly go to the places or to the mountains in which they are found, because all that was in accessible places has been consumed and cut down; . . . because the trees are of such a nature that the roots and stumps of burnt or cut trees do not sprout, and also those which begin afresh take a long time to grow.[39]

Post-medieval

Giuseppe Paulini in 1608 mentions burning as part of the process of winning pasture from forest, or improving existing pasture, in the southern Alps just outside the Mediterranean. He illustrates this with pictures, but spoils the effect by claiming that 'each year we find nearly all the mountains . . . burnt over several times' which cannot be true.[40]

Travellers occasionally mention occupational burning. For example in Valencia (Spain) in the 1790s:

The shepherds burn and destroy the vegetation in one night, mostly to get better pastures, and sometimes by malice.[41]

In Sardinia one of the great eighteenth-century law-codes was eloquent (but ineffective) against fires set by farmers 'greedy to open up and cultivate new lands' and shepherds 'so that the herbage may rise and grow faster for the pasture and sustenance of their herds'. The flocks grazed the burnt areas: evergreen and deciduous oaks did not grow again, pigs ran short of acorns, cities ran short of wood, the King of Spain ran short of revenue from the island, etc., etc. Penalties included fines, compensation and 'penal afflictions of the body'.[42]

Many nineteenth-century travellers in Sardinia complained of too much burning. Tyndale in the 1840s noticed

burning by shepherds and the occasional forest fire; he thought woodcutting was more important in the apparent decline of forest.[43] La Marmora denounced the supposedly immemorial practice of occupational burning. In September 1828, from the summit of Mount Gennargentu, he saw 38 fires burning simultaneously. He listed 26 fires in Nuoro (one of eleven provinces in Sardinia) in two months in the summer of 1849. The law against pasturing newly burnt areas was disregarded.[44] (But the abundance of juniper timber, p. 209, proves that not all of Sardinia has been subject to fire all the time.)

In the south of France, there has long been enough pine to make areas combustible. At Chaume in Vidauban (Provence) the woods of pine, holm-oak, and deciduous oak were twice 'destroyed' by fires in the eighteenth century, but rights of pollarding and coppicing continued to be exercised (pp. 98–9). Coppice-woods were said to be specially combustible because they were 'impenetrable' to anything bigger than a wolf or a wild swine.[45]

In mainland Greece reports begin in the nineteenth century. Pine fires were a feature of the Athenian countryside, as in the much increased pinewoods today. In c. 1847

all Attica was in lurid red smoke for several days; we could not breathe at Athens, for whole sides of forests near shrunk, without a hand to stop them, before the fire.

In 1888 a fire in eastern Attica is said to have burned 750 sq. km; there was another big fire in 1916.[46]

Occupational burning is mentioned in the pinewoods of the Isthmus of Corinth in 1805. It was said in 1830 that arbutus in the Peloponnese

would become a very elegant tree . . . if the shepherds did not have the custom of setting fire to the area where it grows, because of the local [and quite correct] idea that the ashes of burnt plants makes them abundant pasture the following season.[47]

In west Crete in 1612 a 'great and inextinguishable' fire (later re-tellers of the story said it burned for three years) was supposed to have much reduced the accessible woods of cypress and pine.[48] Such conflagrations were evidently rare, not occurring every two or three years as they do now, which suggests that the landscape was not so combustible. We never hear of Venetian or Turkish punitive expeditions setting fire to the countryside to flush out Cretan rebels, which they surely would have done if they could.

Cretan occupational burning is well documented in British and German air photographs of World War II, which show that areas that now appear heavily burnt were already so then. Other areas, especially in east Crete, that now appear unburnt then showed a pattern of occupational burning.

38 E. Bauer, 'Memoria histórica de la legislación de los montes en España hasta finales del siglo XIX', *Ecologia* fuera de serie 1, 1980, pp. 95–111; Fernández & de Mata Carriazo Arroquia 1969, 1, pp. 52 f.

39 ASV: Misti 50 c.131v.

40 See p. 227; Kittredge 1948, p. 6; Perlin 1989, pp. 153 ff.

41 Cavanilles 1795.

42 ASC: MS Ballero 2, p. 77.

43 Tyndale 1849.

44 La Marmora 1860, 2, pp. 565–79; La Marmora 1849.

45 Aubin & others 1980, p. 36.

46 Rackham 1973; T. Wyse, *Impressions of Greece*, Hust & Blackett, London, 1871, p. 251; E. Econo-

midou, 'The Attic landscape throughout the centuries and its human degradation', *LUP* 24, 1993, pp. 33–37.

47 Pouqueville 1820, vol. 5, p. 178; Bory de St Vincent 1836, p. 115.

48 Iseppo Civran, in VAS: Coll Relaz. 80 c.15v–16; F. Cornelius, *Creta Sacra*, Venetis 1754, 1, p. 22.

FIRE AND WEATHER

Fire is not confined to hot countries. In Spain the northern mountains are more fiery than the Mediterranean; their very fire-promoting vegetation outweighs the less hot and less seasonal climate.

In a given climate and vegetation, weather is significant. On a windy day a cigarette, which would go out on a calm day, ignites a small fire; the small fire grows faster and is less easily put out before it becomes a conflagration; once it becomes inextinguishable the wind drives it over ridges or into less flammable vegetation, into which it would not spread on a calm day. Much the same happens in hot weather, or with a late-summer fire when vegetation and soils have been drying for months.

The greatest fire year of the twentieth century in Spain, 1994, followed one of the driest seasons of the century; but the most fiery summers do not simply follow lack of rain in the wet season. Fire statistics for Sardinia, 1950–84, fail to reveal a significant correlation between the area reported as burnt and the previous annual rainfall. The exceptionally fiery year 1983 was preceded by an average winter and a very wet autumn, but came 1½ years after the exceptional drought of 1982. The very non-fiery year 1969 had been rather dry. The relation between fire and rainfall is evidently easily overcome by other factors, such as weather immediately preceding the fire, or varying efficiency of recording fires.

The time of year matters greatly. The later in the dry season the drier the vegetation, the hotter, bigger and more destructive the fire, and the wider the range of plants that get burnt. Fire-sensitive trees such as plane may be little damaged if burnt early; a fire in summer may kill the twigs but allow the tree to recover, pollard-like, from sprouts on the big branches; an autumn fire may kill the tree to ground level.

Fire depends on aspect. In 1980 one of us examined air photographs of fires in the holm-oakwoods of Monte Argentario on the coast of Tuscany. North-facing slopes consistently remained unburnt: presumably the damper vegetation was less flammable. In more arid regions north aspects might be more flammable because there was not enough vegetation on other aspects to support a fire.

An unusually fiery season in SE France and Corsica was 1990. It followed the driest two winters for forty years; drought-killed bushes added to the fuel. July and August were 2°C hotter than average, with a heat-wave in early August. There were three spells of violent *Föhn* winds, each followed by conflagrations.[49]

In Spain, fires are related to weather, but in no simple way: effects of temperature and drought vary from region to region and between large and small fires. Hot dry winds, of the *Föhn* type from the mountains or of the *levante* type from off the Sahara, are liable to turn fires into conflagrations.[50]

Among the fires we have encountered in Crete, the one at Soúyia (Sélinon) in July 1988 was in flammable pines, with a little more than average north wind. It burnt slowly over three days, and all the power of technology failed to extinguish it: the inhabitants told us that a fire in the previous generation of pines had behaved in just the same way. In October 1993 there came a mighty *sirocco* wind off the Sahara whose gusts, like the blast of a furnace, could fling open a locked iron door. This produced a conflagration in the hill country of Apokórona. The landscape, semi-abandoned farmland with maquis and tall phrygana, was not in the first rank of combustibility; but despite the lack of pine the fire swept across hills and ravines and even jumped a big gorge.

FIRE STATISTICS

Official statistics are difficult to interpret. Records are kept of numbers of fires and the area burnt. Sometimes it is recorded whether forest or other land was burnt; this introduces the uncertainty of varying definitions of forest as against maquis or savanna (fires being especially common in lands on the borderline between these). In Italy a cost in money is mysteriously attributed to each fire.

Few statistical analyses can be accepted uncritically. Statisticians rely on people sending in reports, so that fires tend to be under-recorded. The Soúyia conflagration, one of the seven biggest in Crete in the last fifteen years, eluded the statisticians. Fire is complicated by problems of definition and law. A leaf-litter fire in a deciduous oakwood has little practical or ecological consequence, yet (if noticed at all) it counts as one fire, no different from a conflagration in a pinewood. If occupational fires are illegal, this deters people from reporting them, and causes them to be confused with wildfires.

The total number of fires is meaningless, because most fires are small and of little consequence. In the south of France, half the reported fires are of less than 1 ha, yet there is no attempt to define a lower bound below which fires do not count.[51] Numbers of fires depend entirely on the whims of folk deciding whether or not to report a fire of a fraction of a hectare.

In Sardinia the authorities are diligent in looking for fires. They estimate that 18.6 per cent of the island burnt in the calendar years 1980–85, that is to say 3.1 per cent per year, equivalent to a fire cycle of 32 years; this they consider to be too frequent burning and a cause for concern, especially for the forests of the island. (This period included two very fiery years, 1981 and 1983.) However, a visit to Sardinia does not confirm that it is a land of pyromaniacs. Occupational burning is no more than in other pastoral countries. The woods of Sardinia, after centuries of concern about fires, are extensive and increasing.

Something is wrong. Some fires were recorded from all but one of the 287 townships in Sardinia, including many that are entirely arable land. Numbers of fires thus include not merely occupational burning of roughland but even burning of fields after harvest. This is confirmed by an inspection of one year in which we have details of the areas burnt (Table 13.iii). In this averagely fiery year, four-fifths of the area burnt was pasture, which probably represents occupational burning. About one-thirtieth of the pasture in Sardinia was burnt this year; the actual fire cycle in burnt pasture would be shorter

49 Météo France, 'Les conditions météorologiques pendant l'été 90', *FM* 13, 1992, pp. 22–3.

50 Piñol & others 1998.

51 G. Cesti and M. Gouiran, 'Opération Prométhée: indagine sugli incendi boschivi della costa mediterranea francese', *Monti e Boschi* 5, 1992, pp. 37–43.

	Area burnt, sq. km	Percentage of total burnt area
'High forest', conifers	5.0	1
'High forest', broadleaved	19.8	4
Coppice	15.7	3
High maquis	38.0	7
Plantations, mixed	5.0	1
Sown crops	13.9	3
Pasture, bare [i.e. without trees]	172.6	34
Pasture, treed [savanna]	96.9	19
Pasture, bushy	126.7	25
[Seven minor categories]	14.6	3
TOTAL	508.1	100

Table 13.iii. Areas officially recorded as burnt in Sardinia, 1985. [*La Programmazione in Sardegna* 109–11, 1987, p. 107]

than thirty years, since much of the pasture is never burnt. One-sixth of the burnt area was in woods, plantations and tall maquis; much of this represents wildfires. About 1.2 per cent of the total area of such vegetation was burnt. The small remainder was in cultivated ground, and represents intentional burning of fields plus occasional wildfires reaching cultivated land. Sardinia is far from being a land of arsonists: much of the island is burnt on a cycle of ten–twenty years as a regular part of land management. The woods can expect to burn about once in eighty years. Whether this is too frequent burning is a matter for debate based on further evidence, but it is not obviously disastrous.

We have thus been able to reconcile the Sardinian statistics with our own observations, but only through analysis of the details. Crude statistics of area burnt, not distinguishing normal burning from wildfires, are worse than useless.

Analysts are obsessed with 'causes' of fires, meaning the source of ignition. Opération Prométhée studied evidence from some 150,000 fires (including 50,000 forest fires) in the French Mediterranean since 1973.[52] But statistics of causes are incomplete or untrustworthy. The fire itself usually destroys the evidence of its cause, and fire-raising carries powerful legal and political motives for not ascertaining the truth. The French were honest, if unhelpful, in admitting that 70 per cent of fires are of unknown cause.

The statistics, bad as they are, probably permit the following conclusions:
(1) Lightning is not negligible; it accounts for 2–10 per cent of fires in most areas, even though modern landscapes are unattractive to lightning (see p. 221). Artificial lightning – short-circuiting of overhead cables – is also significant.
(2) The distribution of number of fires versus burnt area is very skewed, with most of the burning being accounted for by a few very large fires (Fig. 13.15).
(3) Some years are more fiery than others, with fluctuations of more than tenfold (Fig. 13.17). In Corsica, records from 1973 to 1992 indicate a four-year cycle of fires; half the burning happens in the peak year of the cycle.[53]

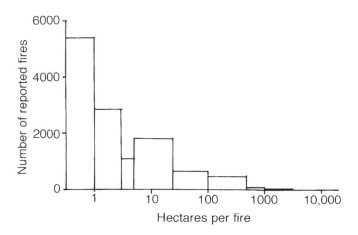

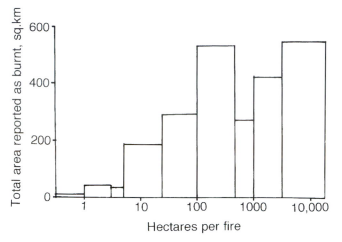

Fig. 13.15. Size *versus* number of fires, Spain 1991. Out of 12,242 reported fires, nearly half the number were of less than one hectare (top diagram). Half the area burnt, however, was by only 70 fires of more than 500 ha each (lower diagram). From official statistics.

(4) Smoking and matches are not all-important as might be expected. Even in Greece in the early 1980s, when we seldom saw any man not smoking, less than 3 per cent of the landscape burnt each year.
(5) The area burnt by reported fires is little different in privately and publicly owned forests, either in France or Spain. We suspect that neglect of fire precautions in private forests offsets the tendency of public foresters to plant fiery kinds of tree.

 EROSION AND FIRE

Burning is said to accelerate erosion. Reports of deluges, for example the Bicorp (Valencia) flood of 1982 (p. 34), often refer to extra erosion on burnt slopes. Supposedly for this reason, laws are said to have been enacted against fires near rivers in the Modego river basin in central Portugal in the fifteenth century.[54]

The theory was reinforced by events in the mediterraneoid San Gabriel Mountains, California, in 1933–4. In November a conflagration (due to forty years of previous fire

52 D. Alexandrian and M. Gouiran, 'Les causes d'incendie – levons le voile', *FM* 13, 1992, pp. 41–48.

53 Conventi 1993.
54 Neves 1981, quoted by Rego & others 1987.

suppression) consumed about 20 sq. km of chaparral. The ground was subsequently soaked by 130 mm of rain. Then came a deluge of over 300 mm at the end of December, which 'produced flood flows bulked with soil, rock fragments, and boulders from the fire-denuded slopes'.[55] As calculated from the volume of material which buried the plain below, the slopes had been denuded by an average of 22 mm. History repeated itself sixty years later: conflagrations in California in November 1993 were followed by great mudflows in February 1994.

This does not always happen. History repeated itself in east Crete with a very different result. The Vasilikí fire, described above, was followed, nine days later, by the Pakhyámmos deluge, 300 mm of rain in 36 hours (pp. 247–8). The effects, five days afterwards, consisted of washing off the ash and some of the charcoal, displacing a few stones up to 10 cm across, and creating the occasional small rill (up to 8 cm deep). Near the base of the slope a cross-slope ditch, alongside a road, acted as a sediment trap; neither then nor in subsequent years has any appreciable sediment collected in it. Even this extreme of fire and deluge failed to destabilize this slope.

Which scenario is more typical, San Gabriel or Pakhyámmos? One would expect fires that increased erosion to attract attention, whereas those that had no effect (or that reduced erosion) might go unreported. There have been many investigations in other countries,[56] but the few in the Mediterranean itself have not included a rigorous comparison of a burnt and an unburnt catchment over one or more whole fire cycles.

Some studies

An opportunity for such a comparison began in August 1990, when a conflagration invaded the Réal Collobrier experimental catchment in the Massif des Maures, Provence. This is a rugged area of deeply dissected crystalline rocks; the mean annual rainfall is about 1,000 mm. Under unburnt forest and maquis the runoff in a year is typically about 550 mm less than the rainfall for the particular year. After the fire there was about 150 mm more annual runoff. Erosion losses in forests, measured before the fire, had been at most 0.002 mm per year. In the first year after the fire, they exceeded 0.6 mm, including much fine material. After three years of increased erosion the site returned to 'normal' in 1993.[57]

Studies begun in NE Spain tend to show greater erosion (as measured by runoff plots and troughs) in burnt than unburnt areas. In one set of plots, for example, the sediment coming from holm-oak woodland was increased by 40 per cent when the woodland was felled, and nearly tenfold when

it was felled and then burned. However, these results were from one season only, lacking heavy rains, with a trifling amount of erosion: even the burnt plot lost only 0.16 ton per ha, equivalent to about 0.006 mm of sediment.[58]

Erosion after fire is soon restrained by the returning vegetation. Another Spanish study, set up after a fire in pinewood and maquis, showed sediment losses in the first year equivalent to 0.14 mm from plots on north-facing slopes, but 0.87 mm on south-facing slopes. This was attributed partly to faster regrowth of the north-facing vegetation.[59] A longer-continued plot study, near Valencia, showed that new undershrubs and pine seedlings, arising in burnt pinewood, reduced sheet-erosion rates to less than half that on bare ground. The heaviest loss, equivalent to only about 0.04 mm in the naturally revegetated plot, occurred when a minor deluge came eight years after the fire. Planted vegetation lost more sediment than natural.[60]

In France, a fire occurred on 28 August 1989 on the south side of the Montagne Sainte-Victoire, Provence, a limestone mountain with intermittent Holocene erosion (p. 304). A sand and silt fill up to 30 cm thick accumulated in stream-channels over the following winter. Plant cover swiftly spread over the burned area; in 1990 and 1991, streams were cutting down through the new fill.[61]

According to our observations, crusts of bryophytes and algae (Fig. 14.12), which hold the soil surface together, are largely killed by a fire, but re-establish themselves in a year or less. Immediately after the fire chemical crusts, or the hard-baked surface of the soil itself, may confer some resistance to water.

Conclusions

Like all aspects of erosion, the influence of fire is enormously variable.[62] It is probably exaggerated by the tendency of erosion specialists to work in erodible areas. In non-erodible areas, erosion remains insignificant despite fire.

We have seen the effects of many fires, and believe that the Vasilikí scenario is more common, though less often reported, than the San Gabriel. For example, the slopes of Monte Porcile are gullied, but most of the gullies are healed, even where now covered by fire-dominated grassland (p. 227); the remaining active badlands are not related to burning.

We do not doubt that fires can provoke serious erosion, but ourselves have only once noticed more than slight rilling. Three further examples are significant. The 1986 deluge in east Crete also fell on a steep young pinewood which had burnt a year before. By then the ground was well covered with grasses and thistles, and no noticeable erosion occurred

[55] J. D. Sinclair, 'Erosion in the San Gabriel Mountains of California', *Transactions of the American Geophysical Union* 35, 1954, pp. 264–68.
[56] G. Giovannini, 'Effect of fire and associated heating wave on the physicochemical parameters related to the soil potential erodibility', *EM* 13, 1987, pp. 113–17; J. A. Vega and F. Diaz-Fierros, 'Wildfire effects on soil erosion', *EM* 13, 1987, pp. 119–25.
[57] Lavabre & others 1993; J. Lavabre and C. Martin, 'Impact d'une incendie de forêt sur

l'hydrologie et l'érosion hydrique d'un petit bassin versant méditerranéen', Walling & Probst 1997, pp. 39–47.
[58] M. Soler and M. Sala, 'Effects of fire and of clearing in a Mediterranean *Quercus ilex* [should be *Q. rotundifolia*?] woodland: an experimental approach', *Catena* 19, 1992, pp. 321–32.
[59] M. A. Marqués and E. Mora, 'The influence of aspect on runoff and soil loss in a Mediterranean burnt forest (Spain)', *Catena* 19, 1992, pp. 333–44.

[60] V. Andreu, J. L. Rubio, R. Cerní, 'Effect of Mediterranean shrub on water erosion control', *EMA* 37, 1995, pp. 5–15.
[61] J. L. Ballais, 'L'érosion consécutive à l'incendie d'Août 1989 sur la Montagne Sainte-Victoire. Trois années d'observations (1989–92)', *BAGF* 5, 1992, pp. 423–27.
[62] M. J. Molina and P. Sanroque, 'Impact of forest fires on desertification processes: a review in relation to soil erodibility', Rubio & Calvo 1996, pp. 145–63.

Fig. 13.16. Sediment deposited on the valley floor by the first heavy rain after a fire. *Soúyia, west Crete, October 1994*

except where there had been bulldozing. Nor have we noticed significant erosion after the series of pine fires around Evróstina in the north Peloponnese. This area is very close to the marl badlands that cluster around the fault-system of the Gulf of Corinth; but here the substrate is compacted gravel, and fires will not cause it to erode.

However, another part of the Soúyia valley (west Crete) burned in August 1994. Heavy rains in September and October left spreads of a few cm of fine sand and silt in the bed of the torrent (Fig. 13.16) and over some hundreds of square metres on the beach at its mouth.

Since fire can sometimes be controlled but seldom eliminated, it would be wise to know the effects of different fire cycles on erodible terrain. To revert to the Californian example: does a conflagration every sixty years (the result of suppressing fires) remove more or less sediment than six modest fires at ten-year intervals (the result of letting fires burn)? Nobody seems to know.

Much erosion after fire is due not to the fire itself but to people 'rehabilitating' the land, especially in Spain. Forestry authorities do not always let well alone: they are pressed by misinformed public opinion to appear to be doing something. They fetch bulldozers to dig access roads into the hillsides, and then grub out the surviving tree-stools and make pseudo-terraces (Chapter 6) for planting new trees. A fiesta of bulldozing will make almost anything erode.[63]

HAVE FIRES INCREASED?

It is thought that fires have increased since World War II. For what the statistics are worth, forest and shrubland fires burnt an average of 1,000 sq. km annually in Portugal, Spain, France, Italy and Greece between 1960 and 1970; 4,800 sq. km between 1975 and 1985; and 5,600 sq. km between 1981 and 1985. This implies that between 2 and 3 per cent of

the Mediterranean 'forest and shrubland' in these countries burned annually between 1975 and 1985. In Spain it is claimed that the area burnt increased from 240 sq. km per year in the 1960s to 900 sq. km per year in 1980–84. Official statistics for the south of France, however, indicate that the area burnt per year fell by more than half between the 1910s and the 1980s, despite a big increase in the area of forests.[64]

Such estimates are of little value: it is rarely possible to distinguish between a real increase in fires, increased efficiency of reporting them, and a widening definition of what counts as a fire. World War II air photographs in Crete suggest that big wildfires were less common when there was more occupational burning. This may be a pattern in other countries.

In Sardinia the annual area officially recorded as burnt increased well over tenfold between 1950 and 1985 (Fig. 13.17). However, as we have seen, this includes occupational burning and stubble-burning as well as wildfires. The reported extent of fires in wooded areas did not increase in proportion – even though the area of woods and plantations

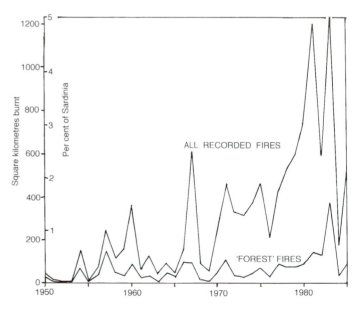

Fig. 13.17. Area burnt each year in Sardinia 1950–85, and area recorded as 'forest' fires, from official statistics.

itself increased in those decades. In most years up to 1958, between half and two-thirds of the burnt area was reported as wooded. After 1972 the proportion of the burnt area reported as wooded was usually between one-tenth and one-eighth; only in 1983 (an anomalous year with a few big fires) was it as much as one-third.[65] We cannot suppose that this is a real change: it is unlikely that the woods of Sardinia are now much less flammable than forty years ago. There are two plausible explanations. One is that the definition of wooded areas has become narrower, so that a burnt savanna counted as wooded in the 1950s but not in the 1980s. The other, more likely, explanation is that record-keepers have become more diligent

[63] For a Spanish critique of this folly, see F. González Bernáldez, 'Consideraciones ecologico-politicas acerca de la conservación y regeneración de la cubierta vegetal en España', *Ecología* fuera de serie 1, 1990, pp. 439–45.
[64] Le Houérou 1987; 'Un millón de hectareas de monte quemadas en diez anos', *Quercus* 19, 1985, pp. 16–24; P. Favre, 'Feux et forêts', *FM* 13, 1992, pp. 31–40.
[65] *La programmazione in Sardegna* 109–111, 1987, p. 15.

and less selective. In the 1950s, when access was difficult, they noted only fires that did significant damage – a large proportion of the woodland fires, and a small proportion of the non-woodland. By the 1980s they were trying to record all fires, whether damaging or not. Whether wildfires or occupational fires have really increased cannot be decided from these statistics.

There is abundant reason for fires and conflagrations to increase. Abandoned farmland has turned into pine forest and other flammable vegetation. Patches of flammable vegetation, formerly separated by fields and vineyards, are now connected by recent woodland, maquis or phrygana. In Greece, pine fires often reveal old terraces under the pines (Fig. 13.18). Sometimes, as at Soúyia, these are the second or third generation of pines to burn on the same site. In the west Mediterranean, not only pines but big undershrubs such as *Cistus monspeliensis* are particularly feared. In Sardinia, Crete and much of Greece, cutting of wood and timber have virtually ceased. Browsing has been concentrated in fewer areas, leaving savanna to be overgrown with *Cistus* and maquis

Fig. 13.18. Burnt pinewood (*Pinus brutia*), revealing old cultivation terraces. *Soúyia, SW Crete, July 1988*

to grow into woodland. Plant material once taken by wood-cutters, cattle and goats now accumulates as fuel.

We believe the increase in fires to be real, although less than the statistics make it. In the Myrtos area of Crete, which we have known for thirty years, there was certainly more burning between 1985 and 1995 than in the previous forty years. A landscape that had been too sparsely vegetated to sustain a fire turned into a peculiarly flammable mixture of undershrubs, grasses and young pines (Fig. 4.25).

FORESTRY AND FIRE

Modern foresters are not content with the natural fire risk. They love to plant fire-promoting species, notably pines and eucalyptus. About 85 per cent of the plantations in Spain are conifers and 13 per cent eucalyptus (Table 13.iv).

Species	Percentage burnt annually 1970–83	Equivalent average interval between fires
Pinus halepensis	4.44	22 years
Pinus pinaster	3.18	31 years
Pinus radiata	3.25	31 years
Pinus sylvestris (non-Mediterranean)	1.60	63 years
Eucalyptus spp	1.02	98 years
Pinus pinea	0.80	125 years
Pinus nigra (mountain)	0.46	217 years

Table 13.iv. Mean percentage of the area of Spanish plantations reported as burnt each year.
[From official statistics quoted in *Quercus* 19, 1985, p. 18.]

Among Spanish foresters' favourite trees, *Pinus pinaster* and *halepensis* can be expected to burn once in 20–30 years. (The Californian *P. radiata*, less grown in Mediterranean Spain, is the favourite in the fiery north.) *P. sylvestris* is not so extremely flammable, but is fire-related in its north European homeland. The lowland *P. pinea* and the mountain *P. nigra* are less fire-promoting, although Italian and Greek experience shows that they too are related to fires. Although in this table *Eucalyptus* is only a moderate risk, it has more recently been catching up on its fires; it is probably as fire-prone as the more flammable pines.

Of the 12,000 sq. km of *P. pinaster* and *P. halepensis* planted in Spain between 1940 and 1981, about 40 per cent were burned in an interval of only eight years. It is claimed that since 1975 the treed area burned in Spain has exceeded the area planted.[66] This was before 1994, the conflagration year when much of the remaining plantation area in Mediterranean Spain burned.

Much of the *Pinus radiata* and *P. pinea* planted in Italy in the 'reafforestation' campaigns of the 1930s was consumed by fire.[67] Not deterred, Italian foresters still plant the four most flammable species in Sardinia, with similar results to Spain.

Foresters have discovered another way to promote fire. They exclude browsing animals, and thus allow flammable vegetation to accumulate. J. R. McNeill noted, as an example of the decay into which (he claimed) the Alpujarra landscape had fallen by 1988, that 'the terraced valley above Válor . . . [has] become goat pasture'.[68] In November 1994 we found two remarkable changes. Almost the entire upper valley had been planted with pines soon after McNeill's visit. In six years these had grown big enough to burn. A fire had consumed all the pines and had spread into the remaining unplanted areas, leaving a blackened landscape as far as the eye could see (Fig. 13.19). It had licked into a live-oak forest (not noticed by McNeill), burning the edges but leaving the core unaffected. Burning revealed that the oaks had previously increased into savanna and old terraces. The entire valley showed no sign of browsing since McNeill's visit, and (to judge by unburnt areas) had accumulated undershrubs (especially the endemic giant spineless gorse *Adenocarpus*

66 P. Martínez Hermosilla, 'Enfoque historico de los trabajos de restauración', *Ecología* fuera de serie 1, 1990, pp. 367–71.

67 For an example, comparing historic with recent photographs, see [D. Moreno and O. Rackham] in *Archeologia preventiva lungo il*

percurso di un metanodotto, ed. R. Maggi, *Quaderni della Soprintendenza Archeologica della Liguria* 4, 1992, pp. 162–3.

68 McNeill 1992, p. 321.

Fig. 13.19. Island of live-oak in the midst of the Válor (Alpujarra) conflagration. The surroundings are burnt pine plantation. The edges of the live-oakwood – mostly recent – are burnt, but the interior escaped. The blackened oaks are burnt by a crown fire and will sprout from their roots; the withered oaks are scorched by a ground fire and will probably grow new twigs. *November 1994*

decorticans) and grasses. No better way could have been devised to set up a conflagration than by mixing these with young pines.

Eucalyptus forestry achieved the same result below Monchique, south Portugal (Fig. 13.20). The infertile schist hills had been poor-quality arable land and cork-oak savanna. In the 1960s they were planted with many sq. km of eucalyptus and some pine. The eucalyptus was felled once and grew again. *Cistus ladanifera* and other undershrubs grew unchecked. This assembly of flammable plants set the scene for a conflagration, which duly happened in 1990. When we were there three years later, *Cistus* and the endemic gorse *Ulex argentea* were growing again from seed. Fuel was accumulating for the next conflagration, which can be expected in about 2005.[69]

These conflagrations have not ruined the fire-adapted native vegetation or the landscape. Live-oak sprouts from its roots; about two-thirds of the eucalypts are coppicing. However, large-scale forestry has been less favourable to tree growth than leaving areas as goat pasture. As a means of conserving soil it has been disastrous. How can erosion be better encouraged than by bulldozing slopes, in preparation for planting trees, and then burning them? These fires, especially if browsing is not resumed, will encourage flammable, fire-dependent wild plants, leading to an endless cycle of future fires.

This has happened elsewhere. The area of *Pinus halepensis* in France is estimated to have risen from 360 sq. km in 1878 (the pines in Cézanne's paintings) to 1,050 sq. km by 1900, partly through deliberate afforestation at the expense of holm-oak, and today is about 1,800 sq. km. We reckon that about three-quarters of the press reports of big fires in Greece mention pine (*P. halepensis* or *brutia*). Five of the seven big roughland fires in Crete were in that small part (some 4 per cent) of the roughland of Crete that is covered by pine.

One estimate claims that of all the forests burned in Greece, Spain, France and Italy about one-third consists of *P. halepensis*, even though it constitutes only 17 per cent of the forests in Greece, 7 per cent of the Spanish, 4 per cent of the French, and 3 per cent of the Italian forests.[70]

A pine-and-fire cycle develops where pines have increased naturally. Areas of young pines repeatedly grow up to burn when they reach the thicket stage, at 15–30 years old (Fig. 13.8). On the fringes of Athens, Mount Hymettus, which probably inspired Plato's famous passage about providing 'only food for bees' (p. 288), is now more wooded than at any time since his day: every time we land at Athens airport nearby, the trees have advanced a little further. Recently it has had its first pine fires. A pine-and-fire cycle has long been established on the island of Chíos, and is unfolding on Rhodes (on both islands among *brutia* pines). One of us, travelling in July 1996 in the pinewoods west of Marmaris (SW Turkey), was surprised that they had got so far without a fire; they had grown up out of savanna as a mixture of pines of different sizes, packed with flammable undershrubs. A conflagration occurred four days later (Fig. 13.9 a,b).

HUMAN RESPONSES TO FIRE

Since 1960, Mediterranean countries have invested strenuously in fire-fighting technology, but without diminishing the area burnt. Spain spends more per head of population on this industry than any other European Community country, but remains one of the most fiery countries in Europe.

As Americans know well, the job of fire-fighting aeroplanes and helicopters is to prevent small fires from growing into middle-sized and large fires. Once a big fire has arisen, in our experience and that of other witnesses, it will burn up to a limit set by weather, topography and vegetation. Fire-fighters can protect villages and small areas of crops. Helicopters and troops make a brave show, and maybe win votes, but we know of no evidence that they influence a big fire.

A remedy is to remove sensitive activities from fiery areas. American and Australian experience shows how extremely dangerous it is to scatter suburban houses in pinewoods or other very combustible vegetation: one's scalp tingles on seeing how carelessly this is done around Athens or on Majorca. After the great fires of 1979 no more building was allowed in forests in SE France. However, Mediterranean countries are not noted for strict observance of building regulations, nor is the death penalty (by fire) sufficient to deter breaches.

Fire and vegetation structure

In olive-growing, fire protection may be one reason for the tradition of ploughing under the trees. If this is neglected or replaced by weedkilling, the dead annual vegetation burns, and the olives are damaged by scorching. We have seen ample evidence in Crete of neglect making olives more combustible. Flammability can depend on the structure of vegetation.

[69] Eucalyptus forestry is less prevalent in Greece, but has been attempted on the island of Kythera, which has an Australian colony. The mixed plantations of Aleppo pine and eucalyptus have already begun to burn.
[70] Le Houérou 1987.

Fig. 13.20. Eucalyptus plantation and neglected cork-oak savanna three years after a conflagration. The cork-oaks have sprouted from their bigger boughs, and the eucalypts from their bases. *Monchique, Algarve, May 1993*

It is the practice in America to alter the structure of forests in the hope either of preventing fires altogether or of keeping a ground fire from reaching the treetops. In Provence, the conflagrations of 1979 concentrated people's minds. Inter-township plans were drawn up for cutting understorey bushes and young trees. This is usually done over 7–10 per cent of the area deemed to be forest, in areas strategically located as firebreaks. Such clearing is required by law (and is sometimes actually done) within 50 m of public roads and habitations. The work is financed partly by the European Community; it has to be repeated every three–four years at the cost of the local inhabitants.[71]

Spanish foresters used to say (before the 1994 conflagrations) that fires in pine plantations were due to bad forest management, not to the inherent flammability of pines. Plans exist for removing undergrowth and leaf-litter from road-sides and other strategic places.[72] On the Cresta del Gallo outside Murcia city, someone has sawn some fifteen lower branches off each of a million or more young pines. No doubt it is possible thus to reduce the risk of an internal fire in a pine plantation. To do so in all the pine plantations in Spain would

be a heroic feat, which could hardly be kept up for long except in very public places. Moreover it is contrary to Spanish foresters' declared purpose of restoring natural forest in the long term.

Managing vegetation structure has its drawbacks. It involves repeated use of machinery and maybe poisons. Fire-fighting roads may do damage in themselves and allow access for less benign activities. A scheme set up in panic after a big fire may be quietly abandoned if there are a few years without big fires. Work calling for annual public expense is at the mercy of economy measures. As the years go by, workmen will require higher wages and may be impossible to find. (In 1969 we noted that the pinewoods of Majorca had been extensively de-branched; we could see no sign in 1994 that this had been done since then.) Removing undergrowth is very damaging to forests as a habitat, and interrupts the self-renewal of the forest or its change to less flammable species. The Spanish state forestry department will not live up to its name, ICONA (Institute for the Conservation of Nature), if it removes undergrowth, which includes young trees.[73]

71 J. C. Valette, E. Rigolot, M. Étienne, 'Intégra-tion des techniques de débroussaillement dans l'aménagement de défense de la forêt contre les incendies', *FM* 14, 1993, pp. 141–54.
72 R. Velez, 'Selvicultura preventiva de incendios forestales', *Ecologia* fuera de serie 1, 1990, pp. 561–71.
73 Another counter-conservation activity of Spanish foresters, at least in the recent past, is planting pines in existing beechwoods, which are of great ecological importance as well as non-flammable; P. Pointereau, E. de Miguel, D. Hickie, *Afforestation of agricultural land – guidelines for environmental assessment*, Commis-sion of the European Communities, 1993, p. 30.

Fire and browsing

Controversy over sheep and goats extends to fire. In Greece, shepherds are blamed by politicians and the public for setting fire to the landscape. Often this is used as an argument against goats; however, Professor V. Papanastasis points out to us that goats, which eat more woody vegetation than sheep, require less frequent burning. Those who would replace goats by sheep, such as R. Tomaselli,[74] must reckon with more occupational burning.

Alternatively, browsing is welcomed, since plant material eaten by livestock is not there to be burnt. The pine fires in the Myrtos area of Crete are largely confined to new pinewoods at low altitudes. The ancient pinewoods in the mountains are much less often burnt: they are severely browsed by goats and contain little undershrubbery or fine fuel. They are about as fire-resistant as pinewoods ever can be. They are used chiefly by beekeepers (p. 51), who have the strongest interest in avoiding a conflagration.

In France it is popularly believed that pasturage is the ideal means of reducing flammable vegetation in forests. It delays the regrowth of bushes and young trees after cutting. The best results have been obtained by short periods of concentrated grazing, to prevent the animals from picking and choosing. It is preferably arranged on a fairly large scale, in those few forests in the south of France where an area of several square km (necessary to sustain at least 300 head of sheep) is in one ownership.

Prescribed burning

Burning to prevent conflagrations was apparently invented in northern Portugal. Frederico Vernhagen reported in 1836 the practice of burning the litter in pinewoods (presumably *Pinus pinea*) annually in winter in order to prevent summer wildfires which would kill the trees. Pines thus burnt, he says, were not damaged in their roots, and grew faster. The choice of weather, even in winter, was critical to prevent the fire from getting out of control.[75] This problem still bedevils prescribed burning: someone has to summon the courage to do it.

'Fuel-reduction burning' was rediscovered in California in the 1880s – it was disparagingly attributed to the Paiute Indians. After long controversy it has become accepted in the United States, Australia and Israel, although it can still be an uncertain and daring operation. In the European Mediterranean, where ownerships are small and nobody dares take responsibility, it is seldom considered.[76]

Firebreaks

Conventional firebreaks are grubbed-out and ploughed strips about 50 m wide, which break up flammable areas. We have never known them stop a big fire: conflagrations have leapt over much bigger obstacles such as the main Athens–Thessaloníki highway and gorges in Crete. Their rational purpose is to provide access for fire-fighters, and as a base for back-burning: when a wildfire is approaching, fire-fighters ignite the vegetation alongside the firebreak to create a wider burnt strip against which the advancing fire-front will go out.

Firebreaks have to be maintained: a single year's neglect produces a flammable crop of dry grass or thistles. Persistent ploughing, especially up and down contours, is as effective a way to promote erosion as human ingenuity can devise. Back-burning requires more organization and resolution than can usually be mustered in the face of a crisis.

Firebreaks are easier to maintain if used for something. Fire protection plans in the south of France include browsed firebreaks, pastured by sheep, cattle, goats, horses, mules or llamas. An alternative, where the terrain permits, is cultivation of non-flammable crops, for example vineyards forming a firebreak within a forest, or the revival of chestnut and olive groves.

In Provence the townships of Hyères, La Londe and Collobrières, after a big fire, combined to divide up their cork-oak forest of 100 sq. km by 100-m-wide corridors of savanna; in all about 10 sq. km, to be browsed by three thousand head of sheep from January to June. The sheep are given fermented molasses, which gives them goatish behaviour, so that they eat cork oak, arbutus, *Phillyrea*, etc. They get a subsidy for their masters according to how much woody vegetation they eat; this saves about 60 per cent of the cost of maintaining firebreaks.[77]

'Green' firebreaks may be problematic. Their awkward shapes are difficult to fence. Browsing to control flammable vegetation will lead to complaints of excessive browsing. Putting firebreaks to use involves reversing the tendency to abandon marginal land, and clashes with the European Community's policy of reducing agriculture and pasturage. The French state subsidizes browsed firebreaks, but the European Community refuses to do so.

CONCLUSIONS

Fire is a natural part of many Mediterranean ecosystems, but people leave it out of their calculations and plans. It comes as a surprise which repetition does little to diminish. It threatens human life when people insist on building houses among fiery vegetation. Foresters grow flammable trees and expect them to reach timber size in the interval between fires.

Experience in California shows that fire cannot be eliminated by any action short of eliminating combustible vegetation. Attempts to suppress fires merely accumulate fuel to cause super-conflagrations later.[78] Although the plants are different, the Mediterranean behaves similarly. Cyprus in the twentieth century, when there was a struggle between shepherds and foresters that the foresters ultimately won,[79] shows that to suppress browsing can result in a fire-dominated landscape.

We are unsure how the present frequencies of fires and conflagrations are related to those of a century ago or of aboriginal times. Fires seldom happen for the first time in long-established vegetation that has never burnt before.

---

74 Tomaselli 1977.
75 Quoted by Rego & others 1987.
76 Pyne 1982; Naveh & Liebermann 1994, chapter 4.

77 A. Challot, 'La place des grandes coupures agricoles et pastorales dans la prévention des incendies de forêt', *FM* 14, 1993, pp. 130–40.
78 The same story of human folly having exactly

the opposite effect to what was intended has been repeated elsewhere in America, notably Yellowstone National Park.
79 Thirgood 1987.

They are either part of an ecological cycle or a consequence of recent change, deliberate (planting flammable trees) or unintentional (abandoning cultivation and pasturage).

Given that no technology can prevent fire, it might seem better to have frequent small fires than rare big ones. But the effects of different frequencies are not simple. In Australia, frequent low-intensity fires result in a more complex ecosystem than infrequent fierce fires. However, frequent burning is unfavourable to the shrubs which are an important part of Australian forests; it tends to favour introduced over native mammals. In some Australian conservation areas there is a deliberately irregular burning policy, involving different intervals, seasons and intensities, as well as areas that are never deliberately burnt but will undergo conflagrations.[80]

Although fire is involved in erosion on certain types of erodible terrain, we doubt whether this matters much. Forest and maquis fires, by definition, do not affect land that by present standards is cultivable. It is often asserted that post-fire erosion removes so much sediment that wild vegetation never returns, but we have not seen this demonstrated in Europe.

It is often claimed that fires cause desertification, but we can name no place where fire has certainly created a new desert. If desertification is progressive, fire could play a part only in the earlier stages: deserts or semi-deserts, by definition, will not burn. We have seen no places where fire damage has begun a progressive decline towards desert. Every fire that we have seen has been followed by vigorous regrowth.

The case against small fires can easily be over-stated. But there are strong reasons for avoiding conflagrations where they have not happened before. In the Samariá Gorge, in Crete, there is a magnificent historic cultural landscape, with ancient cypresses and other fire-sensitive trees. Scars on the trees show that there were periodic small (perhaps occupational) fires that did little damage, but there has been none since the Gorge was made a National Park in 1964. Pines have been invading for decades. The Gorge now risks a tragic conflagration.

Fire 'control' measures often aggravate the risk. The consequences of suppressing fires are not understood. A distant government decides that trees are good and goats are bad, and makes occupational burning a crime. Shepherds' grazing rights are taken away on a legal technicality. Government foresters then add flammable trees to the accumulating flammable wild plants, and are surprised and hurt when a conflagration consumes them. The fires are blamed, not on the vegetation or the tree-planting, but on the source of ignition, the 'notoriously lawless' shepherds. This example of human folly comes from Sardinia,[81] but most Mediterranean countries could tell a similar story.

Lawgivers, in their vanity, like to think they can legislate against fires. In Australia the consequences were spelt out in the inquest on the Victoria conflagration of 1939:

> The law relating to the prevention of fires has failed because it is not fitting for the widely diverse conditions and circumstances which obtain in Victoria. . . .
>
> Settlers and others find it necessary to burn scrub to keep their land clear that their property may be protected from fire, to promote growth . . . [but] it is found impractical to burn in the permitted periods. . . .
>
> Many settlers have not known that the permission of the Forest Officer may be sought. . . .
>
> Many have found that the permission will not be granted as the Forests Officer frequently shuns the responsibility of granting permission or refuses permission at times when experienced people feel that it is safe to burn. . . .
>
> The settler decides to burn in defiance of the law and, not wishing to be detected in the act, leaves the fire untended, either to die out or to rage across the countryside. . .
>
> The law is so notoriously unpopular, because it is unreasonable and inflexible, that there is no public opinion to check an intending law-breaker. . . .[82]

Such stupidities have been endlessly repeated. When, as in Sardinia in 1898, or in West Africa,[83] or in Liguria today, occupational burning has been banned, this does not prevent it from happening, but prevents the authorities from influencing how it is done. People doing the burning lose their flexibility in deciding when and how to burn. The rule is 'light up and get out'; the Forest Service arrests people who stay to keep an eye on the fire. The Law of Unintended Consequences takes precedence over national and European Community law.

Legislators can favour fire-raising directly. Some fires in Greece are attributed to developers who believe that when *dhásos* (land bearing woody plants) is burnt it ceases to be *dhásos* and becomes exempt from laws restricting development on *dhásos*. Whether this is so or not, it is folly to make laws that allow a fire to change the legal status of land.

The anthropological behaviour of bureaucrats conflicts with that of people who burn. Administrators divide responsibility and shun quick decisions,[84] and are not good at handling something so unpredictable as fire. We hesitate to recommend that occupational burning should require official permission. This could result in burning being done too late owing to bureaucratic delay. While the shepherd is waiting for his permit the vegetation dries out; when the permit comes the fire becomes a conflagration.

---

80 P. C. Catling, 'Ecological effects of prescribed burning practices on the mammals of southeastern Australia', *Conservation of Australia's Forest Fauna*, ed. D. Lunney, Royal Zoological Society of New South Wales, Mosman, 1991, pp. 353–64; N. Burrows, B. Ward, A. Robinson, 'Regeneration and flowering responses of plant species in jarrah forest plant communities', *Burning our Bushland: proceedings of a conference about fire and urban bushland*, ed. J. Harris, Urban Bushland Council, W[est] A[ustralia], 1995, pp. 30–33.

81 Cf F. Bocotti, 'Man and forest in Monte Arci', *News of Forest History* 25, 1997, pp. 111–18.

82 [L. E. B. Stretton], *Report of the Royal Commission to inquire into the causes of and measures taken to prevent the bush fires of January, 1939 . . . and the measures to be taken to prevent bush fires in Victoria . . .*, Government Printer, Melbourne, 1939.

83 J. Fairhead and M. Leach, 'Reading forest history backwards: the interaction of policy and local land use in Guinea's forest-savanna mosaic, 1893–1993', *E&H* 1, 1995, pp. 55–91.

84 C. N. Parkinson, *The Law of Delay: interviews and outerviews*, Murray, London, 1970, especially chapter 3, 'Buckmastership'.

Abandonment of land shows no sign of halting. It is probably impractical to prevent the spread of pines and *Cistus* (except in critical areas such as Samariá), but trouble should not be created by planting flammable trees. *Pinus halepensis,* even more flammable than the native *P. brutia,* has been introduced to Crete and thoughtlessly scattered around; by the mercy of God it is not spreading.

Foresters claim that forestry is the ideal use for abandoned land, but choose tree species supposed to be easy to establish and to grow fast. Even in forestry terms, eucalyptus and pines are not ideal. Most of them are destined merely to be ground up as paper-pulp, which may cover the cost of growing them on good terrain. On bad terrain they may never be worth harvesting, even if fire does not get them first. Eucalyptus also depletes water and nutrients, poisons native plants and promotes erosion (p. 258).[85] Native trees, however, are neglected to the point at which people rarely trouble to fell them. Cretan cypress, that noble timber, is unobtainable, although the tree is commoner than for a thousand years.[86] On Sardinia we had great difficulty in finding any stumps to count annual rings.

Mediterranean countries are ill-suited to growing paper-pulp and ordinary timber. Rising costs of labour, and competition from Scandinavia and Russia, will make them even less suitable. The future of Mediterranean forestry should involve more valuable, less flammable native trees, to compete with a diminishing supply of tropical hardwoods. There is much to be said, however, not for afforestation but for developing and extending savanna, especially in Portugal where *montado* might replace some of what is now fiery pine forest.[87]

Traditional campaigns against fire do not work. Anti-fire legislation has been tried for centuries, and the public has been exhorted against sources of ignition for decades. Neither has had any detectable effect on reducing fire (sometimes the reverse), and there are no grounds for hoping that future efforts would succeed where past efforts have failed. The future lies in coming to terms with fire and understanding what makes the landscape combustible. Big fires are part of the price of urbanization, modern forestry, rural depopulation, less intensive land use and the resulting increase in combustible wild vegetation. As we have demonstrated time after time in these pages, these fires are predictable. A young pinewood mixed with cistus is a conflagration waiting to happen: if one source of ignition does not set it off, another will.

Rehabilitation after a fire should usually involve doing nothing and letting regrowth take its course. Trees should not be planted, or land bulldozed, merely because there has been a fire. We cannot offer advice on controlling the pine-and-fire cycle, except to prevent pines from invading in the first place.

SEVEN WAYS TO PROMOTE CONFLAGRATIONS

(1)   Ban occupational burning.
(2)   Ban goats.
(3)   Plant pines and eucalyptus.
(4)   Support the fire-fighting industry.
(5)   Allow land abandonment to join up previously separate tracts of roughland.
(6)   Legislate so that a burnt forest no longer counts for planning purposes as a forest.
(7)   Allow people to build houses in pinewoods.

---

[85] Mabberley & Placito 1993, p. 193.
[86] Rackham & Moody 1996.

[87] A. Souto Cruz and A. A. Monteiro Alves, 'Ecological fire influences on Quercus suber forest ecosystems', *EM* 13, 1987, pp. 69–77.

CHAPTER 14

Current Erosion and its Measurement

Every valley shall be exalted, and every mountain and hill shall be made low: and
the crooked shall be made straight, and the rough places plain.[1]

I cite the following statement by one of our best known foresters: 'the most
permanent, effective and cheapest protection against erosion, however, is a forest
cover. Grass, while effective in preventing erosion, does not diminish the surface
runoff, and serves no other useful purpose.' This statement, of course, is so absurd
that it could not possibly do any harm, but I sometimes wonder whether many
another statement fully as far from the truth as this, and more insidiously deceptive,
has not been made about forest influences. . . . Let us then put the emphasis
where it belongs, not on the trees, but on vegetation of any kind as a means
of preventing erosion.[2] *Carlos G. Bates, experimenter on erosion*, 1924

Erosion has had a bad press. Some ancient Hebrews looked forward with joy to the prospect of a tidier landscape; but for the rest of mankind erosion washes away farmland or covers it with gravel, fills up reservoirs, silts up harbours and is linked to floods. Scholars have thought of more subtle ways in which erosion is bad for the human species: it removes the upper layers of soil in which plant nutrients have accumulated; or it deposits sediment in the lower reaches of rivers, causing (it is said) marshes to extend, mosquitoes to proliferate, malaria to prevail, peoples to flee and empires to fall.[3] These ideas come readily to North Americans, who (with some justification) fear erosion more than Europeans, and have received intense propaganda against it.

For most writers erosion is more than a force of nature; it has been accelerated, if not caused, by human activities, especially the destruction of trees. A central argument of traditional Ruined Landscape theory and of the present debate on desertification is that cultivable land has progressively shrunk. As Pendlebury, the Cretologist of the 1930s, put it:

> Crete, which was once one of the most fertile and prosperous islands in the Mediterranean, is now one of the rockiest and most barren.

For J. R. McNeill, the depopulation of many Mediterranean mountains in recent decades is largely due to the disappearance of cultivable soil rendering human life impossible.[4]

Some such philosophy, we suppose, must inspire authors of documents such as the European Union *Blue Plan* when they make mysterious pronouncements like '16.5 per cent of Spain suffers from severe erosion' – statistics that are meaningless without stating what is meant by 'severe' and how long the erosion has gone on for. They are seldom backed up by any measurement of by how much erosion is supposed to diminish production.

Erosion and Ruined Landscape theory

The critical importance of ploughing in promoting erosion was recognised in Roman times: Pausanias relates siltation by the rivers Akhelöös in Greece and Maeander in what is now Turkey to cultivation in the catchment (p. 342).[5]

However, erosion was incorporated late into Ruined Landscape theory. William Bowles, the Irish geologist who first investigated the badlands of Spain (Chapter 15), appreciated that erosion created cultivable land. Lacking (it seems) any idea of tectonic action as a counterbalance, he foresaw that erosion would ultimately flatten out the earth's surface, so that there would no longer be any gradient to carry rivers away.[6] We cannot find in his writings, nor in those of Cavanilles in the 1790s, any idea that vegetation affected erosion. Antonio Ponz, the passionate Spanish tree-lover and forest-hater of the 1780s, advanced many ingenious reasons for planting trees and grubbing out forests, but not control of erosion.

Meanwhile, the theory that trees prevent floods and erosion had become established in the south of France. By 1760, grubbing of forests was blamed for flooding. The act of the *parlement* of Provence in 1767, requiring new fields to be terraced (p. 115), also says that gullies (*torrens*) have arisen 'when forests have been cut down, or when, by destroying the box [*Buxus*] situated along rivers, one has unclothed the banks of what would serve to prevent their overflowing'. Charles de Ribbe attributed a seventy-year series of erosions in alpine Provence, which (he said) had carried off the best land and driven out the inhabitants, to floods caused by excessive tree-felling. The remaining inhabitants evidently disagreed, for they continued to destroy the woods.[7]

Following deluges in 1840 and 1856, French official foresters devised a grand scheme to prevent flooding in the Pyrenees and Alps. A Service for Re-wooding Mountain

[1] Isaiah 40:4.
[2] C. G. Bates, 'The erosion problem: erosion is the cause of many calamities', *Journal of Forestry* 22, 1924, pp. 499–504 [quoted by Schiff 1962].
[3] A theory publicized, if not invented, by Marsh 1864. For a contrary view, that soil erosion is not

necessarily unnatural or undesirable, see M. S. Boyce, 'Natural regulation or the control of nature?', *The Greater Yellowstone Ecosystem: redefining America's wilderness heritage*, ed. R. B. Keiter and M. S. Boyce, Yale University Press, New Haven, 1991, pp. 183–208.

[4] McNeill 1992 *passim*.
[5] Pausanias VIII.xxiv.11.
[6] bowles 1783.
[7] C. de Ribbe, *La Provence au point de vue des bois*, c. 1776, cited by Marsh 1864, chapter 3. To establish whether he was right would be an

Lands was set up to 'restore' forest cover on a scale that would have wholly disrupted local society and economy. Pyreneans stoutly dissented, to the extent of organizing a Congress Against Reafforestation. Although the foresters had been given special legal powers (by votes of at least 246 to 4), the scheme took sixty years to be partially completed. More recently trees have increased as land has been abandoned for other reasons. Whether any of the original objective has been achieved is doubtful. In the Pyrenees floods have, if anything, increased.[8]

Ruined Landscape theory is thus deeply pessimistic. If it is true, soils are unconservable except by growing trees. All cultivation – even orchards – is ultimately self-destructive; farmers can survive only by going out of business, as was proposed for the Pyrenees. However, there is a curious lack of detail. Are all trees equally useful, and all other vegetation equally useless, at preventing floods, or at preventing the various forms of erosion? Is it only trees in the form of forests that do so? Will savanna or maquis serve? Scholars and propagandists tend to make up their minds on these questions without any studies or measurements to justify their confidence.[9]

Erosion is a natural process, bringing benefits as well as inconveniences. Without it there would be no sedimentation, no jobs for sedimentary geologists, little cultivable land and no fertile river deltas.[10] In Mediterranean countries most cultivation is on sediment accumulated by past erosion; this soil would be too thin and steep to be of much use if returned to the hillsides whence it came.

This still goes on where dam-building has not interrupted it. The traveller Swinburne remarked of the Ebro in the eighteenth century:

> The slime [the waters] leave after great floods is deemed as beneficial to the lands they overflow as those of the Nile are to Egypt.[11]

For millennia the erosion of Ethiopia sent silt and dissolved minerals down the Nile to feed the fertility of Egypt and to top up the Nile delta; today Ethiopia still erodes, but the silt fills up the Roseires, Sennar and Aswan dams instead of fertilizing Egypt, and the Nile delta continues to sink into the sea. The silt of Navarre and Aragon likewise now fills the Ebro dams instead of fertilizing the flood-plain and topping up the delta. In general the bad effects of erosion have been emphasized and its constructive effects overlooked.

Accelerated erosion

Ruined Landscape scholars who accept erosion as a natural process point to *accelerated erosion* resulting from human activities. They apportion erosion into 'natural' and 'anthropogenic' components, of which the latter is supposed to be particularly severe in the Mediterranean.[12] But how is accelerated erosion differentiated? What would be the natural pattern of erosion without accelerating activities?

Here many authors would insert a diagram of erosion rates as a function of annual rainfall. Normal erosion rates are supposed to reach a peak, at about 0.13 mm per year, in regions with 300–350 mm rainfall. With more rainfall, increasing vegetation is supposed to restrain erosion; with less there is not enough rain to wash sediment away.[13] We shall not do this: the Mediterranean has many natural factors predisposing to higher erosion rates, such as the rainfall being concentrated into half the year and the steep slopes resulting from mountain-building.

There is no simple test for accelerated erosion. The dictum of Walter Clay Lowdermilk that 'accelerated erosion . . . proceeds at rates greater than soil formation'[14] is popular but fallacious. No law of nature decrees that soil formation shall outstrip erosion: sometimes it does, sometimes not, for reasons that need have nothing to do with human action.[15] A rough test might be whether landforms are being lowered by erosion faster than they are being raised by tectonic action. Even this is not conclusive, for tectonism and erosion are discontinuous: rapid erosion today may be compensating for a period of rapid uplift millennia ago.

It is easy to assume that well-developed soil profiles with differentiated horizons, like those of northern Europe, are normal, and that something is wrong with the 'truncated', poorly developed soils common in Mediterranean countries. This need not be true: under a violent climatic and tectonic history, sites stable for long enough for soils to develop fully may always have been unusual.

Accelerated erosion might be recognised by comparing existing landforms and soil profiles with those that have existed without human intervention, and determining when that intervention occurred. Soil profiles never disturbed by clearing, occupational burning, grazing or cultivation are rare except in special and inaccessible places, but can sometimes be found buried under later deposits. Conspicuous erosional features, which suggest to the observer that a serious problem exists, often turn out to be natural features of great age, or relics of erosion accelerated by human activity in the distant past.

interesting piece of field and archive research. The Provençal Alps attract deluges (pp. 32–4), are tectonically active and have unstable marl geology (p. 277), so ought to be very erodible whatever the vegetation. Snow-melt is a possible route for forests to influence floods (p. 254), which would operate here but not in most of the Mediterranean proper. De Ribbe mentions great landslides, in which vegetation is unlikely to have been significant (p. 258). The deluges are now thought to have begun in the fourteenth century, not a time of increasing land-use (Bravard 1993). If de Ribbe's floods were unusual, an alternative cause,

unknown to him, might have been the tail of Little Ice Age III (p. 137).
8 J. P. Bravard, 'Approches du changement fluvial dans le bassin du Rhône (XIVᵉ–XIXᵉ siècles)', *Pour une histoire de l'environnement*, ed. C. Beck and R. Delort, CNRS Éditions, Paris, 1993, pp. 97–103; J.-P. Métailié, 'Le fleuve ravageur: risques, catastrophes et aménagement dans les Pyrénées et leur piémont, fin XIIᵉ–XXᵉ siècle', *ibid.*, pp. 105–12; Marsh 1864, chapter 4.
9 Saberwal 1998.
10 Australia, that ancient, stable continent, has little erosion (other than wind erosion) and very infertile soils.

11 Swinburne 1779, p. 81.
12 Lewin & others, 1995, pp. 15 f.
13 J. B. Thornes, 'The ecology of erosion', *Geography* 70, 1985, pp. 222–35.
14 W. C. Lowdermilk, 'Acceleration of erosion above geologic norms', *Transactions of the American Geophysical Union* 15, 1934, pp. 505–9.
15 According to Poesen & Hooke 1997, a soil formation rate of 1 mm per year 'has been accepted for a deep medium-textured soil under a temperate humid climate', but most Mediterranean soils are not like this and form more slowly.

Over what period can accelerated erosion be blamed on deforestation? A powerful influence on scientific thought was an observation by Sir Charles Lyell, the great geologist, in 1846 that forest grubbed out in the south-eastern United States turned into badland within twenty years (p. 272). But, as Lyell pointed out, in this region forest had existed continuously for millions of years, unlike in Europe. Forest, or at least tree-land, may be thought of as the 'normal' state of most Mediterranean countries over the last thirteen thousand years. But on a time-scale of hundreds of thousands of years, the normal state has been 'deforested'; forested interglacials have lasted for less than one-quarter of the time. Those who regard deforestation as an unfortunate artefact, destroying soils, must explain how those soils survived the much longer Pleistocene deforestations.

KINDS OF INLAND EROSION

Sheet erosion

This means the removal of a layer of soil covering a sloping surface. It implies a sheet of water flowing down the slope after heavy rain and carrying solid particles with it. It is difficult to separate from *soil creep*, the gradual downward movement of the top layer of the soil as a whole, which occurs especially on ploughed land. Ploughing and sheet erosion cause soil to accumulate as a *lynchet* on the upper side of any cross-slope barrier – an earthwork, hedge, fence, terrace wall, tree or building (Fig. 6.1i; contrast Plate 14.1).

Fig. 14.1. A stable landscape. Considerable soil is left between the projecting limestone rocks, but it is not moving downslope, as the lack of accumulation against the *kalderími* (constructed mule-track) demonstrates. *Islet of Khálki, Rhodes, May 1986*

Rill erosion

On all but very smooth surfaces, erosion is not uniform. Runoff water is deflected and channelled by small obstacles – stones, grass tussocks, trees – and removes more sediment

Fig. 14.2. Rills in an almond-grove, ploughed on the slope, on schist. *Near Béires, Alpujarra, November 1994*

in some places than others. This can create *rills* which are incipient gullies, usually formed in arable land or other unvegetated surfaces (Fig. 14.2). Rills are defined as gullies on a small enough scale to be obliterated at the next ploughing.

Gullying

A *gully* is a channel big enough to last for years or decades. It may begin as a rill, gradually enlarging through downcutting, side-cutting, head-cutting and branching.[16] Gullies often originate from paths made by persons, cattle and goats.[17] After a time a gully stabilizes and turns into an ordinary ravine. Sands and deep weathered layers of rotten granite are susceptible to gullying. Certain types of soft sediment, especially silts and some clays, develop into whole landscapes of gullies, called *badlands* (Chapter 15). Streams in gullies usually run during storms for only a few hours in the year.

Although gullying is mainly due to water erosion, it can be helped by frost, even at low altitudes. Needle ice is especially effective in lifting a surface layer a few millimetres thick overnight from marl and marly limestone exposures, to sludge downslope as the ice melts in the noonday sun.

Slumps and other mass movements

Hillsides collapse under gravity in various ways. A *slump* or *landslip* is a mass of earth or soft rock that breaks away as a whole from its parent hillside along a *slip plane*. Sometimes it tilts backwards as well as sliding: the breakaway part then comes to rest as a terrace, often crescent-shaped and sloping back into the hillside (Figs. 14.3–14.5). Slumps are determined by geology, often where clay alternates with other strata and is lubricated by springs. They are often set off by deluge, snow-melt or earthquake, or by people digging into unstable slopes. Groups of slumps may merge into *progressive landslides* continuing over many years. Large slumps are important forms of erosion in their own right. Small ones

---

[16] For a detailed study of gullying, see Faulkner 1995.
[17] Gullies are akin to the *holloways* of England,

roads and paths that have sunk several metres below the land surface in the course of centuries (Rackham 1986). We do not understand why

holloways should be rare in most Mediterranean countries.

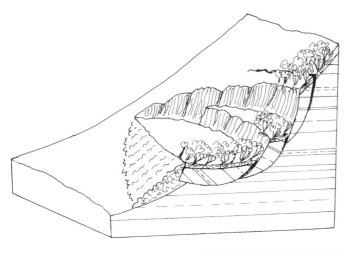

Fig. 14.3. A rotary slump.

Fig. 14.4 Slump in phyllite. *Myrsine in Tourlotí, east Crete, August 1990*

Fig. 14.5a, b. Slumping of an unsupported road-cut in soft clays. *Near Galéni, 20 km SW of Herákleion; April 1988*

expose patches of steep, raw subsoil to other forms of erosion, initiating gullies, as in the badlands of Basilicata and the Píndos Mountains.[18]

Mud-flows and debris-flows occur when extreme rain, perhaps accompanied by earthquake, turns the soil and subsoil temporarily to a liquid. 'A wall of boulders, rocks of all sizes, and oozing mud suddenly appear round the bend in a canyon preceded by a thundrous roar', 'filling houses without pushing in the walls', and leaving a chaotic mixture of clay, silt, sheep, sand, bricks, trees and boulders.[19] California knows them better than the Mediterranean. The southeast corner of France writhes with mud-flows and slumps, with its California-like combination of steep mountains, weak rocks, earthquakes and reckless development.[20] Some creep steadily at a few metres a year; others happen overnight, as when the sea opened its mouth and swallowed the new port of Nice (p. 336). Remains of mud-flows of roughly late Roman date appear to bury earlier deposits on the Khaniá and Frangokástello plains in Crete (p. 296–9).

River erosion

Rivers change through meandering, downcutting and side-cutting.[21] They may side-cut in a small way into flood-plain deposits, but large-scale havoc results when a river eats into the foot of a debris cone or of a steep slope liable to slump (Fig. 14.6). The river may be loaded with more sediment than it can carry away; this accumulates in the channel, causing the water to spill out over the flood-plain so that the stream becomes braided. Later downcutting will convert the deposit into a *river terrace*. These processes are partly due to the inherent instability of rivers, and partly to changes in the river's hydrology or environment from natural or artificial causes.

Wind erosion

Wind picks up and removes silt-sized particles (0.02–0.06 mm diameter). This requires an unvegetated surface and the right kind of soil – especially loess soils, consisting of particles originally brought by wind from somewhere else (p. 29). Stones, potsherds and other coarse fragments are left behind, a process known as *deflation*. Wind erosion produced

[18] Brunsden & Prior 1984; M. Julian and E. Antony, 'Aspects of landslide activity in the Mercantour Massif and the French Riviera, southeastern France', *Geomorphology* 15, 1996, pp. 275–89; Alexander 1982.

[19] A. M. Johnson and J. R. Rodine, 'Debris flow', Brunsden & Prior 1984, pp. 257–361; Jameson & others 1994, pp. 176–85.

[20] Julian & Anthony 1996; Wainwright 1996.

[21] C. W. Thorne, 'Effects of vegetation on river-bank erosion and stability', Thornes 1990, pp. 125–44. [The context is non-Mediterranean.]

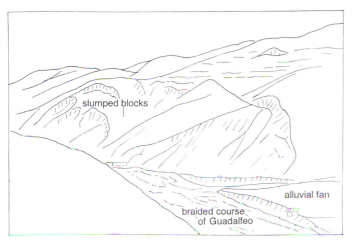

Fig. 14.6a, b. *Left* A side-cutting river: the Río Guadalfeo, Granada. When the river floods it eats into river terraces, causes slumps on valley sides, and erodes alluvial cones. *November 1994*

Fig. 14.7a. *Above* Razor-edged solution features overlying the marks of an ancient quarry. *Ayios Andónios, Varypetro, west Crete, July 1996*

Fig. 14.7b. *Right* After two or three weeks on razor-edged karst, this is what boots look like. *Sphakia, Crete, July 1992*

the 'Great American Dust-Bowl' and the desertification of Soviet Central Asia and western New South Wales (Chapter 1).

Karst erosion

Ordinary rainwater is slightly acid and dissolves limestone, sculpturing the surface into strange landforms and sharp edges (Fig. 14.7). Dissolution rates vary with rock type, precipitation, snow cover and other factors, from less than 20 to more than 100 mm per thousand years; they are not negligible compared to the removal of solid particles.[22] Dissolution bores holes at joints in the rock, and enlarges existing holes and depressions, giving rise to the *karst* landforms characteristic of certain types of limestone. A surface pockmarked with depressions overlies a tangle of unseen fissures and caves (Figs. 4.19, 14.23). These absorb rainwater, and sometimes can drink down even deluges (p. 323).

The insoluble minerals from limestone may be left behind to create a soil, together with extraneous materials such as windblown dust. These are liable to a second process of *underworld erosion*, washing down through fissures and disappearing into the caverns beneath. This is claimed to be hastened by land-uses that diminish the humus content of the soil.[23] Karst landscapes display a complex balance between processes adding to, or subtracting from, their soils.

Scree and gravel

Frost and changes of temperature splinter fragments off cliffs, which accumulate at the bottom of the cliff as a *scree*. A scree may later be eroded by water action and its fragments taken away to form a *gravel*. Stones composing scree are angular; those composing gravel are rounded. They may later be cemented together to form a natural concrete – called a *breccia* if its component stones are angular, and a *conglomerate* if they are rounded.

Many rivers have immense gravel-beds, sometimes 1 km wide, through which the river pursues a shifting, braided course, filling additional channels after heavy rain. The gravels may be bare or more or less covered with vegetation, even woodland, depending on how long ago the last deluge was. These constitute the *fiumari* of southern Italy (Fig. 14.8), or their counterparts in SE Spain, Provence, the Píndos Mountains and Rhodes. To a north European these wonderful, bleached, unused, seemingly desolate river-gravels, close to the buried sites of ancient cities, are the epitome of Ruined Landscape, an affront to civilization. In reality they happen wherever there is a combination of high mountains, short steep river-courses, sudden deluges and a copious supply of non-cemented gravel derived from Pleistocene erosions. They are as prevalent in forested Rhodes (and the very forested northern Alps) as in treeless SE Spain; the Alphéios

---

22 Gams & others 1993; Hempel 1990.
23 D. P. Drew, 'Environmental archaeology and karstic terrains: the example of the Burren, Co. Clare, Ireland', Bell & Limbrey 1982, pp. 115–27.

In the diagram: slumped blocks · alluvial fan · braided course of Guadalfeo

Fig. 14.8. Wide gravel bed of a minor river emerging from mountains. *Rio Trionto, Calabria, May 1993.*

(pp. 291–2) comes from a very forested part of Greece. They are not lifeless; plants such as tamarisks are specially adapted to living in them.

Erosion and soil erosion

It is a common fallacy that all erosion represents the loss of soil. At least one scholarly book has the index entry 'Erosion *see* Soil erosion'. Conservationists measure the volume of sediment deposited in a delta or on a plain, and translate this into so many millions of tons of soil removed from the catchment. Popular writers, quoting such findings, turn the word 'soil' into 'topsoil'. This makes fine propaganda, but whether it is true depends on the kind of erosion. Sheet, rill and wind erosion, working from the top downwards, do indeed remove soil (as long as there is any soil left) but that is not true of the other kinds. With gullying, sidecutting, marine and karst erosion, most of the sediment removed is not soil.

Suppose that some land consists of 50 cm of soil, of which 15 cm can be described as topsoil, overlying soft rock. In it is a gully, 1,000 m long, triangular in section, which tapers evenly from a point at its head to a width of 80 m and a depth of 30 m at its mouth (Fig. 14.9). A moment's calculation shows that the gully has removed 400,000 cubic metres of sediment. The area from which soil has been removed is

Fig. 14.9. Diagram of gully erosion.

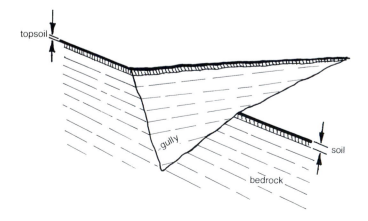

40,000 square metres. The volume of *soil* removed is thus approximately 20,000 cubic metres, or 5 per cent of the total sediment. The volume of *topsoil* removed is about 6,000 cubic metres, or 1.5 per cent of the total sediment. The remaining 95 per cent of the material eroded in forming the gully is soft *rock*. Anyone describing the product of this erosion as 'soil' exaggerates the amount of soil erosion twentyfold; describing it as 'topsoil' commits a 67-fold exaggeration.

This point is illustrated by Metapontum, south Italy (p. 281): a large amount of alluvium has been produced by gully erosion, yet the land surface between the gullies still bears the field ditches of Ancient Greek times, proving that soil erosion has been slight.

DISTRIBUTION OF EROSION

After human activity, climate is next supposed to determine erosion. Maps of 'erosion risk' are published, on which Mediterranean lands, especially those with around 350 mm of annual rainfall, figure as high-risk areas.

Reality is much more complex. Some parts are much more obviously eroding than others, and there is no apparent relation (on a large scale) to climate, vegetation or human history. Erosion is generally obvious in Italy, where gullies and slumps are abundant almost throughout the peninsula. Italians have a rich vocabulary for different kinds of gully, and even a road-sign, FRANA, meaning Slump. Obvious erosion occurs in certain narrowly defined parts of Greece, and is widespread in SE Spain. But in most of Sardinia and Portugal, erosion has to be looked for carefully.

What do the erodible areas of Mediterranean Europe have in common? Not climate: gullying is prevalent in areas of low rainfall (SE Spain), moderate rainfall (north Peloponnese, Rhodes), and high rainfall (Síla Mountains, Italy) and Píndos Mountains, Greece). Nor lack of vegetation: the three most gullied parts of Greece – Píndos, the northern Peloponnese and the island of Rhodes – are among the most wooded parts of Greece. Nor a history of excessive land-use: there is no known difference between the land-use histories of Basilicata and Apulia in Italy, nor of the territories of Corinth and Athens in Greece, to account for one of each pair being more eroded than the other.

By conventional standards Crete should be extremely erodible. It is very mountainous, in places arid, until recently 'barren' with little woodland, and has a history of long civilization, dense population and intensive land-use. It appears in the deepest dye on maps of erosion sensitivity. In reality, modern Crete shows only minor, localized signs of active erosion. It is resistant to all agents of erosion except wind and bulldozer; unlike modern Rhodes, which is less steep, very wooded, has a much more relaxed history of land-use, and yet is very erodible.

Erosion, tectonics and geology

The ultimate driving force of all erosion (except wind erosion) is tectonic action, which lifts up mountains from which material can then be eroded. Without tectonic uplift there would, ultimately, be no erosion.

There is a broad correlation between eroding areas and tectonics in the Mediterranean. Most areas of active erosion are areas of continuing uplift in the Alpine mountain belt: for example, the gullies of the northern Peloponnese are associated with the Corinth fault-zone, notorious for earthquakes today. The Megarid, west of Athens, is said to have been eroded at an average rate of 0.27 mm per year over the last million years in response to the uplift of this fault-zone.[24]

To establish whether uplift is gaining over erosion is difficult. Both vary in space and time, and neither is adequately measured. Among actively eroding regions, long-term vertical movements of 1 to 2 mm per year are known from the northern Peloponnese and 0.25 mm per year from the eastern Pyrenees. In Calabria rates of uplift vary widely over a few kilometres. Archaeological data (over a few thousand years) and tide-gauge data (over a hundred years) tend to give more rapid rates of uplift or sinking than over longer periods; rates of 5 mm per year have been recorded in the Greek island-arc and around volcanoes. At Thermópylae, the site of the battle in 480 BC has since subsided by at least 5 mm per year.[25]

The character of the rock is very important. Well-fissured limestones tend to retain at least a thin cover of soil, because excess rainfall drains into the rock instead of running over the surface. On massive limestones and conglomerates, water is diverted over the surface and prevents soil from lodging, producing a desert (Chapter 17).

Much erosion takes place, not in the bedrock of the mountains, but in secondary deposits. As soon as the mountains began to form, in the Tertiary, they began to erode. Deposits produced by that erosion have been uplifted and tilted by continued mountain-building, and often now lie at angles at which they are not stable. (In the northern Peloponnese, loose marine sands and gravels now lie 1,200 m above sea-level.) If the gravels or screes are bound together by calcium carbonate etc., tilting may not greatly matter; but if not, they are foci for further erosion. *Flysch* is a term for clayey erosion products laid down in the deep sea adjacent to a rising mountain chain and brought above sea-level by continuing uplift. These too are often weakly consolidated and very liable to gullying.

In Basilicata, earlier erosion of the mountains formed deposits ranging in texture from cobble-beds to clay, all of them weakly consolidated; later erosion has attacked these to form spectacular earth-pillars, gullies and cliffs. Dates of marine terraces show that the land has been uplifted at a mean rate of 0.5 mm per year over the last half a million years. That may not seem excessive, but the uplift has not been continuous: phases of sudden movement have been separated by long quiescent periods, and sea-level has fluctuated through more than 100 m several times during the elevation.[26]

Much of the Peloponnese consists of hard limestone ridges overlooking karst basins in which great masses of red sands and gravels were deposited during the Neogene. Some of the basins have outlets down gorges, allowing rivers to carry away more or less of the basin contents and deposit them at the coast.

In Crete the violent tectonic history is counteracted by another peculiarity. Many of the rocks (especially hard limestones) are strong. There was vast erosion in the late Tertiary and Quaternary, much of the material being carried into the sea. If deposited on what is now land or shallow sea, the sediments tend to consolidate into breccias, conglomerates and sandstones. The reason for this may have to do with the climate history, or with the input of Saharan dust to act as a cement. Deposits like those which in Basilicata or Rhodes can be loosened with a finger are concrete-hard in Crete. Even soft materials tend to case-harden. The marls are relatively lacking in the soft clay that generates gullies. Crete, instead of gullying, forms tremendous cliffs and overhangs which require careful examination to tell whether they are of solid limestone, cemented scree, tufa-faced conglomerate (Fig. 3.8) or case-hardened marl. This is, as a colleague put it, 'an amazing tectonic landscape, in which erosion has not caught up with the uplift'.

EROSION AND DELUGES

Many findings point to the importance of deluges, defined roughly as falls of more than 100 mm of rain on one occasion. A deluge can be expected to produce more erosion than ten separate falls of 10 mm of rain. For example, on Greek rivers, when a minor deluge occurs – the sort that happens every five years or so – the river brings down at least twice as much sediment, sometimes twenty times as much, in the month of the deluge as in the rest of the year.[27]

A deluge overwhelms the capacity of the vegetation and soil to divert or absorb the water. Calculations of infiltration and interception are largely irrelevant; most of the water will run off, and nearly all will run off if the vegetation and soil have been soaked by previous rain. Deluges happening early in the season, with minimum plant cover, might be expected to result in gullying and sheet erosion. Late in the season, with the soil near saturation, they ought to promote slumping.

Measurements show, in general, that disproportionately much erosion happens during deluges; it is reasonable to infer that more still has happened during yet heavier falls when no scientist has been on hand to measure it.

What happens in a deluge – the Pakhyámmos event

Oliver Rackham and Jennifer Moody were privileged to be at Pakhyámmos, east Crete, on 23 September 1986. The rainy season opened with a fall of at least 300 mm – half a year's average rain – in 36 hours.[28] It rained a day and a half with the sort of rain that goes through an umbrella, or the canopy of a tree, as if it were not there (Fig. 14.10). The dusty village street looked like a river of blood. The dusty little river, down which no water had come for years, tore away the main road bridge, and great tamarisk trees sailed majestically down it

[24] R. E. L. Collier, M. R. Leeder, J. A. Jackson, 'The impact of Quaternary tectonic activity on river behaviour', Lewin & others 1995, pp. 31–44.
[25] M. Calvet and B. Lemartinel, 'L'expression de l'orogenèse: taux de surrection, mesure de

l'ablation et rythmes d'evolution', *Bulletin de l'Institut Géologique du Bassin d'Aquitaine* 53, 1993, pp. 27–34; N. C. Flemming, 'Predictions of relative coastal sea-level change in the Mediterranean based on archaeological, historical and

tide-gauge data', Jeftic & others 1992, pp. 247–81; Kraft & others 1987.
[26] Brückner 1980.
[27] Poulos & others 1996; cf Wainwright 1996.
[28] For the pressure conditions, see p. 35.

Fig. 14.10. The Pakhyámmos deluge. *24 September 1986*

Fig. 14.11. *Right* Effects of the Pakhyámmos deluge, September 1987.

(a) Sheet-erosion of newly ploughed olive-grove, Spinalónga.

(b) Scouring and deposition in Istron delta, into which millions of tons of water and debris descended from a height.

(c) Oleander in gorge, its bark scoured away by torrent.

(d) Side-cutting of river-bed, a big source of sediment.

(e) Massive sheet-erosion resulting from bulldozing followed by deluge.

(f) Mud-flows from bulldozed slope flowing over patch of phrygana beneath.

(g) Bulldozed false terraces dissolved by deluge.

(h) Mountain road destroyed by gullying.

and out to sea. Rifts in the cloud revealed curtains of water, 300 m high, pouring straight over the cliffs of the great fault-scarp that begins the Sitéia Mountains.

Visible effects were localized to an area 40 by 25 km. The Skinávria Mountains are a rugged little massif 700 m high – a tangle of ravines and small gorges, including hard limestones and conglomerates, hardened marls, river gravels and soft igneous rocks: generally well provided with soil, often terraced, and intermittently populated from Minoan times until the last hundred years. The present cultivation, except on the level delta of the Istron river, consists of olives on false terraces bulldozed into the softer marls. Pasturage at one time covered all the uncultivated land, but has now shrunk to certain limited, often heavily trampled and browsed, areas. Vegetation has increased in less browsed areas, with woods of *Quercus coccifera*, *Juniperus phœnicea* and *Pinus brutia*. Occasional big pollard oaks in jungle-like maquis point to former savanna.

Although rainfall was presumably more than 300 mm in the mountains, the greatest effects were at low altitudes. Juniper-woods, grown up in gorges since the previous deluge, were torn away. Well-rooted oleanders clung on, their bark and wood sandblasted away by sand- and rock-laden water tearing past (Fig. 14.11c). In cultivated basins, tree-trunks and debris were piled high against olive-trees. On hillsides, occasional small gullies (and rare slumps) appeared among phrygana on red limestone soils, marls, soft sandstone and rotten igneous rocks. This was unusual: most new gullying was associated with bulldozing. In the river delta, flooded a metre deep, silt and gravel were deposited in some places; elsewhere, previous sediments were removed (exposing, in one spot, the marks of the plough that had been at work just before the last deluge had covered them).

Most of the mayhem in the valleys was the recycling of deposits already in them. Rivers enlarged their beds sideways, causing slumps (hence the sailing tamarisk-trees; Fig. 14.11d). New cliffs up to 13 m high were created in weak sediments. There was not much input of new sediment, apart from pine-trees and the occasional vehicle. (New sediment came mainly from patches of greenstone conglomerate, one of the few deposits not to have cemented.) Gravel washed down the rivers formed small bars at their mouths, which lasted a few days until a storm at sea removed them.

At intermediate elevations, the roads were destroyed (Fig. 14.11h). Tractors were immobile, but farmers who had not thrown away their mules and donkeys went about their business almost as usual. Fields showed considerable sheet-erosion, but only where newly ploughed. Old-fashioned terraces with dry-stone walls were knocked about, but a winter's work would repair them. Even neglected terrace walls stood up surprisingly well.

On the mountains, the degree of damage lessened. Sheets of water poured over roughland without disturbing the soil. Much-trampled goat-paths were somewhat gullied; occasionally we noted sheet-erosion in sparse phrygana.

Effects of the deluge were not noticeably aggravated by any aspect of traditional Cretan land-use. Damage in the lowlands was an inevitable consequence of so much water pouring into them from a height (Fig. 14.11b). Only bare ground – newly ploughed or severely trampled – showed appreciable loss of soil (Fig. 14.11a), but not to an extent that would matter.

The kind of plant cover made little difference: well-grown pinewoods were no more or less protective than dead grasses on abandoned farmland. Significant protection came from plants so small that few erosion studies trouble to record them: the lichens, mosses, liverworts, *Selaginella* and blue-green algae that encrust otherwise bare ground (Fig. 14.12). These crusts were surprisingly strong, withstanding water-flows, and even trundling boulders, which broke down under-shrubs.

The most potent agent of erosion was the bulldozer. Bull-dozed (like ploughed) fields are left with no surface protection; bulldozing also creates unnaturally steep slopes for water to attack. The 'rivers of blood' came from freshly bull-dozed red soil (Fig. 14.11e). Whole flights of new false terraces, dug out in the latest fashion without retaining walls, melted away like glaciers in Hell (Fig. 14.11g). Roads turned into gullies and diverted gullies into new places. Paved roads were eaten away under the tarmac or burst up from underneath. Occasionally a road-cut slumped. This can happen, even with ordinary heavy rains, all over Crete.

How does the Pakhyámmos deluge stand in history? Old inhabitants told us of a comparable one in 1912. (A bridge was destroyed in 1986 in the Istron delta which had partly collapsed in the 1912 deluge and showed traces of a timber rebuilding, recorded on the British Military Map of c. 1939.) Eighteen months after the 1986 deluge, a lesser, late-season deluge occurred, overlapping part of the same area. The

Fig. 14.12. Ground cover of the clubmoss *Selaginella denticulata* and the lichen *Cladonia foliacea;* both are less than 1 cm high. The projecting teeth of limestone are patinated with minute lichens, which verify that the soil level is stable. *Pakhyámmos, September 1986*

Fig. 14.13. Mass movement in the Tech valley downstream of Prat de Mollo and west of Pas du Loup. It shed large volumes of sediment directly into the river in October 1940, and remains active. *June 1998*

tremendous effects of a deluge funnelled into a great gorge are described on p. 25.

Deluges are normal, rare and irregularly distributed. Trees and bushes in Cretan gorges show sandblasting scars; from the amount of regrowth, the date of the last deluge can be worked out. The Istron delta is filled with deposits from successive deluges at least as far back as the Minoan period. Sedimentation sequences in plains and deltas suggest that deluges have shifted more soil and bedrock than any other agency. Ours, though the biggest in that spot for at least 75 years, was not abnormal. It shifted table- or occasionally car-sized boulders; previous ones have demolished buildings and have shifted boulders as big as houses.

Other reports of deluges

Such were the effects of a deluge on an erosion-resistant landscape. Other accounts, mainly from the south of France, vary with the terrain. When 350 mm of rain fell in a few hours at Montesquieu in October 1986, streambeds were deeply scoured, but even slopes recently burned showed no ravine development – earlier rain had allowed grass to grow since the summer fires. Much the same was true of a storm of 476 mm in 67 hours in autumn 1989; some streams produced sediment from side-cutting their banks.

It was estimated by M. Pardé, a river hydrologist of vast experience, that the '200-year' storm in the basin of the Tech on 17 October 1940 (pp. 32–3, Fig. 14.14) removed the equivalent of 7.65 mm of sediment from the whole catchment, as much as might be expected over a few hundred years without such a deluge. This should not hastily be interpreted as soil erosion: as M. Calvet has pointed out, most of the sediment was derived, not from the slopes of the whole catchment, but from badland erosion, sidecutting, landslides, and gullying in Pliocene and Pleistocene valley fills (Fig. 14.13). Pardé had measurements of sediment in suspension in the rivers: in the

Têt at Prades, for instance, they averaged 41.5 kg per cubic metre over 17–20 October. He calculated, from the thickness of the deposits left, that the Tech had spread between 7½ and 13 million cubic metres of sandy material over the piedmont plain. A somewhat greater quantity, he reckoned, was carried out to sea. A considerably greater volume was dumped on valley floors upstream of Céret, often 5 m, exceptionally 20 m, thick. Over the following winter the rivers cut new channels several metres deep into these fills.[29]

More details have come from a study of applications for compensation by the people affected. Floodwaters, spilling along former distributaries in the river delta and diverted by a railway embankment, scooped out hollows in some places and deposited sediment in others. About 1.3 million cubic metres were brought into the township of Elne and spread over its 700 ha, and 290,000 cubic metres were carried off, giving an average accumulation of 14 cm. The losses included valuable soil; the gains consisted of pale sandy silt in some places and darker, more organic alluvium in others, with coarse sands and gravels near channels and depressions where the floodwater moved fastest.[30] At Bompas, 20 km to the north, about 2 m of sediment like that at Elne covers a Roman road and the remains of medieval buildings; there has therefore either been a deluge much bigger even than that of 1940 or several similar ones.

A deluge on 22 September 1992 drowned 34 people at Vaison-la-Romaine, NE of Avignon. Rainfall totals of between 194 and 448 mm were recorded, mostly in a few hours; in places 30 mm fell in six minutes. The effects were studied by John Wainwright in the following April in an area of 25 by 17 km, north of Mont Ventoux (1,912 m), much of it consisting of marly, well-vegetated badlands. He found 39 landslips, mostly smaller than 100 cubic metres, not counting those in road-cuts. These were often on grassy slopes downslope of treed areas or alongside steep streams. On cultivated areas rills were formed, but most of them had been

---

[29] Pardé 1941; M. Calvet, 'Les inondations d'octobre 1940 en Catalogne: l'aiguat de 40', *Servei Geologie de Catalunya*, 1993; M. Calvet, *Morphogenèse d'une montagne méditerranéenne; les*

Pyrénées orientales (doctoral thesis, Université de Paris I), Presse universitaire de Perpignan, 1994.
[30] Jacob 1997.

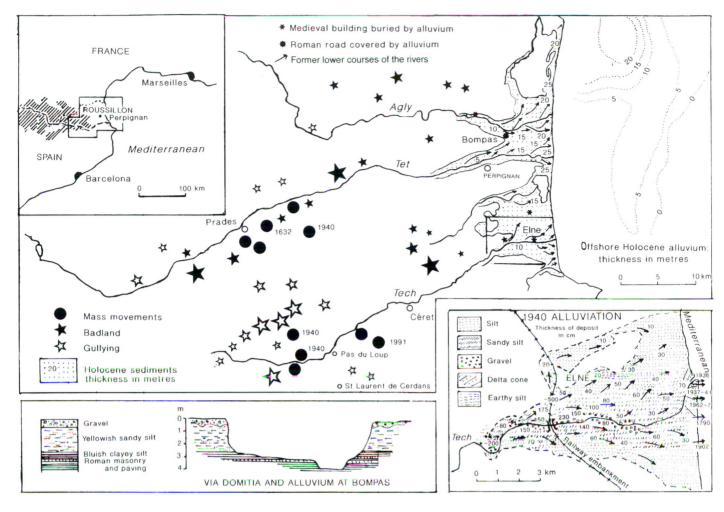

Fig. 14.14. Events in the 1940 deluge in the Roussillon deltas. After Calvet 1993 and Jacob 1997.

ploughed out during the winter. Periglacial deposits at high altitudes were disturbed by small debris flows originating in slumps. The badlands themselves were not much activated (p. 277). In Wainwright's reconstruction most of the flood-water and sediment came from vineyards, orchards and other agricultural land, rather than from badlands. He noted that 'there is stratigraphic evidence along the river Augue Marce for large pulses of sediment from the badlands in the past . . . none were apparent from the current event, which is probably a result of the relatively equilibrium conditions prevailing in these particular badlands'.[31]

Erosion and rainfall intensity

In the Pixinamanna catchment in south Sardinia, O. Seuffert and colleagues set up erosion plots (described below, p. 258) in moderately dense maquis and *Cistus* vegetation. They found that falls of more than 40 mm accounted for up to two-thirds of the erosion, even though they comprised less than 5 per cent of rainfall events and provided little more than 20 per cent of the total rain. Falls of less than 10 mm had little effect.[32]

The erosional effect, they found, is determined not so much by the quantity or intensity of the rain, but by pulses of high intensity (> 0.2 mm per minute, equivalent to 12 mm per hour) within storms. Intense blasts of rain initiate the highly turbulent surface runoff which is mainly responsible for erosion. Topography and wind gustiness influence this short-term rainfall intensity and may be significant in susceptibility to erosion.

In Spain, measurements by Romero Diaz and others indicate that storms of more than 30 mm in an hour are very erosive. They argue that such events, nevertheless, are so rare in their study area that more erosion may be caused by commoner storms of 5–10 mm per hour. However, deluges of more than 200 mm in 24 hours can be expected every few years in most highlands within a hundred km of the Mediterranean (Chapter 2). The importance of bursts of extreme rainfall is confirmed by plot studies in NW Italy: 'practically all soil loss is due to such events [> 0.18 mm per minute] of short duration'.[33]

[31] Wainwright 1996.
[32] O. Seuffert and 5 others, 'Rainfall-runoff and rainfall-erosion-relations on hillslopes. New equations and their experimental background', *Geoökoplus* 9, 1988, pp. 17–40; Seuffert 1992; O. Seuffert, H. Motzer, H. Dieckmann, H. Harres, 'Rainfall and erosion', *Geoökoplus* 3, 1992, pp. 129–37.
[33] Romero Diaz & others 1992; Tropeano 1984.

Adjustment to deluges

Wainwright introduces the idea that landscapes, cultural or natural, are 'adjusted' to rainfall events up to a certain magnitude. Runoff at Vaison-la-Romaine did not rise rapidly until after the first 80–100 mm of rainfall; he inferred that 'the landscape at present is adjusted to the relatively common rainfall events of 100 mm, and that catastrophic events such as that of 22 September 1992 at a frequency of two or three events per century are responsible for more significant changes'. He suggests that landscapes are adjusted to single falls of rain up to about 15 per cent of the annual total.

The Pakhyámmos event produced only a minor catastrophe, which suggests that east Crete is adjusted to deluges of about 200 mm (35 per cent of the annual total), such as occur about twice a century. These storms, however, came early in the season, when the ground was dry. There would probably be a lower threshold for late-season deluges, commoner in Crete than in the south of France (p. 35).

Bulldozing evidently lowers the threshold of adjustment, so that smaller deluges are needed to produce catastrophe.

MEASURING AND ESTIMATING CURRENT EROSION

School textbooks present the total weight of 'topsoil' supposed to be eroded annually over the world's land surface; Heaven only knows how the figure was arrived at, or what the truth is. There is a huge sampling problem. Erosion varies vastly, and a large proportion of the total is concentrated into small areas and short periods. Rates in places 100 m apart can easily differ by a factor of 1,000. (Erosion has been most actively measured in parts of the United States where gentle gradients and uniform soils make sampling easier.[34])

Why measure the current rate of erosion? It is needed to identify areas that are being significantly damaged; to compare different ways of using land, and to determine the effectiveness of conservation measures; and to predict how long a reservoir will last before it fills with sediment.

Erosion specialists usually talk in terms of tons of 'soil' per hectare per annum, as if the weight of soil mattered to anyone. They often fail to make their results meaningful by converting this to the thickness of material lost. We have expressed all the results in terms of mm of soil per year.[35] Moreover, it is often not made clear how far the eroded material is moving: a ton of soil slipping from one terrace to the terrace below is not distinguished from a ton of soil washed into the sea. Some authors, infuriatingly, start the year in January instead of September, which makes it impossible in a Mediterranean climate to draw meaningful comparisons between years.

Empedestalment and patination

An ideal instrument for measuring erosion would be one cheap enough to replicate so that an adequate sample of the local variation could be taken: let us say 100 instruments per sq. km, more in mountainous terrain. It would remain in position for many years without interfering with the normal operation of the landscape, in order to get an adequate sample of variation over time, including exceptional falls of rain. An object approximating to this specification is a tree. The degree to which a tree's roots have been exposed (*empedestalment*), or its root-buttresses buried, is a measure of how much erosion or accumulation has occurred over its lifetime (Figs. 14.15, 14.16).[36] The age of the tree can be estimated from its annual rings, or by comparison with stumps of similar size whose rings have been counted. Pedestals of the dwarf palm *Chamaerops humilis*, which is said to live four hundred years, have been used to estimate erosion rates in Spain.[37] Sheet and rill erosion in natural vegetation can be detected from the empedestalment of undershrubs, left behind as soil is washed away from between them. Most undershrubs, however, live only about twenty years.

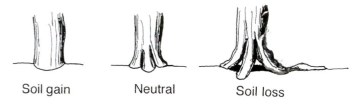

Soil gain   Neutral   Soil loss

Fig. 14.15. Root-buttresses of a tree such as an olive, showing effects of erosion.

High rates of erosion can be estimated by driving pegs into the ground and observing how quickly the sediment recedes around them. Rates of 1.5 to 3.8 mm per year have been measured in Spanish badlands. A variant of this method is to take periodic measurements of the profile of a gully, using as reference points two pegs permanently fixed on either side. This is the method of Rainer Lehmann on Naxos (pp. 264–5); specimen measurements over two years show a mean rate of 1.0 mm per year on ex-farmland on conglomerate, 8.2 mm per year on grano-diorite.[38]

Another observation is on the *patination* of teeth of rock, especially limestone, which project through a thin soil cover. Exposed rock surfaces have a lichen cover which takes many years to form. Where the soil cover is receding, each tooth of rock has an unpatinated zone at its base.[39]

[34] For a list of such studies, see D. J. Mitchell, 'The use of vegetation and land use parameters in modelling catchment sediment yields', Thornes 1990, pp. 289–316.

[35] We use a soil density of 2.5 g per cu. cm, by which 2,500 g per sq. m = 25 t per ha = 22,284 lb per acre = 1 mm thickness. For consistency, we have adjusted the findings of other authors who use different densities, for example 13 t per ha = 1 mm thickness (Poesen & Hooke 1997). Real soils vary in density (a factor commonly left out of

reports of these studies), but there are so many other uncertainties that accuracy on this point is hardly necessary.

[36] In tropical Africa, mounds can build up around tree-bases because termites deposit sediment and loosen soil. We have not observed this in the Mediterranean, but the possibility needs to be watched; Y. Biot, 'The use of tree mounds as benchmarks of previous land surfaces in a semi-arid tree savanna, Botswana', Thornes 1990, pp. 437–50.

[37] A. G. Brown, 'Soil erosion and fire in areas of Mediterranean type vegetation: results from chaparral in southern California, USA and matorral in Andalucia, southern Spain', Thornes 1990, pp. 270–87.

[38] H. M. Scoging, 'Spatial variations in infiltration, runoff and erosion in semi-arid Spain', Bryan & Yair 1982, pp. 89–112; Lehmann 1994.

[39] For weathering patterns on the rock itself, see Fig. 14.7.

Fig. 14.16a. *Left* Old olive-tree empedestalled by the loss of about 1 metre of soil. *Akrotíri Peninsula, west Crete, July 1981*

Fig. 14.16b. *Below* Youngish olive-grove, rapidly empedestalled on loamy clays downslope of an erosion experiment (Table 14.ii). *Guardia Peticara, Basilicata, March 1993*

Fig. 14.16c. *Bottom* Undershrubs empedestalled by rapid erosion from sea spray. *Akrotíri Peninsula, west Crete, July 1981*

Measurement of deposition

Erosion over entire catchments may be estimated from the transport of sediment by rivers or its deposition in reservoirs. This involves measuring sediment concentrations and stream discharge, or the sediment accumulated in a reservoir over a known time. Sediment yield may not be an accurate measure of slope erosion, because sediment travels intermittently; also some material departs in solution or as stones rolled along the river-bed, which is difficult to measure. Some sediment is trapped within the catchment, accumulating in stream channels and on flood-plains, until, decades or centuries later, floods release it by sidecutting river banks. In the Segura catchment in Spain, less than half the material now being eroded from valley slopes gets as far as one of the modern reservoirs; between 54 per cent and 93 per cent is held back at the foot of slopes, in alluvial cones, in valley bottoms and on floodplains.[40]

The amounts of sediment brought to the sea by rivers, divided by the sizes of their basins, shows that the highest erosion rates are in small, mountainous, tectonically active catchments. The highest of all are on volcanic islands such as Taiwan, one of whose catchments is being eroded at 14 mm per year. The highest rates measured in Europe are in Albania: the catchments of the rivers Semani and Shkumbini (a continuation of the badlands of the Píndos Mountains) are being eroded, on average, at 1.7 and 1.4 mm per year. Several Greek and small Apennine rivers have rates equivalent to erosion of more than 0.5 mm per year. Among bigger rivers, the Tiber erodes at only 0.14 mm per year and the Po at 0.11 mm per year. In Spain, the Segura basin, with many badlands, produces an equivalent of 0.08–0.6 mm per year, not counting sediment that fails to reach the reservoirs. A nearby badland river basin, the Guadiana Menor in Granada, was found from a limited number of measurements to be losing only 0.006–0.016 mm per year as suspended sediment, much less than the amount carried away in solution. Among rivers coming out of 'ordinary', not obviously much eroding,

40 Romero Diaz & others 1992.

Fig. 14.17. Flumendosa dam, SE Sardinia. In this catchment, erosion rates are low on account of resistant rocks and soils, quiet tectonics and good vegetation cover. *April 1992*

landscapes, the Uvini, Brimini and Mulargia rivers in Sardinia in 1992–3 gave sediment yields equivalent to 0.0044–0.018 mm per year (cf Fig. 14.17).[41]

The volume of a deposit, of course, *represents total erosion, not soil erosion*. To estimate soil erosion requires knowing what erosion processes were involved, where the sediment came from, and how much of it originated as soil. In deltas, especially, it is dangerous to take the rate of sedimentation as a guide to soil erosion in the whole catchment (Chapter 18).

Instrumental measurements

The above methods are at best semi-quantitative. *Instrumental records* achieve greater precision, but lose representativeness. The instruments are expensive, so cannot be replicated much. They need constant attention and therefore depend on research grants, so are rarely continued long enough to get an adequate sample of variation over time. It is uncertain how far quantitative measurements are representative of an area as a whole. They may overestimate erosion, because scientists are unlikely to 'waste' resources by setting up instruments in sites where erosion is not already thought to be significant. The value of such instruments is in relating the day-to-day march of erosion to weather under different types of vegetation and land-use.

The classic Bates–Henry study of the effect of 'deforestation' on runoff and erosion was done from 1910 to 1926 in the Rocky Mountains.[42] Two adjacent catchments, each of about 80 ha, were carefully selected to be as nearly identical as possible in environment and vegetation, and to be hydrologically watertight. Each was instrumented to record weather, stream-flow and export of sediment in the streams. For eight years, the catchments were treated alike in order to record unavoidable differences between them. Then all the trees on one catchment were felled and removed. Measurements were continued to record differences resulting from the felling and subsequent regrowth of the vegetation.

Felling the trees increased stream-flow by about 15 per cent and loss of sediment from the catchment by about eight-fold. The increase in stream-flow, and most of the increased sediment, were attributable not to any effect on rainwater but to the effect of the trees on melting snow: snow lay until May and was then released more suddenly from the felled catchment. The heaviest single rainfall, of 38 mm (modest by Mediterranean standards), caused the loss of eighteen times as much sediment from the felled catchment as from the intact one: even seven years' regrowth were not sufficient to prevent erosion. However, erosion rates were minute: even the felled catchment lost only the equivalent (on average) of 0.00075 mm per year.

This was an unhurried and exceptionally thorough study. The investigators spent eight years verifying that the sites behaved similarly before beginning the actual experiment. They published the results in great detail, with full accounts of vegetation and methods. The results can be accepted with confidence,[43] but are peculiar (even within the Rocky Mountains) to this combination of geology, climate, and vegetation. The snow-melt factor alone rules out most comparisons with the Mediterranean. They do not reveal how erosion would have been affected by a deluge – even seventeen years were not enough to ensure that one occurred – or by animal browsing after felling.

We know of no such thorough experiment in the Mediterranean basin. This is mainly for anthropological reasons. The equipment needs finance and personnel, which usually limits its life-span to that of a research grant. The work is mostly done on small plots rather than whole catchments; in limestone areas, where catchments are not watertight, it has to be so; this greatly increases the sampling problem. Where sites with different vegetation are compared, the vegetation is usually in place at the start of the observation, which creates doubt as to what exactly is being compared. (If cultivated land has a higher erosion rate than forest, is this an effect of cultivation, or is it because the more erodible soils are those that get cultivated?) Vegetation is often very sensitive to differences of slope and aspect; different erosion rates may then either be directly due to differences of microclimate, or to microclimatic differences acting via the vegetation.[44]

[41] Milliman & Syvitski 1992; S. E. Poulos and others, 'Water-sediment fluxes of Greek rivers . . . annual yields, seasonal variability, delta formation and human impact', *ZG* NF 40, 1996, pp. 243–61; M. A. Romero Diaz and 4 others, 'Variability of overland flow erosion rates in a semi-arid Mediterranean environment under matorral cover, Murcia, Spain', *Catena Supplement* 13, 1988, pp. 1–11; Wise & others 1982; P. Botti, S. Vacca, A. Aru, 'Mass balance of sediment transport and its significance in land evalu-

ation: the example of the tributary basin of the Mulargia reservoir (central-southern Sardinia)', *Land Use and Soil Degradation: MEDALUS in Sardinia: proceedings of the conference held in Sassari, Italy, 27 May* 1994, ed. G. Enne, A. Aru and G. Pulina, *MEDALUS*, 1995, pp. 157–68. [This type of measurement is only approximate, because erosion is increased by deluges and earthquakes, which may not occur within the period of measurement.]

[42] C. G. Bates and A. J. Henry, 'Forest and

stream-flow experiment at Wagon Wheel Gap, Colorado. Final report . . .', *USDA Monthly Weather Review* suppl. 30, 1928.

[43] For reactions to them, see Saberwal 1998.

[44] On the marl of south Crete, north-facing slopes are wooded with pine (*Pinus brutia*), other aspects being phrygana (Rackham 1972). For the inland Ebro valley, see H. C. G Hidalgo and F. Pellicer, 'Spatial distribution patterns of morphogenetic processes in a semi-arid region', Thornes 1990, pp. 399–417.

Publication is limited to the meagre pages of a research report or specialist journal, and it is rare to find a full account of everything that was done.

Critical studies of erosion are seldom done because the administrative organization does not allow it. There is an unsaid assumption that erosion is gradual, the result of ordinary rains rather then deluges. This may not be true, but on a catastrophist theory the observations would not be worth doing, since a deluge, by definition, is unlikely to occur within the duration of a research grant. (If it does occur it may overrun the capacity of the measuring equipment.)

People in a hurry simulate deluges by spraying the soil surface with a known amount of water per hour and measuring runoff and erosion. Simulated rainfall has been found to produce more runoff but less erosion than real rainfall:[45] it does not usually simulate the pulses, gusts and heavy impact of wind-driven raindrops in natural storms.

Modelling and the 'Universal Soil Loss Equation'

Members of MEDALUS have devoted much energy to predicting erosion rates.[46] There already exists a North American model by W. H. Wischmeier and his colleagues. It is not really an equation, but an empirical relationship between observed rates of erosion and various parameters of environment and land use. It begins from a Rainfall Erosion Index, the rate at which soil may be expected to be removed under varying intensities of rainfall and snow-melt, other conditions being standardized. This index depends especially on the maximum rate of rainfall experienced during storms. It is then modified by five coefficients which express the departure from standard conditions of erodibility of a particular soil, length of slope, steepness of slope, vegetation cover and management, and 'support-practice' (a term used to express the kind of erosion-control practices that are usual in the United States). Interactions between these factors appear not to be allowed for.[47]

Settlers in the United States found sheet and rill erosion to be an unexpectedly severe constraint on European-style agriculture. The reason seems to lie in heavy rainfall at the wrong time of year combined with very silty soils, rendering even gentle slopes liable to erosion. Collection of data began in the 1920s and led to a number of regional soil loss 'equations', which were combined in the Universal Soil Loss Equation. The data that generated that Equation were gathered east of the Rocky Mountains. By 1978 an attempt had been made to extend the prediction to the western States, but apparently without additional primary data.[48]

Wischmeier and Smith explain that the term 'Universal' means *universal within the United States,* in contrast to the regional equations. They also declare that their model is

> designed to compute longtime average soil losses from sheet and rill erosion under specified conditions ... it does not predict deposition and does not compute sediment yields from gully, streambank, and streambed erosion.

Students of erosion, disregarding its originators' reservations, have used the model in other parts of the world. Others have constructed their own models for predicting erosion rates; some of these are specific to the Mediterranean, and try to take gullying into account.[49] The primary data are often difficult to collect, especially those relating to rare deluges, which may explain why the Wischmeier–Smith model remains so widely used.

Empirical models will work only within the range of conditions over which the original data were collected. The eastern and middle States, although a large region, are not very diverse: for example, the erodibility factors found ranged only from 0.85 to 1.20. The Mediterranean lies outside the range of climates, soils, crops and cropping practices for which the Universal Soil Loss Equation was constructed. North American soils were generally not cultivated until the nineteenth century, whereas Mediterranean soils have been cultivated and eroded for centuries; many soils now under cultivation are derived from the more resistant illuvial horizons of the original soils.

The Universal Soil Loss Equation fails to predict rates of sheet and rill erosion in the Mediterranean. It is even worse at predicting *total* erosion, which it was never meant to do. For example, the Israel Soil Conservation Service, and Lehmann on Naxos, both found no relation between observed erosion and the prediction.[50] The Equation usually overestimates rates of erosion in the Mediterranean – by a factor of at least 25, according to long-term observations in Portugal (see below). This is responsible for much of the belief in Mediterranean desertification.[51]

Erosion plots

Erosion rates are usually measured on plots of 10–100 square metres in area, in conjunction with measurements of weather, especially rainfall intensity. The results can be regarded as representative of areas with a similar climate, slope, soil and vegetation cover. To take account of all these variables would require measurements on a very large number of plots. It may be possible to reduce the amount of information needed by interpolating the results with statistical models. Much of the modelling activity in MEDALUS is directed to these problems.

[45] Seuffert 1992.

[46] R. P. C. Morgan, 'The art and science of erosion modelling', Rubio & Calvo 1996, pp. 235–49.

[47] W. H. Wischmeier, 'A rainfall erosion index for a Universal Soil Loss Equation', *Soil Science Society of America Proceedings* 23, 1959, pp. 246–49; W. H. Wischmeier and D. D. Smith, *Predicting rainfall erosion losses from cropland east of the Rocky Mountains*, Agricultural Handbook 282, USDA, Washington, 1965; W. H. Wisch-

meier and D. D. Smith, *Predicting rainfall erosion losses. A guide to conservation planning*, Agricultural Handbook 537, USDA, Washington, 1978.

[48] C. H. Lloyd and G. W. Eley, 'Graphical solution of probable soil loss formula for Northeastern Region', *Journal of Soil & Water Conservation* 7, 1952, pp. 189–91; W. H. Wischmeier and D. D. Smith 1978 (see note 47).

[49] See, for example, Gilman & Thornes 1985, chapter 4.

[50] M. Inbar, 'Rates of fluvial erosion in basins

with a Mediterranean type climate', *Catena* 19, 1992, pp. 393–409; Lehmann 1994, p. 148. Lehmann points out the uncertainties involved in estimating the maximum rate of rainfall, and the vegetation and management factors, in Mediterranean conditions.

[51] The Rainfall Erosion Index in mediterraneoid California, as predicted by the model, is less than in any other non-arid part of the States: only one-eighth of that in the very erosion-prone Florida and Georgia.

length of the plot. Plots therefore give lower values for erosion and runoff than expected under natural conditions. They are also too small to give an adequate sample of the patchiness of the landscape: of the rocks, trees, tussocks etc. which divert the flow of water and sediment down a natural slope. It is difficult to scale up plot observations to estimate, even approximately, runoff and erosion from whole catchments. Their usefulness is in comparing the effects of different treatments. These erosion plots form a disorganized series from which it is difficult to draw general conclusions; often the soils are not consistent even within the same group of plots. Few have been set up long enough to experience really heavy rainfall.

Fig. 14.18a. *Left* Erosion plot of standard MEDALUS type, set up in eucalyptus plantation. *Capoterra, Sardinia, May 1993*

Fig. 14.18b. *Below* Larger plot operating since 1961; note the silt-laden runoff water from bare soil. *Vale Formoso, Portugal, April 1993*

An erosion plot is a rectangle laid out up and down slope, contained within sheet-metal walls; runoff from the plot is collected and measured at the bottom (Fig. 14.18). Those used by MEDALUS (Table 14.i) are normally 10 m long and 1 m wide. The plots receive no runoff water or eroded soil from further upslope; *the slope length* (one of the important factors determining the amount of erosion) *is limited to the*

A variant of erosion plots are 'field-boxes' as used by Lehmann on Naxos. A field-box is a container set in the ground so as to trap runoff and sediment coming down a rill or small gully. The catchment, of a few square metres, is estimated from the topography. The field-box, being simpler than the plot, can be better replicated and involves less disturbance.[52]

Table 14.i. Summary of erosion plots. * Set up under the MEDALUS programme; where a date is given, MEDALUS took over existing plots.
[C. Kosmas and 22 others, 'The effect of land use on runoff and soil erosion rates under Mediterranean conditions', *Catena* 29, 1997, pp. 45–59.]

Site	Location	Altitude, m	Annual rainfall, mm	Slope %	Geology	Treatments
Almocreva*	Portugal (Alentejo)	200	560	9	schist + raña	wheat
Vale Formoso (1961–)*	Portugal (Alentejo)	200	583	7–20	schist	wheat, fallow, *Cistus*
El Ardal (1989–)*	Spain (Murcia)	550	276	7–22	limestone + marl	cereals, fallow, undershrubs
Rambla Honda*	Spain (Almería)	690	280	15–42	slate + phyllite	tussock-grass, undershrubs, retama
Canterrane + Réart*	France (Roussillon)	190	582	5–10	Neogene (tends to badland)	vines
Central Piedmont	sub-Mediterranean Italy	100–300	750	31–49	marls, silts, fine sands	vines
Is Olias	SW Sardinia	160	540	12–31	Palaeozoic + colluvium	undershrubs, eucalyptus
Pixinamanna	SW Sardinia	50–700	500	c. 5–20	rotten granite	maquis, bare ground
Petralona*	Greece (NE Macedonia)	220	464	11–20	limstone + marl	maquis, wheat
Spata*	Greece (Attica)	140	496	7–23	Neogene	vines, olives

52   Lehmann 1994, pp. 126–34.

EROSION AND VEGETATION

Europeans like to believe that forests attract water and promote the flow of streams and springs. This is hardly to be expected, since trees lose more water by transpiration than most alternative vegetation. Most observations do not encourage the theory. In the United States, when forests were felled, stream-flow increased by between 34 and 450 mm a year, and gradually diminished as the forest grew again. The results, however, were highly unpredictable.[53] The effect ought to be less in Mediterranean forests, which have no rain for half the year and lack the thick leaf-litter of many American forests.[54]

In the damper Mediterranean climates, up to 600 mm of rainfall are evaporated or transpired per year; the rest goes to runoff and recharging groundwater. Several studies have shown that reduced vegetation leads to less transpiration and more runoff. Under an annual rainfall of 550 to 800 mm, the typical water runoff from grass or bare surfaces may be 120 mm greater than from forest. In one study in Israel, in a year with a rainfall of about 700 mm, maquis contributed 230 mm to ground-water, pine forest 347 mm, and pasture 420 mm.[55]

S. Rambal studied a 35-sq-km karst catchment north of Montpellier with a mean annual rainfall of 1,200 mm. Between 1946 and 1979 forest increased at the expense of 'garrigue'; the leaf area index (ratio of leaf area to ground area) increased from 1.7 to 2.25. The mean annual stream flow decreased from 700 mm to 623 mm. Had rainfall been less than 560 mm there should, in theory, have been no runoff at all except in deluges.[56]

The belief that trees prevent erosion is not quite universal. Mud-flows are an important ecological factor in rain-forests on steep slopes.[57] Archaeological conservationists in England used to recommend the removal of trees from ancient earthworks on the grounds that trees promote erosion. They pointed to disturbance by the growth of roots, the extra eroding action from big drops that accumulate on leaves and hit the soil hard when they fall, and the gross disturbance when a big tree is uprooted in a storm. This last is probably rare in Mediterranean countries.

Trees are supposed to prevent erosion mainly by reducing or delaying runoff. A tree, of course, prevents the whole of a light shower from reaching the ground, and intercepts part of a moderate fall of rain. For a few minutes a tree may retain a few millimetres of rainfall as a film of water on its leaves and twigs. It retains a few millimetres in leaf-mould underneath.[58] Trees may reduce erosion in other ways, for example (in natural or untidy woodland) by logs and debris forming dams in streams which temporarily intercept sediment.[59] There are accounts of vegetation creating changes in soil structure and composition which affect erosion.[60] Much of this work is not directly relevant to the Mediterranean, but a study in the mountains of Catalonia reveals a difference between north- and south-facing slopes; the former have soils with more organic matter and more roots, able to store more water, which may give greater protection from erosion.[61]

Such mechanisms, however, apply to normal rainfall; the more rain falls at any one time, the less effective they are. Most of them have a maximum capacity which is overloaded in times of deluge. If a tree were to retain more than 10 mm or so of precipitation it would collapse under the load.[62] When the leaf-mould is saturated or the log-dams are full, the excess runs off. In a single fall of 100 mm, the tree can have little effect; if the deluge comes when the ground is already soaked by previous rains the tree might as well not be there. As Raphael Zon, dissenting against the American forestry propaganda of the 1920s, put it:

> Soils covered with forests can store up a quantity of water corresponding to a precipitation of . . . in very favorable conditions 0.24 inch [6 mm] at most. A cover of moss can absorb water amounting to from 200 to 900 times their weight; dead leaves . . . 150 to 200 times their weight, and pine needles from 120 to 135 times. These amounts are insignificant when compared to the enormous quantities of precipitation that cause excessive floods . . . the forest floor becomes so saturated with water that it allows any further rain to pass off, just as it would from open ground. Normal rainfall is relevant to such matters as the recharge of aquifers and the supply of ground-water; most erosion, however, is the result of abnormal rainfall.[63]

The only terrain that can swallow an entire deluge is karst with big, gaping dolines and fissures.

Trees thus reduce sheet and rill erosion if the rain is not too heavy. Much of the intercepted rain is evaporated from the tree or infiltrated into the ground.[64] The effect is shared to

53 A. R. Hibbert , 'Forest treatment effects on water yield', Sopper & Lull 1967, pp. 527–43.

54 In tropical and mediterraneoid Australia, felling forests causes the water-table to rise, and is blamed for salinization; D. Sun and G. R. Dickinson, 'Survival and growth responses of a number of Australian tree species planted on a saline site in tropical north Australia', *JAE* 32, 1995, pp. 817–26.

55 A. Shachori and A. Michaeli, 'Water yields of forest, maquis and grass covers in semi-arid regions: a literature review', UNESCO Arid Zone Programme symposium on Methodology of Plant Eco-physiology, Montpellier, 1962; Shachori & others 1967.

56 Rambal 1987; S. Rambal, 'Fire and water yield: a survey and prediction for global change', Moreno & Oechel, 1994, pp. 96–116.

57 J. W. Dalling and E. V. J. Tanner, 'An experimental study of regeneration on landslides in montane rain forest in Jamaica', *JE* 83, 1995, pp. 55–64.

58 'More nearly average figures of 5.5 metric tons [of litter and leaf-mould] per acre dry weight and 200 per cent retention give a depth of 0.1 in. [2.5 mm] of water', Kittredge 1948, p. 195. This is for North American forests, which tend to have much more leaf-mould than those of the Mediterranean.

59 H. A. Viles, '"The agency of organic beings": a selective review of recent work in biogeomorphology', Thornes 1990, pp. 5–24; G. F. Peterken, *Natural Woodland*, Cambridge University Press, 1996.

60 S. W. Trimble, 'Geomorphic effects of vegetation cover and management: some time and space considerations in prediction of erosion and sediment yield', Thornes 1990, pp. 56–65.

61 J. Sevink, 'Soil organic profiles and their importance for hillslope runoff' *Geomorphology* in environments with strong seasonal contrasts, ed. A. C. Imeson and M. Sala, *Catena* suppl. 12, 1988, pp. 31–44.

62 This can happen. On 28–30 December 1995, in the leafless forests of the Ligurian Apennines, heavy rain fell when the air was below freezing. Every twig and bough turned into a cylinder of ice. Whole hillsides of chestnuts and beeches collapsed or fell off the mountains. Such an ice-storm (*galaverna*) is a rare but repeatable phenomenon; surviving trees bear the scars of an event in 1969. (We are indebted to Dr Diego Moreno and villagers of Valletti for demonstrating it.)

63 R. Zon, 'Do forests prevent floods?', *American Forests and Forest Life* 33, 1927, p. 388 [quoted by Schiff 1962].

64 For measurements on maquis in Israel, see Shachori & others 1967.

varying degrees by other vegetation. Maquis, tall dense *Cistus ladanifera* thickets, or a thick mat of grass would be equally effective. The effect is greatest on light rains with little erosive power. *Trees have least effect on the very heavy rains that do most of the erosion.*

Erosion is affected by vegetation less than 5 mm high: crusts of *Selaginella*, mosses, lichens and blue-green algae. These miniature, slow-growing ecosystems indicate that the soil surface is stable. Writers who notice these plants point out that they tend to decrease infiltration and increase runoff. Blue-green algae (and the lichen *Collema* which contains them) secrete mucilage which retains water, and fix nitrogen which may be used by other plants.[65] In a deluge, when most of the rain runs off regardless, crusts hold the surface together and protect the loose soil underneath. They are fragile when dry and easily damaged by trampling; they avoid shade.

Very slight vegetation significantly reduces sheet and rill erosion. Even in the Mediterranean climate, which discourages accumulation of organic matter, topsoil may contain sufficient humus to increase its tenacity. Even straw can reduce soil losses; it matters whether the straw is on the plough-ridges or lies in the furrows.[66]

Trees are said to prevent gullies, slumps and debris flows, but whether they do is doubtful. Gullies are no less abundant in woodland than elsewhere (Fig. 15.14): shallow-rooted trees topple in, while deep-rooted trees are left on pedestals. Trees may encourage slumping if they promote infiltration of water into the soil, where it migrates down to lubricate the slip-planes that initiate a slump. Tree-roots can rarely prevent a slump by holding the soil together, because most slumps are too deep-seated. After deluges, gullies, slumps and debris flows are reported to be at least as prevalent in forests as elsewhere.[67] On steep slopes, the collapse of trees results in avalanches and torrents of debris,[68] but this probably happens in the Mediterranean only on the rare occasion of an ice-storm.

Erosion and cultivation

The MEDALUS erosion plots, taken together, appear to confirm the reputation of vines as a very erosion-promoting crop. Vineyards lost on average 0.06 mm of soil per year, more than any other treatment: vineyards are ploughed at least twice a year and are often on steep slopes. The one site at which olives were studied, however, lost only 0.0003 mm per year, which is surprising considering that well-tended olives, too, are ploughed beneath. (The olive plots were on a different soil from the vines.)

Cereals and natural vegetation were each studied at a larger number of sites. Cereals lost 0.007 mm per year on average, three times as much as semi-natural phrygana.

The Is Olias plots reveal what may be an important finding. Eucalyptus, a favourite foresters' tree, promotes erosion nearly as strongly as cereals. Under each tree the expansion of the roots pushes up a mound of bare soil, on to which rain-water drips. The loose leaf-litter is easily washed away. Although the shade is light, there is little ground vegetation, and in particular no development of moss and lichen crusts.[69]

The age-old practice of picking stones and potsherds off the surface of fields and vineyards often encourages erosion, depending on the size of the stones and the degree to which they are embedded in the soil.[70]

Natural vegetation other than trees

The Pixinamanna erosion plots in Sardinia were run under natural vegetation on rotten-granite soils from autumn 1982 to spring 1985. An area artificially 'cleared' at the start became covered in maquis, *Cistus*, grasses and herbs by April 1984. Erosion from ordinary rains rapidly diminished as the plant cover increased, a cover of 30–35 per cent reducing soil losses by 90 per cent. On soils under maquis up to 3.5 m high, with a ground cover of 35 per cent, only one rainfall event resulted in erosion: a fall of 61 mm in November 1983 yielded the equivalent of 0.008 mm of erosion on a slope of 10°.[71]

In the marl badlands of SE Spain there has been a short-term experimental comparison of runoff and erosion, simulating deluges by high-intensity irrigation sprays. Three sites, some 15 km apart and on somewhat differing soils, were compared: one with low pinewood over undershrubs, one with 'shrub matorral' (continuous cover of undershrubs), one with 'degraded matorral' (low, discontinuous undershrubs). The findings were that there was little difference between pinewood and continuous undershrubs, but erosion from the discontinuous undershrubs was some fivefold greater.[72]

Preliminary results from 18 plots at Rambla Honda, on mica-schist soils on the edge of the Almería desert, show mean annual losses of 0.0138 mm on plots of the tussock-grass *Stipa tenacissima*, 0.0090 mm under the undershrub *Anthyllis cytisoides*, and 0.0024 mm on an alluvial footslope vegetated with the tall switch-plant *Retama sphaerocarpa*. (These figures are from only one year's observation, which, however, included two near-deluges.) The relatively high value from *Stipa* was attributed partly to its tussocky nature, channelling runoff between the tussocks.[73]

[65] R. W. Alexander and A. Calvo, 'The influence of lichens on slope processes in some Spanish badlands', Thornes 1990, pp. 384–98; Williams & others 1995; J. Belknap, 'Surface disturbances: their role in accelerating desertification', *EMA* 37, 1995, pp. 39–57; D. Tongway, 'Monitoring soil productive potential', *EMA* 37, 1995, pp. 303–18.
[66] A. C. Imeson, F. Pérez-Trejo, L. H. Cammeraat, 'The response of landscape units to desertification', Brandt & Thornes 1996, 1996, pp. 446–69; O. R. Stein, W. H. Neibling, T. J. Logan, W. C. Moldenhauer, 'Runoff and soil loss as influenced by tillage and residue cover',

Soil Science Society of America Journal 50, 1986, pp. 1527–31.
[67] Wainwright 1996; F. Gallart and N. Clotet-Pernau, 'Some aspects of the geomorphic processes triggered by an extreme rainfall event: the November 1982 flood in eastern Pyrenees', *Catena suppl.* 13, 1988, pp. 79–85.
[68] N. Garwood, D. P. Janos, N. Brokaw, 'Earthquake-caused landslides: a major disturbance to tropical forests', *Science* 205, 1979, pp. 997–99; A. R. Orme, 'Recurrence of debris production under coniferous forest, Cascade Foothills, northwest United States', Thornes 1990, pp. 67–84.

[69] A. Aru, 'The Rio Santa Lucia site: an integrated study of desertification', Brandt & Thornes 1996, pp. 189–206.
[70] J. Poesen, 'Effects of rock fragments on soil degradation processes in Mediterranean environments', Rubio & Calvo 1996, pp. 185–222.
[71] H. Dieckmann, H. Motzer, H. Harres, O. Seuffert, 'Vegetation and erosion: investigations on erosion plots in southern Sardinia', *Geoökoplus* 3, 1992, pp. 139–49.
[72] C. F. Francis and J. B. Thornes, 'Runoff hydrographs from three Mediterranean vegetation cover types', Thornes 1990, pp. 362–84.
[73] J. Puigdefabrigas and 11 others, 'The Rambla

Erosion and fire

The connection between fire and erosion is reviewed in Chapter 13. The evidence is inconsistent: sometimes massive erosion occurs after a fire, sometimes even a deluge after a fire has little effect. (Or, to put it another way, with some combinations of vegetation and environment erosion, if any, occurs all round the fire cycle. In others it is concentrated into the phase immediately after a fire.)

Erosion and terraces

If erosion control is the motive for constructing terraces, failure to maintain them should cause renewed erosion. It is often claimed that abandonment of terraces results in more erosion than if the terraces had never been built:

> The increased runoff removes in a few years large amounts of soil, which has been stored for centuries behind the terrace walls, and lays it down in the valley bottoms in the form of stream flood deposits.[74]

This is supported by observations in the south Argolid. Air photographs taken in 1961 show intact terraces, but only a few stretches of wall remained in 1979. 'The slopes were covered with scattered boulders, deep gullies had formed, olive trees had fallen, and nearly a foot of new alluvium had accumulated on the valley floor.'[75]

Some authors add that terraces still used but badly maintained result in even more damage. Terrace collapse is part of McNeill's theory of 'undershoot' (p. 91).

The reality, as with everything to do with erosion, is complex. Abandonment of terraces in Spain sometimes increases erosion (until the site has been revegetated), sometimes reduces it.[76] Faint traces of terraces occasionally appear on slopes that now have little soil; the entire contents of the terraces have emigrated. At the other extreme, one more often sees long-abandoned terraces still sound and not eroding, especially where they were well built and are too high for a goat to climb. The very well-built terraces of Majorca, which one of us noted in 1969 as having been abandoned for at least thirty years and grown up to pinewood, were still in good condition in 1994. In Liguria, too, well-built terraces topped with robust grasses remain intact when abandoned. On badland-prone clays and mudstones in Catalonia, abandoned terraces apparently tend to gullying, whereas non-terraced hillsides slump.[77]

More common is an intermediate state, depending on the materials and construction of the terrace wall, the nature of the vegetation, the fire cycle and the degree of browsing. Badly built terrace walls develop breaches, or are swarmed up by goats which knock stones off and create paths. Paths develop into miniature gullies, producing the familiar 'goated' appearance of terraces abandoned for forty years or

Fig. 14.19. Terraces abandoned, breached and gradually reverting to the previous slope. *Ohanes, Alpujarra, November 1994*

Fig. 14.20. Abandoned terraces: their well-built walls are beginning to be colonized by shrubs and trees. *Apéri, Kárpathos, April 1989*

so. As Lehmann described on Naxos, gullies develop miniature fans of debris on the next terrace down. Particularly significant are gullies that descend through more than one successive terrace wall: they widen sideways to involve the whole slope, leaving only scraps of the original walls.[78] We find a common variant in Crete, where the whole wall tumbles on to the terrace below and is buried by the collapse of earth from behind the wall. A typical form in the Alpujarra is for a section of wall to collapse, leaving a triangular breach, behind which is a steep but relatively stable scarp of fill (Fig. 14.19).

Erosion on abandoned terraces can be rapid. Lehmann's measurements indicate losses equivalent to between 0.04 and

Honda field site: interactions of soil and vegetation along a catena in semi-arid southeast Spain', Brandt & Thornes 1996, pp. 137–68.

[74] Van Andel and others, referring to the southern Argolid; Jameson & others 1994, p. 399.

[75] Van Andel & Runnels 1987, pp. 143–50.

[76] Faulkner 1995; cf J. M. Gárcia Ruiz, P. Ruiz Flaño, T. Lasanta, 'Soil erosion after farmland abandonment in subMediterranean mountains: a general outlook', Rubio & Calvo 1996, pp. 165–83.

[77] Haro & others 1992; Gallart & others 1994. The area is only marginally Mediterranean.

[78] J. Arnaez-Vadillo, L. Ortigosa-Izquierdo, M. Oserin, 'Descripción y cuantificación de procesos de erosión en bancales abandonados (Sistema Ibérico, La Rioja)', *Estudios de Geomorfología en España*, ed. F. López Bermúdez and others, Sociedad Española de Geomorfología 1, 1992, pp. 193–201; Lehmann 1994, pp. 150–61.

Fig. 14.21. Nearly everything in this picture is terraced. A few lower terraces are still cultivated; the others are in various stages of abandonment. Except for a few trees and shrubs on the lower terraces, the vegetation is phrygana. Terraces are still visible after decades of disuse, but have not eroded, apart from the contents of one terrace sometimes slipping on to the terrace beneath. *Asphéndou, Sphakiá, Crete, July 1990*

0.8 mm per year, but on only about one-tenth of his study area in Naxos. (His study seasons did not include a deluge.)

Our own observations suggest that it is unusual for erosion to go on until all the terrace fill has been washed all the way down the slope. Most of the material descends no more than one or two terraces. After decades of erosion, the slopes become less steep and re-vegetated. Shrubs on the terrace walls grow into trees; unpalatable trees like cypress may invade from outside (Fig. 4.27); unpalatable undershrubs invade in the face of quite severe browsing. The end result (Fig. 14.21) is a reasonably stable surface on which the entire terrace fill has, in effect, slid down one terrace. Remains of terraces are still visible, and may still reduce erosion. Given enough labour, there would be no great difficulty in restoring them. Undershoot has not happened. This is an 'average' scenario, and examples of more, or less, erosion on abandoned terraces are not far to seek.

In autumn 1995, one of us inspected the terraces at Auriol, between Aix and Marseilles, in the company of Professor Vaudour. He had been concerned about their future thirty-three years earlier, when they were in use. Storm-water was weakening walls and carrying away fine soil, leaving the stones behind. Sheet erosion was followed by rilling, with channels 10 to 15 cm deep and 1 m wide appearing in the fields in the autumn. This erosion, called *eissariado* by the peasants, disappeared with the winter ploughing which seemed to put everything back in place, only to reappear in spring or the next autumn. Incisions a metre deep had developed into ravines within thirty years; streams were entrenching, rows of vines were slipping downhill, and landslides followed heavy rains.[79]

Since 1962 the area under cultivation, including vineyards, has declined by one-third. On terraced slopes the influence of the underlying rock has become more apparent. Thin soils on hard dolomitic limestones have been colonized by pines; frequent fires maintain a *garrigue* of prickly-oak, thyme, rosemary, cistus and arbutus. Deep red soils over softer, bauxitic, marly limestones are now protected by dense maquis and *Ailanthus*. Recent erosion forms vary correspondingly: terraces on dolomites have been disturbed but are largely in place; those on deeper soils are only locally gullied.

79 Vaudour 1962.

Fig. 14.22. Recently abandoned terraces near Auriol, Provence. *November 1995*

(a) Maquis on terraces on dolomitic limestone, showing various stages of pine invasion.

(b) Colonization by pines, preparing the way for the next fire.

DOES EROSION MATTER?

Soil erosion is widely cited as one of the main processes of desertification. To assess its scale critically has been one of the main concerns of MEDALUS. However, experiments initiated under this project have been running for only a few years; this work, like its predecessors, will not yield dependable results unless a long-term basis is found for continuing it.

Most discussions start from the assertion (stated or unstated) that soil is the necessary basis for agriculture and for all plant life. Removing forests to create fields is supposed to set erosion in motion. Erosion removes the soil, leaving either bare rock or badlands in which plant life is impossible. Erosion control may delay, but cannot prevent, this fate. On this plausible but gloomy scenario, nearly all agriculture is, in the long term, unsustainable. Is it true?

Loss of soil – wild vegetation

Despite all that has been written on Mediterranean lands, it is rare to find the whole sequence, from forest to rock, demonstrated on any one piece of land. It is not easy even to show that land, once cultivated, has become uncultivated owing to loss of soil; but it can be done, and here is an example.

On the tremendous south coast of Crete, where high mountains reach the sea, is a limestone plateau at an altitude of 600–700 m, in the township of Ayios Ioánnes, Sphakiá. This is covered with a network of field-walls, probably of Roman date. There can be little doubt that it was once covered with soil and cultivated; remains of a loess-like soil are preserved under the walls.[80] The soil, much of which originated as wind-blown dust, was presumably blown away by wind, as could happen after it was ploughed, or washed down into the limestone underworld. What is left is bare, intractable, razor-edged limestone; but much of it is forest, with big pines growing out of stone-piles and fissures (Fig. 14.23).

The relationship between soils and vegetation is not as simple as geographers and historians would like. Even hard bare rock is seldom utterly barren: hundreds of species of plants are adapted to growing on nothing else. Many hard limestone surfaces are completely covered in differently coloured patches of lichens; there are subsurface lichens which manifest themselves as minute pits. Any Cretan gorge is full of wonderful flowering-plants which grow straight out of holes in the rock.

Fig. 14.23. Recent pinewood on karst limestone which has lost its soil. *Ayios Ioánnes, Sphakiá, Crete, July 1987*

Loss of soil does not preclude trees. At Ayios Ioánnes it has created forest, though probably not the same forest as may have existed before the soil was cultivated. Mount Hymettus (Athens), famous for barrenness since Plato's time, is now rapidly turning into pinewood and burnt pinewood (pp. 236, 288). Where forest is patchy, trees often grow on rock whereas soil-covered areas have grassland (pp. 57, 59).

On soft rocks, gullying makes a habitat for trees. In the badlands of the northern Peloponnese, the lushest vegetation

80 Observations from the Sphakiá Archaeo-logical Survey.

is in long-established gullies, with jungles of arbutus, pines and oaks, tangled with greenbrier (*Smilax*) and other vines. In Bœotia certain undershrubs – *Rhus coriaria*, *Cotinus coggygria*, *Putoria calabrica*, *Daphne oleoides* – specialize in colonizing erosion surfaces, and withstand continuing moderate erosion. Once erosion slackens, raw marl surfaces are often more suitable for tree growth than the uneroded surface. This is probably a matter of root penetration: the original marl has a hard surface layer which roots cannot get into.

Loss of soil – agriculture

An English or American writer naturally assumes that cultivation depends on a well-developed soil: but does it in a Mediterranean environment? The commonest instrument of husbandry in modern Crete and parts of the Peloponnese is the bulldozer, which gouges out false terraces from soft rock in order to grow olives and vines (Figs. 6.9, 6.10). This practice, however destructive, proves that olives and vines grow at least as well in broken-up bedrock as in soil. Even traditional stone-walled terracing does not really conserve soil. Unless the soil is very deep, the making of a terrace on soft rock disperses or buries the original hillside soil and replaces it by broken-up subsoil and bedrock. Because plant growth is so much limited by root penetration, crops grow better in terrace-fill than in the original soil.[81]

Agronomists are exasperated by the 'complacency' of peasants in the face of erosion – meaning the reluctance of peasants to drop whatever else they are doing and turn all their energies to preventing erosion. For example:

> The cultivators know the facts which we have set out, but the majority do not worry. The great ravages, they say, are due to exceptional bad weather, against which one can do nothing; as for the collapses of [terrace-]walls and the gullies, it is adequate, after the rains, to put back the stones and to efface the *eissariados* (rills) by suitable ploughing.[82]

This seems to us to be not so obviously foolish as this author thought; it is the reaction of people who are aware that deluges are the main source of erosion, and have learnt to live with them.

Lost soil (especially topsoil) takes mineral nutrients with it; increased runoff leaches nutrients out of the remaining soil. Lehmann gives figures for both effects on Naxos; they are very variable and almost impossible to translate into kilograms per hectare per year. This probably does not matter much on non-cultivated land. Most remaining wild vegetation is adapted to a nutrient-poor environment, and loss of nutrients may be no disadvantage (except perhaps in places that lack nitrogen-fixing plants). On farmland, losses of nutrients were very important in the past, when fertility was a valuable, inherent property of the soil. They matter less in an age when fertility comes in a sack. Where heavy fertilizing coincides with rapid erosion, as on unterraced vineyards, fertilizer-containing silt may cause trouble when it gets into lakes and reservoirs and eutrophicates them (see the Attica example, pp. 263–4).

Loss of productivity depends on the soil profile. In a field subject to sheet erosion, each year the plough bites a little deeper into the subsoil in compensation for the soil lost on top. For some crops this appears to give a perfectly satisfactory rooting medium. Some of the best commercial wines of Greece grow on rapidly eroding bulldozed marl. Even for cereals, sheet-eroding soft rock can be an adequate rooting medium, given that farmers in southern Europe are not so obsessed with high yields as their modern brethren in NW Europe. A deep sandy soil may be easily erodible but continue to produce if adequately manured. However, if the subsoil contains a layer of dense clay or hard caliche, its productivity will disappear as soon as this layer comes within plough depth.

As pointed out by the Greek soil scientist Nicholas Yassoglou, riverine soils on deltas, drained marshes and flood-plains vary in texture and profile over short distances, according to how they were deposited. They are often exposed to bank and coastal erosion, as a result of the damming of rivers and digging of gravel from river-beds. As these are among the most productive Mediterranean soils, the threat must be taken seriously.

Intensive cultivation of sandy or peaty soils destroys organic matter and harms soil structures, reducing resistance to wind erosion. (The great peat beds of Lake Kopáis in Greece lasted only about fifty years after it was drained). Hedges and windbreaks (of trees or reeds) not only protect growing crops but reduce wind speeds below the critical velocity for deflation. (They are often foolishly removed in order to enlarge 'uneconomic' small fields.) Riverine soils are liable to pollution by heavy metals, and occasionally to being buried by a continuation of the alluvial deposition which originally created them.[83]

Soils on Quaternary river terraces often have sandy surface horizons underlain by heavier material; this favours plant growth, but such soils are susceptible to erosion when irrigated by water-furrows. On older river terraces, erosion may expose dry, indurated horizons that roots are unable to penetrate. River-terrace and alluvial soils may become salty if water-tables rise (p. 257, n. 54).

Regosols, for example deep gravelly and stony soils developed over colluvium, may have no clearly differentiated horizons. If the topsoil erodes away, a stony layer develops on the surface, restricting further erosion. The soil may remain suitable for orchards, which rely on deep moisture. Picking off stones is counter-productive. Such soils, on alluvial fans and cones and on river terraces, are now used for greenhouse cultivation; they have been greatly modified by fertilizers and even by bringing in sand or clay from elsewhere.

We have rarely seen definite examples of fields rendered uncultivable by gullying, or even abandoned in anticipation of gullying; however, we do not doubt that this occasionally happens as part of the normal dynamics of a landscape. We have seen new areas of cultivable land created at a later stage in those dynamics. Fields emerge from the tail-ends of badlands (p. 285); they have stabilized to a degree where even further cultivation does not reawaken gullying.

---

[81] This may explain why terracing is confined to certain regions of the world; in countries such as England, although terracing might be advantageous in controlling erosion, it is not practised because making terraces would damage the soil.

[82] Vaudour 1962, in the south of France.

[83] Yassoglou 1989.

Effects on recharge of ground-water

Eroded sites, especially if bare rock is exposed, tend to increase runoff from ordinary rains at the expense of infiltration. Lehmann regards this as significant (though difficult to measure) in an island such as Naxos, which depends on a limited aquifer. Check-dams to reduce runoff are important in conserving the meagre ground-water of SE Spain (Chapter 19). However, most problems with ground-water result from increased demand rather than diminished supply.

CONTRASTING EXAMPLES OF EROSION

Long-term instrumental measurements – Alentejo, Portugal

At Vale Formoso the problems of replicating and maintaining experimental plots have been surmounted. Fifteen plots, belonging to a permanent institution, have been maintained for forty years under various agricultural treatments. The site, on metamorphic rocks, was chosen because it was thought to be eroding rapidly. Two more plots were set up in 1988, one being 'pasture' or abandoned arable, the other being planted with *Cistus* undershrubs.

Average soil losses in thirty years were equivalent to 0.038 mm per year under wheat, 0.015 mm per year under 'stubble' (fallow?), and 0.058 mm per year under bare soil. Losses month by month were related to the cycle of farming operations. However, the replication problem was not entirely solved. Average soil losses throughout a crop rotation over thirty years ranged from 0.074 mm per year in the most erodible plot to 0.022 mm per year in the least erodible. Variation between plots was greater than between treatments.

For the three years in which the extra plots were observed, the mean losses were 0.016 mm under *Cistus* and 0.007 mm under abandoned arable. These losses include the extremely wet year 1989–90, in which the values for individual treatments were:

Wheat (mean of two plots)	0.389	mm
Stubble (two plots)	0.152 and 0.013	mm
Cistus	0.043	mm
Abandoned arable	0.011	mm

The experiment has continued long enough to give some insight into deluges. Between one-eighth and one-quarter of the erosion in thirty years (varying from plot to plot) took place on the occasion of most active rainfall: that is to say, a single exceptional fall of rain removed from three to seven years'-worth of soil.[84]

Despite the long run of measurements, the Universal Soil Loss Equation predicts rates of erosion 27–40 times those actually measured here.[85]

Crop cover, runoff and erosion in Attica

At Spáta, in the dusty, long-cultivated, now pine-invaded marl hills just outside the fringe of Athens, C. Kosmas and his colleagues have set up runoff plots in a long-established semi-terraced olive-grove on a marly sandstone hillside, and in a vineyard on the plain immediately below.[86]

The site has a series of gravelly, medium- to fine-textured, strongly calcareous soils, varying in depth. Mean annual rainfall is 496 mm. The site had been an olive grove for more than 160 years. Part of it was grubbed out in 1979 and replaced with vines, not terraced and frequently ploughed. Soil surveys made in 1979 and 1991 demonstrated the drastic consequence of the change. Under vines the whole 18-cm thickness of the A soil horizon on slopes of more than 8° was eroded away – at least 15 mm per year – exposing the marl.[87]

Five experimental positions were selected, one with vines, one with olives, and three on ex-cultivated land. The olives, contrary to usual practice, had not been ploughed beneath for more than twenty years, but had semi-natural vegetation of undershrubs and annual and perennial herbs. The vineyard was ploughed as usual, and received fertilizer equivalent to 120 kg per ha of nitrogen each year.

At each position, two erosion plots, each 3 m by 10 m, were observed for five years (1989–94), including the phenomenally dry year 1989–90 and the wet year 1990–1, with near-deluges in January and March. Four ponding rains fell during November 1993 (Table 14.ii).

On the olive plot, vegetation (live and dead) minimized the speed and amount of runoff water: a thunderstorm bringing 98 mm of rain, most of it within one hour, yielded only 2.5 mm of runoff. Under vines, runoff began about 25 minutes later in each rainfall than on the other plots. Presumably this was due to water held up in furrows of the contour ploughing: the time available for infiltration was greater and decreased the total volume of runoff. Bare, smooth, stone-free soil provided the greatest runoff. On the other ex-cultivated soils, stones were thought to have reduced runoff by about 10 per cent. A herbaceous plant cover of 48 per cent was effective, sometimes reducing runoff to a half that from bare soil.

Total sediment losses from bare soil were equivalent to 0.14 mm after the 98-mm storm. On soils with stones on the surface, sediment losses were 32 per cent less. Soil losses from the olive grove were negligible. On land with 48 per cent vegetation, after 1½ years' fallow, soil losses were 74 per cent less than on bare soil. Contour ploughing reduced soil losses from the vines, though the first two rainfalls after the dry season gave high runoff and sediment yields.

Total sediment loss amounted to 0.00010 mm per year under olives and 0.10 mm per year under vines – this despite the olives being on a much steeper slope. The losses measured

[84] M. J. Roxo, 'Lower Alentejo, Bexa and Mertola, Portugal', *MEDALUS I Final Report*, 1993, pp. 406–32; M. J. Roxo and Cortisão Casimiro, 'Human impact on land degradation in the inner Alentejo, Mértola, Portugal', Mairota & others 1998, pp. 106–9.

[85] Seuffert 1992, p. 112, quoting Coutinho & Thomas.

[86] C. Kosmas, 'Spata, Athens, Greece', *MEDALUS I Final Report*, pp. 581–607.

[87] C. S. Kosmas, N. Moustakos, N. G. Danalatos, N. Yassoglou, 'The effect of diminishing soil moisture on soil properties and biomass pro-duction of rainfed wheat', *MEDALUS I Final Report*, C.E.C. EPOCH-CT90-0014-(SMA), 1993; 'The Spata field site: I. The impacts of land use and management on soil properties and erosion. II. The effects of reduced moisture on soil properties and wheat production', Brandt & Thornes 1996, pp. 207–28.

Rainfall event	Olives		Vines		Bare soil		Bare stony soil		Herbaceous vegetation	
mm	runoff	sediment loss	runoff	sediment loss	runoff	sediment loss	runoff	sediment loss	runoff	sediment loss
98.3	98.3	0.16 (0.00006)	9.0	46.4 (0.019)	32.3	344.5 (0.138)	29.9	234.5 (0.093)	25.2	90.4 (0.036)
45.0	45.0	0.02 (0.000008)	15.7	84.4 (0.034)	26.9	72.6 (0.029)	25.4	48.4 (0.019)	24.8	13.0 (0.052)
27.5	27.5	0	0.7	1.7 (0.0007)	12.0	12.8 (0.005)	11.0	13.2 (0.005)	3.9	2.0 (0.0008)
11.8	11.8	0.01 (0.000004)	2.1	4.4 (0.0017)	6.5	16.2 (0.006)	6.3	22.4 (0.009)	4.1	4.8 (0.0019)

Table 14.ii. Runoff and sediment losses resulting from four falls of rain.
Soils strongly calcareous, medium-textured, well-drained, gravelly, slope c. 15 per cent.
Runoff is in mm; sediment losses in grams per sq. m., followed by the rough equivalent in terms of mm of lowering the surface.

under vines were vastly less than had been inferred from observations of changes in the soil profile over the preceding twelve years.

Losses of nitrate from the vineyard, although an insignificant fraction of the nitrate applied, were more than ten times the loss from under olives. Bare soil, with and without stones, lost almost as much nitrate as the soil under vines. The presence of a 48 per cent plant cover reduced the loss of nitrates by about a half.

Arable land on the edge of the Basilicata badlands

In a hilly, clayey area of south Italy runoff and erosion rates were compared on vertic-clay soils with a slope of 14 per cent under direct drilling (no tillage) and three depths of cultivation (Table 14.iii).

Erosion was surprisingly low for arable land, and varied much with year and little with treatment. In most years losses were between 0.04 and 0.08 mm, being on average 25 per cent greater on the untilled soils than on those rotavated or ploughed. The exception was 1984–5, when losses 2½ times the mean were associated with twice the usual rainfall and twice the usual runoff.

In the same area, soil losses were compared under different crops on 0.1-ha plots on a 20 per cent slope between January 1974 and June 1976. They were as follows: horse-beans 0.28 mm per year; durum wheat 0.22; forage crops 0.16; pasture thickened with lucerne 0.10; natural pasture 0.05.[88]

No particular pattern emerges from these observations, except that erosion is much greater in heavy rain. They

confirm that anyone wanting to increase sheet-erosion, especially under vines, should pay attention to ploughing and to removing stones.

Slopes under vines in north Italy

More detail comes from observations in vineyards cultivated on the slope on marls, silts and fine sands in Piedmont in 1981 and 1982. The instruments consisted of silt traps at the foot of slopes 29–58 m long.

Soil losses were closely related to intensity of rainfall. Nearly all erosion occurred in short periods when it rained at more than 0.18 mm per minute. Two or three times each year rainfall exceeded 0.7 mm per minute; on each such occasion over 0.4 mm of soil was eroded.[89] In this very sensitive area, repeated ploughing and weedkilling can so reduce the threshold of adjustment (p. 252) that relatively modest storms have the effect of deluges.

Naxos: collapse of terraces

The middle-sized, mountainous Aegean island of Naxos is perhaps the most delightful of the Cyclades. Its elaborate cultural landscape is well preserved and in part still used. The rugged fertility of the mica-schist valleys, with their terraces and coppiced maple-woods and cliffs bright with the wallflower *Erysimum naxense*, contrasts with the stern, barren marble mountains, dotted with a thin savanna of pollard oaks.

A part of the island with less well-preserved terraces was the scene of the most thorough and comprehensive of erosion

Year	Rainfall total mm	Runoff mm	Soil losses, tons per ha per year			
			Direct drilling	Rotavation to 20 cm	Ploughing to 20 cm	Ploughing to 40 cm
1981–2	467	363	1.23	1.10	1.05	1.00
1982–3	421	368	1.26	1.16	1.09	1.04
1983–4	704	493	1.57	1.43	1.39	1.32
1984–5	1,073	738	6.20	4.56	5.20	5.55
1985–6	606	469	2.00	1.82	1.85	1.72
1986–7	512	345	2.11	1.99	2.12	2.21
1987–8	569	399	1.47	1.44	1.33	1.44
Mean	622	454	2.51	1.93	2.00	2.04

Table 14.iii. Runoff and soil losses from erosion plots in Basilicata. [L. Postiglione, F. Basso, M. Amato, F. Carone, 'Effects of tillage methods on soil losses, on soil characteristics and on crop production in a hilly area of southern Italy', *Agricoltura Mediterranea* 120, 1990, pp. 148–58]

[88] F. Basso and C. Ruggiero, 'Effeti di differenti metodi di lavorazione del terreno sullo sviluppo radicale e sulla produzione di granella di frumento duro in ambiente collinare della Basilicata', *Problemi agronomici per la difesa del fenomeni erosivi* 129, Centro Nazionale di Ricerce, Rome, 1983, pp. 224–37.

[89] Tropeano 1984.

studies, by Rainer Lehmann. The object was to establish the rate and pattern of erosion, and to map the erodibility of a landscape in order to propose anti-erosion measures. Lehmann investigated loss of soil and soil nutrients, and the effects of erosion on the recharge of ground-water; in the Cyclades, as elsewhere, the growth of irrigated agriculture is overtaking the water supply.

The study area – about 8 sq. km in the NW of Naxos – is rugged and geologically varied, including grano-diorite, schists, soft conglomerates and alluvium. It is probably more erodible than the average terraced landscape.

Measurement, from 1989 to 1992, included weather, soil moisture and nutrient content, and the rate at which rainfall infiltrated. The measurements of different types of erosion were simple and easily replicated, involving no expensive construction and no electricity supply. Yet again, alas, the work lasted only three years, during which no deluge occurred. If anything can be learnt about erosion under that restriction, Lehmann learnt it.

Terraces cover about half the study area. Lehmann recognised five stages of deterioration, from Stage 1 (fully maintained) to Stage 5 (reduced to a continuous slope with only scraps of terrace wall remaining). The erosion-measuring bridge and field-box, which we have already mentioned, were devised for measuring the small gullies that occur on goated terraces.

Lehmann tried to estimate how long it takes for terraces to goat and degrade. He calculated the volume of fill removed at each stage, and divided this by an average rate of removal. The latter was, as ever, difficult to ascertain, because most erosion happens over a small proportion of the area and over short periods of time. Measured rates were too low because of the absence of a deluge, and because (as Lehmann had reason to believe) erosion had been more rapid during the Little Ice Age. He thus decided to take the *top* of the range of measured erosion rates as representative of the *mean* historic erosion rate.

To reduce terraces to Stage 5, Lehmann worked out, involves removing 77 cm of terrace fill. At an estimated rate of 1.2 mm per year, this would take six to seven hundred years. On this basis, he concluded, more than half the terraces were abandoned at the latest between the fourteenth and seventeenth centuries AD.

This finding will surprise most students of terraces, who would expect the peak period of terrace cultivation and abandonment to be more recent. We cannot disprove it: each island has its own human history, and periods of decline in the fourteenth and seventeenth centuries are more than likely on Naxos. However, there are vast areas of terraces on Crete approximating to Stages 4 and 5 which, on demographic grounds, supported by travellers' observations,[90] were probably last used in the nineteenth century. Lehmann himself is diffident about the accuracy of his claims. We suspect that he over-estimated the amount of material removed: to judge by his photographs, even at Stage 5 much of the fill has merely tumbled on to the next terrace below, and has not left the slope. Moreover, in estimating the rate of erosion, he may not have allowed sufficiently for unusually little erosion occurring during the years when he was measuring it.

Thessaly – arable land on shallow soils [91]

The modern practice of growing rain-fed autumn-sown cereals year after year appears to be particularly destructive of shallow soils over soft, sandy bedrock. For a month or two after sowing the land is bare and disturbed, and very erodible should intense rain occur. Organic matter, which stabilizes the soil crumb structure, is never high after a hot dry summer, especially if the stubble of the previous crop has been burnt and transhumant sheep (p. 90) have not come.[92]

The great plain of Thessaly and some of its surrounding hills have a long history of pastoralism – earlier in the twentieth century the plain was famous for horses – but are now largely arable land. The soils are derived from shale-sandstone and conglomerates. On the slopes exposure of tree-roots indicates rapid erosion – as much as 1.7 cm per year has been reported. The organic matter content of many soils is said to have halved in forty years.[93]

As part of the MEDALUS programme, erosion rates on the hills were estimated using simulated rainfall. In September 1993, after five dry months, soil surfaces were sprayed with water at a rate of 175 mm per hour for eleven minutes – equivalent to a thunderstorm. The results, like those of other such experiments, indicate much variation in runoff and erosion losses for soils on different parent materials (Table 14.iv) and with different covers. Marls were less absorptive, giving higher runoff rates and sediment losses than other soils. Stubble reduced erosion, probably by protecting the soil structure from raindrop splash.

	Marl	Shale-sandstone	Conglomerate
Percentage runoff	42.5	25.8	32.5
Sediment loss from bare soil, grams per sq.m	491	21	176
equivalent in mm	0.2	0.008	0.07

Table 14.iv. Runoff rates and sediment losses for soils subjected to an artificial thunderstorm at the start of the rainy season.

Marls are most vulnerable to erosion; they remain in use because cultivation can continue into the bedrock as the soil disappears. Shale-sandstones, though less erodible, have a restrictive bedrock; some are being abandoned. They are better protected by straw than other soils. Conglomerate soils, though less productive, have the advantage of better root penetration, which makes their crops less susceptible to drought than on marls. The stony surfaces of shale-sandstone and conglomerate soils help to conserve soil moisture.

[90] For example, D. Grove, J. A. Moody, O. Rackham, 'Excursions in West Crete', *Petromarula* 1, 1990, pp. 77–88.

[91] Most of our information about Thessaly comes from C. S. Kosmas, N. G. Danalatos, N. C. Moustakos, N. Yassoglou, 'Effect of parent material and land use on soil erosion rates under Mediterranean conditions', *MEDALUS Working Paper* 28, Department of Geography, King's College, London, 1994.

[92] Yassoglou 1989.

[93] N. G. Danalatos, *Quantified analysis of selected land use systems in the Larissa region, Greece*, Ph.D. thesis, Agricultural University of Wageningen, Netherlands, 1993.

The Plain of Esparto – erosion on very slight gradients

The rolling country in the south of La Mancha – part of the ancient Plain of Esparto (p. 174) – is now extensively cultivated with cereals and almonds. Fields occupy the wide undulating plains of Hellin and Albacete and smaller, shallow basins on the territory of hill-towns like Liétor and Elche de la Sierra. The hills between are thinly covered with esparto, retama and other semi-desert plants (Chapter 17). These barren hills, haunted by boars and the ghosts of bears, show no sign of former cultivation. To the south-west they rise into the foothills of the Sierra de Segura, now and for centuries one of the great pine-forest areas of Spain. Aleppo pines thin out north-eastwards, ending well before the Hellin plain; they seem to be invading an area that previously had no trees but a very few evergreen oaks.

Fig. 14.24. Erosion on a very gentle slope. Horizontal features are check-dams, built across a barely perceptible valley. Light patches on knolls and at the edges of the plain are where subsoil is exposed and the crop has failed. *West of Hellin, La Mancha, April 1994*

The rocks are commonly deep calcareous marls, not unlike the chalk marls of NW Europe. The hills appear to be highly unstable, especially under ploughing. The shallow rendzina soils of the basins, probably originally washed off the slopes, are in turn washed further down the basins, removing the organic horizon and exposing scalps of white marl on convexities. The underlying material has adequate water-holding, so that acceptable crops can be grown with artifical fertilizers. Sheet and rill erosion, however, continue on slopes of as little as a degree or two. After autumn storms rills deposit white plumes of subsoil over the darker slope-foot colluvium. Ploughing is the critical factor: after a year or two of not cultivating, erosion becomes unnoticeable.[94] Gullying also occurs: the main watercourses are in deep *arroyos*, and some of the clayey marls have developed badlands.

People have built (and successively heightened) dry-stone check-dams across even the shallowest valleys to trap the sediment; some no more than fifty years old are now 2 m high (Fig. 14.24). Even the plain of Hellin is often compartmented by flood-catching banks.

At El Ardal in Murcia are seventeen runoff plots, ten of which have been observed since 1989.[95] The treatments comprise bare land (ploughed downslope), fallow vegetated land treated with 'polymers', fallow land with and without stones, barley, wheat, natural vegetation (mainly *Thymus* and *Rosmarinus*) and cut natural vegetation. There is some replication.

Most erosion has resulted from heavy rain after ploughing. By far the biggest single annual loss was 0.14 mm (360 g per sq. m) from one of the cut-undershrub plots in 1989, but this was not corroborated either by the behaviour of that plot in later years or by the two other plots of that treatment in 1989. The one plot of downslope ploughing produced a loss of 0.07 mm in 1989, and suffered by far the greatest losses of all the plots in the less erosion-prone years 1991 and 1992. The lightest losses in 1989 (0.05 mm or less) came from fallow land with stones and from two of the three natural-vegetation plots.

This is among the few areas where we definitely conclude that cultivation is losing ground to erosion. Cereals and almonds do badly on the scalps of bedrock, probably because of poor root penetration. Check-dams create pockets of good soil, but cannot prevent the loss of the worse soils. The cultivated area (according to the maps we have seen) has changed little within the twentieth century, but cannot go on much longer. Extensive cultivation is possible on these soils for only a limited period of time, at least two-thirds of which has elapsed. How long is that time-span?

In Roman times the Plain of Esparto was a moderately populous and prosperous semi-desert (p. 314). It seems to have depended on intensive cultivation of the best land, the rest being mainly pasture, without much non-irrigated arable. The Tolmo de Minateda is a small mesa with remains of a strongly walled city, the Iberian and Visigothic predecessor of Hellin.[96] In the late middle ages the hill-country is documented in the archives of the Knights of Santiago, lords of Liétor. Much of this vast township was common land and its wealth was in sheep; cultivation was mostly in the river-valleys. The survey of Philip II (sixteenth century) indicates extensive cornland in the plain of Hellin, and even Liétor grew a little grain; but only the less erodible basin-land need have been cultivated, and not necessarily every year.[97]

The present phase of extensive arable cultivation may have begun with an agricultural boom in the mid-eighteenth century.[98] Alternatively it could result from 'land reforms' in the 1810s and 1820s, which privatized common land and

94 Measurements from ex-cultivated erosion plots over marls in Murcia give a rate equivalent to only 0.072–0.13 mm per year; C. F. Francis, 'Soil erosion on fallow fields; an example from Murcia', *Papeles de Geografía* 11, Universidad de Murcia, 1986.

95 F. López Bermúdez, 'El Ardal, Murcia, Spain', *Medalus I Final Report*, 1993, pp. 433–60; F. López Bermúdez and 3 others, 'The El Ardal field site: soil and vegetation cover', Brandt &

Thornes 1996, pp. 169–88; F. López Bermúdez and 3 others, 'El Ardal, Murcia, Spain', Mairota & others 1998, pp. 114–15.

96 On the mesa there is evidence of massive erosion and deposition, dating from around the time when the site was abandoned at the beginning of the Arab period. *Erosion following the sack of a city* may need to be added to the list of types of erosion. We are indebted to Dr Javier Lopez Precioso, Director of the Archaeological

Museum, Hellin, for showing us this site.

97 M. Rodríguez Llopis, *La villa santiaguista de Liétor en la baja edad media*, Istituto de Estudios Albacentenses, Albacete, 1993; Congreso de Historia de Albacete. I. Arqueologia y Prehistoria – Istituto de Estudios Albacentenses, Albacete, 1984 *passim*.

98 G. Lemeunier, 'Crecimiento agricola y roturaciones en el antiguo Marquesado de Villend (s. XVIII)', *A-B* 21, 1987, pp. 5–31.

allowed landowners to cultivate what should not have been cultivated. The hills are dotted with ambitious nineteenth-century farms and derelict mud-brick cottages.

Similar phenomena have created a regular patchwork of coloured soil and pale bedrock, familiar to anyone who has flown over southern Spain or Portugal. Even more susceptible to erosion are the shallow sandy soils developed over Cainozoic sandstones in northern Murcia. On the sandy schists of Alentejo we have seen olive trees, possibly two hundred years old, on soft, puffy, sandy soils, with half a metre depth of roots exposed. These areas, too, are the scene of large-scale cereal cropping or extensive vineyards, in which intense rains cause rill and gully erosion. Behind contour walls across slope depressions, washed-in soil and subsoil can accumulate to a depth of two metres within forty years.

To this extent, erosion is on the point of causing a substantial loss of cultivable land in Spain and Portugal. This can be delayed, though not prevented, by check-dams. For valuable crops like almonds it can be reversed by bulldozing the sediment back whence it came; but even for almonds this is doubtfully worth doing at a time of over-production. The self-destructive theory of cultivation appears to be true for these areas.

In our view this is no great loss. Mechanized agriculture on these soils yields poorly, despite fertilizers. It is economic only at times of under-production, and is now kept going by large subsidies. It is not long established and does not maintain a distinctive ecosystem or a distinctive human culture. This is not an area in which great effort should be put into preventing abandonment.

The Sierra de la Contraviesa and Santopitar – modest erosion on very steep gradients

A very different story is told by these mountain ranges in the extreme south of Spain. The Sierra de la Contraviesa lies between the Alpujarra and the sea. The mountains, up to 1,500 m high, dissected by gorges and ravines, are covered with stony, rather deep soils derived from friable schist. Most of the land is a monoculture of almonds, intensively ploughed underneath. In contrast to the Alpujarra, there is no trace of terraces: slopes are ploughed at angles up to at least 30°. Ravines and gullies too steep to plough are filled with coppice-woods and savannas of live-oak.

This spectacular landscape raises the hair of anyone alive to the evils of ploughing steep slopes. On similarly steep phyllite mica-schist soils in the Sierra de Lujar, further west, almond cultivation has produced dramatic gullies; this however is a recent expansion of cultivation, where natural vegetation has been grubbed out.[99] J. R. McNeill holds up the Contraviesa as a classic example of a landscape ruined by deforestation followed by tree crops which, as he puts it, 'offer minimal obstacles to accelerated erosion'.

We think this is overstated. Rills occur on the steepest slopes, despite the practice of ploughing round the contours; but there is no sign that they remove great quantities of soil. Old almond trees, more than 30 cm in diameter, show no

empedestalment. We relocated the scene of a photograph that McNeill took in 1988 to show how 'an almond grove loses ground to erosion and to encroaching maquis'. Six years on, we could see no sign that this is happening (Fig. 14.25). The almond-grove has not lost ground to maquis – which is indeed increasing, but on rocky, uncultivated parts of the slope. Nor has it lost ground to erosion: the photograph includes a patriarch among almond-trees, 1 m in diameter, which is not empedestalled.

Whether the Contraviesa was ever forested we cannot say; Carvajal's account in 1568 mentions scattered live-oaks much as there are now. Almond cultivation was preceded by figs, vines (Widdrington in c. 1830 praised the wine[100]) and maybe cereals; we saw one threshing-floor.

More hair-raising still is the Santopitar range between Málaga and Vélez Málaga. Cultivation takes the form of vineyards on unterraced 25° slopes, with deep brown soils derived from metamorphic rocks. There is almost no trace of natural vegetation. On these steep, smooth slopes, if anywhere, soil ought to be 'stripped from the hillsides' – as Ruined Landscape writers love to express it – with the greatest ease, especially under vines. In practice, although there is evidently some risk of slumping (which may explain the absence of terraces), we could find no visible sheet or gully erosion; the only rills we could see were where roads had diverted the runoff (Fig. 14.26). Similar slopes to these have apparently been cultivated for at least two hundred and fifty years. For eighteenth-century England they were the source of 'Mountain', then the fashionable dessert wine.[101]

This adds to our impression that the people of Granada province know how to use their complex landscape: which rocks and slopes should be terraced, which cultivated on the slope, and which left uncultivated. Erosion goes on from time to time, as the growth of river-deltas demonstrates (Chapter 18), but the numerous uncultivated badlands, and alluvial cones where tributaries join main rivers, are amply sufficient sources of material. Changes in the cultivated area are due, as elsewhere, to demography, changing markets, and opportunities to emigrate; we doubt whether erosion has played much part.

CONCLUSIONS

Erosion in southern Europe is complex and unpredictable. It is difficult even to measure with sufficient accuracy to compare sites, because of its unevenness over time. If Site A shows more erosion than Site B during a three-year research grant, is this because A was more erodible, or because a storm hit A and missed B? Most measurements are far too small-scale to justify extrapolation to whole landscapes.

We did not produce our last examples to show that erosion does not matter, but to demonstrate that it is not simple. By the standards of Murcia, cultivation in the Contraviesa and Santopitar, with their tenfold steeper slopes, ought not to survive a single winter's rains. Yet south Murcia is losing fertility to erosion despite control measures, whereas the risk in the sierras has not materialized. Their practice is hardly to

---

99 Faulkner 1995.
100 Cook 1834, p. 50.

101 Ponz 1778, XVIII,iv.71, vi.3, vi.12; Swinburne 1779, pp. 204–5.

Fig. 14.25. Almond-grove and old, vegetated gullies in the Sierra de Contraviesa. In our opinion, although management is not ideal, this is a stable landscape. *Between Almegijar and Torviscón, November 1994*

be recommended, but it has not yet brought disaster, despite the 1973 deluge. It appears to be more sustainable than the Murcia practice.

Erosion cannot be assessed by people in distant offices drawing maps of 'erosion hazard' without ever seeing the terrain. The Universal Soil Loss Equation does not work here, and we doubt whether it could be made to work by adding more variables: in the complex environment of Europe enough reliable data are difficult to collect. Modelling can be a means of understanding erosion, but it is still a long way from being predictive.

Our studies and those of others, although often inconclusive or contradictory, suggest the following conclusions.
(1) The primary determinant of erosion is tectonics.
(2) The next determinant is lithology. Particularly liable to erode are certain kinds of clayey marls and massive soft sandstones. Materials deposited in previous (late Tertiary and Quaternary) erosion cycles are very liable to erode again unless, as in Crete, they have been consolidated.
(3) If tectonics and lithology favour erosion, the main determinant of when it happens is weather. Water-erosion depends mainly on deluges rather than ordinary rainfall. If the Sardinian study is generally applicable, it depends mainly on even shorter events: on bursts of ultra-heavy rain within a deluge.

(4) The effect of vegetation is variable and easily exaggerated. Vegetation completely prevents wind erosion; by reducing wind speed near the ground, it reduces (or occasionally increases) the impact velocity of raindrops; it has one form of erosion peculiar to itself (the uprooting of trees); it may slightly increase the risk of slumping. Although extremes can be found, the effect is generally modest and is not limited to trees. *Ploughing, not deforestation, is the critical factor.* There is little evidence on which types of forest best protect against erosion, except that eucalyptus may be counter-protective. Maquis, tall undershrubs and grassland are at least as effective as forest. Felled or burnt forest, stubble, weeds on fallow land, and moss or lichen crusts give very considerable protection. Tussocky vegetation, which channels rainwater between the tussocks, fails to protect. Unterraced vines and almond trees promote erosion.

We have failed to discover a clear distinction between natural and artificially accelerated erosion. Others may be able to draw this line, but they have not explained how it is done. Merely to show that a phase of erosion occurred during human settlement is not enough: there was erosion before human settlement, and there is today in areas that have not been settled.

We are unable to support the view that erosion leads to desertification. Denuded or eroded land rarely becomes desert.

Fig. 14.26. Steep, unterraced, not obviously eroded vineyards on metamorphic rocks. *Cútar, Málaga, April 1994*

How seriously does erosion threaten cultivation? Wind erosion has certainly rendered modest areas incapable of cultivation, though we have not found anywhere where this is still going on. Cultivators have learnt for centuries to live with sheet and rill erosion and to tolerate it on certain types of land. Gullies sometimes advance into fields at their heads, but at their tails they stabilize and allow new fields to be formed; we have no evidence as to which process is gaining.

Erosion in Mediterranean lands should be seen as a limitation on cultivation, rather than a threat. In general it renders poor land uncultivable, rather than spoiling good land. Where farmers think soil is worth conserving, they have usually already protected it. The most threatened lands are areas of rain-fed cereals on thin soils which would no longer be cultivated, at the present time of over-production, but for subsidy. As a threat to cultivation, erosion is much less serious than depletion of ground-water or salinization.

All this, however, applies to traditional forms of land use. Mankind has invented powerful modern means of destabilizing the landscape. The bulldozer is a more effective destabilizer than plough, fire or goat – both by leaving more persistent raw surfaces, and by creating scarps and taluses which are steeper than the normal angle of a hillside (Figs. 14.5, 14.27). The tractor-drawn mouldboard plough, where it can be used, causes more disturbance than the scratchings of an ox-drawn ard.

Four activities especially promote erosion:
(1)   Digging false terraces out of a hillside, and leaving them to stand or fall unsupported. This may be done in order to grow European Community olives or tropical fruit.[102] In Spain the Institute for the Conservation of Nature does it to plant pines: even if the pines survive and do not get burnt, a site so treated appears to lose more sediment than it would if left as *Cistus*.[103]

(2)   Bulldozing land and leaving it for months or years doing nothing with it.
(3)   Joining up fields by removing cross-slope boundaries, thus increasing the slope length.
(4)   Modern road-making. Up to about 1950 roads were carefully built, taking no more land than strictly necessary, supported on at least the downhill side with walls, and properly provided with drains. These fit well into the landscape and have proved to be stable. Since then, roads have vastly increased in some countries; they are no longer built, but merely dug out of the hillside, leaving the scarp and talus unsupported, and compensating for instability by greater width.[104] Roads not only gully and slump themselves but intercept runoff coming down a hillside and channel it into

Fig. 14.27. False terraces bulldozed into Pleistocene gravelly fill in the Gerolákkos basin inland of Mourniés. Runoff from a track running diagonally up the hillside has washed down several of them. *Gerólakkos, west Crete, October 1996*

places where Nature did not mean it to go. This leads to gullying, especially (as Jennifer Moody points out) below hairpin bends where the effects are multiplied (Fig. 14.28). These evils are most prevalent with perfunctorily constructed minor roads.[105]

Bulldozing is probably most prevalent in Crete, and has overcome even the inherent stability of Cretan lithology; most visible erosion in Crete today is associated with (1), (2) or (4). As yet the bulldozer is not so active in more erodible countries. However, especially in Italy, the fashion for consolidating land-holdings encourages erosion. On the marls of Tuscany, the mosaic of small plots of *coltura mista*, broken up by hedges, baulks and terraces, has often been replaced by big fields where runoff can sweep uninterrupted from top to bottom.

[102] A. W. Drescher, 'Impacts of agricultural innovation and transformation of the mountainous hinterland in the Mediterranean – an example from southern Spain (Costa Granadina), *Pirineos* 145–6, 1995, pp. 13–21.
[103] Williams & others 1995.

[104] In Crete, within the last few years, scores of mountain-tops have been disfigured with radio-telephone masts, every one of which has a perfunctorily built road snaking up to it and destabilizing the hillside.
[105] For other countries, see A. D. Ziegler and

T. W. Giambelluca, 'Simulation of runoff and erosion on mountainous roads in northern Thailand: a first look', Walling & Probst 1997, pp. 21–29; W. Froehlich and D. E. Walling, 'The role of unmetalled roads as a sediment source in the fluvial systems of the Polish Flysch Carpathians', *ibid.*, pp. 159–68.

Fig. 14.28. A gully cuts through arbutus maquis on phyllite below a road with hairpin bends. *Ano Sphinári, Kíssamou, west Crete, April 1989*

Management for erosion control

Most reports on erosion end with prescriptions for controlling it. These usually assume that erosion control is all-important and overrides all other considerations: in effect, 'if I were Emperor of the Pyrenees, I would make the Pyreneans do X, Y and Z.' The example of French foresters in the Pyrenees has been followed again and again: the prescriptions have remained harmlessly on the shelf, or have proved to be unenforceable, or if enforced have not produced the effect desired. Consequences and side-effects need to be considered. Is it really worthwhile abolishing ancient common rights, or cluttering the landscape with generations of rusting wire fences, merely in the hope that controlling goats will prevent erosion?

That said, the following considerations are relevant:

(1) *What rates of erosion should be regarded as threatening?* 5 tons per hectare per year, roughly equivalent to a loss of 1 cm in fifty years, is of little consequence if soils are not too shallow and underlain by subsoil capable of cultivation. But this rate might not be acceptable on shallow soils underlain by hard rock.

(2) *What collateral damage is caused?* The most important side-effect is the filling-up of reservoirs.[106] This is more often a problem of a badly placed reservoir than of an unexpected increase in erosion. Those fool enough to site reservoirs downstream of badlands cannot complain if the reservoirs are short-lived.

(3) *What counter-measures are worthwhile?* At present in southern Europe, management proposals are more likely to recommend changes in land use than engineering works. If there is a high erosion risk, the land should be retired from cultivation and put under a protective vegetation cover. Land planning is likely to be the keynote rather than technical soil conservation.

What this means is determined by local circumstances and the Law of Unintended Consequences. Not browsing an area, in order to prevent goating, may instead create a fire-dominated landscape (Chapter 13). Planting trees does more harm than good if slopes have to be bulldozed in order to get the trees to survive. The best place to spend money on erosion control is almost certainly on proper standards of road-building.

EIGHT WAYS TO ENCOURAGE EROSION

(1) Buy a bulldozer and keep it busy.
(2) Grow vines on sloping land and keep them well ploughed.
(3) Grow rain-fed winter cereals.
(4) Do not build stone walls to your terraces.
(5) Rationalize your land-holding and join up small fields.
(6) Pick stones off your land.
(7) Burn your stubble.
(8) Keep more goats than you have pasture to feed.

[106] See Walling & Probst 1997, section 4.

CHAPTER 15

Badlands

Three musket-shots' [distance] are flat and all the rest of great ridges and mountains and ravines, the wildest that I ever went through.[1]

The clay-hills, that encompass it on every side, are the most extraordinary in nature; they are very high, and washed into broken masses, resembling spires, towers, and misshapen rocks. Whole villages are dug in them, the windows of which appear like pigeon, or rather marten [sand-martin] holes. The passage through is remarkably singular, winding for half a mile between two huge rugged walls of earth, without the least mixture of rock or gravel.[2]

Badlands are surrealist places, cut off from the ordinary world; the despair of cartographers and delight of artists, corners of Europe beyond the power of the European Union. Precipitous chasms dissect the land, hundreds of metres deep, separated by narrow remnants of intervening plateau which thin out into knife-edged ridges and pinnacles. The soft rocks give the impression of intensely active erosion. The high cliffs are textured with pigeon-holes and lace-like crusts, crumbling at a finger's touch. Many badlands are in arid areas and give a first impression of unearthly desolation, although (like other deserts) they are full of inconspicuous, strange, specialized plant-life. Others are well vegetated; their land-forms are enhanced by wonderful colours of rocks, trees, flowers and vines. They are not always as unstable as they look, for some badlands in treeless areas are inhabited: people burrow into the cliffs to create good dwellings and wine-cellars without having to get timber. Wooded badlands and their inhabitants inspired romantic painters from Altichiero (1330–85), whose St George at Padua is martyred down in the gullies, to Fragonard (c. 1775), whose dainty lady swings from the root of a great tree overhanging a soft cliff.[4]

Badlands are not just an extension of ordinary erosion, and badland erosion is not soil erosion. They are a separate and localized phenomenon; once seen, they are unforgettable.

The word comes from the *Mauvaises Terres* of North Dakota (United States), named by French travellers in the eighteenth century and revealed in the writings of Catlin and Audubon.[5] These were, presumably, 'bad' lands in the sense of not good for European-style agriculture. The term was taken by American and British writers to other countries. We have found no general word for badlands except in English. Italian has many words for various badland landforms, but apparently no word for badlands as such.

Areas of badland are strung out along the tectonic plate boundary from south and east Spain and SE Italy to mainland Greece, Gávdhos and Turkey (Fig. 15.1).[3] Majorca and Sardinia lack them, and in Crete they are scarce. The total area is a few thousand square km.

Badlands are very photogenic, and spectacular pictures of them adorn the covers of books on desertification. They are held up as land 'degraded' by human activities – the end of a sequence of deforestation, browsing and cultivation which has 'desertified' a once normal landscape and will intrude further on the ordinary world. Those who hold this theory usually present it as a generalization; they seldom say when desertification occurred and why it struck here but not there; still less do they point to remains of past cultivation on the remnants of old land surface between the gullies. They are not put off by the thought that the original *Mauvaises Terres* had never seen cultivation nor any human activity more destructive than Plains Indians hunting bighorn sheep, and yet were fully developed badlands with land-forms like those of Basilicata today.

Badlands occur in unlikely places, for instance near the Niger delta, amongst the oil-palm forests of Awka and the Udi escarpment in SE Nigeria. In 1951 one of us reported on these as being of ancient, largely natural origin.[6] Europe should take warning from the millions recently spent on trying to control them, using concrete structures, with little if any success.

Badlands, despite their beauty, have had a bad press for a century, but this is not based on European experience. The eighteenth-century scientist Cavanilles realized that they were useful. Of the mountains in the Enclave of Ademuz, in eastern Spain, he wrote:

the waters undermine [their bases] without great resistance, and the . . . overlying . . . mass being without cement, it must collapse and be ruined; the same waters must next sweep away the obstacles which the fallen and weakened masses form, and which in process of time go to fertilize the *huertas* [irrigated lands] of Valencia.[7]

[1] F. Colón, *Descripcion y Cosmografia de España*, c. 1518, §6509.
[2] H. Swinburne, *Travels through Spain in the Years 1775 and 1776*, Elmsly, London, 1779, p. 81. Both this and the first quotation are accounts of the badlands of Guadix in Granada, Spain.
[3] Julius Caesar has a possible mention of a badland (*convallem et exesum locum*, 'a raviney and eaten-out place') somewhere in south Spain, where the defeated Pompey had holed up; *Spanish War* 39.
[4] H. Fragonard, 'L'Altalena', in the National Gallery of Art, Washington, D.C.
[5] G. Catlin, *Letters and Notes on the . . . North American Indians*, London, 1844; M. R. Audubon, *Audubon and his Journals*, Nimmo, London, 1898 [Missouri River journals, vol. 1, p. 447–vol. 2, p. 198].
[6] A. T. Grove, 'Land use and soil conservation in parts of Onitsha and Owerri Provinces', *Geological Survey of Nigeria Bulletin* 21, 1951.
[7] A. J. Cavanilles, *Observaciones sobre la historia natural, geografia, agricultura, poblacion y frutos del Reyno de Valencia*, Imprenta Real, Madrid, 1795, vol. II, pp. 70 ff.

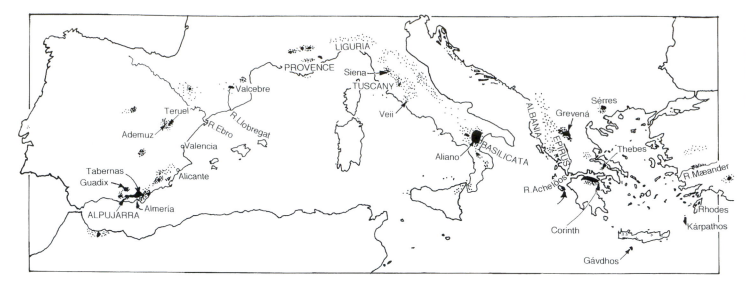

Fig. 15.1. Badlands.

This has happened elsewhere. In Albania badlands have generated the deltas that fringe two-thirds of the coast and provide most of the best farmland in this mountainous, forested country.[8]

American badlands and 'misuse' of the land

Objections to badlands derive largely from American experience in the last century. Lyell the geologist, visiting the Piedmont of Georgia and Alabama in 1846, found that hundreds of gullies had formed in lands converted from forest to tillage twenty years before. He attributed them to the sun's heat opening cracks in the clay which were enlarged by rainwater.[9] (By 1983 the land was again forest and the gullies inactive.[10]) Later in the century gullies developed in Arizona and adjacent states and in south California: pasture-land turned into ravines ('arroyos') within ten years or so. Contemporaries had little hesitation in ascribing this to overgrazing. With these examples before them, and forgetting counter-examples such as the *Mauvaises Terres*, it was easy for scientists to suppose that the badlands of Europe had a similar history, being formed quite suddenly after human misuse of the land.

American badlands have various causes. Many, like the *Mauvaises Terres,* the Dinosaur Badlands and Big Muddy Badlands of Alberta, and those of Utah,[11] have a purely natural origin. Those of California are attributable to the impact of European cattle and methods of cultivation on an ecosystem not adapted to them. The Arizona arroyos were created by a series of unusual deluges, combined with the making of ditches, roads and bridges which channelled the runoff into harmful directions.[12] The Georgia gullies are in a peculiarly unstable micaceous clay formed from rotten granite and gneiss. Here, too, it may be significant that a road appears prominently in Lyell's account. The *Mauvaises*

Terres are the most comparable with southern Europe: in secondary deposits on the flanks of a rapidly upheaving mountain range, far enough north for severe changes of climate and vegetation to have occurred during the Pleistocene.

HOW BADLANDS WORK

Badlands occur where weak, usually Tertiary and Quaternary, deposits lie on or near mountain ranges in a state of active uplift, as in the Rocky Mountains or Basilicata. They avoid relatively stable areas like Sardinia or Australia. They seldom occur in glaciated terrain; presumably ice action and periglacial processes brought down all the unstable sediment in the Quaternary.[13] Badlands occur in the wettest and driest parts of the European Mediterranean, and in all vegetation from forest to desert (Figs. 15.2–15.5).

Badlands are one of the means by which erosion catches up with tectonic uplift. A badland, in principle, begins with a raw surface created by a fault movement, the side-cutting of a river (Fig. 15.3), gully erosion, slumping, the uprooting of a tree, or somebody digging. Usually this scar, left to itself, heals and subsides back to the previous slope of the hill. In badlands material goes on detaching itself, and the debris is washed away instead of piling up at the base of the scar to stabilize it.

Mediterranean badland sediments are of many materials and textures, but commonly contain Pliocene marine blue clays. These clays often have a high sodium content, and readily disperse when exposed at the surface. They may seal the surface against infiltration, so that even light rainfall runs off and attacks the bottoms of the gullies.[14] Gullying in clayey badlands often goes with *piping*, the enlargement of cracks under the surface to form tunnels, which collapse to form

8 Sestini 1992.
9 C. Lyell, *Principles of Geology*, 12th edn, Murray, London, 1875, vol. 1, pp. 338 ff.
10 Oliver Rackham's observation.
11 A. D. Howard, 'Badland morphology and evolution: interpretation using a simulation

model', *ESPL* 22, 1997, pp. 211–27.
12 R. U. Cooke and R. W. Reeves, *Arroyos and environmental change in the American South-West*, Clarendon, Oxford, 1976.
13 The Georgia gullies are an apparent exception, being in a tectonically stable area.

14 Bisci & others 1992; M. D. Soriano, A. Colica, D. Torri, 'Estudio preliminar de la influencia de la estructura y propriedades de los materiales en la evolución de badlands', López Bermúdez & others 1992, pp. 184–91.

Fig. 15.2. Complex badland in south Spain. Note the fields and invading pines. *Alhama de Murcia, SE Spain, April 1994*

Fig. 15.3. Active badland cliffs (despite forest) along a river. *Zákas, Grevená, May 1988*

gullies.[15] Even a thin band of blue clay may create a badland when pipes weaken it, it collapses, and more stable overlying materials slump on top of it.

Badland-forming materials break up when dried or frozen or split by root growth. Summer drought weakens the surface layers, creating shrinkage cracks across and parallel to the surface and detaching polygonal platelets; this produces the friable 'puffy' or 'popcorn' surface familiar in silty or clayey badlands. In silt marls containing salt or gypsum, and montmorillonite clays that swell and shrink, disintegration extends deeper.

Badland material is often shattered by tectonism. Cracks, joints and minor faults form pipes (Fig. 15.7).[16] This can happen, even on nearly horizontal surfaces on ex-arable land and abandoned terraces that were formerly irrigated. Pipes enlarge into tunnels, then into gullies, whose steep sides slump.

Frost, too, is a disruptive agent. Needle-ice dislodges superficial layers of mud-rock. In a cold spell, repeated freezing at night and thawing by day, most effective on bare SW-facing slopes, can shatter the surface to a depth of 5–15 cm.[17] Pipes develop at the base of the disintegrated layer; water penetrates cracks to reach swelling clays in the underlying rock. The next deluge then washes away an entire top layer as

[15] A. Harvey, 'The role of piping in the development of badlands and gully systems in south-east Spain', Bryan & Yair 1982, pp. 317–35.
[16] A. C. Imeson, F. J. P. M. Kwaad, J. M. Verstraten, 'The relationships of soil physical and chemical properties to the development of badlands in Morocco', Bryan & Yair 1982, pp. 47–70; Bull & Kirkby 1997.
[17] Solé & others 1992; D. Torri, A. Colica, D. Rockwell, 'Preliminary study of the erosion mechanisms in a biancana badland (Tuscany, Italy)', *Catena* 23, 1994, pp. 281–94.

Fig. 15.4. *Left* Active though well-vegetated badland. *Samitzar near Ainsa, upper Ebro, Spain, October 1995*

Fig. 15.5. *Below left* Badland in pine forest. *Near Nice, France, Oct 1995*

Fig. 15.6. *Above* End of the gullying process: badland becoming stable and vegetation developing. *Alhama de Murcia, SE Spain, April 1994*

a slurry, delivering it to a stream. This is the likely origin of some valley fill deposits, delta sediments and alluvial spreads, such as the 6 metres that buried the land of the Sybarites on the Basilicata coast.[18]

A common site for badlands is where a meandering river side-cuts the base of an adjacent hillside. Small tributaries coming in at this point are over-steepened and generate badland gullies. The River Maeander itself – the Büyük Menderes in west Turkey – provides examples. The magnificent flysch badlands of SW Turkey are conventionally attributed to deforestation: one of them is called Orman Erosyon, 'Erosion Forest'. Yet a photograph demonstrates that these are long-standing features, often no longer active.[19]

Erosion rates are terrifyingly variable. Spanish investigators stuck 472 metal pins into an Ebro badland. Tertiary clay was found to erode around the pins at 15–20 mm per annum; Holocene sediments, forming another part of the badland,

receded at one-third of this rate. Erosion rates were closely dependent on wet and dry seasons, and there was enormous variation in relation to individual rills and gullies. Rates more than ten times as great have been measured in the Val d'Orcia tributary of the Ombrone in Tuscany.[20] As we shall see, most badlands as a whole appear to erode much less rapidly.

HISTORY AND NATURAL HISTORY OF BADLANDS

Spain

Alpujarra
Badlands are part of the extraordinary Alpujarra landscape. In the west the gullies are in unstable blue and red metamorphic bedrock below the inhabited zone (p. 100). Further east they continue into the yellow Pliocene silts and blue clays that are part of the Tabernas system of badlands.

The big village of Huécija, in the midst of the eastern badlands (Fig. 5.24), is surrounded by immense gullies eroded into a yellow silt some 200 m thick. Cones and mesas between the gullies are capped by remains of a conglomerate bed. Cultivation (now principally table-grapes) is partly on terraces, partly on slopes and partly in the valley floor. Like all the Alpujarra, Huécija has an ingenious system of irrigation canals.

Badlands appear in the Alpujarra's ferocious history. They are not continually changing. As the historian Cabrillana puts it, 'today, as in the sixteenth century, the green of the irrigated fields contrasts with the dazzling white of the "bad lands"'. Huécija was a centre of Muslim revolt in 1569: the Christians barricaded themselves in the church and Augustinian friary, were besieged, burnt out and cut in pieces.[21] The village still

[18] O. H. Bullitt, *Search for Sybaris*, Dent, London, 1971.
[19] H. van Zeist, H. Woldring, D. Stapert, 'Late Quaternary vegetation and climate of south-western Turkey', *Palaeohistoria* 17, 1975, pp. 53–143.
[20] G. Benito, M. Gutiérrez, C. Sancho, 'Erosion rates in badland areas of the central Ebro basin (north-east Spain)', *Catena* 19, 1992, pp. 269–86; Calzolari & Ungaro 1998.
[21] Cabrillana Ciézar 1977, p. 446; Carvajal 1600, V.ix.

clusters round the massive church (with a fire-reddened, blocked doorway) and friary. Some have thought that erosion was set off by neglect of watercourses after the suppression of the revolt,[22] but since Huécija is itself in a gully this cannot be so to any great extent. One of the mesas, Cerro Marchena, was probably chosen for the site of the castle of Marchena mentioned by Al-Edrīsi in the twelfth century. We estimate that Huécija has lost about 4 cubic km of material by erosion; by far the greatest part of this had gone before the Augustinians built their church. A 1930s photograph (cf Fig. 5.25) shows gullies and cultivation terraces exactly as they are now, although there has been at least one deluge in the interval.[23]

After 1569, avenging Christian armies burst into the Alpujarra; deeds of blood and valour were done in the badlands that blocked every road. At Tablate:

> The Moors had demolished the bridge . . . they had left only a few old timbers, which should have formed the centring of the arch, on one side, and on top of these a bit of wall so narrow that it could hardly be crossed by a single man . . . it was undercut and overhung in such a way that if loaded with more than one person it would have come down; and the ravine was so deep that in looking down it turned one's head and disappeared out of sight. . . . the musketeers on both sides opened fire . . . a blessed Friar of the Order of the seraphic Father St Francis, called Brother Cristóbal de Molina, with a crucifix in the left hand and a naked sword in the right, his habit tucked up to his belt . . . calling on the powerful Name of Jesus, came to the dangerous pass . . . and going on, supported in places by the tops of the timbers . . . and in places by the stones and mud-bricks . . . passed over to the side of the enemy, who were looking expectantly for when he would fall. . . . two brave soldiers followed him, but unsuccessfully, for the earth and timber failed and they were . . . dashed in pieces.[24]

The repaired bridge stands yet, dizzily high and narrow, over a tremendous (but vegetated) badland ravine at the west entrance to the Alpujarra. The ravine has apparently not altered since the bridge was built.

Erosion rates are said to range between 0.15 and 0.8 mm per year.[25] As with all badlands, much of the erosion comes from small parts of the area in short periods of time, and cannot be measured with confidence.

The Guadix badlands

Across the Sierra Nevada are the badlands of Guadix, famous for cave-houses dug into the cliffs. Marls, sands and conglomerates accumulated several hundred metres thick in a basin north of the Sierra, and then became unstable when, in the late Tertiary, the Guadiana Menor river breached the ridge that had contained them. Fault movements contribute to the instability.[26] In a semi-arid climate with occasional intense deluges this has resulted in 'some of the most intensive

Fig. 15.7. Row of pipes about to develop into a gully. *Near Sórbas, Almería, Spain, November 1994*

erosional topography in Spain, and, indeed, the world'.[27]

Measurements of erosion in the Guadix area are only about 0.01 mm per year, which Gilman & Thornes regard as incredibly low. However, archaeology confirms that the badlands are not so very unstable. There are many Copper and Bronze Age sites with material, including structures and tombs, that is certainly *in situ*. These are not confined to residual land surfaces on plateaux that gullying has not yet reached. They are located on ridges, hilltops and even on some of the badland slopes. According to the surveyors, they were placed in definite positions in relation to the topography, and are not merely chance survivals on fragments of a once-continuous land surface. The inference is that, notwithstanding the very unstable appearance of the badlands, the present main gullies and ridges already existed 5,000–3,500 years ago and were chosen then as places of burial and settlement. Further east, in the Antas basin, similar evidence goes back into the Neolithic. Gilman & Thornes conclude:

[22] The population of Huécija is given as 100 *vecinos* [households?] in 1568 and 59 in 1586–7; Ortiz & Vincent 1978.

[23] Brenan 1957, opposite p. 167.

[24] Carvajal 1600, V.ix.

[25] J. B. Thornes, *Semi-arid Erosional Systems*, London School of Economics, Geographical Paper 7, 1987; McNeill 1992, p. 320, quoting data

by J. Quirantes Puertas. A much higher figure, 2.4 mm per year, has been claimed for the whole basin of Trevélez, based not on measurement but on the Fournier formula, a kind of European Universal Soil Loss Equation; E. Martin-Vivaldi and Y. Jimenez Olivencia, 'Estudio de la erosion en la cuenca del Río Trevélez (Granada)', López Bermúdez & others 1992, pp. 93–103.

[26] F. J. Martínez Garzon, C. Sierra Ruiz de la Fuente, A. Roca Roca, 'Procesos geomorfoedáficos en la cuenca del Río Huéneja (Guadix, Granada)', López Bermúdez & others, 1992, pp. 251–60.

[27] This account is derived from Gilman & Thornes 1985, chapter 4.

Fig. 15.8. Desierto de Tabernas, SE Spain. *November 1994*

The sparse vegetation cover resulting from a harsh climate and from overgrazing reveals clearly every geomorphic detail and thereby creates a misleading impression of catastrophic erosion. The apparent contrast between the 'hard' Palaeozoic and Mesozoic rocks and the Tertiary deposits reinforces this impression although closer inspection reveals the marls to be well cemented and capable of maintaining nearly vertical slopes. . . .

The argument that widespread erosion was a result of forest clearance and similar early agricultural activities is incompatible with the archaeological evidence . . . more or less the present pattern of slopes, gullies and channels was in existence [in 3000 BC] . . . erosion rates during the last 4000 years in much of the landscape have not been very high.[28]

However, much badland erosion presumably happens in exceptional storms. In 1973 there occurred what was claimed to be a 500-year deluge; the mica-schists, in which some Alpujarrine badlands are developed, were estimated to have lost on average 151 mm of ground (the range of variation being 56 to 420 mm) in this one event.[29]

Desierto de Tabernas

This, perhaps the most famous badland in Spain,[30] is an extension of the eastern Alpujarra in Pliocene clayey silts with some gypsum (Fig. 15.8). Harder beds in the deposits sometimes act as caprocks, but these break up when exposed and are not effective gully-stoppers. As in most badlands, the surface is friable and does not develop a resistant crust.

Vegetation is a sparse version of the local arid phrygana, with esparto grasses and specialized desert plants such as *Launaea lanifera*. The Desierto seems to have been very little used in historic times; it forms part of the vast territory of the town of Tabernas.

These, like other major badlands, are related to tectonic action and river capture. They are difficult to date, but appear to go back through the Quaternary, showing cycles of activity related to the glacial cycles. The landforms are determined by factors such as gypsum content, tendency to piping and the formation of lichen crusts and other vegetation.[31]

[28] Wise & others 1982.
[29] J. R. de la Torre, quoted by Gilman & Thornes 1985, p. 61.
[30] It is used as the setting for Wild West films, which are less exciting than the real history of this part of Spain.
[31] A. M. Harvey, 'Patterns of Quaternary aggradational and dissectional landform development in the Almeria region, southeast Spain: a dry-region, tectonically active landscape', *Die Erde* 118, 1987, pp. 193–215.

South Aragon

Badlands around Teruel attracted early scientific attention. The old city stands on a low, gulley-eaten bluff above the river. W. Bowles described it in c. 1753:

> The surroundings of tervel do not cease to be sand. . . . the said city is on one of these [mountains], which every day continues to disintegrate, and will eventually cause the ruin of it. this defective situation is the only one which i have seen in spain, and proves the poor judgement of the founders. i have also observed that at the rate at which the higher mountains destroy themselves, the middle and lower ones decompose and dissolve into their original soft stone, and thence into earth; and in this way the cliffs sink in some places, raising the earth, leaving sufficient inclination for the springs to be able to have a gradient, and delay the complete equalization and levelling of this terrain. one could apply the same logic to the threatened general levelling of our globe.[32]

Despite Bowles's strictures, Teruel goes back at least to Roman times. Antonio Ponz, a few years later, said:

> The height on which Teruel is founded . . . is hills of white clay, and of a colour tending to red, with a thick cap of hardened gravel, which easily disintegrate by reason of rains.[33]

Bowles understood how badlands function:

> The earth, which i have said is transported from the banks, enters the [River] guadalviar, and goes to discharge itself in the sea of valencia . . .
>
> Albarracin . . . is founded on two limestone rocks, cleft in every part and on every side; so that there is seen to be not more than two inches of solid . . . in which the horizontal clefts begin the destruction, which then the perpendicular ones follow . . . . subdivided into an infinity of other cracks in various directions. this causes a complete daily destruction of the rocks, pieces of which fall every day. this is the natural consequence of the way in which they are cracked, and continue to crack until they fall, disintegrate, and are reduced to cultivable land.
>
> Near these two rocks is another, whose base and top are bedded horizontally and solidly, but the middle is all split obliquely; so that pieces threaten to spall off and fall. albarracin is one of the highest places in spain [1,200 m].
>
> Notice that the words which i use, *clefts* and *cracks*, do not express exactly what i mean. 'cleave' or 'crack' are used, for example, of badly made bricks or floor-tiles which split in the heat of the kiln; of green wood which shrinks, and of the splits which appear in clayey ground in the heat of the sun. all these clefts and cracks arise from the evaporation of water and from the shrinkage of the material. the divisions and gaps in the rocks are not splits or cracks in this

sense; since they come from the decomposition of a part of the mass and the disintegration of its substance, caused by the internal working and movement of the stone, hastened only by cold and heat and water from rain or the river. this is why in this gully gaps are seen from one line to two feet [2 mm to 30 cm] wide; according to the progress of the decomposition and the state which it has reached, the crack is bigger or smaller. the gully itself can now be considered as a big gap. when all the surrounding mountains have thus been decomposed there will remain a great plain of clayey and sandy ground; if by chance there remains in the middle of this plain a great mass of rock 200 or 300 feet high, there will then be many learned articles to explain this phenomenon.

Southern Pyrenees

Badlands are widespread in the Ebro and Llobregat basins. Gullies and mass movement are very active in the Valcebre basin, about 25 km SW of Valcebollère. The rocks are Late Cretaceous mudstones containing susceptible clays. Any disturbance to the soil exposes bedrock which slakes and gullies. Because of the fear of badlands expanding into arable land or pasture, a research project in land conservation has recently been established.[34]

South of France

Jurassic marls, the *Terres Noires* of the Provençal Alps, which underlie about 1000 sq. km, are very prone to gullying. Reports of badland activity in the eighteenth century stimulated the development of the Ruined Landscape idea (p. 81).

In the Büech basin near Sisteron (mean annual rainfall 840 mm) erosion begins in areas devoid of vegetation, extending into vegetated areas in heavy rains. Measurements from silt-traps on three gullies yielded sediment at 190 tons per ha per year – an average lowering of the surface by some 8 mm a year. Gully-slopes, it was estimated, provide 90 per cent of the sediment coming from such landscapes, even though gullies cover less than 10 per cent of the catchment area. The measurements lasted for three years, and were lucky to include one late-season deluge (294 mm), which yielded much of the sediment, and two minor deluges.[35]

Wainwright's study of the Vaison-la-Romaine deluge (p. 250) included marl badlands, generally forested or well vegetated. Badlands shed water earlier during the course of the storm than vineyards or orchards. There were frequent slumps, especially under pine forest. Badlands, however, were not particularly productive of either runoff or sediment in this particular event, although they had been so in past deluges.

[32] bowles 1783, vol. 1, pp. 176 ff.
[33] Ponz 1778, XIII.v.2.
[34] Solé & others 1992; N. Clotet and F. Gallart, 'Sediment yield in a mountainous basin under high Mediterranean climate', *ZG* NF Suppl. 60, 1986, pp. 205–16; P. Llorens and F. Gallart, 'Small basin response in a Mediterranean mountainous abandoned farming area; research design and preliminary results', *Catena* 19, 1992, pp. 309–20; J. C. Balasch, N. Castelltort, P. Llorens, F. Gallart, 'Hydrological and sediment dynamics network design in a Mediterranean mountainous area subject to gully erosion', *Erosion and Transport Monitoring Programmes in River Basins*, ed. J. Bogen and others, IAHS, Wallingford, 1992, pp. 433–42.
[35] M. Bufalo and D. Nahon, 'Erosional processes of Mediterranean badlands', *Geoderma* 52, 1992, pp. 133–47.

Fig. 15.9. Badland forms at Aliano, Basilicata.

(a) *Left Calanchi,* viewed from the middle of the town. *May 1993*

(c) *Right Burroni:* one side of each form is cultivated. *November 1993*

(b) *Biancane:* there is a bush on every mound. *March 1993*

Italy

Badlands derive from the vast quantities of unstable flysch and volcanic rocks in the Apennines, rapidly being uplifted by tectonic action. The Roman land-surveying term *supercilium*, 'eyebrow', would be useful for the fringe of vegetation between the edge of a field and an adjacent gully (Fig. 15.16). A plan of Albisola Superiore (Liguria) in 1719 shows badlands with pine-trees in them, apparently eating into vineyard terraces.[36]

Tuscany and Etruria

Etruscan Veii and medieval Siena both sit on remnant plateaux, eaten into by ancient gullies. The city walls follow the irregular edges of the gullies, which have advanced little since the walls were built; about two-fifths of the wall of Veii survives. Sutri and Nepi, ancient Etruscan towns next to the Ciminian Wood (p. 172), are perched on bluffs at junctions of gullies.

North of Lake Bolsena, badlands are developed in weathered volcanic material overlying flysch and marine Pliocene sediments. The fallout of caesium-137 from atomic bombs and from the Chernobyl nuclear accident of 1986 has been used as a marker for estimating rates of erosion. The results ranged from 1.5 to 8.4 mm per year between May 1986 and February 1991,[37] values 10–100 times the average rate for Italy. Measurements elsewhere, using pins hammered into the surface, have also shown that slopes within badland gullies

[36] Original in Archivio Vescovile, Savona. Oliver Rackham is indebted to the Rev. Sandro Lagomarsini for this allusion.

[37] M. Branca and M. Voltaggio, 'Erosion rate in badlands of central Italy: estimation by radio-caesium isotope ratio from Chernobyl nuclear accident', *Applied Geochemistry* 8, 1993, pp. 437–45.

are lowered at rates of up to 10 mm per year. Slopes at angles of 17–30° can thus retreat at about 2 cm per year, or (if this rate were sustained) 20 m per thousand years.

Basilicata

Among the most magnificent badlands in Europe are those on the Apennine foothills overlooking the Gulf of Táranto. People live precariously in hilltop towns perched among tremendous gullies. The town of Aliano, famous through the writings of Carlo Levi, who was banished there in 1935 (p. 79), is almost cut in two by abysses, from whose depths rise needle-pointed spires of clay, surrounded by a chaos of domes, *calanchi* (ravines), *biancane* (cones of white clay, each crowned with a bush) and *burroni* (forms with a cliff one side and a vegetated slope the other) (Fig. 15.9).

The Basilicata badlands, the acme of Italian instability, lie close to a thrust-front where African and European plates collide. Late Pliocene and Pleistocene erosion deposited several hundred metres of marly and silty clays, sandstones, weak conglomerates and cobbles, seldom interrupted by hard rock. These sediments have been uplifted at a mean rate of about 0.7 mm per year. The Lower Pleistocene surface now stands at a height of about 1,000 m in the west, declining to 350 m towards Apulia (Fig. 15.10). The sediments have become unstable through the downcutting of the Sinni, Agri, Cavone, Basento and Brádano rivers, which now run hundreds of metres below the original land surface. Gullies have eaten back headwards from these rivers and their tributaries.

The badlands vary in texture from cobbles to clay. Ravines around the towns are often in silt or fine sand; clay occurs at lower altitudes, and coarser sediments further inland. The altitude (400–600 m) makes frost a factor, especially during periods of colder climate. Aspect is critical. Looking north-east, one has a vision of clay cliffs; south-westward, the view changes to woods, orchards and fields. South- and SW-facing slopes dry out and have fierce temperature cycles, whereas those facing north and NE are cooler and more vegetated.[38]

Fig. 15.10. Marine terraces by the Gulf of Táranto, the highest and oldest about 730,000 years old, the lowest and youngest about 100,000 years old. They have been uplifted more strongly on the western, Apennine side than on the Apulian side. Badlands are most developed in the Aliano–Pisticci area. (After a drawing by F. Todesco.)
[F. Massari and G. Parea, 'Progradational gravel beach sequences in the hinterland of the Gulf of Taranto', *International Workshop on Fan Deltas 1988, Excursion Guidebook*, ed. A. Colella, Calabria, 1988, p. 104]

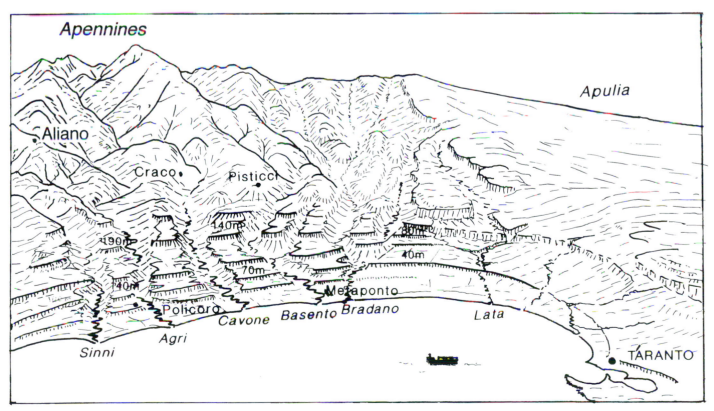

Active badland

Forest 'remaining' in 1950

Forest destroyed between c. 1820 and 1950

Ancient Greek field system

Major Ancient Greek colony

Hill-town with Roman or Greek remains

Hill-town in *-ano*

Medieval hill-town

Fig. 15.11. Part of Basilicata showing:
(1) badlands in 1987–8 (after Del Prete and others);
(2) forest remaining in 1960, and (3) forest disappearing between c. 1830 and 1950 (after Tichy);
(4) ancient hilltop towns;
(5) major Ancient Greek settlements, and regular field systems associated with them.

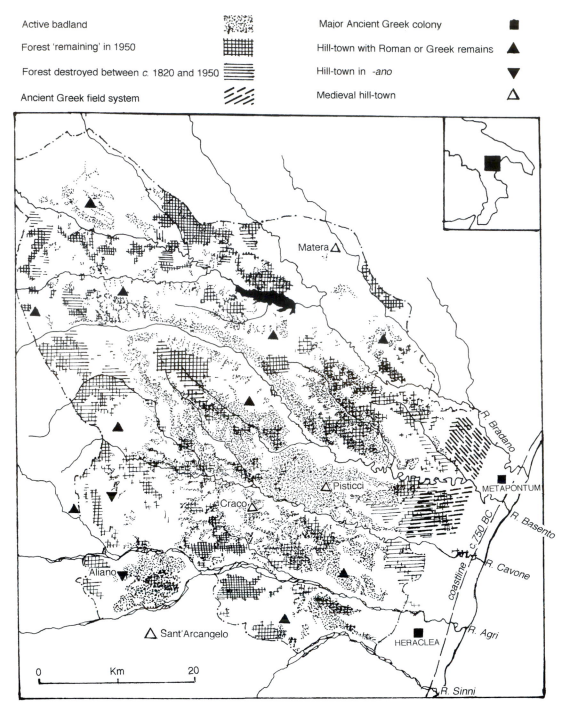

	Non-forest	Forest	Ex-forest	Total
Non-badland	1590 (1583)	236 (270)	113 (85)	1939
Badland	526 (532)	125 (91)	1 (29)	651
Total	2115	361	114	2590

Table 15.i. Area of badlands in part of Basilicata (Fig. 15.3), distributed among forest, non-forest and ex-forest. Areas in sq. km. Figures in brackets are the areas to be expected if there were no association between badland and vegetation.

Calanchi are activated by shallow landslips, but piping is more important in forming *biancane*.[39]

How active are the badlands? Fields are fitted in among the *burroni* and appear to be reasonably stable. Most trees are not empedestalled, although some olives on steep sandy slopes below Aliano are reminiscent of stilt-rooted mangroves. Erosion should be faster around a town: roofs and paving generate extra runoff, which is channelled into paths and tracks leading down to springs and out of the town. In Aliano a house or two is said to fall over the cliff every year; the church joined them after three days of rain in c. 1932.[40] But

[38] M. J. Kirkby, K. Atkinson and J. Lockwood, 'Aspect, vegetation cover and erosion on semi-arid hillslopes', Thornes 1990, pp. 20–40.
[39] Alexander 1982.

the towns are not impermanent; even cellars dug into the soft cliffs are stable if not too near the surface.

Writers tend to attribute the badlands to deforestation, especially in the last century. This lacks conviction: settlement is much older than this. Where *biancane* are separated from the scarps behind by some metres, as they usually are, rates of slope retreat show that gullying must have begun at least several centuries ago. Measurements in Tuscany date individual *biancane*, with a well-developed soil preserved on top, at 300–400 years old.[41]

The theory that badlands were caused by deforestation in the last century can be tested, since Franz Tichy recorded exactly where deforestation occurred. Woods were grubbed out in other parts of Basilicata, but in the 'classic' badland belt through Pisticci and Aliano the woods were located on ridgetops, and most are still there.[42] Higher in the badland zone, for example the middle Agri valley between Grumento and Sant' Arcangelo, woods have increased – partly on badland terrain – in the last hundred years.

Fig. 15.11 superimposes information about badlands[43] and deforestation in the greater part of the Pisticci–Aliano belt. Of this region of 2,600 sq. km, one-quarter is badland. In 1950 (before the main period of recent expansion of forest) 14 per cent was mapped as forest. The area of forest had been about one-third more extensive in the early nineteenth century. There is distinctly *more* badland in forest than would be expected from a random association,[44] but there is almost *no* badland in ex-forest. This disproves the theory that deforestation leads to immediate badland formation. The most reasonable explanation is that badlands occurred randomly between forest and non-forest, but that people seeking new fields avoided grubbing out forests on land already gullied or that was liable to gully in future.

The survival of towns sets an upper limit to the rate of change of landforms. Towns, as we have seen, should be among the first parts of the landscape to be devoured by badland activity. However, the hill-towns are all at least of medieval date, and many have major Roman or Hellenistic antiquities.[45] Only about one town disappears per century: sometimes the collapse was the effect of works intended to buttress the town (Figs. 15.12, 15.13). This indicates that the gross pattern of gullying, although active, is well over two thousand years old. What were hilltops and valleys in Roman times are hilltops and valleys today.

Carlo Levi gave a very depressing picture of Aliano: a political offender would not think kindly of his place of exile. However, there was truth in it. Overcrowding, malaria, flies, the Great Depression, alternate oppression and neglect by a vain and remote government, and the dead hand of bureaucracy had reduced Basilicata to a Third World country in the 1930s. People were even forbidden to emigrate. Such was the lot of many remote parts of southern Europe at that evil time; it had nothing to do with badlands.

When we visited Aliano in 1993 it was very different. Poverty, disease and hopelessness have vanished. The old people, who now preponderate, live sociably, contentedly and hospitably. Their plots of vines and olive trees provide them with wine, fruit and vegetables to supplement their pensions and welfare payments, and give them pleasure and purpose in life. They are much better off than previous generations; Aliano compares favourably with our own homes.

Aliano is not only a place of retirement. Some emigrants have returned with their families, investing foreign earnings in houses, land on the plain, and irrigation equipment. But as with all relatively remote small towns, there is not much employment, and young people seek their fortunes, as their parents did, in cities or abroad.

Aliano's future depends in part on the progress of erosion. The authorities encourage hill-town dwellers to move down to new towns in the valley, but these are hot, unsociable and unpopular, and retain their old unhealthy reputation. The census population has declined by only 30 per cent from its peak in 1951. Aliano is likely to survive amongst the *calanchi*, and even benefit from its beauty and notoriety.

Badlands descend towards the coast in successively younger sediments. In their lowest zone, between 5 and 10 km inland and at 20–100 m elevation, gullies have definitely developed in the historic period. In this zone there are two regular field systems on the land surfaces between the rivers (Fig. 15.11). Straight, parallel ditches, 210 to 240 m apart, form alignments whose remains (interrupted by millennia of later land divisions) cover 48 and 64 sq. km in what is now mainly arable land. They are attributed to country planners among the Greek colonists who founded the city of Metapontum in c. 700 BC.[46] The alignments are fragmented by branching systems of gullies. Although doctrinaire country planners are not necessarily diverted from their geometry by mere practicalities of the terrain, it is hard to believe that the land could have been dissected when the field systems were laid out. Indeed, some gullies follow the alignments and have developed out of ditches. The gullies, to judge from air photographs of c. 1950, are now vegetated and inactive; a few of them have been ploughed out. At some time or times in the past 2,700 years, therefore, climate and land-use have combined to produce a period of intense gully erosion (*not* soil erosion, as the survival of the ditches proves).

The *Tabula Heraclea* leases, of about the 4th century BC (pp. 172–3), are probably the first written record of badland activity. The surveyors had to move one of the boundary markers 'lest it be gravelled under like the previous markers'. The text mentions one of the characteristic islands in the River Agri, where the lessees were forbidden to dig ditches – was this to prevent gullies from being initiated?

40 Levi 1947, p. 49.
41 Calzolari & Ungaro 1988.
42 F. Tichy, *Die Wälder der Basilicata und die Entwaldung im 19. Jahrhundert*, Heidelberger geographische Arbeiten 8, Heidelberg, 1962. [The author repeats the common Italian belief that both gullies and slumps result from destruc-

tion of forests, but points out (p. 124) that most of them are in areas that have not had forest in recent times.]
43 M. Del Prete, M. Bentivegna, V. Summa, *Carta dei Calanchi del Versante Ionico Lucano in Provincia di Matera*, Facoltà di Agraria, Università di Basilicata, 1989.
44 Despite the consideration that badland in

forest is easily overlooked and may be under-recorded.
45 Places ending in -*ano* are ascribed to the early Roman Empire; Racioppi (quoted by Tichy).
46 G. Schmiedt and R. Chevallier, *Caulonia e Metaponto. Applicazioni della fotografia aerea in ricerche di topografia antica nella Magna Graecia*, Istituto Geografico Militare, Florence, 1959.

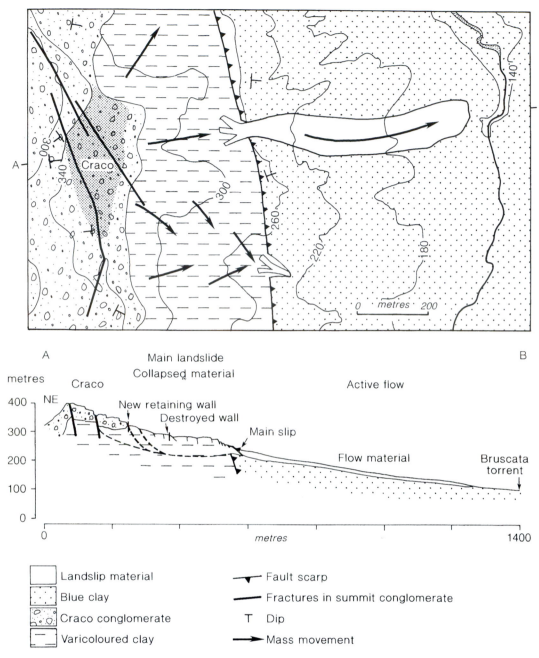

Fig. 15.12. Structures underlying the collapsed town of Craco, Basilicata.

Legend:

- Landslip material
- Blue clay
- Craco conglomerate
- Varicoloured clay

- Fault scarp
- Fractures in summit conglomerate
- T Dip
- Mass movement

Fig. 15.13. Ruins of Craco, abandoned when part of it collapsed. *November 1993*

Greece

Píndos Mountains

Badlands form in clayey marls, weak sandstones, poorly cemented conglomerates and flysch (which often takes the form of clayey shale). Much of the material is derived from ophiolite. A thick mantle of these sediments, dated to the Miocene, covers the mountains in western Grevená province and adjacent parts of Epirus. It continues into Albania, where the rivers have some of the greatest sediment loads in Europe (p. 253).

The landscape is of rolling, vastly wooded hills, with occasional big villages surrounded by wide arable fields. The gullies are predominantly in forest, mountain maquis or savanna, at 500–1,300 m altitude, in a frosty, snowy climate with a short growing season (Fig. 15.14).

Fig. 15.14. Badland in deciduous oakwoods. *Myrsine, Grevená, September 1989*

Typical Grevená badlands are in massive, soft sandstone with occasional harder beds. Scarp slopes tend to be wooded with deciduous oak at lower altitudes and pine at higher; the dip slopes are less vegetated with more cultivation, except where hard beds follow the surface.

We owe the following interpretation to J. A. Moody. Gullying seems to be cyclical. A gully often starts from a small slump or on the outside of a bend in a river (Fig. 15.15). If this is on a scarp slope, the gully extends upslope until it reaches either the ridgetop or a hard bed which acts as a gully-stopper. It then branches and side-cuts until it gradually attenuates into sheet-erosion and then stabilizes. Vegetation makes little difference. Oaks have a shallow root-mat which is undercut, so that the trees topple in, whereas pines are often empedestalled. On dip-slopes the gullies are shallower and less active, but may end by exposing a sheet of hard bedrock.

The gullies look menacing, but arable fields among them are relatively stable (Fig. 15.16). Fields often develop lynchets at their lower edges, but are demarcated by hedges and strips of maquis which act as barriers to further movement. Old gullies become vegetated, and even oaks three hundred years old show no sign of further erosion. The inhabitants have not noticed changes.

Fig. 15.15. Gullies starting from slumps in a steep oakwood. *Pertoúli, Epirus, May 1988*

Fig. 15.16. Mosaic of fields and badlands, with trees at their junctions. *Syndhendhron, Grevená, August 1987*

This is a similar story to Basilicata, except that the sediments here are older and there are no *biancane*. Ophiolite may be a predisposing factor. The gullying process is far older than the lives of men or trees; the entire landscape seems to be formed of gullies superimposed on younger gullies.

A neighbouring badland in Epirus was investigated thirty years ago. The gullies, developed in Older Fill, were very active, to judge by their relation to oak-trees and a Roman aqueduct. They were interpreted as 'a general, climatically induced erosional phase', locally intensified in areas particularly accessible to woodcutters and goats.[47]

North Peloponnese

The foothills bordering the south side of the Gulf of Corinth are eroded into a labyrinth of *burroni* and knife-edged ridges, often 100 m or more high, mantled in pinewoods (which periodically go up in smoke) and maquis (Fig. 15.17). Trees and shrubs cover the old land surfaces and even cling to the cliffs. All this is on Pliocene and later clayey marls and weak sandstones.

[47] D. Harris and C. Vita-Finzi, 'Kokkinopilos – a Greek badland', *GJ* 134, 1968, pp. 537–46.

Fig. 15.17. Well-vegetated, active marl badland, with encroaching pines. *Stímanga near Corinth, August 1989*

The badlands coincide closely with the Corinth fault-zone, almost as wildly unstable as the southern Apennines (p. 43). Late Tertiary or Quaternary marine gravels have been lifted up at least 1,200 m. Gullies appear to begin at seasonal seeps and springs formed during especially wet years.

The amount of vegetation suggests that the badlands are not now very active. Some changes might be expected to reactivate them. Vineyards have been extended on to many less steep slopes, often with gross disturbance from bulldozers. Vineyards reputedly encourage erosion (cf p. 258), and there is little attempt to control it. Although this disturbance – more drastic than anything that prehistoric or historic cultivators could have achieved – has probably increased sheet-erosion, it has not yet resulted in any marked resumption of badland activity. In 1992 a more drastic provocation still, the digging of a new road-cut east of Ayios Geórgios (Old Nemea), did set off a huge slump. On none of our visits have we noticed that recent severe fires had initiated or reactivated gullies.

Bœotia

Around Thebes (modern Thívai) is a small tract of badlands in poorly cemented conglomerate. Only a little (in a sandy facies) is now active, the remainder being in various stages of stabilization, although still poorly vegetated. The badland has been active since Roman times, for it cuts into a Roman building. It probably began in prehistory: the very ancient city of Thebes is defended by two gullies.

Some of the old badland slopes have been ploughed without reactivating them. An air photograph reveals that the present badlands are part of a much larger tract of old gullies, now so completely inactive that they have been ploughed over and made into ordinary fields and vineyards, the gullies being visible as soil-marks.[48] They may still be eroding, but by sheet-erosion and rills which do not prevent cultivation.

Sérres

In NE Greece, tributaries of the Strymon river have cut gullies into soft Mio-Pliocene sedimentary rocks in the foothills of a granite, earthquake-prone mountain range. The gullies provide torrents with heavy loads of coarse debris, which is carried downstream to build up river beds, increasing overbank flooding and reducing the clearance of bridges. The Forestry Department, which has planted these gullies and similar features in northern Greece with *Pinus brutia* and species of *Robinia* and *Fraxinus*, claims some success in reducing the debris supply.

Aegean islands

Rhodes, one of the most wooded and most eroding of islands, has an active badland on the west coast, in marls underlain by clay. It undercuts the cypresses that grow on the slopes. Kárpathos has badlands in flysch under pinewoods.

The islet of Gávdhos has excellent badlands: one enormous gully of blue clay cuts through Roman and earlier archaeology. This remotest European badland is, as far as we know, the earliest identifiable badland to be depicted: the great gully can be identified in Monanni's aerial view of Gávdhos in 1622.[49]

Fig. 15.18. Small badland in the upper layers of the sand and gravel fill of a karst basin. At least fifty years old, it is apparently barely active. *Gerólakkos SE of Khaniá, Crete, October 1994*

Crete, in contrast, is remarkably free of badlands. Only here and there do we find a few hectares of gullies, either in red-beds (Fig. 15.18) or in marls underlain by blue clay. World War II air photographs and the testimony of old men show that these are active for a few decades – sometimes starting from a clay-pit – and then become dormant. One of these badlands was hit by the Pakhyámmos deluge (pp. 247–8) but not reawakened.

THE PATTERN OF BADLANDS

Badlands are ultimately driven by tectonic action. As the mountains around the Mediterranean were uplifted, they were attacked by erosion; the debris of whole mountain-ranges was redeposited lower down the slopes. These deposits have since been rendered unstable by further growth of the mountains, directly as in the Píndos or indirectly as at Guadix. Hence nearly all the present badlands.

---

[48] Rackham 1982, especially Plate 38 f.

[49] ASV: Piante e Disegne: Gozzi. We are indebted to J. A. Moody for comments.

Why does Crete, with its rapid uplift, not run to badlands? The answer lies partly in its steepness – much of the eroded material has been dumped in the sea – and in the scarcity of suitable clays. Another factor is the tendency, for reasons not clear to us, for Cretan deposits to cement. Materials that elsewhere form badlands are in Crete as hard as concrete and form cliffs.[50]

If badlands result from exposing weak sediments to the atmosphere at angles that they cannot sustain, ought not remains of them to exist under the sea, having been formed when the sea-level was lower? Indeed they do. During the Messinian Salinity Crisis, some five million years ago, when the Mediterranean nearly dried up (pp. 40–1), long slopes of Tertiary deposits were exposed to air and rain. The great rivers cut immense canyons in them, with lateral badlands.

Our observations do not allow us to predict where badlands will occur, but the following factors seem to be significant:

	Positive		*Negative*
++++	Lively tectonics	– – – –	Case-hardening
++++	Unconsolidated sediments	– –	Hard beds (which act as crack-stoppers)
+++	Deluges	– –	Strongly bedded or inhomogeneous sediments
+++	Montmorillonite or gypsum-bearing clays	– –	Crusts of mosses, lichens, etc.
+++	Frost action	?	Dense vegetation
++	Bulldozing, Spain		
+	Bulldozing, Greece		
++	Ophiolites		
+	Trampling, digging, road-making		

Badlands show no particular tendency to occur either in dry or wet climates. On the whole, they are probably more vegetated than their surroundings, merely because the landforms discourage people from clearing the vegetation. Ploughing and road-making seem so far to have had surprisingly little effect. We suspect that paths made by people and livestock in ancient or recent times have initiated many gully systems.

Rates of development

Published rates of badland erosion in America and other continents are usually several millimetres a year, sometimes much more. In Europe, individual gullies may erode at this rate, but archaeological evidence shows that many systems, once initiated, are long-standing and much more stable. Erosion rates vary widely from gully to gully and from year to year.[51] This is corroborated by plot measurements near the Basilicata badlands (p. 264).

We know of no extensive badlands in Europe that originated out of ordinary landscapes within historic times. Further investigation may discover some such; but hitherto, wherever there is sufficient evidence, major badlands turn out to be of at least Roman, if not prehistoric, antiquity.[52] Badlands in southern Israel have developed intermittently over an even longer period, beginning some seventy thousand years ago.[53]

ARE BADLANDS BAD?

Uses of badlands

Badlands are little threat to cultivation, except for people who think every hectare of land ought to be cultivated. The quantity of material removed in gullies is sometimes impressive, but does not, on the whole, come from cultivated places. Badlands tend to occupy certain areas and cultivation other areas. Even if fields are lowered by sheet and rill erosion, that does not necessarily make them uncultivable. Gullies occasionally eat up fields, but new fields appear at the tail-end of badlands. In most of the badlands that we have visited the inhabitants seem well aware of what lands are, or are not, safe to cultivate.

The ancient Etruscans were a badland people, and (like modern Italians) loved tunnels. They were forever digging burrows – *cuniculi*, 'rabbits', as the Romans called them – for various purposes: carrying roads or aqueducts through hills; avoiding bridges and viaducts across valleys by taking a road or aqueduct over a dam and diverting the river through a tunnel at the side; diverting streams; perforating valley floors to drain them. Many of these ingenious uses of tunnels can be interpreted as avoiding the troubles that arise with ditches and surface structures in an erodible land. Ploughs were developed to bite into the bedrock and replace lost soil.[54]

In the Alpujarra, J. R. McNeill accuses the colonizers from other parts of Spain, who (allegedly) replaced the Moors after the great revolt, of recklessly expanding cultivation and increasing erosion; he points to the formation of deltas at this time from the resulting silt. We consider this to be overstated. Most erosion in the Alpujarra comes from the badlands which occupy about one-tenth of the area. There is no sign that badland deposits have ever been much cultivated. It is an exaggeration, as McNeill claims, that 'agriculture crept into every nook in every barranca'. Land has been cultivated either on terraces or on the slope, depending on what treatment the rock will withstand. Cultivation remains almost always carefully avoid badlands (Fig. 15.19). Only in the last twenty years has the Forest Authority tried to cultivate badlands by bulldozing false terraces into unstable deposits, with the inevitable consequences (Fig. 15.20).

In south Spain badlands are put to use for fertilizing fields. Terraced fields are constructed and carefully levelled, with stone-built spillways from field to field. A weir on the local

50 There are examples in the Corinth fault-zone, too, such as the intensely hard conglomerate crags of Acrocorinth and south of Nemea.
51 Bull & Kirkby 1997.
52 A first attempt at applying a model of erosion susceptibility to regions such as Spain suggests

that badlands are between 2,700 and 40,000 years old; J. De Ploey, 'Gullying and the age of badlands; an application of the erosional susceptibility model E$_S$' *Catena* Suppl. 23, 1992, pp. 31–46.

53 A. Yair, P. Goldberg, B. Brimer, 'Long term denudation rates in the Zin-Havarim badlands, northern Negev, Israel', Bryan & Yair 1982, pp. 279–91.
54 Potter 1979.

Fig. 15.19. Disused cultivation terraces on metamorphic rocks. They avoid the blue clays that generate badlands. *Vicar, Almería November 1994*

river directs its waters, on the few days each year when it flows, to spread over the fields one by one. This waters the terraces and coats them, each year, with a thin layer of silt from upstream badlands. A ladder of fifteen such fields in the Rambla Honda, in the Tabernas badlands, has built up over centuries to some 3 m high. In the Boloduy valley, threading the Alpujarra badlands, in November 1994, we found people busy repairing the little earth banks round their fields and orchards in readiness for the rain (Fig. 15.21).

Badlands and reservoirs

Badlands shorten the life of reservoirs by filling them with silt. A comic example was a biggish dam at Nijar near Almería, built in 1850 across a *rambla* originating 11 km away and nearly 700 m higher. It filled with sediment in six years. This particular *rambla* had recently been captured and breached by an adjacent, lower, stream. The dam was built across the breach, and the sediments were derived from the reactivated slopes upstream.[55]

In the nearby Segura basin it is estimated that sediment, much of it from badlands, fills up 0.5 per cent of the capacity of the reservoirs per year.[56]

In the eighteenth century silting was not an insuperable problem. Cavanilles describes the periodic scouring of a dam 196 *palmas* high at Tibi near Alicante. A door at the base of the dam would be opened and a tunnel dug into the silt, being gradually enlarged until water began to squirt through the roof. The men would then run for their lives, and the contents of the dam would pour out, silt and all. This desperate

undertaking had to be sanctioned by the local council, whose members once were themselves its victims, when they watched the fun from too low down in the *barranco*.[57]

Such heroics are not for an age of health-and-safety regulations. Silting is a factor determining whether a reservoir site is suitable. The fault is with planners and politicians who love reservoirs, and either ignore this factor or imagine that badlands will obey their commands to stop delivering silt. Planting pines in badlands is supposed to reduce sediment production; but any benefit from afforestation, if not undone at once by the destabilizing effect of the bulldozing necessary to get the pines to grow, will last only until the next deluge or fire. Planting, however, conceals the raw subsoil; the tree cover stops people worrying and is a good advertisement for the forestry department. In reality, reservoirs, though they outlive the governments that built them, have finite useful lives.

Squashing the badlands?

On the marly clays around Siena and in parts of Basilicata, moneyed farmers have bulldozed minor badlands, grinding them down to grow Euro-subsidized durum wheat, to a degree that has begun to attract protests from environmentalists. Badland clays, at least in the laboratory, appear to be stabilized if enough farmyard manure is applied to raise their organic matter content to above 2 per cent. They will not be permanently stable if the farmyards run out of manure or when they encounter very wet winters.[58]

55 E. Roquero, 'Le barrage de Nijar ou d'Isabel II (Almeria, Espagne): une étude de l'influence de la géomorphologie sur l'acceleration de l'ensablement', *ZG* NF Suppl. 83, 1991, pp. 9–16.

56 M. A. Romero, F. Cabezas, F. López Bermúdez, 'Erosion and fluvial sedimentation in the River Segura basin (Spain)', *Catena* 19, 1992, pp. 279–82.

57 Cavanilles 1795, vol. 2, p. 184.
58 C. P. Phillips, 'The badlands of Italy: a vanishing landscape?', *Applied Geography* 18, 1998, pp. 243–57.

Fig. 15.20. False terraces bulldozed into unstable badland as a 'conservation' device. *El Marchal, Vícar, Almería, November 1994*

Fig. 15.21. Silt-gathering irrigated terraces. *Alsodux, eastern Alpujarra, November 1994*

Badlands and deltas

Badlands have the function of creating and maintaining deltas. Many north Mediterranean river deltas, especially those active in historic times, are downstream from badlands: the deltas of SE Spain, the Ebro, many in Italy, in Albania, the Achelöos and others in NW Greece, and the Maeander in Turkey. The future of most deltas is now in doubt because dams below the badlands have cut them off from their supply of silt (Chapter 18).

Conclusion

The Arizona–California model is not a good parallel for Europe. Most, if not all, the larger Mediterranean badlands are ancient and are due to environmental forces outside human control. They are certainly not located in areas that have histories of especially intense human activity. Even modern human activities seldom have much effect on them. They are in a state of dynamic stability, eating into the landscape at one end and gradually ceasing to erode at the other end. They have episodes of activity and of quiescence, determined probably by the frequency of great storms. The present is not a specially active period, but the Little Ice Age may have been.

Badlands are not the result of desertification, nor the end-product of human ignorance and mismanagement. They deserve to be understood and appreciated for what they really are: among the most beautiful and dramatic land-forms that Nature has created.

CHAPTER 16

Erosion in Prehistory and History: Climate, Weather or Man-Made?

If ever there was a soil mantle across the entire Southern Argolid thick enough to support the well-watered forests ancient authors dreamed of, it must have been stripped away by the action of the Ice Age climate long before human beings acquired the ability to damage the land on a grand scale. Extrapolating generously, human activity has aggravated the erosion of the Greek mountains, but the landscape of Classical times was not much richer in soil than it is now.[1]

DATING AND INTERPRETING EROSION

To determine the timing of erosion involves evidence from diverse sources, geological, archaeological and documentary; certainty is elusive. One great obstacle is that much erosion does not go on all the time. Plato, in his famous account of 'Attica', already emphasized the importance of deluges:

> It all lies as a promontory, projecting far from the rest of the continent into the sea. The basin of the sea around it happens to be deep and steep-to. Since many and great deluges (κατακλυσμόν) occurred in the nine thousand years (which years elapsed from that time until now), in the said times and calamities the soil of the earth from the heights did not run down, as in other places, into a noteworthy outpouring, but always flowed round in a circle and disappeared into the deep. There is left from then to now, just as in small islands, only the bones of a sick body, all the fat and soft of the earth having fallen away, only the bare body of the place. But then it was intact and the mountains were high earth-hills, and the plains now called Phelleos were full of fat earth, and there was much woodland in the mountains, of which even now there are visible signs. For some of the mountains now have only food for bees, but not very much time ago trees, from whence roofs were cut for the greatest buildings. And besides, there were many tame trees, and it bore boundless pasturage for herds. And also it garnered the yearly water from Zeus, which was not as now lost by flow from the bareness of the land into the sea; but holding much and receiving it back, and storing it up in the rooftile-producing earth, and drinking down from the heights the water absorbed in the hollows, it provided all the places with plentiful streams of springs and rivers. And even now, on the spots where there formerly were springs, there remain shrines, which are signs that what is now said about this is true.
> . . . The Acropolis was then not as it is now. For now it happened one night waters, carrying away the earth, melted it away and made it bare, when earthquakes happened at the same time as the third extraordinary disaster of water before Deucalion's. . . .[2]

This much-abused passage has passed into the folklore of

erosion as demonstrating the evils of deforestation. But that is not what Plato said. The Greek text is atrocious, and other people's translations smooth over his ambiguities; but there can be no doubt that he meant deluges to be the cause of loss of soil, and loss of soil to be the cause of the loss of trees, not vice versa. Plato, moreover, was not writing about the geographical Attica, but about a fictional Attica which went to war with a fictional Atlantis, and we are not obliged to take what he says literally. If he was inspired by actual observation, he has three important insights. He apparently had in mind the great erosions of the Pleistocene, and made a surprisingly good guess ('nine thousand years') at the date. He appreciated that erosion is more catastrophic than gradual. And he raises the possibility of a yet greater catastrophe if an earthquake coincides with a deluge, the vibration turning to liquid soils and subsoils already soaked by rain – a rare combination which no living European eye has seen.

Interpreting erosion should involve comparing existing land-forms and soil profiles with those that would have existed in the absence of human intervention, and determining when that intervention occurred. But dating erosion is not easy. With or without intervention, nearly all Mediterranean soils are active. Undisturbed ancient profiles can only occasionally be found preserved beneath later deposits.

Evidence for erosion history

Erosion history is a favourite study of archaeologists, who sometimes produce direct information. Fields in Sphakiá (Crete) lost their soil at some time between the Roman period and the age of the oldest pines that now grow on the site (p. 261). In south Attica remains of Classical terraces and farmsteads have been found on limestone on which there is now little or no sediment. This was one of the last areas to be settled in Antiquity, and represents the extension of farming on to land on which agriculture was not sustainable.[3]

Erosion history may be estimated by extrapolating back from present measurements: R. Lehmann made a valiant attempt to do this for the cultivation terraces of Naxos (pp. 264–5). This becomes problematic as uncertainties about past erosion rates mount up with time. One measure of erosion, the empedestalment of trees (p. 252), can be taken back for several centuries.

[1] M. H. Jameson, C. N. Runnels and T. H. van Andel, 1994. [The authors are mistaken in asserting that soil is needed to support forests, see pp. 48–51.]

[2] Plato, *Critias* 111.
[3] H. Lohmann, 'Ein "alter Schafstall" in neuem Licht: die Ruinen von Palaia Kopraisia bei Legrena (Attika)', *Structures rurales et sociétés*

antiques, ed. P. N. Doukellis and L. G. Mendoni, Belles Lettres, Paris, 1994, pp. 81–132; Brückner & Hoffmann 1992.

Fig. 16.1. Post-Roman erosion as demonstrated by the aqueduct of Corinth city. The Roman ground-level is shown by the change from rough to smooth masonry, three courses above the modern concrete underpinning. This side of the ravine, which is loose schist, has lost about 2 metres since Roman times. (The other side, in hard conglomerate, has if anything gained a little sediment.) *June 1985*

Lynchets, where soil has piled up against a cross-slope obstacle, are a useful guide. Objects such as churches, which are datable and have datable alterations, are particularly helpful (Figs. 8.1, 8.2, 16.1). The threshold height of a door inserted into a church, compared with that of the original door, measures the extent to which the ground outside rose between the dates of the doors (p. 131–2). Whether or not sheet erosion is now going on can be detected by looking for fresh sediment deposited in roadside ditches crossing a slope.

Some limestones, when exposed to the air, weather into sharp-edged grooves (Fig. 14.7a). The development of this weathering form (*Rillenkarren*) has been used to measure the time since the rock lost its covering of soil.[4]

Archaeologists, however, use mainly evidence from the sediment derived from erosion. This accumulates at the foot of slopes, on valley floors, in lakes and reservoirs, and at the coast. The sites it came from are seldom worth studying: erosion itself removes the evidence, leaving at best a scatter of potsherds and building stones too heavy to wash or blow away.

Sediment is indirect (proxy) evidence. Research can reveal how and roughly when it was transported to its present site. It is rarely possible (or, maybe, it is rarely done) to establish whence it came or what kind of erosion produced it. Scholars tend to assume that late-Holocene deposits represent soil 'stripped' by sheet-erosion, rather than soft rock eaten out of a small area of gullies. Deposits can consist of existing material re-worked as well as new sediment brought down; this, too, is not adequately taken into account by those who have never witnessed a deluge. Study of sediments focuses attention on the youngest and most easily accessible: older sediments lying underneath may not be exposed.

River terraces

Alluvial terraces on the floors of valleys are much used as proxy evidence for erosion. They record the accumulation or removal of sediments by the river. Accumulation happens when material coming from higher in the catchment is too abundant or too coarse for the river to transport downstream. When the supply is reduced, or the discharge and velocity of the stream increase, the river will excavate material from its bed or its banks and deepen or widen its channel. Downcutting on the lower reaches can be the result of a fall in the level of the river mouth.

Braided streams occupy multiple channels in a wide gravelly bed (p. 245). When more water comes down, they occupy more channels and roll stones along the bed. Streams with a single channel are less variable and carry most of their load in suspension. After heavy (but not exceptional) rain the water level rises above the channel banks, spreads over the floodplain, and deposits sediment. As the river meanders it cuts sideways into the outsides of bends, and builds up low banks of coarser sediment on the insides of bends which later get buried by flood-plain deposits. Any one deposition, therefore, may exhibit coarser and finer facies.

If the sediment supplied to a single-channel stream increases or becomes coarser, it may accumulate on the bed of the channel, reducing its cross-section and its capacity. In times of flood, the river may then distribute splays of coarse material over the flood-plain, turning into a braided stream. It may undercut the confining valley slopes, inducing slumps and gullies which provide still more coarse sediment. The converse may happen when the supply of sediment diminishes in relation to the discharge, allowing one channel to dig its bed deeper and to take over from the other channels.

People disapprove of flood-plains, and regard braided river-beds as waste land which ought to be doing something visibly useful. They build banks or concrete channels to confine the course of the river, often with results very different from those intended. They divert rivers to use the water for irrigation, which may cause sediment to build up in the wrong places.

River terraces may be datable from artefacts or charcoal embedded in them. Their formation and incision are often taken as evidence of periods of rapid erosion and high sediment yield in the catchment, followed by reduced erosion and lower sediment yield. However, they can equally be due to changes in runoff from the catchment through higher rainfall or lower storage capacity. Sediment can come from the river side-cutting into old terraces or the valley sides, as well as from slope erosion. Much alluviation represents the recycling of deposits already on the flood-plain, rather than input of new material. Interest in river history has been revived by new dating methods such as thermoluminescence.[5]

EROSION AND HUMAN ACTIVITY

Postulated Deforestation

Scholars usually attribute historic erosion to deforestation or unspecified 'anthropic activity'. In an ideal world they need to demonstrate:

4 Gams 1993.
5 Lewin & others 1995.

(1)   That erosion occurred in a particular area at a particular time.
(2)   That forests covered that area before the erosion but not since.
(3)   That human activity supplanted the forests shortly before the erosion.
(4)   That subsequent land-use was of a type known to produce erosion of similar terrain in the area today.
(5)   That erosion does not now occur under forest on similar terrain.

Scientists do their best to establish (1), but neglect the other steps in the argument. Evidence is sometimes, but not always, presented for (3), but step (4) is treated perfunctorily (e.g. in 'must have' terms), and (2) and (5) are usually ignored, as are alternative explanations for the erosion phase.[6] Other writers quote the conclusion as further evidence that Erosion is caused by Deforestation, and so the circular argument spins on.

This argument is so common that we need a term for it. When an author asserts that an alluvium or delta was formed as a result of erosion following deforestation in the catchment, but does not give the intermediate steps in the argument, we shall call this *Postulated Deforestation*. The author has not seen fit to present the independent evidence (if any) that a significant part of the catchment changed from forest to non-forest at the appropriate time, and we have failed to discover it for ourselves. This being so, there must be serious doubt whether the deforestation really occurred. (The reader is reminded that cutting down trees does not necessarily constitute deforestation, p. 48.)

Was Paulini right?

Giuseppe Paulini claimed in 1608 that the lagoon of Venice was in danger of getting filled in by debris brought down by floods resulting from recent diminution of forest in the Alps (p. 8). The evidence that deforestation was going on is inconclusive. Since at least the fourteenth century the alpine woods had been managed as a sustainable source of timber. Management may have been strained in the sixteenth century, when the Venetian Senate forbade the growth of big trees (p. 178). However, there were still vast woods, and the alpine economy was still based on exporting timber, as in later centuries and as it is now.[7] Certainly the Venetians were worried about the lagoon filling in, and had just finished a great river-diversion to carry the silt elsewhere; but they had been building such diversions for centuries (pp. 340–1). Paulini writes as though he had never been out in a really big deluge, nor seen what happens when the water-holding capacity of a forest is saturated by the first 30 mm of rain and there are 200 mm more to fall. He notes, however, the fact that forest delays snow-melt – an important and unique feature of forests (p. 254), but one that has little effect in Mediterranean climates.

Whether the years before 1608 were really a period of more than usual flooding at Venice we cannot say; they were so in Crete. Certainly this was a time of peak growth of many deltas (Chapter 18); but for none of those deltas is it certain that there was less forest at this time than before or since. The alternative explanation, which Paulini (as far as he is quoted) does not mention, is that this was the second peak of the Little Ice Age, with an unusual frequency of overwhelming deluges.[8]

Ploughing

Many scholars assume, without further argument, that the arrival of intensive human activity 'strips soil from the hill-sides': that there is an instant change from complete protection before to complete instability after. The assertion derives from colonial experience in North America, and must be applied with caution to the different climates and soils of Mediterranean Europe. In detail it is difficult to visualize. The only historic human activity that destroys nearly all protection is ploughing (or hoeing), and then only for a month or two. Felled or burnt forests retain some protection, which rapidly increases (even under browsing) as the vegetation recovers. Catastrophe might be possible on rare occasions when a deluge – bigger than the Pakhyámmos event – struck immediately after clearing or ploughing. Even then it is hard to imagine a whole hillside losing its soil at once. In prehistoric and historic agriculture hillsides, even if not terraced, would have been divided into little fields separated by baulks, bushes, trees and patches of roughland, and some of the fields would have been fallow. Loss of soil would be patchy at worst. As we saw in the last chapter, some soils fail to erode under the most startlingly risky treatment; others erode under little-disturbed forest.

Ploughing has promoted erosion more efficiently since the nineteenth century. Iron-shod, furrow-turning ploughs then became common, although wooden ards were still used on steep, more erodible terrain. Erosion may have been encouraged by the habit of picking stones off ploughed fields (p. 258). Runoff from town streets, roofs and roadcuts may have encouraged local gullying, especially in badland areas (p. 280).

OLDER AND YOUNGER VALLEY FILLS

The principle of two great ages of deposition (and, by inference, of erosion) was brought into prominence by Claudio Vita-Finzi in 1969 in *The Mediterranean Valleys*.[9] The idea germinated in Tripolitania, where erosional and depositional events in certain wadis could be dated by the remains of Roman dams. Deposits of three main periods were distinguished:

(1)   The earliest contained only Palaeolithic implements and had evidently accumulated in the Pleistocene. It had been cut into by stream action between 9,000 and 3,000 years ago.
(2)   During the Roman Empire, dams were built to store water and to retain sediment. Remains of the dams, founded

---

[6] See M. Inbar, 'Rates of fluvial erosion in basins with a Mediterranean type climate', *Catena* 19, 1992, pp. 393–409. This is but one example of a type of argument that has become a platitude.
[7] M. Agnoletti, 'Gestione del bosco e segagione del legname nell'alta valle del Piave', Caniato 1993, pp. 73–126.

[8] This area is discussed in detail by E. Guidoboni, 'Human factors, extreme events and floods in the lower Po plain . . . in the 16th century', *E&H* 4, 1998, pp. 279–308. She points out that at this time the population was greater than before or since: people were forced to take risks in cultivating slopes and flood-plains, so that floods had greater effects. Tectonic movements may also have had their effects.

[9] A. T. Grove, 'Classic in physical geography revisited', *PPG* 21, 1997, pp. 251–6.

on calcareous crusts, indicate the cross-sectional shape of the wadis at the time of construction. Alluvium accumulated behind the dams until, in the late Empire, floodwaters breached or circumvented the dams, and the Roman alluvium was cut into.

(3) Low river terraces were built up within the downcut wadis. These contained Roman and earlier material, but also pottery and charcoal dated to 1200–1500 AD. This medieval fill had in turn been incised by the current wadis.

Vita-Finzi, examining Mediterranean valleys more widely, found that deposits could be divided into *Older Fill*, of Pleistocene date, and *Younger Fill*, generally dating from 500 to 1500 AD. The latter contains potsherds of various ages down to the time of its deposition, but the former never does. As in Tripolitania, there is often more than one phase of Younger Fill.

Older and Younger Fills were recognised widely (Fig. 16.17a). Vita-Finzi found Younger Fill in all the valleys he visited in Tuscany and Campania (Italy), and in the basins of the Guadalquivir, Tagus and Ebro (Spain). The coastal plains of Valencia and elsewhere round to Gibraltar, he considered, owed much of their substance to this medieval alluvium. In Greece, medieval alluviation was seen as having constructed the Arta plain in Epirus and contributed to the filling of the Kopáis basin in Bœotia, but both were poorly dated.

Older Fill, usually much greater in quantity than Younger, represents erosion over a very long period, especially associated with glacial ages. Hempel dates the huge screes and fans in the mountains of Crete to the Saale glaciation, 350,000 years ago; at lower altitudes, deposits contemporary with the last (Weichsel) glaciation, about 20,000 years ago, predominate.[10]

Younger Fill was originally characterized by containing buff-coloured silty sand with bands of fine gravel, including Roman to medieval potsherds. Everywhere it had been dissected, usually by a single channel, within the last few centuries. It indicated a period when water erosion was more active than before or since. Coming, as it did, in the Dark Ages and medieval period, rather than in Ancient Greek or Roman times, it could not easily be attributed to increased human activity. Although increased pasturage or abandonment of terraces were suggested as factors promoting erosion, both these and what Vita-Finzi called 'the vaguer bogey of medieval devegetation' were strained, unsatisfactory explanations. The apparently synchronous incision and alluviation all round the Mediterranean seemed to eliminate land-use changes as an explanation. Vita-Finzi favoured a climatic explanation: increased erosiveness was associated with either the Little Ice Age or the preceding Medieval Warm Period.

These ideas were controversial. The archaeologist J. Bintliff found supporting evidence; others thought them unacceptable. K. Butzer argued that the fills were much more complex than Vita-Finzi had appreciated. Older Fill included material derived from several different erosional periods. Younger Fill varied in age from one site and region to another. It was post-Roman in parts of Spain around large Roman settlements or where there had been intensive Roman agriculture; it was more recent where there had (supposedly) been post-medieval deforestation and agricultural expansion into marginal environments. Butzer considered it must be the outcome of human activity.[11]

After further investigations in SW France and SE Spain, Vita-Finzi proposed that the age of the fills varied systematically, Older Fill decreasing in age from southerly to more northerly sites as ice-caps retreated, Younger Fill increasing in age with latitude. He later proposed that the pale colour of channel fills deposited in late historical times resulted from iron-rich clays being washed out of them, accounting for this by the 'prevalence of sustained flow in basins previously characterised by ephemeral discharge.' River aggradation, he suggested, shifted south in medieval times as depression belts moved towards the equator under the influence of increasing solar activity and weakened zonal circulation.[12] It is an ingenious argument but still unconfirmed, except that, as we have mentioned (p. 133), many Cretan rivers, now ephemeral, were indeed perennial and flowed continuously during the Little Ice Age.

EXAMPLES OF YOUNGER FILL

Olympia: land of catastrophe

The north-west Peloponnese is wildly unstable. The Corinth fault-zone is one of the liveliest parts of Europe; since the Pliocene the mountains have been uplifted by at least 1,000 m. Erosion during early uplift laid down beds of cobbles, gravels, silts, sands and clays – unconsolidated sediments that later uplift has left at dangerous angles. Hence the gorges and surrealistic badlands, mantled in maquis, pinewoods and burnt pinewoods, which enliven the journey from Corinth to Pátras. Here the factors that promote or hinder erosion should be clearly demonstrated in terms of processes still going on.

In this god-haunted landscape, by the Alphéios River, lies Olympia, seat of the ancient Olympian Games. The area has been block-faulted, tilted and buried in unconsolidated sediments. On the south, the Lapídhas range and Mount Lykoúri (where Zeus turned his priests into wolves) have accumulated more than 300 m of gravels and cobbles on top of 400 m of silty clays and sands. To the north, where Hérakles stuck the Erymanthian boar, uplift and erosion of the mountains in the early Pleistocene deposited very coarse conglomerates and other sediments,[13] in turn eroded by tributaries of the river

[10] Hempel 1990.
[11] J. Bintliff, 'The plain of western Macedonia and the Neolithic site of Nea Nikomedia', *PPS* 42, 1976, pp. 241–62; Bintliff 1977; K. W. Butzer, 'Holocene alluvial sequences: Problems of dating and correlation', *Timescales in Geomorphology*, ed. R. A. Cullingford and others, Wiley, Chichester, 1980, pp. 131–42.

[12] C. Vita-Finzi, 'Age of valley deposits in Périgord', *Nature* 250, 1974, pp. 568–70; Vita-Finzi 1975; C. Vita-Finzi, 'Diachronism in Old World alluvial sequences', *Nature* 263, 1976, pp. 218–9; 'Solar history and paleohydrology during the last two millennia', *Geophysical Research Letters* 22, 1995, pp. 699–702.

[13] J. Hageman, 'Late Cenozoic history of the Pyrgos area, western Peloponnesus', *Proceedings of the VI Colloquium on the Geology of the Aegean Region*, ed. G Kallergis, Institute of Geological and Mining Research, Athens 2, 1977, pp. 667–74; 'Stratigraphy and sedimentary history of the Upper Cenozoic of the Pyrgos area . . .', *Annales Géologiques des Pays Helléniques* 28, 1977, pp. 299–333.

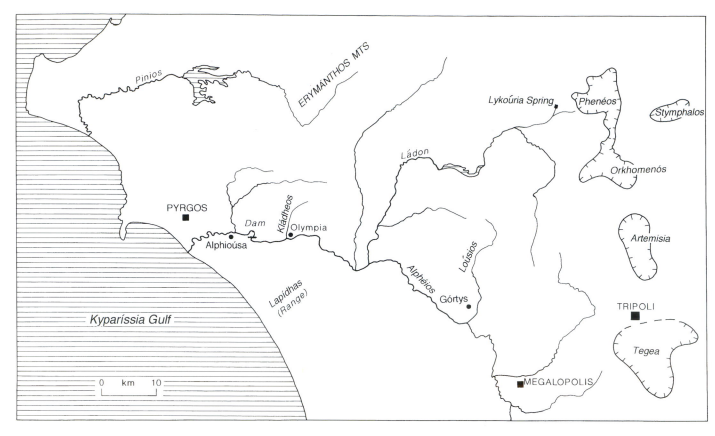

Fig. 16.2. Olympia, the Alphéios catchment, and neighbouring closed limestone basins in the NW Peloponnese.

Alphéios to form deep wooded gullies. There is no lack of all kinds of material – most of it not soil – for further stages of erosional recycling (Fig. 16.2).

Olympia was one of Vita-Finzi's definitive sites for Younger Fill. It attracted his attention through the German excavations of the ancient site. It is close to the Alphéios; the Kladhéos torrent, adjacent on the west, rises in steep, erodible, landslipping hills to the NE. Kronos, a steep little hill, lies immediately north (Fig. 16.3).

From the early Bronze Age there was considerable settlement; a burial mound lay on the later sacred site of Altis. After 1700 BC some stone buildings close to the foot of Kronos were buried by silts, probably from the Kladhéos but possibly washed down from the hill. Definite cult activity at Altis is known from the tenth century BC onwards, well before Hérakles and his descendant Klymenos founded the Games. The temple of Zeus, built in 472–457 BC, was destroyed by an earthquake a millennium later. Subsequently the site was covered by five to six metres of brownish sandy silt, also probably brought down by the Kladhéos (Fig. 16.4); this valley fill extends up affluents rising in hills to the east and north. At Olympia it buried a Byzantine fortress and covered coins which show that it began to accumulate after 600 AD. The Kladhéos later cut down to its original level, breaching a Roman confining wall now mainly on its west side; the

Alphéios shifted north, cutting away or burying the remains of the Hippodrome and forming a cliff in the tail of the Kladhéos sediments. This dissection, which may have begun in the fourteenth or fifteenth century AD, has progressed upstream but has yet to reach the uppermost valleys. There appears to have been little change since before 1766, when the traveller Chandler found and briefly described the temple and stadium. Stanhope's plan of 1811 also shows the remains in much the same relation to the ground levels as they are now.[14]

At first sight this seems to be a counter-example to any theory that erosion results from rising population and more intensive land-use. After being stable through the eleven centuries when Olympia was a major shrine, the catchment then eroded when (to judge by visible remains) civilization was in decline. The claim that immigrant Slav pastoralists 'must have' been damaging the vegetation is possible, but involves a long chain of unverified argument.

But there is more to Olympia than this. About 5 km downstream, near Alphioússa, there are gravel-pits in a river-terrace rising about 10 m above river level. The gravels, some 9 m thick, could have come from the pine-covered gullies on the slopes of the Alphéios's tributaries. They overlie a surface containing tree roots apparently *in situ*, from one of which we have a radiocarbon date indicating that aggradation began after, probably soon after, 2000 BC.[15] Tree-trunks with roots

[14] Huntington 1910; Dufaure 1976; J. Budel, *Climatic Geomorphology*, trans. L. Fischer and D. Busche, Princeton University Press, 1982, pp. 338–46; R. Chandler, *Travels in Greece . . . made at the expense of the Society of Dilettanti*, Oxford, 1776, p. 294; Leake 1830, plan in vol. 1.

[15] 3660±30 bp. The calibration curve is complex at this point, and produces two date ranges at each confidence level: 2110–2090 BC or 2035–1965 BC at 68 per cent, and 2135–2075 BC or 2045–1925 BC at 95 per cent confidence level. The latter of each pair of ranges is the more probable by a factor of about 3. (Dr R. Switsur, Godwin Laboratory, Cambridge, no. Q-2942.)

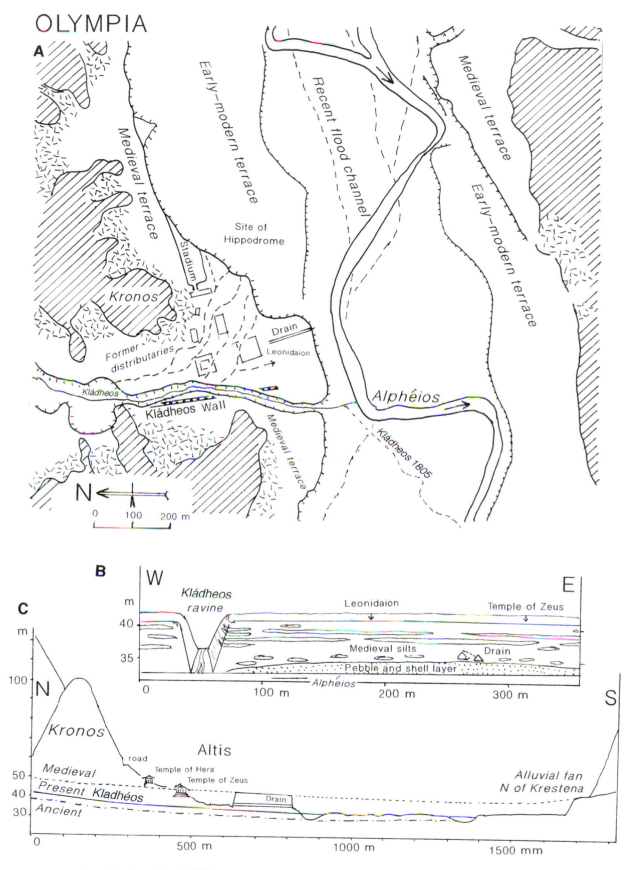

Fig. 16.3. Olympia (adapted from Budel).

A. The archaeological site (Altis), with the Kládheos stream running alongside it to enter the Alphéios from the north.

B. Section, viewed from the south, of the sediments that still cover part of the Altis.

C. North-to-south cross-section of the Kronos hill and the Alphéios terraces and valley floor.

Fig. 16.4. East–west section through the sediments dumped by the Kladhéos over the sacred site of the Altis at Olympia, viewed from the south. A shelly layer of sediment, from the hills overlooking Olympia from the north, is overlain by 1 metre of gravels and then by 5 metres of buff-coloured sandy silt. *September 1994*

attached, radiocarbon dated to within the last three hundred years, lie on the surface of the gravels; others lie in a layer within the gravels at a depth of about 2 m. One of the surface trees was beech, a tree not now recorded in the Peloponnese; others were pine which could have come from the slopes over-looking the Alphéios.

Olympia has one very unusual feature. Phenéos, a large karst basin in the limestone hills 65 km ENE of Olympia, is linked by underground passages to springs in the headwaters of the Ládon, a tributary of the Alphéios.[16] Phenéos had long been feared for its cycles of turning into a lake and then suddenly emptying. The underground channels are blocked and unblocked by earthquakes, or by trees washed down from the surrounding forested and eroding mountains.

In the fourth century BC, Phenéos was usually dry land, but was once inundated for a year before emptying. Next century the lake re-formed until the sinkholes (ζέρεθρα) opened and the lake squirted out into the Ládon, flooding the Alphéios and inundating Olympia. Pliny knew of five such cycles, and another was recorded by Plutarch and Pausanias in the second century AD. There was a particularly high water-level apparently in the eighteenth century, flooding out the monastery of St George below Phenéos town. By 1806 the water had subsided and the basin was again cultivated; Drama–Ali, last of the beys of Corinth, had gratings placed over the three sinkholes on the floor of the Phenéos to keep out driftwood.[17] But the gratings were stolen in the War of Independence of 1821, causing the sinkholes to block and the plain to revert to a lake 40–50 m deep by 1829.[18] On 1 January 1834 it suddenly emptied. By 1838 it was again a lake, and so remained until after 1897.[19] By 1907 it was again land and, apart from one record in the 1930s, it has been dry ever since.

We, like all travellers since Pausanias, have remarked the changes in vegetation at the site of former high-water levels. The nineteenth-century level is marked on the floor by a change from strip fields with many trees above the water-line to a grid of square fields below. On the mountain slopes to the south there is maquis of dense prickly-oak, invaded by occasional fir trees, down to 50 m above the lake; prickly-oak and Spanish broom descend to 40 m, lentisk and Spanish broom to 15 m; below that level is bare rock.

The Phenéos basin lies at about 750 m and covers some 40 sq. km. The exit of the caverns, 5 km south of Lykouriá, is at about 550 m. If the lake reached a depth of 100 m before discharging, about 3 cubic km of water would fall through

Fig. 16.5. Phenéos basin, with old shore-lines of former lake. *May 1998*

(a) SW corner, looking east: prickly-oaks, *Phlomis, Calicotome* and advancing fir-trees in the foreground, below the 50-metre shore-line.

(b) Shore-lines at the south end at 15, 40, 50 and 60 metres above the plain.

[16] Pausanias, *Description of Greece* VIII.xx.1: 'I have heard that the water forming a lake in the Phenéos, which comes down into the sinkholes (βάραθρα) in the mountains, comes up again here and forms the springs of the Ládon.'

[17] For similar gratings to a sinkhole in Antiquity, see *AR* 43, 1996–7, p. 35.

[18] A lake in the Fucino basin, east of Rome, reached its highest level since Neolithic times in the nineteenth century (1800–1820 and again in 1850–61); C. Giraudi, 'Lake levels and climate for the last 30,000 years in the Fucino area (Abruzzo–Central Italy) – a review', *PPP* 70, 1989, pp. 249–60.

[19] Theophrastus, *Historia Plantarum* III.i.2; Pliny, *Natural History* XXXI.xxx.54; Strabo, *Geography* VIII.viii.4, quoting Eratosthenes; P. M. E. Boblaye in Bory de St Vincent 1836, vol. 2, p. 321; Frazer 1898, vol. 4, pp. 231–33, 262–63; Huntington 1910; PW 38, p. 1966.

Fig. 16.6. *Right* Cross-section of the Loúsios valley at Górtys, showing the buried remains of the Hellenistic baths and the half-buried Byzantine chapel in relation to the river channel and terrace deposits. (After Dufaure 1976.)

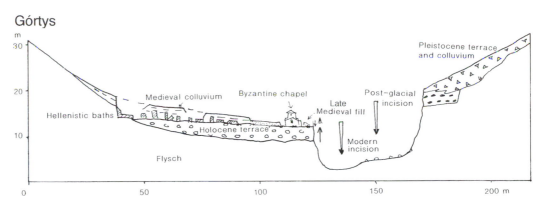

Fig. 16.7a. *Below* Buried bath-house, Górtys. *May 1998*

Fig. 16.7b. *Below right* Buried chapel of St Andrew nearby. *May 1998*

250 m. Such a discharge – thousands of cubic metres per second – could well uproot trees in the Ládon and Alphéios valleys and carry them downstream to be buried (and recently dug up) where the valley floor widens about 5 km downstream of Olympia.[20]

Additional light is thrown on the erosion history by the site of Górtys,[21] 5 km up the Loúsios tributary of the upper Alphéios. French excavators uncovered thermal baths, probably dating to the fourth century BC, built on a Holocene terrace (Figs. 16.6, 16.7a).[22] They had been covered by several metres of sediment, mainly colluvium coming from steep, eroding slopes overlooking the site. The Loúsios had later cut down into this fill to form a lower terrace, on which a Byzantine chapel was built, reusing stone from the Hellenistic buildings. The chapel in its turn was almost buried by slope

and alluvial sediments before the river cut down sharply more than 10 m, through the Holocene terrace and into the underlying rocks. This record indicates three phases of slope and river deposition: (1) prehistoric, (2) early Byzantine, (3) several centuries later.

This complex area illustrates problems of scale in accounting for erosion. On the Alphéios at least two very violent events dislodged huge quantities of gravels, tore out trees, and transported them downstream. The Kladhéos side-valley was little affected by these cataclysms, but had its own drama, probably at another time. A link between them would lie in the gravel deposit on the Alphéios acting as a dam for the Kladhéos, trapping finer sediment which would otherwise have been washed out of the catchment. In the Loúsios tributary there are signs of two historic periods of deposition, each followed by downcutting. No-one who travels in the Peloponnese should be surprised that different areas become ripe for massive erosion at different times. In such unstable terrain it is unrealistic to generalize from one catchment or part of a catchment.[23]

---

[20] Construction work on the Ládon dam would seem to be too recent to provide an alternative explanation, and would have felled rather than uprooted trees.

[21] Not to be confused with Górtyn in Crete.

[22] Dufaure 1976.

[23] Another flood may be imminent. A few km downstream of Olympia a weir is being undermined by flood waters, creating a waterfall over its downstream concrete apron. This is being encouraged by gravel-digging. Fortunately there is (in 1997) no lake waiting to be released from the Phenéos basin, but the Alphéios is notoriously liable to flood and repairs are urgently required.

Date	Olympia	Phenéos basin	Górtys
1766–1990 AD	Archaeological digs 1829, 1875–81, 1936–41 and later	High lake levels in late 18th century, 1821–34, 1838–1907	Archaeological dig 1954
1500–1766 AD	Kladhéos cuts down; Alphéios shifts north		Loúsios cuts down several metres
1200–1500 AD	Kladhéos cuts down		Partial burial of chapel
1000–1200 AD			Chapel of St Andrew built on terrace step
800–1000 AD	Kladhéos river deposits 3–6 m on the Altis site		Loúsios cuts down a few metres
600–800 AD			Thermal baths buried
393–600 AD	Cult and Games abandoned; Zeus temple overthrown; Christian basilica and other buildings		
776 BC–393 AD	Great sanctuary; Olympian Games held regularly; Zeus and other temples and treasuries built	At least 5 cycles of lake filling and emptying	Thermal baths on 'Holocene' terrace
1000–776 BC	Beginnings of Altis sanctuaries		
2000–1000 BC	Bronze Age settlement; Alphióussa gravel terrace begins to form; pre-1700 structures buried		
Earlier prehistoric	Settlement begins		Accumulation of 'Holocene' terrace of Loúsios river

Table 16.i. Chronology of events in the Alphéios basin, Peloponnese.

The Kámpos of Khaniá: a recently fertile plain

A common alluvial situation in the Mediterranean is where a wall of cliff, formed by a fault-line, forms the edge of a plain; a number of streams of small catchment emerge on to the plain through gorges in the fault-scarp; their sediments, fanning out over the plain, form its cultivated surface.

Such a place is the Kámpos behind the city of Khaniá in NW Crete. The plain slopes down about 4 km to the coast from the foot of a scarp 10 km long and 200–600 m high. The scarp is of marls, marly sandstones and conglomerates, usually cemented and case-hardened to give the appearance of hard limestone. From the scarp foot, 60 m above sea-level, the slope gradient diminishes from 10 per cent to less than 3 per cent. Seasonal, single-channel streams and torrents emerge from various gorges and ravines, particularly the long, deep, narrow Thérisso Gorge (Fig. 16.8). Behind the scarp these cut through a hard limestone ridge, with patches of phyllite-quartzite above it. Although as rugged as the Olympia area, this gives no similar impression of wild instability; the sediments of earlier erosional phases seem to be firmly cemented in place, even where they form cliffs. Not even deluges have much effect, except from small-scale slumping on the footslope.

Cultivation extends over the plain and formerly up the terraced base of the scarp. Khaniá city (sited on a small bluff) has been occupied, possibly continuously, since the Final Neolithic, yet there is a curious lack of early major archaeology on the plain. The Kámpos, rich in Turkish and Venetian villas and Byzantine monasteries and once famous for its orchards, is conspicuously lacking in surface archaeological remains of Roman times or earlier. Only one Minoan villa, at Nerokoúrou, has been found by a shallow excavation.

Much of the plain has a unified, planned layout of fields and roads, radiating out from Khaniá.[24] In the fields stand olive-trees of various ages up to about a thousand years; some of these are buried or exposed by up to a metre, but in no obvious pattern.

J. A. Moody found in an archaeological survey that Roman and earlier archaeology exists, but is buried by 2–6 m of more recent sediment. Exposures in ravines and builders' pits sometimes revealed that this sediment lay directly on walls, floors and pottery of Roman and Hellenistic periods, and also had Bronze Age material under it. There are no known pre-medieval structures on top of this fill and no known medieval or later structures buried by it.[25] Massive sedimentation, taking place after Roman times, was sufficiently disruptive to result in the laying-out of new, planned, ownership boundaries. Moody suggested that sedimentation may have been encouraged by the uplift of the western end of the island in the sixth century AD (pp. 43–4).

The matter was reopened by the making of the Khaniá bypass road, which exposes sections up to 4 m deep. In 1994 Holly Raab, a Classical archaeologist, inspected these and other road-cuts to examine the relationship between the age of artefacts and their depth in the ground. She noted the depth of removal or accumulation of soil around trees, especially olives, the ages of which can be roughly estimated from their trunk diameters.[26]

On the low interfluves that constitute most of the kámpos itself, sections show cobble and gravel layers and lenses and the gravelly infills of ancient channels. For example, in a section 1 km north of Mourniés (A, Fig. 16.8), a continuous coarse gravel layer one metre thick, overlain by a metre of red sandy clay, extends 21 m northwards and then thickens and splits into two layers separated by a metre of sandy clay

[24] cf Rackham & Moody 1996, Fig. 12.4.
[25] Moody 1987, pp. 11, 30–31; I. Sanders, *Roman Crete*, Aris and Philips, Warminster, England, 1982, pp. 27–29.
[26] We are grateful to Dr Moody for assistance with interpreting finds.

Fig. 16.8. Plain of Khaniá, Crete.

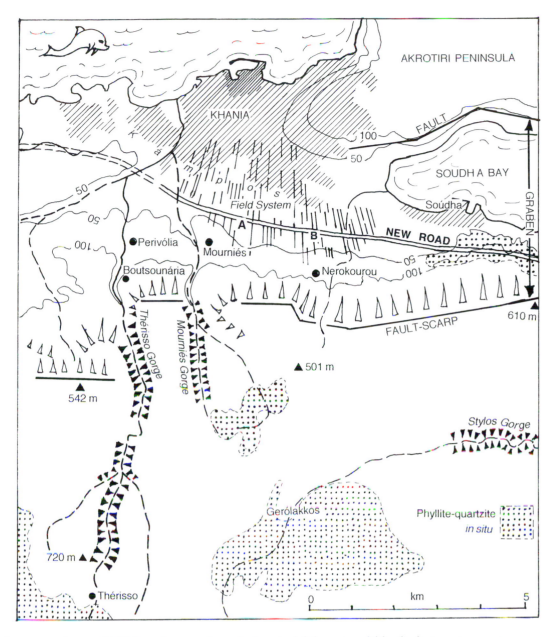

(Fig. 16.9). This is interpreted as a gravel spread laid down by the Mourniés torrent. Sherds within and at the top of this gravel bed are of Hellenistic to Late Roman age, indicating that the gravels accumulated during or after Roman times. A 10-cm layer of fine gravel lies 20 cm above the coarse gravel layer. Just above it, at a depth of 50 to 70 cm, are medieval tiles and sherds. Somewhat thicker layers of similar fine gravel, at a depth of 25–50 cm, exposed by the Nerokoúrou–Mourniés road a kilometre upslope to the south, contain medieval to seventeenth-century potsherds and tiles.

Other road-cuts near Nerokoúrou tell different stories. The gravel beds rise and fall, indicating a rugged topography with many small gullies. Dr Moody points out that in places the sherds associated with them (which are not worn, and have not been transported far) indicate a mid to late Bronze Age deposition covering a Minoan land surface. Bronze Age material sometimes comes within 25 cm of the present (often altered) surface.

In the vicinity of the present streams, exposure and burial of tree roots show that floods have both removed and deposited material down to within the last century or two.

On the upper part of the plain, towards the scarp, the deposit changes to sandy and clayey material that lacks any bedding and is colluvial rather than alluvial (stream-deposited). Slow downslope movement of soil, through creep, rain splash, sheet- and rill-erosion, has resulted in this accumulation at the foot of the scarp. On the lower, gentler slopes, archaeological remains appear, with older artefacts deeper and younger ones near the surface. The higher part of this colluvium has in turn lost sediment through erosion.

On the scarp-foot slopes south of Mourniés, where terraces, irrigation ditches and ploughing complicate soil profiles, 30–50 cm of soil have been removed from the bases of olive trees 2–3 m in diameter and probably 3–4 centuries old. This scarp-foot zone, where water is now within reach of wells 6–7 m deep, has been a favourite place for settlement.

As an example of this footslope, near Boutsounária (west of the deeply incised Thérisso river, but well out of reach of all but its most exceptional floods), a roadside section, 2 m deep, reveals three beds of gravel (< 4 cm diameter), each 10–20 cm

A

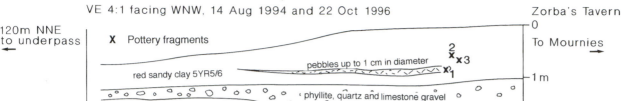

Section to S of Mournies underpass
VE 4:1 facing WNW, 14 Aug 1994 and 22 Oct 1996

Zorba's Tavern

120m NNE to underpass ←

To Mournies →

X Pottery fragments

red sandy clay 5YR5/6

pebbles up to 1 cm in diameter

2
X X 3

X 1

' phyllite, quartz and limestone gravel
up to 10 cm in diameter, sub-rounded
with long axes close to horizontal

grey-brown sand and small pebbles

m 32 30 28 26 24 22 20 18 16 14 12 10 8 6 4 2 0

0

1 m

2

B

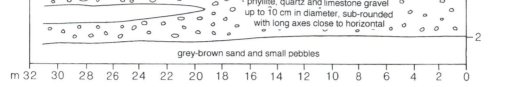

Section to N of Nerokourou underpass
VE 2:1 facing WSW, 15 Sept 1994 and 22 Oct 1996

To Khaniá ←

20m SSE
to underpass →

2-1
X X

X 3

gravel with
some boulders

5
X X 6

7
X

4
X

34m ASL

0 2 4 6 8 10 12 14 16 18 20 22 24 26 28 30 32 34 36 38m

Fig. 16.9. *Above* Road-cuts in the Khaniá plain. A, north of Mourniés; B, north of Nerokoúrou. Fragments of pottery, mainly in water-deposited gravel layers at a depth of a metre or two, provide some indication of the maximum age of deposition. In A, X1, X2 and X3 are probably medieval. In B, X1, X2 and X3 are all small orange fragments of uncertain age; X5, X6 and X7 are probably Bronze Age sherds. Based on a study by H. Raab.

Fig. 16.10. *Left* Section provided by the Khaniá bypass about 1 km north of Mourniés. *October 1994*

thick and, at the bottom of the section, a gravelly bed up to 40 cm thick including cobbles. All these are interbedded with a red to yellow sandy clay. Two medieval to seventeenth-century sherds were found in the top 20 cm, and a Roman sherd was at the top of the middle fine-gravel bed, 90 cm down.

Where did the material come from? Most of it is not the product of soil erosion. The deposits contain rounded or subangular stones of phyllite, quartzite and limestone. Phyllite and quartzite are not local, but came from several km away west or south. They may represent gravels held in an earlier deposit in the Thérisso Gorge. This gorge, unlike nearly all others in Crete, has no gravel beds, but cuts down to solid limestone. However, patches of rounded gravel can be seen cemented to its nearly vertical sides, up to 10 m above the bed (Fig. 16.11); they indicate that the gorge once held several metres of coarse fill, later swept out and down on to the plain

by one or more deluges. A ravine cut in the alluvial fan on which Boutsounária hamlet now sits, at the mouth of the gorge, provided additional sediment. Similar outbursts are likely to have occurred from the Mourniés gorge, where large accumulations of sand and gravel remain in side-valleys and karst hollows. Further east, phyllite-quartzite gravels may have formed a deposit on the less steep slopes of the scarp, held there until a combination of uplift and climate released it down on to the plain. What may be a residual patch of such a deposit still clings to the slope above Nerokoúrou.

The story is unexpectedly complex, and any inferences must be provisional. A systematic archaeological survey, accompanied by trial excavations, is badly needed. However, some conclusions are admissible. For the past fifteen hundred years the plain has had an even, relatively stable surface. The presence of a field boundary every 150 m or so has interrupted soil movement. The streams have incised, and floods have been active only close to them. Only on the footslopes, above the regular fields, has there been much colluvial activity.

In its earlier history the body of the plain has been uneven and less stable than it is now. At least twice, cataclysmic deluges, walls of water crashing out of the gorges, deposited

Catastrophes, therefore, may occur of a much greater magnitude than any that have been scientifically observed. It is also clear that some of the most productive soil in Crete can develop out of debris in only a few centuries.

The Frangokástello plain

This plain, on the south coast of Crete, was studied by the Sphakiá Archeological Survey. More gently sloping than the Khaniá *kámpos*, it is backed by a tremendous fault-scarp split by gorges, whence on rare occasions walls of water may burst out to overwhelm the plain (Figs. 16.12, 16.13). (The gorges still retain large quantities of scree and gravel.) Underlying the present surface is a thin layer of stones and mud spread more or less evenly over the plain, burying the archaeology, except on slightly raised areas where Roman farmsteads (and

Fig. 16.11. Rounded gravels remaining in hollows in a rocky slope overlooking the road up the Thérisso Gorge. The stream-bed must have reached at least to this level (10 metres above the present bed), probably until a few hundred years ago. *October 1994*

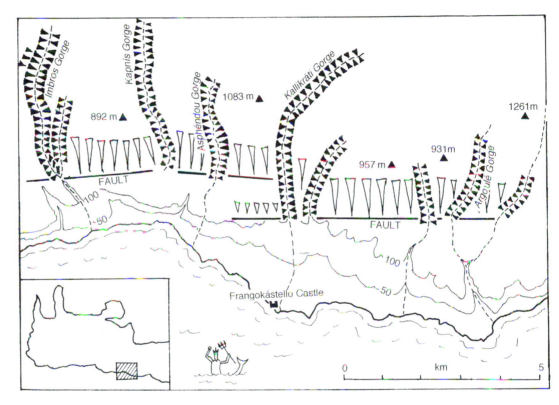

Fig. 16.12. Plain of Frangokástello, Crete.

great masses of coarse and fine material, and left a gullied surface later more or less filled in.[27] The previous settlements vanished. The drainage pattern was probably transformed, and may have been braided for a time. Our own experience of a deluge (p. 248) suggests that much of the upheaval on the plain may have been the recycling of deposits already on it; how much material was brought down by these floods, and how much by deluges further back in prehistory, we cannot say. The last deluge, for some reason, was followed by an infilling phase which left a relatively smooth surface on which some unknown, probably early Byzantine or Arab, authority laid out the regular field system.[28] Later deluges have not been sufficient to have more than local effects.

their carob trees) are still visible. It is later than the earliest Byzantine chapels, which protrude through it. The Roman and Byzantine material itself lies on one or more similar, earlier deposits – they are not stratigraphically distinct – which bury Bronze Age buildings on a still earlier land surface. Here, too, the flood deposits or mud-flows contain phyllite, either from small deposits at the base of the cliffs or washed down through the gorges.

The lack of holes or road-cuts makes dating the Frangokástello events even more problematic than those of Khaniá. Nevertheless, it is possible that the two plains were each covered by alluvium twice by roughly contemporary events.

---

[27] For the testimony of one who witnessed such an event and lived, see p. 25.

[28] A later date for the field system seems excluded by the fact that it transcends ownership and township boundaries. Nobody in the Venetian period or later controlled a large enough tract of land to impose this large-scale geometry.

Fig. 16.13. The plain of Frangokástello and the brooding menace of its gorges. *September 1998*

The southern Argolid: successive Younger Fills

In this eastern, rather arid, peninsula of the Peloponnese the geology is varied, ranging from hard limestones to marls and soft sandstones. The region gives the impression of being moderately unstable, but not so erodible as to form badlands.

Here Tjeerd van Andel and his colleagues have done perhaps the most detailed work on Mediterranean erosion. Fig. 16.14 is a summary diagram of the debris flows and sedimentation which they find in valleys.[29] Older Fill consists of three glacial and interglacial formations. During the Holocene, the area seems to have been more erodible than earlier or at present; they reckon that its various sub-regions have lost between 1.5 and 6.0 m of 'soil' over the past five thousand years. The Holocene erosion phases that they recognise are late Early Bronze Age, Hellenistic, middle

Fig. 16.14. Summary diagram of Holocene valley floor deposits in the Argolid. The debris layers, made up of coarse gravels in a finer matrix, are more resistant to erosion than the intermediate sandy and loamy layers, and form ledges in weathered sections. (Redrawn from van Andel & Zangger 1992.)

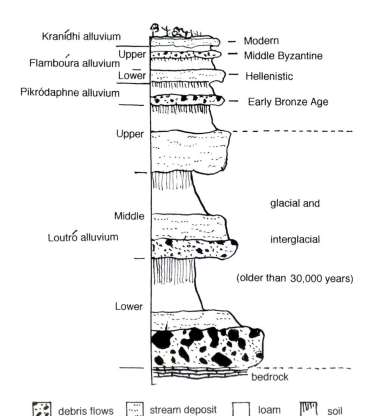

[29] Van Andel & Runnels 1987; Jameson & others 1994; K. O. Pope and T. H. van Andel, 'Late Quaternary alluviation and soil formation in the southern Argolid: its history, causes and archaeological implications', *JAS* 11, 1984, pp. 281–306; van Andel & others 1986.

Byzantine, and post–1700. These phases they associate with different kinds of human activity: rapid expansion of settlement, *or* high population density, *or* collapse of settlement.

By far the greatest Holocene destabilization was in the Early Bronze Age. After a long period of stability, a series of sudden debris-flows laid down sediments on valley floors. These materials, the Pikródhaphne Alluvium, contain Bronze Age sherds and have Bronze Age sites on them; they indicate massive mud-flow erosion between 2300 and 1600 BC.[30] The investigators interpreted this as the collapse of a 'thick soil mantle'. For a cause they considered climatic change or at least 'a few anomalous wet seasons after a long dry period', but decided in favour of a gradual intensification of land use involving a reduction in fallow and expansion of cultivation on to steeper, less stable slopes. However, human economy and population had declined by the time of the collapse, and it is difficult to demonstrate what human activity (short of bulldozing) might have set off the debris-flow mechanism.

diversion channel were built, it is supposed, to prevent a repetition.[31]

These explanations make little reference to tectonic action as a factor. However, this region is not far from the Corinth fault-zone. A recent Greek contribution points to fault movements as an additional source of instability in the late Pleistocene and Holocene.[32]

The volcano island of Mélos

This dry, unstable island in the Cyclades, a famous source of obsidian for stone tools, is described by Davidson and Tasker as 'in the first rank of those areas where erosion is particularly striking'. The associated deposition, on archaeological grounds, began c. 1100 BC and went on until at least 500 AD. Many of the deposits show signs of having been laid down by deluges. Davidson and Tasker conclude that much of the material originated as soil, of which 20–40 cm has gone from many hillslopes.[33]

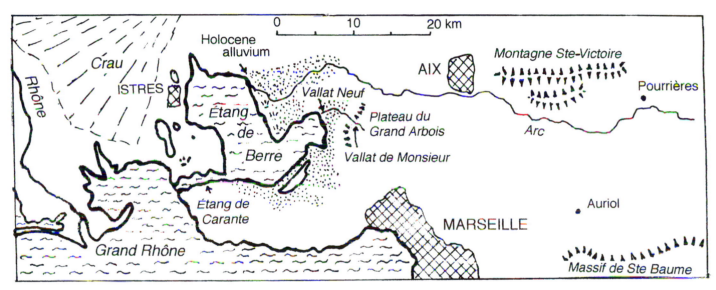

Fig. 16.15. Western Provence. The Étang de Berre lies in a shallow depression, separated from the Rhone delta by the Crau, an immense gravel fan laid down by the river Durance in the Pleistocene. The lagoon is fed by the Arc, which drains a marly depression some 70 km long between limestone ranges, and also by the small Vallat de Monsieur and Vallat Neuf, which come from the sandy and marly slopes of the Plateau du Grand Arbois. Cores from the terraces and deltas of these rivers provide a record of Holocene sedimentation, and hence a record of erosion in their catchments.

On the nearby Argos plain, before 2000 BC, reddish-brown flood-plain deposits 1–3 m thick, containing Early Bronze Age pottery, were spread by the Inachos river across the early Holocene plain. This material is believed to have come from the erosion of brown 'woodland-type' soils overlying Pliocene marls that outcrop around the northern margins of the basin. Since that time no alluviation events have affected all the Argive plain. However, late in the Bronze Age, the lower town of Tiryns was buried under several metres of alluvium by torrential flooding; subsequently a dam and

Debris flows: the Aspromonte, Calabria

The Aspromonte mountains of Calabria, southern Italy, wrapped in a shroud of soft Neogene sediments, are another tectonically lively area, which has risen during the Pleistocene by 1,100–1,300 m.[34] Earthquakes of Richter magnitude 8 recur about once a century; there was a very severe one in 1783 (the 'dry fog' year, p. 137). Erosion in geological times covered the middle slopes with thick, unconsolidated material, setting the scene for earthquakes and deluges to generate slumps and debris flows.

30 Van Andel & others 1990.
31 Van Andel & others 1990; E. Zangger, *The Flood from Heaven*, Sidgwick and Jackson, London, 1992.
32 G. Gaki-Papanastassiou and H. Maroukian, 'Late Quaternary controls on river behaviour

in the eastern part of the Argive plain, eastern Peloponnese, Greece', Lewin & others 1995, pp. 89–95.
33 Davidson & Tasker 1982.
34 M. Sorriso-Valvo, 'Landslide related fans', *Catena* Suppl. 13, 1988, pp. 109–21; P. Ergen-

zinger, 'A conceptual geomorphological model for the development of a Mediterranean river basin under neotectonic stress (Buonamico basin, Calabria, Italy)', *Erosion, Debris Flows and Environment in Mountain Regions*, ed. D. E. Walling, T. R. Davies, B. Hasholt, IAHS Publication 209, 1992, pp. 51–60.

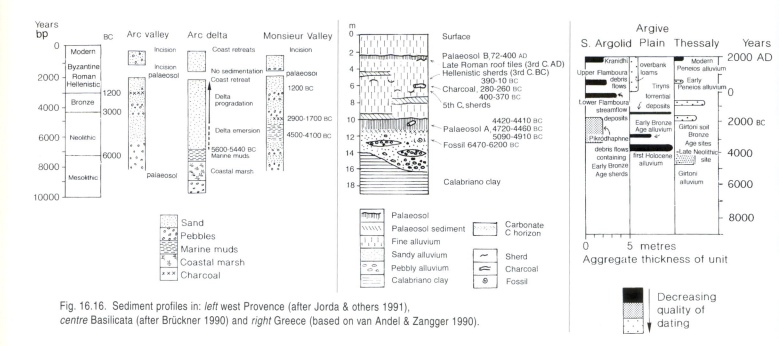

Fig. 16.16. Sediment profiles in: *left* west Provence (after Jorda & others 1991), *centre* Basilicata (after Brückner 1990) and *right* Greece (based on van Andel & Zangger 1990).

At some time in the Holocene a huge debris flow deposited 60 million cubic metres of material on the floor of the Buonamico valley, giving the river a Herculean task of transporting the debris down to the sea. The Constantino landslide, after heavy rains in 1953, seems to have been on a similar scale. A layer of debris 50 m thick was deposited in the narrow mountainous section of the valley; 7 km downstream the width of the river increased to 800 m. A slide in 1972–3 deposited about 7 million cubic metres on the valley floor, where it is now being eaten into by the river.

Here, as in other tectonically very active areas, slopes slump and erode rapidly, and material accumulates on a valley floor. The river carries away the more easily transported finer sediment, incising its channel and creating a flood-plain, with the older material forming a terrace. Such terraces, depending on local events, are unlikely to fit into a regional chronology.

HOLOCENE ALLUVIATION:
A CHRONOLOGICAL SUMMARY

Pre-Neolithic

The final deposits of Older Fill are probably attributable to the Younger Dryas mini-glaciation, about 10,000 BC (p. 22). They may include the 'Upper Fill' in the Odelouca and Enxerim valleys north of Silves in the Algarve, deposited in the Late Pleistocene to heights up to 40 m above river level, and since cut into by as much as 30 m.[35]

In Provence, following a period of downcutting probably in the Allerød warm period (p. 43), a deposit of coarse material has been dated near the top to 8450 ± 450 BC. A later fine-grained 'Holocene principal terrace' was deposited between 7200 and 2500 BC.[36]

In Basilicata, the post-glacial rise in sea-level was accompanied by extensive deposition and accumulation of river alluvium. Radiocarbon dates on an 'embedded fossil' and on underlying buried soils indicate that deposition was in progress around 6000 BC; it continued for at least the next thousand, possibly more than three thousand, years.[37]

Neolithic

Sediments accumulating after 4300 BC in the Étang de Berre, Provence, contain increasing quantities of material interpreted as soil (Figs. 16.15, 16.16). However, in this area, early Neolithic activity does not seem to have resulted in much loss of soil; garrigue was as effective as wildwood in restraining erosion.[38]

In Thessaly, the Girtóni alluvium of the Penéios river is attributed to c. 4500–4000 BC; Bronze Age sites were established on top of it. In general, depositional history in this part of Greece is a function of distant events in the very erodible high Píndos mountains; but van Andel and colleagues point to truncated soil profiles under Neolithic settlements on late Pleistocene to mid Holocene river terraces, and argue that the Girtóni alluvium is the result of soil erosion a thousand years after the Lárissa plain was first settled.[39]

Bronze Age

Deposits in western Provence indicate that soils on the hills to the north were being eroded at the time of the Neolithic to Bronze Age transition, 3500–2900 BC. Water levels in lagoons within closed basins were high, suggesting greater rainfall than at present. Along the middle course of the Vallat Neuf two-thirds of these 'Sub-boreal' sediments,

35 Chester & James 1991.
36 9440±220 bp, 8200–4000 bp, J. Vaudour, 'Introduction à l'étude des édifices travertineux holocènes', *Méditerranée* 57 (1/2), 1986, pp. 3–10; Ballais & Crambes 1992; Jorda & others 1991.
37 Néboit 1984; Brückner & Hoffman 1992.
38 Provansal & others 1994; Leveau & Provansal 1993.
39 Van Andel, Gallis, Toufexis 1995.

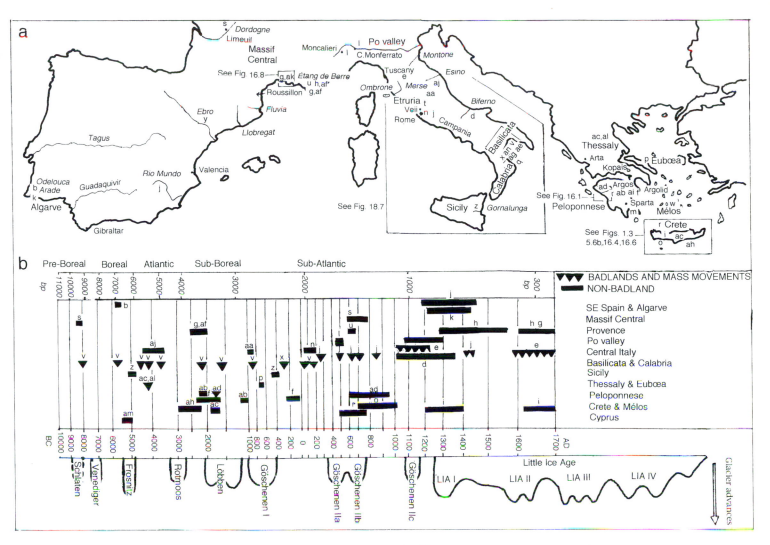

Fig. 16.17. a. Places where Younger Fill has been studied. More detailed maps of West Provence, Italy, the Peloponnese and Crete appear on pp. 301, 340, 292, 297, 299.

b. Dates of Holocene alluvial deposits. Those from badlands, areas of mass movement and unstable volcanics are plotted separately. Also shown are periods of glacial advance and phases of the Little Ice Age. The time-scale is logarithmic, expanding towards the present, to compensate for the generally greater uncertainty of earlier dates. For the key to the letters, see Table 16.iv (p. 310).

apparently the result of torrential floods, accumulated around 2500–2000 BC.[40]

A layer of sediment in Basilicata, deposited later than the main early to mid Holocene terraces, is topped by a buried soil containing charcoal that was dated by Néboit (1984) to c. 2100 BC, but Brückner has obtained dates from two samples of the palaeosol itself of between 5090 and 4410 BC (Fig. 16.16).[41]

Reddish, somewhat clayey, 'palaeoxeralfic' soils in the Mesará in south Crete, locally surmounted by Neolithic implements, were eroded in the middle Bronze Age, about 1900–1700 BC. The river Geropótamos carried the sandy sediment down to the sea and redeposited it on the shore, where winds built up sand-dunes by the river mouth. Locally the dunes covered, and have since protected, early Bronze Age soils.[42]

Iron Age to Early Roman

In the Ebro basin of Spain, van Zuidam found a maximum period of erosion and alluviation between 500 and 100 BC, which he attributed (on not very clear evidence) to the growth of 'Celtic' settlement and cultivation, especially of corn and wine as cash crops. In the same basin, Gutiérrez–Elorsa and Peña–Monné find widespread slope accumulation between 900 and 300 BC, which they attribute to solifluction in a cool wet period. The two interpretations are not necessarily incompatible.[43]

In west Provence, lagoon sediments indicate that erosion was not marked between the Early Bronze Age and late Roman times, although dispersed settlement was quite dense. Cultivation terracing has been invoked to explain the lack of soil losses.[44]

[40] Radiocarbon date 4480±280 bp. Leveau & Provansal 1993; Provansal 1995.

[41] 3760±110 bp, Néboit 1984; 5700±120 bp and 6100±60 bp, Brückner 1990.

[42] Pope 1993.

[43] Van Zuidam 1975; M. Gutiérrez-Elorsa and J. L. Peña-Monné, 'Geomorphology and late Holocene climatic change in Northeastern Spain', *Geomorphology* 23, 1998, pp. 205–17.

[44] Jorda & Provansal 1990; Ballais & Crambes 1992; Provansal & others 1994.

In eastern Sicily, river-terraces began to accumulate about 5000 BC; Bronze Age sites rest on them. Subsequent dissection was followed by further Iron Age alluviation within the resulting valleys. This began before the fifth century BC and continued for more than two centuries, giving a terrace deposit 10 m thick.[45]

In Basilicata, a sediment layer containing sherds from the fifth to fourth centuries BC accumulated in Hellenistic and Roman times. On the Brádano river, this terrace contains charcoal dated 200±160 BC[46] at a depth of 4.2 m. The soil on top has Roman roof-tiles of 70–400 AD. The equivalent terrace in the Cavone valley contains the ruins of a farm destroyed by a flood during the first half of the fourth century BC. The sediment is described as 'bedrock material from the hinterland and the valley slopes (Calabriano clays)', but elsewhere an alternation of thin layers of mineral-rich and organic material within the terrace deposits is held to indicate *soil* erosion. Néboit-Guilhot relates these Iron Age terraces in Sicily and Basilicata to colonialism. Greek immigrants are supposed to have pushed the indigenous people up into the hills, where they cleared land to grow cereals for sale to the Greeks, with, it is conjectured, consequent soil erosion.[47]

Van Andel and Runnels tell of the re-population of the Argolid about 1000 BC, and the expansion of olive-growing on to thinner soils in the fifth century. They depict sedimentation of Hellenistic date (the Lower Flamboúra stream-flood deposits) and write of a gully erosion event 'quite precisely dated to the last few centuries BC (c. 250–50 BC)' that 'took place almost simultaneously over the entire southern Argolid'. They are inclined to attribute it to 'the abandonment of large outlying tracts too extensive to maintain, and perhaps also some coastal ones judged to be too exposed to pirates'. They note a similar decline elsewhere in Greece in the early Roman period.[48]

Late Roman and medieval

Post-Roman river deposits are numerous, and often belong to more than one period. An earlier deposit, about 700 AD, is often followed by one more several centuries later.

In the Algarve, more than 3 m of sediment, containing charcoal and Roman to post-medieval pottery, lie on the floors of the Odelouca and Enxerim valleys. Deposition began around the early Iron Age, to judge by charcoal from a depth of 2.7 m dating to 850±50 BC. It was continued or resumed late in the Roman period, for a mill buried at 1.6 m has been dated to 300±120 AD. The Algarve in late Roman and Moorish times was a prosperous North African colony, and forests – it is said! – were being 'cleared' in the headwaters. The area is supposed to have lost most of its population after the Christian conquest in 1249 AD, but deposition continued, or recommenced, with two dates of 1250±30 AD and 1390±60 AD. The deposit was incised in a subsequent period of relative stability.[49]

In Roussillon, 3 m of river sediment overlie a Roman roadway and medieval masonry at Bompas, 6 km NE of Perpignan. The sediments include 2 m of pale sandy silts, which resemble the 14 cm of deposit left to the SE of Perpignan by the great flood of October 1940 (p. 250). It would have taken many such floods to accumulate 2 m of thickness.[50]

In Provence, Ballais and Crambes describe a river-terrace dated to 625±25 AD. They attribute the onset of sedimentation in valley bottoms to decline of population, abandonment of cultivated land and erosion of slope soils. At the foot of Montagne Ste-Victoire, charcoal from the lower half of another terrace, consisting of 1–3½ m of calcareous pebbles covered by red-brown silty clay, has given dates of 59–429 AD and 438–660 AD. Further deposition occurred between 1300 and 1600 AD. A terrace in the Vallat de Monsieur and on the Arc was initiated in the Middle Ages, but is largely the outcome of deposition between the seventeenth and nineteenth centuries (Fig. 16.16). A thick layer of lagoon sediment also points to rapid erosion in the late Middle Ages, and again especially in the seventeenth to nineteenth centuries.[51]

Vita-Finzi (1975) has dated Younger Fill alongside the upper Po, near Casale Monferrato: wood buried in it at a depth of 11 m dates to 480±60 and 490±60 AD. At Moncalieri, 70 km upstream, wood at 7 m depth has given dates of 1120±80 and 1250±30 AD. These dates allow for two alluvial phases in this part of Piedmont, as in Provence, one post-Roman, the other medieval and later.[52]

In the Arbia tributary of the Ombrone, southern Tuscany, Younger Fill has been dated to 660±120 AD. In a survey of the Merse valley, another tributary of the Ombrone, three stages of erosion (and the corresponding slope deposits and coarse-gravel alluviation) have been dated archaeologically. On the basis of the youngest artefacts in each layer, these date to the eleventh–twelfth, sixteenth–seventeenth, and late nineteenth centuries. The three stages are interpreted as contemporary (within a century or two) with times of economic expansion: agricultural clearances are supposed to have penetrated marginal upland habitats, allowing extreme storms to erode bare ground.[53]

Further south, although some alluvium was laid down seawards of Rome between AD 50 and 200, most deposition appears to have taken place between late Imperial and early medieval times. At various sites around Rome, wood or charcoal specimens in Younger Fill give dates calibrated to 300–800 AD and to the early fifteenth century. There is a similar story nearby in south Etruria adjacent. Valley floors were settled in the first to third centuries AD. Earlier and later, there was too much flooding; deposits include boulders of several tons.[54]

In the Gornalunga valley of east-central Sicily, Judson found a deposit forming a terrace 4–5 m high containing ancient and medieval sherds. In Basilicata a sediment layer 1–2 m thick accumulated after the third century AD to form a terrace 'probably linked to the Middle Ages'. It was incised

[45] Néboit 1984; Néboit-Guilhot 1991.
[46] 2160±160 bp.
[47] Cottechia & others 1969; Néboit 1977.
[48] Van Andel & Runnels 1987; cf S. Alcock, 'Roman Imperialism in the Greek landscape', *Journal of Roman Archaeology* 2, 1989, pp. 5–34.
[49] 2690±70, 1750±90, 780±50, 520±60 bp; Chester & James 1991.
[50] Calvet 1993, 1994; Jacob 1997.
[51] 1300±25 bp. Ballais & Crambes 1992; Jorda & others 1991; Leveau & Provansal 1993; Provansal & others 1994.
[52] 1595±50, 1580±50, 900±50, 790±50, 1400± 100 bp; Vita-Finzi 1975.
[53] 1670±95, 1675±45, 1350±95, 490±30 bp; Hunt & others 1992.
[54] Vita-Finzi 1975; Judson 1963a, 1963b; Potter 1979.

between the eleventh and fifteenth centuries (Fig. 16.16).[55]

In Gidés, Eubœa, after supposedly conservative farming in (drier) Roman times, and before the coming of the Venetians, there was deposition at a rate of one metre per century, although population pressure was low from the seventh to the tenth centuries.[56]

In the southern Argolid, Van Andel and Runnels find 'sheet erosion or catastrophic slope erosion probably during the Middle Byzantine reoccupation' on newly cleared land 'after the empty years of the seventh through the ninth centuries'. Upper Flambóura debris flows were deposited in the valleys below the new settlements.[57]

DATES OF SLUMPS AND LANDSLIDES

In the unstable headwaters of the Llobregat and Fluviá (NE Spain), ten mass movements have been radiocarbon dated: two between 2140 and 1420 BC, four between 1 and 420 AD, four since 1630 AD. Of 66 mass movements since 1880 dated from historical records, 21 occurred between 1907 and 1923, and 15 in 1940. Of 22 since 1940 dated from tree-ring counts, all occurred after 1955, most of them with the deluge of 7–8 November 1982. On that occasion up to 300 mm of rain fell, setting off thousands of shallow landslides, debris flows and rockfalls. Many of these, as is usually the case, reactivated old slumps, such as could have been set in motion originally by great earthquakes such as those of 1755 or 1427–8.[58]

The Florentine Apennines are underlain by a chaotic assemblage of clays, marls and permeable sandstones; the steep valleys that cut into them on both sides are disrupted by giant landslides. On the Adriatic side, the Montone, the river whose sediments bury ancient Ravenna, was dammed by a massive slide, probably between 400 and 750 AD. By 1302 the resulting lake had been filled in, and the river had been diverted into a waterfall which, when Dante saw it, reminded him of the cascade of boiling blood in the eighth circle of Hell. In the upper San Godenzo valley, a headwater of the Arno, a landslide that occurred in 1315 (according to Giovanni Villani's *Istorie Fiorentine*) was reactivated in 1641 and again in 1827. North of Florence, a landslide dammed the Santerno river to form a lake that later drained away; on its bed the town of Firenzuola was founded as a Florentine border post in 1332. Wood in the lake sediments has been dated to 550–660 AD. (There were earthquakes at Ravenna in 492, 501 and 502, and one at Faenza nearby in the first quarter of the sixth century.)[59]

CONCLUSIONS

The original scheme of Older and Younger Fill has to be elaborated. Both of these Fills are multiple, and (like most archaeological phenomena) the Younger Fill turns out on further investigation to have a wider spread of dates than at first supposed. It has to be supplemented by a Bronze Age phase, 2700–1700 BC, and by earlier, less well-defined phases. Where dates can be ascertained precisely enough, alluviation within each phase tends to be local, not synchronous over a wide area.

Attempts at interpreting fills are hampered by vagueness of date. Although many fills are the result of sudden events such as slumps, there is often a long, ill-defined interval between the youngest previous and the oldest subsequent artefact or other datable object. When objects are embedded in the fill itself, it is often unclear what relation they bear to the time-span over which the fill was laid down. For objects dated by radiocarbon, there is the additional uncertainty inherent in the dating process.

The publications of fills from which we work are of variable quality. Many concentrate on the alluvium and say little about its context. They often fail to reveal where the fill came from, how erodible the area is, or even whether badlands are involved. They may be vague about the extent of human activity at the time it was laid down: they tend to be perfunctory with historical documents, often quoting them at second or third hand so that the original cannot be verified. Very rarely do they say anything about climate, unless to assert that there is no reason for supposing it to have been different from today.

In Table 16.iv and Fig. 16.17 we have assembled all the reasonably well-dated Holocene alluvia in Mediterranean Europe known to us, and whether or not (as far as we can discover) they come from badlands or places where mass movements are common. Non-badland deposits bear out the theory of a Younger Fill, dated between 400 and 1400 AD, but with deposits ranging widely on both sides of this period; often there are two Younger Fills in the same place. With badlands, the Younger Fill is earlier and less well marked. Both types of deposit tend to be episodic, with peaks around 2300 and 4300 BC and a marked lull around 3000 BC.

Badlands and the sediments arising from them attract attention out of all proportion to their small extent. They might be expected to erode intermittently most of the time, but Fig. 16.17b suggests that badlands have their episodes of activity like other terrain, but not always at the same time. For example, the badlands of Thebes (Bœotia), active at some time since the Roman period, are now almost dormant (p. 284). The earthquakes, rains, or engineering works that trigger mass movements (p. 243) might be expected to operate on a similar local scale.

This analysis does not bear out Vita-Finzi's theory that the ages of Younger Fill deposits increase with latitude. On the contrary, those from Crete and Greece tend to be earlier than from the south of France.

55 Néboit-Guilhot 1991; Brückner 1990.
56 Genre 1988.
57 Van Andel & Runnels 1987, p. 145; van Andel & others 1990.
58 J. Moya, J. M. Vilaplana, J. Corominas, 'Late

Quaternary and historical landslides in the south-eastern Pyrenees', Matthews 1997, pp. 55–73; F. Gallart, 'Land conservation and hydrological studies in the high Llobregat basin (Barcelona)', Sala & others 1991, pp. 91–96.

59 Dante, *Inferno* xvi.94–105; G. Rodolfi, 'Holocene mass movement activity in the Tosco-Romagnolo Apennines (Italy)', Matthews 1997, pp. 33–46.

Differences in the rates at which evidence is discovered

Reported alluviations become ever more frequent and more narrowly dated as time approaches the present. The more recent an episode, the more likely it is to be noticed, especially as it will have destroyed, reworked or buried the evidence of earlier episodes. Dating methods tend to get more precise, although sometimes (as in Crete) medieval artefacts are difficult to date, and radiocarbon dates are of little use in the last three centuries.

In Crete, any passer-by can notice recent alluviations that have buried churches. Late Bronze Age or Late Roman alluviations can be detected fairly easily by experts, for sites and potsherds of those periods are abundant. In the Classical period, with sparser artefacts, episodes can easily be overlooked or misdated. It is difficult to identify a Neolithic alluviation, because sherds from that period are few and less distinctive. It would be very difficult indeed to recognize one from the early Holocene, since no pre-Neolithic artefacts are known at all from Crete. In general, therefore, greater weight should be given to such earlier alluviations as are known.

Correlation with glacial advances

We established in Chapter 8 that deluges of a sufficient magnitude to move large quantities of sediment, although they can occur at any time, are associated – on the basis of historical records and archaeology – with the Little Ice Age. We should look therefore for evidence of such an association with Little-Ice-Age-like events earlier in the Holocene (Table 16.ii).

Older Fills go with glacial and periglacial conditions, the latest probably attributable to the Younger Dryas period. A cluster of dates from southern France (and several from badland south Italy) could be contemporary with the Venediger and Frosnitz glacial advances in the Alps. Extra weight must be attached to these in view of the difficulty of recognising deposition at that early period.

Dating is seldom precise enough to establish a definite correlation between 'Little Ice Ages' and alluviations. However, Göschenen advances IIb and IIc are clearly to be reckoned with as factors in the main period of Younger Fill, as is Phase I of the Little Ice Age. Phases II and III, although better documented, have less associated alluviation.

The Löbben advance goes with several of the earlier Younger Fills. However, there is no glacial advance to account for the alluviations of around 400 BC to 350 AD and between 4900 and 3350 BC; and the Rotmoos advance comes at a time without known alluviations.

Occasionally climatic change is indicated in the deposits themselves. The Impruneta alluviation (Tiber basin) about 920 BC gave a terrace deposit 7 m thick containing Iron Age pottery. Molluscs in the terrace indicate a cool humid climate, corresponding to the Göschenen I minor glaciation. Incision followed between 250 BC and 150 AD.[60]

Glacial advance	Date	Number of fills known (non-badland + badland)
Last 130 years		3 + 2
19th-cent. advances	1810s and c. 1850 AD	2 + 0
interval		–
Little Ice Age III	1690–1800 AD	3 + 1
interval		*2 + 1*
Little Ice Age II	1570–1640 AD	3 + 1
interval		*1 + 2*
Little Ice Age I	1280–1380 AD	5 + 0
interval		*3 + 1*
Göschenen IIc	1080–1150 AD	2 + 1
interval		*2 + 1*
Göschenen IIb	600–800 AD	5 + 2
interval		*1 + 1*
Göschenen IIa	350–500 AD	1 + 1
interval		*2 + 4*
Göschenen I	1000–400 BC	3 + 1
interval		*1 + 0*
Löbben	2100–1200 BC	4 + 2
interval		*3 + 1*
Rotmoos	3350–2700 BC	0 + 0
interval		*1 + 3*
Frosnitz	5500–4900 BC	2 + 0
interval		*1 + 1*
Venediger	7700–6900 BC	0 + 0
interval		*1 + 0*
Schlaten	c. 8300 BC	1 + 1

Table 16.ii. Correlation between glacial advances and fills. Data for intervals between advances are shown in italics.

Luminescence dating techniques begin to throw light on the problem of early alluviations. In Mediterranean Europe they have first been applied in NE Spain, where dates spanning the last two hundred thousand years have been found for the Pleistocene terraces of the Guadalope, a tributary of the Ebro. The dates of these alluvia have been compared with oxygen isotope records from the Greenland ice and from marine sediment cores, and with European pollen records. It is claimed that river terraces built up during the ice advances of the last two glaciations in the Pleistocene and the Little Ice Age events in the Holocene.[61]

An activity thought to be related to changes of climate, although the connexion is very unclear, is the formation of soft limestone (travertine, tufa) beds in valley bottoms. Calcium carbonate was coming out of solution in the early Holocene; about 2500 BC this process ceased on a large scale, though it still operated with certain springs and leaks from Roman aqueducts. The change coincides roughly with the onset of the Aridization (p. 145). Rivers were becoming more seasonal and more variable. Other factors could include changes in atmospheric carbon dioxide, increasingly muddy waters and declines in water-plants.[62]

[60] Esu & Girotti 1991; Potter 1976.
[61] Fuller & 4 others 1996.
[62] Méditerranée, 57 (1/2), 1992, entire number; A. S. Goudie and others, 'The late-Holocene tufa decline in Europe', The Holocene 3, 1993, pp. 181–86; H. I. Griffiths and H. M. Pedley, 'Did changes in Last Glacial and early Holocene atmospheric CO_2 concentrations control rates of tufa precipitation?', The Holocene 5, 1995, pp. 238–42.

Tectonics

So far neither we, nor our sources, have said much about the links between water erosion and tectonic movement, its ultimate cause. Tectonics can promote erosion by steepening slopes or by earthquakes setting off slumps and mudflows, especially in badlands. These processes tend to happen more at less at random over a longer time-scale than that of the Holocene. However, tectonic action itself, or land-forms catching up with the instabilities that it creates, can be a rapid enough force to be confusable with climatic change.[63]

Human activity and vegetation

The way scholars work inevitably associates erosion and alluviation with human activities. They are often reported as a by-product of investigations by archaeologists whose main interest is in human activities. They are often dated by association with potsherds etc., and are therefore more likely to be assigned to periods from which artefacts are abundant. Alluviations tend to be uninvestigated in areas where there has been little conspicuous human activity.

As a first attempt we have tried to separate those Younger Fills that came at times of intensive or increasing human activity from those associated with recession in human affairs. There are also alluviations at periods of little apparent human activity: among these we include most Neolithic and all earlier alluviations, since Neolithic peoples in the Mediterranean seldom operated on anything like the scale of their Bronze and Iron Age successors. With some fills (especially those that are poorly dated or accumulated over a long period) human activity both increased and decreased. The data are summarized in Table 16.iii. Here, too, the sources are of variable quality: we have done our best to separate those of our informants for whom 'forest clearance' etc. is merely a stock explanation of erosion from those who produce independent evidence for it.

	Human activity				
	High or increasing	Constant	Decreasing	Slight	Variable
Non-badland fills	15	3	7	14	7
Badland fills	8	0	3	13	2

Table 16.iii. Incidence of fills in relation to human activity.

Any theory that alluviation is necessarily caused by increased population and agriculture is thus doomed by many counter-examples. The association with periods of increasing human activity is insignificant; it could even be negative, given the greater attention paid to the study of such periods than of recessions.[64]

Protagonists of the human-activity theory reply by invoking human decline as an alternative cause. Alluviation, they say, is caused not only by ploughing and woodcutting, but also by not ploughing: by abandoning terraces which collapse and shed their contents, or by farmers giving up cultivation and keeping hungry animals with busy hooves. They try to have it both ways. Such an argument needs supporting evidence to be convincing, and this is seldom provided. They seldom demonstrate that these particular farmers had terraces of the kind that erode rapidly when abandoned (p. 259), or that in their environment pasturage is indeed more erosion-promoting than tillage.

Ever since the first ploughman attacked a sloping field, people have been trying to promote erosion by removing vegetation and moving earth. Undeniably they have succeeded at times. Scholars try to attribute any episode of erosion since the Neolithic to human activity; it is easier to make such an attribution than to discover what the activity was or how exactly it caused the erosion, or to eliminate alternative explanations. (Requests for detail on these points tend to go unanswered.) If erosion occurred within five hundred years of people settling in an area, the temptation is almost irresistible to ascribe it to 'expansion of agricultural activity', however tenuous the evidence for that expansion. The reality may be that the greatest deluge in the Holocene happened to come in that interval, and would have caused erosion regardless of what people were doing. To demonstrate clearly that human action was the cause requires the chain of argument which we set out on page 290. Even the most careful scholars, such as van Andel, seldom forge all five links, and many never get beyond stage 1.

Some of the explanations seem rather lame and contrived. Abandonment of terraces and excessive browsing indeed create the occasional gully, but it is hard to see how they could, in ordinary conditions, banish millions of cubic metres of sediment. The range of possible activities is very wide, the connection seldom goes further than an apparent coincidence of date, and the dating itself is often vague. If woodcutting, increasing cultivation, abandoned cultivation, or intensified grazing will all do for an explanation, it would be surprising if one or another of them did not occur almost anywhere during some part of the five hundred years or so within which an alluviation can be dated.

Terracing is a human activity sometimes invoked to explain lack of erosion, for example in Provence, where terrace-building in Roman times and later is supposed to have brought the rapid erosion in prehistory to an end. Here, too, the argument lacks verification, as terraces are very difficult to date.[65]

Two studies involve detailed correlation between Younger Fill and human activity. In Molise (Italy) alluvial deposits have been dated, from associated artefacts, to the mid-Holocene, the Samnite and early Roman period, the high Middle Ages, and the late nineteenth and early twentieth centuries. The prehistoric episode is not dated closely enough

[63] A. E. Mather and 4 others, 'Tectonics versus climate: an example from late Quaternary aggradational and dissectional sequences of the Mula basin, southeast Spain', Lewin & others 1995, pp. 77–88.

[64] Information on the decline of erosion in periods of constant or increasing human activity is more difficult to quantify; an example is the healing of the Theban badlands (p. 284).

[65] Jorda & Provansal 1990.

Table 16.iv. Summary of dated Holocene deposits.

B: deposits likely to be derived from badlands and mass movements.
V: deposits likely to be derived from unstable volcanics.
Sources are listed on p. 310; * contribution to Fig. 16.10.
In giving a range of dates, we cannot usually distinguish between a single event that can be only approximately dated, and a number of events, or a continuous process, that went on for a period of time. Artefacts and charcoal can, of course, be much older than the deposits containing them.

Country and Province	Locality or name of deposit		Date	Relevant human activity	Remarks	Source	Holocene glaciations
It Calabria	Constantino	B	1973 AD	none	great slump	a	
Po Algarve	Odelouca, Enxerim		begins c. 1955 AD	'eucalyptus planting'		b	
It Calabria	Constantino	B	1953 AD	none	great slump	a	
Fr Roussillon	Elne	B	1940 AD	afforestation	great flood deposition	c	
It Molise	Biferno		1850–1930 AD	'prairie-farming'		d	mid 19th-century
It Tuscany	Merse	B	1850–1900 AD	increasing		e	glacier advance
Gr Argolid	Kranídhi		1700–2000 AD	increasing	local gullying	f	
Fr Provence	Étangs		1600–1900 AD	increasing		g*	
Fr Provence	Vallat de Monsieur		1600–1900 AD	increasing		h*	Little Ice Age III, c. 1700
It Tuscany	Merse	B	1600–1800 AD	increasing		e*	
Cr Monofátsi	Axéndi church		1615–c. 1700 AD	constant (late Venetian, early Turkish)		i*	
It Rome	Treia	B?	early 15th cent.	high		j*	Little Ice Age II, c. 1600
Fr Provence	Mont Ste-Victoire		1290–1570 AD	roughly constant?		h*	
Po Algarve	Boi		1220–1440 AD	sharp decline (post-Moorish)	2 radiocarbon dates	k*	
Sp Murcia	Rio Mundo (Liétor Alluvium)		1060–1280 until 1470–1600 AD	roughly constant? (late Arab, Christian)	4 radiocarbon dates	j*	Little Ice Age I, c. 1320
Cr Kydhonía	Aikyrgiánnis church		1200–1400 AD	probably increasing (early Venetian)	High Medieval	i*	
It Piedmont	Moncalieri		1040–1300 AD	expansion	2 radiocarbon dates	l*	
It Molise	Boiano		1000–1350 AD (weak)	expansion (medieval)		d*	Göschenen IIc c. 1100
It Tuscany	Merse	B	1000–1200 AD	'expansion' (medieval)		e*	
Po Algarve	Odelouca, Enxerim		c. 700–1500 AD	expansion (Arab) followed by decline (Christian)		b	
Gr Argolid	Upper Flamboúra		400–1700 AD	variable (Byzantine, early Turkish)	localized sheet erosion and debris flows	f	
Gr Peloponnese	Sparta		c. 500–1500 AD	variable (Byzantine)		m	
It Latium	around Rome		c. 400–1400 AD	variable (medieval)	major alluviation	n	
Cr Sphakiá	Frangokástello		c. 700–1000 AD	decline (Byzantine)		o*	
Gr Eubœa	Gidés		c. 500–1200 AD	low (Byzantine)		p	
It Rome	Crescenza		9th cent.	relatively low		j*	
It Basilicata	Brádano	B	c. 400–1200 AD	decline? (early medieval)		q	
Gr Peloponnese	Olympia (Kladhéios)		600–c. 1000 AD	probably low (early medieval)	5 m of fine sand	see p.292*	
It Etruria	Arbia	B	centred on 7th cent.	decline		j*	
Cr Kydhonía	Khaniá plain		500–c. 800 AD	high (early Byzantine), declining		r*	Göschenen IIb c. 600–800 AD
Fr Dordogne	Limeuil–Manaurie		600–800 AD	post-Roman 'abandonment'	7.5 m of silty sand over early Holocene	s*	
It Rome	Fiano	B?	7th or 8th cent.	recession		j*	
It Etruria	Arbia	B	centred on 7th cent.	recession		j*	
It Etruria	various	B	300–c. 1000 AD	recession (late Roman to early medieval)		t	

Country and Province	Locality or name of deposit		Date	Relevant human activity	Remarks	Source	Holocene glaciations
Fr Provence			600–650 AD	'abandonment' (post-Roman)		u*	↑
It Rome	Fiano	B?	c. 500 AD	decline		j*	Göschenen IIa
It Piedmont	Casale Monferrato		c. 500 AD	decline (post-Roman)		l*	c. 350–500 AD
Fr Provence			59–429 AD to 440–660 AD	variable (late Roman)		h	↓
It Rome	Treia	B?	c. 4th cent.	high		j*	
It Latium	around Rome		50–209 AD	very high (Roman)	minor alluviation	n*	
It Molise	Biferno		300 BC–200 AD	agricultural expansion (Hellenistic)	Iron Age & Roman	d	
It Basilicata	Lower Cavone	B	110 ± 70 AD (? too old)	high (Roman)		v*	
Gr Mélos	Phylakopí etc.	V	c. 600 BC– c. 500 AD	variable (early Classical to early Byzantine)	unstable volcanic island	w	
Gr Argolid	Lower Flamboúra		c. 300–50 BC	agricultural abandonment (Hellenistic)	gully erosion	f*	
Gr Mélos		V	c. 1100 BC–500 AD (intermittent)	variable (Iron to early Byzantine)		w	
It Basilicata	Brádano	B	around 360–340 BC; before 70–400 AD	constant high (Classical), then decline (Roman)?		x*	
It Etruria	various	B	c. 500–100 BC	variable (Etruscan)		t	
Sp Aragon	Ebro	B	(700–)500–100 BC (–100 AD)	expansion (Celtiberian)	supposed to have ended because soil had all gone	y	
It Basilicata	Cavone	B	around 300–350 BC	constant high (Classical)		x*	
Sicily	Gornalunga		550–350 BC	expansion (Greek)	10 m of deposit	z*	
Gr Eubœa	Eretria		720–680 BC	low? (Iron)		p*	↑
Po Algarve	Arade		after 900 BC (and into Roman and beyond)	probably increasing	1 radio-carbon date	b	
It Calabria	Middle Cacchiavia	B	850 ± 60 BC	low? (Iron)		v*	Göschenen I
It Umbria	Impruneta valley		c. 920 BC	low? (Iron)	incised between 250 BC and 150 AD	aa*	1000–500 BC
Gr Peloponnese	Argos plain		1100–1000 BC	decline (Late Bronze)	torrential flood	ab*	↓
Cr Kydhonía	Khaniá plain		1400–c. 800 BC	constant high (Late Bronze), then decline (Iron)		r	↑
Cr Sphakiá	Frangokástello		1500–c. 800 BC	constant high (Late Bronze), then decline (Iron)		o	Löbben 2000–1250 BC
It Basilicata	Lower Cavone	B	1530 ± 60 BC (? too old)	low? (Early Bronze)		v*	
Cr Mesará			1900–1700 BC	increasing (Middle Minoan I, II)		ac*	
Gr Peloponnese	Olympia (Alphéios)	B	beginning c. 2000 BC	low?		ad* (p.292)	
Gr S. Argolid	Pikródhaphne		2300–1600 BC	high (late Early Bronze), then declining (Mid Bronze)	sheet erosion & debris flows	f*	↓
It Basilicata		B	up to c. 2100 BC	low? (late Neolithic)		ae	
It Basilicata	Middle Cacchiavia	B	2180 ± 90 BC	low? (late Neolithic)		v*	
Gr Argolid	Argos plain		2300–2000 BC	low? (late Early Bronze)		ab*	
Fr Provence	Vallat Neuf		2500–2000 BC	(late Copper, Early Bronze)		g, af*	
It Basilicata		B	before 2450–2000 BC	low? (late Neolithic)		ag	

(continued on p. 310)

Country and Province	Locality or name of deposit		Date	Relevant human activity	Remarks	Source	Holocene glaciations
Cr Kainoúrgi	Ayiofárango		2900–2200 BC	increasing (Early Bronze)		ah*	glacial advance 3350–2700 BC
Gr Peloponnese	Argos Plain		before 3000 BC	increasing? (Early Bronze)	buried a Middle Neolithic site	ai	↓
It Basilicata	Middle Cacchiavia	B	3580 ± 80 BC	(mid Neolithic)		v*	
It Marche	Esino		c. 4500–3500 BC	(early–mid Neolithic)		aj	
It Basilicata	Basento	B	4040 ± 870 BC	(early–mid Neolithic)		v	
Fr Provence	Étang de Berre		4300 BC onward	Neolithic	deposits of 'soil'	g, ak	
It Basilicata	Lower Basento	B	4260 ± 80 BC	(early Neolithic)		v*	
Gr Thessaly	Girtoni Alluvium	B	4500–4000 BC	Neolithic	'soil erosion'	ac, al*	
It Basilicata	Lower Brádano	B	4420 ± 70 BC	early Neolithic		v*	
Sicily			5000 BC	early Neolithic		z*	↑
Cyprus	Vasilikos		5540–5010 BC	early Neolithic		am*	
It Basilicata	Brádano, Basento, Cavone	B	before 6100 to at least 5090–4410 BC	slight (pre-Neolithic)		an	Frosnitz 5500–4900 BC
Fr Provence	Holocene Principal Terrace		7000–4000 BC	slight (pre-Neolithic, early Neolithic)		h, af	↓
Po Algarve	Odelouca, Enxerim		c. 5800 BC	slight (Mesolithic)	end of Older Fill	b*	
It Basilicata	Lower Cavone	B	5840 ± 60 BC (? too old)	slight (Mesolithic?)		v*	
							Venediger 7700–6900 BC
It Basilicata	Middle Cacchiavia	B	8170 ± 220 BC (too old)	slight (Upper Palaeolithic)		v*	↑
Fr Dordogne	Montignac		8600–8200 BC	slight (Upper Palaeolithic)	scree, slope wash, valley fill 25 m thick	s*	Schlaten c. 8300 BC
Fr Provence			before 8900–8000 BC	slight (Upper Palaeolithic)		u	↓
It Basilicata	Middle Cacchiavia	B	10,250 ± 210 BC	slight (Upper Palaeolithic)		v	

Sources

a	Ergenzinger 1992	i	O. R. field observation	q	Néboit-Guilhot 1991	x	Cottechia and others 1969;	af	Provansal 1995
b	Chester & James 1991	j	Vita-Finzi 1976	r	H. Raab & J. A. Moody		Néboit 1977	ag	Brückner 1980
c	Calvet 1994: Jacob 1997	k	Devereux 1989		field observation	y	van Zuidam 1975	ah	Doe in Bintliff 1977
d	Barker & Hunt 1995	l	Vita-Finzi 1975	s	Vita-Finzi 1974	z	Néboit 1984	ai	van Andel & Zangger 1990
e	Hunt and others 1992	m	Bintliff (1977) 398	t	Potter 1979	aa	Esu & Girotti 1991; Potter 1976	aj	Cilla and others 1994
f	van Andel and others 1986;	n	Judson 1963a	u	Ballais & Crambes 1992	ab	van Andel and others 1990	ak	Provansal and others 1994
	Jameson and others 1994 ch. 3	o	J. A. Moody field observation	v	Abbott & Valastro 1995	ac	Pope 1993	al	van Andel and others 1995
g	Leveau & Provansal 1993		(Sphakiá Survey)	w	Davidson & Tasker 1982	ad	A. T. G. field observation	am	Gomez 1986
h	Jorda and others 1991	p	Genre 1988			ae	Brückner 1983	an	Brückner & Hoffman 1992

to be linked to either Neolithic or Bronze Age activity, but the other three fit in very well with known expansions of settlement. However, even here there was a phase of expansion in the post-medieval period which is not so far known to have left datable sediment (even though it coincided with the second peak of the Little Ice Age).[66]

The second such correlation is summarized by T. W. Potter from south Etruria. This correlation is negative. The only period when valley floors were free enough from flooding and alluviation to be settled was the early Roman Empire, when land use was at a maximum. Potter rejects the conventional explanation in terms of 'clearance of forest' and accelerated erosion:

> . . . it is hard to esplain why a period of intensive land-utilization, such as the first and second centuries AD in south Etruria . . . should coincide with a phase of very low silt deposition in the valleys.[67]

In this badland environment, continued alluviation is evidently the normal state of affairs. Only in unusual periods,

---

[66] Barker & Hunt 1995.   [67] Potter 1979.

of which the early Roman Empire seems to be one, was flooding a sufficiently rare occurrence for alluviation not to occur.

We propose one additional human activity that has been largely overlooked: the channelling of water flow by bridges, culverts and drains, and also by the overflowing of irrigation ditches. All these concentrate the water flow, often in places where Nature did not intend it to go. This is strongly implicated in the formation and extension of gullies (arroyos) in the SW United States (p. 272).[68] May not the building of bridges and water-mills, the runoff from towns and the draining of fields, have had a similar effect in Europe?

The causes of Younger Fill

Like our predecessors, we have failed to achieve a clear separation between the effects of tectonics, climate, weather and human activity. This is partly because of the rarity of good dates, and partly because the four factors are likely to interact. In our view, the weight of evidence is against human activity as the principal cause, and towards an explanation in terms of weather.

The last five thousand years were different from the earlier Holocene both in climate and in cultural history. If, as we argued in Chapter 9, the later Holocene – the typical Mediterranean climate today – is drier and more seasonal than the earlier Holocene, it can also have more deluges of the massively earth-moving kind. Moreover, weather and human activity are unlikely to be independent: Little-Ice-Age-like periods probably encourage some kinds of human activity and discourage others.

The pattern in time of Younger Fill tells against human activity being the main cause. It is associated with a particular phase of the Bronze Age in Greece, but not with the corresponding period in Crete; it did not, as far as we know, follow the Bronze Age around as that culture spread through southern Europe. There is not a plethora of reports from areas of intensest human activity. The Classical, Hellenistic and Roman periods, in one or other of which there was unusually intense human activity in most parts of southern Europe, are marked by few occurrences of Younger Fill except associated with badlands – in contrast to the succeeding 'Dark Ages'. It is difficult to resist the conclusion that the primary cause was climate or weather. From the little detail yet available, it appears that the Classical-to-Roman period was of generally stable climate, and that instability set in during the Dark Ages. The connexion becomes somewhat more easily verified with the Little Ice Age.

The pattern in time correlates better with episodes of abnormal weather than with human activity. Alluviating events such as deluges were more prevalent in some centuries than in others; but any individual deluge was local in extent, so that Younger Fills in different areas are not synchronous. That is not to deny that climate, human activity and earth-quakes may be contributory factors. It is unreasonable to expect the same combination of causes to have operated everywhere. Soft, collapsing Basilicata is very different from hard, cliff-bound Crete.

A final argument tells against human activity as the prime cause. The steady increase in the incidence of Younger Fill came to an end a hundred years ago. It was still being formed locally in the nineteenth century, often until just before the time of maximum population. This is no longer so, except in such very unstable areas as Calabria. Almost all historical cultural activities, however, can still be observed, at least on a small scale. People still cultivate sloping fields, burn forests, grub out maquis, cultivate terraces in bad condition and abandon terraces.

But the destabilizers now have power at their elbow. In effect, a vast experiment has been set up. Cretan farmers, Greek road-builders and Spanish foresters have been given bulldozers and tractor-drawn ploughs with which to create bare slopes and soft cliffs on a scale that dwarfs the efforts of their Bronze Age and medieval predecessors. If the 'human activities' interpretation is right, the greatest phase of erosion and alluviation in the Holocene should be the present. The landscape should be collapsing year by year – except in old-fashioned regions, like Majorca, Corsica, and Amári in Crete, which have so far opted out of the earth-moving festival. It is not. As we argued in Chapter 14, too much bulldozing is asking for trouble, but trouble has not yet materialized on the gigantic scale that archaeological precedent would lead us to expect. It is still possible for dam-builders to be unmindful of alluviation. Specialists in Younger Fill are not hastening to record the last remaining historic deposits before they are overwhelmed by new ones. The twentieth century is likely to go down as a relatively quiet period in the Younger Fill story.[69] The conventional interpretation of post-Neolithic erosion is therefore exaggerated.

This general explanation can be applied also to particular areas. Any archaeologist who claims that Bronze or Iron Age farmers 'stripped the soil' from a survey area should state whether or not the stripping has now been resumed, as the remaining soil is disturbed by machines. If the answer is yes, the question of prehistoric activity remains open. If the answer is no, it is hard to believe that ards and harrows destabilized the soil where tractors and bulldozers have failed.

Future research should concentrate on getting better-dated alluvial deposits from which to establish correlations both with climate and with human activities. More information, too, is needed about the incidence of Younger Fill and where it does not occur: for example, why is it poorly developed in Bœotia,[70] although this is not far from, nor dissimilar to, nor historically unlike, the Argolid? Unfortunately, scholars turn their attention to places where they know there is Younger Fill: it is difficult to publish a learned article on something that did not happen.

---

[68] Cooke & Reeves 1976.
[69] One exception is the Algarve, where renewed sedimentation in the mid-twentieth century is associated with planting eucalyptus (an erosion-promoting tree, p. 258) and the accompanying bulldozing; Chester & James 1991.
[70] Oliver Rackham's experience with the Bœotia Survey, 1979–81.

CHAPTER 17

Euro-Deserts and Karst

It is formed of nothing except piled-up rocks and sharp, hard stones;
it has no plant-earth, as if the land had been passed through a sieve
with the intention of keeping nothing but the stones.[1]

Two areas of southern Europe verge on being dry deserts by reason of low rainfall alone – SE Spain and SE Crete, where annual rainfall is less than 300 mm. In others low rainfall combines with poor water retention or poor root penetration to produce a semi-desert with sparse vegetation and no native trees: for example, La Mancha in Spain, the low Aegean islands, the marls of middle Crete, and probably Malta and SE Sicily. There are also rock deserts, especially of hard limestone, even in high-rainfall areas, and cold deserts at high altitudes (Fig. 17.1).

Deserts are defined, roughly, as areas that lack indigenous trees except in favoured spots such as spring-lines, rock fissures and ancient ruins. We are concerned not with cut-down or bitten-down trees, but with places where there is no sign that trees have existed at all.

especially if there are cliffs or other sites excluded from browsing.

Tracing the history of deserts is difficult. Self-evidently they preserve no pollen, and contribute little to any nearby pollen-bearing deposit. They seldom generate income, and hence are excluded from many types of historical document. However, at La Cova de l'Esperit, a limestone semi-desert in the Corbières (south France), there is a charcoal record. Juniper was dominant throughout the early and middle Holocene, indicating a landscape hardly more wooded than today. Only in the late Neolithic do charcoals of holm-oak, wild-olive and box become significant. Deciduous oak never appears.[3]

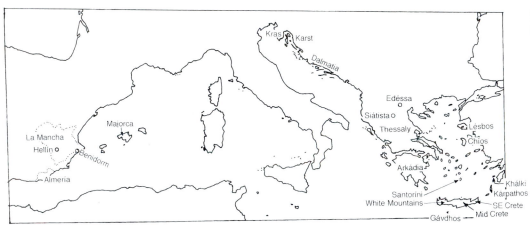

Fig. 17.1. Deserts and semi-deserts in southern Europe.

Existing deserts and semi-deserts as a test of desertification

The idea of desertification was based on observations (real or mistaken) of the expansion of the world's deserts. An early sign of desertification in southern Europe should be the enlargement of existing deserts and semi-deserts. The vegetation zones round their edges ought to be moving outwards. There ought to be a transition zone where trees grew in the recent past but not now: a zone of dead trees and old, now drought-stunted, trees that are not being replaced.[2] The picture may locally be confused by browsing – replacement trees may exist as shrubs, but be unable to make the transition to the tree form because they get bitten down. It should be possible to distinguish this from lack of replacement at all,

COLD DESERTS

An example is the alpine desert of west Crete. The cypress-woods that ring the White Mountains (Fig. 4.27) give out at altitudes ranging from 1,500 m on the periphery to 1,800 m in the interior. The trees get gradually shorter, reaching finally the last massive, surrealistically twisted ancient cypresses (Fig. 4.22), many of them reduced to scraps of foliage on bleached skeletons. Above them stretches the High Desert, vast cones and funnels of karstic limestone, black and pink and glaring white, with less than 1 per cent plant cover; but those plants are bizarre and specialized, and many of them (like *Anchusa cespitosa*) are endemics confined to this desert.

The High Desert, the most extraordinary landscape that

---

[1] Al-Edrīsi, twelfth-century geographer, on the immediate environs of Almería, SE Spain; p. 189.

[2] This is familiar in SW Australia where salinization is increasing.

[3] Vernet 1997, p. 121.

Fig. 17.2. High Desert east of Pákhnes, the highest mountain in west Crete. *July 1987*

Fig. 17.3. Alpine desert, Mount Galátzo, Majorca. *March 1969*

Majorca, although only 1,445 m high, has a high desert (Fig. 17.3). Here too the reason is probably the lack of an island tree tolerating both cold and drought. The highest trees are *Quercus ilex*, which like Cretan cypress also descends to sea-level. As in Crete, the desert has existed for long enough to have acquired endemic and specially adapted plants, such as *Smilax balearica* and a ground-oak-forming variety of *Quercus ilex*.[4]

DRY DESERTS

Eastern Spain – La Mancha

The arid eastern 200 km of the Meseta is a plateau 600–700 m high, sprinkled with small sierras. To the Romans it was the Plain of Esparto (p. 174); they probably established the traditional pattern of well-irrigated gardens round the towns, barley on the better patches of dry land and vast, meagre pastures for transhumant sheep. To the medieval Arabs it was a scatter of cities, the home of prudent and virtuous sages.[5] In the sixteenth century Don Quixote would have found windmills, mud villages at long intervals, castles and sheepfolds on rare eminences, mysterious lonely *ventas* or inns, and dusty towns haunted by plague and the Inquisition. Later travellers, such as Ford, hurried through the wide and boring steppes (Fig. 17.5); they did not stay to discover the wetlands and other natural delights.

Over a wide area the rainfall is less than 450 mm (Fig. 17.4), falling below 150 mm in many individual years. Red soils, often with a caliche barrier at a depth of about half a metre,

even Crete can show (Fig. 17.2), is determined by climate and lithology. Precipitation, although great, falls as snow, whose meltwater disappears into the karst. Although flocks pass through the Desert on the way to pasture, browsing is not the determining factor: cypress is an unpalatable tree and does not go higher on cliffs. The Desert existed long before Crete had human inhabitants, as shown by the evolution of plants specialized to its unique environment.

The tree-limit is probably determined by frost: we observed further dieback of the cypresses after the hard winter of 1992. No new trees have appeared at the upper limit. The High Desert has extended downward by about 50 metres altitude in the last few hundred years. The cause of this minor desertification presumably lies in fluctuations of climate and has nothing to do with human activity, nor is it linked to any extension of dry deserts at low altitudes.

Other islands have alpine deserts. Kárpathos reaches only 1,215 m – half the height of west Crete – and yet has a zone above the tree-limit. The island lacks cypress; the highest trees are *Pinus brutia*, less tolerant of cold and of the extreme winds on this island.

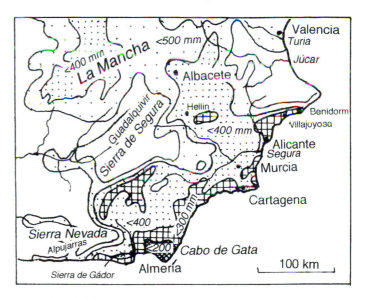

Fig. 17.4. Rainfall map of SE Spain.

are underlain by Tertiary and Jurassic limestones which store water. The rivers Júcar (famous for floods) and Guadiana are sunk in cañons tens of metres deep.

The dry zone is marked by lack of native trees, and by drought-adapted plants and plant communities. To a European it comes as a surprise to find whole mountains with not

---

4 For its unique plant communities, see Folch i Guillèn 1981.

5 J. A. Pacheco Paniagua, 'Chinchilla en las fuentes árabes', *A-B* 13, 1984, pp. 13–23.

Fig. 17.5. Plain of Esparto: steppe and modern almond cultivation. *Near Hellín, La Mancha, April 1994*

Fig. 17.6. *Retama (Lygos sphaerocarpa).* Note the savanna-like spacing of the bushes. *Liétor, La Mancha, April 1994*

Fig. 17.7. *Right* Esparto-grass *(Stipa tenacissima).* Plantations, as here, with the grass tussocks in rows, are not uncommon. *Elche de la Sierra, La Mancha, April 1994*

even the occasional bitten-down tree or trees on cliffs. Vegetation cover, however, rarely falls below 10 per cent, and there is a rich, specialized flora and fauna, extending even to wild boar. The coastal desert (Lygeo-Scorpiuretum sulcatae) extends north well into Catalonia.[6]

The characteristic plants are *retama* and *esparto*. Retama, *Lygos sphaerocarpa*, is a big leguminous undershrub some 2 m high, similar to English broom (Fig. 17.6); it is common in the semi-deserts of the southern Mediterranean but in Europe only in Spain and Portugal. Its name is Hebrew and Arabic: the prophet Elijah, when Jezebel wanted his blood, camped in the Judaean 'wilderness' under a *rothem* bush.[7] Its enormously deep roots reach groundwater. It is long-lived, and coppices if cut down; it never grows into a tree.

There are two esparto-grasses, *Stipa tenacissima* and *Lygeum spartum*, of which *Stipa* is the commoner in Spain. This big, tussocky, long-lived perennial grass grows over the mountains where even retama cannot reach water. It has been a fibre-plant since the Neolithic, being gathered, and often planted, for commerce (Fig. 17.7). It is an ideal arid-land cash crop: when all else fails through drought people can earn a little money from esparto.

The limit of the arid zone is defined by the absence of live-oak, normally confined to areas with at least 500 mm annual rainfall. Aleppo pine, *Pinus halepensis*, can live (at least for a time) on 400 mm. If lack of trees were due to browsing or woodcutting, rather than climate, they ought to occur more widely on cliffs than in accessible places. According to our own observations, this is not so.

The Spanish forestry authorities have planted trees on a vast scale, under the guise of 'reforestation', where trees would not grow otherwise. Hillsides are carved into pseudoterraces by courageous bulldozer drivers and planted with *Pinus halepensis*. The bulldozing breaks up bedrock and allows the pines sufficient root penetration to get started. However, even this drought-resistant tree has fared badly; if the trees live they are consumed by wildfires (p. 235), as happened on a large scale in the Sierra de Gádor in 1994.

On the edge of this region, between it and the Almería desert, are charcoal sites, notably the Cova de Cendrès near Teulada. These show a change from oak (deciduous and evergreen) to wild-olive and Aleppo pine, dated between 5000 and 3000 BC. We interpret this as showing that the Aridization

(p. 145) took effect here, a little later than usual. By the Bronze Age the vegetation had been reduced to the phrygana plus pines that predominate today.[8]

The Plain of Esparto was probably bigger in Roman times (p. 174). Its being a desert did not prevent it from having a fairly populous and wealthy culture, with big Iron Age sites, Roman towns, villas, bridges, and Visigothic and Arab cities.[9] Irrigated *huertas* and orchards (p. 78) and their aqueducts still

[6] Folch i Guillèn 1981.

[7] 1 Kings 19:4, 5 ('juniper' in King James's Bible).

[8] Vernet 1997, pp. 129 ff., summarizing work of Badal Garcia.

[9] L. Roldan Gomez, 'La investigación arqueologica de la epoca romana en Albacete', *A-B* 20, 1987, pp. 37–66; J. López Precioso and F. Sala Sellés, 'La nécropolis del Bancal del Estanco Viejo (Minateda-Hellín, Albacete)', *Lucentum* 7, 1988,

pp. 133–59; G. Carrasco Serrano, 'Comunicaciones romanas de la provincia de Albacete en los itinerarios de época clásica', *A-B* 23, 1988, pp. 35–42; J. J. Baquero Aguilar and others, 'Los puentes romanos de Isso (Hellín)' *A-B* 12, 1983, pp. 47–79.

Fig. 17.8. Brackish lake in La Mancha. *Sierra near Hellin, April 1994*

remain, as in the gorge of the Río Mundo around the wonderful mountain-town of Ayna.

The travels of Cavanilles in the Kingdom – the present Province – of Valencia in c. 1795 define the limits of the semi-desert. Cultivation depended largely on Roman irrigation systems, fed by springs or rivers coming from the interior. Outside the *huertas* or irrigated areas, roughland was of two kinds. The northern and western mountains were well vegetated, with woods, chiefly of pines, where there was enough root penetration; orchards of carobs, almonds and olives were widespread. For example, the Desierto de las Palmas was a well-vegetated area, a 'desert' only in the sense of a land inhabited by Discalced Carmelite monks (cf p. 14n.). In contrast, the south of Valencia was a land of esparto, whose gathering and processing supported eight townships and the city of Alicante. At the border with Murcia to the west

> the arid plains of . . . Murcia [give way to] the greenwoods of the valley. To a harsh soil, almost uncultivated for more than two hours [distance], there succeed thick woods of pines and deciduous oaks, planted by nature to serve as a boundary between the kingdoms of Murcia and Valencia.

Cavanilles claimed – but without giving evidence – that for a century both woodland and esparto had been diminishing through over-use and lack of protection. Since his time the limits of the semi-desert seem to have changed little. In some places we saw a slight increase of vegetation compared to his description, in others a slight decrease. In most places, including the Murcia border, Cavanilles's accounts and maps agree closely with the modern map.

An area of de-desertification is between Benidorm and Villajoyosa, on the coast near Alicante. Cavanilles described this as 'loose sterile sands without trees'. Those parts that have escaped being built on are now either orange-groves or pinewoods and savanna. Although some of the latter is new plantation (with the accompanying fires) much of it appears to be long-established, partly after an attempt at cultivation. The marl hills further west, in contrast, are still semi-desert, covered with the esparto that in Cavanilles's time supplied a factory at Villajoyosa.[10]

A century ago much of the pasture area, where soils permitted, was converted to dry-farming cereals and vines. Wheat and barley continue, though they depend on subsidies and their days are numbered from soil erosion (p. 266). Almonds, another twentieth-century crop, are threatened by drought, erosion, disease and competition from America. The next fashion was for vines, said to have covered one-quarter of La Mancha by 1985. The craze for irrigating with ground-water – even vines! – came in the 1960s; by 1987 the area thus watered had risen to 1,300 sq. km (4.6 per cent of La Mancha), using twice as much water as the aquifer will sustain (Chapter 19).

La Mancha used to be celebrated for oases, lagoons and wetlands: some fresh, some salt, some containing gypsum (Fig. 17.8); some temporary, some permanent. Many of the fresh-water wetlands were destroyed by drainage, and others by over-pumping and the retreat of the water-table. A notable, though very precarious, survivor are the springs of the Guadiana (p. 354).

Almería and the eastern Alpujarra

This is the core of the dry part of Spain, the classic European desert (Fig. 17.9). For centuries travellers have marvelled at its romantic sterility: esparto-scattered hills, retama-lined

Fig. 17.9. Treeless desert mountains (of schist) and abandoned floodwater-irrigated fields. *Rambla Honda, Tabernas, Almería, April 1994*

ravines, wondrous badlands, drought-bitten almonds in the plains, bizarre spiny plants outlying from the Sahara (e.g. *Launaea arborescens* and *L. lanifera*), and relics of attempts at cultivating arid-land crops such as sisal.

The limits of the arid zone vary with the geology (Fig. 17.10). The hard limestone of the Sierra de Gádor is mostly too dry even for retama. Very rarely, however, there are well-grown elms. This paradox – that the only tree in an arid landscape should be a water-demanding (and very palatable)

10 Cavanilles 1795, vol. 2, p. 243.

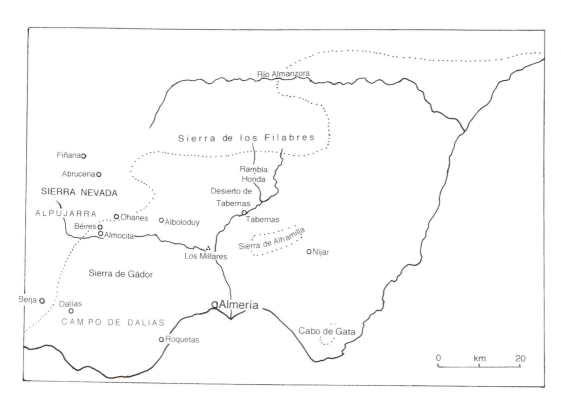

Fig. 17.10. The arid zone of Almería.

species – is explained by small aquifers that occasionally seep to the surface. The inhabitants tap these by 'water-mines' (*minas de agua*), adits driven into the hillsides, and use the water for small-scale irrigation. We have searched the great north-facing cliffs of the Sierra de Gádor with binoculars without finding a single tree other than small hackberries (*Celtis australis*) and willows, both of which depend on seeps. There can be little doubt that the absence of trees is not due to browsing or woodcutting.

The nearest wild trees to Almería are in two regions of different character. About 40 km to the north, in the upper valley of the Río Almanzora, is an area of pines (*Pinus halepensis* and *pinaster*), forming patches of woodland, mainly on the surrounding mountains above 600 m; there is some live-oak and a few deciduous oaks and cork-oaks; scattered pines descend into the lowlands. About 30 km NW of Almería, around Ohanes (Fig. 5.26), there begin the live-oak woods and savannas that form a zone along the whole south-facing side of the Sierra Nevada (Fig. 5.22). On the Sierra de Gádor we have seen only a few scattered live-oaks at about 1,000 m, the nearest being only 14 km from Almería.

Settlement began in the Neolithic (Almería Culture), and by the third millennium BC had produced the Los Millares Culture, one of the most flourishing of Chalcolithic Spain (p. 102). This might be taken to indicate that dryness is no barrier to civilization, but it began before the Aridization (p. 145), when the region was less dry; it had sporadic occurrences of northern trees such as hazel. The presence at metalworking sites of charcoal and pollen of pine, olive, evergreen

and deciduous oak, juniper and 'mastic' suggests that even in the second millennium the lowlands were not quite as arid as they are now. In our view, the oaks were probably relicts from the pre-Aridization climate. For domestic fuel, however, people burnt shrubs and undershrubs – lentisk, rosemary, *Lycium intricatum* – indicating a locality not much less dry than now.[11]

The copious archives of the 1520s reveal this as a prosperous area, well adapted to its natural limitations. It had suffered little from the transition from Islam to Christianity; the ex-Moorish population was increasing, happily ignorant of their coming fate (p. 103). They lived in small, intensively cultivated, terraced oases among the steppe and badlands (pp. 285–6). The principal crops in the records were olives and mulberries, so that they were not short of wood.[12] Clauses in leases state whether the rent was to be remitted 'for drought or for much water'. The impression is of a landscape like parts of the eastern Alpujarra today, not yet reached by the craze for over-pumping ground-water (Chapter 19).

Andrés Sánchez Picón and his colleagues have shown a retreat of *wild* trees between 1750 and 1850, which they claim as an example of desertification.[13] This important finding is worth examining.

Historical records and place-names, from 1511 to 1901, show pines and pinewoods to have been generally present in the Almanzora, especially in the mountains, in the same area as pine is now. There was, and is, a small outlier on the mountainous headland of Cabo de Gata.[14] The surviving stands are regarded by the authors as relicts, but they do not say on what

[11] Chapman 1978; Vernet 1997, pp. 148 ff., quoting work by M. O. Rodriguez and others; Gilman & Thornes 1985, p. 99, quoting W. von Scoch and F. H. Schweingruber, *Archäol. Korrespondenzblatt* 12, 1982, pp. 451–55.
[12] Cabrillana Ciézar 1977.

[13] J. G. and J. G. Latorre, 'Los bosques ignorados de Almería. Una interpretación histórica y ecológica', *Historia y medio ambiente en el territorio almeriense*, ed. A. Sánchez Picón, Universidad de Almería, 1996, pp. 99–126; A. Sanchez Picón,

'La presión humana sobre el monte en Almería durante el siglo XIX', *ibid.*, pp. 169–202.
[14] We often find it difficult in Spain to distinguish surviving native pines from early introduced trees and their offspring.

basis nor how old the trees are. There has been some diminution of pine, but it is not simple. The Ensenada survey (c. 1750) enumerated 50,000 pines on the territory of Cuevas, where there are now very few, but also very few historical records. There were '167,830' pines in the vast territory of Vélez Blanco, where (it seems) there are no earlier historical records and no present occurrence. In some of the mountains a recent expansion of pine was reported.

At higher altitudes in the pinewood area, there were (and sometimes still are) live-oaks and occasional less drought-tolerant maples, deciduous oaks, mountain pines and cork-oaks. Small stands of some of these have been reported on the higher elevations of the Sierra de Alhamilla and Cabo de Gata.[15]

The live-oaks of the eastern Alpujarra definitely diminished between 1750 and 1850. In Béires, for example, it was said in c. 1850 that

> the famous live-oakery which there used to be in the boundary sierra has been consumed in the lead and iron works which are nearby; there exist only a few live-oaks and oak bushes on the south-facing side of the river of Ohanes; even the *monte bajo* [here meaning phrygana or steppe] which the land produces in such abundance has grown scarce.[16]

Evidence of this type might be suspect, on the grounds that people notice felling but not regrowth; they exaggerate the past woods that they remember, and play down the woods of the present. However, there is confirmation from the Enseñada Survey of c. 1750, which reports 51,112 live-oaks in Béires. In Alboloduy, further east, there had been '654,033' stems of live-oak in 1750, but only 15,540 in 1850; so great a change can hardly be due to a mere difference in how big an oak trunk had to get before it was counted.

There had been live-oaks even in the Sierra de Gádor. The biggest report is of 500,000 stems in c. 1750 in Roquetas, which seems scarcely possible for a lowland, south-facing township, and is an unusually round number for the Ensenada Survey; we suspect that the township boundaries have changed. Lesser stands were reported or remembered in Dalías, Berja and Almocita.

There is thus general agreement on a drastic decline of live-oak between 1750 and 1850.[17] The commonest cause given at the time was consumption by the newly expanded lead-mines of the Sierra de Gádor, followed by domestic consumption, ironworking and grubbing-out for farmland.

Lead-mining was insignificant up to 1822. From then until 1840 the Sierra de Gádor was one of the world's chief lead producers. Lead was smelted at relatively low temperatures using esparto as fuel. This competed with other uses of esparto, especially paper-making; there were ordinances against pulling up esparto plants, which fuel-gatherers presumably did to avoid the trouble of cutting the tough leaves. However, from 1824 onwards, furnaces were installed

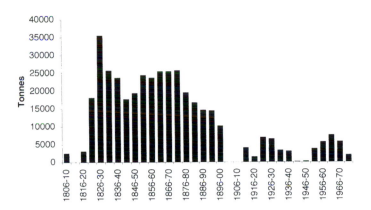

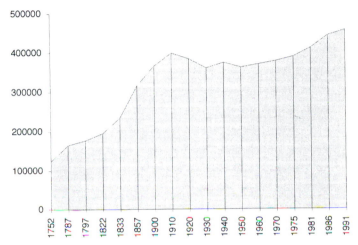

Fig. 17.11a. Reported lead production in Almería province, 1806–1975.

Fig. 17.11b. Census population in Almería province, 1752–1991.

to recover more lead from the slag of previous smeltings, which required a higher temperature than an esparto fire would produce. Although at first they used imported coke, they took to consuming live-oak charcoal. Sánchez Picón considers fuel for re-smelting to have been the chief consumer of trees, although mining timber was also needed. He calculates that the lead recovered by secondary smelting, although only a few per cent of the total, could easily have accounted for half a million live-oaks.

Fuel was needed by the growing population locally and in Almería city, and by tens of thousands of miners. Growing markets among villagers, miners and townspeople called for more arable land, which encroached on woods and savannas. Sánchez Picón and colleagues base their argument for desertification on increased consumption of wood and decreased area of trees.

However, the matter is not so straightforward. Live-oaks sprout, and should not ordinarily be killed by cutting them down. They might be grubbed out for expanding cultivation, but we would expect trees to avoid easily cultivable sites; only about one-sixth of Almería province was cultivated by 1850.

[15] J. G. and J. G. Latorre, 'Primeros datos sobre la presencia de un roble marcescente (*Quercus faginea* Lam.) en un medio arido', *Real Sociedad Española de Historia Natural, Tomo Extraordinario*, 1996, pp. 355–57; 'Alcornocales en zonas áridas. El uso de información histórica al servicio de la Ecología', *ibid.*, pp. 358–60; 'Los pinares invisibles del sureste árido español. Ecología e historia de unos ecosistemos ignorados', *ibid.*, pp. 361–63; Gilman & Thornes 1985, pp. 110, 115, 122.

[16] Bauer Manderscheid 1991.

[17] Except in Fiñana and Abrucena on the north side of the Sierra Nevada, where more stems were reported in c. 1850 than in c. 1750 (although it was claimed in Abrucena at the latter date that the woods were 'going to destruction').

If the trees were destroyed by felling followed by browsing the shoots, there ought to be survivors on cliffs, which we have looked for in vain.

We doubt whether the Sierra de Gádor has ever been much forested. The account of the Alpujarrine war of the 1560s (p. 103) does not suggest that there was more forest than there is now. There are few if any *encina* (live-oak) place-names. The claim that there were more than 100,000 live-oaks in the eighteenth century (in 700 sq. km of mountains) seems to rest on the one, suspicious, record for c. 1750. Widdrington in c. 1830 describes the Gádor as a 'high and perfectly bare limestone ridge'; the eight thousand miners lived in huts built of stone 'with the least possible quantity of wood'.[18] This is a significant observation because large-scale mining had not then reached its peak; nearly all smelting up to that time had been with esparto or coke.

A detailed comparison is needed between historical documents and the present state of the trees. We have not been to see whether the live-oakery of Berja in fact has disappeared; but its eastward extension into Ohanes is still very much alive (Fig. 5.26). Are there yet live-oakeries in hidden valleys of the Sierra de Gádor, as there are in the Sierra de Contraviesa (p. 101)?

We infer that trees indeed pass their climatic limit (500 mm rainfall for live-oak, 400 mm for Aleppo pine) in the Almería lowlands. Historic records are in and around the mountains, where the rainfall is greater – even though the mountains have a history of denser population than the lowlands.

Trees diminished (on the whole) during the mining period, and have recovered some of the lost ground since. The change was mainly a depletion of the numbers of trees in areas where trees existed, rather than an extension of the area wholly without trees. However, there is evidence for a retreat of pine and live-oak from lower elevations, especially in Alboloduy where (as far as we know) there are no live-oaks now. Deciduous oak and cork-oak, more moisture-dependent, have declined in their relict stations in the Sierra de Filábres.

Although it is easy to attribute these changes to obvious forms of human activity, the questions remain of why these trees escaped felling before c. 1830, and whether, had they not been felled then, they would still be alive. In a region with little woodland and millennia of moderately dense population, it is inconceivable that wildwood still persisted as late as the eighteenth century: people must long before have come to terms with trees and achieved a balance between felling and regrowth. (They would take notice when foreign businessmen bought up their wood supplies at prices that they could not afford, or took away the permanent bases of pollards.) Had there been no felling since 1830, many of the trees would by now have died from other causes; whether or not they were replaced would depend on other factors. Aleppo pines, especially, are short-lived and have to start again from drought-sensitive seedlings. They would be expected to shift within a pine area, as seems to have happened here, in response to fires promoting germination or dry seasons hindering it.

Although Sánchez Picón rejected the hypothesis of climatic change, on the grounds that the climate had remained relatively constant, we incline to a different view.

We would expect the Little Ice Age to cause a contraction of the Almería desert. The marginal woods and savannas of the eighteenth century would have arisen in a period of generally less arid, more variable climate. When the trees were felled for the last time in the nineteenth century the climate was unfavourable for replacing them. The rare survival of deciduous oaks and cork-oaks, trees of a wetter climate that are not very sensitive to woodcutting, confirms this view.

Almería is thus an example of the expansion of a desert, but on a modest scale – kilometres rather than tens of kilometres. Whether it is an example of human desertification is doubtful. There was abnormally much tree-felling for a short period, but this happened a hundred and fifty years ago and there has been ample opportunity for recovery. Felling may have hastened, and drawn attention to, a change that would otherwise have happened through less favourable climate. Even in what might be supposed to be a very sensitive area, this is not an unequivocal confirmation of Ruined Landscape theory.

South-east Crete

The other area that might be called the European Sahara is the SE of Crete, extending round the coast from Myrtos to Káto Zákro and beyond (Fig. 17.12).

The eastern peninsula, like all of Crete, is a land of contrasts. On its NW side, the towering fault-scarp of the Thryphtí Mountains (Fig. 3.1) attracts rain; on metamorphic rocks, this creates patches of relatively lush vegetation including deciduous oaks. Further east is a high arid limestone plateau with mountain-plains, a marginal land for settlement, with shadeless villages now lapsing into dusty decay. Dryness increases south-eastwards towards vast, crumbling, treeless, gorge-slotted cliffs on the coast (Fig. 17.13). This rain-shadow is the desert: annual rainfall is less than 400 mm, falling probably to 200 mm at the SE corner.

The principal trees are pine, prickly-oak and carob. *Pinus brutia*, rather less drought-tolerant than *P. halepensis*, occurs in two parts of east Crete and in two other principal areas in Crete (Fig. 17.14); in each of these it is increasing. The evergreen oak, *Quercus coccifera*, here usually appears as a ground-oak form; it seldom grows within 1 km of the sea.

Pine and prickly-oak give out in the arid region, but in different patterns. Oak is less gregarious than pine, and outliers occur on cliffs and other favourable spots. Where oak and pine fail there may be the occasional carob, more drought-resistant and perhaps a relic of ancient cultivation. The gorges, almost uniquely in Crete, are treeless except for occasional wild pear; this indicates that climate rather than browsing is the limiting factor. They are rich in curious semidesert plants such as *Camphorosma lessingii*, an outlier from central Asia.

In the second millennium BC Káto Zákro boasted a great palace and town; another town lay at Paláikastro further north on the east coast. Káto Zákro is an oasis, where much of the ground-water in the peninsula emerges (Fig. 17.13b). In Roman times there was a 'city' at Ampelos within the present

---

[18] Cooke 1834, p. 48.

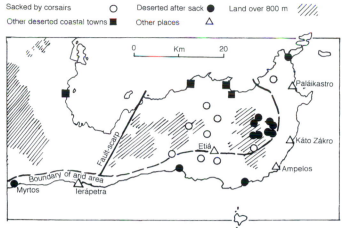

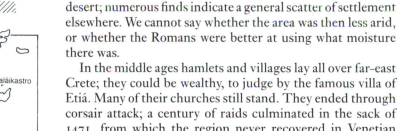

desert; numerous finds indicate a general scatter of settlement elsewhere. We cannot say whether the area was then less arid, or whether the Romans were better at using what moisture there was.

In the middle ages hamlets and villages lay all over far-east Crete; they could be wealthy, to judge by the famous villa of Etiá. Many of their churches still stand. They ended through corsair attack; a century of raids culminated in the sack of 1471, from which the region never recovered in Venetian times. It was re-settled after the Turkish conquest in c. 1650.[19] For a time it prospered; but the drought and winter cold are too much for modern Cretans. The southern and eastern fringes have probably not been inhabited since Roman times except by hermits, although they are intermittently culti-vated near springs.

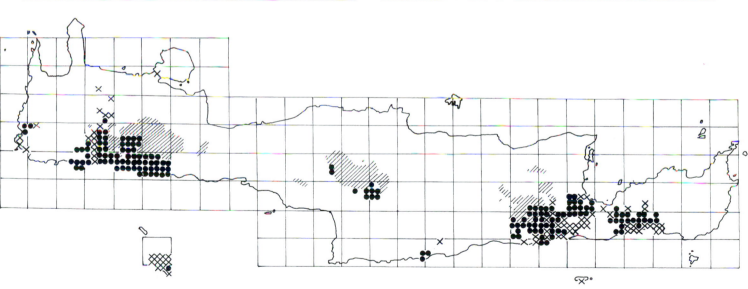

Fig. 17.12. *Top* The dry zone of east Crete.

Fig. 17.13a. *Centre left* Cape Goudhoúra, SE Crete. *August 1978*

Fig. 17.13b. *Centre right* Oasis of Káto Zákro, east Crete. *May 1990*

Fig. 17.14. *Above* Distribution of *Pinus brutia* in Crete, by 2-km squares. Crosses represent recent spread. Land over 1,200 metres is hatched.

Records of vegetation begin with Buondelmonti in 1415. He says that some of the mountains were treeless, but he records pines in three places, for example:

> we went towards Mount Dictano, well furnished with pines, from which no small amount of pitch is prepared, which merchants retail in various parts of the island.

He noted all the major pine areas still existing in Crete, and one small outlier on the south coast: 'I saw Zephirus facing Damietta [in Egypt], in whose summit a thick grove of pines is visible on the top'.[20] This could be the extreme SE corner of Crete, where there are now no pines, or one of the mountains east of Ierápetra, where pines are still visible from offshore.

The pines of east Crete are mentioned by other travellers, especially Raulin in 1845 and Trevor-Battye in 1909. Some of the present trees are at least two hundred years old. Pines seem always to have been present, but were never a big source of timber. In the Myrtos area, air photographs of World War II show pinewoods on north-facing slopes, evidently limited by drought.

Since 1945 agriculture has been exchanged for market gardening in plastic greenhouses on the coastal plains and foothills, wherever water can be tapped. Dry-land cultivation is abandoned, and pasturage has much declined. When one of us visited Myrtos in 1968 the change had begun. In the next twenty years pines greatly increased; they were no longer confined to north-facing slopes, but had invaded other aspects, growing very slowly. This gave rise (as we predicted, p. 217) to a conflagration in 1994. Undershrubs much increased, especially the browsing-sensitive *Atriplex halimus* and *Ebenus cretica* (Fig. 4.25).[21]

Similar changes have happened in other fringe areas of the semi-desert. A symptom is the spread of *Ebenus* off the cliffs where it had previously grown; whole hillsides in May are crimson with this magnificent endemic (Fig. 4.37). As elsewhere (Fig. 4.35), lack of browsing turns phrygana into taller phrygana, not into maquis. The pine belt has expanded round its edges, reaching the south coast at one point. A new equilibrium is developing, this time with enough vegetation to support fire. A browsing-dominated landscape is turning into a fire-dominated landscape.

These changes constitute de-desertification. The desert area has shrunk at its edges, but its core is little altered. In the extreme SE there is little browsing – there is little to browse or burn – and few signs of change. This is probably too dry for *Ebenus* or pine.

Santoríni[22]

There is more wine on this island than water![23]

Small islands, without mountains big enough to attract rain, tend to be drier than the mainland. The surrealist Aegean island of Santoríni or Théra is one such: its highest point, though reaching 566 m, is an isolated spike of limestone.

Santoríni is the half-submerged caldera of a great volcano (Fig. 17.15), whose explosion in the seventeenth century BC failed to destroy Minoan Crete, and probably did not alter the world's weather. This eruption, together with its Pleistocene predecessors, left a sea-filled crater 6 by 10 km, surrounded by two islands. The islands slope up from the outer coast to the edge of the crater and there plunge into the sea in cliffs up to 350 m high above the water-line and continuing far below it. The rocks are volcanic, apart from the small limestone

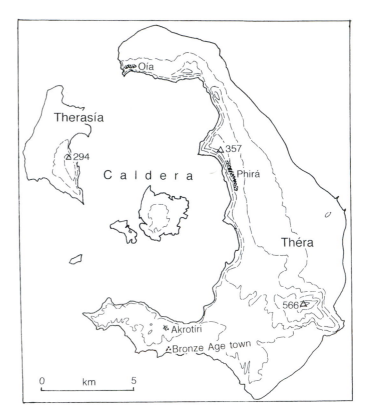

Fig. 17.15. Santoríni, with its submerged crater and the recent volcano in the middle.

peak. Hard lavas and pyroclastics from the 'Minoan' and previous eruptions, of brilliant colours and bizarre shapes, are overlain by thick layers of tephra.

Rainfall is about 380 mm a year, but additional moisture (and reduced evaporation) come from fog, even in summer. Soils, consisting mostly of volcanic ground glass, afford poor water retention but excellent root penetration.

Santoríni is virtually treeless. Oaks, junipers and even the drought-defying wild pear are absent; lentisk only just occurs. Introduced pines and cypresses barely survive; olives usually die. The only tree much cultivated is carob. Santoríni, with its vine-growing history, is probably less browsed than other islands, but even the cliffs lack trees.

The volcanic soils yield passable crops if tended to increase root penetration and conserve moisture. Barley – the most drought-resistant cereal – used to be much grown. The characteristic crop is vines, widely spaced in a very distinctive low-growing style (Fig. 17.16a) on terraces of powdered glass walled with volcanic bombs.

The island has been prosperous and well-populated at times, though ever threatened by earthquake and eruption. In recent centuries its prosperity came from seafaring and trade. The 'Minoan' eruption ended a wealthy period: the magnificent town, now revealed by the Akrotíri excavations, was but one among many settlements destroyed.

As in other Euro-deserts, some of the houses (including quite grand ones) are burrowed into soft cliffs. There is also a unique form of vernacular architecture that avoids using

[20] Buondelmonti (van Spitael edn), lines 98, 818, 846.

[21] Rackham 1972, Rackham 1990b.

[22] Abridged from O. Rackham, 'Observations on the historical ecology of Santorini', Hardy & others 1990, pp. 384–91.

[23] Christos Doumas, archaeologist.

Fig. 17.16a. Vine in volcanic soil, Santoríni. *April 1977*

Fig. 17.16b. Cave-house dug into tephra, with shell-vaults above. *Oía, Santoríni, September 1989*

timber by using the shell-vault for roofs and upper floors: a thin, rigid shell made of small boulders stuck together with a strong cement made from tephra (Fig. 17.16b). Bronze Age people used neither of these, but had plenty of timber – maybe imported – in their buildings.

Excavations have shown that the island was less desert-like at the time of the 'Minoan' eruption. It grew olives as well as vines, and there are reports of tamarisk and an oak tree; we have seen remains of tree-roots under soils buried by the last eruption. Soils were better developed; there had been 14,500 years since the previous eruption. Santoríni, however, cannot have been more than ordinarily fertile, and its prosperity must have depended either on some special, high-value crop or on non-agricultural activity. The island was little, if any, higher before the eruption than it is now; it would therefore not have attracted more rain. We are inclined to attribute the change in vegetation to the tail-end of the Aridization (p. 145).

LIMESTONE DESERTS

Even in relatively high-rainfall areas patches may lack vegetation because of lack of moisture: either the soil or rock do not retain moisture, or they cannot be penetrated by roots. Lack of soil precludes the growth of most crops but not of wild plants, especially trees.

The desert island of Khálki

Limestones vary in the extent to which they form and retain soil, and the degree of fissuring. On Rhodes there are a dozen outcrops of Mesozoic limestones of very different properties. Some are very hard and desert-like, such as the great hog's-back of Mount Atavyrios which dominates the island. Others are abrupt, cliffy, and wooded, such as Mounts Prophétes Elías and Akramyti. Further north, the glaring white marble of the high peak of Kerketévs on Sámos contrasts with the pinewooded marble of the Samsun Dağ on the Turkish coast.

The islet of Khálki,[24] 10 by 3 km (Fig. 17.18), is Greece at its sternest and most desolate. The visitor, leaving Rhodes and passing arid but relatively green Alímnia, reaches a bay beneath frowning mountains of grey and orange limestone, almost without visible plant life. This first impression is moderated by a small fertile valley with great terebinth trees, but Khálki shows every sign of extreme depopulation. If anywhere is a Ruined Landscape, this should be.

About 10 per cent of Khálki consists of basins and plains containing Older Fill. A further 30 per cent is hillsides with

soil; the rest is bare rock and cliffs. There are very few trees or shrubs, mainly lentisk and juniper; pine is extinct (except as an introduction), but the place-name Pefkió indicates its former presence; prickly-oak and carob are very rare. There are cliff endemics, but no trees on cliffs.

Khálki's magnificent antiquities imply dense population and intensive land-use over a long period. Place-names such as Andromássos, in the language of Minoan Crete, seem to derive from Bronze Age habitation. In the island basins are stumps of masonry from Archaic Greek (c. 600 BC) onwards. The largely deserted inland town of Khorió clusters around these remains, with a medieval castle of the Knights Hospitallers towering above it. There are sixth- to ninth-century churches. All possible cultivable land has been terraced and enwalled; tens of thousands of tons of stones have been picked off fields and made into massive 'consumption dykes' or field-walls (Fig. 6.7). The houses contain great timbers of cypress and deciduous oak, brought laboriously from Rhodes or Turkey. On this waterless island every house has a capacious cistern for storing rainwater.

The Pleistocene ecology, as on other small islands, is unknown. Khálki would have been either 'excessively' browsed or unbrowsed, depending on whether or not it had native mammals.

---

[24] This section is derived from observations by Oliver Rackham (with very much kind help from Professor Nicolas Vernicos) during the European Economic Community and UNESCO seminar, 'Measures against Desertification in Southern Europe', held on Khálki in 1986.

There are scraps of written history. Theophrastus says Khálki had two crops of wheat a year. Wey, the fifteenth-century traveller, said that 'they do not till the ground with beasts because of the multitude of stones, but with iron tools'.[25]

Historic prosperity did not derive from agriculture, but from sponging and trade. Khálki's excellent harbour was used to transfer goods brought by caique from Turkey and the other islands to ships for more distant voyages. This would explain the large population and the leisure that people had for shifting stones.

Depopulation is explained by distant political events. Italy, having defeated the Ottoman Empire in the Libyan war of

Fig. 17.18. The island of Khálki. *March 1990*

have no cultivable soil within them. The effect would be to lose some of the worse soils, and slightly to improve the better soils. However, the patination of limestone teeth does not indicate soil movement at present (cf Fig. 14.1). It would be physically possible to restore much of the cultivated land.

Khálki cannot be held up as an example of desertification. For centuries it was renowned as ultra-stony. It is a semi-desert because of poor water retention and poor root penetration, and there is no sign that it has been otherwise.

Thessaly

The hills surrounding the great plain of Thessaly are some-times cited for desertification induced by overgrazing and unwise land-use within the last hundred and sixty years.[26] They are diverse: some, of hard limestone and marble, are barren except of phrygana and steppe; some are more vege-tated; some are dissected by minor badlands.

We have compared the travels of Leake (a meticulous observer) in 1805 with what is there now.[27] The plain was open country with scattered trees around towns and villages. Now it is almost all arable land, but then most of it was pastured: shepherds brought their flocks in winter from distant mountains. Leake found in November 'a fine turf, now covered with sheep and cattle, with the exception of a few patches of arable land'.[28]

The hills around the Thessaly plain were as diverse as they are now. On the north

> The river of Krátzova in the lower part of its course flows through a thick wood of large planes. Further up, its valley is well cultivated, and on the heights still further, at the distance of six or eight miles . . . is seen the large village of Mirítza [now Oxíneia?]. On the northern side of this valley the heights are covered with oaks . . . but where the forest is not so thick, the trees being in general intermixed with vineyards around several villages: few of the oaks are of any considerable size.[29]

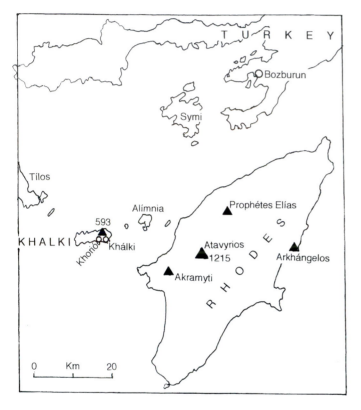

Fig. 17.17. Rhodes and Khálki.

1912, seized Rhodes, and incidentally Khálki, as booty. The island had been faring very well up to that time, and the town was being shifted down to the coast. Cut off from the main-land, the economy collapsed; the people emigrated to produce sponges at Khalki's colony in Florida; and when we were there in 1986 the port town was still half-built.

The determinant of Khálki is evidently the hard, massive Mesozoic limestone, like that of other barren massifs on Rhodes; even Alímnia is different. Browsing, although severe, seems not to be the primary factor. Within enclosures phrygana is taller but otherwise little different from outside (Fig. 4.35). Cliffs are not refugia for trees. There are signs of historic or prehistoric soil movements: some field-walls now

[25] Theophrastus, *Historia Plantarum* VIII.ii.9; G. Williams, ed., *The Itineraries of William Wey, Fellow of Eton College*, Roxburghe Club, London, 1857.

[26] M. D. Argyropoulos and S. P. Paraskevo-poulos, 'Desertification in Thessaly (Central Greece): a socio-economic perspective', MEDA-LUS working paper 39, 1994.

[27] Oliver Rackham visited the area with Dr J. A. Moody in spring 1990.
[28] Leake 1835, vol. 1, p. 455.
[29] Leake 1835, vol. 4, p. 262.

This area is much the same now, except that there are few vineyards.

In another place Leake records 'the mountain of Vlokhó, which, by its abruptness, insulated situation, and white rocks, attracts the spectator's notice from every part of the surrounding country'. He appreciated the correlation between vegetation and geology:

> The mountains to the northward and eastward [of Agrapha, adjoining the Thessalian plain on the SW] are of dark-coloured rock, and covered with woods of pine and oak: in the opposite direction the rocks are white, bare, and full of caverns.[30]

The only obvious change in natural vegetation since Leake's time is a modest increase in plane-woods along the rivers. The hard limestone hills have apparently been desert-like for at least two hundred years; any study of their desertification must begin by demonstrating that they have not always been desert-like.

A wonderful marble mountain lies 30 km further north at Siátista. Its glaring white slopes are seen from all over south Macedonia. Its desert-like state is partly the result of severe browsing, from which there is some sign of present recovery, but there is no sign or record that the vegetation has ever been radically different. The limestone is massive, and there is evidently not enough root penetration. In the midst is the historic town of Siátista, prosperous in Leake's time; its timber-framed buildings are of oak from the vast oakwoods on the surrounding sandstones, and pine from the distant Píndos Mountains.

Middle Crete

Extending SE from Herákleion, a belt of marls and soft limestones straggles across Crete to the Mesará plain and the south coast. This land of rolling hills, vineyards and dry villages is often strikingly lacking in trees, whether upstanding or bitten-down. It can hardly be called a desert, but its roughlands, instead of maquis–phrygana–steppe mosaic, are often monotonous tracts of the chicken-wire undershrub *Sarcopoterium spinosum*. It is unloved by the modern traveller, except briefly in May when (since the decline of browsing) the hills turn crimson with *Ebenus* blossom. This was the scene of Knossós, Arkhánes and other illustrious Minoan sites in the second millennium BC, and of their Roman and medieval successors.

Why is it so arid-looking? The rainfall of about 600 mm is well within the limit of tree growth. Marly limestone has good moisture retention. The clue probably lies in poor root penetration. In a frost-free climate marl does not weather easily; it often develops a hardened surface, and roots do not bore into its compact structure. Only undershrubs specialized for this habitat do well. The few trees tend to be either around springs and streams, or where a boulder or a geological fault allows roots to penetrate. If broken up by terracing or deep ploughing, however, marl supports crops, especially vines, reasonably well.

KARST

Karst landscapes form over certain types of hard limestone, and are defined by their bizarre solution features. Sinkholes, caves and many kinds of hole and pit, each with a book-learned name in an arbitrarily chosen language, are generated by the patterns in which rainwater dissolves the limestone (p. 245). More than half the Mediterranean hard limestone is karstic; non-karst kinds include the hard, unfissured desert limestones, and the platey limestones in which many Cretan gorges are formed. In this region marls and soft limestones seldom karstify.

Fig. 17.19. Limestone desert in central Chíos. *January 1988*

The name comes from the Kras, the hill-country behind Trieste. Until the mid-nineteenth century this was a treeless land, generally held to have been made barren and stony by overexploitation.[31] The Austrians extended the term *karst* to include the limestone belt, 600 by 100 km, along the Dalmatian coast. A century ago Cvijic and other geographers promoted it as a term for similar limestones elsewhere in the world. Karst landscapes include the Causses of the Massif Central and much of the Maritime Alps of southern France. Much of mainland Greece, the limestone Cyclades, Chíos (Fig. 17.19), Karpathos and Crete are karstic. There are karsts in southern Italy, Sardinia, eastern and southern Spain (though not most of the driest areas) and the Balearics.

Karsts range from the High Desert of west Crete to the deep sheltered bays and promontories of SW Turkey which are submerged karst. Their vegetation varies from desert to savanna to forest. Even the driest, most desertic karst landscapes are enormously important as water stores. Karstic limestones can take in over half the rainfall, even where, as in the Apuan 'Alps' of Tuscany, mean annual values exceed 2,500 mm.[32] Sometimes they funnel down an entire deluge (p. 257). Rivers and runoff disappear through sinkholes into an underworld of caverns. They emerge at great springs, often where a geological fault disrupts a cavern, to create oases (as at Káto Zákro above). In the Peloponnese the waters of Lake Stymphalos (of the brazen-billed birds) emerge not

30 Leake 1835, vol. 4, pp. 323, 273 ff.
31 Gams 1993.

32 P. Forti, L. Micheli, L. Piccini, G. Pranzini, 'The karst aquifers of Tuscany (Italy)', Günay & others 1993, pp. 341–49.

Fig. 17.20. Lake Stymphalos, northern Peloponnese. *May 1998*

The bigger karst depressions, like the mountain-plains of Crete, are formed by blocks of the earth's crust subsiding between faults; they may involve other rocks as well as limestone; they have rivers, which disappear into sinkholes. Such depressions enlarge through continued fault-movement, solution of the limestone, and sediment getting carried into the underworld. They fill with scree and gravel during glacial periods, with sediment eroded off the surrounding hillsides, and with wind-blown dust. The state of any depression results from the balance between these processes.[35] Part of the floor may collapse, taking masses of soil and sediment down into the underworld. On the floor of the Omalós Plain in Crete, a small badland eats into the accumulation of tens of thousands of years of limestone residues, rotten remains of vanished phyllites, and Saharan dust (p. 29). Karst processes still go on: shell-holes of World War I, for example on Monte Grappa (Italy), have grown into small dolines.

Vegetation is not confined to areas with soil. Many karsts are clad in maquis, forest or savanna: evergreen oaks and firs root deep into fissures. Trees tend to grow on bare rock and grassland on areas that have soil (Fig. 4.19). Olive will grow on nearly bare rock, but is a tree of soil-pockets rather than fissures; hence the pocket-terraces (p. 107) often built around each tree.

Depressions with accumulated soil are often the only cultivable land (Fig. 6.18). They are surrounded by walls, reinforced with spiny boughs to keep out animals grazing outside. The floor is usually not quite level, and fields on it form low terraces. In the Aegean, edges of depressions are favourite sites for Minoan settlements and Roman farms. In Dalmatia the oval floor of a doline is usually divided into two halves to allow a biennial rotation. In Crete and Apulia, terraces ringing a doline or extending out of a mountain-plain show that the hillsides, as well as the floor, were cultivated in times of greater pressure on land. The ultimate in karst cultivation is reached in tiny cultivated plots between limestone teeth, as in the *jardins du karst* near Nîmes and their counterparts in the Akrotíri Peninsula (Crete) and the Bozburun Peninsula (SW Turkey).

Slovenia

It has been claimed that the Slovenian karst, though inhabited since at least late Palaeolithic times, remained covered by forest, mainly of sub-Mediterranean trees, from the end of the last glaciation until about three thousand years ago. Soils were thicker and more continuous than now. The forest was destroyed by Illyrian cattle-breeders who arrived about 500 BC and began occupational burning; this gradually reduced the humus content of the soil, weakening its structure and allowing the mineral soil to be sucked into the underworld (p. 245). The development of the type of groove called *Rillenkarren* is held to indicate the former existence of a soil, lost not more than three thousand years ago.[36] An indicator of

in the nearby Gulf of Corinth but on the shore beyond Argos, 50 km away; those of Lake Tákka, the outlet of the central depression of Arkádia, emerge under the sea.[33]

Lakes form where karst outlets are not enough to carry away either normal rainfall or deluges, occasionally (as in the sixteenth and seventeenth centuries in Cretan mountain-plains, p. 133) or regularly. They often fluctuate. Lake Kopáis in Bœotia, before its destruction in the last century, filled and emptied on an annual cycle; its input came from a biggish river. Lake Phenéos in the Peloponnese is notorious for filling and suddenly emptying (p. 294). Lake Stymphalos behaves more gently, and not in sympathy with Phenéos: for most of the twentieth century Phenéos has been dry and Stymphalos has been quite full (Fig. 17.20).

Karst soils are derived from the residue left where limestone has dissolved, from wind-blown dust from distant deserts, and from the remains of other deposits formerly overlying the limestone, together with organic matter. There is a sharp boundary between mineral soil and limestone.

Insoluble residues vary from one limestone to another, typically forming less than 5 per cent of the rock. They are commonly a red silty clay, consisting of quartz, kaolinite, haematite and manganese oxides. The amount released by solution, a thickness of a millimetre or two per thousand years, accumulates in hollows in the surface rock to form the classic *terra rossa* (red earth) in places where there has not been much input from other sources. This varies in quantity, depending on how much limestone was dissolved and how much residue it left. In Istria and Apulia it can form a thick continuous layer.[34] Often it has been concentrated by erosion into the floors of basins, leaving fragments in clefts on the hillsides.

[33] A. Morfis and H. Zojer, 'Karst hydrogeology of the Central and Eastern Peloponnese', *Steirische Beiträge zur Hydrogeologie* 37–8, 1986, pp. 1–301.
[34] Nicod 1990.

[35] Rackham & Moody 1996, pp. 27–8.
[36] Gams 1993; Gams & others 1993. The evidence for the existence of forest and the manner of its disappearance is not stated. Pollen deposits on this coast indicate deciduous oaks and

other trees in the early Holocene, though the Non-Forest Index was not particularly low and may mean savanna (Table 10.ii). It is also unclear where the soils came from and how they survived the 'deforestation' of the last glaciation.

surface erosion is the thick colluvium that has accumulated at the foot of slopes in large dolines, burying buildings or date-able archaeological remains.[37]

Pasturage for sheep and goats reached a maximum by the early nineteenth century; cultivated land was limited to doline floors where eroded soil had accumulated. When dairy cattle later replaced shepherding, stones were picked off the grassland, and limestone teeth hammered away, to avoid damaging scythes when hay was cut. Hundreds or thousands of tons of stones per hectare were piled up or built into walls.

Woodland, of very small extent in historic times, began to increase in the mid-nineteenth century through land aban-donment. Plantations were initiated in the 1880s. Tito, like other aegophobic tyrants, banned goats in 1953. When we passed by in 1967 there was extensive maquis. By 1990 there were 10,000 ha of plantations, mainly *Pinus nigra*, which were extending by natural seeding; wide areas of abandoned pastures and hayfields were still being overgrown.

The Kras is thus turning from semi-desert into forest. This is not altogether to the good. It has lost its character as a distinctive landscape. The fire risk has vastly increased. The water that once percolated underground, feeding springs and boreholes, is now transpired by trees.

Dalmatia

In the Dinaric Karst the limestones are nearly pure carbon-ate, leaving little insoluble residue. Soils, never very thick, were thinned by periglacial processes during the last glacia-tion. This is the wettest part of the Mediterranean, with 1,000–4,000 mm annual rainfall, which leaves a large surplus for runoff, percolation and springs. It is an area of great contrasts, especially in islands such as Rab and Krk, where coppice-woods and maquis adjoin naked, pink rock. This is partly due to differences in the qualities of limestones, but doubtless also to human history; boundaries between forest, maquis, savanna and desert coincide both with boundary walls and with changes in rock.

Some of the walls and terraces are ancient. On Hrvar island, rectangular Graeco-Roman systems of large fields were subdivided in the Middle Ages. Mixed cultivation developed in the Italian manner. As in Slovenia, stones were cleared to ever-increasing depths in ploughland, meadows and orchards; fresh stones and outcrops were exposed as soil trickled down into the underworld.[38]

After World War II, agronomists encouraged farmers to plant vines and olives within thick walls to protect them from the *bora*, the terrible north wind. This fashion did not last. More recently, bushes have encroached on grazing areas; black pine has greatly increased, and with it the fire risk. The hydrology has deteriorated. Where the geology allows, karst hollows have been plugged to store water for hydro-electric schemes. Many karst springs are undrinkable because of pollution from rubbish dumps and agricultural chemicals.

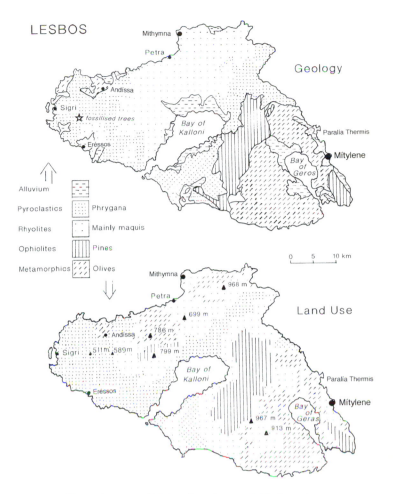

Fig. 17.21. Vegetation map of Lésbos (based on a satellite image) compared with a simplified geological map. The plant cover is very closely related to the underlying rocks.

OTHER ROCK DESERTS

Lésbos

Most of the East Aegean island of Lésbos consists of igneous rocks from the Miocene, 18–15 million years ago. Pyroclastic eruptions buried a 'petrified forest' of *Sequoia* and other Tertiary trees. Their products now cover the west of the island and a belt extending across the middle. These areas have been described as severely desertified: the story is the usual one of severe browsing of shrubs in late summer, or burning followed by browsing of the regrowth.[39]

Another supposedly degraded area is a south-facing head-land at Paralía Thermís, about 12 km north of Mytilíni town. This is barren at the seaward end, while the landward end is terraced with olives. The barren eastern end is underlain by a hard marbley limestone (quarried on the northern side); the olive-terraces are underlain by quite deep soils over a friable schist. A photograph of about a century ago demonstrates that little has changed since then (Fig. 17.23).

[37] Sauro 1993.

[38] I. Gams, 'Classical Karst: Dinaric Karst from Postojna to Capljina', *Man's Impact in Dinaric Karst: Guidebook, International Geographical Union Study Group on Man's Impact in Karst*, ed. I. Gams and P. Habic, Ljubljana, 1987.

[39] E. Koutsidou, C. Giourga, A. Loumou, N. S. Margaris, *Vegetation recovery in grazing exclosures in Greece*, MEDALUS working paper 18, 1994.

Fig. 17.22. Andíssa Bay, Lésbos, (a) in the 1930s, (b) in September 1993. Woodland has grown denser and olive-groves have increased.

Fig. 17.23. Headland at Paralía Thermís, Lésbos, (a) early in the twentieth century, (b) in September 1993.

Land-use on Lésbos is largely explained by relief and geology. Vegetation is sparse on hard silicic volcanic rocks such as rhyolites, andesites, pyroclastics, and welded tuffs or ignimbrites. The area above c. 700 m (the climatic limit for olives) is largely bare rock; the pyroclastics are mainly under phrygana; the pine forests between the two great gulfs are on ophiolites. Irrigated cultivation lies on alluvium on the north side of the central gulf. Olives are widespread over metamorphic rocks. 'Degradation' is largely to be explained by lack of fissures in the pyroclastics, into which trees might root, and thin, poor soils overlying them; there is no real evidence that vegetation has ever been abundant. Cultivators have chosen to build terraces, to plant and cultivate where the soils will reward their efforts. In Lésbos, where large areas are under olives, most of them pollards a century or more old, pocket-terraces have been built round individual trees. Unsuitable sites, as on pyroclastic rocks, are used for grazing or not used at all.

Similar but younger pyroclastics near Edéssa, northern Greek Macedonia, have the same 'desertic' appearance as the pyroclastics on Lésbos.

Arkadian schists

Metamorphic rocks, too, are diverse in their vegetation. Phyllite-quartzites in west Crete, for example, are often densely wooded, but variants where hard quartzite predominates appear arid even in a high-rainfall area.

Hermas remarked on the diverse vegetation of Arkádia in the Peloponnese (p. 167). Some schist areas, such as the northern Párnon Mountains, are as ruggedly fertile as anywhere in Greece, and are very wooded despite centuries of dense population. Only a few km away, there is near-desert (Fig. 17.24). The ground is covered with loose flakes of mica-schist, with the most meagre steppe and occasional bushes of deciduous oak and hawthorn. In valley bottoms, however, are occasional full-sized white poplars: as in Almería, the only trees in a semi-desert are moisture-loving and very palatable.

The desert apparently depends on this particular schist rock, which crumbles away into scree-like rubble. Root penetration is excellent for those few plants that can retain a footing on the loose surface, but there is no moisture retention. Rainfall goes straight through the rock and down to a water-table, which is reached by the poplars in the valleys.

Philippson visited the site in 1888 and drew these conclusions; it appears then to have been even less vegetated than it is now.[40]

CONCLUSIONS

We began by asking whether desertification can be detectable through existing deserts and semi-deserts expanding. To summarize our findings:

West Crete (cold) ..slight contraction in last few centuries.

La Mancha (dry) ..contraction since Roman times; now physically stable but over-exploited.

Almería (dry) ......modest expansion early last century, probably not now continuing.

SE Crete (dry)......steady contraction over sixty years.

Santoríni (dry) ....less arid in the seventeenth century BC; no evidence of recent change.

Khálki (rock) ......slight expansion through erosion over two and a half thousand years; loss of pines.

Thessaly (rock) ....no evidence of change.

Mid Crete (rock) ..no evidence of change.

Kras (karst)..........loss of soil (and by inference of vegetation) over three thousand years; rapid de-desertification since land abandonment.

Dalmatia (karst) ..some evidence of desertification in historic times; rapid de-desertification since land abandonment.

Lésbos (rock) ....no good evidence of change.

Arkádia (rock)......no change over 110 years.

Gávdhos islet........almost treeless a century ago, now (dry, rock): very wooded (p. 61).

The answer is clear: the history of deserts does not convey a message of alarm. The Sahara is not expanding northward: the two places in the front line of such an expansion, Gávdhos and SE Crete, are definite examples of de-desertification. Deserts may have transition zones at their edges which can fluctuate at times of greater or less pressure on land. These marginal zones have sometimes been heavily used in the recent past, but at present there are easier ways of making a living. Examples of contraction of deserts are clearer and more numerous than instances of expansion. Where desertification can be demonstrated it has proved reversible.

Deserts are regarded as useless land and are much abused. The abuse of karst is the subject of a special commission of the International Geographical Union, headed by Professor U. Sauro. Limestone is saleable, and industrial quarrying has produced enormous, arbitrary excavations. Depopulation has been more than counterbalanced by urbanization and tourism in Provence, the Venetian Prealps, Attica and the Akrotíri Peninsula on Crete. The nature of karst protects it

Fig. 17.24. The crumbly schist desert of Arkádia. The shredded trees are white poplar. *Near Vérvena, September 1984*

from dams, but water supplies are over-used and polluted. Pig, human and chicken sewage drains into sinkholes; rubbish is cast into dolines; noxious gases find their way into caves, sinkholes and springs.[41]

Some karst landscapes, as in low-altitude Majorca, are easy to re-wood. If grazing is restricted, karst turns into maquis or forest within a few decades. As in the Mediterranean generally, this can replace a distinctive cultural landscape with the all-too-common patchwork of forest and burnt forest. On the island of Chíos there is a different story. Much of the hard limestone mountains is covered with phrygana. For many years local groups have been trying to extend the forests of *Pinus brutia*. Sometimes this succeeds: Chíos is locked into a particularly vicious pine-and-fire cycle. However, in some of the experimental plots established since 1977, from which browsing animals have been excluded, the pines have died and other changes have happened. Sometimes, phrygana has merely grown into tall phrygana. In other exclosures, native trees, bushes and other plants have appeared, notably wild pear, lentisk and prickly-oak. So impressive has been this effect that planting is no longer done and exclosure alone is left to establish native vegetation.[42]

Deserts are among the wonders of the world, full of beauty and delight and biodiversity. The few in Europe should be appreciated and cherished. They may not be as grand as those of Australia or Arabia, but Europe would be duller without them. There is no longer any need to make them 'blossom as the rose'. The most effective conservation measure is merely not to abuse them. Mostly they are not cultivable by modern standards, although where old-fashioned cultivation does exist it should be encouraged. Increase in vegetation is probably inevitable, but should not unthinkingly be promoted. Nature did not intend these special places to be like everywhere else.

40 Philippson 1892, p. 165.

41 Nicod 1990; D. C. Ford, 'Environmental change in karst areas', *EG* 21, 1993, pp. 107–9;

E. Custodio, A. Bayo, M. Pascual, X. Bosch, 'Results from studies in several karst formations in southern Catalonia (Spain)', Günay & others

1993, pp. 295–326.

42 Koutsidou & others, see note 39.

CHAPTER 18

Deltas and Soft Coasts

Torre de Mare is supposed to be placed in the precise site of the ancient city,
which must have extended all round this spot. It is a square tower, built by the
[thirteenth-century] Angevin kings, and it is probable that the sea washed its
foundations at the time of construction, but the accumulation of soil produced by
the irregular and variable courses of the rivers Bradano and Bassento between
which it is situated have left an intervening space of nearly a mile.[1]

EXPANSION OF THE LAND

On most Mediterranean coasts today, long monotonous
sandy beaches alternate with wild beautiful cliffs and
promontories. Inlets and good harbours are rare, except on
islands. But this has not always been so. In Roman times there
were many inlets and lagoons; rivers such as the Ebro had
harbours at their mouths instead of deltas. The loss of
harbours is one reason why these coasts 'contain veritable
graveyards of towns and port-cities defunct in the last two or
three thousand years and maybe earlier', as Catherine Delano
Smith puts it. The growth of deltas has also given southern
Europe some of its best farmland. These tendencies have
been appreciated for many centuries, notably at Venice where
medieval hydrologists schemed and laboured to avert the fate
that overtook the Roman cities in this area.

Low alluvial coasts shift rapidly with the supply and
removal of sediment. In the near-tideless Mediterranean,
most of the sediment comes not from coast erosion, but from
inland erosion via rivers. The growth of deltas and infilling of
bays should therefore be related to rates of erosion in the river
basins. Historians assume that delta material represents soil
eroded from the catchment. Time after time, scholars have
jumped to the conclusion that rapid coastal accretion is the
immediate result of population pressure, deforestation and
soil erosion. (But not everyone: Spanish geomorphologists
have explained the coastal changes inside and outside the
Straits of Gibraltar entirely in terms of changing sea-level,
tectonics, wind and weather.[2])

The reality is not simple. The visible Ebro delta is the top
metre or two of a huge truncated pyramid, rising out of water
up to 1,200 metres deep. If it was not visible six hundred years
ago, that does not mean that all the hundreds of cubic kilometres of the cone represent sediment brought down since
then. The cone has been growing for millions, not hundreds,
of years; sediment added since the Middle Ages forms the
veneer on the surface that has brought it above present sea-
level.

Coasts are well documented, especially by seafarers. Maps
and sea-charts can be used for determining shoreline changes
since the sixteenth century. There is abundant archaeological
material from the excavation of deserted coastal cities.
Tracing the prehistory of sedimentation entails coring to
depths of tens of metres, analysing the layers of material, and
dating organic layers in the cores.

Fig. 18.1. Deltas discussed in this chapter. Arrows indicate the dominant wave direction.

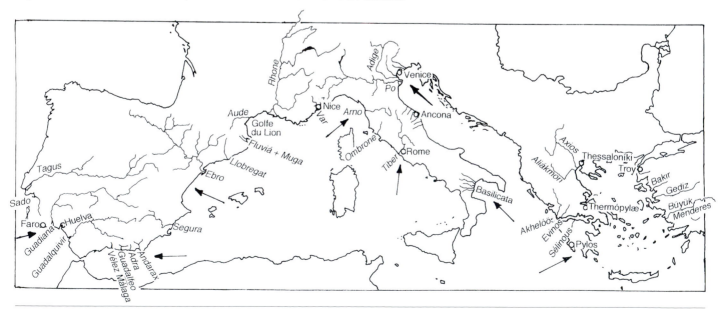

---

[1] K. Craven, on the site of Metapontum, Basilicata; *A tour through the southern provinces of the*
Kingdom of Naples, Rodwell and Martin, London,
1821.

[2] Goy & others 1996.

Fig. 18.2. Delta of the River Thyamis, NW Greece. Note the discoloured plume, which may be of sediment originating from badlands, and the modernistic grid of fields. *September 1998*

HOW COASTAL CHANGES WORK

Clay and silt carried downstream by a river after heavy rain are normally either deposited over its floodplain and delta or transported out to sea, giving a plume at the surface discernible from the air (Fig. 18.2).[3] They may feed a submarine delta fan. Sand is carried along the coast to one side or the other of the river mouth by longshore drift; it may be thrown up by the waves to form a barrier beach, possibly with a lagoon behind. Or it may blow onshore to form dunes; since sandbanks are not exposed at low tide, Mediterranean dunes tend to be meagre. Gravel, transported as bed-load, usually accumulates at the river mouth.

With little tidal current, sediments are moved along the shore mainly by waves. Movement is determined largely by the orientation of the shore to the direction of approach of the dominant waves; this is usually the angle of longest fetch, the direction of the longest stretch of water crossed by onshore winds (Fig. 18.1). Beaches tend to align themselves at right angles to that direction.

Delta fronts soon respond to reductions in sediment supplies; they retreat within a few years of rivers being dammed. Advances of delta fronts in the past, in response to increases in erosion rates upstream, are likely to have been less immediate. Not all the material eroded from the slopes of headwater valleys is carried directly downstream to the coast; much of it is stored in alluvial fans, floodplains and river terraces (p. 253).

People affect these changes in various ways by intention, default or carelessness. Tectonic instability complicates the task of linking coastal changes with erosion inland, both by altering local relative sea-level and by increasing the supply of sediment.

COASTS AND GLACIAL CYCLES

The behaviour of deltas and coasts must be seen against glacial history. During much of the last glaciation the present coasts were well above sea-level. Deltas then laid down are now about 100 metres under the sea, where some of them form the unseen foundations of their present successors. After glaciation ended, absolute sea-level rose rapidly to reach roughly its present level about 7,400 (real) years ago, about the time of the early Neolithic. This formed drowned coastlines, with complicated bays, branched inlets at the mouths of rivers and jagged promontories. Since that time, sediment supply, wave action and currents have been working to bring coasts into balance with the new sea-level. This has

---

[3] Some of these 'sediment plumes' are really plankton blooms: J. D. Milliman, 'Flux and fate of fluvial sediment and water in coastal seas', *Ocean Margin Processes in Global Change*, ed. R. F. C. Mantoura and others, Wiley, Chichester, 1991, pp. 69–89.

happened regardless of how intensive human activity has been: it applies to the Orinoco delta as much as to the Ebro. In the Mediterranean, radiocarbon dates from the earliest sediments in various deltas vary from about 7000 to 5000 BC, much the same range as in the world as a whole.[4]

On the mainland, deltas have formed and coasts have been straightened, often via an intermediate phase with lagoons. Bays remain where no large rivers flow into them (e.g. La Spezia in Liguria, Kotor in Montenegro). On islands and narrow peninsulas, the changes were slower and are often not complete. Coasts may still be jagged (e.g. Sorrento near Naples); bays often remain; Sardinia is remarkable for the many surviving lagoons. Crete, with its cliff-bound, sea-notched coasts, has a relatively stable outline despite being violently uplifted and depressed. Coastal springs have armoured the coast with jagged ledges and reefs of beach-rock, adding to the navigational terrors of the island.

Local uplift or subsidence particularly affect old deltas from previous glacial cycles. Where tectonic uplift has been rapid, c. 1 mm per year, sediments from the height of the last glaciation but one, now at least 120,000 years old, are by now uplifted to near present sea-level, or were eroded away as they were exposed. At the other extreme, big deltas subside under their own weight, and near them sediments from previous glaciations are presumably buried beneath sediments from the last glaciation.

It is hardly surprising that these adjustments should have continued into historic times. To catch up with a 100-metre rise in sea-level takes several thousand years; this does not in itself prove that either human action or unusual climate has been at work.

The Late Quaternary

This period is illustrated in the Baie des Anges, near Nice. Pleistocene gravels underlying the floors of the Var and adjacent valleys, and extending over the narrow shelves at the river mouths, were deposited mainly when sea-level was lower in the Last Glaciation. They are overlain by beds of end-glacial peats, dated to about 11,000 bp. These in turn lie under early-Holocene river sediments, whose accumulation more or less kept pace with the rise of sea-level. Organic remains indicate wildwood inland, with alder, oak (probably *Quercus pubescens)* and pine in various habitats. Above these Quaternary river-terraces, Neolithic artefacts appear in buried soils dated to 4000–5000 BC.[5]

After sea-level stopped rising, gravel barrier-beaches developed around river mouths and in embayments. Rivers such as the Var deposited fine-grained sediments on floodplains and marshes, but flushed coarse as well as fine material to their mouths. This change of sediments has been interpreted as denoting a change in climate, with 'Atlantic' wet-temperate conditions preceding aridization.[6]

REGIONAL CHANGES IN THE HOLOCENE

Atlantic coasts

The Tagus is the only big river of southern Europe not to have filled in its mouth. The cause is unlikely to be lack of human activity: its basin is not significantly different from, and certainly no less deforested than, the rest of Spain. Aided by its tributaries from the east, the Tagus has only partially progressed to filling up the great gulf that once extended from Setúbal to Santarém.

The Guadiana and Guadalquivir, the other great rivers draining the Meseta, end in a wide, wave-smoothed bay. The tidal range is modest, only 2 m; westerly currents and prevailing winds from the SW shift sediment along the shore mainly eastwards, as the eastward-pointing sand-bars testify. The Guadiana has formed only a modest delta (73 sq. km): most of its sediment has gone to create barrier beaches, dunes and lagoons stretching as far as Cadiz. A spit projecting SE across the river mouth would be cut off by floods and storms from time to time and flung on to the eastern shore. Most of the accumulation happened in the middle Holocene, to judge by sediments, still only 5 km from the coast, that go back to before 2000 BC. There was a pause in growth between 1200 and 900 BC, attributed to a local rise in sea-level, and then the deposition of spits and bars was resumed. Growth was again interrupted between 870 and 1000 AD, during the main part of the Medieval Warm Period. Sediment accumulation resumed in the opening phase of the Little Ice Age, was interrupted by the *tsunami* accompanying the Lisbon earthquake of 1755, and continued into the twentieth century. Some of the sixteenth-century coastal watch-towers (*torres de mar*) are now up to 600 metres inland; others have been ruined by encroachment. A jetty was built on the Portuguese side of the mouth in 1974, in an effort to stabilize the shipping channel. It remains to be seen how it will affect the eastward transport of Guadiana sediment.[7]

The Guadalquivir, in contrast, has formed a delta inside a huge shallow embayment, still partly open sea in Roman times. By the twelfth century AD the *marismas* of the delta, like many marshes, were valuable land for dairy cattle.[8] Much of this area now forms the Doñana National Park. Its future is threatened by erosion since the building of a long wall, intended to maintain the navigation channel to Huelva, which has interrupted the eastward movement of sediment.

Granada

The rivers between Algeciras and Cartagena (Figs. 5.22, 18.3) drain the great mountains in the south of Spain. They have steep gradients, and from time to time receive vast deluges from autumn storms, bringing down enormous loads of debris. (One storm in October 1973 added 40 ha to the small

---

4 D. J. Stanley and A. G. Warne, 'Worldwide initiation of Holocene marine deltas by deceleration of sea-level rise', *Science* 265, 1994, pp. 228–31.
5 C14 date 5620±200 bp.
6 M. Dubar and E. J. Anthony, 'Holocene environmental change in river mouth sedimentation in the Baie des Anges, French Riviera', *QR* 43, 1995, pp. 329–43.
7 C. Zaso, C. J. Dabrio, J. L. Goy, 'The evolution of the coastal lowlands of Huelva and Cadiz (south-west Spain) during the Holocene', Tooley & Jelgersma 1992, pp. 204–17; J. A. Morales, 'Evolution and facies architecture of the meso-tidal Guadiana river delta (S.W. Spain–Portugal)', *MG* 138, 1997, pp. 127–48.
8 Butzer 1988.

Fig. 18.3. Growth of the Guadalfeo, Adra and Andarax deltas (after Hoffmann).

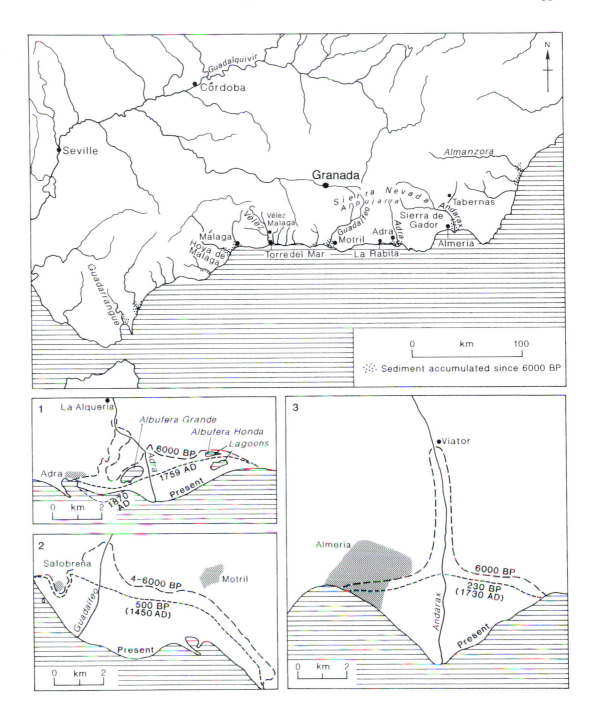

Adra delta.[9]) The mountains, partly of hard rocks, also contain badlands in soft sandstones and marls (pp. 274–5). So far there is no evidence for significant tectonic or sea-level changes in the historical period.

When sea-level stopped rising, all the rivers reached the sea via embayments and estuaries. They have since filled in their estuaries and gone on to generate protruding deltas. The development of these river mouths and deltas has been traced from radiocarbon dates of sediment cores plus written records.[10] Sedimentation is conventionally attributed to upheavals in the catchment resulting from ethnic cleansing in the late sixteenth century, the expulsion of the 'Moors' and

their replacement by colonists ignorant of how to control erosion, followed by mining and overpopulation in the nineteenth century (pp. 103–5).

The Río de Vélez basin includes much of the Santopitar mountains. Their fertile, soft but apparently stable slopes have been densely populated at least since medieval times, to judge by the many Arab settlement-names. The estuary originally extended 7 km inland. Over the last five millennia BC, sediment accumulated in it at a rate of about 2 metres per thousand years. The river went on to form a cusped delta. The rapid siltation, supposedly in the sixteenth century,[11] which deposited 30 metres of sediment at the mouth of the

9 Hoenerbach 1980.
10 Hoffmann 1987.
11 This is based on the belief that Christian galleys sailed up to Vélez Málaga during the siege

of 1497, but contemporary accounts are vague on this point (M. Diego de Valera, *Cronaca de los Reyos Católicos*, ed. J. de M. Carriazo, Molina, Madrid, 1927, pp. 212–37; Fernández & de Mata

Carriazo Arroquia 1969, vol. 1, pp. 685 ff.). Torre del Mar, successor to a coastguard tower built in the sixteenth century, is almost on the present coast.

Vélez, has been attributed to 'massive deforestation which caused extensive soil erosion',[12] but the evidence for this is not stated (Postulated Deforestation, p. 290). An alternative explanation is increasingly frequent deluges, such as occurred in the 1540s, 1617 and 1626 (Chapter 8).

The caption on a map of 1785 is said to comment on the great growth of the delta since 1750, long after the Moors had gone but before any effects of mining or over-population could have made themselves felt.[13] (But a supporter of the Ruined Landscape hypothesis might have an explanation ready in the growth of the Mountain wine trade, p. 267.)

A dam has been built on one of the tributaries, and the tip of the delta is now rapidly eroding; the medieval lighthouse west of the river mouth is now half encircled by the sea.

The next river, the Guadalfeo, rises in the Sierra Nevada. It falls nearly 3,000 metres in 8 km, and then enters a broad lower valley floor, across which the river shifts its course from one flood to another. The sediments are largely derived from schists, with finer greyish sands at and below sea-level, and brown-grey, platy gravels on top of them. Much of the material comes from alluvial cones, laid down by tributaries and cut away by the main stream when it floods; from masses of soft rock sliding directly into the river bed; and from the badlands of the western Alpujarra.

The Guadalfeo delta has advanced seawards by about 3 km over the last five hundred years. Much of the sediment appears to come from badlands and other uncultivated terrain. The expulsion of the Moors coincided with the second phase of the Little Ice Age; deluges from the 1590s onwards are a cause that better fits the evidence.

The local water board evidently disapproves of the Guadalfeo delta: when we visited it a large dam was being built below its chief source of sediment.

The Río Grande de Adra, rising in the middle Sierra Nevada, has a similarly steep gradient, and a catchment that includes badlands in the middle Alpujarra. The original estuary may have penetrated about 4 km inland; it still existed about 800 BC, to judge by a Phœnician settlement that would have overlooked it. By Arab times, two *albufera* lagoons had come into existence behind a barrier beach east of the Adra mouth. Radiocarbon dating of peat gives an age of c. 1150 AD.

There was probably not much change at the time of ethnic cleansing and the Little Ice Age: the main growth of the Adra delta came later. Maps of 1759, 1784 and 1786 show a small asymmetrical delta and the two lagoons. In the nineteenth century the delta grew so fast that farmers were induced to buy the shore and sea-bed in anticipation of dry land. The increasing wetlands supposedly encouraged mosquitoes and malaria. The port of Adra was saved by diverting the river in 1870 through the Albufera Grande. This lagoon soon filled with sediment, and a new delta rapidly developed. Meanwhile, the old western delta near the town was eroded, and buyers of the sea-bed lost their money. By 1931 a barrier beach had cut off a new lagoon. The building of a harbour mole west of Adra town in 1908 impeded eastward longshore drift, resulting in some erosion of the eastern delta.

Mining and metal-smelting in the Sierra de Gádor (about one-quarter of which is in the catchment of the Adra), mainly between 1818 and 1890 (Fig. 17.11), are said to have destroyed forests, whence the increased erosion and deposition. The coincidence of date would make this a neat example of the workings of a Ruined Landscape. It is more than likely that mining spoil-heaps contributed to the contents of the delta. However, we are not aware that anyone has done the detailed fieldwork needed to establish where the sediment came from, or whether forests – had there been any (pp. 317–8) – would have prevented erosion on this mountain. And there is an alternative explanation. Fourteen deluges swept the valley between 1823 and 1910; given the steepness and unstable rocks of much of the catchment, they would have brought down sediment whatever the vegetation.

Fig. 18.4. Attempt to retain the Andarax delta, whose silt supply has been cut off, by means of concrete walls and gabions. *November 1994*

The Adra delta, too, is probably doomed by the building of the Benínar dam (p. 357) below its main source of silt. This dam contrives to retain silt but to leak water; sinkholes in a hidden limestone bed under the river alluvium have prevented it from filling properly.[14]

The Río Andarax has a bigger but very arid catchment, including the SE slopes of the Sierra Nevada, half the Sierra de Gádor and some of the finest badlands in Europe (pp. 274–6). Its steep longitudinal gradient continues down to the mouth; the river falls 100 metres in its last 10 km. It emerges near the city of Almería, which has been a port for over two thousand years; the port remains in existence because the growing delta has bypassed it.

The original Andarax estuary extended some 8 km above the present mouth. Boreholes have revealed marine and brackish-water lagoon sediments, just below present sea-level, representing an accumulation of about 1 metre thickness per thousand years. They are covered by coarse sediments brought down by floods within the last few

[12] Lumsden 1992, pp. 182–3, quoting from a publication of the Instituto Tecnológico Geominero de España, in *Riesgos Geológicos*, November 1987.

[13] Lario quotes F. Llobet, *Mapa del Reyno de Sevilla*, 1767; T. López, *Atlas Geográfico de España*, 1810. (We have failed to verify these comments on copies of the maps in Cambridge University Library.)

[14] J. Benavente and 4 others, 'Karstic hydrogeological investigations for the construction of the Rules reservoir (Granada, Spain)', Günay & others 1993, pp. 25–31.

millennia. A Punic sherd dating from about 600 BC was found in them at a depth of 2.4 metres. Since the mid-eighteenth century the river – fed by mines? or deluges? – has built a protruding delta about 6 sq. km in extent. It is now being eroded near the river mouth (Fig. 18.4).[15]

South-east Spain

The Río Segura, much bigger than the preceding rivers, comes from soft rocks and badlands, including most of the arid Plain of Esparto (p. 174). In Roman times it flowed into the Sinus Illictanus, the bay of *Illici* (modern Elx), which is now infilled apart from some lagoons. *Torres de mar* show that there has been no significant advance since the sixteenth century.

This coast is moderately stable. A huge lagoon, the Mar Menor, remains behind a barrier beach; no large river flows into it. Even the Valencia delta, fed by the badlands of Teruel (p. 277), has not grown much: the Roman city of Valencia, founded in 138 BC at the mouth of the flood-prone Río Túria, is still only 4 km from the sea. The map published by Cavanilles in 1795 shows that since then the great lagoon (La Albufera) has much silted up; although there has been little change at Valencia city, the coast in the south of the delta has advanced by about 1 km.

The Ebro

This longest river in Spain rises only 30 km from the Atlantic. Its catchment of 86,000 sq. km includes the southern slopes of the Pyrenees, areas of flysch and Miocene gypsum and marls, and extensive badlands. The river's mean annual discharge in the first half of the twentieth century was about 20 cubic km. Its visible delta occupies 320 sq. km; this continues underwater as a 'pro-delta', a mud belt extending 25 km out over the continental shelf, 100 km along the shelf to the SW, and 28–60 metres thick.[16]

The Ebro delta is a unique landscape, one of the few places where a genuine fenland agriculture has developed. Ricefields, divided by low banks and flooded for much of the year, are tilled by huge tractors towing aquatic ploughs (Fig. 18.5). About one-quarter of the delta is reed-beds, lagoons, and natural vegetation, one of the great European localities for flamingos and other delta birds.[17] All this is protected from the sea by a tenuous line of dunes.

In Roman times the delta was submerged: the coastline ran straight from Ampolla to Amposta (Fig. 18.7).[18] There was an estuary up to Tortosa, where the Llotja or market hall of 1377

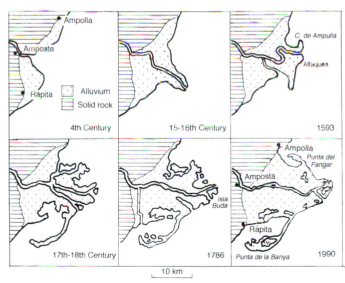

Fig. 18.5. *Above left* Ebro delta: a flock of egrets follows the underwater plough. *November 1994*

Fig. 18.6. *Top* Low, receding dunes near the main mouth of the Ebro. *November 1994*

Fig. 18.7. *Above* Evolution of the visible Ebro delta over the last fifteen hundred years, showing its rapid expansion in the sixteenth and seventeenth centuries. (After Esteban & Puigserver and others.)

[15] Hoffmann 1987.
[16] A. Palanques, F. Plana, A. Maldonado, 'Recent influence of man on the Ebro margin sedimentation system, northwestern Mediterranean Sea', *MG* 95, 1990, pp. 247–263; J. I. Díaz and 3 others, 'Morpho-structure and sedimentology of the Holocene Ebro prodelta mud belt . . .', *Continental Shelf Research* 16, 1996, pp. 435–56.
[17] R. M. Comes, *El Delta de l'Ebre*, Terra Nostra, Barcelona, 1989.
[18] N. Dupré, 'Évolution de la ligne de rivage à l'embouchure de l'Ebre', Trousset 1987, pp. 25–34.

is a testimony to the prosperity of the port. Navigation by sea-going ships declined as the visible delta grew rapidly from the fifteenth to the seventeenth centuries. It is recorded by successive maps from 1585 onwards, and by the position of coastguard towers. Great floods were recorded in 1325, 1329, 1380, 1448, 1466, 1488, 1518, 1609 and 1617. By the late eighteenth century the delta had reached nearly its present extent, except that the snout along the main distributary has since lengthened by about 6 km.

For a long time the delta was uninhabited and used mainly for seasonal pasturage and salt-making, but from the mid-eighteenth century planned settlements were established. These ran the risk of floods, political troubles and malaria; only in the twentieth century has the delta become prosperous.[19]

The Ebro delta has been studied in three dimensions, tracing the development of its submarine cone since the Messinian, about five million years ago. It grew faster in glacial than in interglacial periods. In the Pliocene, during previous interglacials and in the early Holocene, sediment was added at about 6 million tons a year, equivalent to lowering of the surface of the Ebro basin by 0.028 mm per year.[20] When sea-level was lower during glaciations, deposition was three to four times faster. The higher rates of erosion have been attributed to 'deforestation', meaning reduced tree cover during glacial periods.[21] This is possible, but the erosion-restraining qualities of glacial steppe are unknown. Alternative reasons for faster erosion in cold periods include larger glaciers in the Pyrenees, more frost action, and more wind-deposited dust providing a sediment source.

In the decades before 1940 the Ebro carried at least 22 million tons of sediment annually, about the same as during glacial maxima.[22] Rates of deposition in the pro-delta over the last hundred years were more than double those of the early Holocene; they represent the erosion of 0.05–0.1 mm thickness per year, averaged over the whole catchment. These higher rates of erosion have been attributed to Postulated Deforestation, cultivation and over-grazing.[23] However, it may be no coincidence that the visible delta grew most actively during the Little Ice Age.

After some insignificant eighteenth-century dams on tributaries, the Ebro was dammed in many places between 1940 and 1970. Although the lowest dams are supposed to let through at least 150 cubic metres of water per second (equivalent to 4.7 cubic km per year), lest a nuclear power station boil, they trap 99 per cent of the sediment; less than 0.2 million tons per year of clays and muds get through to

Tortosa. The delta is starved, especially of the sand that provided its infrastructure. Between 1946 and 1990, the snout regressed by 1.5 km. The southern and northern ears, Punta de la Banya and Punta del Fangar, have retreated and moved towards the coast.[24]

To the visitor from northern Europe the great delta seems desperately fragile. Only the thinnest line of dunes, which in England would not survive one winter's storms, separates some of the most productive farmland in Spain from the Mediterranean. A slight rise in relative sea-level could return it to the expanse of shallow sea which it was in Roman times. The current subsidence rate is put at 3 mm a year.[25] Already in the nineteenth century its growth-rate began to decline. The silt that created and maintained the delta now fills dams instead. Will it be there a hundred years from now?

Catalonia

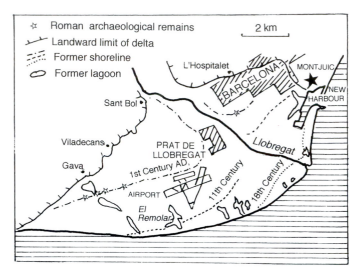

Fig. 18.8. Evolution of the Llobregat delta (after Marqués & Julia).

The river Llobregat rises in the eastern Pyrenees; badlands in Eocene marls and Upper Cretaceous clays provide much of its sediment. The Roman shoreline, marked by pottery from 200 BC–100 AD when the area exported wine, now lies 3 km inland (Fig. 18.8). The eleventh-century shoreline, dated by documents said to mention the Remolar lagoon, had moved 2–3 km seawards. Since then, the south side of the delta has suffered from erosion, while the mouth of the river has

[19] V. Tofiño de San Miguel, *Atlas Maritimo de España*, Madrid, 1789; W. Barentsz, *Caertboeck van de Midlandtsche Zee*, Amsterdam, 1595 (facsimile, Philip, London); J. P. López, *La colonització del delta de l'Ebre i la seva reglamentació en el darrer terç del s. XVIII*, Institut d'Estudis Rapitencs, San Carlos de la Ràpita, 1993; J. Bayerri i Raga, *Les terres del'Ebre, encara un futur . . . ?*, Cooperativa Gràfica Dertosense, Dertosa, 1985; L. Esteban and S. Puigserver, *El delta del Ebro y su Parque Natural*, Sendai, Barcelona, 1993; A. Maldonado, 'Dinamica sedimentaria y evolution litoral reciente del delta del

Ebro', *Sistema Integrado del Ebro*, ed. M. Mariño, Madrid, 1983, pp. 157–63; Sánchez-Arcilla & others 1998.

[20] Erosion of catchments in Roussillon (eastern Pyrenees) in Plio-Pleistocene times lowered the surface, on average, by between 0.020 and 0.200 mm per year: Calvet 1993.

[21] A. Maldonado, *El delta del Ebro, Estudio sedimentológico y estratigráfico*, Departamento de Estratigrafia, Barcelona, 1972; Nelson 1990.

[22] Nelson 1990, pp. 410–11.

[23] J. Guillén and A. Palanques, 'A historical perspective of the morphological evolution in the

lower Ebro river', *EG* 30, 1997, pp. 174–80. (They produce no independent evidence for the intense deforestation invoked in the catchment between 1500 and 1650.)

[24] J. Guillén and A. Palanques, 'Sediment dynamics and hydrodynamics in the lower course of a river highly regulated by dams: the Ebro River', *Sedimentology* 39, 1992, pp. 567–79; Sánchez-Arcilla & others 1998.

[25] A. Sánchez-Arcilla and 6 others, 'Impacts of sea-level rise on the Ebro Delta: a first approach', *Ocean & Coastal Management* 30, 1996, pp. 197–216.

Fig. 18.9. Muga delta: a lagoon and hedged fields in the Aiguamolls de l'Empordá, with one of the tower blocks of Empuriabrava; in the background are the easternmost Pyrenees, source of the delta sediments. *June 1998*

shifted 2–3 km further east. Apparently there was not much growth during the Little Ice Age. From the 1950s the growth of Barcelona city, airport and harbour began to overtake the delta.[26]

At the east end of the Pyrenees, renowned for deluges (Chapter 2), the combined deltas of the Fluviá and Muga are among the most active and temperamental in the Mediterranean (cf p. 305). They occupy 120 sq. km, coming from a catchment of only 2,000 sq. km, which extends 30–50 km up into the Pyrenees at well over 1,000 m. The underlying prodelta of silts and fine sands, which began to build up before the mid-Holocene rise in sea-level was completed, now extends 23 km offshore to occupy 365 sq. km of the seabed. Channel-like features in it are thought to result from Pleistocene river floods.

Since Roman times the shore has pushed forward 2–5 km. In the late Middle Ages there was fenland farming, with seasonally flooded pastures and ricefields. Castelló d'Empúries, shown at the mouth of the Muga on early eighteenth-century maps, is now 3 km inland, separated from the sea by floodplains, wetlands and brackish lagoons.[27] In the nineteenth century the fenland economy began to break down. A huge lagoon was drained for growing maize and fruit trees; rice growing ceased in the 1950s, and much of the wetland was brought under cultivation. The coast north of the Muga mouth was developed for tourism, including the monstrous 'residential marina' of Empuriabrava with its own airport (Fig. 18.9).

By 1976 tourist development was threatening to devour the delta south of the Muga too, which became a conservation issue. In 1983, the Natural Park of Aiguamolls de l'Empordá was set up – 50 sq. km of fen pastures, wetlands and the remains of the lagoon. As a wetland reserve for birds, fishes, reptiles and potentially otters it is second in Catalonia only to the Ebro delta.

The wetland is now threatened from a different direction. Owing to dam-building in the headwaters of the Muga and Fluviá, the coast is receding. Floods did damage in 1987, 1992, 1993 and 1994. The Natural Park can probably adapt to these events more readily than farmers and hoteliers.[28]

The Rhone

The Rhone delta is five times as big as the Ebro's. Most of its water comes from outside the Mediterranean region. The terrain varies from vineyards at the edges, through rice-fields and rough grassland (the Camargue), to reed-beds, salt-marshes, artificial saltpans and huge lagoons. Like the Ebro delta, it still contains first-class natural ecosystems. As in other fenlands, roads run on roddons (the raised beds of abandoned distributaries). It ends in a feeble line of dunes.

The delta's history is somewhat controversial. A complicating factor – and an additional menace to its future – results from its great size. The weight of the 1–2 km thickness of silt, accumulated over five million years, is enough to depress the earth's crust. Subsidence takes several thousand years to be completed; the present rate (2 mm per year relative to local sea-level[29]) may depend on the increased rate at which silt was added during the last glaciation. Subsidence is shown by the high proportion of lagoons to land.

The changing area of the Rhone delta cannot be used as an indicator of long-term erosion in its catchment: too little is known of the three-dimensional changes in its volume and its sedimentary stratigraphy. Assuming subsidence to have been slow and steady, the stratigraphy might provide some guide to short-term variations in sediment yield from the catchment.

The delta front near the mouth of the Grand-Rhône distributary has advanced by about 10 km since sea-level reached its present height. Archaeological materials suggest that this advance was completed by Classical times, and that the delta is not very much bigger today. Silt addition has been roughly balanced by subsidence. A Roman tiled floor at Tour du Vallat is under 2 metres of silt, and land that grew cereals in Roman times is now waterlogged.[30]

The Rhone delta was well supplied with sediment from the Alps. Until the mid-fourteenth century the Alpine streams seem to have been 'well behaved', driving mills and crossed by many bridges; then, quite suddenly, they became much more torrential, carrying a heavy load of sediment and acquiring multiple channels.[31] This change in character could

[26] Delano Smith 1979, p. 329; M.-A. Marqués and R. Julia, 'Données sur l'evolution du littoral dans le nord-est de l'Espagne', Trousset 1987, pp. 15–23; F. Breton and P. Esteban, 'The management and recuperation of beaches in Catalonia', Dooly & Doody 1995, pp. 511–17.

[27] J. I. Diaz and G. Ercilla, 'Holocene depositional history of the Fluviá-Muga prodelta, northwestern Mediterranean Sea', *MG* 111,

1993, pp. 83–92.

[28] D. Sauri-Pujol, A. Ribas-Palom, D. Roset-Pages, 'From hazards to resources: assessing the role of floods in the lowlands of the Costa Brava (Girona, Spain)', Dooly & Doody 1995, pp. 169–74.

[29] S. Suanez and M. Provansal, 'Morpho-sedimentary behaviour of the deltaic fringe in comparison to the relative sea-level rise on the

Rhone delta', *QSR* 15, 1996, pp. 811–18.

[30] Delano Smith 1979, pp. 328–9; G. Arnaud-Fassetta and M. Provansal, 'Étude géomorphologique du delta du Rhône: l'évolution des milieux de sédimentation fluviatiles au cours de l'Holocène récent', *Méditerranée* 78, 1993, pp. 31–42.

[31] Bravard 1993.

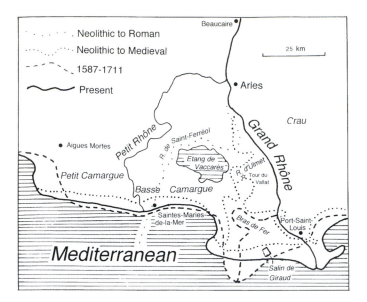

Fig. 18.10. Changes in the Rhone delta (after Arnaud-Fassetta & Provansal).

have been associated with the early stages of the Little Ice Age. Increased land-use is unlikely in these decades after the Black Death, but was thought to be the cause of the continuing instability in the eighteenth century (p. 241).

Visible expansion of the delta came later (Fig. 18.10). Between 1587 and 1711, at the mouth of the Bras de Fer, the shore advanced at about 50 metres annually.[32] Erosion in the basin may then have been at a maximum: Alpine glaciers were advanced, people and livestock were multiplying in the foothills, cultivation was extending in the mountains, and badlands were notoriously active (p. 277). The greatest well-authenticated Rhone floods, those of 1840 and 1856, which led to afforestation programmes (p. 242), came also at a time when glaciers were advanced.

This delta has been starved of silt by sixteen dams built since 1950. The sediment now regularly reaching the mouth of the Grand-Rhône is estimated at no more than 2.2 million tons a year, compared with 5.5 million in 1957 and more than 40 million (21 million cubic metres) in the nineteenth century.[33] The Petit-Rhône carries about a quarter as much. In spite of (or because of?) protective groynes, the delta west of the Grand-Rhône and adjacent coasts is retreating, especially during great storms as in November 1982.[34] Especially at risk are the saltpans which supply a quarter of France's salt. A history of malaria caused the ancient towns to be built well inland, safe from any likely rise in sea-level; but the same cannot be said of new tourist installations occupying 40 per cent of the Gulf of Lions shoreline.

Gulf of Lions

Along the Gulf of Lions, most of the embayments that still existed in Roman times have been turned into lagoons or filled up altogether. Of the Gulf of Capestang all that remains is the Étang de Vendres, and of the Gulf of Narbonne only the *étangs* of Bages, Sigean, Ayrolle and Gruisan remain. The river Aude was deliberately diverted in 1398 to a new mouth some 20 km to the north on account of flood damage around Narbonne.[35]

The Étang de Berre is an ancient lagoon separated from the sea. Sedimentation occurred during the Neolithic and Bronze Age; after a gap in the Iron Age and Roman period it resumed in medieval and early modern times (pp. 302, 304). The delta of the River Arc in the lagoon has advanced about a kilometre.[36]

Further east the rivers are among the most dramatically steep in Europe. Rising in a part of the Alps prone to earthquakes and deluges, they bring down copious sediments from soft marls, moraines, badlands and landslips. At Fréjus the Emperor Augustus constructed a port in the first century BC: 50-ton ships could still use it until the mid-seventeenth century, but it is now 2 km inland.

Nice lies near the mouth of the river Var, which rises on the Rock of Three Bishops, 3,000 metres high and only 80 km inland. This middle-sized river, still undammed, brings down nearly a half as much sediment as the great Ebro and a quarter of that of the mighty Rhone in the days of their glory. However, it has only a small delta; excess sediment slumps over the narrow continental shelf into deep water. Successive main roads and a railway are crammed between the mountains and the sea.

The events of 16 October 1979 illustrate what modern earth-moving, unassisted by earthquake, can achieve on such unstable terrain. The citizens of Nice had taken early to the air; in 1910 they built a modest airstrip on the delta. This grew into an international airport, which involved extending the delta into the sea by 3½ sq. km. Not content, they must needs have a new harbour as well: they added landfill to landfill until the delta collapsed. One wet afternoon a submarine landslide swallowed up half a cubic kilometre of miscellaneous debris, piled up by Nature and Man; the nearly finished port slid into the depths of the Bay of Angels, never to be seen again; a tsunami 4 metres high ravaged the coast; 120 km out to sea and 2,500 metres deep a turbidity current swept away submarine cables.

The disaster was preceded by a deluge of the size to be expected every few years (p. 32). Subsequent analysis showed that the delta had been recklessly overloaded and steepened by 'reclamation' works. The deluge produced underwater jets of heavy, silt-laden river water, a turbidity current that swept down the submarine Var Canyon at 40 km per hour, tearing away the base of the delta slopes and precipitating collapse.[37]

[32] R. J. Russell, 'Geomorphology of the Rhone Delta', *AAAG* 32, 1942, pp. 149–254.

[33] Floods in October 1993 and January 1994, on a scale similar to those of the nineteenth century, broke through dykes and laid down thick splays over the adjacent marshes.

[34] Arnaud-Fassetta & Provansal 1993 (see note 30); J.-J. Corre, 'The coast line of the Gulf of Lions: impact of a warming of the atmosphere in the next few decades', Tooley & Jelgersma 1977, pp. 152–69.

[35] P. Ambert, 'Modifications historiques des paysages littoraux en Languedoc central: état actuel de connaissances', Trousset 1987, pp. 35–43.

[36] Leveau & Provansal 1993.

[37] D. J. W. Piper and B. Savoye, 'Processes of late Quaternary turbidity current flow and deposition on the Var deep-sea fan, north-west Mediterranean Sea', *Sedimentology* 40, 1993, pp. 557–82; E. J. Anthony and M. Julian, 'Aspects of landslide activity in the Mercantour Massif and the French Riviera, southeastern France', *Geomorphology* 15, 1996, pp. 275–89; 'The 1979 Var delta landslide on the French Riviera: a retrospective analysis', *JCR* 13, 1997, pp. 27–35.

Sardinia

The story of Sardinia is little known. Lagoons survive to a degree unique in the European Mediterranean. To judge by abundant stromatolites – a kind of tropical shallow-water algal concretion – in stone used in the Roman buildings at Nora, they were also well developed in earlier interglacials. For centuries the lagoons were most valuable sources of fish and salt.[38]

Delano Smith has written of 'the growing amount of sedimentary evidence for a period of exceptional erosion and, by implication, climatic wetness not more than a couple of centuries ago'.[39] She suggests that the deposition of coarse sediments on the Gulf of Oristano could be associated with seventeenth- and eighteenth-century land hunger, when forest exploitation and land clearance coincided with the Little Ice Age. But there is not much sign of land hunger in Sardinia at that time, even though the population was beginning to increase, and signs of severe erosion are few.

Western Italy

The Apennines include extremely unstable rocks (Chapters 3, 17). The rivers draining these regions have some of the heaviest sediment loads in the world; but even these are now sometimes exceeded by the rate at which sand and gravel are excavated.

The River Arno has a steep, rainy basin, with extensive but now not very active badlands and landslips, and also great forests. For seven hundred years it has contained the greatest concentration of human activity in southern Europe. The river mouth has advanced seaward by 7 km over the last two thousand years, forming a delta which has filled in the bay of Pisa but not protruded beyond it (Fig. 18.11).

There is some evidence for a retreat of the shore in the fourteenth century, despite the great floods of that period (Fig. 8.4c). Some attribute the retreat to the effects of the Black Death on land use; others to coastal 'reclamation' near Pisa. In 1558 and 1568 canals were dug to lead the river round Pisa and divert floodwater into coastal marshes. From then onwards, in the mid and late Little Ice Age, the delta front advanced more rapidly, at 11 metres per year on average; each successive kilometre of advance represented more sediment than the previous kilometre. Commentators are divided in opinion between the effects of felling (or rather, grubbing out) forest and of the Little Ice Age.[40]

In 1957 two dams were built about 60 km upstream of Florence for hydro-electricity and to lessen flooding, and since 1976 a canal has been dug to divert part of large floods directly to the sea. Sediment deposition is filling up the dams and starving the river channel downstream. In addition enormous amounts of sand and gravel have been excavated: since 1850 the river bed above Florence has been lowered 2 to 4 m, while between Florence and Pisa the bed of the Arno has cut down 5 to 9 m, mainly since 1954.

The volume of sediment reaching the mouth annually is estimated to have diminished from 5.15 million cubic metres between 1500 and 1800 to 1.91 million cubic metres over the last fifty years. Billi and Rinaldi assert that recent afforestation has cut off supplies of sediment to the river, and conclude that much of the sediment comes from bed and bank erosion – though the latter must have been reduced by enwalling.[41] Another factor is that in the twentieth century the mean rainfall and the peak discharges of the river have declined (though the flood of 1966 was as big as any recorded). The amount of sediment now reaching the mouth is too little to prevent the delta front and its beaches from retreating.

The Ombrone, the next river southwards, drains a smaller, less mountainous but still steep basin with a greater abundance of landslips and badlands. Its soft siltstones and mudrocks form excellent farmland, are very susceptible to sheet-erosion when ploughed in large fields, and continue to be farmland despite erosion. The area now losing soil greatly exceeds the area gaining soil.[42]

With all this sediment, the Ombrone has a remarkably large delta despite rapid subsidence of the coast (about 3 mm per year 1891–1951). When sea-level ceased to rise, the river ended in a lagoon behind a sand-spit. This persisted for some four thousand years, but silting finally gained over subsidence and the lagoon filled up by about 50 BC. The harbour was abandoned and the coast depopulated, supposedly because of malaria. There is some evidence for growth, shifting and recession of the coast during the Roman period and after; commentators have linked these to the rise and fall of population, but the changes seem to be poorly dated and there is no supporting evidence for the link.[43]

After the thirteenth century the Ombrone delta gained consistently. By the eighteenth century the mouth had advanced by 2 km (4 metres a year on average). The rapid sedimentation has been attributed, as ever, to tree grubbing and cultivation.[44] In 1830, the flow of the river was diverted to deposit its sandy load in a coastal lagoon, thus creating the Grosseto Plain, an area of about 70 sq. km. The accumulation is estimated at 6.5 million cubic metres of sediment per year between 1830 and 1834. By the twentieth century (1924–1973) only about 1.0 million cubic metres per year was coming down.[45] It has been argued that the slackening of deposition since the mid-nineteenth century is the result of

[38] Pinna 1903, *passim*.

[39] Delano Smith 1979, p. 352.

[40] V. Alessandro, C. Bartolini, C. Caputo, E. Pranzini, 'Land use impact on Arno, Ombrone, and Tiber deltas during historical times', Quélennec 1990, pp. 261–65; C. Caputo and 4 others, 'Present erosion and dynamics of Italian beaches', ZG NF Suppl. 81, 1991, pp. 31–39; Pasquinucci & Mazzanti 1987.

[41] P. Billi and M. Rinaldi, 'Human impact on sediment yield and channel dynamics in the Arno basin (central Italy)', Walling & Probst 1997, pp. 301–11.

[42] D. D. Gilbertson and C. O. Hunt, 'Geoarchaeology and palaeoecology of late[-] and post-medieval slope and fluvial deposits in Tuscany: a field trip commentary', *IV° Ciclo di Lezioni sulla Ricerca Applicata in Campo Archeologia: Siena*, 1991 [no pagination].

[43] L. Innocenti and E. Pranzini, 'Geomorphological evolution and sedimentology of the Ombrone River delta, Italy', *JCR* 9, 1993, pp. 481–93.

[44] Pasquinucci & Mazzanti 1987; P. L. Aminti and E. Pranzini, 'Variations in longshore sediment transport rates as a consequence of beach erosion in a cuspate delta', Quélennec 1990, pp. 130–34.

[45] Frangipane & Paris 1994. Sediment production varied more than tenfold from year to year.

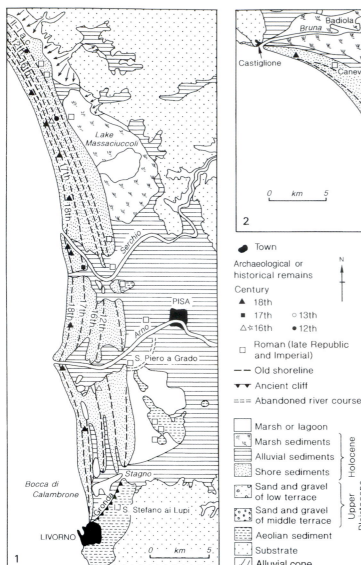

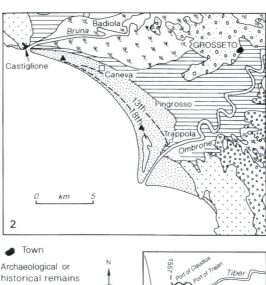

Fig. 18.11. Growth of (1) the Arno and (2) the Ombrone deltas (after Pasquinucci & Mazzanti), and (3) the Tiber delta (after Le Goff & Bradford).

slackening land use in the catchment.[46] It may, however, be that the mid-nineteenth century had been a period of unusually rapid deposition owing to deluges. We would ask whether sediment production has again increased since 1973 owing to the increase of arable land, bulldozing and amalgamation of fields, in the hills around Siena.

Away from rivers this coast is stable. The rocky island of Monte Argentario is joined to the mainland by two *tomboli* or barrier beaches enclosing a lagoon. Into this lagoon projects a peninsula on which stands Orbetello, a town of Etruscan origin. Because it belonged to Spain, and because of the impregnable military defences afforded by enclosed shallow water, it has an incomparable series of early maps going back to 1472. The geography has not significantly altered for at least five hundred years.[47]

The River Tiber has a catchment of similar character to the Arno, but with a longer history of intensive human activity

owing to the mega-city of Rome. To Horace and Virgil it was the Yellow River (*flavus Tiberis*), from its suspended sediment. Its delta is rather less active. The structure is well known to archaeologists, and the stratigraphy is known in three dimensions from the logs of well-borers. At the beginning of the Holocene the Tiber emptied into a lagoon behind barrier islands, lying within its great low-level Pleistocene delta. During the sea-level rise, the lagoon remained in existence but moved inland and to a higher level, and began to be filled in by flood sediments. Once sea-level had ceased to rise, a chain of islets bounding this peaty lagoon continued the line of the coasts on either side. Infilling continued until, by the Roman period, the lagoon had been cut in two, and the delta had begun to protrude into the sea.[48]

The city of Ostia Antica, near one of the main mouths of the Tiber, continued as the port of Rome for several centuries; we are tempted to connect this stability with the slightly

---

46 The area under forest is said to have increased between 1829 and 1991 from 29 per cent to 40 per cent, and the area under grazing to have declined from 37 per cent to 7 per cent. These figures should be accepted with caution, since they are highly sensitive to changes of definition

(p. 19). There is, however, no doubt of the reality of the changes.
47 L. Rombai and G. Ciampi, *Cartografica storica dei Presidios in Maremma*, Consorzio Universitario della Toscana meridionale, Siena, 1979.

48 Bradford 1957, pp. 237–56; P. Bellotti, S. Milli, P. Tortora, P. Valeri, 'Physical stratigraphy and sedimentology of the Late Pleistocene–Holocene Tiber delta depositional sequence', *Sedimentology* 42, 1995, pp. 617–34.

earlier period of quiescence in the sedimentation history of south Etruria (p. 304). In the first century AD the emperors Claudius and Trajan built the greatest artificial harbour in the Empire (Fig. 18.11) for the shipping that fed Rome.

Despite resumed flooding and sedimentation in the hinterland, the delta apparently paused in its progradation until the late Middle Ages. It then advanced rapidly, especially after a deluge in 1557 had permanently altered the lower river. There are said to have been ten floods in the second half of the fifteenth century, and thirteen in the seventeenth, compared with no more than six in all the preceding centuries AD.[49] The advance continued until, by 1900, Trajan's harbour was 5 km inland. There has been the usual debate between those who assign the advance of the delta to (mostly Postulated) deforestation and those who point to increased rainfall as the cause.[50] (There is evidence for the conversion of forest to savanna in this part of the Apennines, but several centuries earlier, p. 210). Another supposed cause was the embanking of the Tiber higher up, which caused the channel to deepen and more silt to come down. By the end of the nineteenth century the river was carrying down 6.8 million tons of silt to the sea annually – less than the Ombrone was bringing from one-tenth of the catchment area. Since the 1940s most of the silt goes to filling a dam, and since the 1950s the coast has been eroding ever faster.[51]

Basilicata

The Gulf of Táranto, beneath the arch-badlands of Basilicata, is a classic site of coastal civilizations buried beneath the combined deltas of five parallel rivers. Archaeologists commonly have to dig through 6–12 metres of sediment to reach Classical sites.[52]

The coast of the Gulf has been tilted eastwards since Early Pleistocene times, with uplift of 700 metres near the lower Sinni and slight subsidence of the Otranto peninsula (Fig. 15.10). Throughout the Pleistocene, the rivers have carried very large loads of sediment to the coast in times of flood. The surfaces of the confluent deltas are graded to successive interglacial sea-levels, giving a flight of marine terraces, the oldest being some 720,000 years old.[53]

Historic deposits, though notable, are small compared to those of the Quaternary. Despite the assistance of uplift, the shore-line has advanced only 1.2 km since 700 BC. A zone 4 km wide behind this new land has been covered with alluvium. Immediately behind this zone, the land surface of c. 700 BC still survives, bearing the field systems that went with Metapontum and other ancient cities. At some unknown date in the last 2,700 years this surface was dissected with gullies. Much of the material deposited on the plain presumably comes from these local badlands rather than from soil

erosion over a wide area (p. 281).

Even in this actively eroding area dams were built in the 1950s. Although they are not low enough on the rivers to cut off all the silt supply, the advance of the coast has been halted or reversed in the last fifty years.

East coast of Italy

The coast either side of Ancona, underlain by Plio-Pleistocene clays and sands, is notoriously unstable and subject to landslipping. Rivers rising 1,000 metres or more up in the limestones and marly limestones of the Umbria-Marche ridge, about 60 km inland, carry great quantities of sediment to the coast. In the 1930s and 1940s annual sediment yields from the Metauro, Cesano, Esino, Chienti, Aso and Tronto valleys (Fig. 18.12) were around 1,000 tons per sq. km, equivalent to a lowering of the surface of about 0.2–0.5 mm per year. Most of the sediment came from very heavy storms, as in October 1940 when 200–300 mm of rain fell in 48 hours. Over that year, the Foglia carried almost three times and the Tronto over five times its mean suspended load.[54]

During the Holocene, meandering rivers in Marche cut down through Older Fill to form terraces, which now slope away downwards beneath coastal plain deposits. Starting between 1100 and 1400 AD, incision gave way to sedimentation. Braided rivers deposited some 10 metres of sediment containing abundant trunks of elm and poplar uprooted from riverine woods, which vary in date from the Middle Ages to the nineteenth century. The deposition seems to mark the onset or reactivation of the Marche badlands. After 1900, the rivers began to cut down again, going through the sediments that had accumulated over six to nine hundred years, and even, in their lower courses, another 10 metres or more into Older Fill.[55]

These events affected the coastline around Ancona. When the sea had reached its present level, the coast was characterized by unstable cliffs and sandy pocket beaches. By the third century BC its inlets had been filled with river-borne sediment and were fronted by barrier beaches and lagoons. When sediment was coming down, between 1100 and 1900 AD, deltas formed at river mouths and sandy beaches accumulated in front of the cliffs, protecting them against sea erosion. This changed in the course of the twentieth century, especially when gravel was dug between 1940 and 1960, and when check-dams were built below bridges to protect their foundations. Now the coast is slumping, and attempts are being made to protect it by artificial barriers.

M. Coltorti is inclined to explain the medieval-to-modern deposition period by the 'clearing' of woodland, to which Little Ice Age historical documents are said to provide

[49] D. Camuffo, 'Reconstructing the climate and the air pollution of Rome during the life of the Trajan Column', *The Science of the Total Environment* 128, 1993, pp. 205–26.

[50] G. Barker and A. Grant, eds., 'Ancient and modern pastoralism in central Italy: an interdisciplinary study in the Cicolano mountains', *Papers of the British School at Rome* 59, 1991, pp. 15–88; J. Le Gall, 'Le fleuve, la mer et les hommes aux bouches du Tibre', Trousset 1987,

pp. 89–93; Segre 1967, quoted in Alessandro & others, note 38.

[51] Milliman & Syvitski 1992.

[52] Delano Smith 1979, p. 330; Brückner 1980; H. A. Meyerhoff, 'Geologic factors affecting the search for Sybaris', *The Search for Sybaris 1960–65*, ed. F. G. Rainey and C. M. Lerici, Lerici, Rome, 1967, vol. 1, pp. 250–311; J. B. Ward-Perkins, *Landscape and History in Central Italy*, Blackwell, Oxford, 1964.

[53] P. J. Hearty and G. D. Pra, 'The age and stratigraphy of Middle Pleistocene and younger deposits along the Gulf of Taranto (southeast Italy)', *JCR* 8, 1992, pp. 882–905; Bruckner 1980; Néboit 1984.

[54] AQUATER, *Regione Marche. Studio generale per la difesa della costa: primera fase*, Rapporti di Settore 2, San Lorenzo in Campo, 1982.

[55] M. Coltorti, 'Human impact in the Holocene fluvial and coastal evolution of the Marche region, central Italy', *Catena* 30, 1997, pp. 311–35.

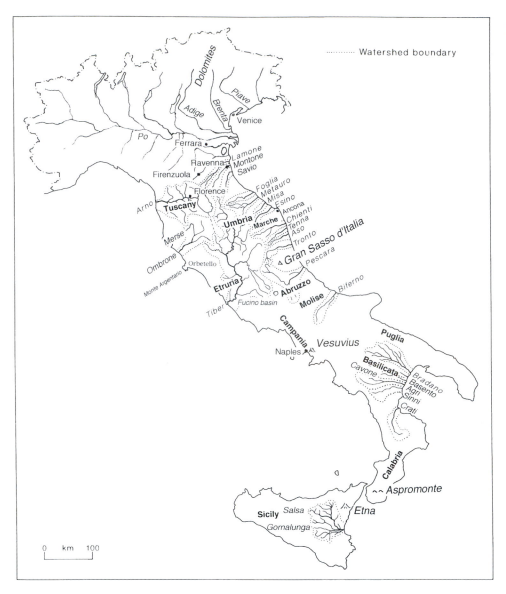

Fig. 18.12. Italian rivers mentioned in the text.

testimony. The landscape has experienced many such vicissitudes over the last two thousand years; it is surprising that he does not attribute the river and coastal changes, as we would, to more frequent deluges during the Little Ice Age.

The North Adriatic

The most brilliant example of delta technology is Venice. The city and its outlying islands are the remaining fragments of the delta of the Brenta, drowned by rising sea-level but protected from wave-action by the barrier beach of the Lido. The city was built from the sixth to the twelfth centuries AD on about two hundred islets. These are associated with the roddon-like levées of sinuous distributaries later to be rejuvenated as the Grand and Giudecca Canals. The lesser canals, however, have straight lengths and sharp corners, and are not like the organic growth of meandering salt-marsh creeks. Nor

do they exhibit the regular narrow rectangles of planned delta towns such as Chioggia.[56] We suspect that they represent the ditches of a drowned Roman field-system.

Had events been left to Nature, siltation would probably have overtaken subsidence in the Middle Ages, and Venice would have become just another semi-coastal town like Ravenna. The Venetians intervened to prevent this, and to preserve their peculiar way of life and their very effective defences against attack. They got rid of the rivers. In successive canalizations, the Brenta was diverted southward outside the lagoon altogether, the Sile and Piave to the eastward. The system, assisted by continuing subsidence of the land,[57] has been remarkably effective; it still functions despite two hundred years of comparative neglect.

The Venetian lagoon, which covers 550 sq. km, is part of the confluent system of a dozen river deltas that borders the Adriatic for 230 km. The biggest delta is that of the great river

[56] R. J. Goy, *Chioggia and the Villages of the Venetian Lagoon*, Cambridge University Press, 1985, chapters on Chioggia and (by G. Marzenin) on Venice.

[57] There are about seven steps up to the front doors of the palaces on the Grand Canal, now visible only at unusually low water; they suggest rapid subsidence of about a metre from the sixteenth century onwards. The bridges on the lesser canals, now humpbacked to let barges pass under, corroborate this.

Po. By about the fifth century BC the river had filled in its drowned valley; the coast was defined by an arc of barrier beaches, of about that date, continuing the curve of the Venetian Lido round to Ravenna (Fig. 18.13). The Po went on to develop a sequence of protruding deltas. The Roman delta formed a little to the north of the port of Ravenna, which it cut off from the sea by its growth. Other outlets at this time were the Padus Vetus ('Old Po') and the Po di Volano. The Padus Vetus built a delta and silted up in the early Middle Ages, and the Po di Primaro came into existence. The Volano, by then the main stream, built a delta in its turn. In the twelfth century it was diverted above Ferrara, after a major flood, into the Po di Ficarolo with its mouth some 20 km northwards.[58] In the fourteenth century, the river migrated further north still, forming a fourth delta which threatened to encroach on the lagoon of Venice. Over the next four hundred years the Venetians successively diverted the rivers away from the lagoon.

The story is told in maps from 1470 onwards. In the late sixteenth century the two southern of the five distributaries, with their feeders, were cut off from the Po itself, reducing the amount of silt getting into the Po delta from the Apennines. The remaining distributaries began to build lobes rapidly seawards. The northern one, the Po di Ficarolo, threatened to impinge on the Venice lagoon. Between 1604 and 1607 it was diverted southward by a 5-km cut, the Taglio di Porto Viro, which has formed the fifth, modern delta. Similar diversions protected the lagoon of Comácchio.[59]

Of the successive Po deltas, the second and third (medieval) are the smallest. The modern is the biggest in area and probably in volume; it formed in only three hundred years. In the seventeenth century the NE lobe of the delta, starved of sediment, ceased to advance, but the SE lobe advanced 4–9 km along a front of 30–35 km – a rate of deposition, as P. Fabbri has noted, unparalleled in historical times. Such an acceleration, he continues, has been recorded at most river outlets in Italy – at most large river outlets in southern Europe, we might add – 'and may be related to the extensive deforestation in this period to meet increased needs for farmland and the high demand for timber all over Europe'. After 1897, with dam-building and sand- and gravel-digging in river beds, the Po delta advanced more slowly. Since the 1950s the southern delta has been retreating at a rate of 5–10 metres a year, and subsiding at 5–20 mm a year, calling for protective works in an attempt to preserve tourist developments on the coast.

The Venetian lagoon has been preserved – the earliest map of Venice and its saltmarshes, of the fourteenth century, is remarkably like the present – but its ecology has changed. Natural subsidence is no longer compensated by sediment inflow, resulting in coast erosion and increased salinity. Subsidence has been augmented by pumping of water, extraction of methane and compaction of drained marshland. An area of 2,400 sq. km, extending 40 km inland to the gates of Ferrara

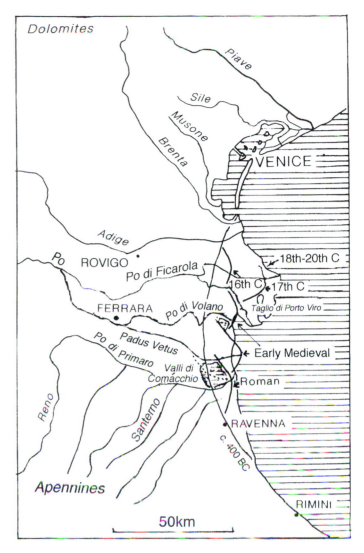

Fig. 18.13. Northward migration of the Po delta (after Fabbri). The old shorelines are marked by dunes raised a metre or so, on which modern roads tend to run, with settlements strung out along them. One such shoreline bisects the lagoon called Valli di Comácchio, presumably formed later by subsidence.

and Rovigo, is now between 2 metres below and 2 metres above sea-level. In Venice, by 1970, the pumping of groundwater had drawn down the aquifer level by 20 metres, causing the land surface to subside 20 cm more than it would have done naturally. Excessive pumping was stopped, and by 1990 ground-water levels had largely recovered, but there was only a 2-cm recovery of the land level.[60]

Greece

The biggest delta on the west coast is that of the Akhelóös, a steep river from extremely erodible mountains which debouches into a shallow sea (Fig. 18.14a).[61] This delta

[58] Sestini 1992; C. Cencini, 'Physical processes and human activities in the evolution of the Po delta, Italy', *JCR* 14, 1998, pp. 774–93.

[59] P. Fabbri, 'Coastline variations in the Po delta', *ZG NF* Suppl. 57, 1985, pp. 155–67; M. Bondesan and 6 others, 'Coastal areas at risk from storm surge and sea-level rise in northeastern Italy', *JCR* 11, 1995, pp. 1354–79.

[60] Lumsden 1992, p. 183, reporting results by the Istituto per lo Studio della Dinamica delle Grandi Massi; Bondesan (see note 59); Sestini 1992.

[61] D. J. W. Piper and A. G. Panagos, 'Growth patterns of the Acheloos and Evinos deltas, western Greece', *Sedimentary Geology* 28, 1981, pp. 111–32.

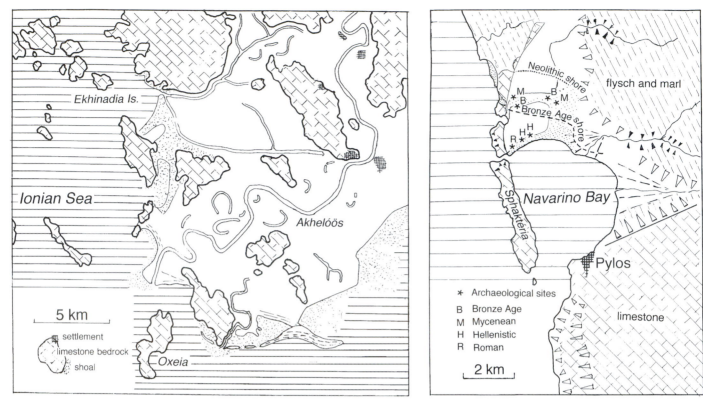

Fig. 18.14a. *Above* Akhelóös delta (after Piper & Panagos). Fig. 18.14b. *Above right* Pylos (after Kraft and Bousquet & others).

formed mainly before written records. Herodotus in the fifth century BC records that it had joined half of the Ekhinades islands to the mainland. Thucydides, a little later, expected that the remaining islands would be joined up, but this had not happened by the time of Strabo (first century BC), who gives the distances of the islands from the mainland. Pausanias (second century AD) remarks that the Akhelóös was bringing down less mud than it used to, because Ætolia had been depopulated and its lands were uncultivated.[62] Since Strabo there has been little net change, except that the main mouth has advanced 2 km since 1830. Half the Ekhinades are still free.

The coast is subsiding, as witness the large lagoons. The growth of the delta depends on whether accretion gains on subsidence. Growth seems to have been rapid in and around the fifth century BC, and then to have stopped. Pausanias's interpretation, however interesting, is probably wrong. The Ætolia Survey has not confirmed that settlement was lacking in the Roman period;[63] and even if it was, the delta evidently did not resume growth as Ætolia recovered from depopulation. We cannot say whether the growth in the Classical period was due to a burst of accretion or to a pause in subsidence. The two processes stayed otherwise roughly in balance – or did, until the inevitable dam-building stopped the accretion.

Peloponnese

The north coast of the Peloponnese, where the Corinth fault-zone plunges beneath the Gulf, was sacred to Poseidon, god of earthquakes and the sea. The short steep rivers, fed abundantly with sediment from the badlands on the uplifted side of the fault, have only small deltas much as does the Var. These grow fast, assisted perhaps by coastal uplift, but on reaching a certain size slump off the edge into the abyss – often triggered by an earthquake. In 373 BC the city of Helike, on the Sélinous delta, behaved rudely to some visitors who had asked for a cult statue of Poseidon. One night in the following winter the offended god swallowed the delta; Helike, like Nice harbour, was never seen again, except for the drowned tree-tops of Poseidon's sacred grove. Herakleides Pontikos, whose contemporary account is preserved by Strabo and Pausanias, gives a convincing description of a tsunami. The delta grew again and fell off again in 1748 AD, again on 23 August 1817, again in 1861, and presumably will soon do so yet again.[64] Some geophysicists put these events in the context of the general widening of the Gulf of Corinth by 4–6 mm a year.[65]

The bay of Pylos (Fig. 18.14b), although it has no large river, was partly filled during the middle Holocene. By 2500 BC about half the bay was filled in. Since then there has been only a minor advance, in Hellenistic or Roman times,

[62] Herodotus II.10; Thucydides II.cii.3–5; Strabo, *Geography* X.ii.19; Pausanias VIII.xxiv.11.

[63] S. Bommeljé and 6 others, *Aetolia and the*

Aetolians, Parnassus Press, Utrecht, 1987.

[64] Strabo, *Geography* VIII.vii.2; Pausanias VII.xxiv.4–6; Leake 1830, vol. 3, p. 402; B. Bousquet, J. J. Dufoure, P. Y. Pechoux, 'Temps

historiques et évolution des paysages égéens', *Méditerranée* 48 (2), 1983, pp. 3–25.

[65] N. Mouyaris, D. Papastamatiou, C. Vita-Finzi, 'The Helice Fault?' *Terramotae* 4, 1993, pp. 124–29.

Fig. 18.15. Thermópylae and the Sperkhéios delta. The sea is shown as in 1938, with earlier and later shorelines added. (Adapted after Kraft & others).

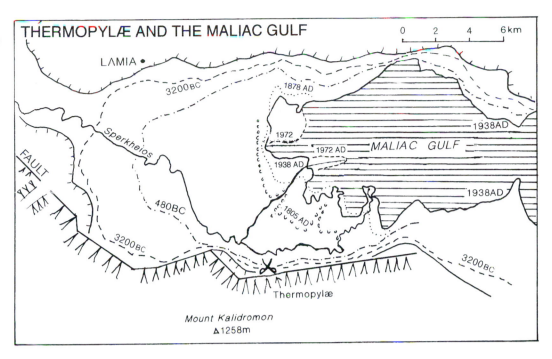

cutting off part of the bay as the Osmanaga lagoon. The topography of the battle of Sphaktéria in 425 BC was much the same as that of the battle of Navaríno in 1827 AD or of today.[66]

Thermópylae

A famous locality for coastal change is the Pass of Thermópylae, where three hundred Spartan warriors held back the entire Persian army in 480 BC, and scene of many other battles down to World War II. The site is on the south side of the Maliac Gulf, a sheltered bay formed by a graben, where a fault-cliff rises on one flank of the delta of the River Sperkhéios. At the time of the battle there was barely room for an army to squeeze past the base of the cliff, but the advancing delta has now formed a plain 5 km wide (Fig. 18.15).

The bay has probably been filling in throughout prehistory, but at times the sea has risen to lap against the cliff and the pass has been unusable. Between 3200 BC and 480 BC the delta advanced by 3 km, and continued at about the same rate until the nineteenth century. Sedimentation was gaining on the very rapid subsidence: the site of the battle is said to be buried by 20 metres of silt. When W. M. Leake mapped the scene in 1805, some 3 km of delta had come to separate the sea from the cliff. The delta reached nearly its present extent by the end of the nineteenth century. It has advanced very little since, despite the absence of dams on the river.[67]

Thermópylae is a most complex site, the result of the changing balance between fault movements, variations of sea-level and the varying activity of the badlands that supply much of the sediment. The river may also have been artificially diverted southwards in order to create the 'Rice Grounds' and 'Salt works' shown on Leake's map.

Thessaloníki

This city stands on what is left of the Thermaic Gulf, a bay nearly filled in by two rivers, the Aliákmon and the Axios (Vardar). Their catchments include badlands. The area has greatly subsided, to judge by the site of Mekyberna, 50 km SE, where the harbour has gone at least 10 metres under the sea in 2,400 years.[68] Pélla, the capital of Macedonia at the time of Philip and Alexander (fourth century BC), lies beside a wide alluvial plain. Until about 500 BC the plain had been a shallow embayment, extending west of Thessaloniki for about 60 km, with Pélla close to its north shore. By 100 AD the two deltas had met, pinching off the bay into a lagoon. Growth since then has been much slower. The late-Roman Klidhí bridge on the Thessaloníki–Athens road is not far from the modern bridge, only a few km inland (Fig. 18.16).[69]

West coast of Asia Minor

Troy is another famous instance of the infilling of a bay. It was surrounded by the sea on three sides five thousand years ago, and could still be approached from the sea at the end of the Bronze Age. Deposits by the River Scamander (Kara Menderes) have been steadily filling up the bay. This process was completed at some time after the Roman period, leaving Troy 7 km inland (Fig. 18.17). Swift currents in the Dardanelles prevent the delta from advancing further.[70]

[66] J. C. Kraft, G. Rapp, S. E. Aschenbrenner, 'Late Holocene paleogeographical reconstructions in the area of the Bay of Navarino: Sandy Pylos', *JAS* 7, 1980, pp. 187–210.
[67] Leake 1835, plate at end of vol. 2; Kraft & others 1987.

[68] F. K. Chaniotes in *AR* 43, 1996–7, p. 69.
[69] J. Bintliff, 'The plain of western Macedonia and the Neolithic site of Nea Nikomedia', *PPS* 42, 1976, pp. 241–62.
[70] J. C. Kraft, 'Geological reconstructions of coastal morphologies in Greece and Turkey:

Troy, the Gulf of Messenia, and the embayment at Methoni', Trousset 1987, pp. 155–57; D. Eisma, 'Stream deposition and erosion by the eastern shore of the Aegean', Brice 1978, pp. 67–81.

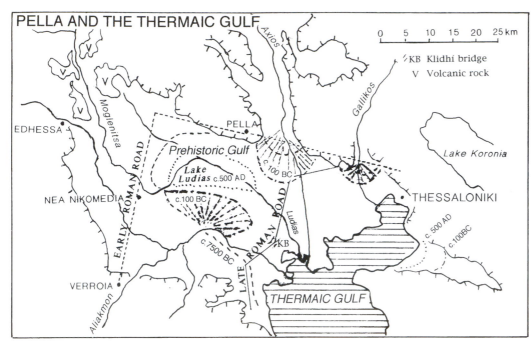

PELLA AND THE THERMAIC GULF

Fig. 18.16. *Left and below* Possible stages in infilling the Thermaic Gulf (after Bintliff).

Fig. 18.17. *Right* Coastal infilling at Troy and in the bays of west Anatolia (based on Kraft & others).

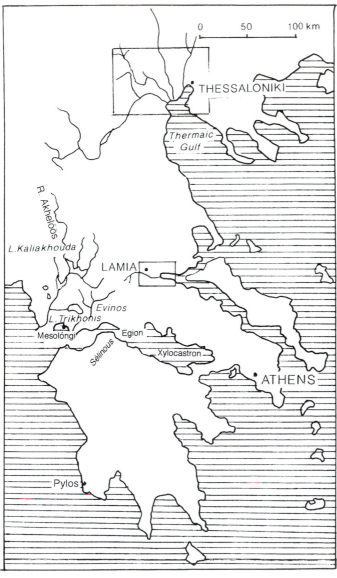

The great rivers of western Anatolia – Bakır, Gediz, Little (Küçük) and Great (Büyük) Menderes – flow down to the Aegean in east–west grabens, areas of subsidence between parallel faults that are still active. During glacial periods of low sea-level the grabens formed fjord-like inlets as much as 100 km long and 10 km wide. The very active hinterland has produced vast quantities of sediment from volcanics, badlands, etc. Alluvial cones up to 100 metres high, formed by Pleistocene torrents emerging from short, steep side-valleys, delivered coarse sediments to the valley floors and acted as dams to form short-lived lakes. The grabens were thus filled with great thicknesses of lake and river sediments.[71]

This set the scene for the Holocene, in which the valleys became the setting for the brilliant, unstable ancient cities of Pergamon, Smyrna, Ephesus and Miletus. The rise in sea-level had converted the valleys into long, shallow embayments overlooked by hills 600–800 metres high. Sediments brought by the tributaries filled these up, leaving some residual lagoons. The sea retreated at rates of as much as 2 km a century. Pliny has much to say of cities left inland and islands annexed.[72] Ephesus and Miletus, founded as ports at one stage of this process, were later abandoned far inland.

Only the Gediz graben still retains a fragment of its bay and a big port. Subsidence – said to be 1–2 mm per year – has been exceeded by siltation. Seismic profiles have revealed some evidence of the three-dimensional structure of this delta and of its submerged pro-delta. The town of Temnos, now Emiralem 30 km up the river, may have been near the sea in 1000 BC. The Gediz delta extended rapidly west and then

[71] G. Evans, 'The recent sedimentation of Turkey and the adjacent Mediterranean and Black Seas: a review', *Geology and History of Turkey*, ed. A. S. Campbell, Petroleum Exploration Society of Libya, Tripoli, 1971, pp. 385–406; R. J. Russell, 'Alluvial morphology of Anatolian rivers', *AAAG* 44, 1954, pp. 363–91.

[72] Pliny, *Natural History* V.xxxi.

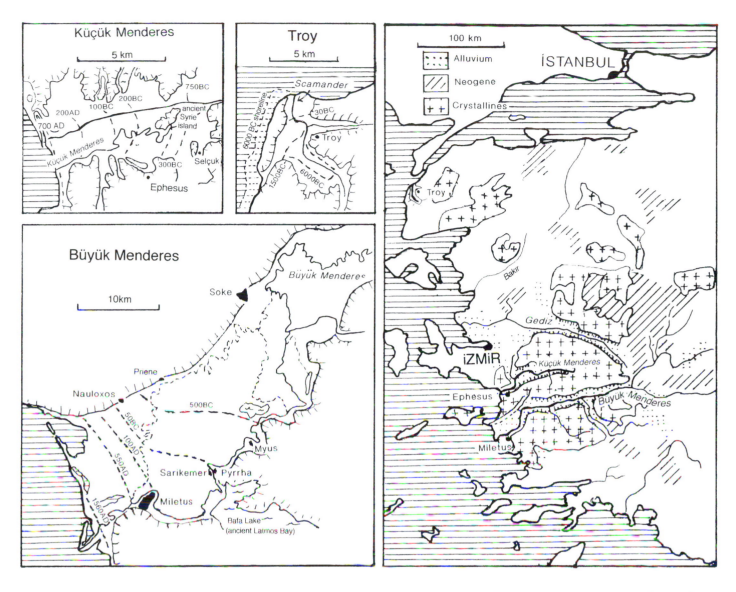

south, reaching nearly its present extent by the first century AD. Pliny recalled that the town of Leucae (now 3 km inland) had once been an island.[73] Since then there has not been much addition, but the lowermost course of the river has shifted. Biblical Smyrna, well out of the way of the main delta, took on a new lease of life as Turkish İzmir; it is protected (rather like Venice) by an artificial diversion of the main river made in 1886 to prevent silting of the shipping channel. The annual sediment discharge was about 11 million tons before the river was dammed in 1960; between maps of 1887 and 1944 the plain at the new mouth extended about 5 km. But the visible delta has not changed much over the last thousand years.[74]

The fate of western Turkey was for long held up as an example of the far-reaching consequences of Ruined Landscape. Mismanagement of land in the interior (especially, it was said, by the carefree and nomadic Turk) led to harbours being replaced by malarial fens and to a graveyard of Greek cities. It is now perfectly clear that alluviation began long before either Turks or Greek cities. The driving force is the tectonic instability of Anatolia, the most seismically active part of the Alpine mountain-building belt. The graben faults continue to deliver sediment: for example the Gediz earthquake of 1970 set off earthflows, slumps and large-scale landslides especially at the margins of the grabens.[75]

Growth-rates are not consistent. The Gediz delta extended apparently evenly from 1000 BC to 1000 AD, and has since not changed much. The Küçük Menderes delta advanced most rapidly between 300 and 100 BC, in the earlier part of the heyday of Ephesus (which moved in 286 BC to a less silted-up site). The Büyük Menderes delta may have grown most rapidly between 500 BC and 700 AD, isolating Miletus from the sea; here too there is some sign of less rapid growth in the last few centuries.[76] Most of the rivers have recently cut down several metres into their valley infills.

[73] Pliny, *Natural History* V.xxxi.119.

[74] A. E. Aksu and D. J. W. Piper, 'Progradation of the Late Quaternary Gediz delta, Turkey', *MG* 54, 1984, pp. 1–25.

[75] S. Erinç, 'The Gediz earthquake of 1970', Campbell 1971, pp. 443–51.

[76] S. Erinç, 'Changes in the physical environment in Turkey since the end of the Last Glacial',

Brice 1978, pp. 87–110; Eisma (see note 70), pp. 70–72; M. and R. Higgins, *A Geological Companion to Greece and the Aegean*, Duckworth, London, 1996.

SUMMARY TABLES

Delta	5000–1500 BC	1500 BC–500 BC	500–1 BC	1–500 AD	500–1200 AD	1200–1500 AD	1500–1750 AD	1750–1900 AD	1900–1996 AD	Badlands (•) and other special features
Po Tagus										estuarine deposition
Po, Sp Guadiana	+++	++	+++	++	+	++	+ –	+ –	– –	
Sp Río de Vélez						?	++++	?	– – –	•
Sp Guadalfeo							+++?	?	?	••
Sp Adra			+++?	+++?				++++	–	•• mining
Sp Andarax								+++	– –	••• mining
Sp Segura				+++	?	?	0	0	0	•••
Sp Ebro	slight	slight	slight	slight	slight	+	++++	++	– –	•••
Sp Llobregat	?	?	?	?	+++	?	?	?	– –	••
Fr Rhone	?	?	?	++	stable	stable	stable	+	–	comes from Alps; delta subsides under own weight
Fr Arc (Provence)	+++	slight	slight	slight	?	+++	+++	?	?	•
It Arno				?	++	–	++++	+++	?	• dense population since c. 1200
It Ombrone	?	?	++	+	–?	+++	++++	+++	– –	•••? subsidence
It Tiber	++	++?	+	+	+	++	++++	+++	– – –	• dense population since 100 BC
It Basilicata	?	?	+++?	++++	?	+++	++++	+++	++ –	•••
It Po	?	?	+	++	++	+++	++++	+++	0	• comes from Alps and Apennines; subsidence
It Venice	?	?	?	+++?	– –	stable	stable	stable	stable	comes from Alps; subsidence; artificial control
Gr Akhelóös	++++	+++	+++	stable	stable	stable	stable	+	stable	•••
Gr Pylos	++++	++	+?	stable	stable	stable	stable	stable	stable	
Gr Thermópylae	+++	+++	+++	++?	++?	++?	+++	++++	0	•• tectonics; artificial embanking?
Gr Thermáic Gulf	+++	+++	++++	+	+	+	+	+	+	••?
Tk Troy	+++	++++	+++	+++	+++?	stable?	stable?	stable?	stable?	restrained by sea-current
Tk Gediz	++	++++	++++	+++	+++	+	+	+	+	•?
Tk Ephesus	+++	+++	++++	++++	+++	+	+	+	+	••?
Tk Miletus	+++?	+++	+++	+++	+++?	?	?	?	?	•?

Table 18.i. Apparent advance (+) or retreat (–) of deltas at different periods.

Date	Climate	Culture	Catchment	Sea-level	Coast
1960–2000 AD	Warm	European Union Bulldozer Plastic pipes	Bulldozing Over-pumping of aquifers Rivers dammed and excavated Urban growth, rural decline	Stable with local variations	Deltas retreat Tourist development Coast defences
1900–1960	Warming	Internal combustion engine End of malaria	Land abandonment Mechanized farming Urban growth, rural decline	Stable with local variations	Rivers dammed Deltas stop growing
1750–1900	Little Ice Age IV	Steam engines Bureaucratic culture	Maximum cultivated area Urban growth	Stable with local variations	Coasts repopulated Many deltas drained and cultivated
1650–1750	Little Ice Age III		Local expansion of cultivated area	Stable with local variations	Delta growth continues
1500–1650	Little Ice Age II	Early–modern 'peasant' and commercial economies	Expansion of cultivated area Younger Fill continues	Stable with local variations	Peak growth of some deltas
1200–1500	Little Ice Age I	Late medieval	Maximum population, then Black Death, then slow repopulation Coasts abandoned Third phase of Younger Fill	Stable with local variations	Deltas slowly expand
800–1200	Medieval Warm Period	Medieval	Slow increase of population	Stable with local variations	Deltas slowly expand
500–800 AD	Göschenen II	Byzantine, Arab	Land abandonment Second phase of Younger Fill	Stable with local variations	Deltas slowly expand
100 BC–500 AD		Roman, Byzantine	Peak of population in some areas Pause in Younger Fill accumulation	Stable with local variations	Rivers build deltas and block harbours
5000–100 BC	Göschenen I Löbben Rotmoos	Neolithic, Bronze, Iron Age	Expansion of browsing, cultivation, ?agricultural terracing First phase of Younger Fill	Stable with local variations	Estuaries fill with sediment River muds accumulate as pro-deltas Some deltas become visible Barrier beaches and dunes form
9000–5000 BC	Warm, rainy (pre-Aridization)	Mesolithic to early Neolithic Herding and cultivation begin	Rivers cut channels in Older Fill Trees spread Soil formation Islands colonized	Rises to present level	Estuaries form and begin to be filled with sediment Barrier beaches form Cliffs cut
10,000–9000 BC	Cold (Younger Dryas)	Mesolithic	Final deposit added to Older Fill	Minor fall	Minor fall in sea-level
14,000–10,000 BC	Warming	End of Palaeolithic	Mega-fauna exterminated on mainland Trees increase Rivers cut channels in Older Fill	Begins to rise	Lower valleys flooded Sedimentation moves up
40,000–14,000 BC	Cold (+ warm interval)	Late Palaeolithic	Steppe, savanna, little forest; abundant large mammals Periglacial activity Older Fill accumulates	110 m below present	Rivers cut down near their mouths Deltas accumulate on continental shelf; pro-delta deposits spill over the edge

Table 18.ii. Chronology of delta growth in relation to environment and human cultures.
This is presented as a *rough* generalization for the *west and middle Mediterranean*. In Greece and Turkey there has been much less growth of deltas since AD 500.

RELATIVE SIZES OF DELTAS

On average the area of a visible delta is about 1.8 per cent of the area of its catchment. Individual deltas can vary widely from this ratio, but the relationship holds over a thousandfold range in the sizes of deltas and of catchments (Fig. 18.18). Smaller catchments tend to have slightly bigger deltas in proportion than larger catchments.

Unless affected by uplift or subsidence, every delta represents the accumulation of about 110 metres thickness of deposit in eight thousand years. The average thickness will probably be about half this, depending on the profile over which the deposit was laid down. The accumulation of 55 metres of deposit in eight thousand years over 1.8 per cent of the catchment is equivalent to the removal of 0.12 mm of material per year from the whole catchment – a figure within the usual range of erosion rates.

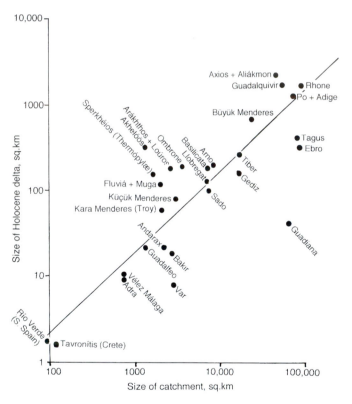

Fig. 18.18. Sizes of deltas in relation to size of their catchments. Both scales are logarithmic.

What factors, other than the size of the catchment, determine the size of a delta? Table 18.iii lists deltas in the order in which they exceed, or fall short of, the size appropriate to their catchments. At one extreme, the vast extent of the Po–Adige delta is presumably due to a favourable topography – a shallow bay at the end of a sheltered sea, already largely filled with the delta deposits of previous interglacials – combined with ready sources of sediment from the Alps and the erodible flanks of the Apennines. Quite as big, in relation

to its much smaller catchment, is the Akhelóös delta, whose special feature is a vast supply of sediment from the uplifting, unstable flysch rocks of the southern Píndhos mountains. At the other extreme, the great Guadiana has a tiny delta, truncated by the currents, tides and surf of the Atlantic. The Var has a very small delta because the steep-to coast leaves no room for a bigger one.

The most consistent determinant of the relative sizes of deltas is thus the topography of the estuary or continental shelf covered by them. The Ebro, for example, has a relatively small delta because it rises from deep water. Other important factors are the effects of exposure to waves and currents, especially combined with tides. Uplift or subsidence have less effect, other than on lagoons within the delta. Climate is not very significant, although (other things being equal) small deltas are more frequent in drier regions.

Copious supply of sediment appears to be responsible for the large size of some deltas, such as the Akhelóös, Sperkhéios, Fluviá and Ombrone. However, it is easily overcome by other factors: the Granada rivers have below-average deltas despite coming out of badlands.

Human activities are difficult to assess. Some deforestation and some cultivation have occurred in all catchments, although it is difficult to establish that they occurred at the specific times needed to support claims about the histories of particular deltas, and not at other times. It is even more difficult to establish whether these activities took the particular forms needed to cause erosion (e.g. ploughing a smooth unbroken slope). The best we can do in Table 18.iii is to point out those basins that have contained concentrated civilizations – an Ephesus or a Florence – or where mining has been linked to the growth of deltas. There is clearly no tendency for these catchments to be the ones with the biggest deltas. Apart from the Po with its special circumstances, the biggest deltas in relation to their catchments tend to be those of obscure Greek, Catalan and Tuscan rivers in which, as far as we can tell, there has been no more than average – often less than average – cultivation and deforestation.

CONCLUSIONS

Deltas are more difficult to study than river terraces. They are three-dimensional objects, on which information is usually accessible about only two dimensions at best. Evidence often exists about the growth of a delta in length (although even this may not be closely dated), less often in area, very seldom about growth in volume. Many factors are involved. In the short term, beach regimes vary from decade to decade and from one stretch of coast to another,[77] and these fluctuations are difficult to distinguish from systematic changes. Even neighbouring deltas can have very different histories.

Scholars, unwilling to admit ignorance, love simple explanations on the lines of 'the delta advanced rapidly after 1600 because people had been cutting down trees in the catchment' – without troubling to give the evidence that tree-felling was indeed more active at this time than before and since, or, if it was, that it did accelerate erosion. In reality, the increased

Delta	Country	CATCHMENT				DELTA			
		Human activity	Climate (+ wet, – dry)	Erodibility	Steepness	Underwater topography (+ favourable, – unfavourable to extension of delta)	Exposure	Tide	Uplift (– subsidence)
Po + Adige	It	+	+++	++ (Alps)	++	+++		+	––
Akhelóös	Gr	+	+++	+++	++	++			––
Sperkhéios	Gr	+	+	+++		+++			––––
Arákhthos + Loúros	Gr	+	++	+++	++	+++			++?
Fluviá + Muga	Sp	+	+++		+++				
Ombrone	It	++	++	+++	+				–––
Axios + Aliákmon	Gr	+	+	++		+++			–––
Guadalquivir	Sp	+++ (Cordova)	–			+++	+++	+++	
Büyük Menderes	Tu	++	++	++		++			––
Basilicata	It	+	+	+++	++		++		++
Küçük Menderes	Tu	++++ (Ephesus)	++	+++		++			––
Arno	It	++++ (Florence, Pisa)	++	++	++				
Rhone	Fr	+	+++	++ (Alps)					––
Tiber	It	++++ (Rome)	++	++	+				
Guadalfeo	Sp	+	++	+++	+++	–	++		
Río Verde	Sp	++	++	+	++		+		
Sado	Po	–	++				+++		
Vélez Málaga	Sp	++ (unterraced vines)	++	++	+++	––	++		
Tavronítis	Cr	+	++	+++	++		++		++
Adra	Sp	++ (mining)	+	+++	+++	––	++		
Gediz	Tu	+++	+	+++		++			––
Andarax	Sp	++ (mining)	––	+++	++	–	++		
Tagus	Po	+	+			++		++	
Ebro	Sp	+	–	++		–	+		
Var	Fr	+	+++	+++	+++	––––	++		
Guadiana	Sp, Po	+	–				+++	+++	

Table 18.iii. Some Holocene deltas arranged in descending order of size relative to that of the catchment, in relation to certain factors of the catchment and of the delta

supply of sediment could have been due to a great slump having arrived on the valley floor (p. 302); or to great floods releasing material eroded thousands of years before and stored on river terraces. Or the supply of sediment may have remained constant but its effects have become more noticeable, for example by previous sedimentation having filled up a basin nearly to the water surface.

It is rarely known from whereabouts in a catchment the deposits in a delta came. For most Mediterranean deltas it is reasonable to suppose that much of the material came from badlands, mass movements or glacial formations; without such sources there would be much less development of deltas.

Little is known of how deltas behaved before human civilization. For the early Holocene the evidence is hidden under the sea. In the Ebro delta, where deposits from before the sea-level rise have been measured, siltation was about one-third of the average rate for the last two thousand years. Sea-level rise stopped only a thousand years or so before the onset of civilization; the deposits for this critical interval are difficult to date and mostly inaccessible beneath later deposits.

Deltas and coast deposits, so far as we can tell, fail to conform to the pattern in time of erosion history, with various phases of Younger Fill activity in the Bronze Age and the historic period. This is to be expected for those whose sediment comes from badlands, which display some activity for most of the time, and landslips and slumps, which tend to happen at random. Basilicata, with its abundant archaeological sites buried by delta deposits, would seem to be an excellent place to test the links between delta growth, human population and climate; we know of no serious attempt to do this.

In the west Mediterranean the indications, for what they are worth, are that visible growth of deltas was most active after the Roman period, especially during the second and third phases of the Little Ice Age. This is also true of Italy, but with some evidence of Roman and earlier growth. In the eastern deltas, however, the advance was mainly in prehistory or down to the early Middle Ages; little extension can be attributed to the Little Ice Age.

What causes can be assigned? There was deliberate intervention at Venice (to divert the growth of the delta) and perhaps at Thermópylae (to take advantage of it). The exceptional growth of the Adra and Andarax in the nineteenth century can plausibly be linked to mining; the Vélez and Guadalfeo deltas, adjacent but with less mining, grew earlier. However, the link is not proved: the badlands in these catchments are much more extensive than any mine-dumps, and the deltas are not unduly large.

The sixteenth- and seventeenth-century growth of deltas came during the Little Ice Age and at a time of some increase in intensive land-use. As far as can be judged, deltas grew irrespective of the degree of that increase. Delta growth affected Spain, then in severe recession of human affairs. In Italy it affected the Arno, where land-use had been intensive for centuries already; it affected the Tiber, where the intensive land-use of ancient Rome had, it seems, failed to add to the delta to the same extent. Explanations in terms of changes in land-use are somewhat lame: various different changes have to be invoked to be consistent with different sets of observations.

Delta expansion provides indirect evidence of erosion rates over the last seven thousand years. There is no consistent support for the idea that it has been caused by artificially accelerated erosion, still less by soil erosion. Where rates of sediment supply have recently been high they have always been high. Accelerated deposition can be as readily attributed to more severe weather in the Little Ice Age as to deforestation and soil erosion.

RECENT CHANGES AND THE FUTURE

The one very consistent feature of Table 18.i is that deltas and coasts are now no longer advancing; most of them are retreating. This is not due to a rise in absolute sea-level, which at about 1 mm per year during the twentieth century is barely significant.

The main cause is undoubtedly damming of rivers. Neo-Castorization, as Vörösmarty and colleagues call it – the propensity of contemporary mankind to copy the beaver, *Castor fiber*, and build dams – has become a worldwide geological force. Dams are built partly to 'control' floods (or, rather, to encourage people to move into areas within reach of floods). They thus cut off the transport of sediment by floods. The sediment that does get through is mainly silt or clay, which is carried out to sea. This has happened, for instance, with all the big rivers of Greece.[78] Even the Catalan coast, with its hyper-active feeders, is now receding after dam-building.

Contributory factors are: people digging gravel out of river beds; harbour works trapping sediment migrating along the shore, and starving the coast on the downdrift side; sea-defence works, which prevent (or delay) marine erosion in one place and accelerate it in another; building hardened sea-fronts, which cause the beach to disappear; imprisoning dunes under development; and pumping out water or gas with accompanying subsidence.[79]

The North Adriatic deltas are at risk from sea surges. In the Gulf of Venice these are caused by south-easterly *sirocco* winds combined with high tides – the normal tidal range in the northern Adriatic is as much as a metre. The greatest surge in the twentieth century was in November 1966.[80] Matters are expected to get worse, by more than half a metre, in the twenty-first century: further subsidence, about half of it natural, of at least 20 cm, plus a sea-level rise of about 35 cm (at the present best guess).

Deltas and coastal plains include the best farmland, sites for hotels and airports, and some of the most famous wildlife habitats in Europe. Erosion and floods kept them alive; without sediment supplies, they will be eaten away and revert to the shallow seas which they were not long ago. Their future does not always attract the attention that it should. Discussions of the Ebro delta's problems centre on urbanization and administration, rather than on getting rid of the dams that threaten its very existence.[81]

[78] C. J. Vörösmarty, M. Meybeck, B. Fekete, K. Sharma, 'The potential impact of neo-Castorization on sediment transport by the global network of rivers', Walling & Probst 1997, pp. 261–73; Poulos & others 1996.

[79] C. Caputo and 4 others, 'Present erosion and dynamics of Italian beaches', *ZG* NF Suppl. 81, 1991, pp. 31–39.

[80] M. Bondesan (see note 59), p. 1360 and Fig. 1.

[81] Bayerri i Riga (see note 19).

CHAPTER 19

Over-Use of Ground-Water

And Isaac's servants digged in the valley, and found there a well of springing water.
And the herdmen of Gerar did strive with Isaac's herdmen, saying, The water is
our's; and he called the name of the well Esek . . . And they digged another well,
and strove for that also . . . And he removed from thence, and digged
another well; and for that they strove not.[1]

All sorts of commons problems are readily and frequently managed in sensible,
virtuous, sustainable ways by local people who entirely lack the pretensions to be
trained economists. Conversely, it becomes obvious that it is the very trained
experts who often undo, destroy and wreck sensible arrangements for
managing commons.[2]

The idea of this book grew out of a conference which one of us attended in 1986 on the desert island of Khálki (pp. 321–2) in the SE Aegean. For thousands of years the people had lived frugally using rainwater hoarded in cisterns. In the 1970s somebody discovered ground-water, and the islanders looked forward to a life of luxury, turning on hot and cold taps instead of pulling buckets. The supply, however, was non-renewable. By 1986 the water in the tap – if any – might be fresh or salt, and the half-life of an electric water-heater was nine months. The ancient cisterns were neglected and confused with cesspits. This story is a miniature of what happens on a grander scale all over the Mediterranean; elsewhere crops and money are involved, not merely people.

Traditional Mediterranean agriculture involves crops such as wheat and barley, grown over the winter and harvested in early summer, plus olive and almond trees, vines and figs, which grow during summer drought. Crops needing moisture all the year – citrus and other fruit trees, salad plants, vegetables – are normally grown on small areas of irrigated, intensively cultivated land. This division between *regadio* (irrigated) and *secano* (dry-farming) areas is most marked in Spain (pp. 77–8).

Traditional irrigation depends on springs, rivers and occasionally ground-water – usually on a small enough scale for the participants to get together and share out the water by turns. It could, however, involve quite large canals, tunnels, and even dams, as in Roman Spain; later the kings of Spain organized the building of bigger dams.

Dependence on ground-water irrigation

Since the 1950s irrigation has grown through the successive coming of concrete aqueducts, deep boreholes, electric pumps, plastic pipes, storage tanks on high ground and long-distance canals. Irrigated crops are grown all over valley floors and coastal plains, except where limited by frost. Though of relatively small extent, these now supply much of the salad and fruit of northern Europe as well as that of local consumers, including tourists. The produce is of high value per area cultivated: in 1988 some 13,000 ha in the Campo de Dalías (SE Spain) produced crops that sold for 325 million ECU, about 25,000 ECU per ha.[3] Moreover, some dry-land crops are now irrigated: in Crete even olives are now treated as an irrigation-demanding crop.

By far the most costly farmland is under plastic greenhouses. Fifty years ago, before local ground-water had been exploited and before there were links with distant urban markets, much of the greenhouse country was ex-arable land worth almost nothing. Changes came swiftly after 1955. While woodland and maquis have encroached on terraced hillsides, and people have left the inland countryside, plastic has spread over once-barren coastal terraces and alluvial fans. These newly valuable lands, so recently 'reclaimed', are the areas of Mediterranean Europe most threatened by what some would call desertification. Of 30,000 sq. km of irrigated land in Spain, about 30 per cent are supplied with ground-water, and the aquifers are beginning to run out.

Other uses

Coastal plains also have industry, tourism and expanding cities, which compete with agriculture for water. To put this in perspective, we were told in Spain that people use about 0.3 cu. m per day and hotels 0.1 cu. m. per occupant per star per day. Golf – which should be the royal and ancient sport of waterless sand-dunes – has come to 'need' 100 cu. m per day per hole.

The water drunk by agriculture dwarfs nearly all other uses. A basin of 100 sq. km might be irrigated with 20 M cu. m of water per year: a modest amount, equivalent to 200 mm of additional rainfall – many irrigators apply much more.[4] The produce might be worth something like 100 million ECU to its three thousand growers. The same quantity of water should provide for the drinking and washing of a city of about 200,000 people, or for a ten-star hotel with 100,000 guests, or for a thousand-hole golf-course.

[1] Genesis 26:18–22.
[2] Matt Ridley, *The Origins of Virtue*, Viking, London, 1996.
[3] M. A. Garcia-Dory, 'Agricultura intensiva y explotación de los recursos naturales en el Campo de Dalias', *Quercus* 59, 1991, pp. 43–5. The value per hectare is probably underestimated, since not every hectare is cultivated each year.
[4] 1 M cu. m = 1 million cubic metres = 0.001 cu. km = 1 hm3 (cubic hectometre, a Spanish measure) = 1 m on 1 sq. km.

Drying-out of aquifers

The input to many aquifers from rainfall, flooding rivers and mountain snow is not enough to replace the water pumped for irrigation. In the whole Guadiana basin in Spain, it is claimed, pumping takes 92 per cent of the available water, and in the Segura 85 per cent;[5] in parts of these basins, therefore, more water is taken out than comes in, the deficit being supplied from storage. (The deficit is not always so bad as the figures make out, because irrigators often apply water so profusely that some of it percolates back into the aquifer, taking pollutants with it.)

The level of ground-water is measured either by the *water-table* (the level below which soil or rock pores are full of water) or the *piezometric level* (the height to which water rises in a well). If the water beneath a water-table is under artesian pressure, the piezometric level will be higher than the water-table. Beneath any one place there are often several aquifers, each with its own water-table and piezometric level, separated by impermeable strata.

Water tables and piezometric levels commonly fall, especially in dry years; declines of one metre per year are not unusual, and 10 metres per year not unknown. Falling ground-water levels increase costs of pumping, but as this is usually insignificant in relation to total production costs it is not a deterrent to over-use.

Pollution

If the water-table declines too far, a much greater danger arises. Once a coastal water-table falls below sea-level, sea-water percolates into the ground-water and then the soil. Once salt water has got in, it is very difficult to get it out (with one remarkable exception, p. 353); ground-water becomes useless for most purposes.

Concentrations of dissolved salts in general can be measured in terms of the electrical conductivity, expressed in scholastic units called microsieverts per centimetre (mS cm⁻¹). These values for conductivity are very roughly similar to values for dissolved salts in parts per million or milligrams per litre. Sea-water has a conductivity of about 30,000 mS cm⁻¹. Water with conductivity exceeding 600 mS cm⁻¹ (2 per cent sea-water) is generally too brackish for human consumption; at 1,200 mS cm⁻¹ (4 per cent sea-water) cattle refuse it; above this value it begins to damage crops, some being more sensitive than others.

Ground-water can be polluted by industrial waste, sewage from people and pigs, or sulphuric and nitric acids washed by rain out of a polluted atmosphere. In southern Europe many industrial plants are sited on top of, or to windward of, the aquifers of irrigated coastal areas. Pollutants are carried down through the soil and by streams and via karstic caverns to reach ground-water.

Pollution by heavy metals (lead, mercury, etc.) may come from natural ores in rocks, or from current, disused or ancient mineral workings. Old workings are often especially dangerous because much of the metal was left behind in the waste. Examples are the great copper mines in SW Spain and adjacent Portugal.[6] These often lie upstream of modern dams. Ancient and modern disused mines in SW Sardinia have tailing ponds, precariously held up by frail earthen dams, perched above the reservoir that supplies Cagliari. Such pollution problems are not special to the Mediterranean, although they may be worsened when water gathered from extensive semi-arid catchments is concentrated and applied to small irrigated areas.

Irrigated land is farmed intensively, with much more fertilizer and pesticide applied per hectare than in dry-land farming. The Mediterranean has nearly caught up with the rest of Europe. If the figures are to be believed, application of nitrogenous fertilizers to cultivated land in Spain, Portugal and Greece doubled in twenty years, from 20 kg per ha in 1970 to 40 kg per ha in 1990. Much of this percolates into the ground-water. Contamination is now very high along the east coast of Spain, and is increasing in the Guadiana basin, the plains of La Mancha and in Majorca.[7]

These problems are aggravated by drought, which (depending on the time of year) increases consumption or reduces recharge. They are already aggravated by the erosion of delta fronts (Chapter 18), and would be made much worse by global warming or rising sea-level.

Surface water supplies are often administered by government agencies, which in theory can conserve supplies. Ground-water, however, is treated as if it were private property, and is often abstracted by syndicates or individual farmers who hire contractors to sink wells. If water is found, they hope yields will be sustained long enough to pay for the investment. Authorities have little influence over the depth of the wells and the amount extracted. Problems of sharing out water, which might be solved if only the environment and the users were concerned, get worse when politicians, creditors and shareholders have interests as well.

There can also be competition, both for the water itself and for investment in using it, between users of surface water and ground-water. Ground-water users claim more water on the grounds that they achieve more agricultural production per cubic metre.

THE TROUBLES OF SPAIN

We shall deal especially with Spain, where irrigation is on the largest scale and encounters the most complex practical and political problems. We summarize some of the many books and articles that have been written, often in connexion with MEDALUS programmes (pp. 16–17).

It is said (we pass over the question of whether the figures have any meaning) that roughly 37 per cent of all Spanish agricultural production comes from ground-water irrigation, which uses about 4,000–5,000 M cu. m of water per year. Surface-water irrigation yields some 25 per cent of total

5 B. López-Camacho, A. Sánchez-González, A. Batlle, 'Overexploitation problems in Spain', *Selected papers on Aquifer Overexploitation*, ed. I. Simmers and others, *Hydrogeology* 3, 1992, pp. 363–71.

6 This was written before the collapse of the modern tailing pond at Aznalcóllar near Seville in April 1998, which let mercury, arsenic, lead and zinc into the rivers Guadiamar (on which the Coto Doñana nature reserve depends) and Guadalquivir and over surrounding farmland.

7 Lumsden 1992, pp. 276–80.

Fig. 19.1. Ground-water troubles in Spain and the Balearics. Most of the permeable Mediterranean coastal areas are affected by sea-water intrusion. (After Lopez-Camacho & others).

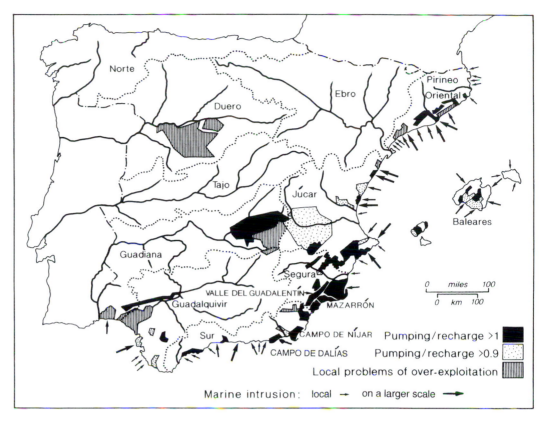

production, using about 20,000 M cu. m per year. Dry-land farming yields the remaining 37 per cent of production.

About one-third of the irrigated land in Spain is within reach of sea-water, especially in the very arid coastal areas of the south and south-east (Fig. 19.1).

Over-pumping: the Guadalentín basin and the south coast

MEDALUS has given much attention to the basin of the Guadalentín, lower tributary of the Río Segura. The Alto Guadalentín aquifer occupies 236 sq. km of a 3,394 sq. km catchment, with an average annual rainfall (1968–87) of about 300 mm. Until recently its water could be drawn from a shallow depth, even from artesian wells. Expanding irrigation increased the rate of extraction from 24 M cu. m in 1973 to 69 M cu. m in 1987; the input from rainfall is estimated at only 60 M cu. m. The water-table in this aquifer dropped by 200 metres between 1970 and 1995.[8]

Even in this inland location, retreat of the water caused much more trouble than merely the extra costs of pumping, and the drying out of farms that could not afford extra pumps. When water is drawn from near the base of the aquifer, where the geology changes, sulphates and chlorides get in. In 1991 the water pumped out contained dissolved salts roughly equivalent to at least 4 per cent sea-water, on occasion reaching 16 per cent sea-water.[9] Its use for irrigation risks poisoning many crops, and also permanently alkalizing the soils, when calcium ions are replaced by sodium. Moreover, when

water is sucked up from more than 150 metres down, the release of hydrostatic pressure brings carbon dioxide out of solution. Wells and pumps corrode, carbonate deposits block pipes, and iron carbonate and sodium bicarbonate are deposited in soils. All of these are harmful, costly and difficult to remedy.[10]

In coastal basins sea-water adds to these threats. In the Segura basin (other than the Guadalentín), where 140 M cu. m are being extracted annually to irrigate more than 25,000 ha, the deficit amounts to about 50 M cu. m per year, and three-quarters of the aquifer area is feeling the effects of sea-water intrusion.[11]

The small aquifers in the Holocene deltas on the south coast (p. 331) are all individual. The Río Vélez generally manages at least a trickle throughout the year. Its delta is not much affected by sea-water: a barrier of fine sediments at the delta front appears to keep out the sea during those months each year when the water-table is low. The Rio Verde delta, however, is recharged by torrents, and is pumped down mainly in winter; it gets salinized each year and recovers. The delta sediments are coarse, and a large volume of salt water enters the aquifer during the dry season. However, clays are lacking to bind the sodium ions, and in autumn the incoming fresh water washes out the salt. At Castell de Ferro the water is pumped mainly in spring and autumn for drip-irrigation in greenhouses; it is perpetually brackish, probably through sea-water getting in through karst caverns in a coastal outcrop of limestone.[12]

---

[8] J. C. Céron and A. Pulido-Bosch, 'Groundwater problems resulting from CO₂ pollution and overexploitation in Alto Guadalentín aquifer (Murcia, Spain)', *EG* 28, 1996, pp. 223–28.

[9] Measured conductivity values were between 1120 and 4860 mS cm⁻¹.

[10] A. Pulido-Bosch and 6 others, 'Field Trips and Excursion Guide', *Proceedings of XXIII International Congress of Hydrogeologists*, Puerto de la Cruz, Canary Islands, 2, 1991, p. 136.

[11] Lumsden 1992, pp. 268–70, 276–80.

[12] M. L. Calvache and A. Pulido-Bosch, 'Effects of geology and human activity on the dynamics of salt-water intrusion in three coastal aquifers in southern Spain', *EG* 30, 1997, pp. 215–22.

Collateral damage to wetlands: La Mancha

Irrigation in the arid plains of La Mancha (p. 313) has been increasing for a century. The earlier water-demanding crops such as maize and lucerne (alfalfa) have decreased in recent years. They have been more than made up for by increasing irrigation of vines, of which La Mancha had 7,500 sq. km in 1985 (supposedly the greatest concentration of vines in the world) and still 6,000 sq. km in 1995. Forgetting both the European Union's over-production and the principle that good wine should not be watered, vine-growers by 1987 were irrigating 1,300 sq. km of vineyards. (Much of the wine goes down the throats of industrial-alcohol stills.)

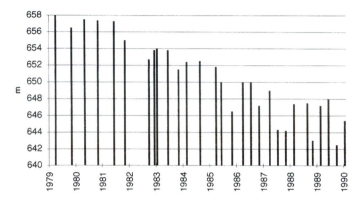

Fig. 19.2. Falling water-table at Cerro Verdiscos (714 m above sea-level). In the wetter mid-1990s it largely recovered.

The volume of irrigation water reached 500 M cu. m per year by the 1990s, compared with only about half that amount getting into the aquifer from the annual rainfall. Piezometric levels were falling at about a metre a year (Fig. 19.2), which diminished the flow of the rivers Guadiana and Júcar during the drought years of the 1980s and early 1990s.

La Mancha has (or had) many oases, lakes and wetlands, fresh or salt, seasonal or permanent (Fig. 17.8). The Lagunas de Ruidera and Tablas de Daimiel survived into the 1990s as two of the most famous wetlands in Spain, both being National Parks.

The Lagunas de Ruidera (Fig. 19.3) are on the valley floor of a headwater of the Guadiana. They form a staircase of fifteen beautiful bluish-green lakes, each lake held back by a barrier of travertine built up by filamentous algae growing in the carbonate-rich river water. (This is said to be comparable in Europe only with the Plitvice lakes in Croatia.) Elms and poplars grow at the water's edge. Steep valley sides are clad in live-oaks, junipers and the broom *Genista scorpius*, interrupted by holiday homes and plantations of exotic trees. The reedy pools are excellent habitats for aquatic birds.

The Tablas de Daimiel, about 70 km westward, occupy nearly 2,000 ha at the confluence of the Cigüela and Guadiana. The Cigüela is a salty seasonal stream; the Guadiana is fresh and normally perennial, being fed by great springs from western La Mancha's 'Aquifer 23', so that the wetland is unusually diverse. To the north lies a live-oak savanna, to the south an agricultural landscape, now irrigated. Open water, reedbeds and tamarisk-fringed islands are internationally renowned bird habitats.

Already in the 1960s the Tablas were threatened by further drainage for agriculture, a campaign against which led to their being scheduled a National Park. Nevertheless, pumping from Aquifer 23 for irrigation lowered the local water-table by 10–20 m, and a dam cut off some of the Guadiana supply. Only in 1987 were steps taken to prevent final desiccation by building a dam 1.9 km long on the down-valley side of the wetland, and diverting some water from the Tagus-Segura Aqueduct into the Cigüela.[13]

By 1994 the wetlands were in a sorry state. The level of the Lagunas had fallen steadily since 1988, and many of them were dry. In the Tablas reeds had spread into the little remaining open water, and pollution of the Cigüela was increasing. However, by spring 1997, when one of us first visited the area, the scene had changed. The Lagunas were full following two wet winters, and the Tablas were watery once more. It remains to be seen whether this is merely a temporary remission of desiccation. Much depends on the future of the climate – and of the National Water Plan.

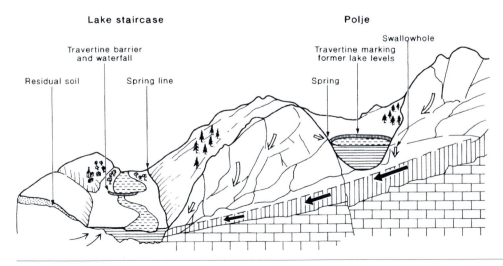

Lake staircase

Residual soil   Travertine barrier and waterfall   Spring line

Polje

Swallowhole

Travertine marking former lake levels   Spring

Fig. 19.3. Composite diagram of Mediterranean karst terrain. On the left, pools like the Lagunas de Ruidera are held back by travertine barriers. The *polje* to the right is comparable to the flooding basins in the Peloponnese and the mountain-plains of Crete (Chapters 16 and 17).

13 P. Molina Vicente, 'Restauración de ecosistemas singulares y frágiles. Zonas humedas', *Ecología* fuera de serie 1, 1990, pp. 331–40.

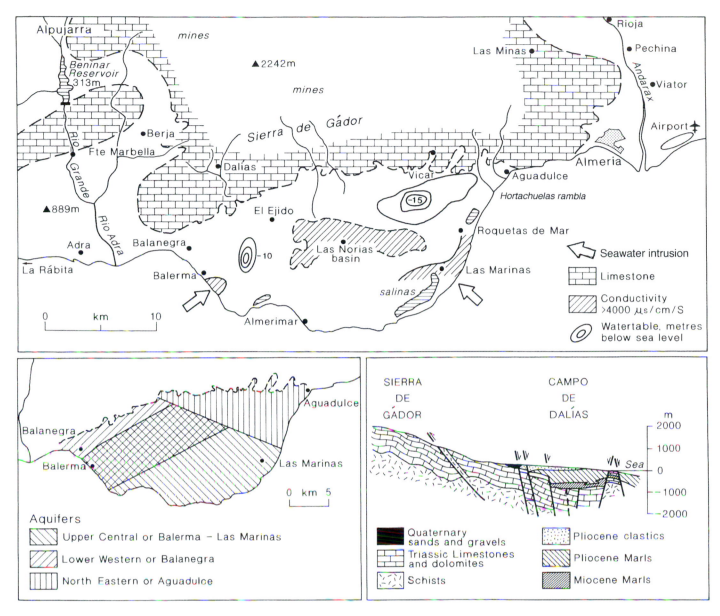

Fig. 19.4. Campo de Dalías and the lower Andarax.

The complexities of the Campo de Dalías

The Campo de Dalías (Fig. 19.4) is excellent for growing tropical crops in winter. It is a fault-bounded coastal plain, such as we have already encountered many times. The towering Sierra de Gádor at the back shelters it from north winds. January mean temperature is 12°C, frost is almost unknown, and the sun shines for 2,800 hours out of 4,380. But it is very dry: mean annual rainfall diminishes from 500 mm at the base of the Sierra to less than 225 mm on the coast. The Campo competes for water with the growing city of Almería and with a muddle of tourism (Fig. 19.5). It relies entirely on ground-water. This grasped our own attention when in 1994 we discovered the extraordinary qualities of coffee made with the brackish and gypsiferous water of a place called Aguadulce,

'Fresh Water'. All this is a new development, which did not exist at all half a century ago.

The Campo comprises most of the medieval *ta'a* or district of Dalías. This was modestly prosperous, with arable cultivation and orchards, but chiefly known for its 'most beautiful plains for grazing animals in winter'. It was wrecked by the Christian conquest of 1569; four of the six settlements were deserted, and the remaining two merged as the inland village of Dalías. In the mid-eighteenth century the plain had a sparse cover of juniper, lentisk, euphorbia and rosemary.[14] In 1924 Gerald Brenan walked across it:

> For fifteen miles the road ran in a perfectly straight line across a stony desert without, so far as I could make out, passing a single house or tree on the way. One could see it

[14] Carvajal 1600, p. 204; D. Hurtado de Mendoza, *Guerra de Granada*, ed. B. Blanco-González, Castalia, Madrid, 1970, Anejo III; J.-P. Thauvin, *Etude hydrologique, modélisation et gestion des aquifères du Campo de Dalias*, Ph.D. thesis, University of Nice, 1986.

appearing and disappearing into slight hollows in the whitish ground till it merged into the skyline. This desert is known as the Campo de Dalías. Actually it is a delta of stone and rubble pushed eight miles out into the sea by the erosion of the Sierra de Gádor, and running down to it at a slight incline.[15]

In the late 1930s it was one of the poorest places in Spain, with only a few hectares of barley, figs and vegetables.

Here was one of the latest and (so far) most successful achievements of the 'internal colonization' movement for settling inexperienced peasants on poor-quality land. In the 1950s Brenan reported that

> its aspect has changed. The underground springs that in past times fed the Roman station of Murgi have been tapped, and the once-arid plain is dotted with white farm-houses and green with corn and fruit trees.

By 1950, 500 ha of vines, maize, sugar beet, potatoes, cotton and peas and 770 ha of barley were being irrigated. About 4,500 people lived on the plain – and one tourist, namely Brenan.

Irrigation had begun when the Instituto Nacional de Colonización bored some wells in the 1950s. Colonists were each 'given' 3.5 ha of land with some seed, a little tractor and a dwelling. Yields were poor because of salty and infertile soils; the enterprise would have failed, as many such schemes did, had somebody not remembered in 1957 a traditional technique of making artificial soils.

Cultivation beds, *enarenados*, had been made for a century or more around Adra and at the western end of the Campo de Dalías.[16] Organic manure, spread 2–10 cm thick over a layer of sandy clay 20–30 cm thick, is in turn covered with 10–12 cm of coarse beach sand. The sandy clay retains water in the organic layer where it is needed by the roots of the crop. The coarse sand stops capillary rise to the surface where the water would evaporate. The dry sand warms in the sun, the heat being transmitted to the manure and causing rapid root growth. The clayey layer absorbs excess salts. An artificial soil of this kind is usually replaced after 3–5 years, and is thus suitable only for high-value crops.

Plastic greenhouses began in 1961. The area increased to 1,000 ha by 1971, 6,000 by 1979, 12,000 by 1984, and 17,000 ha by 1995 – the greatest concentration of crops under plastic in Europe, and the main source of wealth in Almería province. Three crops of fruit, vegetables or flowers could be grown in two years.

Early success attracted immigrants from Alpujarra (contributing to its decline, p. 100), Málaga and more distant parts of Spain. Family smallholdings were well adapted to greenhouse cultivation. The proprietors had little education or training; they had learnt their jobs on a family farm or by working for someone else. By 1986 one-third of the farms had a mechanical cultivator; more than half had water-carriers and storage tanks. Over half owned a well or a share in a well; some got their water from the Instituto Andaluz de Reforma Agraria.

Fig. 19.5. Muddled coexistence of greenhouses, tower blocks and hotels at the east end of the Campo de Dalías. *November 1994*

In 1991, about 85 per cent of the holdings were of less than 3 ha, in one or two parcels usually close to one another. Each hectare called for 500–800 days of work annually, rising to a peak in April and May, when three or four people were needed. Labour is largely provided by the family, although even for 3 ha help is called in from friends, neighbours or hired workers.[17]

At first, produce was consumed in Spain; as exports increased, packaging, presentation, quality control and transport came to encroach on the gross profit. Most of the marketing is now done through a dozen big companies. Commercial handling is said to account for 45 per cent of the total value of sales. Much of this expenditure is local: about half the packaging is made in Almería province. Many other Almerian industries and services depend on irrigated agriculture. Over a hundred enterprises supply seeds, manures, implements, transport, irrigation equipment, plastic sheeting, sand and clay, or refrigerated storage. Factories recycle some of the 13,000 tons of scrap plastic generated each year. Other enterprises dispose, by goatodegradation or otherwise, of some of the half a million tons of plant remains arising annually.[18]

By 1986 some 75,000 people lived on the Campo, almost all of them directly or indirectly from agriculture. At the same time, despite the lack of space and drinkable water, a grey, cramped, gritty form of tourist development began to invade the east corner of the plain. Later *urbanización* has brought the outer suburbs of Almería within the Campo.

Like agricultural colonists all over the world, these farmers got into debt. To finance *enarenados* and greenhouses, they borrowed money from the colonization agency or from private banks. By the mid-1980s the 120 sq. km of the Campo under plastic was said to represent an investment of something like 500 million ECU, of which the land was worth about 100 million, and debts amounted to 200 million. The total profit in the mid-1980s from agricultural produce amounted to about 100 million ECU annually, plus 70 million paid to local industry and 50 million to hired labour. By 1988, foreign companies were beginning to appear, although as yet they have not taken over much land.

[15] Brenan 1957, p. 192.
[16] Thauvin 1986 (see note 14).

[17] M.-C. Hernadez Porcel, 'Agriculture and tourism in the Campo de Dalías: possible conflicts', unpublished paper, 1991.
[18] Pulido Bosch & others 1996.

	Upper Central	Upper North-East	Lower Western	Lower North-East
Extent	225 sq. km	150 sq. km?	195 sq. km	> 150 sq. km?
Depth	Shallow	Shallow	Shallow to > 500 m	Deep
Geology	Unconsolidated Pleistocene sediments	Plio-Pleistocene sediments	Limestones and dolomites underlain by phyllites; supposed to be isolated from sea by impermeable barrier	Dolomites, calcareous sandstones
Wells	Hundreds of shallow private wells	Mainly large bores	> 50 large commercial bores	100 bores near Aguadulce
Annual input, M cu. m	24 (rain) Sewage of El Ejido town used to be a major recharge	5–14 (rainfall) 5–10 from Upper Central 4 returning from excess irrigation	Unknown amount from Upper Central 15? from sea	Unknown
Annual output, M cu. m	15 (extraction) 1–2 (leakage to sea) Unknown amounts to Upper North-East and Lower Western	26 (extraction; partly to city of Almería)	34 in 1984	50 (extraction) At least 3 to sea
Quality	Increasingly brackish away from mountains; up to 25% sea-water	Worsening; 30% sea-water in places	Originally very good; mostly still drinkable	Sea-water intrusion beginning
Changes	Many wells now abandoned	Piezometric level falling at c. 1 m per year	Piezometric level has fallen at c. 1 m per year; salinization still localized	Piezometric level falling at c. 1 m per year

Table 19.i. Principal aquifers of the Campo de Dalías.

Brenan was wrong about the Campo being a delta. It results from a huge tectonic downthrust of the Triassic dolomites, conglomerates and phyllites that form the Sierra de Gádor. Over these have accumulated successive deposits of conglomerates, calcareous sandstones, blue clays and marls of Miocene to Pliocene dates, Pleistocene marine sediments, and finally screes and fans from the fault-scarp of the Sierra. The bedrock has been shattered by further faulting, and planed off in times of higher relative sea-level. The subdued surface relief thus hides a most complex buried structure.

Ground-water presumably comes from the Sierra de Gádor, percolating either directly through the rocks or from mountain torrents: the route is not always known. It may be contaminated by minerals from the rocks, by brine from higher sea-levels now trapped between faults, or by sewage from the 75,000 people.

An average hectare of the plastic-covered Campo is said to be irrigated with 7,800 cu. m of water per year (in addition to 1,000–5,000 cu. m. of rainfall). This is more than necessary, as nitrate getting into the ground-water proves. With drip irrigation, and a sandy clay base to reduce losses by percolation, about 5,000 cu. m should suffice. The total volume of water required annually to irrigate 17,000 ha is thus about 85 M cu. m. There are four major aquifers, whose characteristics are summarized in Table 19.i.

After twenty years it was noticed that levels in wells were falling at around a metre per year. In 1977 the Geological and Mining Institute warned of the risk of seawater intrusion, and some attempt was made to regulate the sinking of new boreholes. By 1982 the water deficit in the preceding year alone had reached 30 M cu. m, piezometric levels had fallen locally below sea-level, and areas were being lost to salinization. Yet the Council of Ministers, evidently hating the Campo de Dalías, subsidized to the extent of 15 million ECU a Dutch initiative to irrigate a further 20 sq. km in an area that had already proved difficult.[19] The maximum extraction of which we have record was 122 M cu. m in 1986.

A desperate attempt has been made to get more water from a dam at Benínar on the River Adra, which flows out of the middle Alpujarra (p. 332). The river was supposed to discharge 35 M cu. m annually, of which 13 M were to be diverted to Almería city and 12 M to the Campo de Dalías, leaving 10 M for users in the Adra delta. Prognostications were made of the effect on the salt content of springs and aquifers below the dam.[20] However, the reservoir sits on leaky limestone, and when we were there in 1994 it had never filled.

Short of channelling huge quantities of water from distant parts of Spain, or introducing desalinization on a heroic scale, water use will have to diminish. A decline in the irrigated area is probably inevitable, but could be mitigated by growing less thirsty crops, varying the cropping season and not over-irrigating.

Some palliative works have been done. Trickle irrigation has reduced evaporation. To promote infiltration to the Upper Central aquifer and to remove the risk of ordinary floods from the plain, more than 170 check-dams have been constructed in the gullies on the south side of the Sierra de Gádor. They will not last long in this unstable terrain unless the sediment they trap is excavated.[21]

Water quality could be improved by recycling the sewage of the towns on the Campo (at least 5 M cu. m per year) via proper treatment, instead of informally as at present.

There is some scope for keeping up production by methodical organization. Now that the aquifers are better

[19] M. A. Garcia-Dory, 1991 (see note 3).
[20] El Amrani-Paaza, J. Benavente and J. J. Cruz-Sanjulián, 'Modélisation hydrogéochimique de l'aquifère du delta du río Adra (Andalousie, Espagne)', *Hydrogéologie* 3, 1995, pp. 47–58.
[21] Pulido Bosch & others 1996.

understood, water pumped from threatened parts of them could be replaced by supplies piped from less threatened places. Underground water barriers could be created by stopping extraction – indeed by pumping in fresh water – in zones next to the sea and around parts of aquifers already contaminated. A report in 1992 recommended thus:[22]

(a)   Protect the Lower Western aquifer by recharging it along the coast with water [should there be any] from the Benínar reservoir, and control use of water in the coastal zone by requiring an annual licence to pump.
(b)   Compensate for reduced extraction from the Lower Western aquifer by increasing annual extraction (of poorer quality water) from the Upper Central by 5–8 M cu. m.
(c)   Keep the Upper North-East aquifer at its current levels by forbidding extraction from boreholes less than 100 metres deep within 500 metres of the coast near Aguadulce. No new exploitation to be allowed NE of the Hortachuelas rambla.

All this depends on social factors. Water barriers would involve far more careful measurement and control of pumping and land use than in the past. Big landowners, with land in different places, have already begun to do this. A general plan would involve getting the cooperation of a multitude of smallholders. Traditional private rights to groundwater were in theory superseded by legislation in 1986. Local legislation might prevent new installations, but cannot readily control the replacement of existing equipment by more powerful pumps which enable an individual to beggar his neighbours. The social responsibility that has worked so well elsewhere in Spain cannot be created by legislation nor in a hurry.

Falling water levels and sea-water intrusion may not even be the most urgent of the Campo de Dalías's problems. Others concern supplies of sand, clay and organic manure, and dependence on outside sources of seed and pesticides. All these are manageable at a price. A greater problem still is likely to be production costs. Fruit and vegetable growing is a competitive industry. Irrigation enterprises are profitable through family or cheap labour. Families may not always be willing to work hard in unpleasant conditions for low and risky incomes, when costs may rise and prices fall for reasons outside their control.

Much of the labour is already provided by Africans, especially from northern Morocco, Senegal and Gambia, who will work (illegally) for little more than their keep. There are more African immigrants to Almería than any other region, drawn by job opportunities in the plastic greenhouses of the Campos de Dalías and Níjar.[23]

The Campo de Dalías is a dangerously over-developed area, whose future would lie in diversifying out of agriculture. Tourism is a possible alternative, although poor-quality development has already taken much of the coast.

The dry corner of Spain: Almería and the Andarax

The Campo de Dalías forms less than 4 per cent of the area of Almería province, which includes the growing city of Almería, the eastern Alpujarra (pp. 274–5), mountains, badlands and deserts, traditional irrigation, and modern irrigated cultivation. The deserts have long been thought a reproach to civilization and are littered with the remains of attempts to get them to blossom. Latterly there has been tourist development, mainly for Spanish holidaymakers, on the coast.

The mean annual discharge of the rivers in the province, equivalent to only 11 per cent of precipitation, is said to amount to 320 M cu. m., of which about 220 M cu. m are taken. Annual water use by towns and industry is 57 M cu. m, and by agriculture (excluding the Campo de Dalías) 250 M cu. m.[24] If the latter figure is true, some 80–100 cu. m per year must come from ground-water.

The lower Andarax valley (Fig. 19.4) is an area of modern intensive cultivation. During the twentieth century people and agriculture have moved down towards Almería city as new sources of water have been tapped. Prosperity has proved short-lived, as pumping has sucked salts into the aquifers from hidden evaporite deposits. Citrus groves, once covering 145 sq. km, are the most sensitive crops and the first to be abandoned.

Three main aquifers can be distinguished. The uppermost 'detritic aquifer', in Pliocene and Quaternary alluvial beds, is the most used. It is badly contaminated with gypsum and salt from evaporites (up to about 17 per cent sea-water) and with nitrate from fertilizers and stockyards. At the base of the Sierra de Gádor is an aquifer of relatively pure water, but charged with carbon dioxide which clogs or dissolves equipment as in the Guadalentín. A deep aquifer is contaminated with boron, which is poisonous especially to citrus, as well as ordinary salts. Water problems here are more of quality than quantity: only in dry years is there a deficit of as much as 10 M cu. m in the detritic aquifer, and sea-water intrusion is localized to the delta.[25] In future Almería city could provide at least 10 M cu. m of recycled water from sewage.

CRETE[26]

Crete, like many limestone areas, is much richer in water than its mainly arid appearance would suggest. Total precipitation is said to be about 7300 M cu. m per year on the island's 8,300 sq. km,[27] of which between 25 per cent and 60 per cent percolates into the aquifers. This water passes through limestone fissures and caverns, often emerging where one of the many faults intersects a cave. The resulting springs can discharge as much as 70 M cu. m per year each. Crete still has a few permanent rivers, and used to have many (p. 133). Coastal aquifers – where the sea forms a barrier – give rise to

22   Technographic, Poligono Industrial Calonge, *Recursos Naturales y Crecimiento Económico en el 'Campo de Dalías'* [commercial report], 1992.
23   F. Checa, 'Oportunidades socioeconómicas en el proceso migratorio de los immigrantes africanos en Almería', *Agricultura y Sociedad* 77, 1995, pp. 41–82.

24   M. Arenas, J. Gomez de las Heras, A. González, *Desertificacion por escasez del recurso agua en Murcia, Almeria y vertiente Mediterranea de Granada*, Instituto Technologico Geominero de Espania, Le de acuíferos, Almeria, 1989.
25   A. Pulido Bosch, F. Sanchez Martos, J. L. Martínez Vidal, F. Navarrete, 'Groundwater

problems in a semi-arid area (Low Andarax river, Almería, Spain)', *EG* 20, 1992, pp. 195–204.
26   See M. Knithakis, 'The hydro-geology of Crete', *Petromarula* 1, 1990, pp. 65–67; E. D'Assiras, 'Aquifers and irrigation water in the Khaniá area', *Petromarula* 1, 1990, pp. 68–9.
27   An accurate figure is impossible because there are no weather stations above 900 m.

freshwater springs on the beach or in the sea itself, and to the famous *álmyroi*, salt-water springs arising inland. Domestic water has seldom been a difficulty, except in the district of Sphakiá which lives out of cisterns, and the city of Herákleion where there is only wine and *rakí* to drink.

Cretan traditional irrigation is second only to Spanish. Springs and rivers are used to irrigate orchards, terraced polyculture and meadows. In the seventeenth century Venetian villas had magnificent garden waterworks.[28] Some aquifers were shallow enough to be reached by dipping-wells, such as may still be seen on the remote Katharó Plain. In the twentieth century there appeared metal, cloth-sailed wind-pumps, adapted from American windmill technology, which still operate on a reduced scale in the Ierápetra and Lassíthi Plains. Small concrete-lined channels, fragile and inflexible, were built from the 1920s onwards.

New technology in Crete is in the hands of public agencies. Water is pumped out of an aquifer to a tank on a hill, whence it flows under gravity through successively smaller plastic pipes to reach the nozzles that supply individual trees. Each farmer has a meter and pays a small fee per cubic metre used. What was left of the traditional multiple cultivation in the plains has been replaced by oranges and other tropical fruits. Olive cultivation has vastly increased, on top of centuries of previous increases.

This began in the west part of the Mesará, the southern plain, in the 1970s (p. 96). The water-table was 2–5 metres down, within reach of dipping-wells. The economy of the western Mesará soon came to depend on irrigated fruit and vegetables: the NW corner of the plain is now the proverbial 'sea of plastic' (Fig. 5.17). By the late 1970s, following a run of dry years, the water-table had fallen to 5–20 m. It quickly recovered with above-average rainfall in the years around 1980, but went down even faster in the late 1980s and early 1990s. It is not yet clear whether this was due to lack of rain or to over-extraction.[29]

Since the 1980s, the Organization for the Development of West Crete (OADyK), a semi-autonomous body (what the English would call a quango) financed by the European Community, has been concerned especially with supplying water to farmers and to towns.

Irrigation in Crete has brought unheard-of prosperity to many lowland townships. But it is beginning to get out of hand, and we sense a spirit of not knowing when to stop. Development creeps up to unreasonable heights in the foothills and gets to waterless and uninhabited coastal plains. Bulldozers carve false terraces out of unstable hillsides, destroying maquis and heath (and the archaeology) for yet more olives. We hear of this being done with money allocated by the European Union to support grazing. Non-limestone gorges (one of the rarest and least well-known habitats in Crete) are threatened by dams; we hear of a threat even to the Omalós mountain-plain, one of the great botanical and cultural treasures of the island. Even Crete shows signs of running out of water. Rivers dry up, and those remaining are polluted by the processing residues of the ever increasing quantity of olives. Although sea-water intrusion is not yet serious, sewage and agricultural chemicals begin to get into the aquifers.

As in Spain, the new prosperity is based on only a few crops and is unstable. Much of it is already unviable in economic terms. In most years a large part of the orange crop is unsold, cast over a cliff or fed to sheep: orange-fed mutton ought to be a Cretan speciality. The new olives can succeed only by putting someone else's olives out of business. Prosperity in the lowlands has, if anything, hastened the decline of mountain villages and hamlets. However, once over-development has got thus far, retreat causes hardship or worse. Bonuses harden into necessities; subsidies become treated as permanent income. New olives are of special varieties that depend on being irrigated for ever. And Crete is a fiercely independent island: more than once the central government offices have been ignited for failure to maintain European Union subsidies or to banish plant disease.

CONCLUSIONS

Irrigation was once a blessing. The arrangements which groups of cultivators made to share out surface water are a splendid counter-example to the doctrine of the Tragedy of the Commons (see quotation, p. 351). In the anciently irrigated areas of Spain, they are elaborate, flexible and self-policing; they survived the vicissitudes and disasters of centuries, and were exported to Spanish colonies (p. 78). This happens when resources are exploited gradually by people whose livelihoods depend on them and who fully understand their limitations. It begins to break down with new irrigation, when boreholes are drilled into aquifers whose behaviour is not properly understood. When, additionally, there is intervention by creditors with duties to their shareholders, governments answerable to their electors, and savants with theories to put into practice, the complexities and conflicts of interests become unresolvable; a Tragedy of the Commons turns irrigation into a curse.

Irrigation used to be integrated with dry-land farming. With the coming of a long-distance economy, as V. N. Rao has pointed out, new irrigation tends to encourage neglect of nearby dry-land farming.[30] Examples can be seen all over the Mediterranean.

With new irrigation, as in SE Spain, Parkinson's Second Law – Expenditure rises to meet Income – applies to water.[31] However much water is provided, use will rise to meet the supply, and there will always be a shortage.

The Spanish and Cretan scenarios are at the ends of a spectrum of over-use, examples of which occur in many lowlands of southern Europe. In Roussillon, even the drenching Pyrenean rainfall barely suffices for 300,000 residents (plus tourists) in an area of 850 sq. km, besides irrigation; there are risks of contamination both from sea-water and from pollution in an overlying shallow aquifer.[32] In the

[28] M. Boschini, *Isola et Regno di Candia*, 1625, BNP: Ital. 383 f.7–10.

[29] Information from B. Devereux.

[30] V. N. Rao, 'Linking irrigation with develop-ment', *Economic & Political Weekly* 13, 1978, p. 24.

[31] C. N. Parkinson, *The Law and the Profits*, John Murray, London, 1960, chapter 2.

[32] M. Chabart, J. J. Collin, J. P. Marchal, 'Modelling short-term water resource trends in the context of a possible "desertification" of southern Europe', Brandt & Thornes 1996, pp. 389–430.

Argolid plain (NE Peloponnese, Greece), water use has increased beyond all reason since 1963. About 120 M cu. m are said to be used annually on an irrigated area of 180 sq. km, equivalent to nearly 700 mm of extra rainfall; most of the cultivation is citrus orchards, on to which water is sprayed to prevent frost as well as drought. Here the water does not exist to irrigate on such a prodigal scale: boreholes have descended from 200 to 400 metres over the last twenty years, chasing falling piezometric levels. By 1963 some of the farmland reported a salt content as high as 0.4 per cent in the soil; even distant springs, from which water was piped to wash down the salt, had a salt content of 2 per cent sea-water, only just fresh enough for irrigation.[33]

Surface water is relatively easy to manage. A king or state or quango builds a reservoir, usually with big subsidies, and sells the water to cultivators who are captive customers. Ground-water, on the other hand, is tapped by private operators, who look to God or their neighbours (via sideways percolation) or the state to replenish their supplies. They may recognise that the water-table is falling, but it happens slowly, and there may seem to be no sense in restraint while the going is good, so bores are deepened and extraction continues.

Some water 'shortages' in south Spain have been met by canals bringing in water from the great rivers of central Spain, notably the Tagus. This may cause more political trouble than it is worth, as people on the way demand a share (and then a bigger share) of the water passing through their territory. The people of La Mancha, for example, are not easily persuaded that a cubic metre of water would be better spent in distant Almería or Murcia than in their own province. An international conflict of interest arises when the Portuguese have to make do with what little of the Tagus is left to cross the border.

The West Cretan model is perhaps the least unsatisfactory. Most ground-water is controlled by a development authority which can, in principle, force the users to share out a finite resource. Even here, irrigation on the plains, by attracting cultivators, has probably hastened depopulation in the mountains. And there is collateral damage. Irrigation has encouraged the destruction of soils, natural vegetation and archaeology, and the uglification of the landscape by bulldozing hills and siting tanks in prominent positions. Rivers and springs dry up or are polluted. The relentless search for yet more water leads to proposals for dams which are not only destructive but dangerous in an earthquake-prone land.

Remote, previously innocuous plains are covered in plastic and rotten plastic – at least one-quarter of the greenhouses are derelict at any one time – which blows about until it blows into the sea. Greenhouses are a very uncomfortable working environment, unpopular because of the chemicals used. We sense an urge to pursue development for its own sake; some irrigation specialists think it a reproach to their profession if any rivers reach the sea at all. All this is for a new way of life with an uncertain future: people have trusted their livelihoods to an industry that may be short-lived for reasons over which they have no control.

The Campo de Dalías situation can be regarded in different ways. At one level, it is a problem of industrialization, of boom and bust such as Almería has seen several times before. If agriculture becomes a modern industry, it becomes mortal like other modern industries, especially if it uses a finite resource. The Campo is likely to enjoy fifty years of prosperity which, by the standards of mining or car-manufacturing, is a good life-span. It may well fail from other causes, as mines often do, before its finite resource is used up. The tragedy is that people have committed their lives and their credit to an activity that they hoped would last out their time. Those who encourage industrialization rarely look forward to the declining phase of the industrial cycle.

The Campo might also be thought of as the Tragedy of the Commons come true: of what happens when people are let loose on an unfamiliar new resource and have no time to understand it, or to work out the social and legal system needed to use it properly. Many scholars interpret this as impending desertification, but it cannot be so: the worst that can happen is that the plain reverts to being the desert that it was in the 1940s.

In most areas the situation is still just retrievable. An obvious shortcoming is that water is usually free to the user, or charged at no more than the cost of delivery. If users had to pay for the water itself they would be more economical. Proprietors in areas greatly threatened by pollution or salinization might be made, or paid, not to irrigate their land. Woodland or pasture into which rainfall infiltrates to replenish the catchment should be preserved: further irrigation or development should not be allowed to occupy even more of the catchment than it does already. Aquifers adjoining the sea should be protected by artificial recharge. Physical planning is inevitably involved. The longer conservation measures are delayed, the more painful and expensive they will be.

33 R. A. F. Seaton, M. Lemon, S. E. van der Leeuw, 'Agricultural production and water quality in the Argolid valley, Greece', *Understanding the natural and anthropogenic causes of soil degradation and desertification in the Mediterranean basin*, ed. S. E. van der Leeuw, Synthesis of the European Community Archaeomedes Project 6, 1996, pp. 281–326.

CHAPTER 20

Desertification or Change: what to do about it?

The nation, its palate jaded by a surfeit of verse, tragedies, comedies, opera, novels and
theological disputes over the nature of grace, finally settled on the subject of cereals . . .
Many eminently useful things were written on farming; everyone read them, except,
unfortunately, the farmers themselves.[1]

There is no desire here to pretend that degradation is not a problem in the Mediterranean and
Near East. At the same time, much of the vegetation evidence for degradation is problematic at
best, in part because Mediterranean and Near East vegetation is more resilient than even
[Z.] Naveh . . . has recognised. One wonders, in fact, whether humid temperate realms are not in
some respects more sensitive to human impacts, contrary to the widespread belief that semi-arid
regions are especially fragile. Regardless, it seems that in the past negative feedbacks operated to
constrain human ability to destroy the Mediterranean and Near East environment, while today
the dramatic increase in inputs of energy and capital from elsewhere create a potentially
more dangerous situation.[2]

Late September – early in the new year – is a good time to be
in the Mediterranean. The last charter flights take off; the
beaches empty; departing tourists are sped northward by the
first storms. Parasols are stacked away, kiosks shuttered, boats
pulled high up the shore. Grey-green waves grind up
cigarette-ends, and wash plastic bottles and plastic sheeting
out to sea for currents to carry to the next seaside resort. In
plastic greenhouses, winter salad crops are watered and
sprayed. The first heavy rain, always a joyous event, refreshes
city and village, sweeping away the small rubbish. At last
there is room to park a car. The local language reasserts itself.
Sheep trot along the village street: in coffee-houses wood-
stoves and spirit-stills are reassembled from storage and fired
up; young men load their shotguns and set out for the bush.

Old human ways are resumed, so far as they can be. Natural
systems, likewise, struggle to resume their accustomed cycles
in spite of modern interventions. Life returns to farm and
maquis: green miraculously replaces desert fawn. Sheep
disperse from the hot concrete sheds where they have
languished on imported fodder through the summer. Olives
ripen on venerable trees in ancient groves, and also on dwarf
trees, irrigated and aligned down steep hillsides or along bulldozed false terraces. Reservoirs begin to fill with water, and
also with the sediment that would once have gone to replenish the sandy shores at river mouths, now showing signs of
wear and tear.

This book began with the nature of desertification and
whether it is affecting Mediterranean Europe. We pointed
out that this concept is often confused and confusing, covering a miscellany of unwanted changes in the landscape. In
trying to understand the situation and provide an answer to
the question 'desertification or change?' we have described
some of the changes, natural and man-made, in southern
Europe during the Holocene.

Mediterranean Europe combines winter rainfall and
summer drought with tectonic instability, characteristics that
it shares only with southern California and middle Chile. In
these three regions there is nothing abnormal about violent
natural events – deluges, fires, earthquakes, volcanic eruptions – which can change the landscape overnight. There are,
besides, longer-term instabilities resulting from crustal
movements, Pleistocene glaciations and coastal changes. The
Mediterranean climate has existed for far too short a time for
plants to have become fully adapted to it. The sea and desert
barrier reduced the scope for plants and animals to avoid
changes by migrating; many endemics have had to sit out
glacial cycles on their own particular cliffs and islands.

Because of this combination of natural instabilities,
perhaps greater than in any other part of the world that has
not actually been glaciated, one might expect Mediterranean
lands to be relatively robust in the face of human activities.
The activities that Europeans think of as civilization have a
much longer and more complex history in southern Europe
than in other countries with similar climates; their effects are
still not well understood despite centuries of study. Human
activities cannot be distinguished from natural changes on
time-scale alone. Research is barely beyond the stage where
fresh information creates more problems than it solves.

It is likely that people had already transformed the landscapes of most parts of Mediterranean Europe by the Bronze
Age, four thousand years ago. This 'humanization' of landscape and vegetation overlapped in time with the appearance
of something like the present Mediterranean climate.
On grounds of chronology and of how climates work, humanization cannot have been the cause of the change of climate.
To what extent changing climate helped the humanization
has still to be understood.

Since then, the ups and downs of human history and of

---

[1] Voltaire, *Dictionnaire philosophique*, c. 1750
(one might say much the same today about
desertification). Quoted by Annie Moulin,
Peasantry and Society in France since 1789, trans.
M. C. and M. K. Cleary, Cambridge, p. 10. We are
unable to find this aphorism in its supposed
source, but let it stand on its own merit.
[2] Mark A. Blumler 1993.

natural events have both had effects on plant cover, land use and landscape. There have been tectonic happenings such as the eruption of Santoríni and the uplift of the west end of Crete. Of wider significance have been variations from decade to decade in the occurrence of extreme weather events, about which we still have little information except for the Little Ice Age.

Climate

Throughout the second half of the Holocene the climate has shown variations on a Mediterranean-type theme. Whether volcanoes or sunspots or both be the cause, there have been many periods of violent and changeable weather.

We have so far failed to demonstrate effects of global warming in the Mediterranean. Changes in climate within the twentieth century are inconsistent, and no greater than those of previous centuries. Attempts to predict effects of global warming founder on the complexity of the Mediterranean.

The twentieth century has been relatively stable, leading people to expect a not too variable climate; they have exploited water and other resources to the full, leaving no reserves against future instability. The twenty-first century may not be so kind.

Vegetation

We have, we hope, thoroughly demolished the myth of the Ruined Landscape, at least as a general proposition. Ruined Landscapes may yet be found in odd corners, although Mediterranean vegetation is so resilient that they would require some special combination of circumstances. Mediterranean vegetation should be understood on its own terms and using its own categories, rather than misinterpreted as degraded forms of a once-universal forest of tall, timber-quality trees. It makes no sense to reproach Greece for not being like northern Europe.

Desertification, likewise, is a rare phenomenon, greatly outweighed in the twentieth century by examples of de-desertification. Even if the term is widened to include land becoming unusable for agriculture, the amount of land that has become physically uncultivable is a very small fraction of that which could be cultivated but has been abandoned for other reasons.

Whether the ecology was ever ruined by early human action is still difficult to say. An apparent example is the extermination of the endemic faunas of the islands; however, the endemic island floras still survive, as far as we know almost unscathed. Early human intervention often took relatively subdued forms, such as converting savanna to farmland with trees, or altering the frequency of fires. This was a peaceful change, compared to the course of events in England, New England, New South Wales or the island of St Helena; to natural upheavals in the Mediterranean during glaciations; or to the treatment of coastal Spain in recent decades.

Human wealth and mobility

Written records, archaeological survey and excavation, and the study of the landscape itself give information about settlement and economy, and less securely about population. We have drawn inferences about landscape history in particular areas with which we have gained some familiarity.

Until very recently, the rural landscapes of Mediterranean Europe resulted from the activities of people whose livelihoods depended on the produce of their fields, pastures and trees. After World War II this suddenly ceased to be so. Instead of 30 per cent of the working population being 'employed in agriculture', the figure is supposed to be about 3 per cent. The change, however, is not quite as great as this: the 3 per cent probably excludes many city or coastal dwellers who still own a bit of land, as well as fertilizer-manufacturers and tractor-menders. At the same time people have moved from what was often a period of unusual poverty to unprecedented wealth. Their children, comparing their own wealth with their contemporaries' rather than with their grandparents', forget that it is unprecedented, and expect wealth to go on for ever.

There has always been a mobile population in southern Europe; traders by sea and land, armies and pirates, scholars and administrators going about their business. But most men and nearly all women were settled in their towns, hamlets or farms, at least for part of the year. Now most families have a car. People move between cities and the country daily or weekly. A hundred million holidaymakers throng the shores of southern Europe every year. A landscape is now not so much the home and surroundings of the community that created it; it is but one of a host of scenes that the traveller speeds past in a journey.

Erosion

Erosion is a natural, often episodic, process, proceeding at rates that vary greatly in space and time. Its ultimate cause is tectonic action, and many of the more erodible areas are associated with great fault zones.

We have collated data on the age of alluvial accumulations in river terraces and deltas to show that erosion in river catchments increased in particular periods. We would attribute this mainly to a greater frequency of deluges, probably associated with the weather changes that caused glaciers to advance for decades at a time, especially during the Little Ice Age between the thirteenth and nineteenth centuries.

To identify accelerated erosion resulting from human activity is more difficult than is usually assumed. The dates of both are seldom precise enough to establish correlations, and there is a dearth of information on how different types of vegetation or of human intervention promote or restrain erosion on different kinds of terrain. On present evidence, variations in erosion are explained less fully in terms of human activity than of climate, weather, geology, and tectonics. This could easily be changed by more rigorous and better-dated searches for links, but we doubt if it will again be possible simply to postulate human activities (or inactivities, if they serve better) as the generalized cause of erosion.

We argue that the threat of soil erosion is exaggerated. Much erosion is not soil erosion. In the Mediterranean, most soil erosion has traditionally resulted from ploughing. In the eighteenth and nineteenth centuries this was locally important where erodible soils were being brought under the plough. Most such have now been abandoned. The remaining erodible ploughlands are not very productive, and could easily be withdrawn from cultivation: they would already have been abandoned but for subsidies.

Deserts, badlands and barren karst, scarcely mentioned by travellers in the past who were more interested in human landscapes, now attract attention and comment. They are regarded as indicators that there is something wrong with the environment, and held up as the end-product of desertification. We have tried to show that these three kinds of landscape are usually ancient and determined by their geology. It can seldom be shown that their origination followed, still less was caused by, human activities.

Thus far for the period up to 1950. Now the diffusion of the bulldozer creates a new and more powerful method of promoting erosion and other degradation. Those who fear erosion would do well to attend to the excesses of earth-moving: to the creation of yet more false terraces for vines and olives in an age of over-production; to the amalgamation and 'rationalization' of fields; to digging up hillsides and then growing nothing; to digging out yet more unsupported minor roads; and to the bizarre practice of bulldozing slopes and planting pines (in the name of erosion control) which turn into burnt pines.

Land abandonment

Terrace cultivation is adapted to the power of men and beasts, or the small machines, 'mechanical donkeys', that served farmers well in the mid-twentieth century. Nearly all terraced areas have been abandoned for cereal cultivation, and in some areas for olive- and vine- growing too: the terraces are still usable, but people have found easier ways to make a living.

Agricultural production has been concentrated in gentle terrain, where large machines can be used, and moderately high yields can be got with fertilizers and pesticides and water. Where irrigation is available in plains it tends to hasten the abandonment of nearby mountain agriculture.

At the same time pasturage has become concentrated and separated from other activities. Where roads, shelters and waterholes have been constructed, flocks congregate and over-graze the remaining wild vegetation or ex-agricultural land. In other, more remote areas browsing has declined to the point at which sensitive plants have moved down from their usual cliff habitats. The fashion for private enterprise, at the expense of family and communal practices, tends to promote both over- and under-grazing, as well as littering the landscape with generations of decrepit fences.

The contraction of cultivation in Europe contrasts with the African Mediterranean, where a growing population still pursues self-sufficiency in cereals. In that generally less mountainous terrain, cultivation is still spreading into semi-arid phrygana; it is also being intensified, with mechanization, substitution of wheat for barley, and cropping every year instead of having fallow years. All these changes increase the perils of drought.[3] Self-sufficiency has been pursued in much of the European Mediterranean at various times in the past, usually with similar results in terms of scarcity in dry years. The retreat from marginal land, whatever its other consequences, at least puts an end to that hazard.

Much of southern Europe, especially SW Iberia, has a history of savanna. Earlier transhumance practices tend to be replaced by sedentary livestock, fed on cereals in the summer hungry season when herbage and acorns are scarce. Livestock diseases and increased labour costs have undermined these new practices. Attempts to increase cereal production risk degrading soils and trees; but less intensive use of the land allows cistus and other flammable bushes to occupy increasing areas. As a result, one of Europe's prime wildlife habitats – underrated by ecologists trained to appreciate only forests – is deteriorating.

Land abandonment, neglect of traditional uses of roughland, increase of flammable vegetation (aggravated by tree-planting), and mindless suppression of ordinary fires are among the reasons for wildfires becoming more conspicuous and destructive, to the point at which they threaten lives and property and create ugly landscapes. This risk could be reduced – given political pertinacity – by sensible measures to reduce fuel and to keep people and valuable property away from likely conflagration sites.

Future land-use

The scarce factor of production in agriculture in Mediterranean Europe is now water, not land. Irrigation has been extended to take full advantage of the water available under average conditions. When rainfall is below average (as it is in most years and may be for several successive years!), there are calls for increased supplies, requiring transfer from a distance, more storage in reservoirs, and more pumping of groundwater. Water transfer is objected to by people in the donor areas or on the route. Reservoirs in semi-arid areas fill rapidly with sediment because of the large catchment area in relation to reservoir capacity and the high sediment yields. They trap sediments that would otherwise keep coastal beaches in existence. Reservoirs drown natural features; over-pumping dries up rivers and wetlands. Finally, over-exploitation of coastal ground-waters leads to sea-water intrusion and the threatened collapse of industrial greenhouse production.

Olive growing is a classic example of people taking a fashionable activity beyond reasonable limits and not knowing when to stop. As an investment, planting olives takes several years to produce a return, and the grower does not know until later whether it will be profitable. But without target prices well above market prices, special subsidies for small and middle-sized producers (of less than 500 kg), and consumption aid as well, much of the crop would hardly be worth picking. When subsidies decline, as one day they must, there

---

3 W. Swearingen, 'Northwest Africa', Glantz
1994, pp. 117–33.

will be a competition between new and old trees. We cannot be sure of the outcome – in west Crete old trees are probably more productive than new ones – but there is a risk that sooner or later ancient trees will be abandoned or even destroyed. Ancient trees, especially olives, are important historic and landscape features and an important habitat for wildlife, which should be maintained. There is more to olives than just another agri-business crop.

Subsidies, as a means of promoting proper land-use, have proved clumsy and sometimes perverse. They have failed to keep people on the land, and (like land reform before them) have often benefited those who least needed them.

What does the future hold for the economics of ordinary agriculture in Mediterranean Europe? Governments' restrictions on trade have declined and are likely to decline further.[4] In consequence, prices for agricultural products are expected to rise on the world market, and may decline within the European Union. At the same time, the decline of subsidized 'support' prices will result in a reduction of the returns to European Union farmers, though this may be offset to some extent by direct area payments. Southern Europe will meet more competition from imported produce, especially as eastern Europe recovers from the ravages of Communism. Total farm output in Mediterranean Europe is likely to fall, the price of land is likely to decline further, and capital will be taken out of agriculture.

In some areas there may be a tendency to revert to less capital-intensive methods: to substitute extra land for costly inputs such as chemicals, bought-in animal feedstuffs, and even irrigation water (though this last is still very cheap). Plant breeding could help in producing crop varieties that do well in somewhat adverse conditions.

The future demography of farming is unclear. North Europeans notice the high proportion of Mediterranean farmers 'reaching the age of retirement'. In terms of Mediterranean realities this may always have been so: farming is a job that people retire into, rather than out of (for example, they may take up the family farm, and officially become farmers, on the death of a parent, or on retiring from foreign employment). We foresee that farming as a full-time livelihood will continue to contract, and will become dependent either on irrigation or on having a big farm. The future, especially on difficult terrain, lies in part-time farming.

However, the tendency for land to go out of production, or to be downgraded to a lower category of production, shows no sign of slackening, nor can we see why it should. First-class land still sells for high prices, especially if it can be irrigated, but it was said ten years ago of southern France that 'you would practically have to pay people to accept a poorer category'.[5]

Future landscape

We are in no position to advise on agricultural policy. Nor would we trespass into the town planner's territory, the future of the Euro-Mediterranean's urban scene. We conclude with observations on landscape conservation, a complex subject where aesthetics, land economy and ecological well-being overlap.

Not all recent developments are bad. Production of harmful effluents from intensive stock-rearing may be reduced if there is less grain feeding and more grazing of pastures. Such trends may be encouraged by the growing demand for 'organic' foodstuffs.

However, there appear to us to be two fundamental problems. The first is 'old-fashioned modernism': the urge, still irresistible in some Mediterranean countries, to pursue some fashionable technology such as bulldozing or piped irrigation to its physical limit (or beyond), without considering any but the most immediate consequences. The second is the lack of any clear understanding of what to do with roughland that has lost its traditional uses. Such land is regarded as expendable, and frittered away on low-density housing, golf-courses, pine and eucalyptus plantations, quarries, rubbish dumps, superfluous olives and the like. Here, too, only the most immediate consequences are considered: when fire, starting in the rubbish, consumes pines, houses and olives this is thought of as a crime or a tragedy, rather than the foreseeable effect of the aforesaid actions.

Near-natural landscapes tend to be reproached for being unusual, or misinterpreted as the results of maltreatment, rather than appreciated as beautiful and extraordinary. Gorges, cliffs, deserts, deltas, karst and high mountains tend to survive (where they do) more through being difficult to destroy than because they are valued as wonders of Europe. Even the coast, source of modern wealth, is maltreated in ways that will reduce its earning power or destroy it altogether. (What sadder sight is there than an expensive seaside hotel with an expensive swimming-pool because the sea is unfit to swim in?)

Most Mediterranean landscapes, however, are blends of the natural and man-made. The human contribution to their meaning and beauty has been in part conscious and deliberate, in part the inadvertent product of people living in particular ecological settings. They have long been a source of joy to visitors and, we believe, to many of the local inhabitants. As precious and as vulnerable as other art forms, they deserve to be protected from being desecrated or frittered away.

The economic incentive is not to be gainsaid; it has to be turned to account. Landscapes that are the product of ancient economies and societies cannot be preserved in a non-functional state: to survive they must be kept in working order. National Parks and other protected areas have increased, and will probably increase further as more land becomes available. This is good, but we deplore the philosophy, still prevalent in some countries, of treating national parks as if they were wilderness and trying to banish all human activities other than tourism. Old-fashioned human activities have shaped the character of virtually every national park in Europe, and need to be understood and upheld.

Mass tourism will continue to call for sandy beaches and swimming pools, until the wheel of fashion rolls on in some

---

4 General Agreement on Tariffs and Trade treaty.

5 H. Clout, 'France', *Policies and Plans for Rural People: an International Perspective*, ed. P. Cloke, Unwin Hyman, London, 1988, pp. 98–119.

other direction. But other visitors, often coming 'out of season' and therefore doubly valuable, are attracted by traditional and more remote landscapes, their people, plants and animals. It is ironic that most people visit Crete for the seaside, which is little different from other seasides except that it suffers from having no tide. Fewer come for the unique antiquities; far too few come for the glamorous mountains which are the defining feature of Crete, or to understand a different way of life.

The accommodation of more discriminating needs can respect local styles of buildings and behaviour; it might provide employment for young people through the year, help to support communities with their children and schools, and thereby indirectly assist in the protection of the landscapes around them.

Until twenty years ago the ecological imperative was of concern mainly to people like us, the authors. Biodiversity has now become a mysterious keyword, a quality to be maintained. Most changes in the Mediterranean landscape have reduced diversity as well as promoting uglification. Monotonous, large-scale formations have replaced the symbiotic, harmonious arrangements of the small landscape units that landscape ecologists call ecotopes.[6] Only continued occupation by people gaining their livelihoods locally can maintain the man-made diversity typical of Mediterranean Europe.

Zev Naveh, the leading exponent of landscape ecology, to whom we owe a deep debt for his stimulus, sees the need for direct intervention to prevent what he calls the neotechnological degradation of the landscape. We have pointed to the bulldozer, the electric pump and the land consolidator as agents of such degradation, needing to be tamed and trained. Naveh argues that generally 'threatening trends can only be reversed by more powerful negative inputs';[7] he advocates the development of comprehensive methods for assessing biodiversity as a basis for planning and managing landscapes.

While we agree with Naveh's diagnosis, we would be wary of following this treatment. We see dangers in centralized landscape planning and management. Experts and politicians have repeatedly shown themselves to be fallible. It is better to encourage people to live and work amongst the small-scale, historically diversified, patterns of farmland, orchard, grassland, shrubland, savanna and woodland that are in need of protection, and to help and advise communities to make decisions for themselves about their own land. The Age of Enlightenment has been an age of scientific achievement and of science-based mistakes; some of the latter would have been averted had the experts listened to local people before shouting at them. It will not end with the millennium.

THE END

---

6 Naveh 1994b.
7 Naveh 1994b, p. 190.

Bibliography

Abbreviations of archives and titles

AAAG	*Annals of the Association of American Geographers*	E&H	*Environment and History*	NSMS	*Nimbus:* Società Meteorologica Subalpina, Torino
A-B	*Al-Basit: Revista de Estudios Albacetenses*	EG	*Environmental Geology,* later *Environmental Geology and Water Science*	PBN	Paris, Bibliothèque Nationale
ABN	*Acta Botanica Neerlandica*			PG	*Physio-Géo* (CNRS Meudon)
AR	*Archaeological Reports* (Society for the Promotion of Hellenic Studies)	EM	*Ecologia Mediterranea*	PKF	*Paläoklimaforschung – Palaeoclimate Research* [occasional publications by the Akademie der Wissenschaften und der Literatur, Mainz; some of them form a series, *European Palaeoclimate and Man,* edited also by B. Frenzel and the European Science Foundation]
		EMA	*Environmental Monitoring and Assessment*		
ASC	Archivio di Stato, Cagliari	ESPL	*Earth Surface Processes and Landforms*		
ASPG	Archivio di Stato, Palermo, S. Maria della Gancia (VTLO: Vice Portolano di Termini, Lettere e Ordini; VPT: Vice Portolano di Termini)	FM	*Forêt Méditerranéenne*		
		GJ	*Geographical Journal*		
		GR	*Geographical Review*		
ASPTPR	Archivio di Stato, Palermo, Tribunale del Real Patrimonio (Audit Office)	IAHS	International Association for Hydrological Sciences	PPG	*Progress in Physical Grography*
				PPP	*Palaeogeography, Palaeoclimatology, Palaeoecology*
ASV	Archivio di Stato, Venice (PTM: Provveditor di Terra e Mare)	IJC	*International Journal of Climatology*		
		JAE	*Journal of Applied Ecology*	PPS	*Proceedings of the Prehistoric Society*
BAGF	*Bulletin de l'Association des Géographes Français*	JAS	*Journal of Archaeological Science*	PW	Pauly-Wissowa, *Real-Encyclopädie der Classischen Altertumswissenschaft,* 2nd edn, Metzler, Stuttgart
		JB	*Journal of Biogeography*		
		JC	*Journal of Climatology*		
BARIS	*British Archaeological Reports International Series*	JCR	*Journal of Coastal Research*	QR	*Quaternary Research*
		JE	*Journal of Ecology*	QS	*Quaderni Storici*
BARSS	*British Archaeological Reports Supplementary Series*	JFA	*Journal of Field Archaeology*	QSR	*Quaternary Science Reviews*
		JGR	*Journal of Geophysical Research*	SA	*Scientific American*
BCH	*Bulletin de Corréspondance Hellénique*	JH	*Journal of Hydrology*	USDA	United States Department of Agriculture
BNP	Bibliothèque Nationale, Paris	LUP	*Landscape & Urban Planning*		
BSA	*Annual of the British School at Athens*	MG	*Marine Geology*	WE	*The World Economy*
CC	*Climatic Change*	MOB	Meteorological Office, Bracknell, England archive	ZG	*Zeitschrift für Geomorphologie*
CNRS	Centre National de Recherches Scientifiques				

Abbott, J. T. and Valastro, S. T. 1995, 'The Holocene alluvial records of the chorai of Metapontum, Basilicata, and Croton, Calabria, Italy', Lewin and others 1995, pp. 195–205

Agulhon, M. 1970 (trans. J. Lloyd, 1982), *The Republic in the Village: the people of the Var from the French Revolution to the Second Republic,* Cambridge

Alarcón, Antonio de, D. P. 1892 (3rd edn), *La Alpujarra,* Sucesores de Rivadeneyra, Madrid

Alexander, D. 1982 'Differences between "calanchi" and "biancane" badlands in Italy', Bryan and Yair 1982, pp. 71–87

Al-Edrīsi (Abu-Abd-Alla-Mohamed Al-Edrīsi) (twelfth century), *Descripción de España,* trans. A Blázquez,: Déposito de la Guerra, Madrid 1901

Alfonso XI, *Libro de la Monteria . . .,* ed. D. P. Seniff, Hispanic Seminary of Medieval Studies, Madison 1983

Allbaugh, L. G. 1953, *Crete: A Case Study of an Underdeveloped Area,* Princeton, Princeton University Press

Allen, H. 1997, 'The environmental conditions of the Kopais basin, Boeotia during the post glacial with special reference to the Mycenaean period', *Recent Developments in the History and Archaeology of Central Greece,* ed. J. Bintliff, *BARIS* 666, pp. 39–58

Ambroise, R., Frapa, P., Giorgis, S. 1989 *Paysages de terrasses,* Édisud, Aix-en-Provence

Atherden, M. A. and Hall, J. A. 1999, 'Human impact on vegetation in the White Mountains of Crete since AD 500', *The Holocene* 9, pp. 183-93

Aubin, M. C. and 13 others 1980, *Évolution de l'interface hommes-milieu en zone bioclimatique mediterranéenne; le cas de Vidauban et de son terroir viticole,* Centre National de Recherches Scientifiques, A.S.P. PIREN 1, Université de Nice

Bailey, G. N. and 5 others 1986, 'Palaeolithic investigations at Klithi [Epirus]: preliminary results of the 1984 and 1985 field seasons', *BSA* 81, pp. 7–35.

Balabanian, O. 1984, *Problemas agrícolas e reformas agrárias no Alto Alentejo e na Estramadura Espanhola,* Lisboa

Ballais, J.-L. and Crambes, A. 1992, 'Morpho-genèse holocène, géosystèmes et anthro-pisation sur la montagne Sainte-Victoire', *Méditerranée* 1.2, pp. 29–41

Baraldi, E., Calegari, M., Moreno, D. 1992, 'Ironworks economy and woodmanship practices: chestnut woodland culture in Ligurian Apennines (16–19th c.)', *Proto-industries et histoire des forêts, Cahiers de l'ISARD* 3, Toulouse, pp. 135–49

Barker, G. W. and Hunt, C. O. 1995, 'Quaternary valley floor erosion and alluviation in the Biferno Valley, Molise, Italy: the role of tectonics, climate, sea level change, and human activity', Lewin and others 1995, pp. 145–57

Barriendos Vallvé, M. and Martín-Vide, J. 1998, 'Secular climatic oscillations as indicated by catastrophic floods in the Spanish Mediterranean coastal area (14th–19th centuries)', *CC* 38, pp. 473–91

Bauer Manderscheid, E. 1991, *Los montes de España en la Historia,* Fundación Conde del Valle de Salazar, Madrid

Behnke, R. H., Scoones, I., Kerven, K., eds. 1993, *Range Ecology at Disequilibrium: new models of natural variability and pastoral adaptation in African savannas,* Overseas Development Institute, London

Behre, K.-E. 1981, 'The interpretation of anthropogenic indicators in pollen diagrams', *Pollen et Spores* 23, pp. 225–45

Bell, M. and Boardman, J. eds. 1992, *Past and Present Erosion,* Oxbow, Oxford

Bell, M. and Limbrey, S. 1982, *Archaeological aspects of woodland ecology, BARIS* 146

Beug, H.-J. 1967, 'On the forest history of the Dalmatian coast', *Review of Palaeobotany and Palynology* 2, pp. 271–79

Beug, H.-J. 1977, 'Vegetationsgeschichtliche Untersuchungen im Küstenbereich von Istrien (Jugoslawien)', *Flora* 166, pp. 357–81

Billi, P. and Rinaldi, M. 1997, 'Human impact on sediment yield and channel dynamics in the Arno River Basin (central Italy)', Walling and Probst 1997, pp. 301–11

Bintliff, J. L. 1977, *Natural Environment and Human Settlement in Prehistoric Greece: BARSS* 28

Bisci, C. and 5 others 1992, 'The Sant' Agata Feltrea landslide (Marche region, central Italy): a case of recurrent earthflow evolving from a deep-seated gravitational slope deformation', *Geomorphology* 15, pp. 351–61

Blanchemanche, P. 1990, *Batisseurs de Paysages: terrassement, épierrement et petite hydraulique agricoles en Europe XVIIᵉ–XIXᵉ siècles*, Maison des Sciences de l'Homme, Paris

Blumler, M. A. 1993, 'Successional pattern and landscape sensitivity in the Mediterranean and Near East', *Landscape Sensitivity*, ed. D. S. G. Thomas and R. J. Allison, John Wiley, Chichester, pp. 287–305

Bonatti, E. 1970, 'Pollen in the lake sediments [of Monterosi and Baccano]', *Transactions of the American Philosophical Society* NS 60/4, pp. 26–31

Borrow, G. 1842, *The Bible in Spain* [many editions]

Bory de St Vincent, J. B. G. M., ed. 1836, *Expedition Scientifique de Morée*, Levrault, Paris

Bottema, S. 1979, 'Pollen analytical investigations in Thessaly (Greece)', *Palaeohistoria* 21, pp. 20–39 and diagrams

Bottema, S. 1980, 'Palynological investigations on Crete', *Review of Palaeobotany and Palynology* 31, pp. 193–217

Bottema, S. 1985, 'Palynological investigations in Greece with special reference to pollen as an indicator of human activity', *Palaeohistoria* 24, pp. 257–89

Bottema, S. and others, eds. 1990, *Man's role in the shaping of the Eastern Mediterranean landscape*, Balkema, Rotterdam

Bottema, S. and Woldring, H. 'Anthropogenic indicators in the pollen record of the eastern Mediterranean', Bottema and others 1990, pp. 231–65

bowles, g. [i.e. w.] 1783, *Introduzione alla storia naturale e alla geografica fisica di spagna* (trans. f. milizia), parma

Bradford, J. S. P. 1957, *Ancient Landscapes: studies in field archaeology*, Bell, London

Bradley, R. S. and Jones, P. D., eds. 1992, *Climate since A.D. 1500*, Routledge, London

Brandt, C. J. and Thornes, J. B., eds. 1996, *Mediterranean Desertification and Land Use*, Wiley, Chichester

Braudel, F. 1966, *The Mediterranean and the Mediterranean world in the age of Philip II [of Spain]* (trans. S. Reynolds 1972–3), Collins, London

Bravard, J.-P. 1993, 'Approches du changement fluvial dans le bassin du Rhône (XIV-XIX siècles)', *Pour une histoire de l'environnement*, ed. C. Beck and R. Delort, CNRS Éditions, Paris, pp. 97–103

Brenan, G. 1957, *South from Granada*, Hamish Hamilton, London

Brice, W. C., ed. 1978, *The Environmental History of the Near and Middle East*, Academic Press, London

Bruce-Chwatt, L. J. and de Zulueta, J. 1980, *The rise and fall of malaria in Europe: a historico-epidemiological study*, Oxford

Brückner, H. 1980, 'Marine Terrassen in Süd-italien: eine quartärmorphologische Studie über das Küstentiefland von Metapont', *Düsseldorfer geographische Schriften* 14

Brückner, H. 1983, 'Holozäne Bodebildungen in den Alluvionen süditalienischer Flüsse', *ZG NF Suppl.* 48, pp. 99–116

Brückner, H. 1990, 'Changes in the Mediterranean ecosystem during antiquity - a geomorphological approach as seen in two examples', Bottema and others 1990, pp. 127–37

Brückner, H. and Hoffmann, G. 1992, 'Human-induced erosion processes in Mediterranean countries', *Geoökoplus* 3, pp. 97–110

Brunsden, D. and Prior, D. B., eds. 1984, *Slope Instability*, Wiley, Chichester

Bryan, R. and Yair, A., eds. 1982, *Badland geomorphology and piping*, Geo Books, Norwich

Buffoni, L., Chlistovsky, F., Maugeri, M. 1996, *1763-1995: 233 anni di rilevazioni termiche giornaliere a Milano-Brera*, CUSL, Milan

Bull, L. J. and Kirkby, M. J. 1997, 'Gully processes and modelling', *PPG* 21, pp. 354–74

Buondelmonti, C. c. 1415, *Descriptio Insule Crete*, ed. M.-A. van Spitael, 2nd edn, Spanakis, Herákleion, 1981

Burgess, C. 1987, 'Fieldwork in the Évora district, Alentejo, Portugal, 1986–1988: a preliminary report', *Northern Archaeology* 8, pp. 35–63

Butzer, K. W. 1988, 'Cattle and sheep from Old to New Spain: historical antecedents', *AAAG* 78, pp. 29–56

Butzer, K. W., Mateu, J. F., Butzer, E. K., Kraus, P. 1985, 'Irrigation agrosystems in eastern Spain: Roman or Islamic origins?', *AAAG* 75, pp. 479–509

Cabrillana Ciézar, N. 1977, 'Aportación a la historía rural de Almería en el siglo XVI', *Andalucia de la Edad Medievale y Moderna*, ed. M. M. Ladera Guesaden, *Cuadernos de Historía Anexos de la Revista Hispana* 7, pp. 41–74

Calvet, M. 1993, *Crues catastrophiques et vitesse de l'érosion dans les Pyrénées Orientales*, Servei Geologie de Catalunya, pp. 93–101

Calvet, M. 1994, *Morphogenèse d'une montagne méditerranéenne; les Pyrénées orientales*, Presse Universitaire de Perpignan

Calzolari, C. and Ungaro, F. 1998, 'Geomorphic features of a badland (biancane) area (Central Italy): characterisation, distribution and quantitative spatial analysis', *Catena* 31, pp. 237–56

Caniato, G., ed. 1993, *La Via del Fiume dalle Dolomiti a Venezia*, Cierre, Verona

Carletti, F., ed. 1993, *Demani Civici e Risorse Ambientali*, Jovene, Napoli

Carvajal, del Marmol, L. 1600, *Historia del rebelión y castigo de los Moriscos del Reino de Granada*, Málaga [page-numbers refer to the 1852 Madrid edition]

Casimiro Mendes, J. 1993, 'Local scale micrometeorology and time series analysis', *MEDALUS Final Report*, pp. 17–38

Castex, J.-M. 1980, *L'aménagement des pentes et des sols dans les Alpes-Maritimes et le Var*, Laboratoire de Géographie Raoul Blanchard, Université de Nice

Cavaciocchi, S., ed. 1996, *L'uomo e la Foresta Secc. XIII–XVIII*, Le Monnier, Prato

Cavaillès, H. 1905, 'La question forestière en Espagne', *Annales de Géographie* 14, pp. 318–31

Cavanilles, A. J. 1795, *Observaciones sobre la historia natural, geografía, agricultura, poblacion y frutos del Reyno de Valencia*, Imprenta Real, Madrid

Chabal, L. 1997, *Forêts et Sociétés en Languedoc (Néolithique final, Antiquité tardive): l'anthracologie, méthode et paléocologie*, Maison des Sciences de l'Homme

Chapman, R. W. 1978, 'The evidence for prehistoric water control in south-east Spain', *Journal of Arid Environments* 1, pp. 261–74

Cheddadi, R., Rossignol-Strick, M., Fontugne, M. 1991, 'Eastern Mediterranean palaeoclimates from 26 to 5 ka B.P. documented by pollen and isotopic analysis of a core in the anoxic Bannock Basin', *MG* 100, pp. 53–66

Chester, D. K. and James, P. A. 1991, 'Holocene alluviation in the Algarve, southern Portugal: the case for an anthropogenic cause', *JAS* 18, pp. 73–87

Cilla, G. and others 1994, 'Holocene fluvial dynamics in mountain areas: the case of the river Esino (Apennino Umbro-Marchigiano)', *Geografia Fisica Dinamica Quaternaria* 17, pp. 163–74

Clark, P. ed. 1985, *The European Crisis of the 1590s*, Allen and Unwin, London

Colón, F., *Descripcion y Cosmografía de España [1517-]* Patronato de Huérfanos de Administración Militar, Madrid 1910

Conte, M. and Sorani, R. 1993, 'Some severe drought events in the central Mediterranean', *MEDALUS Working Paper* 8

Conventi, Y. 1993, 'Désertification potentielle: impact anthropique et évolution de la mosaique des paysages en Corse', *MEDIMONT contributions to preliminary report, first version*, pp. 22–44

Cook, S. E. [alias Widdrington] 1834, *Sketches in Spain during the years 1829, 30, 31 and 32*, Boone, London

Cooke, R. U. and Reeves, R. W. 1976, *Arroyos and Environmental Change in the American South-West*, Oxford

Cottechia, V., Dai Pra, G., Magri, G. 1969, 'Oscillazioni tirreniane e oloceniche del livello mare nel golfo di Taranto, correlate da datazioni col metodo del radiocarbonio', *Geologica Applicata e Idrogeologia* 4, pp. 93–148

Còveri, L. and Moreno, D. eds., 1983, *Studi di etnografia e dialettologia ligure in memoria di Hugo Plomteux*, SAGEP, Genoa

Crutzen, P. J. and Goldhammer, J. G., eds. 1993, *Fire in the environment: the ecological, atmospheric, and climatic importance of vegetation fires*, Wiley, Chichester

Dallman, P. R. 1998, *Plant life in the world's Mediterranean climates*, Oxford

Davidson, D. and Tasker, C. 1982, 'Geomorphological evolution during the late Holocene', *An Island Polity [Mélos]*, ed. C. Renfrew and M. Wagstaff, Cambridge, pp. 81–94

De Reparaz, A. 1990, 'La culture en terrasses, expression de la petite paysannerie méditerranéenne traditionnelle', *Méditerranée* 71, pp. 23–29

Delano Smith, C. 1979, *Western Mediterranean Europe: a historical geography of Italy, Spain and southern France since the Neolithic period*, Academic Press, London

Di Castri, F. and Mooney, H.A. 1973, *Mediterranean Type Ecosystems: origin and structure*, Chapman and Hall, London

Di Martino, P. 1993, 'Deforestation and natural regeneration of woodland: the forest history of Molise, Italy, over the last two centuries', Watkins 1993, pp. 69–92

Di Martino, P. 1996, *Storia del Paesaggio Forestale del Molise (sec. XIX–XX)*, Lampo, Campobasso

Dooly, M. G. and Doody, J. P., eds. 1995, *Directions in European Coastal Management*, Samara, Cardigan (Wales)

Driver, T. S. and Chapman, G. P., eds. 1996, *Time-scales and Environmental Change*, Routledge, London

Dufaure, J. J. 1976, 'La terrace holocène d'Olympie et ses équivalents méditerranéens', *Bulletin de l'Association Géographique Française* 433, pp. 85–94

Edrīsi *see* Al-Edrīsi

Embleton, C., ed. 1984, *Geomorphology of Europe*, Macmillan, London

Esu, D. and Girotti, O. 1991, 'Un terrazzo dell' Eta del Ferro nel bacino del Tevere presso Attigliano (Umbria)', *Geografia Fisica e Dinamica Quaternaria* 14, pp. 119–22

Eyre, S. R. 1963, *Vegetation and soils: a world picture*, Edward Arnold, London

Fantechi, R. and Margaris, N. S., eds. 1984, *Desertification in Europe*, Reidel, Dordrecht

Faulkner, H. 1995, 'Gully erosion associated with the expansion of unterraced almond cultivation in the coastal Sierra de Lujar, S. Spain', *Land Degradation & Rehabilitation* 6, pp. 179–200

Fernández, L. S. and de Mata Carriazo Arroquia, J. 1969, *Historia de España: La España de los reyes Católicos (1474–1516)*, Espasa-Calpe, Madrid

Finley, M. I. and others 1986, *A History of Sicily*, Chatto and Windus, London

Flohn, H. and Fantechi, R., eds. 1984, *The Climate of Europe: Past, Present and Future*, Reidel, Dordrecht

Folch i Guillèn, R. 1981, *La vegetació dels països catalans*, Ketres, Barcelona

Folland, C. K. and others 1990, 'Observed climate variations and change', *Climate Change; the IPCC Scientific Assessment*, ed. J. T. Houghton and others, Cambridge, pp. 195–238

Font Tullot, I. 1988, *Historia del Clima de España*, Instituto Nacional de Meteorologia, Madrid

Forbes, H. 1997, 'A "waste" of resources: aspects of landscape exploitation in lowland Greek agriculture', *Aegean Strategies: studies of culture and environment on the European fringe*, ed. P. N. Kardulias and M. Shutes, Rowman and Littlefield, Lanham (Maryland), pp. 187–213

Ford, R. 1845, *A Hand-book for Travellers in Spain and Readers at Home . . .*, Murray, London

Frank, A. H. E. 1969, 'Pollen stratigraphy of the lake of Vico (central Italy)', *PPP* 6, pp. 67–85

Frangipane, A. and Paris, E. 1994, 'Long-term variability of sediment transport in the Ombrone River basin (Italy)', *Variability in Stream Erosion and Sediment Transport*, ed. Olive, L. J., Loughran, R. J., Kesby, J. A., IAHS Publ. 224, pp. 317–24

Frazer, J. 1898, *Pausanias's Description of Greece*, Macmillan, London

Frenzel, B., ed. 1994a, *Evaluation of Land Surfaces Cleared from Forests in the Mediterranean Region during the Time of the Roman Empire*, PKF 10

Frenzel, B., ed. 1994b, *Climatic Trends and Anomalies in Europe 1675–1715*, PKF 13

Fuller, I. C. and 5 others 1996, 'Geochronologies and environmental records of Quaternary fluvial sequences in the Guadelope basin, northeast Spain, based on luminescence dating', *Global Continental Changes; the Context of Palaeohydrology*, ed. J. Branson, A. G. Brown, and K. J. Gregory, *Geological Society Special Publication No. 115*, Geological Society, London, pp. 99–120

Gallart, F., Llorens, P., Latron, J. 1994, 'Studying the role of old agricultural terraces on runoff generation in a small Mediterranean mountainous basin', *JH* 159, pp. 291–303

Gams, I. 1993, 'Origin of the term "karst" and the transformation of the classical Karst (kras)', *EG* 21, pp. 110–14

Gams, I. and others 1993, 'Environmental change and human impacts on the mediterranean karsts of France, Italy and the Dinaric region', *Catena Suppl.* 25, pp. 59–98

Garnsey, P. D. A. 1988, *Famine and Food Supply in the Graeco-Roman World: responses to risk and crisis*, Cambridge

Genre, C. 1988, 'Les alluvionnements historiques en Eubée, Grèce', *Études Mediterranéennes* 12, pp. 229–58

Gilman, A. and Thornes, J. B. 1985, *Land-use and prehistory in south-east Spain*, Allen and Unwin, London

Glantz, M. H. 1994, *Drought follows the plow*, Cambridge

Gomez, B. 1986, 'The alluvial terraces and fills of the Lower Vasilikos Valley, in the vicinity of Kalavasos, Cyprus', *Transactions of the Institute of British Geographers* NS 12, pp. 345–59

Gómez-Moreno, M. 1951, 'De la Alpujarra', *Al-Andalus* 16, pp. 17–36

González Bernaldez, F. 1991, 'Ecological consequences of the abandonment of traditional land use systems in central Spain', *Options Méditerranéennes* 15, pp. 23–9

Goy, J. L. and 6 others 1996, 'Global and regional factors controlling changes of coastlines in southern Iberia (Spain) during the Holocene', *QSR* 15, pp. 773–80

Grove, A. T., Moody, J., Rackham, O. 1991, *Crete and the South Aegean Islands: effects of changing climate on the environment*, European Community Contract EV4C-0073-UK, final report

Grove, J. M. 1988, *The Little Ice Age*, Methuen, London

Grove, J. M. and Conterio, A. 1992a, *Reconstruction of past Mediterranean climate*, European Community Contract EV4C-0044-UK(H), final report 2

Grove, J. M. and Conterio, A. 1992b, 'Little Ice Age Climate in the Eastern Mediterranean', *Proceedings of the International Symposium on Little Ice Age Climate*, ed., T. Mikami, Tokyo Metropolitan University, pp. 221–26

Grove, J. M. and Conterio, A. 1994, 'Climate in the eastern and central Mediterranean, 1675 to 1715', *PKF* 13, pp. 275–85

Grove, J. M. and Conterio, A. 1995, 'The climate of Crete in the sixteenth and seventeenth centuries', *CC* 30, pp. 223–47

Grove, R. H. 1995, *Green Imperialism*, Cambridge University Press

Grüger, E. 1996, 'Vegetation changes', *Dalmatia: archaeological and ecological studies in a Mediterranean landscape*, ed. J. Chapman, R. Shiel, Š. Batovič, Leicester University Press, pp. 33–43.

Guidoboni, E., Comastri, E., Traina, G. 1994, *Catalogue of ancient earthquakes in the Mediterranean area up to the 10th century*, Istituto Nazionale di Geofisica, Rome

Guilaine, J. 1993, 'Six millénaires d'histoire de l'environnement: étude interdisciplinaire de l'abri sous roche de Font–Juvénal (Conquessur-Orbiel, Aude)', *Pour une histoire de l'environnement*, ed. C. Beck and R. Delort, CNRS Éditions, Paris, pp. 165–71

Günay, G., Johnson, A. I., W. Back, eds. 1993, *Hydrological Processes in Karst Terranes [sic]*, IAHS Publication 207, Wallingford

Gutiérrez, C., Prieto, F., Garcia-Dory, M. A. 1985, 'Evolución del encinar en España', *Quercus* 16, pp. 4–9

Hardy, D. A. and others, eds. 1990, *Thera and the Aegean World III*, Thera Foundation, London

Haro, S., Fernandez, J. F., Josa March, R., Gallart, F. 1992, 'Papel hidrologico y geomorfologico de las propriedades del suelo en una zona pirenaica de campos abandonados (Cal Parisa, Vallcebre)', *Estudios de Geomorfologia en España*, ed. P. López-Bermúdez and others, pp. 243–50

Hempel, L. 1990, *Forschungen zur Physischen Geographie der Insel Kreta im Quartär*, Vandenhoeck and Ruprecht, Göttingen

Hoenerbach, V. 1980, 'Cultivos Enarenados', eine Sonderanbauform an der andalusischen Mittelmeerküste, dissertation, University of Bonn

Hoffmann, G. 1987, 'Holozänstratigraphie und Küstenlinienverlagerung an der andalusischen Mittelmeerküste', *Berichte au dem Fachbereich Geowissenschaften der Universität Bremen* 2, pp. 45–62

Holzhauser, H. and Zumbühl, H. J. 1996, 'The history of the Lower Grindelwald Glacier during the last 2800 years – paleosols, fossil wood and historical pictorial records – new results', *ZG* NF Suppl. 104, pp. 95–127

Hunt, C. O., Gilbertson, D. D. and Donahue, R. E. 1992, 'Palaeoenvironmental evidence for agricultural soil erosion from Late Holocene deposits in the Montagnola Senese, Italy', Bell and Boardman 1992, pp. 163–74

Huntington, E. 1910, 'The burial of Olympia', *GJ* 36, pp. 657–86

Huntley, B. and Birks, H. J. B. 1983, *An atlas of past and present pollen maps for Europe: 0–13000 years ago*, Cambridge

Jacob, N. 1997, 'La crue d'octobre 1940 dans la basse vallée du Tech (Roussillon) d'après les dossiers des sinistrés', *Annales de Géographie* 596, pp. 414–24

Jameson, M. H., Runnels, C. N., van Andel, T. H. 1994, *A Greek Countryside: the southern Argolid from prehistory to the present day,* Stanford

Jeftic, L., Milliman, J. D., Sestini, G., eds. 1992, *Climatic change and the Mediterranean,* Edward Arnold, London

Joffre, R., Vacher, J., de los Llanos, C., Long, G. 1988, 'The dehesa: an agrosilvopastoral system of the Mediterranean region with special reference to the Sierra Morena area of Spain', *Agroforestry Systems* 6, pp. 71–96

Jones, W. H. S. 1907, *Malaria: a neglected factor in the history of Greece and Rome,* Macmillan and Bowes, Cambridge

Jorda, M. and Provansal, M. 1990, 'Terrasses de culture et bilan érosif en région méditerranéenne', *Méditerranée* 71, pp. 55–62

Jorda, M., Parron, C., Provansal, M., Roux, M. 1991, 'Érosion et détritisme holocènes en Basse-Provence calcaire: l'impact de l'anthropisation', *PG* 22–23, pp. 37–47

Judson, S. 1963a, 'Erosion and deposition of Italian stream valleys during historic time', *Science* 140, pp. 898–9

Judson, S. 1963b, 'Stream changes during historic time in east-central Sicily', *American Journal of Archaeology* 67, pp. 287–89

Julian, M. and Antony, E. 1996, 'Aspects of landslide activity in the Mercantour massif and the French Riviera, southeastern France', *Geomorphology* 15, pp. 275–89

Katzoulis, B. D. and Kabetzidis, H. D. 1989, 'Analysis of the long-term precipitation series at Athens, Greece', *CC* 14, pp. 263–90

Kelly, M. G. and Huntley, B. 1991, 'An 11 000-year record of vegetation and environment from Lago di Martignano, Latium, Italy', *Journal of Quaternary Science* 6, pp. 209–24

Kittridge, J. 1948, *Forest Influences: the effects of woody vegetation on climate, water and soil, with applications to the conservation of water and the control of floods and erosion,* McGraw-Hill, New York

Klein, J. 1920, *The Mesta: a study in Spanish economic history 1273–1836,* Harvard University Press

Kraft, J. C., Rapp, G. R. Jr., Aschenbrenner, S. E. 1980, 'Late Holocene palaeogeomorphic reconstructions in the area of the bay of Navarino: Sandy Pylos' *JAS* 7, pp. 187–210

Kraft, J. C. and 4 others, 1987, 'The pass at Thermopylae, Greece' *JFA* 14, pp. 181–98

La Marmora (A. della Marmora) 1849, *Sopra il taglio di centomila alberi di quercia da farsi in Sardegna.* Cagliari [a rare pamphlet; copy in ASC: SE 256¹²]

La Marmora, A. 1860, *Itinéraire de l'Île de Sardaigne,* Bocca, Turin

Lane, F. C. 1934, *Venetian Ships and Shipbuilders of the Renaissance,* Johns Hopkins University Press, Baltimore

Lautensasch, H. 1959, 'Maurische Züge in geographischen Bild der Iberischen Halbinsel', *Bonner geographische Abhandlungen* 28, pp. 1–98

Lavabre, J., Sempre Torres, D., Cernesson, F. 1993, 'Changes in the hydrological response of a small Mediterranean basin a year after a wildfire', *JH* 142, pp. 273–99

Lea, H. C. 1901, *The Moriscos of Spain: their conversion and expulsion,* Bernard Quaritch, London

Leake, W. M. 1835, *Travels in Northern Greece,* London [travels 1804–5]

Leake, W. M. 1830, *Travels in the Morea,* Murray, London [travels 1805]

Lehmann, R. 1994, *Landscape Degradation, Soil Erosion and Conservation on the Cycladic Island of Naxos, Greece,* Geographisches Institut, University of Basel

Le Houérou, H. N. 1987, 'Vegetation wildfires in the Mediterranean basin: evolution and trends', *EM* 13 (4), pp. 13–24

Le Play, M. F. 1834, 'Itinéraire d'un voyage en Espagne', *Annales des Mines* 3ᵉ sér. 5, pp. 175–208, 209–36

Leporati, E. and Mercalli, I. 1994, 'Snowfall series of Turin, 1784–1992: climatological analysis and action on structures', *Annals of Glaciology* 19, pp. 77–84

Leveau, P. and Provansal, M. 1993, 'Systèmes agricoles et évolution du paysage depuis le néolithique au nord-est de l'Étang de Berre', *Pour une Histoire de l'Environnement,* ed. C. Beck and R. Delore, CNRS Paris, pp. 173–99

Levi, C. 1947, *Christ stopped at Eboli,* Farrar, Strauss (Penguin edn., Harmondsworth 1982)

Lewin, J., Macklin, M. G., Woodward, J. C., eds. 1995, *Mediterranean Quaternary River Environments,* Balkema, Rotterdam

Link, H. F. 1801, trans. J. Hinckley, *Travels in Portugal, and through France and Spain . . . ,* Longman, London

Lionetti, M. 1996, 'The Italian floods of 4–6 November 1994', *Weather* 51 (January), pp. 18–27

Livet, R. 1962, *Habitat rural et structures agraires en Basse-Provence,* Faculté de Lettres, Aix-en-Provence

Llasat, M.-C. and Rodriguez, R. 1997, 'Towards a regionalization of extreme rainfall events in the Mediterranean area', *Regional Hydrology: Concepts and Models for Sustainable Water Resource Management,* IAHS 246, pp. 215–22

López-Bermúdez, P. and others 1992, *Estudios de Geomorfologia en España 1, 2,* Sociedad Española de Geomorfologia, University of Murcia

Lowe, J. J., Davite, C., Moreno, D., Maggi, R. 1993, 'Stratigrafica pollinica olocenica e storia delle risorse boschive dell'Appennino settentrionale', *Rivista Geografica Italiana* 102, pp. 267–310

Lowe, J. J., Davite, C., Moreno, D., Maggi, R. 1994, 'Holocene pollen stratigraphy and human interference in the woodlands of the northern Apennines, Italy', *The Holocene* 4, pp. 153–64

Lumsden, G. I., ed. 1992, *Geology and the Environment,* Clarendon, Oxford

Lyrintzis, G. 1996, 'Human impact trend in Crete: the case of Psilorites Mountain', *Environmental Conservation* 23, pp. 140–48

Mabberley, D. J. and Placito, P. J. 1993, *Algarve Plants and Landscape: passing tradition and ecological change,* Oxford

Mack Smith, D. 1968, *Modern Sicily after 1713,* Chatto and Windus, London

McNeill, J. R. 1992, *The mountains of the Mediterranean world,* Cambridge

Mairota, P., Thornes, J. B., Geeson, N., eds. 1998, *Atlas of Mediterranean Environments in Europe: the desertification context,* Wiley, Chichester

Mancebo, J. M., Molina, J. R., Camino, F. 1993, 'Pinus sylvestris L. en la vertiente septentrional de las Sierra de Gredos (Avila)', *Ecologia* 7, pp. 233–45

Mancini, F. 1986, 'Soil conservation problems in Italy after the Council of Research finalized project', Fantechi and Margaris 1986, pp. 147–51

Marsh, G. P. 1864, *Man and Nature: or, physical geography as modified by human action,* Scribner, New York

Martín Vide, X. 1985, *Pluges i Inundacions a la Mediterrània,* Ketres, Barcelona, pp. 42–45

Matthews, J. A. and others, eds. 1997, *Rapid mass movement as a source of climatic evidence for the Holocene, PKF* 19

Meiggs, R. 1982, *Trees and Timber in the Ancient Mediterranean World,* Clarendon, Oxford

Menéndez Amor, J. and Florschütz, F. 1961, 'Resultado del análisis polínico de una serie de muestras de turba recogidas en la Ereta del Pedregal (Navarrés, Valencia)', *Archivo de Prehistoria Levantina* 9, pp. 97–99

Météo-France [1995], *Inventaire des situations à précipitations diluviennnes sur le Languedoc-Roussillon, la Provence-Alpes, Côte d'Azur et la Corse. Période 1958–1994,* Service Central d'Exploitation de la Météorologie [Aix-en-Provence]

Milliman, J. G. and Syvitski, J. P. M. 1992, 'Geomorphic/tectonic control of sediment discharge to the ocean: the importance of small mountainous rivers', *Journal of Geology* 100, pp. 525–44

Miro-Granada y Gelabert, J. 1974, *Les crues catastrophiques sur la Méditerranée occidentale,* Flash Floods Symposium, IAHS 112, pp. 119–32

Miro-Grenada y Gelabert, J. 1983, 'Consideraciónes generales sobre la meteorología de las riadas en el Levante español', *Estudios Geográficos* 44 (170–71), pp. 31–54

Moody, J. A. 1987, *The environmental and cultural prehistory of the Khania region of west Crete,* PhD dissertation, University of Minnesota, Minneapolis

Moody, J., Rackham, O., Rapp, G., Jr. 1996, 'Environmental archaeology of prehistoric NW Crete', *JFA* 23, pp. 273–97

Moreno, D. 1990, *Dal documento al terreno: storia e archeologia dei sistemi agro-silvo-pastorali,* Il Mulino, Bologna

Moreno, D., Croce, G. F., Guido, M. A., Montanari, C. 1993, 'Pine plantations on ancient grassland: ecological changes in the Mediterranean mountains of Liguria, Italy, during the 19th and 20th centuries', Watkins 1993, pp. 93–110

Moreno, D. and Poggi, G. 1996, 'Storia delle risorse boschive nelle montagne mediterranee: modelli di interpretazione per le produzioni foraggere in regime consuetudinario', Cavaciocchi 1996, pp. 635–54

Moreno, J. M. and Oechel, W. C. 1994, *The Role of Fire in Mediterranean-Type Ecosystems,* Springer, New York

Naveh, Z, 1994a, 'The role of fire and its management in the conservation of Mediterranean ecosystems and landscapes', Moreno and Oechel 1994, pp. 163–85

Naveh, Z. 1994b, 'Biodiversity and landscape management', *Biodiversity and Landscapes*, ed. Ke Chung Kim and R. D. Weaver, Cambridge, pp. 187–209

Naveh, Z. and Liebermann, A. 1994, *Landscape Ecology: theory and application*, Springer, New York, 2nd edn

Néboit, R. 1977, 'Un exemple de morphogenèse accélérée dans l'antiquité: les vallées du Basento et du Cavone en Lucanie (Italie)', *Méditerranée* 4, pp. 39-27

Néboit, R. 1984, 'Érosion des sols et colonisation Grecque en Sicile et en Grande Grèce', *BAGF* 499, pp. 5-21

Néboit-Guilhot, R. 1991, 'Critères d'identification des facteurs d'orientation du sens de la morphogénèse fluviale en Italie du Sud et en Sicile', *PG* 22–3, pp. 61–66

Nelson, C. H. 1990, 'Estimated post-Messinian sediment supply and sedimentation rates on the Ebro continental margin, Spain', *MG* 95, pp. 395–418

Nicod, J. 1990, 'Murettes et terrasses de culture dans les régions karstiques Méditerranéennes', *Méditerranée* 71 (3/4), pp. 43–50

Nicol-Pichard, S. 1987, 'Analyse pollinique d'une séquence tardi et postglaciaire à Tourves (Var, France)', *EM* 13 (1/2), pp. 29–42

Ortiz, A. D. and Vincent, B. 1978, *Historia de los Moriscos: vida y tragedia de una minoría*, Biblioteca de la Revista de Occidente, Madrid

Palmieri, S., Siani, A. M., D'Agostino, A. 1991, 'Climate fluctuations and trends in Italy within the last 100 years', *Annales Geophysicae* 9, pp. 769–76

Palutikof, J. P., Conte, M., Casimiro Mendes, J., Goodess, C. M., Espiritu Santo, F. 1996, 'Climate and Climate Change', Brandt and Thornes 1996, pp. 43–86

Pardé, M. 1941, 'La formidable crue d'octobre 1940 dans les Pyrénées-Orientales', *Revue Géographique des Pyrénées et du Sud-Ouest* 12, pp. 237-79.

Parsons, J. J. 1962, 'The acorn-hog economy of the oak woodlands of southwestern Spain', *Geographical Review* 52, pp. 210–35

Pashley, R. 1837, *Travels in Crete*, Murray, London

Pasquinucci, M. and Mazzanti, R. 1987, 'La costa tirrenica da Luni a Portus Cosanus', *Déplacements . . .*, ed. P. Trousset, pp. 96–106

Perlin, J. 1989, *A Forest Journey: the role of wood in the development of civilization*, Norton, New York

Philippson, A. 1892, *Der Peloponnes*, Friedländer, Berlin [travels 1888]

Pichard, G. 1995, 'Les crues sur le bas Rhône de 1500 à nos jours: pour une histoire hydro-climatique', *Méditerranée* 80 (3/4), pp. 105–16

Pinna, M. 1903, *Indice dei Documenti Cagliaritani del Regio Archivio de Stato, dal 1323 al 1720*, Cagliari

Piñol, J., Terradas, J., Lloret, F. 1998, 'Climate warming, wildfire hazard, and wildfire occurrence in coastal eastern Spain', *CC* 38, pp. 345–57

Planchais, N. 1982, 'Palynologie lagunaire de l'Étang de Mauguio. Paléoenvironnement végétal et évolution anthropique', *Pollen et Spores* 24, pp. 93–118

Poesen, J. W. A. and Hooke, J. M. 1997, 'Erosion, flooding and channel management in Mediterranean environments of southern Europe', *PPG* 21, pp. 157–99

Pons, A. and Reille, M. 1988, 'The Holocene and Upper Pleistocene pollen record from Padul (Granada, Spain): a new study', *PPP* 66, pp. 243–63

Ponz, D. Antonio 1778, *Viaje de España*, Ibarra, Madrid

Pope, K. O. 1993, 'Geology and Soils [of the western Mesara plain in Crete], *Hesperia* 62, pp. 197–204

Potter, T. W. 1979, *The changing landscape of South Etruria*, Elek, London

Potter, T. W. 1976, 'Valleys and settlement, some new evidence', *World Archaeology* 8, pp. 206-19

Poulos, S. E., Collins, M., Evans, G. 1996, 'Water-sediment fluxes of Greek rivers: annual yields, seasonal variability, delta formation and human impact', *ZG* NF 40, pp. 243–61

Pouqueville, F. C. H. L. 1820, *Voyage de la Grèce*, Paris [travels 1805]

Provansal, M. 1995, 'The role of climate in landscape morphogenesis since the Bronze Age in Provence, southeastern France', *Holocene* 5, pp. 348–53

Provansal, M., Bertucchi, L., Pelissier, M. 1994, 'Les milieux palustres de Provence occidentale, indicateurs de la morphogénèse holocene', *ZG* NF 38, pp. 185–205

Pulido Bosch, A. and 6 others 1996, 'La contaminación en los acuíferos del Campo de Dalías y delta del Andarax (Almería)', *Recursos Naturales y Medio Ambiente en el Sureste Península*, ed. L. García-Rossell and A. Navarro Flores, Instituto de Estudios Almerienses, Almería, pp. 363–81

Pyne, S. J. 1982, *Fire in America: a cultural history of wildland and rural fire*, Princeton University Press

Quélennec, R. E., ed. 1990, *Littoral 1990*: EUROCOAST Symposium, Marseille

Raban, A. 1996, *Bad Land*, Vintage Books, New York

Rackham, O. 1972, 'The vegetation of the Myrtos region', in P. M. Warren, *Myrtos: an Early Bronze Age Settlement in Crete*, Thames and Hudson, London, pp. 283–304

Rackham, O. 1980, *Ancient woodland: its history, ecology and vegetation in England*, Edward Arnold, London

Rackham, O. 1983, 'Observations on the historical ecology of Boeotia', *BSA* 78, pp. 291–351, pl. 34–38

Rackham, O. 1986a, *The History of the [British and Irish] Countryside*, Dent, London

Rackham, O. 1986b, 'Charcoal', Renfrew and others 1986, pp. 55–62

Rackham, O. 1989, *The Last Forest: the story of Hatfield Forest*, Dent, London

Rackham, O. 1990a, *Trees and Woodland in the British Landscape*, 2nd edn, Dent, London

Rackham, O. 1990b, 'The greening of Myrtos', Bottema and others 1990, pp. 341–48

Rackham, O. 1998, 'Savanna in Europe', *The Ecological History of European Forests*, ed. K. J. Kirby and C Watkins, CAB International, Wallingford, pp. 1–24

Rackham, O. [forthcoming], 'Observations on the historical ecology of Laconia', in W. Cavanagh and others, *The Laconia Survey; continuity and change in a rural Greek landscape*, 2nd edn, BSA Suppl. Vol. 1

Rackham, O. and Moody, J. A. 1996, *The Making of the Cretan Landscape*, Manchester University Press

Rambal, S. 1987, 'Évolution de l'occupation des terres et ressources en eau en région méditerranéenne karstique', *JH* 93, pp. 339–57

Raulin, V. 1858, 1860, 'Déscription physique de l'île de Crète', *Bulletin de la Société Linnéenne de Bordeaux* 22, pp. 109–213, 307–426, 491–584; 23, pp. 1–50, 70–157, 321–444 [travels 1845]

Rego, F., Botelho, H., Bunting, S. 1987, 'Prescribed fire effects on soils and vegetation in *Pinus pinaster* forests in northern Portugal', *EM* 13, pp. 189–95

Reille, M. 1992, 'New pollen-analytical researches in Corsica: the problem of *Quercus ilex* L. and *Erica arborea* L., the origin of *Pinus halepensis* Miller forests', *New Phytologist* 122, pp. 359–78

Renfrew, C., Gimbutas, M., Elster, E. S. 1986, *Excavations at Sitagroi, a prehistoric village in northeast Greece, Volume 1*, Institute of Archaeology, Los Angeles

Rodrigo, F. S., Esteban-Parra, M. J., Castro-Diez, Y. 1994, 'An attempt to reconstruct the rainfall regime of Andalusia . . . from 1601 A.D. to 1650 A.D. using historical documents', *CC* 27, pp. 397–418

Rodrigo, F. S., Esteban-Parra, M. J., Castro-Diez, Y. 1995, 'The onset of the Little Ice Age in Andalusia . . . : detection and characterization from documentary sources', *Annales Geophysicae* 13, pp. 330–38

Romero Diaz, M. A., Cabezas, F., and López Bermúdez, F. 1992, 'Erosion and fluvial sedimentation in the River Segura basin (Spain)', *Catena* 19, pp. 379–92

Röthlisberger, F. 1986, *10,000 Jahre Gletschergeschichte der Erde*, Sauerländer, Aarau

Rubio, J. L. and Calvo, A., eds. 1996, *Soil degradation and desertification in Mediterranean environments*', Geoforma, Logroño

Russo, G. and Sacchini, A. 1994, 'Brief survey on recurrences of extreme rainfalls in Genoa, Italy', *Időjárás* [Quarterly Journal of the Hungarian Meteorological Service] 98, pp. 251–60

Saberwal, V. L. 1998, 'Science and the desiccationist discourse of the 20th century', *E&H* 4, pp. 309–43

Sala, M. and others, eds. 1991, *Soil Erosion Studies in Spain*, Geoforma, Logroño

Sallares, R. 1991, *The Ecology of the Ancient Greek World*, Duckworth, London

Sánchez-Arcilla, A., Jimenez, J. A., Valdemoro, H. I. 1998, 'The Ebro delta: morphodynamics and vulnerability', *JCR* 14, pp. 754–72

Sauerwein, F. 1969, 'Das Siedlungsbild der Peloponnes um das Jahr 1700', *Erdkunde* 23, pp. 237–44 + suppl.

Sauro, U. 1993, 'Human impact on the karst of the Venetian Fore-Alps, Italy', *EG* 21, pp. 115–21

Schiff, L. S. 1962, *Fire and Water: scientific heresy in the Forest Service*, Harvard University Press

Schweizer, P. 1988, *Shepherds, Workers, Intellectuals: centre–periphery relationships in a Sardinian village*, University of Stockholm Studies in Social Anthropology

Sestini, G. 1992, 'The impact of climatic changes and sea-level rise on two deltaic lowlands of the eastern Mediterranean', Tooley and Jelgersma 1977, pp. 170–203

Seuffert, O. 1992, 'The project "Geoöko-dynamik" (Geoecodynamics) in southern Sardinia', *Geoökoplus* 3, pp. 111–28

Shachori, A., Rosenzweig, D., Poljakoff-Mayber, A. 1967, 'Effect of Mediterranean vegetation on the moisture régime', Sopper and Lull 1967, pp. 291–311

Shipley, G. and Salmon, J., eds. 1996, *Human Landscapes in Classical Antiquity: environment and culture*, Routledge, London

Solé, A. and 5 others 1992, 'How mudrock and soil physical properties influence badland formation at Vallcebre (Pre-Pyrenees, NE Spain)', *Catena* 19, pp. 287–300

Sopper, W. E. and Lull, H. W., eds. 1967, *Forest Hydrology*, Pergamon, Oxford

Spratt, T. A. B. 1865, *Travels and Researches in Crete*, van Voorst, London

Stavrinidhos, N. S. (Ν. Σ. Σταυρινίδου) 1985–6, Μεταφράσεις Τουρκικῶν Ἱστόρικων Ἔγγραφων, Vikelia, Herákleion

Stevenson, A. C. and Harrison, R. J. 1992, 'Ancient forests in Spain: a model for land-use and dry forest management in south-west Spain from 4000 BC to 1900 AD', *PPS* 58, pp. 227–47

Swinburne, H. 1779, *Travels through Spain in the Years 1775 and 1776*, Elmsly, London

Terral, J.-F. and Arnold-Simard, G. 1996, 'Beginnings of olive cultivation in eastern Spain in relation to Holocene bioclimatic changes', *QR* 46, pp. 176–85.

Thirgood, J. V. 1987, *Cyprus: a chronicle of its forests, land, and people*, University of British Columbia Press, Vancouver

Thomas, D. S. G. and Middleton, N. J. 1994, *Desertification: exploding the myth*, John Wiley, Chichester

Thomas, R. G. 1993, 'Rome rainfall and sunspot numbers', *Journal of Atmospheric and Terrestrial Physics* 55, pp. 155–64

Thornes, J. B., ed. 1990, *Vegetation and erosion: processes and environments*, Wiley, Chichester

Tichy, F. 1962, *Die Wälder der Basilicata und die Entwaldung im 19. Jahrhundert*, Heidelberger geographische Arbeiten 8, Heidelberg

Tooley, M. J. and Jelgersma, S., eds. 1992, *Impacts of Sea-level Rise on European Coastal Lowlands*, Blackwell, Oxford

Tomaselli, R. 1977, 'Degradation of the Mediterranean maquis', *Mediterranean forests and maquis: ecology, conservation and management*, UNESCO, Paris, pp. 33–72

Tournefort, P. de 1741, *A Voyage into the Levant*, Midwinter and others, London [French original 1717; travels 1700]

Trevor-Battye, A. 1913, *Camping in Crete*, Witherby, London [travels 1909]

Triandafyllidou-Baladié (Τριανταφυλλίδου-Baladié), G. 1988, Τὸ Εμπόριο καὶ ἡ Οἰκονομία τῆς Κρήτης (1669-1795), Vikelia, Herákleion

Tropeano, D. 1984, 'Rate of soil erosion processes on vineyards in central Piedmont (NW Italy)', *ESPL* 9, pp. 253–66

Trousset, P., ed. 1987, *Déplacements des lignes de rivage en Méditerranée d'après les données de l'archéologie*, CNRS, Paris

Tykot, R. H. and Andrews, T. K., eds. 1992, *Sardinia in the Mediterranean: a footprint in the sea*, Sheffield Academic Press

Tyndale, J. W. 1849, *The island of Sardinia*, Bentley, London

Tzedakis, P. C. 1993, 'Long-term tree populations in northwest Greece through multiple Quaternary climatic cycles', *Nature, London* 364, pp. 437–40

van Andel, T. H. 1998, 'Middle and upper Palaeolithic environments and the calibration of ^{14}C dates beyond 10,000 BP', *Antiquity* 72, pp. 26–33

van Andel, T. H. and Runnels, C. 1987, *Beyond the Acropolis: a rural Greek past*, Stanford University Press

van Andel, T. H., Gallis, K., Toufexis, G. 1995, 'Early Neolithic farming in a Thessalian river landscape, Greece', Lewin and others 1995, pp. 131–43

van Andel, T. H., Runnels, C. N., Pope, K. O. 1986, 'Five thousand years of land use and abuse in the Southern Argolid, Greece', *Hesperia* 55, pp. 103–28

van Andel, T. H. and Zangger, E. 1990, 'Land-scape stability and destabilization in the prehistory of Greece', Bottema and others 1990, pp. 139–57

van Andel, T. H., Zangger, E., Demitrack, A. 1990, 'Land use and soil erosion in prehistoric and historical Greece', *JFA* 17, pp. 379–96

van Zuidam, R. A.1975, 'Geomorphology and archaeology: evidences of interrelation at historic sites in the Zaragoza region, Spain', *ZG* NF 19, pp. 319–28

Vaudour, J. 1962, 'L'érosion des sols à Auriol (B[ouches]-du-Rh[ône])', *Méditerranée* 3 (1), pp. 73–80

Vernet, J.-L. 1997, *L'Homme et la Forêt Méditerranéenne de la Préhistoire à nos jours*, Errance, Paris

Vita-Finzi, C. 1969, *The Mediterranean Valleys*, Cambridge University Press

Vita-Finzi, C. 1974, 'Age of valley deposits in Périgord', *Nature* 250, pp. 568–70

Vita-Finzi, C. 1975, 'Late Quaternary alluvial deposits in Italy', *Geology of Italy*, ed. C. Squyres, Tripoli, Libya, pp. 329–40

Vita-Finzi, C. 1976, 'Diachronism in Old World alluvial sequences', *Nature* 263, pp. 218–9

Wace, A. J. B. and Thompson, M. S. 1914, *The Nomads of the Balkans*, Methuen, London

Wainwright, J. 1996, 'Hillslope response to extreme storm events: the example of the Vaison-La-Romaine event', *Advances in Hillslope Processes*, ed. M. G. Anderson and S. M. Brooks, Wiley, Chichester, pp. 998–1026

Walling, D. E. and Probst, J.-L. 1997, *Human Impact on Erosion and Sedimentation*, IAHS publication 245, Wallingford

Walter, H. and Lieth, H. 1960– , *Klimadiagramm-Weltatlas*, Fischer, Jena

Watkins, C., ed. 1993, *Ecological effects of afforestation: studies in the history and ecology of afforestation in Western Europe*, CAB International, Wallingford

Watts, W. A., Allen, J. R. M., Huntley, B., Fritz, S. C. 1996, 'Vegetation history and climate of the last 15,000 years at Laghi di Monticchio, southern Italy', *QSR* 15, pp. 113–22

Wells, B., ed. 1992, *Agriculture in Ancient Greece*, Svenska Institutet i Athen, Stockholm

Wells, B., Runnels, C., Zangger, E. 1990, 'The Berbati–Limnes archaeological survey. The 1988 season', *Opuscula Atheniensia* 15, pp. 207–38

Widdrington *see* Cook

Wigley, T. M. L., Ingram, M. J., Farmer, G., eds. 1981, *Climate and History: studies in past climates and their effect on man*, Cambridge

Wijmstra, T. A. 1969, 'Palynology of the first 30 metres of a 120 m deep section in northern Greece', *ABN* 18, pp. 511–27

Wijmstra, T. A. and Smit, A. 1976, 'Palynology of the middle part . . . of the 120 m deep section in northern Greece (Macedonia)', *ABN* 25, pp. 297–312

Williams, A. G. and 4 others 1995, 'A field study of the influence of land management and soil properties on runoff and soil loss in central Spain', *EMA* 37, pp. 333–45

Wise, S. M., Thornes, J. B., Gilman, A. 1982, 'How old are the badlands? A case study from south-east Spain', Bryan and Yair 1982, pp. 259–77

Yassoglou, N. 1989, 'Desertification in Greece' *Strategies to combat desertification in Mediterranean Europe*, ed. J. L. Rubio and R. J. Ricon, Commission of European Communities Report, EYR 11175 E/ES, pp. 148–62

Zielinski, G. A. 1995, 'Stratospheric loading and optical depth estimated of explosive volcanism over the last 2100 years derived from the Greenland Ice Sheet Project 2 ice core', *JGR* 100, pp. 20937–55

Index

Picture Credits

The publishers would like to thank the following for the illustrations used in this book. Errors or omissions are inadvertent and will be corrected in subsequent editions if notification is given in writing to the publisher. References are to plate numbers.

Cambridge University Library: 6.14a
Geographical Handbook Series:
    Greece 3, p. 503: 17.22a
A. T. Grove: 2.5; 3.3; 5.8, 5.9a, b, 5.11, 5.13, 5.14, 5.16, 5.17, 5.18, 5.21, 5.26; 6.5, 6.6, 6.9, 6.10, 6.14b, 6.15, 6.17;

13.16; 14.6, 14.13, 14.16b, 14.17, 14.18b, 14.20, 14.22a, b, 14.25, 14.27; 15.4, 15.5, 15.8, 15.9b, c, 15.13, 15.18, 15.20; 16.4, 16.5a, b, 16.7a, b, 16.10, 16.11; 17.19, 17.20, 17.22b, 17.23b; 18.4
W. Grove: 7.4, 18.9
N. Margaris: 17.23a
Oliver Rackham: 3.1, 3.2, 3.5, 3.6, 3.7, 3.8, 3.10, 3.11; 4.1, 4.5a–f, 4.6, 4.7, 4.8, 4.9, 4.10, 4.11, 4.12a, b, 4.13, 4.14, 4.15, 4.16, 4.18a, b, 4.19, 4.20, 4.21, 4.22, 4.24b, 4.25a, 4.27, 4.28, 4.29, 4.30, 4.32, 4.33b, 4.34, 4.35,

4.37; 5.2, 5.5, 5.12, 5.20, 5.23, 5.24, 5.25; 6.2, 6.3, 6.4, 6.7, 6.8, 6.11, 6.13, 6.18; 8.1, 8.2; 9.1; 10.1a, b, 10.3, 10.4; 11.1, 11.2, 11.4; 12.3a, b, 12.5, 12.6, 12.7, 12.10a, b, 12.11, 12.12, 12.13, 12.14, 12.16, 12.17a, b, 12.18, 12.19, 12.20, 12.21, 12.22, 12.23, 12.24a, b, 12.25, 12.26, 12.27, 12.28, 12.29; 13.1, 13.2, 13.3, 13.4, 13.5, 13.6, 13.7, 13.8a, b, 13.9a, b, 13.10, 13.11, 13.12, 13.13, 13.14a–e, 13.18, 13.19, 13.20; 14.1, 14.2, 14.4, 14.7a, b, 14.8, 14.10, 14.11a–h, 14.12, 14.16a, c, 14.18a, 14.19, 14.21, 14.23,

14.24, 14.26, 14.28; 15.2, 15.6, 15.7, 15.9a, 15.14, 15.15, 15.16, 15.17, 15.19, 15.21; 16.1, 16.13; 17.2, 17.3, 17.5, 17.6, 17.7, 17.8, 17.10, 17.14a, b, 17.16a, b, 17.18, 17.24; 18.2, 18.5, 18.6; 19.5
Reproduced by permission of the Board of Trustees of the National Museums and Galleries on Merseyside (Walker Art Gallery, Liverpool): 1.1
J. A. Moody: 14.11d, 15.3
M. Young (after photograph by A. T. Grove): 14.5